## Recommended Dietary Allowances (RDA) and Adequate Intakes (AI) for Vitamins

| Age (yr) | Thiamin RDA (mg/day) | Riboflavin RDA (mg/day) | Niacin RDA (mg/day)[a] | Biotin AI (µg/day) | Pantothenic acid AI (mg/day) | Vitamin B6 RDA (mg/day) | Folate RDA (µg/day)[b] | Vitamin B12 RDA (µg/day) | Choline AI (mg/day) | Vitamin C RDA (mg/day) | Vitamin A RDA (µg/day)[c] | Vitamin D AI (µg/day)[d] | Vitamin E RDA (mg/day)[e] | Vitamin K AI (µg/day) |
|---|---|---|---|---|---|---|---|---|---|---|---|---|---|---|
| **Infants** | | | | | | | | | | | | | | |
| 0–0.5 | 0.2 | 0.3 | 2 | 5 | 1.7 | 0.1 | 65 | 0.4 | 125 | 40 | 400 | 5 | 4 | 2.0 |
| 0.5–1 | 0.3 | 0.4 | 4 | 6 | 1.8 | 0.3 | 80 | 0.5 | 150 | 50 | 500 | 5 | 5 | 2.5 |
| **Children** | | | | | | | | | | | | | | |
| 1–3 | 0.5 | 0.5 | 6 | 8 | 2 | 0.5 | 150 | 0.9 | 200 | 15 | 300 | 5 | 6 | 30 |
| 4–8 | 0.6 | 0.6 | 8 | 12 | 3 | 0.6 | 200 | 1.2 | 250 | 25 | 400 | 5 | 7 | 55 |
| **Males** | | | | | | | | | | | | | | |
| 9–13 | 0.9 | 0.9 | 12 | 20 | 4 | 1.0 | 300 | 1.8 | 375 | 45 | 600 | 5 | 11 | 60 |
| 14–18 | 1.2 | 1.3 | 16 | 25 | 5 | 1.3 | 400 | 2.4 | 550 | 75 | 900 | 5 | 15 | 75 |
| 19–30 | 1.2 | 1.3 | 16 | 30 | 5 | 1.3 | 400 | 2.4 | 550 | 90 | 900 | 5 | 15 | 120 |
| 31–50 | 1.2 | 1.3 | 16 | 30 | 5 | 1.3 | 400 | 2.4 | 550 | 90 | 900 | 5 | 15 | 120 |
| 51–70 | 1.2 | 1.3 | 16 | 30 | 5 | 1.7 | 400 | 2.4 | 550 | 90 | 900 | 10 | 15 | 120 |
| >70 | 1.2 | 1.3 | 16 | 30 | 5 | 1.7 | 400 | 2.4 | 550 | 90 | 900 | 15 | 15 | 120 |
| **Females** | | | | | | | | | | | | | | |
| 9–13 | 0.9 | 0.9 | 12 | 20 | 4 | 1.0 | 300 | 1.8 | 375 | 45 | 600 | 5 | 11 | 60 |
| 14–18 | 1.0 | 1.0 | 14 | 25 | 5 | 1.2 | 400 | 2.4 | 400 | 65 | 700 | 5 | 15 | 75 |
| 19–30 | 1.1 | 1.1 | 14 | 30 | 5 | 1.3 | 400 | 2.4 | 425 | 75 | 700 | 5 | 15 | 90 |
| 31–50 | 1.1 | 1.1 | 14 | 30 | 5 | 1.3 | 400 | 2.4 | 425 | 75 | 700 | 5 | 15 | 90 |
| 51–70 | 1.1 | 1.1 | 14 | 30 | 5 | 1.5 | 400 | 2.4 | 425 | 75 | 700 | 10 | 15 | 90 |
| >70 | 1.1 | 1.1 | 14 | 30 | 5 | 1.5 | 400 | 2.4 | 425 | 75 | 700 | 15 | 15 | 90 |
| **Pregnancy** | | | | | | | | | | | | | | |
| ≤18 | 1.4 | 1.4 | 18 | 30 | 6 | 1.9 | 600 | 2.6 | 450 | 80 | 750 | 5 | 15 | 75 |
| 19–30 | 1.4 | 1.4 | 18 | 30 | 6 | 1.9 | 600 | 2.6 | 450 | 85 | 770 | 5 | 15 | 90 |
| 31–50 | 1.4 | 1.4 | 18 | 30 | 6 | 1.9 | 600 | 2.6 | 450 | 85 | 770 | 5 | 15 | 90 |
| **Lactation** | | | | | | | | | | | | | | |
| ≤18 | 1.4 | 1.6 | 17 | 35 | 7 | 2.0 | 500 | 2.8 | 550 | 115 | 1200 | 5 | 19 | 75 |
| 19–30 | 1.4 | 1.6 | 17 | 35 | 7 | 2.0 | 500 | 2.8 | 550 | 120 | 1300 | 5 | 19 | 90 |
| 31–50 | 1.4 | 1.6 | 17 | 35 | 7 | 2.0 | 500 | 2.8 | 550 | 120 | 1300 | 5 | 19 | 90 |

NOTE: For all nutrients, values for infants are AI.

[a] Niacin recommendations are expressed as niacin equivalents (NE), except for recommendations for infants younger than 6 months, which are expressed as preformed niacin.

[b] Folate recommendations are expressed as dietary folate equivalents (DFE).

[c] Vitamin A recommendations are expressed as retinol activity equivalents (RAE).

[d] Vitamin D recommendations are expressed as cholecalciferol.

[e] Vitamin E recommendations are expressed as α-tocopherol.

## Recommended Dietary Allowances (RDA) and Adequate Intakes (AI) for Minerals

| Age (yr) | Sodium AI (mg/day) | Chloride AI (mg/day) | Potassium AI (mg/day) | Calcium AI (mg/day) | Phosphorus RDA (mg/day) | Magnesium RDA (mg/day) | Iron RDA (mg/day) | Zinc RDA (mg/day) | Iodine RDA (µg/day) | Selenium RDA (µg/day) | Copper RDA (µg/day) | Manganese AI (mg/day) | Fluoride AI (mg/day) | Chromium AI (µg/day) | Molybdenum RDA (µg/day) |
|---|---|---|---|---|---|---|---|---|---|---|---|---|---|---|---|
| **Infants** | | | | | | | | | | | | | | | |
| 0–0.5 | 120 | 180 | 400 | 210 | 100 | 30 | 0.27 | 2 | 110 | 15 | 200 | 0.003 | 0.01 | 0.2 | 2 |
| 0.5–1 | 370 | 570 | 700 | 270 | 275 | 75 | 11 | 3 | 130 | 20 | 220 | 0.6 | 0.5 | 5.5 | 3 |
| **Children** | | | | | | | | | | | | | | | |
| 1–3 | 1000 | 1500 | 3000 | 500 | 460 | 80 | 7 | 3 | 90 | 20 | 340 | 1.2 | 0.7 | 11 | 17 |
| 4–8 | 1200 | 1900 | 3800 | 800 | 500 | 130 | 10 | 5 | 90 | 30 | 440 | 1.5 | 1.0 | 15 | 22 |
| **Males** | | | | | | | | | | | | | | | |
| 9–13 | 1500 | 2300 | 4500 | 1300 | 1250 | 240 | 8 | 8 | 120 | 40 | 700 | 1.9 | 2 | 25 | 34 |
| 14–18 | 1500 | 2300 | 4700 | 1300 | 1250 | 410 | 11 | 11 | 150 | 55 | 890 | 2.2 | 3 | 35 | 43 |
| 19–30 | 1500 | 2300 | 4700 | 1000 | 700 | 400 | 8 | 11 | 150 | 55 | 900 | 2.3 | 4 | 35 | 45 |
| 31–50 | 1500 | 2300 | 4700 | 1000 | 700 | 420 | 8 | 11 | 150 | 55 | 900 | 2.3 | 4 | 35 | 45 |
| 51–70 | 1300 | 2000 | 4700 | 1200 | 700 | 420 | 8 | 11 | 150 | 55 | 900 | 2.3 | 4 | 30 | 45 |
| >70 | 1200 | 1800 | 4700 | 1200 | 700 | 420 | 8 | 11 | 150 | 55 | 900 | 2.3 | 4 | 30 | 45 |
| **Females** | | | | | | | | | | | | | | | |
| 9–13 | 1500 | 2300 | 4500 | 1300 | 1250 | 240 | 8 | 8 | 120 | 40 | 700 | 1.6 | 2 | 21 | 34 |
| 14–18 | 1500 | 2300 | 4700 | 1300 | 1250 | 360 | 15 | 9 | 150 | 55 | 890 | 1.6 | 3 | 24 | 43 |
| 19–30 | 1500 | 2300 | 4700 | 1000 | 700 | 310 | 18 | 8 | 150 | 55 | 900 | 1.8 | 3 | 25 | 45 |
| 31–50 | 1500 | 2300 | 4700 | 1000 | 700 | 320 | 18 | 8 | 150 | 55 | 900 | 1.8 | 3 | 25 | 45 |
| 51–70 | 1300 | 2000 | 4700 | 1200 | 700 | 320 | 8 | 8 | 150 | 55 | 900 | 1.8 | 3 | 20 | 45 |
| >70 | 1200 | 1800 | 4700 | 1200 | 700 | 320 | 8 | 8 | 150 | 55 | 900 | 1.8 | 3 | 20 | 45 |
| **Pregnancy** | | | | | | | | | | | | | | | |
| ≤18 | 1500 | 2300 | 4700 | 1300 | 1250 | 400 | 27 | 12 | 220 | 60 | 1000 | 2.0 | 3 | 29 | 50 |
| 19–30 | 1500 | 2300 | 4700 | 1000 | 700 | 350 | 27 | 11 | 220 | 60 | 1000 | 2.0 | 3 | 30 | 50 |
| 31–50 | 1500 | 2300 | 4700 | 1000 | 700 | 360 | 27 | 11 | 220 | 60 | 1000 | 2.0 | 3 | 30 | 50 |
| **Lactation** | | | | | | | | | | | | | | | |
| ≤18 | 1500 | 2300 | 5100 | 1300 | 1250 | 360 | 10 | 13 | 290 | 70 | 1300 | 2.6 | 3 | 44 | 50 |
| 19–30 | 1500 | 2300 | 5100 | 1000 | 700 | 310 | 9 | 12 | 290 | 70 | 1300 | 2.6 | 3 | 45 | 50 |
| 31–50 | 1500 | 2300 | 5100 | 1000 | 700 | 320 | 9 | 12 | 290 | 70 | 1300 | 2.6 | 3 | 45 | 50 |

## Tolerable Upper Intake Levels (UL) for Vitamins

| Age (yr) | Niacin (mg/day)[a] | Vitamin B6 (mg/day) | Folate (µg/day)[a] | Choline (mg/day) | Vitamin C (mg/day) | Vitamin A (µg/day)[b] | Vitamin D (µg/day) | Vitamin E (mg/day)[c] |
|---|---|---|---|---|---|---|---|---|
| **Infants** | | | | | | | | |
| 0–0.5 | — | — | — | — | — | 600 | 25 | — |
| 0.5–1 | — | — | — | — | — | 600 | 25 | — |
| **Children** | | | | | | | | |
| 1–3 | 10 | 30 | 300 | 1000 | 400 | 600 | 50 | 200 |
| 4–8 | 15 | 40 | 400 | 1000 | 650 | 900 | 50 | 300 |
| **Adolescents** | | | | | | | | |
| 9–13 | 20 | 60 | 600 | 2000 | 1200 | 1700 | 50 | 600 |
| 14–18 | 30 | 80 | 800 | 3000 | 1800 | 2800 | 50 | 800 |
| **Adults** | | | | | | | | |
| 19–70 | 35 | 100 | 1000 | 3500 | 2000 | 3000 | 50 | 1000 |
| >70 | 35 | 100 | 1000 | 3500 | 2000 | 3000 | 50 | 1000 |
| **Pregnancy** | | | | | | | | |
| ≤18 | 30 | 80 | 800 | 3000 | 1800 | 2800 | 50 | 800 |
| 19–50 | 35 | 100 | 1000 | 3500 | 2000 | 3000 | 50 | 1000 |
| **Lactation** | | | | | | | | |
| ≤18 | 30 | 80 | 800 | 3000 | 1800 | 2800 | 50 | 800 |
| 19–50 | 35 | 100 | 1000 | 3500 | 2000 | 3000 | 50 | 1000 |

[a]The UL for niacin and folate apply to synthetic forms obtained from supplements, fortified foods, or a combination of the two.

[b]The UL for vitamin A applies to the preformed vitamin only.
[c]The UL for vitamin E applies to any form of supplemental α-tocopherol, fortified foods, or a combination of the two.

## Tolerable Upper Intake Levels (UL) for Minerals

| Age (yr) | Sodium (mg/day) | Chloride (mg/day) | Calcium (mg/day) | Phosphorus (mg/day) | Magnesium (mg/day)[d] | Iron (mg/day)[b] | Zinc (mg/day) | Iodine (µg/day) | Selenium (µg/day) | Copper (µg/day) | Manganese (mg/day) | Fluoride (mg/day) | Molybdenum (µg/day) | Boron (mg/day) | Nickel (mg/day) |
|---|---|---|---|---|---|---|---|---|---|---|---|---|---|---|---|
| **Infants** | | | | | | | | | | | | | | | |
| 0–0.5 | —[e] | —[e] | — | — | — | 40 | 4 | — | 45 | — | — | 0.7 | — | — | — |
| 0.5–1 | —[e] | —[e] | — | — | — | 40 | 5 | — | 60 | — | — | 0.9 | — | — | — |
| **Children** | | | | | | | | | | | | | | | |
| 1–3 | 1500 | 2300 | 2500 | 3000 | 65 | 40 | 7 | 200 | 90 | 1000 | 2 | 1.3 | 300 | 3 | 0.2 |
| 4–8 | 1900 | 2900 | 2500 | 3000 | 110 | 40 | 12 | 300 | 150 | 3000 | 3 | 2.2 | 600 | 6 | 0.3 |
| **Adolescents** | | | | | | | | | | | | | | | |
| 9–13 | 2200 | 3400 | 2500 | 4000 | 350 | 40 | 23 | 600 | 280 | 5000 | 6 | 10 | 1100 | 11 | 0.6 |
| 14–18 | 2300 | 3600 | 2500 | 4000 | 350 | 45 | 34 | 900 | 400 | 8000 | 9 | 10 | 1700 | 17 | 1.0 |
| **Adults** | | | | | | | | | | | | | | | |
| 19–70 | 2300 | 3600 | 2500 | 4000 | 350 | 45 | 40 | 1100 | 400 | 10,000 | 11 | 10 | 2000 | 20 | 1.0 |
| >70 | 2300 | 3600 | 2500 | 3000 | 350 | 45 | 40 | 1100 | 400 | 10,000 | 11 | 10 | 2000 | 20 | 1.0 |
| **Pregnancy** | | | | | | | | | | | | | | | |
| ≤18 | 2300 | 3600 | 2500 | 3500 | 350 | 45 | 34 | 900 | 400 | 8000 | 9 | 10 | 1700 | 17 | 1.0 |
| 19–50 | 2300 | 3600 | 2500 | 3500 | 350 | 45 | 40 | 1100 | 400 | 10,000 | 11 | 10 | 2000 | 20 | 1.0 |
| **Lactation** | | | | | | | | | | | | | | | |
| ≤18 | 2300 | 3600 | 2500 | 4000 | 350 | 45 | 34 | 900 | 400 | 8000 | 9 | 10 | 1700 | 17 | 1.0 |
| 19–50 | 2300 | 3600 | 2500 | 4000 | 350 | 45 | 40 | 1100 | 400 | 10,000 | 11 | 10 | 2000 | 20 | 1.0 |

[d]The UL for magnesium applies to synthetic forms obtained from supplements or drugs only.
[e]Source of intake should be from human milk (or formula) and food only.

NOTE: An Upper Limit was not established for vitamins and minerals not listed and for those age groups listed with a dash (—) because of a lack of data, not because these nutrients are safe to consume at any level of intake. All nutrients can have adverse effects when intakes are excessive.

SOURCE: Adapted with permission from the *Dietary Reference Intakes* series, National Academies Press. Copyright 1997, 1998, 2000, 2001, by the National Academy of Sciences. Courtesy of the National Academies Press, Washington, D.C.

## Acceptable Macronutrient Distribution Ranges (AMDR)

| Macronutrient | Range (percent of energy) | | |
|---|---|---|---|
| | Children, 1–3 years | Children, 4–18 years | Adults 19+ years |
| Fat | 30–40 | 25–35 | 20–35 |
| ω-6 polyunsaturated acids[a] (linoleic acid) | 5–10 | 5–10 | 5–10 |
| ω-3 polyunsaturated fatty acids[a] (linolenic acid) | 0.6–1.2 | 0.6–1.2 | 0.6–1.2 |
| Carbohydrate | 45–65 | 45–65 | 45–65 |
| Protein | 5–20 | 10–30 | 10–35 |

[a]Approximately 10% of the total can come from longer-chain ω-3 or ω-6 fatty acids.

SOURCE: Adapted from Institute of Medicine, *Dietary Reference Intakes for Energy, Carbohydrate, Fiber, Fat, Fatty Acids, Cholesterol, Protein, and Amino Acids*, National Academies Press; Washington, D.C., 2005.

# Nutritional Sciences

## From Fundamentals to Food

### MICHELLE McGUIRE, PhD
### KATHY A. BEERMAN, PhD
WASHINGTON STATE UNIVERSITY

THOMSON
™
WADSWORTH

Australia • Brazil • Canada • Mexico • Singapore • Spain • United Kingdom • United States

**NUTRITIONAL SCIENCES: FROM FUNDAMENTALS TO FOOD**
Michelle McGuire and Kathy A. Beerman

PUBLISHER    Peter Marshall
DIRECTOR, MARKET DEVELOPMENT    Star MacKenzie Burruto
DEVELOPMENT EDITOR    Elizabeth Howe
ASSISTANT EDITOR    Elesha Feldman
EDITORIAL ASSISTANT    Lauren Vogelbaum
TECHNOLOGY PROJECT MANAGER    Donna Kelley
MARKETING MANAGER    Jennifer Somerville
MARKETING ASSISTANT    Catie Ronquillo
MARKETING COMMUNICATIONS MANAGER    Jessica Perry
PROJECT MANAGER, EDITORIAL PRODUCTION    Sandra Craig
CREATIVE DIRECTOR    Rob Hugel
ART DIRECTOR    Lee Friedman
PRINT BUYER    Barbara Britton
PERMISSIONS EDITOR    Sarah D'Stair
PRODUCTION    Martha Emry
TEXT AND COVER DESIGNER    Diane Beasley
ART CONSULTANT    Carolyn Deacy
PHOTO RESEARCHERS    Myrna Engler, Stephen Forsling
COPY EDITOR    Linda Purrington
ILLUSTRATIONS    Dragonfly Media Group
COVER IMAGE    © Peter Holmes/age fotostock/SuperStock
COMPOSITOR    Graphic World
PRINTER    CTPS

Printed in China

2  3  4  5  6  7  09  08  07

For more information about our products, contact us at:
Thomson Learning Academic Resource Center
1-800-423-0563
For permission to use material from this text or product, submit a request online at
http://www.thomsonrights.com.
Any additional questions about permissions can be submitted by e-mail to thomsonrights@thomson.com.

Thomson Higher Education
10 Davis Drive
Belmont, CA 94002-3098
USA

Library of Congress Control Number: 2006921376

ISBN-13: 978-0-534-53717-3
ISBN-10: 0-534-53717-0

# Contents in Brief

# Focus On . . .

# Contents

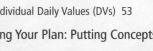

## CHAPTER 6 Protein 172

## CHAPTER 7 Lipids 220

## CHAPTER 8 Energy Metabolism 268

## CHAPTER 9 Energy Balance and Body Weight Regulation 318

## CHAPTER 10 Water-Soluble Vitamins 376

## CHAPTER 11 Fat-Soluble Vitamins 434

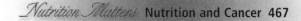

## CHAPTER 12 The Trace Minerals 480

## CHAPTER 13   The Major Minerals and Water 514

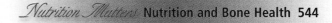

## CHAPTER 14   Life Cycle Nutrition 554

# About the Authors

*To the three people who have most shaped my life–giving it character, meaning, and texture–and have supported me in my dreams and goals. To my mom, Carol Brown, and in memory of my dad, Gene Brown, whose love for me has never wavered and whose never-ending belief in me has provided strength and resolve even when life has taken an uncertain turn. Thank you. To my husband, Mark McGuire, for his ability and willingness to listen and advise, be critical without being hurtful, love, soothe, nurture, and comfort. Thank you.*

*Shelley*

*This book is dedicated to those who give my life meaning. To my children, Anna and Michael, who make everything worthwhile. To my husband, Steven, for his unconditional love, patience, support, comfort, and ongoing encouragement—thank you for all that you do. To my father, Morris Beerman, who continues to amaze me with his intellect, wisdom, compassion, and kindness. In memory of my mother, Zenda, who taught us all the meaning of the word determination.*

*Kathy*

**MICHELLE "SHELLEY" McGUIRE** is associate professor of nutrition and teaches in the Department of Food Science and Human Nutrition at Washington State University. She earned a bachelor's degree in biology from the University of Illinois, a master's degree in nutritional sciences from the University of Illinois, and a doctorate in human nutrition from Cornell University. Since completing her doctorate in 1995, she has received several young investigator awards for her research and leadership skills in the area of human nutrition including the International Life Sciences Institute's Future Leader Award and the International Society for Research in Human Milk and Lactation's Ehrlich-Koldovsky Award. She was also selected to participate in the Dannon Institute's Nutrition Leadership Institute. Dr. McGuire is the author of numerous articles in prominent journals, specializing in research related to understanding how breast feeding and lactation influence maternal and child health. Her recent research has focused on investigating how maternal diet influences milk fat content and thus energy intake and fatty acid consumption by the breast-feeding child. She also has a major research interest in understanding the interactions among maternal nutritional status, breast-feeding characteristics, and duration of lactational amenorrhea. She is an active member of the American Society for Nutrition and the International Society for Research in Human Milk and Lactation. She has taught Washington State University's introductory nutrition course for nonscience majors for more than ten years. She has a strong appreciation for the goals and objectives of this course as well as the wide variety of student needs associated with it. Dr. McGuire also teaches an undergraduate course on life cycle nutrition. This teaching experience, coupled with her strong writing skills and research background, has helped her create an exciting and refreshing text for the introductory nutrition course.

**KATHY BEERMAN** is professor of nutrition and teaches in the Department of Food Science and Human Nutrition at Washington State University. She earned bachelor's and master's degrees from the State University of New York at Brockport, Department of Health Sciences, and a doctorate from Oregon State University, Department of Health. The author of several published articles, she specializes in research focusing on dietary practices of college students and the effects of isoflavones on health parameters in postmenopausal women. Dr. Beerman has also published several articles assessing the effectiveness of computer technology in the college classroom. She is a member of the American Society for Nutrition. Dr. Beerman teaches the introductory nutrition course for health majors, as well as courses in nutrition education, life cycle nutrition, and pathophysiology. Since joining the faculty at Washington State University in 1989, she has taught more than 10,000 students, and has been the recipient of several college and university teaching awards, including the Burlington Northern Faculty Meritorious Achievement in Teaching Award, the R. M. Wade Foundation Award for Excellence in Teaching, and the WSU Mortar Board's Distinguished Professor Award. Dr. Beerman and Dr. McGuire coordinately oversee the undergraduate nutritional sciences option for students preparing for health professions (such as medicine) or graduate school. Dr. Beerman's years of teaching experience combined with her wide knowledge base in nutrition and health sciences has helped create this innovative introductory nutrition text.

# Preface

$\mathcal{W}$e wrote this introductory nutrition textbook to explain nutrition concepts clearly and completely in a way that all students, regardless of level or background, can understand. Fundamental science concepts are carefully described and then applied to nutrition in a stepwise, student-friendly manner. The accompanying artwork provides beautiful examples and illustrations that clarify the science. It is our goal to show all students, even those with little or no science background, that the science of nutrition is approachable, understandable, and useful.

## The Fundamentals Are Important

To our knowledge, this is the first introductory nutrition textbook to present a "primer" chapter (Chapter 3) on biology and chemistry at a level appropriate for both science and nonscience students. This chapter provides the background necessary to master nutrition concepts. Without knowledge of fundamental science principles, many capable students in introductory nutrition courses are lost from the very first class. For example, when teaching nutrition we often refer to chemical bonds—how else do we distinguish between saturated and unsaturated fatty acids? When teaching the importance of ATP we discuss the breaking of phosphate bonds. But how can a student grasp these concepts without knowing what a chemical bond is? To help the student succeed, we explain the basics of chemical bonding as well as many other fundamental science concepts. Students enrolled in introductory nutrition for professions other than health care (often to fulfill a science requirement) may not have taken any science coursework. Thus we provide students with the chemistry, physiology, and anatomy terminology necessary to really learn nutrition science.

## An Integrated Approach

As the field of nutrition grows, so does the amount of information we must learn and teach. To present the content efficiently, and to reinforce the most important concepts, we have organized the text in an integrated way. First, we integrated many important concepts throughout the book. For example, the concept of active and passive transport mechanisms is carefully described in Chapter 4, and it is returned to numerous times throughout Chapters 5 through 13. Students can easily refer to the **Connections**, such as the one on the right, found on many pages for a quick reminder of these important topics.

Second, to present information more efficiently, we developed **Focus On . . .** boxes that integrate issues related to scientific innovation, diet and health, food, clinical applications, nutrition through the life cycle, and sports nutrition.

CONNECTIONS Hydrolytic reactions break chemical bonds by the addition of water (Chapter 3, page 77).

*Focus On* THE PROCESS OF SCIENCE

*Focus On* DIET AND HEALTH

*Focus On* FOOD

*Focus On* CLINICAL APPLICATIONS

*Focus On* LIFE CYCLE NUTRITION

*Focus On* SPORTS NUTRITION

Each chapter contains several of these featured topics, which elaborate on information presented in the chapter. For example, *Focus on Diet and Health: Lactose Intolerance* accompanies the section on carbohydrate digestion. *Focus on Sports Nutrition: Protein and Amino Acid Supplements* accompanies information on dietary requirements for protein. During the testing phase, this feature received high praise from students.

We also strived to improve the integration of important nutrients as they are related to maintenance of health or risk for chronic disease. Our approach was to address these topics in a more comprehensive way by including what we call **Nutrition Matters** pieces, which follow most of the nutrient chapters. These "minichapters" deal with important nutrition-related issues—such as food safety, nutrition and cancer, and dysfunctional eating. They are up-to-date, comprehensive, yet clear and student-friendly.

We have also integrated basic knowledge with application. For example, many of the chapters conclude with a section titled **Putting Concepts into Action** in which nutrition concepts presented in the chapter, such as nutrient density and protein and fiber recommendations, are integrated into to a real-life scenario. In these sections, hypothetical individuals apply and integrate the chapter concepts to solve a nutrition-related problem or to better understand their own health. For example, in Chapter 2 the fictional character

*Nutrition Matters*

**Nutrition, Impaired Glucose Regulation, and Diabetes**

**What Is Diabetes?**

**Discovery of Insulin**

Jodi implements some of the diet assessment tools described in the chapter, such as the online MyPyramid Tracker tool, to discover how her nutrient intake compares to recommendations. Jodi then makes dietary changes. Thus **Putting Concept into Action** sections relate and integrate science concepts into student-friendly scenarios.

## Illustrations

All students benefit from seeing science concepts articulated in clear, well-organized illustrations. The figures in the text were designed using a unique captioning system, and the consistent use of blue text boxes and lines quickly identifies the key points of each figure. Many visual summaries take students step-by-step through complex processes, from the whole-body "big picture" to the details. Effort was made to ensure that components found in many figures are displayed in the same way. For example, the anatomical drawings of various organs such as the liver, pancreas, and stomach have a consistent appearance, glucose is always colored blue when it appears in a figure, and phospholipids are always drawn and colored the same way.

## Book Length

Most introductory nutrition courses are taught over one semester or quarter, yet textbooks written for these courses contain more information than can be covered within a 10- to 15-week period. By presenting concepts concisely, we created a slimmer, trimmer textbook that could more easily be covered in a single semester. Additional resources that might be found in an appendix are provided to students via the book's website or as a bundled extra such as the food composition table, a booklet provided free with every new text and available online.

## Application of Nutritional Science to Health and Healthy Eating

This book provides both a serious review of the science of nutrition while also including its application to health and making good food choices. The **Food Matters** sections help interpret the USDA Dietary Guidelines for Americans into ideas for choosing and preparing foods in the most health-conscious manner. As teachers, we want our students to be equipped with the knowledge and skills to make good decisions about the foods they eat. As nutrition scientists, we want our students to be equipped with the knowledge and skills to think analytically. We wrote this book in such a way that students will take a good, hard look at the scientific endeavor of nutrition research in Chapter 1, which contains information concerning scientific method, study design, interactions, confounding variables, and sorting nutrition fact from fiction. This information is then integrated into the remaining chapters so that students can apply these concepts to nutrition claims that bombard them on a daily basis.

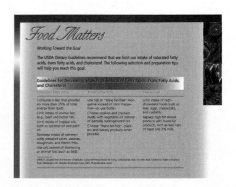

## Pedagogical Tools

Each chapter contains a number of helpful tools to help the student master the chapter content. **Essential Concepts** are summaries of each major heading. Students can use the summaries to preview the chapter material, as well as review the concepts presented. In classroom testing, students found these features especially helpful in learning and retaining concepts. Each chapter also contains a margin **glossary** defining key terms on each page. To help students prepare for tests, a set of **Review Questions** is provided at the end of each chapter. The questions are multiple-choice and essay questions with answers given in the Appendix of the text. **Practice Calculations** provide hands-on exercises to build critical thinking and quantitative skills. The calculations give examples of mathematics-related problems encountered in nutrition. Answers are provided in the Appendix. Last, chapter **Notes** include specific references and provide additional sources of information on topics covered in the chapter.

## Supplements

Because teaching an introductory nutrition course effectively requires many instructor resources, we have developed a package of tools to make the teaching and learning experience a positive one.

A **Multimedia Manager** contains electronic versions of every figure in the text, using a unique "stepped" format so that figures can be broken down and reassembled to make custom lecture presentation slides. Also included on the Multimedia Manager are hundreds of animations and video clips. Other helpful resources include a test bank, instructor's manual, transparency acetates, Web Tutor and Web CT access, and the popular diet analysis tool, Diet Analysis+. The complete supplements package is described in detail in the front of the Instructor's Edition of this text and on the book's website at **www.thomsonedu.com/nutrition/mcguire.**

# Acknowledgments

*M*uch goes into writing and publishing a textbook—more than we ever could have imagined. We are fortunate to have teamed up with the experienced and skillful professionals at Thomson Wadsworth. Their guidance was invaluable, enabling us to transform our vision into a quality textbook. We will forever be grateful for their encouragement and continued support throughout this entire process. Our special thanks is extended to Peter Marshall, who both encouraged us to go forward with our ideas and who then put together a team that could make those ideas happen. In the world of publishing, he is a true visionary. Much appreciation goes to Elizabeth Howe, who continued to amaze us with her fresh ideas and her ability to assemble all the pieces while coordinating the many reviews of each chapter. Thank you to Star MacKenzie Burruto, who served as the book's enthusiastic ambassador by communicating her ideas and suggestions to us—her work has greatly improved this book. A million thanks are extended to Martha Emry, for patiently orchestrating the players who magically transformed all the various pieces into a quality book, and to Sandra Craig, who continued to skillfully juggle no matter how many balls were seemingly in the air. We are also grateful to Carolyn Deacy for assisting us with the art manuscript. Her insights and ideas helped make the illustrations in this book truly instructive. We are also indebted to the talented team of artists at Dragonfly Media Group who took our sometimes sketchy ideas and developed them into artful illustrations; to our creative photo researcher Myrna Engler, whose untimely death saddened us and whom we dearly miss; to Matthew Farruggio for his stunning chapter opening photographs; to Linda Purrington, who read each chapter with her keen and discerning eyes, making countless suggestions that greatly improved the text; and to Graphic World for skillfully bringing the pages to life. Thank you to Jennifer Somerville and the entire marketing team for effectively taking this beautiful creation to the academic world, to Elesha Feldman for preparing the student and instructor resources that accompany the text, and to Lauren Vogelbaum for her coordination of reviews. We would also like to recognize Josh Fletcher, our Thomson sales representative, for encouraging us to pursue this project.

We also want to express our sincere appreciation to Washington State University, the College of Agriculture, Human, and Natural Resource Sciences, and the Department of Food Science and Human Nutrition for allowing us to pursue this effort. Jodi Anderson and Carolee Armfield deserve special thanks for their steadfast encouragement and support. Finally, and perhaps foremost, we wish to thank our friends and families, who helped in more ways than they will ever know. Thank you for being fellow travelers on this journey.

We would also like to thank the student reviewers at the following schools for their insightful feedback: Arizona State University; Brown University; Clayton College and State University; Delaware State University; Florida State University; Iowa State University; Johnson County Community College; Louisiana State University; Miami University; Missouri State University; Northeastern University; Pepperdine University; State University of New York College at Oneonta; Texas State University; University of Houston; University of Oklahoma; University of Tennessee, Martin; University of Wyoming; Washington State University; and West Virginia University.

Our heartfelt thanks are also extended to the expert reviewers who found time in their busy schedules to read initial drafts of several of the chapters in this book and help us craft them into up-to-date and accurate pieces. In this regard, we acknowledge Drs. Jean-Pierre Habicht and Kenneth Carpenter, who critically reviewed Chapter 1; Dr. Peter Garlick, who reviewed Chapter 6; Drs. Dale Bauman and Darlene Berryman, who reviewed Chapter 7, Dr. Ron Brosemer, who reviewed Chapter 8; Drs. Grace Marquis and Mary Penny whose knowledge of community and global nutrition issues was most helpful; Dr. David Kritchevsky, who reviewed the coverage of cardiovascular disease; Dr. Alan McCurdy, who reviewed food safety; Drs. Heidi Kalkwarf and Karen S. Wosje who reviewed coverage of osteoporosis and Dr. Edward Frongillo and Claire M. Horan who collaborated with us to write the Nutrition Matters on food insecurity. Our husbands, Drs. Steven McGeehan and Mark McGuire, also provided much appreciated and critical feedback for many of the chapters. In addition, we acknowledge the many other academicians who each carefully reviewed portions of this book. Your comments and contributions have been invaluable.

## Reviewers, Editorial Board Workshop Participants, and Class Test Participants

Linda Johnston Lolkus*
*Indiana University–Purdue University, Fort Wayne*

Clara Lowden
*Riverside Community College*

Mary Mead
*University of California, Berkeley*

Myrtle McCulloch
*Georgetown University*

Glen F. McNeil
*Fort Hays State University*

Kathleen J. Melanson
*University of Rhode Island*

Juliet Mevi-Shiflett
*Diablo Valley College*

Anahita Mistry*
*Eastern Michigan University*

Huanbiao Mo
*Texas Woman's University*

Gaile Moe
*Seattle Pacific University*

Mohey Mowafy
*Northern Michigan University*

Kathy D. Munoz
*Humboldt State University*

Judy Myhand*†
*Louisiana State University*

Steven Nizielski
*Grand Valley State University*

Amy Ozier*
*Northern Illinois University, DeKalb*

Anna M. Page*†
*Johnson County Community College*

Kathleen Page
*Bucknell University*

Eleanor B. Pella
*Harrisburg Area Community College, Gettysburg*

Susan Polasek
*University of Texas, Austin*

Mercy Popoola†
*Clayton College and State University*

William R. Proulx
*State University of New York College at Oneonta*

Elizabeth Quintana
*West Virginia University*

Jenice Rankins*†
*Florida State University*

Maureen Reidenauer
*Camden County College*

Robert D. Reynolds
*University of Illinois, Chicago*

Judy Richman†
*Northeastern University*

Chris Roberts*
*University of California Los Angeles*

Carmen R. Roman-Shriver*
*Texas Tech University*

Andrew Rorschach*†
*University of Houston*

Connie S. Ruiz
*Lamar University*

Ross Santell*
*Alcorn State University*

Linda Sartor
*University of Pennsylvania*

Peter Schaefer*
*Hudson Valley Community College*

Kevin Schalinske*
*Iowa State University*

Claudia Schopper
*West Virginia University*

Neil Shay
*University of Notre Dame*

Adria Sherman
*Rutgers University*

Sarah Short
*Syracuse University*

Sandra S. Shortt
*Cedarville University*

Brent Shriver*
*Texas Tech University*

Deborah Silverman
*Eastern Michigan University*

Joanne Slavin*
*University of Minnesota*

Margaret K. Snooks
*University of Houston, Clear Lake*

LuAnn Soliah
*Baylor University*

Arlene Spark*
*Hunter College*

Diana-Marie Spillman*†
*Miami University of Ohio*

Catrinel Stanciu
*Louisiana State University*

Carol Stinson
*University of Louisville*

Jon A. Story*
*Purdue University*

Maria Sun
*Southwest Tennessee Community College*

Melanie Taylor
*University of Utah*

Forrest Thye*
*Virginia Tech University*

Carol Turner
*New Mexico State University*

Dhiraj Vattem*†
*Texas State University*

Priya Venkatesan*
*Pasadena City College*

Eric Vlahov
*University of Tampa*

Janelle Walter
*Baylor University*

Dana Wassmer
*Cosumnes River College*

M. K. (Suzy) Weems
*Baylor University*

Lavern Whisenton-Davidson*
*Millersville University*

Mary W. Wilson
*Eastern Kentucky University*

Stacie L. Wing-Gaia
*University of Utah*

Ira Wolinsky
*University of Houston*

Jane E. Ziegler
*Cedar Crest College*

Donna L. Zoss
*Purdue University*

*Denotes editorial board participation.
† Denotes class test participation

# Chapter 1

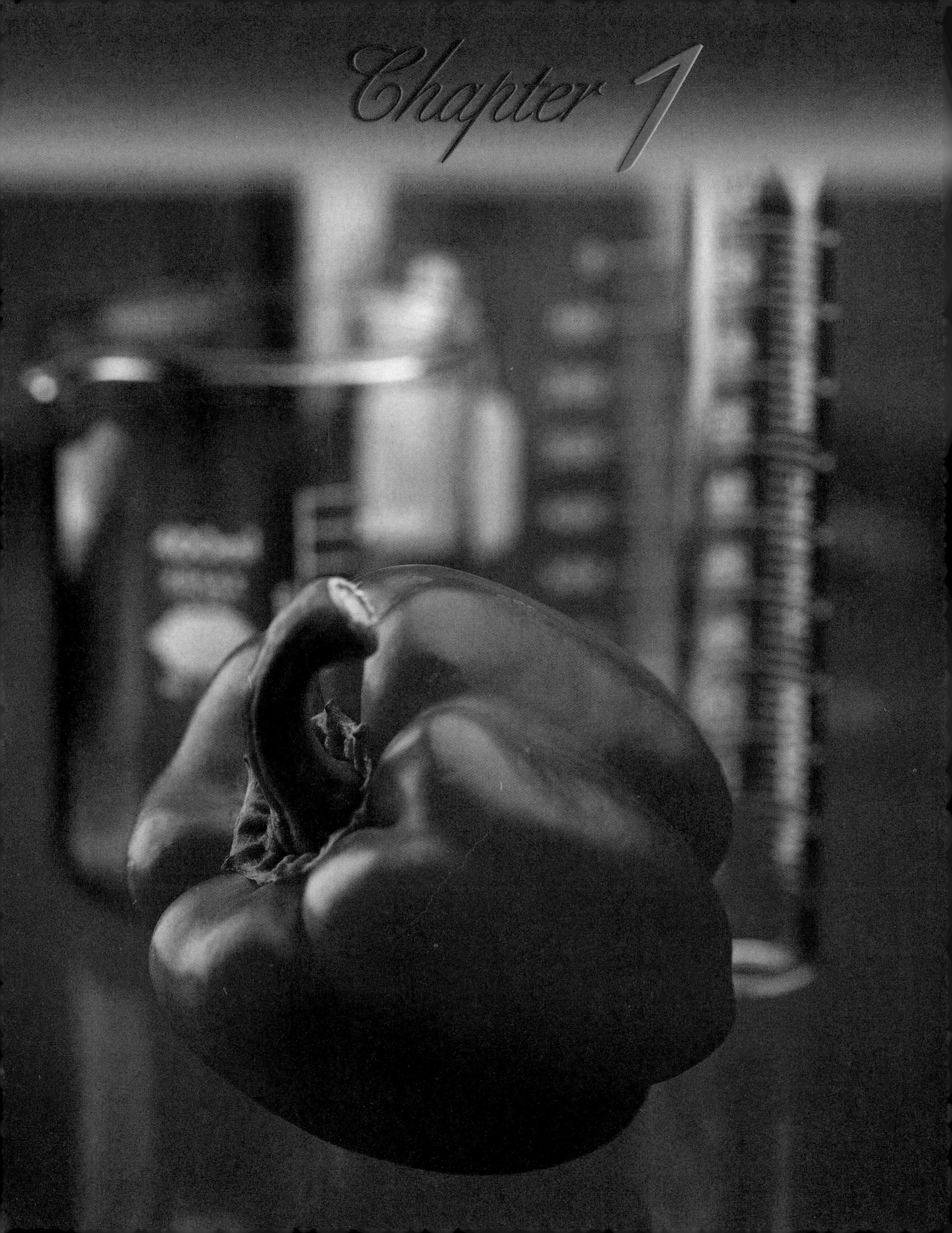

# The Science of Nutrition

*Let thy food be thy medicine and thy medicine be thy food.*
—Hippocrates (460–364 B.C.)

cience is powerful. It helps explain our world, makes it a better place to live, and helps keep us healthy. Yet scientific progress almost always generates considerable debate. Not surprisingly, nutritional discoveries, which have helped prevent and cure disease for centuries, are often met with both excitement and skepticism. For example, one day a newspaper headline announces, "Vitamin A decreases risk of heart disease." A year later, another headline claims, "Vitamin A increases risk of cancer." "You should eat more fish" is followed by "Fish contains dangerous heavy metals." While nutritional science offers great hope for improving health, it also generates intense debate.

Nutrition and its impact on human health are of crucial importance. Whereas nutritional deficiencies were once a major health challenge in the United States, nutritional abundance contributes to many of today's health problems.[1] For example, dietary practices play a large role in predisposing people to obesity, cancer, heart disease, osteoporosis, and type 2 diabetes.[2] This chapter discusses fundamental concepts necessary to understand how good nutrition is basic to health. You will also learn how scientists study nutrition. Without this knowledge, making sound decisions about selecting a healthy diet—based on scientific reason, not rumor—is difficult.

Photos © Matthew Farruggio

1

# Introduction to Nutrition and the Nutrients

The term **nutrition** refers to the science of how living organisms obtain and use food to support all the processes required for their existence. The study of nutrition encompasses a wide variety of scientific disciplines. Some nutritionists are interested in food production and availability, whereas others study why people choose to eat certain things. Still others study cultural influences on food intake and nutritional status. In addition, some scientists study how the body breaks down food and uses it for all the chemical and mechanical processes important for health.

Scientists who study nutrition, called nutritional scientists, can be found in many disciplines, including immunology, medicine, genetics, biology, physiology, biochemistry, education, psychology, and sociology. **Dietitians** are nutrition professionals who help people make dietary changes and food choices to support a healthy lifestyle. Thus the science of nutrition, collectively called the **nutritional sciences,** reflects a broad spectrum of academic and social disciplines.

## Nutrients and Their Classifications

Several basic concepts are central to nutritional science. First is the concept of a **nutrient,** which is a substance in foods used by the body to serve one or more of the following purposes: provide a source of energy, provide structure, or regulate chemical reactions.

It is important to recognize that not all compounds in food are nutrients. To convince yourself of this, you need only examine almost any food label (Figure 1.1). Many of these compounds are not nutrients, because they are not used by the body to support its functions. Foods that contain high levels of nutrients are more nutritious than foods that do not.

**nutrition** The science of how living organisms obtain and use food to support processes required for life.

**dietitian** A nutritionist who works as a clinician, assisting people in making healthy dietary choices.

**nutritional sciences** A broad spectrum of academic and social disciplines related to nutrition.

**nutrient** A substance in foods used by the body for energy, maintenance of body structures, or regulation of chemical processes.

## FIGURE 1.1   Not Everything in a Food Is a Nutrient

Most foods, especially those that are highly processed, contain both nutrients and nonnutrients. Some nonnutrients are added to provide color, flavor, texture, and/or freshness.

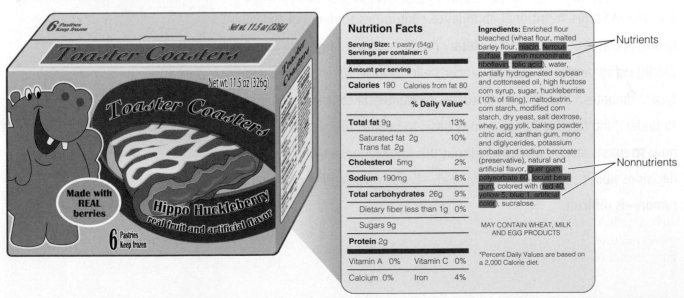

## FIGURE 1.2 Micronutrients versus Macronutrients

Vitamins and minerals are micronutrients, whereas water, carbohydrates, proteins, and lipids are macronutrients.

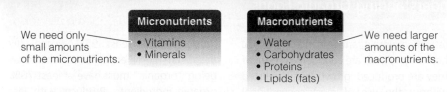

We need only small amounts of the micronutrients.

**Micronutrients**
- Vitamins
- Minerals

**Macronutrients**
- Water
- Carbohydrates
- Proteins
- Lipids (fats)

We need larger amounts of the macronutrients.

## Essential, Nonessential, and Conditionally Essential Nutrients

Although all nutrients in foods *can* be used by the body, we only need to consume some of them. **Essential nutrients** must be obtained from the diet, because the body needs them and either cannot make them at all or cannot make them in adequate amounts. **Nonessential nutrients** are ones the body can make in sufficient amounts when they are needed. We do not need to consume nonessential nutrients. Most foods contain a mixture of essential and nonessential nutrients. For example, milk contains a variety of essential vitamins and minerals as well as several nonessential nutrients. Also, at times a normally nonessential nutrient becomes essential. During these times, it is called a **conditionally essential nutrient.** For example, older children and adults have two essential lipids—lipids that must be obtained through the diet—whereas babies are thought to have at least four. The additional lipids are "conditionally essential" during early life. Certain diseases also cause normally nonessential nutrients to become conditionally essential.

## Macronutrients versus Micronutrients

Nutrients are classified into six categories based on their chemical structure and composition (Figure 1.2). These categories include carbohydrates, proteins, lipids, water, minerals, and vitamins. Water, carbohydrates, proteins, and lipids are called **macronutrients,** because they are needed in large quantities. Vitamins and minerals are called **micronutrients,** because we need only very small amounts of them. For example, a typical adult requires about 2,726 pounds (1,239 kilograms) of the macronutrient protein over the course of a lifetime, but only about 0.3 pound (0.14 kg) of the micronutrient iron.*

## Organic versus Inorganic Nutrients

Nutrients can also be classified as being organic or inorganic. By definition, molecules that contain carbon atoms bonded to hydrogen atoms or other carbon atoms are called **organic compounds.** Carbohydrates, proteins, lipids, and vitamins are chemically organic. Water and minerals are **inorganic** because they do not contain carbon-carbon or carbon-hydrogen bonds. In this way, all foods are considered "organic"—at least in the chemical sense of the term. The term *organic* is also used to describe how a food (plant or animal) is grown and harvested. **Organic foods** are grown without using pesticides, herbicides, or synthetic growth promoters. When a food is labeled as being "certified organic," it has been grown and processed according to U.S. Department of Agriculture (USDA) national organic standards (Figure 1.3). You can read more about organic foods in the Focus on Food feature.

**essential nutrient** A substance that must be obtained from the diet, because the body needs it and cannot make it in required amounts.

**nonessential nutrient** A substance found in food and used by the body to promote health but not required to be consumed in the diet.

**conditionally essential nutrient** Normally nonessential nutrient that, under certain circumstances, becomes essential.

**macronutrients** The class of nutrients that we need to consume in relatively large quantities.

**micronutrients** The class of nutrients that we need to consume in relatively small quantities.

**organic compound** A substance that contains carbon–carbon bonds or carbon–hydrogen bonds.

**inorganic compound** A substance that does not contain carbon–carbon bonds or carbon–hydrogen bonds.

**organic foods** Plant and animal foods that have been grown, harvested, and processed without conventional pesticides, fertilizers, growth promoters, bioengineering, or ionizing radiation.

## FIGURE 1.3 The USDA's "Certified Organic" Seal

---

* Pounds can be converted to kilograms by dividing by 2.2; kilograms can be converted to pounds by multiplying by 2.2.

# Focus On FOOD
## Understanding Organic Foods

With an annual industry growth rate of around 20%, "organic foods" are increasing in popularity. Organically produced foods are now found in most supermarkets. But what makes a food organic, what standards are used to define the term *organic*, and how are these standards enforced? And are there benefits of consuming organic foods?

A chemist understands the term *organic* to generally mean a carbon-containing compound. In this sense, all foods are chemically organic. However, the food industry uses the term *organic* to mean something quite different. In 1992, the U.S. federal government established the National Organic Standards Board (NOSB) to help develop standards for substances to be used in organic production. The NOSB developed the following definition: "Organic agriculture is an ecological production management system that promotes and enhances biodiversity, biological cycles and soil biological activity . . . based on management practices that restore, maintain, and enhance ecological harmony." Foods are considered organic if they are produced, grown, and harvested without the use of most conventional pesticides, fertilizers made with synthetic ingredients, bioengineering, or ionizing radiation. Furthermore, meat, eggs, and dairy products must come from livestock raised without growth hormones and antibiotics.

What assurance do consumers have that foods labeled as "organic" are actually produced according to these guidelines? In 2000, the USDA released new national standards for the production, handling, and processing of organically grown agricultural products. These practices, which were implemented in 2002, ensure that foods labeled as "organic" meet national standards. To learn more about these standards, visit the USDA's National Organic Program website at http://www.ams.usda.gov/nop/indexIE.htm.

Foods with the USDA Organic Seal and labeled as being "100% organic" must have at least 95% organically produced ingredients. Foods labeled as being "organic" must have at least 70% organic ingredients. Products with less than 70% organic ingredients may list specific organically produced ingredients on the side panel of the package but may not make any organic claims on the front of the package. An example is shown in Figure 1.4.

Whether organic foods offer nutritional or safety advantages over other foods remains controversial. In fact, the USDA makes no claims that organically produced food is safer or more nutritious than conventionally produced food, and the labeling of foods as "organic" is not meant to suggest enhanced nutritional quality or food safety. Rather, the difference between organic foods and conventionally produced foods involves the methods used to grow, handle, and process them. Whether these alternative agricultural practices promote enhanced environmental integrity and balance is an area of active debate.

## FIGURE 1.4 Understanding Food Labels of "Organic Products"

| Must have 95–100% certified organic ingredients. | Must have at least 70% certified organic ingredients. | Organic ingredients can be listed on side panel. | No organic claim is being made. |

## Phytochemicals, Zoonutrients, and Functional Foods

As scientists learn more about the relationship between diet and health, they are learning that foods contain more than the traditional nutrients that influence our health. When health-promoting compounds such as these are found in plants, they are called **phytochemicals.** Others, called **zoonutrients,** are found in animal foods. Although phytochemicals and zoonutrients are not essential nutrients, researchers think that many are beneficial to health.

### Phytochemicals

The term *phytochemical* refers to a wide variety of substances found in plants.[3] Although these compounds are not considered essential nutrients, evidence is growing that they can help reduce the risk of developing certain diseases. Many "health claims" on food packaging labels refer to phytochemicals. For example, consuming phytochemicals found in some breakfast cereals may decrease the risk of heart disease. Grapes and wine contain phytochemicals that may reduce the risk of heart disease, and garlic contains phytochemicals that seem to inhibit cancer. You will learn more about these and other phytochemicals throughout this book.

### Zoonutrients

Zoonutrients are compounds uniquely present in animal foods that provide health benefits beyond the provision of essential nutrients and energy.[4] Examples of zoonutrients include a variety of nonessential lipids, found in fish and dairy products, that are thought to decrease risk for heart disease. Another example of a zoonutrient is found in the larval jelly produced by honey bees. This substance is antimicrobial and may reduce the risk of infection.

### Functional Foods

**Functional foods** are those that contain one or more substances such as an essential nutrient, phytochemical, or zoonutrient thought to influence health.[5] Thus consuming these foods may result in more optimal health, often above and beyond what can be attributed to the essential nutrients in them. For example, soy milk is considered a functional food, because it contains a variety of phytochemicals thought to decrease risk of some cancers and osteoporosis. Other examples are milk, which has been shown to be rich in zoonutrients that may lower risks of cancer and high blood pressure, and tomatoes, which may be heart healthy. Although consuming functional foods may improve health, the mechanisms by which this occurs are often not well understood.

Why might scientists consider this plate of spaghetti and glass of wine *functional foods?*

## ESSENTIAL *Concepts*

The study of nutrition represents a wide spectrum of scientific disciplines. This is because what we eat influences every physiological system in the body. The term *nutrition* refers to how the body uses substances in food to promote and sustain life. Some nutrients must come from our foods; these are called essential nutrients. Other nutrients—the non-essential nutrients—can be used by the body, but we can make them when needed. When the body cannot make a typically nonessential nutrient in adequate amounts, the nutrient is considered conditionally essential. Nutrients can also be classified as macronutrients or micronutrients depending on the amount the body needs. Water, carbohydrates, proteins,

**phytochemical** (phy – to – CHEM – i – cal) A substance found in plants and thought to benefit human health (above and beyond the provision of essential nutrients and energy).

**zoonutrient** (zo – o – NU – tri – ent) A substance found in animal foods and thought to benefit human health (above and beyond the provision of essential nutrients and energy).

**functional food** A food that contains an essential nutrient, phytochemical, or zoonutrient and that is thought to benefit human health.

and lipids are macronutrients; vitamins and minerals are micronutrients. The term *organic* is used by chemists to describe most substances that contain carbon, whereas "organic foods" are those that are produced without using synthetic fertilizers, hormones, or other drugs. Phytochemicals and zoonutrients are compounds found in plant and animal foods, respectively, that are not essential nutrients but may improve health. Foods that contain essential nutrients, phytochemicals, or zoonutrients are called functional foods.

# General Functions of the Nutrients: Structure, Regulation, and Energy

Nutrients play numerous roles in the body related to structure, regulation, and energy production. Each class of nutrient consists of many different compounds and contributes to most of these roles in one way or another. These functions are summarized in Table 1.1 and are described briefly next.

## Carbohydrates

Carbohydrates consist of carbon, hydrogen, and oxygen atoms and serve a variety of purposes in the body. Among the many different types of carbohydrates, of primary importance is the role that glucose plays in the body. Most cells use glucose as their primary source of energy, and some rely exclusively on glucose as their source of energy. Carbohydrates are used for many other purposes as well. For example, some are needed to manufacture the genetic material (DNA) that is in every cell of the body. Other carbohydrates play roles in maintaining health of the digestive system and may help decrease risk of certain conditions, including heart disease and type 2 diabetes. Carbohydrates are also important regulatory components in the membranes that surround our cells.

Grains and cereals provide most of the carbohydrates in the diet.

© Polara Studios Inc.

## Protein

Proteins consist primarily of carbon, oxygen, nitrogen, and hydrogen atoms. Some proteins also contain nitrogen atoms, and some contain sulfur atoms. The thousands of proteins in the body play many roles, including serving as

## TABLE 1.1 General Functions of the Major Nutrient Classes

| Nutrient Class | Function[a] | | |
| --- | --- | --- | --- |
| | Structure | Regulation | Energy |
| Carbohydrates | X | X | X |
| Proteins | X | X | x |
| Lipids | X | X | X |
| Water | X | X | x[b] |
| Vitamins | | X | x[b] |
| Minerals | X | X | x[b] |

[a] "X" indicates primary function; "x" indicates secondary function.
[b] These are not energy-yielding nutrients but are involved in the use of carbohydrates, proteins, and lipids for energy (ATP) production.

a source of energy. Proteins also compose the major structural material in many parts of the body, including muscle, bone, and skin. Proteins let us move, allow our complex internal communication system to function, keep us healthy by fulfilling their roles in the immune system, and drive the many chemical reactions needed for life.

## Lipids

Lipids, which include a variety of oils and fats found in foods and the body, generally consist of carbon, oxygen, and hydrogen atoms. They provide large amounts of energy and are important for the structure of cell membranes and for the development and structure of the nervous system and reproductive system. Lipids also regulate a variety of cellular processes.

Meats, dairy products, legumes, eggs, and nuts are good sources of protein.

## Water

Water, made of oxygen and hydrogen atoms, makes up approximately 60% of our total body weight. Its functions are varied and vital. Water acts as the medium in which the body transports nutrients, gases, and waste products. It also serves as the environment in which chemical reactions occur, and is itself involved in many of these reactions. Water is also important in regulating body temperature, protecting our internal organs from damage, and insulating the body.

Lipids found in olives are thought to impart important health benefits.

## Vitamins

The vitamins have a variety of different chemical structures. Although they all contain carbon, oxygen, and hydrogen atoms, some vitamins also contain other substances such as phosphorus and sulfur atoms. Vitamins are involved in regulating body processes as well as in promoting growth and development. Some vitamins also protect the body from the damaging effects of toxic compounds. Vitamins themselves are not used for structure or energy. However, they play important roles in building and maintaining tissues as well as in using the energy in carbohydrates, lipids, and proteins. Vitamins can be classified, based on how they interact with water, as either water soluble (vitamin C and the B vitamins) or fat soluble (vitamins A, D, E, and K). Much research today is focused on the roles of vitamins in preventing and treating diseases such as heart disease and cancer.

Consuming a balanced diet is important when it comes to getting enough of the micronutrients.

## Minerals

Technically speaking, minerals are substances that occur naturally in the earth, such as iron, selenium, and sodium. In the human body, all inorganic substances with the exception of water are called minerals. At least 16 minerals are considered essential nutrients, and each serves its own specific purpose. Some minerals, such as calcium, provide the matrix for various major structural components in the body (such as bone), whereas others, such as sodium, help regulate a variety of body processes (such as water balance). Still other minerals, such as selenium, facilitate chemical reactions. Minerals are not used directly for energy, although many are involved in energy-producing reactions. Scientists are still discovering the importance of many of the minerals in preventing and treating disease.

## Energy

All the nutrient classes are involved in "energy" production in some way. But what do we actually mean by energy? **Energy** is defined as the capacity of a physical system to do work. Thus, if something has energy, it can cause something else to happen.

In terms of nutrition, carbohydrates, proteins, and lipids all contain chemical energy. Cells can transfer the energy from these nutrients into a special substance called **adenosine triphosphate (ATP).** The energy in ATP can then be used to power all the processes needed by the body. Thus carbohydrates, proteins, and lipids are called **energy-yielding nutrients.**

The ability of the body to use energy in nutrients can be compared to the use of wind power to light a room. Wind has mechanical energy. This is evident during a windstorm, when the energy moves and breaks things. However, the wind's energy in this form cannot run an electrical appliance. Its mechanical energy must first be transformed into electrical energy. Similarly, the energy in energy-yielding nutrients must first be transformed into ATP before the body can use it. It is important to recognize that energy per se is not a nutrient. Instead, the body uses energy found in foods to grow, develop, move, and fuel the many chemical reactions required for life.

### What Is a Calorie?

The amount of energy in foods varies and is measured in units called **calories**—the more calories a food has, the more energy (ATP) the body can get from it. Because 1 calorie represents a very small amount of energy, the energy content of foods is typically expressed in units of 1,000 calories or **kilocalories.** This is often abbreviated as "kcalories" or "kcal." In addition, a kilocalorie is sometimes referred to as a **Calorie** (note the capital "C"), as on food labels. One Calorie is equivalent to 1,000 calories or 1 kilocalorie.

### The Caloric Content of Macronutrients and Foods

Carbohydrates and proteins provide approximately 4 kcal per gram (g), whereas lipids provide approximately 9 kcal per gram. Thus, 10 g of a pure carbohydrate or protein would contain 40 (4 × 10) kcal, whereas 10 g of a pure lipid would contain 90 (9 × 10) kcal. Although alcohol is not considered a nutrient, it provides 7 kcal per gram.

The caloric content of a food can be measured in the laboratory using a **bomb calorimeter.** As shown in Figure 1.5, using a bomb calorimeter involves placing the food in an airtight chamber, which is surrounded by water. Oxygen is then pumped into the chamber, and the food is ignited. The energy initially contained in the food's chemical bonds is released as heat, and the heat given off during combustion raises the temperature of the surrounding water. The change in temperature represents the amount of energy originally in the food. A calorie is defined as the amount of heat required to raise the temperature of 1 g of water one degree Celsius. A kilocalorie is the amount of heat required to raise the temperature of 1 kilogram (kg) of water one degree Celsius.

Obviously, people do not use a bomb calorimeter to determine the caloric content of their food. Instead, it is usually estimated mathematically. In the Focus on Food feature you can practice estimating the caloric content of foods.

**energy** The capacity to do work.

**adenosine triphosphate (ATP)** (a – DEN – o – sine tri – PHOS – phate) A chemical used by the body when it needs to perform work.

**energy-yielding nutrient** A nutrient that the body can use to produce ATP.

**calorie** A unit of measure used to express the amount of energy in a food.

**kilocalorie** (kcal, or Calorie) 1,000 calories.

**bomb calorimeter** (cal – o – RIM – e – ter) A device used to measure the amount of energy in a food.

## FIGURE 1.5    Bomb Calorimeter

Bomb calorimetry is used to determine the energy content of a food.

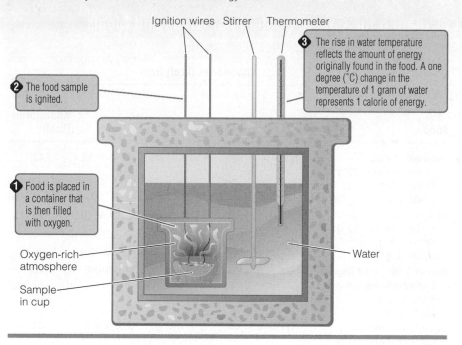

Ignition wires    Stirrer    Thermometer

**3** The rise in water temperature reflects the amount of energy originally found in the food. A one degree (°C) change in the temperature of 1 gram of water represents 1 calorie of energy.

**2** The food sample is ignited.

**1** Food is placed in a container that is then filled with oxygen.

Oxygen-rich atmosphere

Sample in cup

Water

## ESSENTIAL *Concepts*

Carbohydrates, proteins, and lipids all provide energy and are referred to as energy-yielding nutrients. Energy is defined as the capacity to perform work, and the energy found in foods is transformed to usable energy (ATP) within the body. Carbohydrates, proteins, and lipids provide 4, 4, and 9 kcal per gram, respectively. Macronutrients perform a variety of other functions in the body, including those related to structure, movement, growth, and development. Water makes up about 60% of total body weight and serves as the medium in which all chemical reactions occur, helps eliminate waste products, regulates body temperature, and provides insulation and protection. Vitamins are classified as being either water or lipid soluble and serve many purposes. Similarly, many essential minerals have specific functions regarding structure, regulation, and energy use.

# Scientific Method in Nutrition Research

Scientists test theories (including those related to nutrition) in many ways, most of which involve a series of steps collectively called the **scientific method.** Scientific method involves three steps: making an observation, proposing a hypothesis, and data collection.[6] Although the use of scientific method in nutrition research by no means guarantees that what a scientist concludes is the truth, its careful application provides a safeguard that the conclusions are likely valid. For scientific method to be conducted properly, each step has additional conditions that must be considered. These are described next.

**scientific method** Steps used by scientists to explain observations.

## *Focus On* FOOD

### Calculating the Caloric Contents of Foods and Meals

Almost all foods contain energy that the body can use to grow, develop, move, and function. The amount of energy in a food or a meal can be calculated by knowing how many grams of each energy-yielding nutrient is present in each food. To determine the caloric content of a food, you need only multiply the number of grams of carbohydrates, proteins, and lipids by 4, 4, and 9, respectively.

For example, consider the caloric content of a breakfast consisting of oatmeal, low-fat (1%) milk, brown sugar, raisins, and orange juice. The amounts of each energy-yielding nutrient in these foods can be found in the food composition table that accompanies this text and are provided here. The total caloric content of this breakfast is 496 kcal, or 496 Calories. Note that, because of rounding errors and other factors, the total number of kilocalories (or Calories) listed for a food in a food composition table or a food label may differ slightly from the value obtained from calculations. However, these differences are very small.

You can also calculate what percentage of energy comes from each of the energy-yielding nutrient classes. In our example of the oatmeal breakfast, these values can be calculated as follows:

Percent kilocalories from carbohydrates:
    $(396 \div 496) \times 100 = 80\%$

| Food | Kilocalories (kcal) from | | | Total Kilocalories (kcal) |
| --- | --- | --- | --- | --- |
| | Carbohydrates | Protein | Lipids | |
| Oatmeal, 1 cup<br>  Carbohydrates: 25 g<br>  Protein: 6 g<br>  Lipids: 2 g | $25 \times 4 = 100$ | $6 \times 4 = 24$ | $2 \times 9 = 18$ | 142 |
| Milk, 1 cup<br>  Carbohydrates: 12 g<br>  Protein: 8 g<br>  Lipids: 2 g | $12 \times 4 = 48$ | $8 \times 4 = 32$ | $2 \times 9 = 18$ | 98 |
| Brown Sugar, 2 tablespoons<br>  Carbohydrates: 24 g<br>  Protein: 0 g<br>  Lipids: 0 g | $24 \times 4 = 96$ | $0 \times 4 = 0$ | $0 \times 9 = 0$ | 96 |
| Raisins, ½ ounce<br>  Carbohydrates: 11 g<br>  Protein: 0 g<br>  Lipids: 0 g | $11 \times 4 = 44$ | $0 \times 4 = 0$ | $0 \times 9 = 0$ | 44 |
| Orange Juice, 1 cup<br>  Carbohydrates: 27 g<br>  Protein: 2 g<br>  Lipids: 0 g | $27 \times 4 = 108$ | $2 \times 4 = 8$ | $0 \times 9 = 0$ | 116 |
| Totals | 396 | 64 | 36 | 496 |

Percent kilocalories from protein:
    $(64 \div 496) \times 100 = 13\%$
Percent kilocalories from lipids:
    $(36 \div 496) \times 100 = 7\%$

Thus, this meal provides 80, 13, and 7% of its energy from carbohydrates, protein, and lipids, respectively. Note that these percentages should total 100 $(80 + 13 + 7)$. Currently, it is recommended that 45 to 65% of a day's energy come from carbohydrates, 10 to 35% from proteins, and 20 to 35% from lipids.

## Step 1: The Observation

Making an appropriate and accurate observation (Step 1) about an event serves as the framework for the rest of scientific method. If the observation is flawed, the resulting conclusion will likely be flawed as well. Take the observation that there has been an alarming rise in childhood obesity during the past few decades.[7] Where did this information come from? Were the data collected appropriately, are obesity rates the same among all socioeconomic groups, are girls more likely to be obese than boys, and at what age do the rates of obesity increase? Answering these questions helps assure that the observation is complete and accurate.

A careful researcher interested in studying childhood obesity using scientific method must consider all these questions before moving on to the next step of developing an explanation for the observation. This is because the explanation, or hypothesis, for the observation depends on the answers to these questions. Although it is tempting to think that an observation is simply a statement of what is observed, scientists must carefully consider all the available facts and data to assure that their observation is complete and correct.

## Step 2: The Hypothesis

Once an observation has been made, the next step is to explain why the event occurred. Scientists propose **hypotheses** to explain their observations. For example, someone might hypothesize that the increase in childhood obesity is due to a lack of exercise. Another hypothesis might be that childhood obesity is caused by consuming too much fat. Or someone could hypothesize that childhood obesity is caused by both a lack of exercise and too much fat. Regardless as to which is correct, all three are reasonable hypotheses (or explanations) as to why childhood obesity has increased.

### Hypotheses: Predicting Causal Relationships versus Correlations

Scientists can make two different types of hypotheses: those that predict **cause-and-effect relationships** (also called causal relationships) and those that predict **correlations** (also called associations). It is important to understand the difference between these kinds of hypotheses, because they require different kinds of studies to test them. A cause-and-effect relationship describes a situation when we know that one factor (such as excess energy intake) causes another factor (such as weight gain). In other words, factor A (such as excess energy intake) causes factor B (such as weight gain). When the relationship between A and B is causal, we can say that an alteration in A will *cause* an alteration in B.

When two factors are correlated (or associated) with each other, we can say that a change in A is related to a change in B. For example, television watching is associated with weight gain in children. However, it is not the television watching, per se, that causes children to become obese. Instead, other factors associated with television watching (such as snacking and being inactive) actually cause weight gain. Thus, changing television-watching behavior will not necessarily change obesity rates. Understanding the difference between causal relationships and correlations is important in all scientific disciplines, including nutrition. Although many studies are designed to test for correlations, their results are unfortunately interpreted or overstated as proving causal relationships.

### Hypotheses: Predicting Simple Relationships versus Interactions

It is also helpful to differentiate between a **simple relationship** and one that involves an **interaction**. A relationship between two variables is considered a simple relationship if it cannot be altered by other factors. For example, we know that consuming inadequate iron results in iron-deficiency anemia. This fact is not altered by any other variable, such as age, activity level, sex, or other dietary factors.

Most relationships between diet and health are not this simple. For example, high dietary intakes of certain fats can increase the risk of heart disease in some, but not all, people. This relationship is much stronger in individuals with a family history of heart disease. The relationship between the intake of these fats and heart disease is not simple, because it involves an interaction with genetic factors. Fat intake and genetics *interact* to influence risk of heart

**hypothesis** (hy – PO – the – sis) A prediction about the relationship between variables.

**cause-and-effect relationship** (also called causal relationship) When an alteration in one variable causes a change in another variable.

**correlation** (also called association) When a change in one variable is related to a change in another variable.

**simple relationship** A relationship between two factors that is not influenced or modified by another factor.

**interaction** When the relationship between two factors is influenced or modified by another factor.

## FIGURE 1.6  Simple Relationships versus Interactions

Most relationships between nutrition and health are not simple but instead involve interactions with other factors.

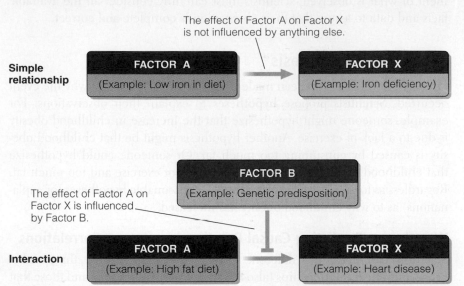

disease. Similarly, the relationship between energy intake and risk of childhood obesity is not a simple relationship, because activity level interacts with energy intake to influence obesity risk. High caloric intake may lead to obesity in inactive, but not physically active, children. Thus it would be inappropriate for researchers to recommend that all children eat less to decrease their risk of obesity. The difference between a simple relationship and an interaction is shown in Figure 1.6.

There are many examples of how **lifestyle factors** (such as diet and exercise), **environmental factors** (such as exposure to pollutants), and **genetic factors** interact to influence health. Typically, scientific research first focuses on simple relationships and then moves on to explore the many interactions that exist among factors. An example is the long-held advice that all people should limit their intake of salt to prevent hypertension (high blood pressure). However, scientists now know that high salt intake only increases the risk of high blood pressure in some people with certain genetic makeups.[8] High salt intake does not influence blood pressure in most people. Current recommendations have been adjusted to reflect this information. It is important to remember that relationships initially thought simple often turn out to involve more interactions than originally believed. This process often leads to changes in dietary recommendations over time.

## Step 3: Data Collection

Although making an accurate observation (Step 1) and developing an appropriate hypothesis (Step 2) are critical to scientific method, these are not nearly as complex as the final step—data collection. Without supporting data, a hypothesis is simply an unproven conjecture—not a scientific finding.

Scientific studies require the scientist to design an appropriate study, carefully conduct the study, and interpret the data correctly. Certain kinds of hypotheses can only be tested using certain kinds of study designs. Furthermore, how the data are collected, analyzed, and interpreted can greatly influence the correctness of the conclusions the investigator draws.

**lifestyle factor** Behavioral component of our lives over which we may or may not have control (such as diet and tobacco use).

**environmental factor** An element or variable in our surroundings that we may or may not have control over (such as pollution and temperature).

**genetic factor** An inherited element or variable in our lives that cannot be altered.

If the study design is flawed, a good observation and hypothesis can be completely wasted.

A basic understanding of experimental practices can help you discern which nutrition claims are bogus and which ones are reasonable. For example, scientists conduct many kinds of studies, and the choice of which design to use depends on the type of hypothesis. In addition, for widely accepted ethical reasons, some studies cannot be conducted on humans. Instead, experimental animals or cell cultures are used. Each of these kinds of studies has its strengths and weaknesses. Although a thorough discussion of experimental design is beyond the scope of this book, a general understanding of the appropriate use of study designs can help you sort nutrition fact from fiction.

## Epidemiologic Studies: Testing for Correlations

When the hypothesis suggests that one factor is correlated or associated with another, an **epidemiologic study** is often conducted (Figure 1.7). In these studies, scientists make observations and record information; they do not actually ask people to change their behaviors, alter their food intake patterns, or undergo any sort of treatment. Epidemiologic studies are used to determine the relationships or correlations among various factors within a large group of people and are not used to test hypotheses predicting causal relationships.

### FIGURE 1.7 Epidemiologic versus Intervention Studies

Epidemiologic studies are used to investigate correlations, whereas intervention studies are used to investigate causal relationships.

Epidemiologic studies are used to examine trends in populations and to test for correlation.

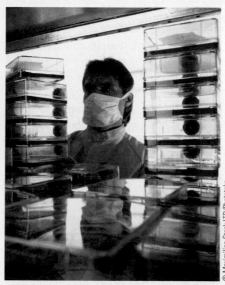

Intervention studies are used to examine causal relationships and include human studies, animal studies, and cell culture studies.

**epidemiologic study** (e – pi – de – mi – o – LO – gic) A study in which data are collected from a group of individuals who are not asked to change their behaviors in any way.

For example, if the hypothesis states, "Dietary fat intake is related to obesity in American children," an epidemiologic study could be conducted to estimate fat intake and body weight in a group of American children. The data would then be analyzed to determine the relationship between these two factors (fat intake and obesity). Nutritional scientists often use epidemiologic studies when there is not sufficient knowledge to put forth a cause-and-effect hypothesis or when conducting another type of study would be impractical or unethical.

***Advantages and Limitations of Epidemiologic Studies***   As with all study designs, there are both positive aspects and limitations to epidemiologic studies. Epidemiologic studies can be used to explore the complex interactions that exist among genetic, environmental, and lifestyle factors. Often these interactions cannot be studied adequately in other types of experiments. An example of an important epidemiologic study is the Framingham Heart Study, which was initiated in the 1940s to investigate the relationship between nutrition and heart health. This study has resulted in the publication of hundreds of scientific articles and continues to provide important information about how genetic, lifestyle, and environmental factors influence long-term health. You can read more about this study in the Focus on Diet and Health feature.

Limitations of epidemiologic studies include the fact that their results cannot be interpreted as proving a relationship to be causal. In addition, it is important to recognize that subjects studied in a particular epidemiologic study may not be representative of the entire population. For example, were only men studied? Were any minority groups included in the study? These considerations are important when determining whether the study's findings are applicable to the entire population or only a portion of it.

## *Focus On* DIET AND HEALTH
### The Framingham Heart Study

An example of an important, large-scale epidemiologic study that has changed how we view the relationship between lifestyle factors and heart disease is the Framingham Heart Study. With growing evidence that heart disease was becoming a major public health problem in the 1940s, the U.S. government began this large-scale study to investigate what factors were related to risk of this disease. In 1948 the U.S. Public Health Service (PHS) together with the Boston University School of Medicine began to study 5,209 healthy residents of Framingham, Massachusetts, to document lifestyle, environmental, and genetic factors that might lend insight as to how heart disease could best be prevented and treated.

The importance of the information gained from the Framingham Heart Study cannot be overstated. Before Framingham, scientists did not recognize the relationship between serum cholesterol and risk of heart disease. Now all programs designed to prevent and treat heart disease include a major emphasis on reducing serum cholesterol. In addition, before Framingham people did not know cigarette smoking was a risk factor in developing heart disease. However, the study demonstrated that smokers were at increased risk of developing this disease and that the risk was related to the number of cigarettes smoked each day. Thus, data from this epidemiologic study have changed how we think about the relationship between lifestyle factors (such as nutrition and smoking) and risk of heart disease.

Over 1,000 peer-reviewed publications have resulted from the Framingham Heart Study, which was expanded in 1971 to study children of the original study participants. This continuation will for years to come enable the medical community to gain valuable clues as to how many diseases can be prevented and treated.

## Intervention Studies: Testing for Causal Relationships

When a hypothesis suggests a causal relationship, the scientist has several options for the type of experiment to use. In most cases, the researcher does not design an epidemiologic study but instead conducts an **intervention study.** These studies can be carried out using humans, experimental animals, or cell culture systems, as illustrated in Figure 1.7. When the study is conducted on people, it is called a human intervention study.

***The Use of a Control Group*** In contrast to epidemiologic studies, intervention studies require that participants be exposed to some sort of treatment or intervention. Usually, some participants receive the treatment, and others do not. Participants who do not receive the treatment or intervention are said to be in a **control group.** A control group is needed to determine whether the effects seen in the treatment group are actually caused by the treatment or are caused by some other aspect of the study. For example, researchers may test the hypothesis that nutrition education decreases obesity in children. They may have some children attend a nutrition education class (the treatment group), whereas others do not (the control group). The researchers would then measure whether the children receiving the treatment gained less weight than children in the control group.

Intervention studies comparing a treatment group to a control group are scientifically powerful, because their results can provide evidence that the relationship between two factors is causal in nature. A major strength of a human intervention study is that the results can be directly applied to humans. For example, if nutrition education in the previously described study resulted in weight loss, we can assume this effect will be true for other children as well. However, the same treatment may not have similar effects in adults. Furthermore, if participants in this study were all enrolled in a private sports academy, we might question whether the same results would be found in other children. Therefore, as with epidemiologic studies, before extending the conclusions to the general population one must be sure the study participants are representative of the population of interest.

***Controlling for Hawthorne Effect, Placebo Effect, and Researcher Bias***
Although conducting a human intervention study has many benefits, researchers must consider several factors. For example, just being in a study can influence a person's behaviors. This phenomenon in which study participation causes a person to behave differently is called the **Hawthorne effect.** Another phenomenon, called the **placebo effect,** occurs when an observable effect of the treatment seems to arise just because the individual *expects* or *believes* the treatment will work. If the Hawthorne effect or placebo effect occurs in a study, the conclusions researchers draw may not be accurate. In addition, sometimes the scientist conducting the study can inadvertently influence the outcome by knowing which subjects are receiving the treatment and which ones are not. This is called **researcher bias.** Scientists use several techniques to decrease the chances of Hawthorne effect, placebo effect, and researcher bias, as described next.

***Random Assignment to Group and Confounding Variables*** To conduct a valid study, the researcher must minimize the Hawthorne effect, the placebo effect, and researcher bias. One way to avoid these problems is by **random assignment** of participants to the treatment or control groups. This is important, because it distributes possible **confounding variables** equally among study

**intervention study** An experiment in which something is altered or changed to determine its effect on something else.

**control group** A group of people, animals, or cells in an intervention study that does not receive the experimental treatment.

**Hawthorne effect** Phenomenon in which study results are influenced by alteration of something that is not related to the actual study intervention.

**placebo effect** (pla – CE – bo) The phenomenon in which there is an apparent effect of the treatment because the individual expects or believes that it will work.

**researcher bias** When the researcher influences the results of a study.

**random assignment** When study participants have equal chance of being assigned to each experimental group.

**confounding variable** A factor, other than the one of interest, that might influence the outcome of an experiment.

groups. Confounding variables are factors other than the ones of interest that might influence the outcome of the study. For example, consider our study designed to test whether nutrition education influences childhood obesity. Random assignment of children to either treatment or control group would help ensure that confounding variables (such as activity level) are equally distributed in both groups. All research reports should state whether participants were randomly assigned to study group.

The unintentional influence of confounding variables can also be minimized by excluding certain study participants from being in an experiment. For example, smoking can influence the birth weight of a baby. Therefore, if investigators wanted to study the relationship between caffeine intake during pregnancy and birth weight, they might want to exclude smokers from the study. This is especially important in nutrition studies, because of the many interactions among genetic, lifestyle, and environmental factors.

***Blinded and Placebo-Controlled Studies*** Another important technique that minimizes the Hawthorne effect, placebo effect, and researcher bias is the "blinding" of the experimenter and/or participant so that neither knows to which group the participant has been assigned. One way to "blind" the study is to give the participants in the control group something that looks, smells, and tastes just like the treatment. This "fake" or imitation treatment is called a **placebo.** For example, a sugar pill could be given instead of a vitamin supplement. The use of a placebo helps assure that changes seen in the treatment group are actually due to the treatment and not to the placebo effect.

When the researchers, but not the participants, know who is in the treatment and placebo groups, the study is said to be a **single-blind study.** It is best if neither the scientist nor the participant knows to which group the participant has been assigned. When neither knows, the study is called a **double-blind study. Randomized, double-blind, placebo-controlled** intervention studies are considered to have the ideal experimental design to test a hypothesis about a causal relationship (Figure 1.8).

***Animal and Cell Culture Studies*** **Animal studies** can provide important data that cannot be collected from humans. Using animals in nutrition research has many advantages. For example, the genetic variability among laboratory

**placebo** A "fake" treatment given to the control group that cannot be distinguished from the actual treatment.

**single-blind study** A human experiment in which the participants do not know to which group they have been assigned.

**double-blind study** A human experiment in which neither the participants nor the scientists know to which group the participants have been assigned.

**randomized, double-blind, placebo-controlled study** The type of experiment that is considered to be the ideal research design for testing a cause-and-effect hypothesis.

**animal study** The use of experimental animal subjects such as mice, rats, or primates.

## FIGURE 1.8  The Ideal Nutrition Intervention Study

Randomized, double-blind, placebo-controlled intervention studies are considered the "gold standard" in nutrition research.

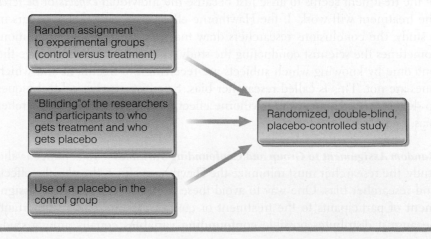

animals is much less than the genetic variability among people. In addition, scientists can more easily study interactions in animal studies, as compared to human studies. This is because researchers can carefully control all aspects of an animal's environment, as compared to a human's environment. Although animal studies have advantages, there are also limitations. For example, one must always question whether the data collected from these studies are applicable to humans. It is not always possible to test whether the results of an animal study are valid in humans. However, sometimes a series of convincing animal studies can lay the foundation for scientists to move forward and conduct appropriate follow-up human intervention trials.

Sometimes scientists want to study causal relationships at the cellular level. In these situations, researchers can use **cell culture systems,** which are specific types of cells that can be grown in laboratory conditions. For example, certain breast, adipose (fat), skin, and muscle cell culture systems can be grown in the laboratory to study specific diseases. In this way, scientists worldwide can use identical cell culture systems.

Using cell culture systems provides scientists with a powerful tool to study what is happening inside a cell in response to a particular nutrient or treatment. Cell culture systems are examples of what scientists call *in vitro* **systems,** meaning that the scientists are studying natural phenomena in an environment that is outside of a living organism. In contrast, human and animal studies are considered *in vivo* **systems,** meaning that they involve the study of natural phenomena within living organisms. However, applying data from cell culture systems to human health has limitations. First, cells that grow readily in the laboratory are usually not representative of normal, healthy cells. In addition, cells within the body interact with other cells; these interactions cannot be studied with a cell culture system made up of a single kind of cell.

## ESSENTIAL *Concepts*

Most research is conducted using a three-step process called the scientific method. Scientific method involves making an observation, generating an explanation (or hypothesis), and testing the explanation by conducting a study. With each of these steps, several considerations must be made. For example, one must be careful that the observation is valid, the hypothesis should be well grounded in scientific evidence or theory, and the study should be appropriate for the hypothesis. Epidemiologic studies are conducted to investigate associations or correlations, whereas intervention studies can test causal relationships. Scientists use many techniques to decrease bias in studies. These include randomization, control groups, placebos, and blinding. Sometimes it is not possible or practical to test a hypothesis using humans as participants. In these cases, researchers turn to animal models or cell cultures.

## Making Sound Decisions About Nutrition Claims

Our understanding about the relationship between nutrition and health is continually changing. It is the nature of science to be a long and winding road of discoveries. Therefore, you should not be surprised when dietary recommendations change over time—this is expected. Nevertheless, it is important to be able to evaluate nutritional claims. Use the checklist in Figure 1.9 to

**cell culture system** Specific type of cells that can be grown in the laboratory and used for research purposes.

*in vitro* Involving the use of cells or environments that are not part of a living organism.

*in vivo* Involving the study of natural phenomena in a living organism.

**FIGURE 1.9** Evaluating Nutrition Claims

**Checklist for evaluating nutrition claims**

| YES | NO | |
|---|---|---|
| ✓ | ☐ | Published in peer-reviewed journal? |
| ✓ | ☐ | Research conducted by a scientist? |
| ☐ | ✓ | Did funding source bias results? |
| ✓ | ☐ | Was experiment appropriate? |
| ✓ | ☐ | Are nutrition claims supported by reputable organizations? |

**primary (information) source** The publication in which a scientific finding was first published.

**peer-reviewed journal** A publication that requires a group of scientists to read and approve a study before it is published.

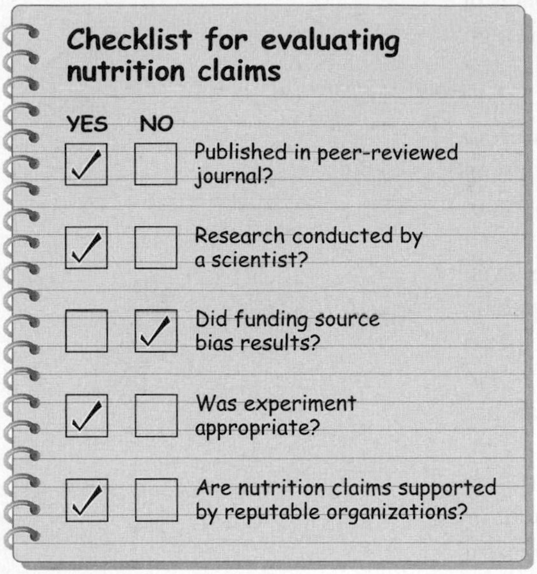

help you in this process, and keep these important questions in mind when you make nutrition-related decisions.

## Where Did the Information Come From?

One consideration in determining whether nutrition claims are reputable is the source from which the claims are obtained. As you can imagine, not all sources of information have the same credibility, and it is often difficult to judge the validity of nutrition-related claims.

Instead, you must determine the **primary source** of the information; in other words, where it was first reported or published. In general, believable primary sources of information are **peer-reviewed journals** or publications put out by reputable organizations. Having a paper published in a peer-reviewed journal means that the paper was read and "approved" by a group of scientists knowledgeable in that area of study. Table 1.2 provides a list of some reputable peer-reviewed journals that publish nutrition-related articles. In addition, many private and governmental organizations publish credible information concerning diet and health. These agencies are considered some of the most reliable and unbiased sources of information available to the public and to the scientific community. Table 1.2 provides contact information for some of these groups as well.

When you are evaluating whether nutrition claims are valid, it is important to find out where the information was originally published. As a rule, question new nutrition claims that have not been published in a peer-reviewed journal or other reputable publication. Often you need not go any further than this step to determine that a nutrition claim is not even worth considering.

**TABLE 1.2** Some Reliable Sources of Nutrition Information

| Peer-Reviewed Journals | Government and Private Agencies |
|---|---|
| American Journal of Clinical Nutrition | American Cancer Society (http://www.cancer.org) |
| Annals of Nutrition and Metabolism | American Diabetes Association (http://www.diabetes.org) |
| Annual Review of Nutrition | American Dietetic Association (http://www.eatright.org) |
| Appetite | American Heart Association (http://www.americanheart.org) |
| British Journal of Nutrition | American Medical Association (http://www.ama-assn.org) |
| Clinical Nutrition | Centers for Disease Control and Prevention (CDC; http://www.cdc.gov) |
| European Journal of Nutrition | Mayo Clinic (http://www.mayoclinic.org) |
| Journal of Human Nutrition and Dietetics | National Academy of Sciences (http://www.nas.edu) |
| Journal of Nutrition | National Institutes of Health (NIH; http://www.nih.gov) |
| Journal of the American College of Nutrition | NIH Office of Dietary Supplements (http://www.ods.od.nih.gov) |
| Journal of the American Dietetic Association | U.S. Department of Agriculture (USDA; http://www.usda.gov) |
| JAMA (Journal of the American Medical Association) | U.S. Food and Drug Administration (FDA; http://www.fda.gov) |
| Journal of Pediatric Gastroenterology and Nutrition | |
| Lancet | |
| Nature | |
| New England Journal of Medicine | |
| Nutrition | |
| Nutrition Research | |
| Public Health Nutrition | |
| Science | |
| Scientific American | |

## Who Conducted the Research?

Asking "Who conducted the research?" is the next step in evaluating a nutrition claim. In general, most reputable nutrition research is conducted by researchers at universities or medical schools. Researchers at a variety of private and public institutions and organizations conduct reliable nutrition research as well. It is important that the individuals conducting the research were qualified and knowledgeable, and finding out where they work and what their qualifications are can help you make this determination.

## Who Paid for the Research?

There are many ways that scientists receive money to fund their research, but most involve applying for grant money from private companies or state or federal agencies. In most cases, the source of the money does not influence the outcome of the study, even if the funding agency has something to gain or lose by the study results. For example, consider research designed to study the effect of milk consumption on weight gain in children. The dairy industry might be interested in funding such a study. This arrangement poses no ethical problems whatsoever, as long as the results are not biased by what the dairy industry would like the research to conclude. However, in some circumstances the funding agency may be able to influence the findings. It is very difficult to determine if this is a problem in a particular study, but it is something to keep in mind when evaluating the validity of a nutrition claim.

## Was the Experimental Design Appropriate?

Once you have determined that a nutrition claim (1) has been published in a reputable journal or report, (2) was conducted by a qualified researcher, and (3) was not biased by the funding source, you are ready to consider the research itself. In other words, was the research conducted in a way that was appropriate to test the hypothesis? And do the conclusions fit the study design?

One of the best ways to find out details about a study comes from the U.S. National Library of Medicine, which supports a searchable biomedical database called **PubMed.** This database allows easy access to over 11 million biomedical journal citations and can be found at http://www.ncbi.nlm.nih.gov/PubMed. You can use details found at PubMed to ask important questions about a study. Was it an epidemiologic study or a human intervention trial? Do the results suggest an association or cause-and-effect relationship? Was an appropriate control group used? Was the study double-blinded? Was a placebo used in the control group? With your knowledge of study design, you can evaluate the research itself and determine whether the nutrition claim is likely to be valid.

## Do Major Public Health Organizations Agree with the Conclusions?

Because even the best experiment does not always provide conclusive evidence that a particular nutrient influences health in a certain way, public health experts usually wait for the results of several studies before they begin to make overall claims about the effect of a certain nutrient on health. Thus, it is advisable to determine whether a claim is supported by major public

**PubMed** A computerized database that allows access to approximately 11 million biomedical journals.

health organizations. For example, you could check whether the American Heart Association supports a claim concerning nutrition and heart disease, or whether the American Cancer Society supports a claim concerning nutrition and cancer.

## ESSENTIAL *Concepts*

When it comes to nutrition claims, separating fact from fiction can be difficult. However, by doing some investigative work, it is possible to largely determine the validity of any nutrition claim. Publication in a peer-reviewed journal indicates the information is probably reliable, and most reputable research is conducted by people affiliated with a research university or health-related agency. Although most research is not influenced by the source of funding, consider whether the funding agency might have biased the conclusions made by the investigators. Also, it is important to consider whether the design of the study was appropriate and whether major public health groups support the study conclusions.

# Nutrition and Health: An Overview

Nutrition plays an important role in health and disease. Indeed, both consuming too little or too much of a nutrient can cause illness. Although nutrient deficiencies remain serious public health issues in developing parts of the world, the major nutrition issues in the United States have shifted from concern about not getting enough nutrients to concern about overconsumption and poor food choices. At the same time, the leading illnesses have shifted from infectious disease and nutritional deficiencies to those related to overnutrition and old age. Nutrition is considered by many health professionals to be one of the most important lifestyle factors influencing our nation's health.

## Assessing Overall Health of a Population

To study the relationship between nutrition and health, it is important to understand some basic concepts concerning how a population's "health" is assessed. In other words, what do scientists and other health professionals measure when they want to determine whether a population is becoming more or less healthy?

There are many measures of a population's health, but the most frequently used are morbidity and mortality rates. A **rate** is a measure of some event, disease, or condition in relation to a specific population group, along with some specification of time. For example, speed is expressed as a rate (such as miles per hour). **Morbidity rates** assess illness over time, whereas **mortality rates** assess deaths over time.

### Mortality Rate

An example of a mortality rate is **infant mortality rate,** defined as the number of infant deaths (<1 year of age) per 1,000 live births in a given year. Infant mortality rates are often used to assess the health and well-being of a society, as they reflect a complex web of environmental, social, economic, medical, and technological factors that interact to influence overall health. Infant mortality rates in the United States have dropped dramatically in the past century, as shown in Figure 1.10.[9] Many factors influence infant mortality rate, including

**rate** A measure of something within a specific period of time.

**morbidity rate** (mor – BID – i – ty) The number of illnesses in a given period of time.

**mortality rate** (mor – TAL – i – ty) The number of deaths in a given period of time.

**infant mortality rate** The number of infant deaths (< 1 year of age) per 1,000 live births in a given year.

**FIGURE 1.10** Changes in Life Expectancy and Infant Mortality Rate from 1900 to 2000

Since 1900, life expectancy has increased while the infant mortality rate has decreased. These shifts indicate greater health in the society.

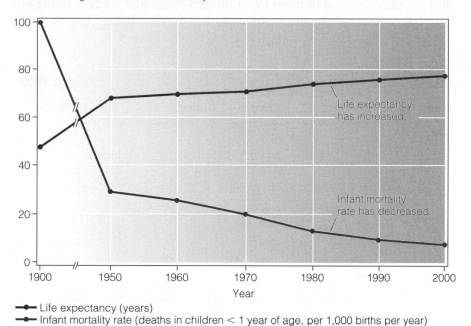

Life expectancy has increased.

Infant mortality rate has decreased.

→● Life expectancy (years)
→● Infant mortality rate (deaths in children < 1 year of age, per 1,000 births per year)

SOURCES: Adapted from Centers for Disease Control and Prevention. Health, United States, 2004. Available from: http://cdc.gov/nchs/data/hus/hus04trend.pdf. And from Centers for Disease Control and Prevention. Achievements in Public Health 1990–1999: Healthier Mothers and Babies. Morbidity and Mortality Weekly Report. 1999;48:849–58.

genetic factors, access to medical care, substance abuse, adequate maternal nutrition, and weight gain during pregnancy.

Mortality rates can also be assessed for various diseases, such as heart disease, cancer, stroke, and diabetes. These rates are expressed as the number of deaths from a particular disease per 100,000 people in a given year. Because of a combination of factors, U.S. mortality rates from various diseases have shifted dramatically over the past century.[10] This shift—and especially how it relates to nutrition—is described in more detail in the next section.

## Morbidity Rates: Incidence and Prevalence

Whereas mortality rates assess the number of *deaths* in a given period of time, morbidity rates reflect *illness* in a given period of time. Morbidity rates can be expressed as either incidence or prevalence. The **incidence** of a disease is the number of people who are newly diagnosed with the disease in a given period of time, whereas the **prevalence** is the total number of people with a particular disease in a given period of time. Like mortality rates, morbidity rates (incidence and prevalence) in the United States have changed drastically during the last few decades. For example, in 2002 tuberculosis incidence had declined for the 10th consecutive year, whereas prevalence of diabetes had increased.[11]

## Life Expectancy

Another indicator of societal health is **life expectancy,** which is defined as the average number of years of life remaining to a person at a particular age. The life expectancy of a person at birth is also called **longevity.** Average life expec-

**incidence** The number of people who are newly diagnosed with a condition in a given period of time.

**prevalence** The total number of people who have a condition in a given period of time.

**life expectancy** A statistical prediction of the average number of years of life remaining to a person at a specific age.

**longevity** Life expectancy at birth.

tancy in the United States has increased dramatically over the past century.[12] Today, an American baby can expect to live an average of 77 years, whereas a 75-year-old person can expect to live another 12 years (to the age of 87). There are many reasons for this, one of which is better nutrition.[13] This aging trend also has important implications for nutritional issues that are now facing our population. For example, the major public health issues facing our population today are diseases that take long periods of time to develop.

**FIGURE 1.11  Five Leading Causes of Death in 1902, 1950, and 2002**

Chronic diseases such as heart disease, cancer, and stroke have replaced infectious diseases such as tuberculosis, pneumonia, and diarrhea as our leading causes of death.

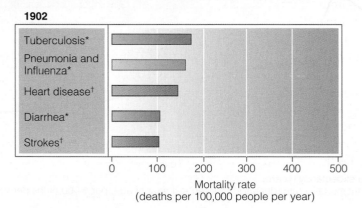

**1902**

Mortality rate
(deaths per 100,000 people per year)

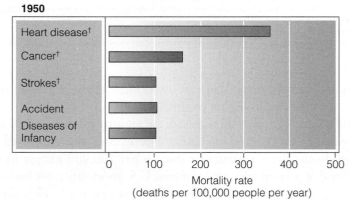

**1950**

Mortality rate
(deaths per 100,000 people per year)

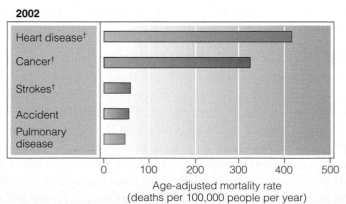

**2002**

Age-adjusted mortality rate
(deaths per 100,000 people per year)

*Infectious disease
†Chronic disease

SOURCES: Data for 1902 and 1950: Centers for Disease Control and Prevention. Leading Causes of Death, 1900–1998. Available from: http://www.cdc.gov.data/dus/lead1900_98.pdf. Data for 2002: Centers for Disease Control and Prevention. Health, United States, 2004. Available from: http://cdc.gov/nchs/data/hus/hus04trend.pdf.

## The Shift from Infectious Disease to Chronic Disease

During the early 1900s, the major causes of death in the United States were infectious diseases (Figure 1.11).[14] **Infectious diseases** are caused by pathogens (such as bacteria, viruses, fungi, parasites, or other microorganisms) and are contagious. Infectious diseases such as pneumonia, tuberculosis, and influenza were rampant during the early part of the 20th century, accounting for one third of the nation's deaths. In addition to infectious disease, nutritional deficiencies contributed to the high morbidity and mortality rates in the early 1900s.[15]

Public health efforts such as clean water, sewage disposal, the discovery of antibiotics, and childhood vaccination programs helped to reduce the incidence of many infectious diseases.[16] Figure 1.12 shows the decrease in mortality from infectious disease in relation to the timing of some of these accomplishments in health promotion. Improved nutritional status during the past century has also led to better health. As a combined result of decreased infectious disease, better nutrition, and other advances in health care, infant mortality rates have steadily decreased and life expectancy rates have increased faster than any other time in U.S. history. Often referred to as the "graying of America," increasing longevity greatly influences our nation's health.[17]

For example, because of the graying of America the United States is faced with the rising incidence of what are called chronic diseases. **Chronic diseases,** which are noninfectious illnesses that develop slowly and persist for a long time, include such conditions as heart disease, cancer, osteoporosis, diabetes, and stroke. These diseases are now the nation's major causes of disability and death.[18]

## Risk Factors Related to Health and Disease

Unlike infectious diseases, chronic diseases are not prevented by vaccines or cured by medication. Although many chronic diseases can be treated, they are largely caused by what people do, and in some cases, what people do not do. Lifestyle and environmental factors as well as genetic characteristics that

Today, people can expect to live longer than ever before.

---

### FIGURE 1.12  Infectious Disease and Public Health Accomplishments in the United States from 1900 to 2000

Many public health accomplishments have led to decreased rates of death from infectious disease in the past century.

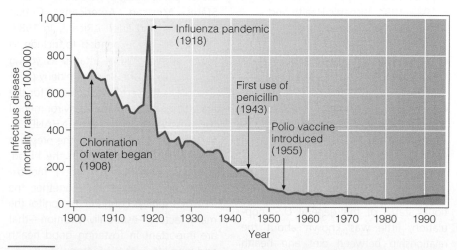

SOURCES: Adapted from Armstrong BL, Conn LA, Pinner RW. Trends in Infectious Disease Mortality in the United States During the 20th Century. JAMA (Journal of the American Medical Association). 1999;281:61–6. And from Centers for Disease Control and Prevention. Achievements in Public Health, 1900–1999: Control of Infectious Diseases. Morbidity and Mortality Weekly Report. 1999;48:621–9.

**infectious disease** A contagious illness caused by a pathogen such as a bacteria, virus, or parasite.

**chronic disease** A noninfectious disease that develops slowly and persists over time.

are related to a person's health and risk of developing chronic diseases are called **risk factors.** For example, the major lifestyle risk factors associated with heart disease, cancer, and stroke—the top three leading causes of death in the United States—include tobacco use, lack of physical activity, and a range of poor dietary habits. There are also risk factors related to genetics and environment that can influence health. These include things such as sex (gender) and age as well as pollution, ambient temperature, and exposure to sunlight.

One of the major risk factors for many of the chronic diseases is being overweight or obese. Nationwide and worldwide obesity is becoming a health crisis of epidemic proportions.[19] The shift from undernutrition to overnutrition or unbalanced nutrition as societies adopt more industrialized economies—called the **nutrition transition**—is strongly related to many of the chronic diseases facing us today.[20]

Because research strongly suggests that poor nutrition (such as overeating) is a major risk factor for chronic disease, the U.S. government has invested significant funds to monitor and evaluate the relationship between diet and health. In particular, the **National Health and Nutrition Examination Survey (NHANES)** has provided vast amounts of information concerning nutrition and health trends for over 30 years. The NHANES study is described in more detail in the Focus on Diet and Health feature.

## Why Studying Nutrition Is More Important Than Ever

In the United States, chronic disease and health problems caused, at least in part, by poor diet represent one of the most serious public health issues. Researchers estimate that 65% of adults ($\geq$20 years of age) are either over-

**risk factor** A lifestyle, environmental, or genetic factor related to a person's chances of developing a disease.

**nutrition transition** The shift from undernutrition to overnutrition or unbalanced nutrition that often occurs simultaneously with the industrialization of a society.

**National Health and Nutrition Examination Survey (NHANES)** A federally funded epidemiologic study begun in the 1970s to assess trends in diet and health in the U.S. population.

---

# *Focus On* DIET AND HEALTH
## The National Health and Nutrition Examination Surveys

By the late 1960s, public health professionals were becoming concerned with the growing number of people suffering from chronic diseases such as cancer, heart disease, and osteoporosis. Researchers were beginning to discover links between dietary habits and disease. In response to this, the National Center for Health Statistics (NCHS)—a branch of the U.S. Public Health Service—initiated a large, ongoing epidemiologic study to simultaneously monitor nutrition and health in the U.S. population. This was called the National Health and Nutrition Examination Survey (NHANES), and five such surveys of this type have been conducted.

- 1971–1975: National Health and Nutrition Examination Survey I (NHANES I)

- 1976–1980: National Health and Nutrition Examination Survey II (NHANES II)
- 1982–1984: Hispanic Health and Nutrition Examination Survey (HHANES)
- 1988–1994: National Health and Nutrition Examination Survey III (NHANES III)
- 1999–present: National Health and Nutrition Examination Survey IV (NHANES IV)

Although both the NHANES I and NHANES II studies provided extensive information about the health and nutritional status of the general U.S. population, little was known about the relationship between diet and health among the many ethnic groups. In response to this, HHANES was con-

ducted in the 1980s to study the health and nutritional status of the three largest Hispanic subgroups in the United States—Mexican Americans, Cuban Americans, and Puerto Ricans. In 1988, NHANES III was expanded to include all major ethnic groups. Infants as young as 1 month of age and the elderly (>74 years of age) were also included for the first time. In addition, environmental concerns were addressed in NHANES III, such as pesticide exposure, the presence of certain trace elements in the blood, and the amounts of carbon monoxide present in the blood. Together, the NHANES surveys continue to monitor the many factors—especially nutrition—that are important in fostering good health and long life in the U.S. population.

weight or obese and that over 280,000 deaths each year are due to obesity alone. More than 64 million Americans now have cardiovascular disease (the leading cause of death), 50 million have high blood pressure, and over 1 million have type 2 diabetes. In addition, cancer accounts for 25% of all deaths in the United States annually. Consuming a healthy balance of nutrients plays a major role in preventing each of these conditions. As the incidence and prevalence of these diseases increase, it is becoming more important than ever that we pay attention to what we eat throughout our lifetimes. As the old saying goes, "An ounce of prevention is worth a pound of cure."

# ESSENTIAL *Concepts*

Nutrition is an important factor in assuring optimal health. Over the past century, the primary public health concerns have shifted from infectious disease and nutritional deficiencies to chronic disease and overnutrition. These shifts are reflected in morbidity and mortality rates of various diseases as well as life expectancy. Whereas infant mortality rates and mortality from infectious disease have decreased, life expectancy has increased. This "graying of America" has been accompanied by a phenomenon called the nutrition transition, which is the shift from undernutrition to overnutrition or unbalanced nutrition as a society becomes more industrialized. The National Health and Nutrition Examination Surveys (NHANES) are a series of federally funded, epidemiologic studies begun in the 1970s and designed to collect information related to nutrition and health in the United States. Many studies suggest that poor dietary practices are associated with a growing prevalence of chronic disease. Thus understanding the relationship between diet and health has never been more important.

## Review Questions    Answers are found in Appendix G.

1. Which of the following is *not* considered an essential nutrient?
   a. Lipids
   b. Water-soluble vitamins
   c. Phytochemicals
   d. Minerals

2. Which macronutrient constitutes the largest proportion of body weight?
   a. Carbohydrates
   b. Proteins
   c. Lipids
   d. Water

3. In which order, from left to right, does scientific method proceed?
   a. Hypothesis, observation, data collection
   b. Observation, data collection, hypothesis
   c. Data collection, observation, hypothesis
   d. Observation, hypothesis, data collection

4. Epidemiologic studies are best used to test hypotheses related to _____.
   a. associations
   b. animal studies
   c. causal relationships
   d. chemical mechanisms

5. A factor other than the one of interest that might influence the outcome of a study is called a _____.
   a. placebo
   b. researcher bias
   c. confounding variable
   d. *in vitro* factor

6. Infant mortality rate is defined as the number of infant deaths occurring in the first
   a. two years of life.
   b. year of life.
   c. year of life per 100,000 live births.
   d. year of life per 1,000 live births.

7. The _____ of a disease is the number of new cases in a given year.
   a. mortality rate
   b. prevalence
   c. incidence
   d. longevity

8. Life expectancy has _____ in the United States during the past century.
   a. increased
   b. decreased
   c. not changed

9. Poor nutrition is considered a(n) _____ risk factor for many chronic diseases.
   a. lifestyle
   b. environmental
   c. genetic
   d. all of the above

10. Which of the following is considered a chronic disease?
    a. Pneumonia
    b. Osteoporosis
    c. Tuberculosis
    d. Influenza

11. Describe what is meant by an essential nutrient, a nonessential nutrient, and a conditionally essential nutrient.

12. What is a randomized, double-blind, placebo-controlled study, and why is this considered the "gold standard" in nutrition research?

13. Explain what is meant by "nutrition transition." What nutrition implications are related to the nutrition transition? What might happen if they were *not*?

## Practice Calculations   Answers are found in Appendix G

1. If a serving of food contains 50 g of carbohydrate, 25 g of protein, and 5 g of lipid, how many kilocalories does the serving contain? What are the percentages of calories from the three macronutrient classes for this food?

2. Using a food composition table, calculate the percent kilocalories (% kcal) from the three macronutrient classes for your favorite meal. In this exercise, include a beverage, a fruit, a vegetable, and a meat, dairy, or bean product.

## Media Links   A variety of study tools for this chapter are available at our website, www.thomsonedu.com/nutrition/mcguire.

Prepare for tests and deepen your understanding of chapter concepts with these online materials:
- Practice tests
- Flashcards
- Glossary
- Student lecture notebook
- Web links
- Animations
- Chapter summaries, learning objectives, and crossword puzzles

## Notes

1. Carpenter KJ. A short history of nutritional science: Part 1 (1785–1885). Journal of Nutrition. 2003;133:638–45. Carpenter KJ. A short history of nutritional science: Part 2 (1885–1912). Journal of Nutrition. 2003;133:975–84. Carpenter KJ. A short history of nutritional science: Part 3 (1945–1985). Journal of Nutrition. 2003;133:3023–32. Carpenter KJ. A short history of nutritional science: Part 4 (1945–1985). Journal of Nutrition. 2003;133:3331–42.

2. Allison DB, Fontaine KR, Manson JE, Stevens J, VanItallie, TB. Annual deaths attributable to obesity in the United States. JAMA (Journal of the American Medical Association). 1999;282:1530–38. Bray GA. Medical consequences of obesity. Journal of Clinical Endocrinology and Metabolism. 2004;89:2583–9.

3. Liu RH. Potential synergy of phytochemicals in cancer prevention: mechanism of action. Journal of Nutrition. 2004;134:3479S–85S.

4. Ward RE, German JB. Zoonutrients and health. Food Technology. 2003;57:30–36.

5. Lock AL, Bauman DE. Modifying milk fat composition of dairy cows to enhance fatty acids beneficial to human health. Lipids. 2004;39:1197–206.

6. Carey SS. A beginner's guide to scientific method, 3rd ed. Belmont, CA: Wadsworth/Thomson Learning; 2004.

7. Hedley AA, Ogden CL, Johnson CL, Carroll MD, Curtin LR, Flegal KN. Prevalence of overweight and obesity among US children, adolescents, and adults, 1999–2002. JAMA (Journal of the American Medical Association). 2004;291:2847–50. Wang Y, Monteiro C, Popkin BM. Trends of obesity and underweight in older children and adolescents in the United States, Brazil, China, and Russia. American Journal of Clinical Nutrition. 2002;75:971–7. National Center for Health Statistics (NCHS). Health, United States, 2005. Hyattsville, MD: Public Health Service; 2005. Available from: http://www.cdc.gov/nchs/hus.htm.

8. Beeks E, Kessels AG, Kroon AA, van der Klauw MM, de Leeuw PW. Genetic predisposition to salt-sensitivity: a systematic review. Journal of Hypertension. 2004;22:1243–9.

Healthier mothers and babies. Morbidity and Mortality Weekly Report. 1999;48:849–58.

9. Centers for Disease Control and Prevention. Achievements in public health, 1900–1999. Hyattsville, MD: Public Health Service; 2004. Available from: http://www.cdc.gov/nchs/hus.htm.

10. Centers for Disease Control and Prevention. Leading causes of death, 1900–1998. Available from: http://www.cdc.gov/nchs/data/dvs/lead1900_98.pdf.

11. National Center for Health Statistics (NCHS). Health, United States, 2005. Hyattsville, MD: Public Health Service; 2005. Available from: http://www.cdc.gov/nchs/hus.htm.

12. National Center for Health Statistics (NCHS). Health, United States, 2005. Hyattsville, MD: Public Health Service; 2005. Available from: http://www.cdc.gov/nchs/hus.htm.

13. Perls T, Terry D. Understanding the determinants of exceptional longevity. Annals of Internal Medicine. 2003; 139:445–9.

14. Centers for Disease Control and Prevention. Achievements in public health, 1900–1999; Control of infectious diseases. Morbidity and Mortality Weekly Report. 1999;48:621–9. Armstrong GL, Conn LA, Pinner RW. Trends in infectious disease mortality in the United States during the 20th century. JAMA (Journal of the American Medical Association). 1999;281:61–6.

15. Carpenter KJ. A short history of nutritional science: Part 2 (1885–1912). Journal of Nutrition. 2003;133:975–84.

16. Centers for Disease Control and Prevention. Achievements in public health, 1900–1999: Control of infectious diseases. Morbidity and Mortality Weekly Report. 1999;48:621–9.

17. Anderson RE, Smith RD, Benson ES. The accelerated graying of American pathology. Human Pathology, 1991; 22:210–4.

18. National Center for Health Statistics (NCHS). Health, United States, 2005. Hyattsville, MD, Public Health Service; 2005. Available from: http://www.cdc.gov/nchs/hus.htm.

19. Hedley AA, Ogden CL, Johnson CL, Carroll MD, Curtin LR, Flegal KN. Prevalence of overweight and obesity among US children, adolescents, and adults, 1999–2002. JAMA (Journal of the American Medical Association). 2004;291:2847–50. Wang Y, Monteiro C, Popkin BM. Trends of obesity and underweight in older children and adolescents in the United States, Brazil, China, and Russia. American Journal of Clinical Nutrition. 2002;75:971–7. National Center for Health Statistics (NCHS). Health, United States, 2005. Hyattsville, MD: Public Health Service; 2005. Available from: http://www.cdc.gov/nchs/hus.htm. James PT, Leach R, Kalamara E, Shayeghi M. The worldwide obesity epidemic. Obesity Research. 2001;9:2228S–33S.

20. Popkin BM. The nutrition transition and obesity in the developing world. Journal of Nutrition. 2001;131:871S–3S. Popkin BM, Gordon-Larsen P. The nutrition transition: worldwide obesity dynamics and their determinants. International Journal of Obesity and Related Metabolic Disorders. 2004;28:S2–9.

**Check out the following sources for additional information.**

21. American Cancer Society. Cancer facts and figures, 2006. Atlanta, GA: American Cancer Society; 2006. Available from: http://www.cancer.org.

22. American Heart Association. Heart and stroke statistics—2006 update. Dallas, TX: American Heart Association: 2006. Available from: http://www.americanheart.org.

23. Cordain L, Eaton SB, Sebastian A, Mann N, Lindeberg S, Watkins BA, O'Keefe JH, Brand-Miller J. Origins and evolution of the Western diet: health implications for the 21st century. American Journal of Clinical Nutrition. 2005;81:341–54.

24. Kleiber M. The fire of life—an introduction to animal energetics, 2nd ed. Huntington, NY: Robert E. Krieger Publishing; 1975.

25. Willett, W. Lessons from dietary studies in Adventists and questions for the future. American Journal of Clinical Nutrition. 2003;78:539S–43S.

## Chapter 2

# Assessing Nutritional Status and Guidelines for Dietary Planning

ood and eating make up a large part of the cultures in which we live. For some, food is plentiful and associated with good times—family gatherings, social occasions, and special events. For others, food is limited, and getting enough to eat is a great concern. In either case, food and the rituals around eating are integral to the fabric of our cultures.

On a basic level, food is simply a vehicle for nutrients, which provide the building blocks and energy for all the body's structures and functions. Our health deteriorates when we consume too little or too much of certain nutrients for long periods of time. Getting appropriate amounts of each nutrient is important for health and well-being. The decisions we make about the food we eat every day are important, and we need to consider our choices thoughtfully. How do we know if we are consuming enough, but not too much, of all the required nutrients? How can we choose foods to meet our nutrient needs?

Learning the answers to these questions requires a basic understanding of several important concepts, such as nutritional adequacy, nutritional status, dietary assessment, and diet planning. In this chapter, you will learn the fundamentals of these concepts and will learn about federal and private programs that have developed regulations and guidelines to help us stay healthy by choosing foods wisely.

Photos © Matthew Farruggio

# Nutritional Status and Malnutrition

Poor nutrition can lead to suboptimal physiological function and poor health. Consuming too little of a nutrient is called **undernutrition,** and can cause a **nutritional deficiency,** which can be serious and sometimes fatal. Conversely, consuming too much of some nutrients—called **overnutrition**—can also be unhealthy. Overnutrition of energy can lead to obesity and its related health consequences, whereas overconsumption of some vitamins and minerals, called a **nutritional toxicity,** can be fatal. Undernutrition and overnutrition make up the extreme ends of what is called the **nutritional status** continuum, and both are examples of **malnutrition.** The relationship between dietary intake and physiological function (or health) is illustrated in Figure 2.1.

## Primary versus Secondary Malnutrition

There are two types of malnutrition, which are differentiated by cause—primary malnutrition and secondary malnutrition. **Primary malnutrition** is due to inadequate diet, whereas **secondary malnutrition** is caused by other factors. For example, a person may be deficient in one of the B vitamins because he or she consumes too little vitamin-rich fruits and vegetables (this is primary malnutrition) or may be deficient in the B vitamins because an illness interferes with its absorption (secondary malnutrition).

## Nutritional Adequacy

**Nutritional adequacy** occurs when a person consumes the amount of a nutrient that is required to meet physiological needs. However, what is adequate for one person may be inadequate for another. Although the science is inexact,

**undernutrition** (or **nutritional deficiency**) Inadequate intake of one or more nutrients and/or energy.

**overnutrition** Overconsumption of one or more nutrients and/or energy.

**nutritional toxicity** Overconsumption of a nutrient resulting in dangerous (toxic) effects.

**nutritional status** The health of a person as it relates to how well his or her diet meets that person's individual nutrient requirements.

**malnutrition** Poor nutritional status caused by either undernutrition or overnutrition.

**primary malnutrition** Poor nutritional status caused strictly by inadequate diet.

**secondary malnutrition** Poor nutritional status caused by factors such as illness.

**nutritional adequacy** The situation in which a person consumes the required amount of a nutrient to meet physiological needs.

**FIGURE 2.1  Nutritional Status as Related to Nutrient Intake and Level of Function**

Dietary intake can determine nutritional status. Both overnutrition and undernutrition can cause suboptimal physiological functioning.

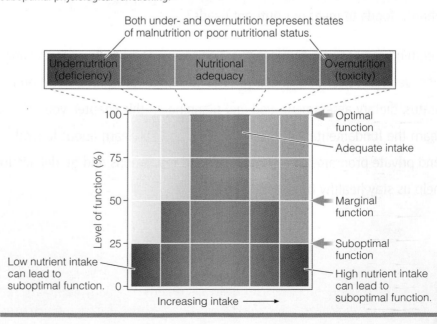

there are many techniques, guidelines, and tools that are used to determine an individual's nutritional status and dietary adequacy and to help people choose foods wisely. These are described in the next section.

## ESSENTIAL *Concepts*

A person's nutritional status depends on whether sufficient amounts of nutrients are available to support optimal physiological function. Nutritional adequacy occurs when nutrient intake is adequate (but not too much) to provide for nutrient needs. Nutritional adequacy supports optimal physiological function, whereas under- and overnutrition do not. Both undernutrition and overnutrition are forms of malnutrition. Primary malnutrition is due to inadequate diet, whereas secondary malnutrition may be caused by other factors such as illness.

# The ABCDs of Nutritional Status Assessment

There are many ways to evaluate nutritional adequacy—for example, dietary assessment. The other components of nutritional status assessment include anthropometric measurements, biochemical measurements, and clinical assessment. Collectively, these are the "**ABCDs**" of nutritional status assessment, and each of these assessment methods is described next.

## Anthropometric Measurements

**Anthropometric measurements** assess physical dimensions and body composition. Examples of anthropometric measurements include height, weight, bone mass, and head circumference. These measurements vary with age (and sometimes with sex and race) and nutritional status. Anthropometry is particularly useful in assessing nutritional status in circumstances when chronic malnutrition is likely, providing important information about overall nutritional status.

### Height and Weight

Height and weight are the most common anthropometric measures. Weight is often used to assess a person's risk for certain chronic diseases, such as heart disease and type 2 diabetes. This is because being overweight or obese can lead to many health problems, especially as we age. Weight is also used to assess nutritional adequacy during pregnancy as well as during the progression of disease. Changes in height can indicate nutritional status in the elderly, especially with regard to bone mass.

Height and weight are also used to assess nutritional status in infants and children, and are typically measured shortly after birth and throughout childhood. Health care workers compare a particular anthropometric measurement at a given time point, to what is called a reference standard. **Reference standards** are values that reflect the normal range for a particular parameter in a healthy population. In Chapter 14 you can learn more about the use of reference standards during infancy and childhood.

### Body Composition

Body weight is the sum of the lean mass (protein), body fat, water, and bone mass. The distribution of these components is known as **body composition,** which can be indicative of a person's nutritional status. Many methods are

**"ABCDs" of nutritional status assessment** Four components of assessing nutritional status: anthropometric measurements, biochemical measurements, clinical assessment, and dietary assessment.

**anthropometric measurements** Measurements or estimates of physical aspects of the body such as height, weight, circumferences, and body composition.

**reference standard** A value that represents what would be expected in a healthy population of similar age and sex.

**body composition** The distribution of fat, lean mass (muscle), and minerals in the body.

used to assess body fatness, but the most common of these is **body mass index (BMI).** BMI is calculated as weight (in kilograms) divided by height squared (in square meters).* BMI values are highly correlated with body fat content, and both very low and high BMIs are related to malnutrition and increased risk for disease.

Clinicians and researchers use many other methods to estimate body fat as well. Another common measurement is **skinfold thickness.** Because much of the body's fat lies directly below the skin in subcutaneous fat, the thicker the skinfold, the more fat is present. Note that many of these methods can be used to estimate lean body mass as well. Other indicators of body composition are waist circumference, hip circumference, and waist-to-hip ratio (calculated as waist circumference divided by hip circumference). Increases in these measurements typically indicate increased body fat content. These anthropometric measurements can also show where fat is *distributed* in the body. Because they are easy to measure, circumferences are often assessed in the clinical setting. Chapter 9 provides more detail concerning these and other indicators of body fat and lean mass.

Bone mineral content and bone mineral density are additional anthropometric measurements used to assess nutritional status—especially in regard to minerals, such as calcium. One instrument used to measure bone mineral content is the **dual-energy x-ray absorptiometer (DEXA or DXA),** which uses very low amounts of x-rays to produce a detailed picture of the various components of the body, including the bone. DEXA machines can also be used to determine body fat and lean mass. The use of DEXA is described in more detail in Chapter 9 and in the Nutrition Matters following Chapter 13.

### Advantages and Disadvantages of Anthropometry

Many anthropometric measurements are easy and inexpensive to obtain and give important information concerning nutritional status. Thus anthropometry is routinely used in both clinical and research settings. If you use a bathroom scale, you are taking an anthropometric measurement.

However, using anthropometry alone to assess nutritional status has limitations. For example, many nonnutritional factors such as disease, genetics, and physical activity level can make it difficult to interpret these measurements. An athlete with increased muscle mass may have higher than "normal" body weight and BMI because of long hours of physical training—this is not from overnutrition. Even when nutrient deficiency is suspected to be causing an abnormal anthropometric measure, it is virtually impossible to determine which nutrient is problematic. Thus anthropometric measurements alone are not considered diagnostic and must be supported by other measures of nutritional status.

## Biochemical Measurements

To assess health and nutritional status, **biochemical measurements** are often used. These involve laboratory analysis of a biological sample, such as blood or urine. In some cases the sample is analyzed for a specific nutrient. In others, it is analyzed for an indicator, called a **biological marker** or biomarker, that reflects the nutrient's function. Often more than one biochemical measurement is used to assess a person's nutritional status.

**body mass index (BMI)** An indicator of body fatness calculated as weight (kg) divided by height squared (m²).

**skinfold thickness** An anthropometric measurement of body fatness.

**dual-energy x-ray absorptiometry (DEXA or DXA)** A method of measuring body composition.

**biochemical measurement** Laboratory analysis of biological samples, such as blood and urine, used in nutritional assessment.

**biological marker (biomarker)** A measurement in a biological sample such as blood or urine that reflects a nutrient's function.

---

* To calculate weight in kilograms from weight in pounds, divide by 2.2. To calculate height in meters from height in feet, divide by 3.3.

As with anthropometric measurements, there are advantages and disadvantages of using biochemical measurements. On the positive side, biochemical measurements can help diagnose a specific nutrient deficiency or excess, whereas anthropometric measurements cannot. However, collecting and analyzing biological samples often require technical expertise and costly procedures. In addition, many factors such as time of day, age, sex, activity patterns, and use of certain drugs can influence the level of nutrients or their biomarkers in blood or urine. Nutritionists and other health care providers must carefully consider these factors when interpreting biochemical measurements.

## Clinical Assessment

Nutritional status is often evaluated during a clinical assessment. During this procedure, the clinician or researcher takes a **medical history** to obtain information about previous disease, weight loss and/or gain, surgeries, medications, and other relevant information such as family history. Specifically, the health care provider asks the patient whether he or she has been experiencing any unusual **symptoms** such as lack of energy, blurred vision, or loss of appetite. Symptoms are phenomena an individual experiences as being somewhat abnormal. Because they are not observable by others and can only be reported by the patient, symptoms are often overlooked.

The medical history is typically followed by a physical examination to determine whether there are visible signs of health problems. **Signs** of illness are different from symptoms, because signs can be seen by other people. Some signs may indicate poor nutritional status, such as skin rashes, swollen ankles (edema), or bleeding gums. Physical examinations almost always include anthropometric measurements such as height and weight as well.

Clinical assessment has advantages; for example, it is the only way that health care providers can ask questions concerning symptoms of malnutrition. Furthermore, signs of some extreme forms of malnutrition are very distinct, and observation of them can make clinical diagnosis of a particular nutrient deficiency or toxicity quite accurate. However, because the signs and symptoms of many nutrient deficiencies are not apparent until they become severe, clinical assessment may not be adequate when malnutrition is more moderate in nature.

## Dietary Assessment

Although anthropometric, biochemical, and clinical assessments can be used to assess nutritional status, it is also important to examine what an individual eats. There are several **dietary assessment methods** used to collect information about a person's nutrient intake. **Retrospective methods** require the person to remember what he or she ate in the past. **Prospective methods** require a person to keep track of what he or she eats for one or more days. These methods are described next.

### Retrospective Methods: Recalls and Questionnaires

The two main retrospective dietary assessment methods are the 24-hour recall method and the food frequency questionnaire. In the **24-hour recall method,** a trained person interviews the participant and records everything that person has eaten or drunk in the previous 24 hours. Food models and standard household measuring tools are often used to help achieve accuracy. However,

**medical history** Questions asked to assess overall health.

**symptoms** Manifestations of disease that cannot be seen by others, such as stomach pain or loss of appetite.

**signs** Physical indicators of disease that can be seen by others, such as pale skin and skin rashes.

**dietary assessment** The evaluation of a person's dietary intake.

**retrospective dietary assessment** Type of dietary assessment that assesses previously consumed foods and beverages.

**prospective dietary assessment** Type of dietary assessment that assesses present food and beverage intake.

**24-hour recall** A retrospective dietary assessment method that analyzes each food and drink consumed over the previous 24 hours.

because 24-hour recalls are based on a single day and require the person to remember specific information, they often do not represent the person's usual food intake.

Another retrospective dietary assessment method is the **food frequency questionnaire,** which typically elicits information concerning food intake *patterns*, especially over an extended period of time. For this method, a person completes a questionnaire that includes a list of foods and questions about frequency and amount consumed. The person completing the questionnaire reads through the list, indicating the frequency at which he or she consumes each food and what the typical serving size is. This is illustrated in Figure 2.2. However, it is not possible to list all foods, and it is sometimes difficult to construct questionnaires that are culturally appropriate for particular ethnic groups. It is also difficult to quantify nutrient and energy intakes using food frequency questionnaires. Thus information from a food frequency questionnaire, although useful, is limited in accuracy and completeness.

## Prospective Methods: Diet Records

Although retrospective methods of dietary assessment are relatively simple, their usefulness depends on a person's memory. To more accurately assess a person's diet, it is better to record foods and beverages while the person is consuming them. This method is called keeping a **diet record,** or food record. Portion sizes are either estimated using standard household measurements (such as tablespoon or cup) or by weighing the food. Clinicians consider the diet record to be one of the most accurate methods of dietary assessment. The steps involved in conducting a diet record are outlined in the Focus on Clinical Applications feature.

**food frequency questionnaire** A retrospective dietary assessment method that assesses food selection patterns over an extended period of time.

**diet record** A prospective dietary assessment method that requires the individual to write down detailed information about foods and drinks consumed over a specified period of time.

---

### FIGURE 2.2 Sample Questions on a Food Frequency Questionnaire

Food frequency questionnaires can be used to assess nutritional status. This example might be used as part of a questionnaire designed to assess dairy foods intake.

| Hot chocolate made with lowfat/reduced fat milk: | Evaporated milk: | Half-and-half: |
|---|---|---|
| **Times per day** | **Times per day** | **Times per day** |
| ☐ Never or almost never | ☐ Never or almost never | ☐ Never or almost never |
| ☑ Less than once | ☐ Less than once | ☐ Less than once |
| ☐ 1 | ☑ 1 | ☐ 1 |
| ☐ 2–3 | ☐ 2–3 | ☑ 2–3 |
| ☐ 4–5 | ☐ 4–5 | ☐ 4–5 |
| ☐ Over 5 | ☐ Over 5 | ☐ Over 5 |
| **Usual amount consumed** | **Usual amount consumed** | **Usual amount consumed** |
| ☐ 1/2 cup | ☑ 1–2 tbsp. | ☑ 1–2 tbsp. |
| ☑ 1 cup | ☐ 3–4 tbsp. | ☐ 3–4 tbsp. |
| ☐ 1 1/2 cup | ☐ 1/2 cup | ☐ 1/2 cup |
| ☐ 2 cups | | |
| ☐ 2 1/2 cups or more | | |

## *Focus On* CLINICAL APPLICATIONS

### Tips for Compiling an Accurate and Representative Diet Record

Keeping and analyzing a diet record is one of the most accurate ways to determine the adequacy of one's diet. However, if not compiled correctly the diet record can be inaccurate and misleading. Some basic techniques and tips can help assure that a diet record is as accurate as possible.

- *Detail is important.* It is important to include as much detail, such as brand names and preparation methods, as you can for all foods and beverages that are consumed. For mixed dishes (also called composite foods), estimate the amount of each ingredient. An example is "salad," which might include lettuce, tomatoes, eggs, and so on. The more detail that is included in the diet record, the more accurate the analysis can be.

- *Estimate or measure serving sizes.* Ideally, weigh or measure foods using standard household devices such as measuring cups or spoons. If this is not possible, estimate serving sizes as carefully as possible. Some restaurants provide detailed information concerning the amounts of food served, and such detail can be very helpful.

- *Record dietary information for several days.* Because what we eat varies from day to day, it is important to keep diet records for more than a single day. Preferably, compile food records for three days, including one weekend day.

- *Choose representative, "normal" days.* Choose days that are representative of typical eating patterns, and avoid "special" days such as holidays and birthdays.

- *Do not change your normal eating patterns.* When keeping a diet record people often alter normal eating patterns to be more convenient and/or healthful. Resist this temptation, as it is especially important that diet records reflect normal intake.

- *Avoid times of illness or stress.* Being sick or under unusual stress can influence food preferences and overall intake. It is best to avoid these times when collecting a diet record.

- *Be prepared.* Carrying your food record with you at all times lets you more accurately record everything you eat and drink. You never know when you are going to want to stop for a cup of coffee or have a snack with a friend.

## Food Composition Tables and Computerized Dietary Analysis Software

The next step in conducting a dietary assessment is to determine the nutrient and energy content of the diet. Information concerning the nutrient content of thousands of foods is readily available in two basic forms: **food composition tables** and **computerized nutrient databases.** Food composition tables, such as the one that accompanies this book, can be accessed on the U.S. Department of Agriculture (USDA) website (http://www.ars.usda.gov/nutrientdata).

Using food composition tables can be time consuming and tedious. Fortunately, easy-to-use, computerized nutrient databases are available. For example, the USDA offers a dietary analysis tool online with its MyPyramid food guidance system. This feature, called MyPyramid Tracker, can be accessed on the Internet (http://www.mypyramid.gov) and is described in more detail later in this chapter. Commercial dietary assessment software programs are also available. Computerized nutrient databases make it relatively easy for anyone to analyze his or her diet.

## ESSENTIAL *Concepts*

Nutritional status can be assessed in several ways, including anthropometric measurements, biochemical measurements, clinical assessment, and dietary assessment. Anthropometric measurements include measures of body dimensions and composition. However, biochemical analyses of blood and/or urine samples can provide more detailed information about

**food composition table** Tabulated information concerning the nutrient and energy contents of foods.

**computerized nutrient database** Software that provides information concerning the nutrient and energy contents of many foods.

nutrient status. Clinical assessment involves conducting a medical history and physical examination to check for signs and symptoms of malnutrition. In addition, dietary assessment is often used to estimate nutrient and energy intake. Such methods include diet recalls, food frequency questionnaires, and diet records. After collecting dietary data, a person can estimate nutrient and energy intake using food composition tables or dietary assessment software.

# Dietary Reference Standards: Dietary Reference Intakes (DRIs)

It is not enough to simply *know* how much nutrients and energy a person consumes; dietary assessment must go one step further and determine whether these amounts are likely to be adequate. To this end, the Institute of Medicine has developed a set of nutritional standards to assess the adequacy of a person's diet.

The Institute of Medicine is a division of the National Academy of Sciences, which is a private (nongovernmental) organization of distinguished scholars dedicated to furthering science and general welfare. The Institute of Medicine has established a set of dietary reference standards called the **Dietary Reference Intakes (DRIs)**. This umbrella term is used to describe a set of four groups of values called the Estimated Average Requirements (EARs), Recommended Dietary Allowances (RDAs), Adequate Intake levels (AIs), and the Tolerable Upper Intake Levels (ULs). These are illustrated in Figure 2.3. The DRIs also include calculations for Estimated Energy Requirements (EERs), which can be used to assess whether energy intake

**Dietary Reference Intakes (DRIs)** A set of four types of nutrient intake reference standards used to assess and plan dietary intake; these include the Estimated Average Requirements (EARs), Recommended Dietary Allowances (RDAs), Adequate Intake levels (AIs), and the Tolerable Upper Intake Levels (ULs).

## FIGURE 2.3 Dietary Reference Intake (DRI) Standards

There are four sets of DRI standards: EARs, RDAs, AIs, and ULs. Each set of DRI standards is used for a different purpose.

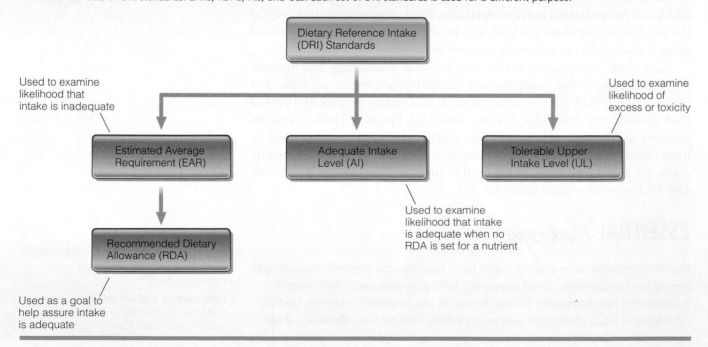

is sufficient, and Acceptable Macronutrient Distribution Ranges (AMDRs), which provide a recommended distribution of the macronutrients in terms of energy consumption. The DRIs and AMDRs are provided inside the covers of this book and are discussed in detail here. The EERs are also discussed.

## A Brief History of Nutrient Recommendations

Before examining each of the DRIs in detail, it is useful to learn about the history of nutrition recommendations and standards in the United States. The dietary standards, published in 1943 by the National Academy of Sciences, were called the RDAs and were used as a goal for good nutrition. At that time, malnutrition in the United States was generally due to undernutrition and nutritional deficiencies were common. Thus the RDAs were designed primarily to prevent nutrient deficiencies. The RDAs were revised approximately every five years, with the last (10th) edition being released in 1989. However, as knowledge of nutrition improved it became clear that the RDAs had some limitations. Thus, in 1994 nutritional scientists proposed revising the "former RDAs" into the DRIs.

Like the former RDAs, the currently used DRIs were established by highly qualified scientists and represent our best knowledge of recommended intake for all the essential nutrients. Because nutrient requirements differ by sex, age, and life stage (such as pregnancy and lactation), DRI values are differentiated by each of these variables. Note that both Canada and the United States recognize the DRIs as their "official" set of dietary reference standards.

To understand how the DRIs were established, it is important to first look at the term **nutrient requirement.** A nutrient requirement is the amount of a nutrient that must be consumed to prevent deficiencies. This amount results in nutritional adequacy. Many factors influence a person's nutrient requirements, including age, sex, genetics, medications, and many other environmental influences. Note that the DRIs are only estimates of average nutrient requirements in a healthy population—your level may be less than or greater than the average. The following sections discuss each of these dietary reference standards in detail and how to use them in dietary assessment.

## Estimated Average Requirement (EAR)

The **Estimated Average Requirement (EAR)** for a particular nutrient is the intake value thought to meet the requirement of half the healthy individuals in a particular life stage and sex. In other words, if a woman consumes the EAR value for a particular nutrient, she is consuming the amount that meets the requirement of about half (50%) of all women in the same age group. This concept is illustrated in Figure 2.4.

### Establishing an EAR

To see how the EARs were set, consider an example, such as iron. The EARs for iron were determined by a panel of micronutrient experts who first reviewed the functions of iron in the body as well as the potential outcomes of poor iron status throughout the life cycle. Then they selected the most appropriate measure that could be used to assess iron status during each life stage. When appropriate, they also distinguished between males and females. For example, they estimated the average iron requirements for older infants by taking into account increasing blood volume and increasing iron storage,

**nutrient requirement** The lowest chronic intake level of a nutrient that supports a defined level of nutritional status in a particular individual.

**Estimated Average Requirement (EAR)** The amount of a nutrient that meets the physiological requirements of half (50%) the healthy population of similar individuals.

**FIGURE 2.4** **Estimated Average Requirements (EARs) and Recommended Dietary Allowances (RDAs)**

EARs represent the intake values needed by about half of a healthy population, whereas RDAs represent the intake values needed by about 97% of a healthy population.

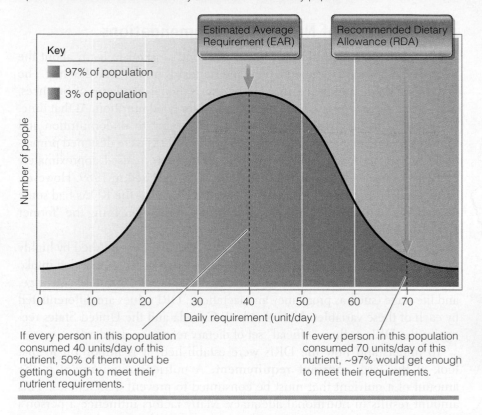

If every person in this population consumed 40 units/day of this nutrient, 50% of them would be getting enough to meet their nutrient requirements.

If every person in this population consumed 70 units/day of this nutrient, ~97% would get enough to meet their requirements.

which are important during this period of life. For adult women, the panel took into account iron loss from menstruation. Then the members established EARs, using mathematical equations that accounted for all known factors and predicted the amount of iron needed to meet the requirements of half the population. The EARs are very useful in research settings to evaluate whether a group of people is likely to be consuming adequate amounts of a nutrient. However, it is inappropriate to use the EAR values as recommended dietary intake goals for dietary intake of individuals.

## Recommended Dietary Allowance (RDA)

The **Recommended Dietary Allowances (RDAs)** are used as nutrient intake goals for individuals. The RDA values were derived directly from the EARs using mathematical equations, and Figure 2.4 illustrates the relationship between the EARs and the RDAs. The RDA for a particular nutrient is the daily dietary intake level considered sufficient to meet the nutrient requirements of nearly all (about 97%) healthy individuals in a particular life stage and sex. In other words, if all the people in a certain population consumed the RDA for a nutrient, 97% would be getting sufficient amounts of the nutrient. In this way, the RDAs include a built-in safety margin to help assure adequate nutrient intake in the population. Unless specifically noted, RDAs do not distinguish between whether the nutrient is found in foods, is added to foods, or is consumed in supplement form.

**Recommended Dietary Allowance (RDA)** The average chronic intake level of a nutrient thought to meet the nutrient requirements of nearly all (97%) healthy people in a particular physiological state and age.

## Adequate Intake Level (AI)

When scientific evidence was insufficient to establish an EAR and thus to accurately set an RDA, the scientists derived an **Adequate Intake (AI) level** instead. In other words, the establishment of an AI instead of an RDA for a nutrient means that more research is needed. Like the RDAs, AIs are meant to be used as nutrient intake goals for individuals. The AIs were based on experimentally documented intake levels of a nutrient that seemed to maintain adequate nutritional status in healthy people, and in general AI levels are thought to exceed true nutrient requirements. An example of a nutrient with an AI instead of an RDA is calcium. More research is required to be able to establish RDAs for this mineral. In addition, there are no RDAs for young infants (0–6 months of age), and AIs for this group were based on the daily mean nutrient intake supplied by human milk for breastfed babies. You can see which nutrients have AI versus RDA values by examining the DRI tables inside the front cover of this book.

## Tolerable Upper Intake Level (UL)

The RDA and AI values have been established to prevent deficiencies and decrease risk of chronic degenerative diseases. However, avoiding the other end of the nutritional status continuum—nutrient overconsumption or toxicity—is also important. As such, the **Tolerable Upper Intake Levels (ULs)** have been established as the highest level of usual daily nutrient intake likely to pose no risk of adverse health effects.

ULs are not to be used as target intake levels or goals. Instead, the ULs provide limits for those who take supplements or consume large amounts of fortified foods. Some nutrients are harmful at very high intakes and thus are toxic. Note that scientific data are insufficient to allow the establishment of UL values for all nutrients. The lack of ULs for a particular nutrient indicates the need for caution in consuming high intakes of that nutrient; it does not mean that high intakes pose no risk.

## Using DRIs in Dietary Assessment

Even after completing a dietary assessment and identifying the appropriate DRI values, it is nearly impossible to determine what your actual personal nutrient requirements are. Remember that the EARs are simply *Estimated* Average Requirements. Thus your individual requirement may be more or less than the EAR. Given these inherent limitations, some inferences about your diet can be made by comparing the results of your dietary assessment with your DRI values.[1] These concepts are illustrated in Figure 2.5 and are listed as follows.

**When EARs and RDAs have been established:**
- If intake of a nutrient is much less than the EAR, then it is likely to be inadequate—resulting in increased risk of nutrient deficiency.
- If intake of a nutrient is between the EAR and the RDA, then it should probably be increased.
- If intake of a nutrient is between the RDA and the UL, then it is probably adequate.

**When only AIs are available:**
- If intake of a nutrient equals or exceeds the AI, then it is probably adequate.
- If intake of a nutrient falls below the AI, no conclusion can be made concerning the adequacy of the diet.

**Adequate Intake (AI) level** Nutrient intake of healthy populations that appears to support adequate nutritional status set as goals of dietary assessment and planning when RDAs cannot be established.

**Tolerable Upper Intake Level (UL)** The highest level of chronic intake of a nutrient thought to be not detrimental to health.

## FIGURE 2.5   Using Dietary Reference Intake (DRI) Standards

The DRI values can be used as goals for dietary intake. In general, intakes are considered adequate if they are between the RDA and the UL.

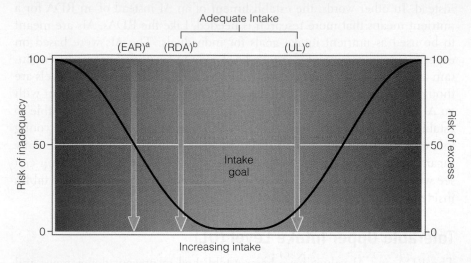

a Estimated Average Requirement
b Recommended Dietary Allowance
c Tolerable Upper Intake Levels

SOURCE: Adapted from the Institute of Medicine. Dietary reference intakes for energy, carbohydrates, fiber, fat, fatty acids, cholesterol, protein, and amino acids. Washington, DC: National Academies Press; 2005.

Although using dietary assessment in combination with DRIs is an excellent way to assess dietary adequacy, it should not be used as the sole method of nutritional status assessment. To get a more complete view of a person's nutritional status, information from all four assessment methods—anthropometric, biochemical, clinical, and dietary—should be considered.

## Estimated Energy Requirements (EERs) and Physical Activity Patterns

Establishing recommendations for energy intake presented a significant challenge to the DRI committee. This is because energy requirements (like nutrient requirements) are influenced by many factors, including age, sex, weight, height, and level of physical activity. For example, it is easy to imagine that a 23-year-old sedentary person may require less energy than a 23-year-old marathon runner—even if they are both women weighing 130 pounds. Therefore scientists established equations to calculate what are called the **Estimated Energy Requirements (EERs)**,[2] which represent the average energy intakes needed to maintain weight in a healthy person of a particular age, sex, weight, height, and physical activity level, consistent with good health. The EER equations for adults of healthy weight are provided next, and others can be found in Appendix B.

$$Adult\ man:\ EER = 662 - [9.53 \times age\ (y)] + PA \times [15.91 \times wt\ (kg) + 539.6 \times ht\ (m)]$$

$$Adult\ woman:\ EER = 354 - [6.91 \times age\ (y)] + PA \times [9.36 \times wt\ (kg) + 726 \times ht\ (m)]$$

In this equation, PA refers to physical activity level, which is categorized as sedentary, low active, active, or very active. Table 2.1 shows examples of these

**Estimated Energy Requirement (EER)** Average energy intake required to maintain energy balance in healthy individuals based on sex, age, physical activity level, weight, and height.

| TABLE 2.1 | Physical Activity (PA) Categories and Values | |
| --- | --- | --- |

| Activity Level Category | Physical Activity (PA) Value | | Description |
| --- | --- | --- | --- |
| | Men | Women | |
| Sedentary | 1.00 | 1.00 | No physical activity aside from that needed for independent living |
| Low active | 1.11 | 1.12 | 1.5–3 miles/day at 2–4 miles/hour in addition to the light activity associated with typical day-to-day life |
| Active | 1.25 | 1.27 | 3–10 miles/day at 2–4 miles/hour in addition to the light activity associated with typical day-to-day life |
| Very active | 1.48 | 1.45 | 10 or more miles/day at 2–4 miles/hour in addition to the light activity associated with typical day-to-day life |

SOURCE: Institute of Medicine. Dietary Reference Intakes for energy, carbohydrate, fiber, fat, fatty acids, cholesterol, protein, and amino acids. Washington, DC: National Academies Press; 2005. Please note that these values only apply to normal weight, nonpregnant, nonlactating adults. Values for children, pregnant or lactating women, and overweight or obese individuals are different.

activity categories and their corresponding values. A sample practice calculation for determining an EER follows.

***Calculating Estimated Energy Requirements (EERs)***   Kyung-Soon is a 38-year-old woman who weighs 115 pounds, is 5 feet 4 inches tall, and has a low activity level. Her Estimated Energy Requirement (EER) is calculated as follows:

Age = 38 years

Physical activity (PA) level = 1.12

Weight (wt) = 115 pounds = 52.3 kg (115 ÷ 2.2)

Height (ht) = 5 feet 4 inches = 5.3 feet = 1.6 m (5.3 ÷ 3.3)

$$EER = 354 - [6.91 \times age\ (y)] + PA \times [9.36 \times wt\ (kg) + 726 \times ht\ (m)]$$
$$= 354\ (6.91 \times 38) + 1.12 \times (9.36 \times 52.3 + 726 \times 1.6)$$
$$= 354 - 262.6 + 1.12 \times (489.5 + 1161.6)$$
$$= 91.4 + 1.12 \times 1651.1$$
$$= 1,941\ kcal$$

## Acceptable Macronutrient Distribution Ranges (AMDRs)

Scientists are also interested in understanding the best distribution of energy sources in the diet. Thus the DRI committee established the **Acceptable Macronutrient Distribution Ranges (AMDRs).** The AMDRs reflect the ranges of intakes for each class of energy source that are associated with reduced risk of chronic disease while providing adequate intakes of essential nutrients. The AMDRs, which are expressed as percentages of total energy intake, are listed next and inside the front cover of your book for easy future reference.

### Acceptable Macronutrient Distribution Ranges (AMDRs)
Carbohydrates: 45 to 65% of total energy
Protein: 10 to 35% of total energy
Fat: 20 to 35% of total energy

Scientists have also established AMDRs for the essential fatty acids and the polyunsaturated fatty acids. These are described in more detail in the

**Acceptable Macronutrient Distribution Ranges (AMDRs)** Recommendations concerning the distribution or percentages of energy from each of the macronutrient classes.

appropriate chapters. Note that several chapters in this book provide practice calculations regarding the AMDR values. You can find them at the ends of these chapters.

## ESSENTIAL *Concepts*

Dietary reference standards are estimated levels of nutrients that meet basic biological needs and promote optimal health. The Institute of Medicine has formulated a set of such standards, called the Dietary Reference Intakes (DRIs), that include the Estimated Average Requirements (EARs), Recommended Dietary Allowances (RDAs), Adequate Intake levels (AIs), and Tolerable Upper Intake Levels (ULs). The EARs estimate actual nutrient requirements in various population groups; RDAs are based on EARs and are considered to be recommended intakes for individuals. When the Institute of Medicine could not establish EARs and RDAs, AI levels were set as guidelines for intakes. The ULs are not recommendations but are levels that should not be exceeded. In general, dietary intake values that fall between the RDA and the UL levels are considered adequate. The DRI committee has also established Estimated Energy Requirement (EER) equations and Acceptable Macronutrient Distribution Ranges (AMDRs) to provide guidance as to total energy intake and distribution of energy intake from the macronutrients.

# Planning a Nutritious Diet

Knowing whether your diet provides all the essential nutrients in adequate amounts is an important first step to assure health and reduce risk of chronic disease. However, being able to choose appropriate amounts of nutritious foods on a daily basis is key to this process. Because the link between nutrition and health is so strong, the U.S. government invests significant resources in developing tools to help people choose and plan healthy diets. This section discusses these tools to help people choose foods wisely.

## A Brief History of the Federal Food Guidance Systems

Several government agencies are involved in assuring the nation's health. One is the U.S. Department of Agriculture (USDA), which works both to optimize the nation's agricultural productivity and to promote a nutritious diet in the United States. In 1894 the USDA published its first set of nutritional recommendations for Americans.[3] Although these recommendations were limited and referred mostly to total calorie and protein intake, they provided the groundwork from which all other food guides would later be developed.

Since this publication, there have been a succession of versions—all designed to translate nutrient intake recommendations into guidelines for dietary planning.[4] Generally known as the **USDA Food Guides,** these guidance systems have classified foods into food groups and have made recommendations on the relative amounts of each food group to consume (see Figure 2.6). In 1980 the U.S. Department of Health and Human Services (DHHS) and the USDA issued a new form of dietary recommendations called the **Dietary Guidelines for Americans,** which provided specific advice about how good dietary habits can promote health and reduce risk for major chronic disease. These guidelines are revised about every five years. The current version of the federal food guidance system, called **MyPyramid,** was released in 2005.

**USDA Food Guides** Dietary recommendations developed by the USDA based on categorizing foods into "food groups."

**Dietary Guidelines for Americans** Dietary recommendations, developed by the USDA and DHHS, that give specific nutritional guidance to individuals as well as advice about physical activity, alcohol intake, and food safety.

**MyPyramid** A graphic and accompanying interactive website developed by the USDA to illustrate the recommendations put forth in the Dietary Guidelines for Americans.

## FIGURE 2.6 Development of USDA Food Guides

The U.S. Department of Agriculture (USDA) has published many food guides during the past century.

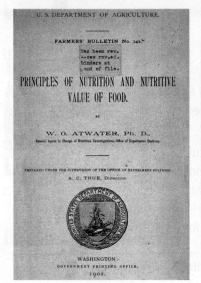

1902

Depression era

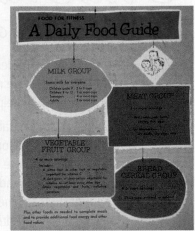

Basic 4

SOURCE: USDA, at http://www.usda.gov.

Because MyPyramid was specifically developed to translate the principles of the USDA 2005 Dietary Guidelines for Americans into practice, these will be discussed together.

## Dietary Guidelines for Americans

The 2005 Dietary Guidelines for Americans differ from previous versions in that their development relied heavily on mathematical equations that predicted the "best diet" for reducing risk of the major chronic diseases—overweight and obesity, hypertension, type 2 diabetes, heart disease, cancer, and osteoporosis. The recommendations were also devised to help boost several nutrients that are often low in the American diet, including vitamin E, vitamin A, calcium, magnesium, potassium, and dietary fiber. The current edition of the Dietary Guidelines also includes advice related to alcohol intake and food-borne illness.

The 2005 USDA Dietary Guidelines give nine key recommendations to improve health. These are summarized next and described in more detail in Appendix C. These guidelines are designed to evolve as our understanding of nutrition and health promotion advances. In years to come, you can follow these changes on the USDA website (http://www.usda.gov).

### The 2005 USDA Dietary Guidelines for Americans
- Consume a variety of foods within and among the basic food groups while staying within energy needs.
- Control calorie intake to manage body weight.
- Be physically active every day.
- Increase daily intake of fruits and vegetables, whole grains, and nonfat or low-fat milk and milk products.
- Choose fats wisely for good health.

## Food Matters

### Working Toward the Goal

The USDA Dietary Guidelines recommend that we choose a variety of fruits and vegetables each day, while not consuming too many calories. The following selection and preparation tips will help you meet these goals.

- Buy fruits and vegetables that are easy to prepare. For example, buy precut packages of fruit (such as melon or pineapple chunks) for a healthy snack in seconds.
- Consider frozen juice bars (100% juice) as healthy alternatives to high-fat snacks.
- Make a Waldorf salad, with apples, celery, walnuts, and dressing.
- Add fruit such as pineapple or peaches to kabobs as part of a barbecue meal.
- Keep a package of dried fruit in your desk or bag.
- As a snack, spread peanut butter on apple slices or top frozen yogurt with berries or slices of kiwi fruit.
- Plan some meals around a vegetable main dish, such as a vegetable stir-fry or soup.
- Shred carrots or zucchini into meatloaf, casseroles, quick breads, and muffins.
- Include chopped vegetables in pasta sauce or lasagna.
- Keep a bowl of cut-up vegetables in a see-through container in the refrigerator.

- Choose carbohydrates wisely for good health.
- Choose and prepare foods with little salt.
- If you drink alcoholic beverages, do so in moderation.
- Keep food safe to eat.

To help you translate the Dietary Guidelines for Americans into healthy eating habits, most chapters in this textbook contain a Food Matters feature that is based on the Dietary Guidelines. This feature provides food selection and preparation tips. The first Food Matters is provided here.

### Key Concepts: Variety, Balance, Moderation, and Nutrient Density

The Dietary Guidelines for Americans emphasize four important concepts: variety, balance, moderation, and nutrient density. Consuming a **variety** of foods within each food group is important, because this helps assure that a person is getting adequate amounts of all the essential nutrients. **Balance,** or proportionality, indicates the relative amount of calories obtained from each of the food groups. A balance among the macronutrients can assure adequate micronutrient intake and helps prevent excessive weight gain.

**Moderation,** or not consuming too much of a particular food, is also important—especially when it comes to controlling caloric intake and maintaining a healthy body weight. The Dietary Guidelines recommend that we limit calorie intake, especially from added sugars, solid fats, and alcoholic beverages—sources of calories that are very poor sources of essential nutrients. Another important concept in the Dietary Guidelines is **nutrient density,** which is defined as the amount of nutrients that are in a food relative to its

**variety** A concept that emphasizes the importance of eating different varieties of a food type.

**balance** (or proportionality) A concept that emphasizes eating appropriate relative amounts of foods from each food group.

**moderation** A concept that emphasizes not consuming too much of a particular type of food.

**nutrient density** The relative ratio of nutrients in a food in comparison to total calories.

energy content. Foods with high nutrient densities (nutrient-dense foods) provide high amounts of essential nutrients relative to the amount of calories. Nutrient density is often expressed as the amount of a particular nutrient per 1,000 kcal of a given food. Foods with low nutrient densities are high in calories relative to the amount of micronutrients. You can read more about nutrient density in the Focus on Food feature.

**discretionary calorie allowance** The amount of calories that can still be eaten, without promoting weight gain, after all the essential nutrients have been consumed.

# *Focus On* FOOD
## Taking a Closer Look at Nutrient Density

© Myrna Engler

Nutrient density is a key concept in current nutritional guidelines. This is because consumption of nutrient-dense foods can help a person meet nutritional requirements while preventing weight gain and, ultimately, obesity. However, the concept of nutrient density is sometimes difficult to grasp.[5] After all, foods are not labeled for their nutrient densities, nor can you always know by looking at them which foods have high nutrient densities.

The concept of nutrient density is probably best understood by simply comparing the nutrient densities of several somewhat similar foods. For example, let us assess the nutrient densities for vitamin A and calcium in premium ice cream, sherbet, frozen nonfat yogurt, and a slushy. For each of these foods, the nutrient density with respect to calcium and vitamin A is calculated as follows: Nutrient density = (amount of nutrient per serving) ÷ (kcal per serving) × 1,000.

As you can see, ice cream has the highest nutrient density for vitamin A (589 RAE per 1,000 kcal), whereas fro-

zen yogurt has the highest nutrient density for calcium (1,955 mg per 1,000 kcal). The slushy has the lowest nutrient densities of all. Of course, most foods offer a variety of different nutrients, and just because a food may be a good source of one nutrient, it may be a poor source of another. In other words, there is no perfect food. This is why the Dietary Guidelines recommend that we eat a variety of foods, especially those that are nutrient-dense. This approach to eating makes it easier to consume enough of all the essential nutrients without consuming too many calories.

Another concept related to nutrient density is called **discretionary calorie allowance,** which is the amount of calories that can be consumed beyond what is needed to obtain required amounts of energy and essential nutrients. For example, a person requiring 2,000 kcal could meet his or her nutrient needs by consuming 1,740 kcal from a nutrient-dense diet. This person would have a discretionary calorie allowance of 260 (2,000 minus 1,740) kcal. This discretionary calorie allowance would give this person the opportunity to eat a slice of cheesecake (about 250 kcal) for dessert without gaining weight. However, if the nutrient density of this person's diet had been low, he or she might not have any discretionary calories left after obtaining all the essential nutrients. If so, the extra calories in the cheesecake might lead to weight gain. Choosing foods with high nutrient densities helps increase a person's discretionary calorie allowance, making it easier to maintain a healthy body weight.

| Dessert | Vitamin A (RAE[a]/serving) | Calcium (mg/serving) | Energy (kcal/serving) | Nutrient Density | |
|---|---|---|---|---|---|
| | | | | Vitamin A (RAE/1,000 kcal) | Calcium (mg/1,000 kcal) |
| Ice cream | 156 | 169 | 265 | 589 | 638 |
| Sherbet | 85 | 206 | 235 | 362 | 877 |
| Frozen yogurt | 4 | 436 | 223 | 18 | 1,955 |
| Slushy | 0 | 0 | 220 | 0 | 0 |

[a] RAE, retinol activity equivalent.

### Alcoholic Beverages

Of particular interest to some people are the Dietary Guidelines recommendations regarding alcohol consumption. Alcohol provides very few essential nutrients and therefore is of low nutrient density. Excess alcohol consumption also can alter judgment and can lead to addiction, serious health concerns, and societal problems. Thus consuming large amounts of alcohol is never recommended. However, moderate alcohol consumption (1 to 2 drinks each day) may have beneficial health effects in some, especially middle-aged and older adults. This does not hold true for young adults, and it is not recommended that anyone begin drinking or drink more frequently on the basis of potential health benefits. More information concerning the relationship between alcohol and health is provided in the Nutrition Matters following Chapter 10.

### Establishment of the Current USDA Food Guide

Aside from making overall recommendations concerning the optimal diet for achieving health, the Dietary Guidelines also establish the current USDA Food Guide. This guide includes seven food groups: fruits, vegetables, grains, lean meats and beans, dairy products, oils, and discretionary calories. The recommended number of servings from each food group is based on a person's calorie requirements. The more calories a person needs, the more food that person should eat. This Food Guide formed the basis of the more consumer-friendly MyPyramid. Although the Food Guide and related MyPyramid are recognized as the government's "official" dietary recommendations, the committee also acknowledges that other eating plans such as the Dietary Approaches to Stop Hypertension (DASH) diet may be just as appropriate. The DASH diet is described in more detail in the Nutrition Matters following Chapter 7.

## MyPyramid: Putting Recommendations into Practice

To help people put the recommendations of the Dietary Guidelines into practice, the USDA has established its newest food guidance system called MyPyramid, which replaces the USDA Food Guide Pyramid published in 1992. Shown in Figure 2.7, the MyPyramid symbol of a person climbing a multicolored pyramid is simple. Although each color on the pyramid represents a different food group, there is no mention of numbers of servings. This is quite different from previously published USDA food guides, including the 1992 Food Guide Pyramid. This change in the food guidance system was implemented because it is recognized that food intake recommendations must be individualized, depending on a person's age, sex, and physical activity level. Table 2.2 outlines the ranges of intake recommended by MyPyramid.

To find out how much of each food group is recommended, a person must visit the MyPyramid website (http://www.mypyramid.gov) and provide personal information (age, sex, and physical activity level) under the section called "MyPyramid Plan." This generates a "personal pyramid" with specific recommendations on food intake patterns, serving sizes, and menu selection. In all, there are 12 different sets of recommended dietary patterns. Examples of two personalized pyramids are provided in Figure 2.8 (page 48). You can see from these pyramids that the recommended amounts of food groups, oils, and discretionary calories (called "extras") depend on estimated energy requirements. This website also provides in-depth descriptions of types and amounts of foods that fit into each food group, as well as specific information

## FIGURE 2.7 USDA MyPyramid Icon

MyPyramid graphically illustrates the dietary recommendations outlined in the USDA's 2005 Dietary Guidelines for Americans.

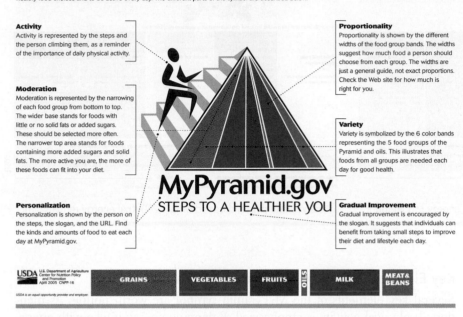

### Anatomy of MyPyramid

**One size doesn't fit all**
USDA's new MyPyramid symbolizes a personalized approach to healthy eating and physical activity. The symbol has been designed to be simple. It has been developed to remind consumers to make healthy food choices and to be active every day. The different parts of the symbol are described below.

**Activity**
Activity is represented by the steps and the person climbing them, as a reminder of the importance of daily physical activity.

**Moderation**
Moderation is represented by the narrowing of each food group from bottom to top. The wider base stands for foods with little or no solid fats or added sugars. These should be selected more often. The narrower top area stands for foods containing more added sugars and solid fats. The more active you are, the more of these foods can fit into your diet.

**Personalization**
Personalization is shown by the person on the steps, the slogan, and the URL. Find the kinds and amounts of food to eat each day at MyPyramid.gov.

**Proportionality**
Proportionality is shown by the different widths of the food group bands. The widths suggest how much food a person should choose from each group. The widths are just a general guide, not exact proportions. Check the Web site for how much is right for you.

**Variety**
Variety is symbolized by the 6 color bands representing the 5 food groups of the Pyramid and oils. This illustrates that foods from all groups are needed each day for good health.

**Gradual Improvement**
Gradual improvement is encouraged by the slogan. It suggests that individuals can benefit from taking small steps to improve their diet and lifestyle each day.

**MyPyramid.gov**
**STEPS TO A HEALTHIER YOU**

USDA U.S. Department of Agriculture Center for Nutrition Policy and Promotion April 2005 CNPP-16
USDA is an equal opportunity provider and employer.

GRAINS  VEGETABLES  FRUITS  OILS  MILK  MEAT & BEANS

about meal planning and special subgroups (such as vegetarians). Also available is the MyPyramid Tracker computerized nutrient database. In this way, a person can assess whether a diet meets the recommendations put forth by the USDA as well as by the Institute of Medicine.

The slogan of the MyPyramid food guidance system—"Steps to a Healthier You"—is reflected in the image of a person climbing the pyramid. This is the first federal food guidance system that has included a physical activity component, and its developers hope it can help consumers choose the right amount and kinds of food to balance their daily physical activity.

## TABLE 2.2 Amounts and Servings of Each Food Group Recommended by the USDA MyPyramid

| Food Category | Amounts Recommended [a,b] | Dietary Significance |
|---|---|---|
| Grains | 3–10 oz. per day (or equivalent) | Major sources of B vitamins, iron, magnesium, selenium, energy, and dietary fiber |
| Vegetables | 1–4 cups per day | Rich sources of potassium, vitamins A, E, and C, folate and dietary fiber; high-nutrient-density foods |
| Fruits | 1–2½ cups per day | Rich sources of folate, vitamins A and C, potassium, and fiber |
| Dairy products | 2–3 cups per day | Major sources of calcium, potassium, vitamin D, and protein |
| Meats, poultry, eggs, fish, nuts, seeds, and dry beans | 2–7 oz. per day (or equivalent) | Rich sources of protein, magnesium, iron, zinc, B vitamins, vitamin D, energy, and potassium |
| Oils | 3–11 tsp. per day | Source of essential fatty acids and vitamin E |

[a] Amounts depend on age, sex, and physical activity level. Personalized recommendations can be generated by going to the MyPyramid website (http://www.mypyramid.gov).
[b] Examples of serving sizes and equivalents can be found on the MyPyramid website.

## FIGURE 2.8 USDA Personalized Food Pyramids

There are 12 different pyramids in the USDA MyPyramid food guidance system, each based on a particular energy requirement.

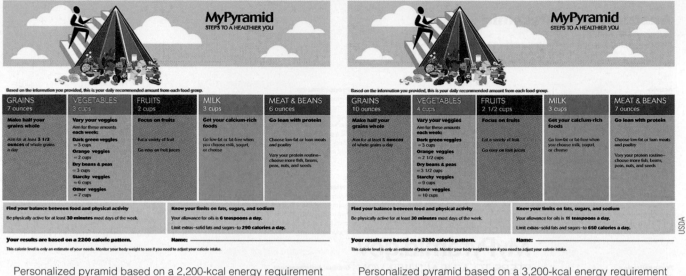

Personalized pyramid based on a 2,200-kcal energy requirement          Personalized pyramid based on a 3,200-kcal energy requirement

### Key Elements of the MyPyramid Symbol and Message

Several key elements of the MyPyramid symbol are worth noting. As mentioned, physical activity is emphasized. Balancing energy intake with energy expenditure is a major component of MyPyramid. In addition, six of the food groups (grains, vegetables, fruits, oils, dairy products, and meat and beans) outlined in the Dietary Guidelines are represented in different colors, each with different widths on the MyPyramid symbol. The colors represent the relevance of consuming foods from each food group, and the varying widths of the colored bands at the base of the pyramid represent the idea of proportionality (or balance) among food groups. The MyPyramid guide recommends that we choose foods *in approximate proportion* to the base widths of the bands. For example, we should consume more servings of dairy products (blue band) than meat and beans (purple band).

Moderate intake of solid fats and added sugars is represented by the narrowing of each food group's stripe from bottom to top. Solid fats include saturated fats and *trans* fats, such as those found in beef and vegetable shortening, respectively. Similarly, the widening of each stripe at the bottom of the pyramid is meant to encourage the consumption of more nutrient-dense foods. In other words, although it is acceptable to consume some higher-fat and higher-sugar foods from the grain group (such as donuts), most foods from this group should be low-fat with no added sugar.

### *Healthy People 2010*: Assessing our National Goals for Healthy Living

The U.S. Department of Health and Human Services developed a comprehensive document called ***Healthy People 2010*** as a set of overall health objectives for the nation to achieve by the year 2010. Unlike the Dietary Guidelines and MyPyramid, which provide information to individuals, the goals and objectives put forth in *Healthy People 2010* are meant to be used by govern-

***Healthy People 2010*** A publication that sets long-term national goals for improving overall health of Americans.

## TABLE 2.3  *Healthy People 2010* Nutrition-Related Goals

| Healthy People 2010 Focus Areas | General Goals | Examples of Nutrition-Related Indicators |
|---|---|---|
| Arthritis, osteoporosis, and chronic back conditions | Prevent illness and disability related to arthritis and other rheumatic conditions, osteoporosis, and chronic back conditions. | Increase the proportion of persons who meet dietary recommendations for calcium from 46 to 75%. |
| Cancer | Reduce the number of new cancer cases as well as the illness, disability, and death caused by cancer. | Increase the proportion of persons who consume at least 2 daily servings of fruit from 28 to 75%. |
| Diabetes | Reduce the prevalence and economic burden of diabetes, and improve the quality of life for all persons who have or are at risk for diabetes. | Reduce the proportion of adults who are obese from 23 to 15%. |
| Food safety | Reduce food-borne illness. | Reduce outbreaks of infections caused by key food-borne bacteria by 50%. |
| Heart disease and stroke | Improve cardiovascular health and quality of life through the prevention, detection, and treatment of risk factors; early identification and treatment of heart attacks and strokes; and prevention of recurrent cardiovascular events. | Increase the proportion of people who consume less than 10% of calories from saturated fat from 36 to 75%. |
| Maternal, infant, and child health | Improve the health and well-being of women, infants, children, and families. | Reduce iron deficiency among young children and females of childbearing age. |
| Nutrition and overweight | Promote health and reduce chronic disease associated with diet and weight. | Reduce the proportion of children and adolescents who are overweight or obese from 11 to 5%. |

SOURCE: U.S. Department of Health and Human Services. *Healthy People 2010: Understanding and Improving Health,* 2nd ed. Washington, DC: U.S. Government Printing Office; 2000.

ment agencies, communities, and professional organizations to help develop programs to improve the health of our communities. *Healthy People 2010* is designed to achieve two overarching goals: (1) increase quality and years of healthy life and (2) eliminate health disparities among different segments of the population. Within these goals, there are 28 different "focus areas," such as cancer, diabetes, food safety, maternal health, child health, nutrition, and obesity. In addition, a set of health indicators was identified to help track the nation's progress in meeting these objectives, such as physical activity and obesity. Using these health indicators, progress toward meeting the nation's health goals can be tracked and assessed in years to come. Some of the focus areas as well as examples of nutrition-specific goals and health indicators are listed in Table 2.3.

Although it is beyond the scope of this book to discuss each objective and indicator, many of the diet-related goals are consistent with the recommendations put forth in the 2005 Dietary Guidelines. In fact, *Healthy People 2010* often refers to the Guidelines for specific detail concerning nutrition recommendations. In this way, the Guidelines help individuals meet their specific nutrition goals, whereas *Healthy People 2010* helps the nation assess whether its national health goals are being met.

# ESSENTIAL *Concepts*

For over a century, the USDA has put forth dietary plans to help people stay healthy. The current version of the USDA dietary guidance system, called MyPyramid, emphasizes the concepts of variety, moderation, proportionality, and physical activity. MyPyramid includes a graphic to illustrate concepts of dietary planning as well as an interactive website designed to help people obtain personalized dietary guidance and conduct dietary assessments. MyPyramid is based on the 2005 Dietary Guidelines for Americans. These Dietary Guidelines provide science-based nutritional recommendations to promote optimal health and reduce the risk of chronic disease in the United States. Although these recommendations focus on increasing intakes of important nutrients, they also focus on moderation. In addition, they stress the importance of physical activity in promoting a healthy body weight. The concepts of increased nutrient density and discretionary calorie allowances are central to the Dietary Guidelines' recommendations and thus to the key messages of MyPyramid. Dietary goals for supporting overall health of the nation are also part of the *Healthy People 2010* report.

# Using Food Labels to Plan a Healthy Diet

How can we use the Dietary Guidelines and MyPyramid to help us choose appropriate amounts of nutritious types of foods? The answer to this lies in the vast amount of information on a food's label. Thus it is important to understand how to read a food label and then apply that knowledge to making healthful food choices.

## Food Labels and Nutrition Facts Panels

The regulation of nutrition labeling for foods was established in 1973 by the U.S. Food and Drug Administration (FDA) as a way to help consumers become aware of the nutrient content of foods. The FDA requires most packaged foods with more than one ingredient to have certain information printed on their labels. Required components of a food label include the following information:

- Product name and place of business
- Product net weight
- Product ingredient content (from most abundant to least abundant ingredient)
- Company name and address
- Product code (UPC bar code)
- Product dating if applicable
- Religious symbols if applicable
- Safe handling instructions if applicable (such as for raw meats)
- Special warning instructions if applicable (such as for aspartame)
- Nutrition Facts panel outlining specified nutrient information

Of special interest to many people is the **Nutrition Facts panel,** which is a required component of most food labels. An example of a food label—including a Nutrition Facts panel—is provided in Figure 2.9.

The FDA requires several elements on a Nutrition Facts panel. First, the manufacturer must include information concerning the serving size of the food. Serving sizes have been standardized so that the nutrient contents of similar foods can be easily compared. For example, you can compare the amount of iron in a single serving of two similar breakfast cereals by simply comparing Nutrition Facts panels. In addition, information concerning total energy (listed as Calories), total carbohydrates (including dietary fiber), sugar, and protein must be provided.

**Nutrition Facts panel** A required component of food packaging that contains information about the nutrient content of the food.

## FIGURE 2.9 Understanding Food Labels and Nutrition Facts Panels

The FDA requires food labels to have specific information such as manufacturer's name, net contents, and nutrition information. Nutrition Facts panels provide a variety of information useful in dietary planning.

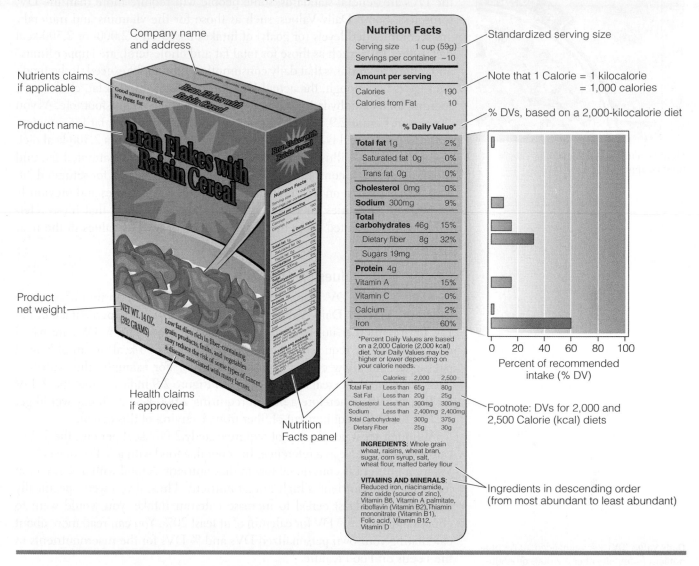

Nutrition Facts panels must also provide information concerning specific nutrients that the Dietary Guidelines either limit or promote. Nutrients that the Dietary Guidelines suggest we limit include total fat, saturated fat, *trans* fat, cholesterol, and sodium. Conversely, we are encouraged to increase dietary fiber, vitamins A and C, and the minerals calcium and iron. The FDA requires that information concerning all these nutrients be included on Nutrition Facts panels. In this way, food labels can help consumers choose foods that specifically meet the goals of the Dietary Guidelines. You can learn more details about food labels by visiting the FDA website at http://www.cfsan.fda.gov/~dms/foodlab.html.

### Daily Values (DVs)

Although Nutrition Facts panels provide an impressive amount of nutrition information, using this information for diet planning can be somewhat challenging. For example, a Nutrition Facts panel might state that a food has 20 mg

Health claims such as those shown above must be approved by the FDA.

of calcium, but the consumer reviewing the food label may not know whether 20 mg of calcium is a little calcium or a lot of calcium. Thus the **Daily Values (DVs)** were created to give consumers a benchmark for comparison. However, the DVs are general standards; some people will require more than the DV, some less. Some Daily Values, such as those for the vitamins and minerals, are recommended levels (or goals) of intakes based on a 2,000- or 2,500-kcal diet. Other DVs, such as those for total fat and cholesterol, are "upper limits" and represent amounts that daily consumption should not exceed. If the food's package is big enough, the actual DVs for total fat, saturated fat, cholesterol, sodium, total carbohydrate, and fiber intake are provided in a footnote. As you can see in Figure 2.9, the DV (or upper limit) for saturated fat for someone requiring 2,000 kcal is 20 g/day; this increases to 25 g/day for a 2,500-kcal diet. In fact, the cereal illustrated in this figure provides 0 g of saturated fat, and therefore does not contribute at all to a person's upper limit for saturated fat. The DVs are based on a combination of "former RDA" values and previously established guidelines for macronutrient intake. It is likely that these references will be updated using the current DRI and AMDR values in the near future.

### Percent Daily Values (% DVs)

In addition to the DVs, the Nutrition Facts panel provides a unit of measure called the **Percent Daily Value (% DV)**. The % DV is the percentage of the DV for a particular nutrient in 1 serving of the food. The % DVs are based on a daily energy requirement of 2,000 kcal (not 2,500 kcal) to simplify food packaging and allow comparison among foods. For example, the Nutrition Facts panel on the cereal box illustrated in Figure 2.9 indicates that the % DV for fiber is 32%. Therefore a person requiring 2,000 kcal each day would get 32% of the daily requirement of fiber from 1 serving of this cereal.

Although most people do not require exactly 2,000 kcal per day, the % DV values can still serve as a reference. In general, a food with a % DV of less than 5% for a nutrient is considered low in that nutrient. A food with a % DV of at least 20% for a nutrient is high in that nutrient. Thus, if you were specifically choosing a breakfast cereal to increase calcium intake, you would want to choose one with a % DV for calcium of at least 20%. You can read more about calculating your own personalized DVs and % DVs for the macronutrients in the Focus on Food feature.

## Nutrient Content Claims and Health Claims

Food packaging can also contain additional nutrition-related information. For example, **nutrient content claims** describe the level of a nutrient (or its "content") in a food. These include phrases such as "sugar free," "low sodium," and "good source of." The use of these terms is regulated by the FDA, and some of the approved definitions are provided in Table 2.4.

Some food manufacturers also include **health claims** on their packaging. For example, a food manufacturer might claim that consumption of its product decreases risk for heart disease. This is quite different from a nutrient content claim, which simply states that a food contains or does not contain a particular nutrient. Like other parts of a food's package, health claims must be approved by the FDA.

There are two kinds of health claims: **unqualified** and **qualified.** (To qualify something is to indicate some kind of limitation on it.) Both types of health claims describe a relationship between a specific food component or whole

**Daily Value (DV)** Recommended intake of a nutrient based on either a 2,000- or 2,500-kcal diet.

**Percent Daily Value (% DV)** The percentage of the recommended intake (DV) of a nutrient provided by a single serving of a food.

**nutrient content claims** FDA-regulated phrases and words that can be included on a food's packaging to describe its nutrient content.

**health claims** FDA-approved statements relating a food or food component to health benefits.

**unqualified health claims** Health claims that can be included on a food's packaging to describe a specific, scientifically supported health benefit.

**qualified health claims** Statements concerning less well established health benefits that have been ascribed to a particular food or food component.

# Focus On FOOD

## Calculating Individual Daily Values (DVs)

The Daily Values (DVs) and percent Daily Values (% DVs) help consumers choose and compare foods. As described, DVs are formulated for a reference individual requiring 2,000 or 2,500 kcal daily. Some DVs (such as cholesterol) do not depend on calorie requirements, because the recommended upper limit for intake is the same for all adults. In these cases, there is no need to make adjustments in the DV if your calorie requirement is not 2,000 or 2,500 kcal. However, energy requirements can influence the DVs of the macronutrients. Thus, if your caloric requirement differs much from 2,000 or 2,500 kcal, you may want to calculate these DVs for yourself. You can do this by using the following "calculation factors," which are those used currently by the FDA in its food-labeling regulations. Remember that these values were established using older guidelines and that in the near future they will likely change to reflect the current DRI and AMDR values.

- 60% of kilocalories should come from carbohydrates.
- 10% of kilocalories should come from protein.
- <30% of kilocalories should come from fat.
- <10% of kilocalories should come from saturated fat.
- At least 11.5 g fiber should be consumed per 1,000 kcal.

Keep in mind that carbohydrates and proteins have 4 kcal per gram, whereas fats have 9 kcal per gram. Examples of practice calculations related to DVs are provided here.

***Practice Calculation*** If a person's energy requirement (EER) is 2,200 kcal, what are the Daily Values (DVs) for carbohydrates, protein, and fat? What are the % DVs for a serving of a hypothetical food containing 200 g carbohydrate, 18 g protein, and 5 g fat?

### Carbohydrate

- 60% of 2,200 kcal = $0.60 \times 2,200$ = 1,320 kcal
- 1,320 kcal of carbohydrate = $(1,320 \div 4)$ or 330 g carbohydrate
- DV for carbohydrates is 330 g
- % DV for carbohydrate from this food is about 61% ($200 \times 100 \div 330$)

### Protein

- 10% of 2,200 kcal = $0.10 \times 2,200$ = 220 kcal
- 220 kcal protein = $(220 \div 4)$ or 55 g protein
- DV for protein is 55 g
- % DV for protein from this food is about 33% ($18 \times 100 \div 55$)

### Fat

- 30% of 2,200 kcal = $0.30 \times 2,200$ = 660 kcal
- 660 kcal fat = $(660 \div 9)$ or 73 g fat
- DV for fat is 73 g
- % DV for fat from this food is about 7% ($5 \times 100 \div 73$)

## TABLE 2.4    FDA-Approved Nutrient Content Claims

| Wording | Description |
| --- | --- |
| "Light" or "Lite" | If 50% or more of the calories are from fat, fat must be reduced by at least 50% as compared to a regular product. If less than 50% of calories are from fat, fat must be reduced at least 50% or calories reduced at least 1/3 compared to a regular product. |
| "Reduced Calories" | At least 25% fewer calories per serving compared to a regular product. |
| "Calorie Free" | Less than 5 kcal (Calories) per serving. |
| "Fat Free" | Less than 0.5 g fat per labeled serving. |
| "Low Fat" | 3 g fat or less per serving. |
| "Saturated Fat Free" | Less than 0.5 g saturated fat and less than 0.5 g *trans* fatty acids per serving. |
| "Low in Saturated Fat" | 1 g saturated fat per serving and containing 15% or less of calories from saturated fat. |
| "Cholesterol Free" | Less than 2 mg cholesterol per serving. Note that cholesterol claims are only allowed when food contains 2 g or less saturated fat per serving. |
| "Low in Cholesterol" | 20 mg cholesterol or less per serving. |
| "Sodium Free" | Less than 5 mg sodium per serving. |
| "Low in Sodium" | 140 mg or less sodium per serving. |
| "Sugar Free" | Less than 0.5 g sugars per serving. This does not include sugar alcohols. |
| "High," "Rich in," or "Excellent Source of" | Contains 20% or more of the Daily Value (DV) to describe protein, vitamins, minerals, dietary fiber, or potassium per serving. |
| "Good Source of," "Contains," or "Provides" | 10–19% of the Daily Value (DV) per serving. |
| "More," "Added," "Extra," or "Plus" | 10% or more of the Daily Value (DV) per serving. May only be used for vitamins, minerals, protein, dietary fiber, and potassium. |
| "Fresh" | A raw food that has not been frozen, heat processed, or otherwise preserved. |
| "Fresh Frozen" | Food was quickly frozen while still fresh. |

SOURCE: Adapted from U.S. Department of Health and Human Services and U.S. Food and Drug Administration. *A Food Labeling Guide—Appendix B*. Available at: http://www.cfsan.fda.gov/~dms/flg-6b.html.

food and a disease or health-related condition. Whereas unqualified claims are supported by considerable research, qualified health claims have less scientific backing and must be accompanied by a disclaimer (or "qualifier") statement. You can usually tell if a health claim is qualified, because it contains a statement such as "However, FDA has determined that this evidence is limited and not conclusive." The health claims currently permitted by the FDA are listed in Appendix D and are mentioned throughout this text.

## ESSENTIAL *Concepts*

Food labels give consumers useful information to help in making food choices. Most packaging contains information about the manufacturer of a food as well as the ingredients. In addition, Nutrition Facts panels provide information that can help us choose healthful foods. Using the Daily Values (DVs) and % DVs makes it relatively easy to choose foods to specifically address potential nutritional inadequacies and to help prevent chronic diseases. In addition, % DVs can be used to quickly and easily compare foods. Nutrient content claims and health claims are also valuable tools when planning a healthy diet.

# Planning Your Diet and Assessing Your Plan: Putting Concepts into Action

One benefit of studying nutrition is that much of what you learn can be applied directly to your own life. You now have the information needed to assess your diet and choose foods to improve it, and a hypothetical example is provided here. Remember, however, that dietary assessment is only one component of conducting a complete nutritional assessment.

## Step 1: Setting the Stage and Setting the Goals

For this hypothetical case, consider a 21-year-old student named Jodi who is interested in making sure that her diet is adequate. Jodi is in good health, weighs 135 pounds (61.4 kg), and attends aerobics classes twice a week at the university. Lately, however, she has felt tired and is wondering if her diet might be deficient. She has heard that iron and vitamin deficiencies can make a person feel tired and would like to know more about her nutritional status to determine whether poor diet might be influencing her health.

## Step 2: Assessing Nutritional Status

First, Jodi examines her anthropometric measurements to determine whether there is any evidence of overall malnutrition. Given that Jodi is 5 feet, 3 inches tall (1.6 m), her BMI is 23.9 kg/m². As a BMI of 18.5 to 24.9 kg/m² indicates adequate weight status, this measure suggests that she is at a healthy body weight. However, because BMI is not a good indicator of overall dietary adequacy, Jodi decides to conduct a dietary assessment using the three-day diet record method. She carefully records everything she eats and drinks for three days, paying close attention to portion sizes and describing all components of more complex foods such as the lasagna served in the cafeteria.

Next Jodi visits the MyPyramid website and enters all the information into its database. Using this software, she is able to find out how her dietary intake compares to the DRI values for all the essential vitamins, minerals, and macronutrients as well as energy. In addition, she is able to compare her dietary intake of certain food groups to that recommended in the USDA Dietary Guidelines and MyPyramid.

The results of her dietary analysis indicate that almost all the necessary nutrients are supplied in her diet at levels that fall between her RDAs and ULs. In addition, her total energy intake is acceptable. However, the percentage of calories coming from fat is 40%—higher than recommended—and fiber intake is below the recommended 20 to 25 g/day. Further, her intake of iron is only 13 mg/day—72% of the RDA. This suggests she may have inadequate intakes of fiber and iron. Similarly, her vitamin $B_{12}$ intake is only 1.9 micrograms ($\mu$g)/day—representing only 79% of the RDA for this nutrient. Jodi does some research on why the body requires iron and vitamin $B_{12}$ and is surprised to learn that both are needed for energy production. Jodi thinks she should increase intakes of both these nutrients and try to decrease fat and increase fiber.

## Step 3: Setting the Table to Meet the Goals

Using nutrient databases on the USDA website, Jodi learns that lean meats and fortified breakfast cereals are good sources of iron but do not supply high amounts of fat. She also learns that meat and dairy products supply vitamin $B_{12}$ and that fiber is found in whole-grain products, peas and lentils, and some fruits and vegetables. Jodi begins to specifically choose these foods while limiting her total fat intake.

She begins reading Nutrition Facts panels on packaged foods, and when possible she chooses foods with lower total fat contents. In the cafeteria, Jodi looks at cereal box labels and begins to eat fortified breakfast cereals instead of unfortified ones. In addition, when possible she selects foods labeled as providing, in each serving, at least 20% of the DV for iron. She also looks for products that contain relatively large amounts of whole-grain components, and makes sure she eats enough lean meat and low-fat dairy products.

## Comparing the Plan to the Assessment: Did It Succeed?

Jodi wants to know whether her alterations in food choices and diet planning improved her nutrient intake. Therefore, she does another three-day dietary assessment as she had done previously, and learns that the simple changes she has made to her diet have resulted in more adequate intakes of iron, vitamin $B_{12}$, and fiber while simultaneously decreasing her total fat intake. She has succeeded! However, Jodi also decides to make an appointment with her health care provider to make sure her overall health is good.

This example of using dietary assessment and planning techniques can be easily applied to any person's life. You now have all the tools needed to do each of these steps and the knowledge to make real changes in your diet and health. Some introductory nutrition classes require students to undertake this exercise as a class assignment. If this is the case for you, then you will get firsthand experience at these procedures. If not, consider doing this for your own benefit. Whether required or optional, conducting a personal dietary assessment will make the rest of this course more meaningful, because you will be able to apply the knowledge to your own diet and overall health.

# ESSENTIAL *Concepts*

It is relatively easy to conduct some portions of a nutritional status assessment, such as measuring weight and height and calculating BMI. In addition, using diet records in conjunction with computerized dietary assessment programs allows any person to compare his or her dietary intake with nutrient recommendations. For example, intakes can be compared to DRI values as well as recommendations put forth in the Dietary Guidelines and MyPyramid. Using this information in conjunction with that found in food composition tables and food labels, food choices can be made to improve the diet.

## Review Questions  Answers are found in Appendix G.

1. Which of the following situations is an example of malnutrition?
   a. When a person consumes too much of a particular nutrient
   b. When a person consumes spoiled food and becomes ill
   c. When a person does not eat enough of one or more essential nutrients or energy
   d. a and c

2. Body mass index (BMI) is often used to assess nutritional status. What kind of measurement is BMI, and how is it calculated?
   a. Clinical symptom — Height divided by weight
   b. Anthropometric measure — Height divided by weight squared
   c. Clinical sign — Weight divided by height
   d. Anthropometric measure — Weight divided by height squared

3. Which of the following represents one challenge in using a 24-hour recall to assess nutritional status?
   a. A person must write down everything he or she eats while consuming it.
   b. A biological sample, such as blood, must be collected.
   c. It is difficult to remember the portion sizes that were consumed.
   d. A person must collect samples of everything he or she eats for 24 hours.

4. Which of the following definitions describes the Recommended Dietary Allowances (RDAs)?
   a. The intake value thought to meet the requirement of half the healthy individuals in a population
   b. The average daily dietary intake level considered to meet the nutrient requirements of nearly all healthy individuals in a population
   c. Intake levels of a nutrient that seem to maintain adequate nutritional status in a population of healthy individuals

   d. The highest level of continuing nutrient intake likely to pose no risk of adverse health effects

5. Which of the following is the Acceptable Macronutrient Distribution Range (AMDR) for dietary fat?
   a. 20 to 35% of calories
   b. 10 to 35% of calories
   c. 45 to 50% of calories
   d. 45 to 65% of calories

6. Which of the following terms refers to consuming the relatively appropriate amount of calories from protein, carbohydrate, and fat?
   a. Variety
   b. Balance
   c. Moderation
   d. Both a and c

7. A person who chooses highly nutritious, low-fat food would likely have a diet that is
   a. high in nutrient density and low in discretionary calories.
   b. low in nutrient density and low in discretionary calories.
   c. high in nutrient density and high in discretionary calories.
   d. low in nutrient density and high in discretionary calories.

8. Which of the following was developed as a set of overall health objectives for the nation?
   a. *Healthy People 2010*
   b. The Dietary Guidelines for Americans
   c. The Dietary Reference Intakes
   d. The U.S. Department of Agriculture Food Guide

9. Which of the following is *not* required to be included on a food label?
   a. Product ingredient content
   b. Nutrient content claims
   c. Company name and address
   d. Nutrition Facts panel

10. Which of the following represents the type of claim with the greatest scientific evidence, and who approves these claims?

   a. Qualified health claim     U.S. Department of Agriculture

   b. Unqualified health claim     U.S. Department of Agriculture

   c. Qualified health claim     U.S. Food and Drug Administration

   d. Unqualified health claim     U.S. Food and Drug Administration

11. There are many ways to assess nutritional status. List the four basic categories of methods and when each type might be used. Also, describe one advantage and one disadvantage of each type of method.

12. Describe the four categories of Dietary Reference Intake (DRI) values. Which of these are meant to be used as dietary intake goals? How are these sets of standards related to each other?

13. Describe how the Dietary Reference Intakes (DRIs) can be used for diet planning and how they are important for establishing the Daily Values (DVs) listed on Nutrition Facts panels.

## Practice Calculations   Answers are found in Appendix G.

1. The Acceptable Macronutrient Distribution Range (AMDR) for carbohydrates is 45 to 65% of calories. If your estimated caloric requirement is 2,000 kcal/day, how many calories should come from carbohydrate? What dietary changes might you make if your carbohydrate intake is too high? Would there be any potential negative outcomes of decreasing carbohydrate intake?

2. Using the following equation, please calculate the Estimated Energy Requirement (EER) for a 21-year-old person weighing 150 pounds and being 5 feet, 10 inches, tall. Assume an active physical activity level with a Physical Activity (PA) factor of 1.25.

$$EER = 662 - 9.53 \times age\ (y) + PA \times [15.91 \times wt\ (kg) + 539.6 \times ht\ (m)]$$

3. Consider a relatively healthy woman who consumes 50 mg of vitamin E each day. The Estimated Average Requirement (EAR) is 12 mg/day, the Recommended Dietary Allowance (RDA) is 15 mg/day, and the Tolerable Upper Limit (UL) is 1,000 mg of vitamin E/day. What percentage of her RDA is she consuming? Should this woman be concerned about her vitamin E intake? Explain.

4. Using the food composition table that accompanies this text or other sources, calculate the vitamin $B_{12}$ content of a lunch containing a hamburger on a white bun (with ketchup and mayonnaise), a small order of french fries, and a large cola soft drink. What percentage of your Recommended Dietary Allowance of vitamin $B_{12}$ does this represent? What foods could be added to this lunch to increase the vitamin $B_{12}$ content?

5. If an individual requires 2,500 kcal each day and must eat 2,400 kcal to get his or her Recommended Dietary Allowance (RDA) of all essential nutrients, what amount of calories makes up his or her "discretionary calorie allowance"?

## Media Links   A variety of study tools for this chapter are available at our website, www.thomsonedu.com/nutrition/mcguire.

Prepare for tests and deepen your understanding of chapter concepts with these online materials:

- Practice tests
- Flashcards
- Glossary
- Student lecture notebook
- Web links
- Animations
- Chapter summaries, learning objectives, and crossword puzzles

# Notes

1. Institute of Medicine. Dietary Reference Intakes: applications in dietary assessment. Washington, DC: National Academies Press; 2000.

2. Institute of Medicine. Dietary Reference Intakes for energy, carbohydrate, fiber, fat, fatty acids, cholesterol, protein, and amino acids. Washington, DC: National Academies Press; 2005.

3. Atwater WO. Foods: nutritive value and cost. Farmers' Bulletin. No. 23. Washington, DC: U.S. Government Printing Office; 1894.

4. Haughton B, Gussow JD, Dodds, JM. An historical study of the underlying assumptions for United States food guides from 1917 through the basic four food group guide. Journal of Nutrition Education. 1987;19:169–76. Hertzlet AA, Anderson HL. Food guides in the United States. Journal of the American Dietetics Association. 1974;64:19–28.Welsh S, Davis C, Shaw A. A brief history of food guides in the United States—Food Guide Pyramid. *Nutrition Today.* 1992 (Nov.-Dec.):6–11. Welsh S, Davis C, Shaw A. Development of the Food Guide Pyramid—Food Guide Pyramid. *Nutrition Today.* 1992 (Nov.-Dec.):12–15.

5. Drewnoski A. Concept of a nutritious food: toward a nutrient density score. American Journal of Clinical Nutrition, 2005;82:721–32. Drewnowski A, Darmon N. The economics of obesity: dietary energy density and energy cost. American Journal of Clinical Nutrition. 2005;82 (Suppl 1):265S–73S.

**Check out the following sources for additional information.**

6. Butler D, Pearson H. Flash in the pan? Nature. 2005;433:794–6.

7. Butler D, Schneider A. Food FAQs. Nature. 2005;433:798–7.

8. Devaney B, Barr S. DRI, EAR, RDA, AI, UL: Making sense of this alphabet soup. Pediatric Basics. 2002;97:2–9.

9. Gibson RS. *Principles of nutritional assessment,* 2nd ed. New York: Oxford University Press; 2005.

10. Institute of Medicine. Dietary Reference Intakes: applications in dietary planning. Washington, DC: National Academies Press; 2003.

11. Institute of Medicine. Dietary Reference Intakes: guiding principles for nutrition labeling and fortification. Washington, DC: National Academies Press; 2003.

12. U.S. Department of Health and Human Services and U.S. Department of Agriculture. Dietary Guidelines for Americans 2005. Washington, DC: U.S. Government Printing Office; 2005. Available from: http://www.healthierus.gov/dietaryguidelines.

13. U.S. Department of Health and Human Services. Healthy People 2010: understanding and improving health. 2nd ed. Washington, DC: U.S. Government Printing Office; 2000.

14. U.S. Food and Drug Administration. A food labeling guide—Appendix A (Oct. 2004). Available from: http://www.cfsan.fda.gov/~dms/flg-6a.html.

15. U.S. Food and Drug Administration. A food labeling guide—Appendix B (June 1999). Available from: http://www.cfsan.fda.gov/~dms/flg-6b.html.

16. U.S. Food and Drug Administration. A food labeling guide—Appendix C (Nov. 2000). Available from: http://www.cfsan.fda.gov/~dms/flg-6c.html.

17. U.S. Food and Drug Administration. How to understand and use the nutrition facts label (Nov. 2004). Available from: http://www.cfsan.fda.gov/~dms/foodlab.html.

18. U.S. Food and Drug Administration. Summary of qualified health claims permitted (Sept. 2003). Available from: http://www.cfsan.fda.gov/~dms/qhc-sum.html.

19. Wyse BW, Windham CT, Hansen RG. Nutrition intervention: panacea or Pandora's box? Journal of the American Dietetic Association. 1985;85:1084–90.

Chapter 3

# Chemical and Biological Aspects of Nutrition

*T*he human body consists of carbon, hydrogen, oxygen, nitrogen, and a few other assorted elements. When joined together, these elements are transformed into large, life-sustaining compounds, or molecules, such as proteins, carbohydrates, lipids, and nucleic acids. Cells carry out the vital functions of biological life. Our bodies are made of trillions of cells, which differ vastly in size, function, and shape. Cells of similar structure and specialization form tissues. The four different types of tissues comprise over 40 organs, which make up 11 unique organ systems. This chapter discusses the chemical compounds found in food and their many roles in the biological processes of life. These concepts are fundamental to the study of nutrition. The science of biological chemistry, or biochemistry, is central to understanding how the body uses foods to support optimal health.

**FIGURE 3.1    Levels of Organization in the Body**

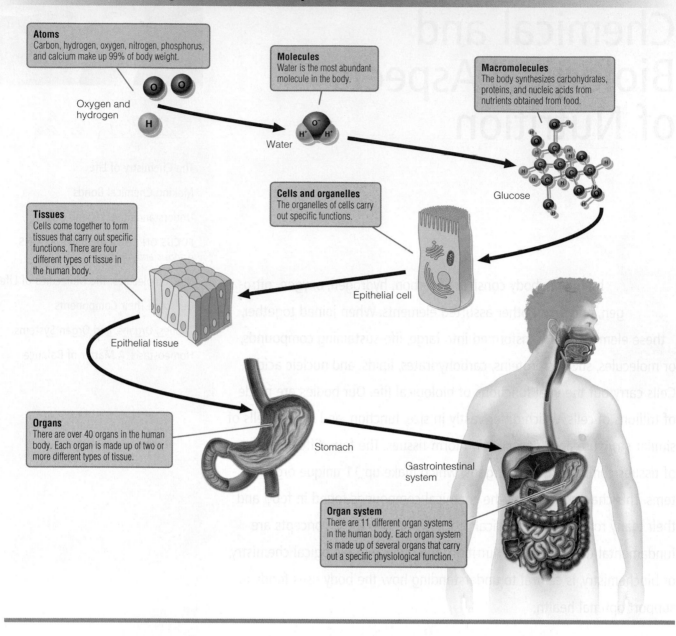

**Atoms**
Carbon, hydrogen, oxygen, nitrogen, phosphorus, and calcium make up 99% of body weight.

Oxygen and hydrogen

**Molecules**
Water is the most abundant molecule in the body.

Water

**Macromolecules**
The body synthesizes carbohydrates, proteins, and nucleic acids from nutrients obtained from food.

Glucose

**Cells and organelles**
The organelles of cells carry out specific functions.

Epithelial cell

**Tissues**
Cells come together to form tissues that carry out specific functions. There are four different types of tissue in the human body.

Epithelial tissue

**Organs**
There are over 40 organs in the human body. Each organ is made up of two or more different types of tissue.

Stomach

Gastrointestinal system

**Organ system**
There are 11 different organ systems in the human body. Each organ system is made up of several organs that carry out a specific physiological function.

# The Chemistry of Life

The organization of atoms into molecules, molecules into macromolecules, macromolecules into cells, cells into tissues, tissues into organs, and organs into organ systems is indeed remarkable (Figure 3.1). This entire circuitry is made of and fueled by the nutrients contained in food. To appreciate these life-sustaining functions, it is important to have a basic understanding of chemistry—the science that deals with **matter.**

## Atoms, Elements, and Molecules

It is hard to imagine the existence of something that we cannot see, taste, touch, or hear. Yet submicroscopic particles, called **atoms,** are the fundamental units that make up the world around us. The atom itself, illustrated

**matter**  Material that has mass and occupies space.

**atom**  The smallest portion that an element can be divided into and still retain its properties.

## FIGURE 3.2  A Model of an Atom

Atoms consist of subatomic particles—protons, neutrons, and electrons. Electrons orbit the nucleus of an atom in spaces called shells. The shell closest to the nucleus holds up to 2 electrons. Subsequent shells hold up to 8 electrons. The last, or outermost shell of an atom is called the valence shell.

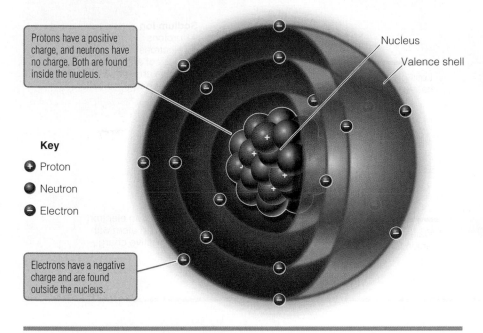

Protons have a positive charge, and neutrons have no charge. Both are found inside the nucleus.

Nucleus

Valence shell

**Key**
- Proton
- Neutron
- Electron

Electrons have a negative charge and are found outside the nucleus.

in Figure 3.2, consists of still smaller (subatomic) units—uncharged **neutrons** and positively charged **protons,** both housed in the center or nucleus of the atom. **Electrons,** which have a negative charge, orbit the nucleus of an atom in spaces called shells. In most cases, the net positive charge of protons is balanced by an equal number of electrons.

The shells surrounding the nucleus of an atom hold a certain number of electrons. For example, the shell closest to the nucleus holds up to 2 electrons, whereas the next shell holds up to 8 electrons. Atoms with more than 10 electrons (2 in the first shell and 8 in the next) have additional shells, each adding another layer much like the layers of an onion. The last, or outermost, shell of an atom is called the **valence shell.** The electrons in the valence shell play an important role in forming chemical bonds.

## Ions

Most atoms have equal numbers of protons and electrons. When the number of protons having positive charges equals the number of electrons having negative charges, the atom is neutral. However, it is possible for atoms to gain or lose electrons, as shown in Figure 3.3. When this occurs, the numbers of protons and electrons are no longer equal. As a result, an atom has a net positive or negative charge. Atoms that have an unequal number of protons and electrons are called **ions.** However, it is important to note that molecules can also be ions. For example, the hydroxide ion ($OH^-$), which consists of a hydrogen and oxygen atom, has an overall negative charge.

An ion with a net positive charge is called a **cation,** and an ion with a net negative charge is called an **anion.** For example, a hydrogen atom has 1 proton and 1 electron. The loss of an electron results in a hydrogen ion with a net positive charge—in other words, a cation. This can be written as a simple equation: $H \longleftrightarrow H^+ + e^-$, where $H^+$ refers to a positively charged hydrogen

**neutron** A subatomic particle in the nucleus of an atom with no electrical charge.

**proton** A subatomic particle in the nucleus of an atom that carries a positive charge.

**electron** A subatomic particle that orbits around the nucleus of an atom that carries a negative charge.

**valence shell** (VA – lence) The outermost orbital of an atom.

**ion** An atom that has acquired an electrical charge by gaining or losing one or more electrons.

**cation** (CAT – i – on) An ion that has a net positive charge.

**anion** (AN – i – on) An ion that has a net negative charge.

## FIGURE 3.3 Ions

Atoms with unequal numbers of protons and electrons are ions. An ion can have a positive charge (cation) or a negative charge (anion).

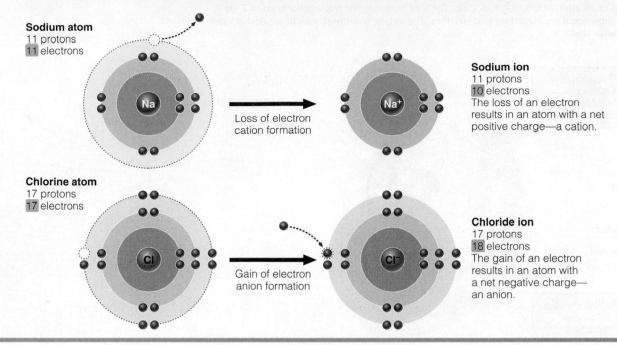

**Sodium atom**
11 protons
11 electrons

**Sodium ion**
11 protons
10 electrons
The loss of an electron results in an atom with a net positive charge—a cation.

Loss of electron cation formation

**Chlorine atom**
17 protons
17 electrons

**Chloride ion**
17 protons
18 electrons
The gain of an electron results in an atom with a net negative charge—an anion.

Gain of electron anion formation

ion (cation) and e⁻ refers to a negatively charged electron. The double arrow indicates that this is a reversible reaction, meaning that uncharged hydrogen can be formed from the joining of a H⁺ ion and an electron.

Important ions found in the human body include sodium (Na⁺), potassium (K⁺), calcium (Ca²⁺), chloride (Cl⁻), iodide (I⁻), and fluoride (F⁻). Note that when atoms such as chlorine, iodine, and fluorine gain an electron, the resulting anions undergo a name change; the suffix *–ine* becomes *–ide*.

## Elements

An **element** is defined as a pure substance made up of only one type of atom. Elements, and their characteristics, are listed in the periodic table, shown in Figure 3.4. There are approximately 92 naturally occurring elements, 20 of which are essential for human health. In fact, just 6 elements—carbon, oxygen, hydrogen, nitrogen, calcium, and phosphorus—account for 99% of total body weight. These few basic elements provide the raw materials needed to form large molecules, or macromolecules, such as proteins, carbohydrates, lipids, and nucleic acids, found in living systems.

## Molecules

We often abbreviate water as $H_2O$, because when 2 hydrogen atoms and 1 oxygen atom are chemically joined together, a molecule of water is formed. A **molecule** is defined as 2 or more atoms joined together by chemical bonds. A **molecular formula,** such as $H_2O$, is used to describe the number and type of atoms present in a molecule. For example, glucose, an important source of energy in the body, has a molecular formula of $C_6H_{12}O_6$. These numbers and letters tell us that a molecule of glucose consists of 6 carbon, 12 hydrogen, and 6 oxygen atoms. When more than one molecule of a substance is present, a number is placed before the molecular formula. For example, three molecules of water is written as 3 $H_2O$. Some molecules, such as oxygen ($O_2$), consist of only

**element** A substance made up of only one type of atom.

**molecule** A substance held together by chemical bonds.

**molecular formula** (mo – LEC – u – lar) Indicates the number and types of atoms in a molecule.

**FIGURE 3.4    The Periodic Table of Elements**

| | | | | | | | | | | | | | | | | | |
|---|---|---|---|---|---|---|---|---|---|---|---|---|---|---|---|---|---|
| 1<br>H<br>Hydrogen | | | | | | | | | | | | | | | | | 2<br>Helium<br>He |
| 3<br>Lithium<br>Li | 4<br>Beryllium<br>Be | | | | | | | | | | | 5<br>Boron<br>B | 6<br>Carbon<br>C | 7<br>Nitrogen<br>N | 8<br>Oxygen<br>O | 9<br>Fluorine<br>F | 10<br>Neon<br>Ne |
| 11<br>Sodium<br>Na | 12<br>Magnesium<br>Mg | | | | | | | | | | | 13<br>Aluminum<br>Al | 14<br>Silicon<br>Si | 15<br>Phosphorus<br>P | 16<br>Sulfur<br>S | 17<br>Chlorine<br>Cl | 18<br>Argon<br>Ar |
| 19<br>Potassium<br>K | 20<br>Calcium<br>Ca | 21<br>Scandium<br>Sc | 22<br>Titanium<br>Ti | 23<br>Vanadium<br>V | 24<br>Chromium<br>Cr | 25<br>Manganese<br>Mn | 26<br>Iron<br>Fe | 27<br>Cobalt<br>Co | 28<br>Nickel<br>Ni | 29<br>Copper<br>Cu | 30<br>Zinc<br>Zn | 31<br>Gallium<br>Ga | 32<br>Germanium<br>Ge | 33<br>Arsenic<br>As | 34<br>Selenium<br>Se | 35<br>Bromine<br>Br | 36<br>Krypton<br>Kr |
| 37<br>Rubidium<br>Rb | 38<br>Strontium<br>Sr | 39<br>Yttrium<br>Y | 40<br>Zirconium<br>Zr | 41<br>Niobium<br>Nb | 42<br>Molybdenum<br>Mo | 43<br>Technetium<br>Tc | 44<br>Ruthenium<br>Ru | 45<br>Rhodium<br>Rh | 46<br>Palladium<br>Pd | 47<br>Silver<br>Ag | 48<br>Cadmium<br>Cd | 49<br>Indium<br>In | 50<br>Tin<br>Sn | 51<br>Antimony<br>Sb | 52<br>Tellurium<br>Te | 53<br>Iodine<br>I | 54<br>Xenon<br>Xe |
| 55<br>Cesium<br>Cs | 56<br>Barium<br>Ba | 71<br>Lutetium<br>Lu | 72<br>Hafnium<br>Hf | 73<br>Tantalum<br>Ta | 74<br>Tungsten<br>W | 75<br>Rhenium<br>Re | 76<br>Osmium<br>Os | 77<br>Iridium<br>Ir | 78<br>Platinum<br>Pt | 79<br>Gold<br>Au | 80<br>Mercury<br>Hg | 81<br>Thallium<br>Tl | 82<br>Lead<br>Pb | 83<br>Bismuth<br>Bi | 84<br>Polonium<br>Po | 85<br>Astatine<br>At | 86<br>Radon<br>Rn |
| 87<br>Francium<br>Fr | 88<br>Radium<br>Ra | 103<br>Lawrencium<br>Lr | 104<br>Rutherfordium<br>Rf | 105<br>Dubnium<br>Db | 106<br>Seaborgium<br>Sg | 107<br>Bohrium<br>Bh | 108<br>Hassium<br>Hs | 109<br>Meitnerium<br>Mt | | | | | | | | | |

| 57<br>Lanthanum<br>La | 58<br>Cerium<br>Ce | 59<br>Praseodymium<br>Pr | 60<br>Neodymium<br>Nd | 61<br>Promethium<br>Pm | 62<br>Samarium<br>Sm | 63<br>Europium<br>Eu | 64<br>Gadolinium<br>Gd | 65<br>Terbium<br>Tb | 66<br>Dysprosium<br>Dy | 67<br>Holmium<br>Ho | 68<br>Erbium<br>Er | 69<br>Thulium<br>Tm | 70<br>Ytterbium<br>Yb |
|---|---|---|---|---|---|---|---|---|---|---|---|---|---|
| 89<br>Actinium<br>Ac | 90<br>Thorium<br>Th | 91<br>Protactinium<br>Pa | 92<br>Uranium<br>U | 93<br>Neptunium<br>Np | 94<br>Plutonium<br>Pu | 95<br>Americium<br>Am | 96<br>Curium<br>Cm | 97<br>Berkelium<br>Bk | 98<br>Californium<br>Cf | 99<br>Einsteinium<br>Es | 100<br>Fermium<br>Fm | 101<br>Mendelevium<br>Md | 102<br>Nobelium<br>No |

Just 6 elements account for 99% of body weight in humans.

| Element | Atomic symbol | % of humans by weight | Functions in life |
|---|---|---|---|
| Oxygen | O | 65 | Found in water and other organic molecules |
| Carbon | C | 18 | Found in all organic molecules |
| Hydrogen | H | 10 | Found in organic molecules and water |
| Nitrogen | N | 3 | Component of proteins |
| Calcium | Ca | 2 | Component of bone, teeth, and body fluids |
| Phosphorus | P | 1 | Found in cell membranes and bone matrix |

one type of atom. However, most molecules are made up of different atoms. Molecules composed of 2 or more different atoms, such as water ($H_2O$) and glucose ($C_6H_{12}O_6$), are called **compounds.** This is illustrated in Figure 3.5.

# ESSENTIAL *Concepts*

Matter is made up of submicroscopic particles called atoms. Atoms consist of subatomic units—neutrons, protons, and electrons. Whereas neutrons and protons are found in the nucleus, electrons orbit the nucleus of an atom in spaces called shells. Each shell holds a certain number of electrons. The outermost shell of an atom is called the valence shell and plays a role in forming chemical bonds. When the number of protons with positive charges equals the number of electrons with negative charges, an atom is neutral. Atoms that have unequal numbers of protons and electrons are called ions. Ions with positive charges are

**compound** A molecule made up of two or more different types of atoms.

## FIGURE 3.5 Understanding Molecular Formulas

Molecular formulas are used to indicate the number and types of atoms in a molecule.

Some molecules such as hydrogen are made up of only 1 type of atom.

The subscript indicates there are 2 hydrogen atoms in this molecule

$H_2$

A compound is a molecule that is made of 2 or more types of atoms.

Glucose is made of 6 carbon, 12 hydrogen, and 6 oxygen atoms.

$C_6H_{12}O_6$

A number in front of the molecule indicates the number of molecules.

There are 3 water molecules. Each molecule is made of 2 hydrogen and 1 oxygen atom.

$3H_2O$

called cations, and ions with negative charges are called anions. Like atoms combine with each other to form elements, whereas molecules result when chemical bonds join together 2 or more atoms. Molecules composed of 2 or more different atoms are called compounds. A molecular formula describes the number and type of atoms in a molecule. The number in front of the molecular formula indicates the number of molecules present.

# Making Chemical Bonds

The atoms in a molecule are held together by **chemical bonds.** Chemical bonds enable a relatively small number of atoms to form millions of different molecules. Without chemical bonds, the molecular world would fall apart. As mentioned earlier, the valence or outer-shell electrons are important in chemical bonding. In fact, the valence electrons that form chemical bonds can be thought of as the "glue" that holds atoms together in a molecule.

## Satisfying the Octet Rule

The valence shell of *most* atoms holds a maximum of 8 electrons, and atoms strive to have this shell completely filled. When an atom has a completely filled valence shell, it is said to contain an octet of electrons. Because an atom is very stable when its valence shell is full, atoms go to great lengths to achieve this chemical arrangement.

The need to have a completely filled valence shell, called the **octet rule,** causes atoms to interact in a variety of ways. Many atoms fill their outer valence shell by a process called **electron transfer.** Electron transfer occurs when atoms accept electrons from, or donate electrons to, other atoms. Another way atoms satisfy the octet rule is by **electron sharing.** Electron sharing involves the ability of atoms to share unpaired electrons in order to completely fill the valence shell. The tendency of atoms to gain, lose, or share electrons to satisfy the octet rule is the basis of chemical bonding.

In many cases, the arrangement of the valence electrons lets us predict an atom's behavior (Figure 3.6). Atoms with nearly complete shells satisfy the octet rule by accepting additional electrons from other atoms. For example, a chlorine atom (Cl) contains 7 valence electrons. Thus, chlorine needs only

**chemical bonds** Electric forces that hold atoms together in a molecule.

**octet rule** (OC – tet) The "desire" of an atom to have 8 electrons in its outer valence shell.

**electron transfer** The transfer of 1 or more electrons between atoms.

**electron sharing** The sharing of 1 or more valence electrons between atoms.

## FIGURE 3.6 Electron Transfer and the Octet Rule

Chlorine, with a nearly complete valence shell, satisfies the octet rule by accepting an electron from sodium, which has only 1 electron in its valence shell. By accepting an electron, chlorine becomes chloride.

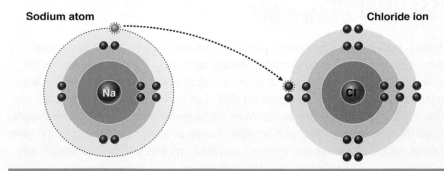

1 additional electron to satisfy the octet rule. When chlorine gains an electron, it becomes the anion chloride ($Cl^-$). Likewise, atoms with only 1 or 2 valence electrons can easily satisfy the octet rule by donating their electrons. For example, sodium (Na) contains 1 valence electron, so it is unlikely that sodium will find an atom to donate the 7 electrons needed to complete its outer shell. Sodium can more easily achieve the octet configuration by donating its single valence electron. When this happens, sodium becomes a cation ($Na^+$). The tendency for sodium to donate 1 valence electron and the tendency for chlorine to accept 1 electron is the driving force behind the formation of the molecule sodium chloride (NaCl), otherwise known as table salt.

The transfer of electrons between atoms is a very important chemical event. The loss of 1 or more electrons is called **oxidation.** Atoms that lose one or more electrons are said to be oxidized, and become more positively charged. For example, when iron (Fe) loses 2 electrons the net charge changes to $Fe^{2+}$. When Fe is fully oxidized, it loses 3 electrons, becoming $Fe^{3+}$. Therefore, the oxidation of iron increases the net positive charge (Fe to $Fe^{3+}$), reflecting the loss of 3 negatively charged electrons. In fact, when iron-containing proteins in red meat are exposed to air, they become oxidized, causing the meat to turn brown.

Conversely, the gain of 1 or more electrons is called **reduction,** and atoms become more negative during this process. **Reduction-oxidation (or redox) reactions** often take place simultaneously and are said to be coupled reactions (Figure 3.7). That is, the loss of an electron by one atom (oxidation) results in the gain of an electron by another (reduction). As you will later learn, redox reactions enable some molecules to transfer energy to other molecules, an important process in the body's use of nutrients for energy (ATP) production.

## Chemical Bonds

The transfer or sharing of electrons between atoms leads to the formation of chemical bonds. The two major types of chemical bonds, called **ionic bonds** and **covalent bonds,** are illustrated in Figure 3.8. These bonds, along with a third kind of chemical bond called a hydrogen bond, are discussed next.

### Ionic Bonds: A Forceful Attraction

Atoms with positive charges (cations) are naturally drawn to atoms with negative charges (anions). This force of attraction between ions results in the formation of an ionic bond. Common table salt, which is made up of sodium and chloride ions, provides a good example of ionic bonding. Table salt and other similar **ionic compounds** are made up of thousands of oppositely charged ions that are held together by ionic bonds to form a crystalline structure. This arrangement increases the force of ionic attraction even more.

Under the influence of water, some ionic bonds weaken. In fact, ionic compounds such as table salt dissolve when placed in water, meaning that the positively charged cations and negatively charged anions separate. Substances that produce ions when dissolved in fluids such as water are called **electrolytes.** However, the terms *electrolyte* and *ion* are often used interchangeably. In the body, salts that dissolve into ions serve a variety of important functions. For example, ions such as sodium, potassium, and calcium play major roles in blood pressure regulation, the activity of nerve cells, and muscle contraction.

**FIGURE 3.7    Electron Transfer in Reduction-Oxidation (Redox) Reactions**

Redox reactions are coupled reactions in which 1 or more electrons are transferred between molecules.

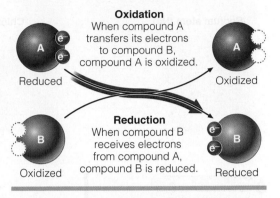

**Oxidation**
When compound A transfers its electrons to compound B, compound A is oxidized.

Reduced    Oxidized

**Reduction**
When compound B receives electrons from compound A, compound B is reduced.

Oxidized    Reduced

**oxidation** (ox – i – DA – tion) The loss of one or more electrons.

**reduction** The gain of one or more electrons.

**reduction-oxidation (redox) reactions** Chemical reactions that take place simultaneously whereby one molecule gives up 1 or more electrons (is oxidized) while the other molecule receives 1 or more electrons (is reduced).

**ionic bond** (i – ON – ic) The force of attraction between ions with opposite charges.

**covalent bond** (co – VA – lent) A chemical bond created by the sharing of one or more electrons.

**ionic compound** A substance, such as table salt, composed of ions.

**electrolyte** (e – LEC – tro – lyte) A chemical compound that separates into ions when in solution.

## FIGURE 3.8 Formation of Ionic and Covalent Chemical Bonds

Ionic bonds do not involve the sharing of electrons, whereas covalent bonds do.

### A. Ionic bond formation
Ionic bonds are formed when ions of opposite charge are attracted to each other.

Sodium atom (Na)

Chlorine atom (Cl)

**1** Atoms become ions when electrons are transferred.

Sodium ion (Na⁺)

Chloride ion (Cl⁻)

**2** Ions with positive charges (cations) are naturally drawn to atoms with negative charges (anions).

**3** This force of attraction results in the formation of an ionic bond.

Na⁺    Cl⁻

**4** Ionic compounds are held together by ionic bonds.

Salt crystals

**Ionic compound (NaCl)**

### B. Covalent bond formation
Covalent bonds are formed when electrons are shared between atoms.

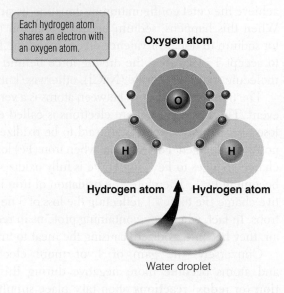

Each hydrogen atom shares an electron with an oxygen atom.

Oxygen atom

Hydrogen atom    Hydrogen atom

Water droplet

## Covalent Bonds: An Electron Tug-of-War

Whereas ionic bonds result from the forceful attractions between oppositely charged ions, covalent bonds are the result of electron sharing. In fact, you can think of a covalent bond as 2 atoms in a game of electron "tug-of-war." Each atom holds on tightly to its electrons while also pulling on the other's electrons. If each atom pulls with similar strength, neither wins the tug-of-war. Therefore, a compromise is reached—electron sharing, the basis of the covalent bond.

The number of electrons in the outer valence shell of an atom determines the number of electrons that it can share (Figure 3.9). For example, a carbon atom has 4 electrons in its outermost shell, and because of the octet rule, seeks to share 4 electrons with other atoms. In other words, carbon atoms can form 4 covalent bonds. A carbon atom can share its electrons in a variety of configurations, making carbon atoms quite versatile. In fact, carbon atoms are found in a wide variety of molecules and therefore play an important role in all living organisms. Many examples of carbon-containing "organic" molecules are found in the body and in food, including proteins, lipids, carbohydrates, and vitamins. However, not all carbon-containing molecules are organic compounds. Carbon-containing molecules such as carbon dioxide, carbon monoxide, and calcium carbonate are considered inorganic molecules.

***Polar and Nonpolar Covalent Bonds***   There are two types of covalent bonds, and the type of bond in a molecule is determined by an atom's ability to share its electrons. A nonpolar covalent bond is formed when atoms share their electrons equally. The simplest example of a nonpolar covalent molecule is hydrogen ($H_2$). Because the hydrogen atoms are identical, neither atom can dominate in controlling the shared electrons. As such, the electrons are shared equally, and the molecule is **nonpolar.**

Conversely, electrons in polar covalent bonds are not shared equally. In the case of a polar covalent bond, one atom has a greater tendency to keep its own electrons as well as to draw away the other atom's electrons.* In polar covalent bonds, one atom in a molecule has a stronger attraction for the shared electrons but does not completely "steal" them. As a result, one atom in the molecule becomes partially positive, because it has lost control of its electron(s) some of the time. The other atom in the molecule becomes partially negative, because it gains an electron (or electrons) some of the time. The unequal distribution of shared electrons results in a covalent bond that has a more positive charge at one end and a more negative charge at the other. Molecules exhibiting this type of unequal charge distribution are called **polar molecules.** The illustration of polar and nonpolar covalent bonds in Figure 3.10 emphasizes these important differences.

Polar molecules are attracted to other polar molecules and ions. This is why water, which is one of the most polar compounds in nature, is so important in the human body. When a compound such as table salt (NaCl) is put into water, the polar water molecules are strongly attracted to the charges on sodium and chloride ions, and the table salt quickly dissolves into its constituent ions ($Na^+$ and $Cl^-$). As shown in Figure 3.11, these attractions are so strong that the water molecules surround the sodium and chloride ions, which keeps them dispersed. Substances such as table salt that dissolve in liquids are called **solutes,** and liquids that dissolve them are called **solvents.** In this example, water is the

*Recall that ionic bonds are formed when electrons are completely transferred from 1 atom to another.

### FIGURE 3.9   The Octet Rule and Carbon Atoms

To satisfy the octet rule, a carbon atom can share an additional 4 electrons with another atom or atoms.

**Carbon atom**

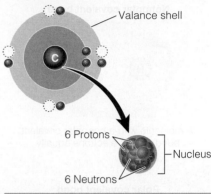

Valence shell

6 Protons

6 Neutrons

Nucleus

**CONNECTIONS** Organic molecules contain a carbon atom bonded to a carbon or hydrogen atom, whereas inorganic molecules do not (Chapter 1, page 3).

**nonpolar** Molecules that are held together by covalent bonds that share electrons equally.

**polar molecule** Molecules that have both partial positively and negatively charged portions.

**solute** (SOL – ute) A substance that dissolves in a solvent.

**solvent** The component of a solution in which a solute dissolves.

## FIGURE 3.10 Nonpolar and Polar Covalent Bonds

Molecules with nonpolar covalent bonds do not have charges, whereas those with polar covalent bonds have partial charges.

**Nonpolar covalent bond**

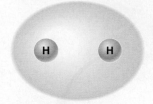

A molecule with a nonpolar covalent bond shares electrons equally.

**Polar covalent bond**

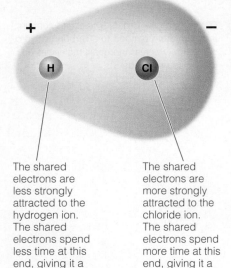

The shared electrons are less strongly attracted to the hydrogen ion. The shared electrons spend less time at this end, giving it a partial positive charge.

The shared electrons are more strongly attracted to the chloride ion. The shared electrons spend more time at this end, giving it a partial negative charge.

Molecules with a polar covalent bond do not share electrons equally.

## FIGURE 3.11 Water as a Solvent

Because water is polar, when table salt (NaCl) is put in water, the salt dissolves. A solute dissolved in solvent together form a solution.

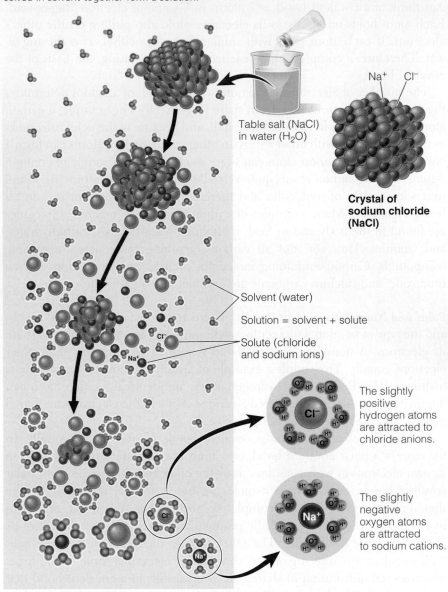

Table salt (NaCl) in water ($H_2O$)

**Crystal of sodium chloride (NaCl)**

Solvent (water)

Solution = solvent + solute

Solute (chloride and sodium ions)

The slightly positive hydrogen atoms are attracted to chloride anions.

The slightly negative oxygen atoms are attracted to sodium cations.

**solution** A mixture of two or more substances that are uniformly dispersed.

**hydrophilic** (hy – dro – PHIL – ic) A substance that dissolves or mixes with water.

**hydrophobic** (hy – dro – PHO – bic) A substance that does not dissolve or mix with water.

**hydrogen bond** A force of attraction between hydrogen atoms and atoms such as oxygen or nitrogen.

solvent, whereas table salt is the solute. A **solution** forms when a solute is dissolved in a solvent. Water is well known for its versatility as a solvent.

In general, polar molecules dissolve in water and are considered **hydrophilic,** meaning "water-loving." In comparison, most nonpolar compounds do not dissolve in water and are considered **hydrophobic,** meaning "water-fearing." Now you know why oil (hydrophobic) and water (hydrophilic) do not mix. In simple terms, they are not attracted to each other.

## Hydrogen Bonds: A Weak But Stabilizing Force

Whereas ionic and covalent bonds hold atoms together in molecules, **hydrogen bonds** are weak electrical attractions that form between hydrogen atoms and atoms with negative charges, such as oxygen or nitrogen. Hydrogen bonds form between atoms within the same molecule or between different mol-

## FIGURE 3.12  Hydrogen Bonds

Hydrogen bonds provide structure and stability to molecules. Hydrogen bonds form between water molecules and are also found in large molecules such as proteins and DNA.

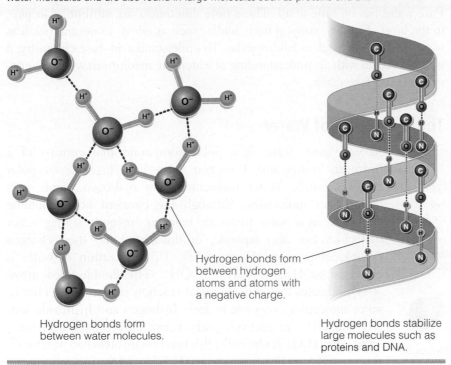

Hydrogen bonds form between water molecules.

Hydrogen bonds form between hydrogen atoms and atoms with a negative charge.

Hydrogen bonds stabilize large molecules such as proteins and DNA.

ecules. Although hydrogen bonds are weak attractions, they play an important role in helping large molecules retain their shape and structure (Figure 3.12). For example, the intricate shapes of proteins and the genetic material DNA are maintained, in part, by numerous hydrogen bonds within these molecules. Similarly, the presence of hydrogen bonds between water molecules makes water unique in terms of its chemical and physical properties.

# ESSENTIAL *Concepts*

The valence shell of most atoms can hold a maximum of 8 electrons, or an octet. An atom attempts to fill its valence shell by electron transfer or sharing. Atoms with nearly complete valence shells tend to satisfy the octet rule by accepting electrons from other atoms. Atoms with only 1 or 2 valence shell electrons tend to donate electrons to other atoms. Atoms that gain electrons become reduced, whereas those that lose electrons become oxidized. The transfer or sharing of electrons is the basis of chemical bonding. Chemical bonds are the forces that hold atoms together. Ionic and covalent bonds are the two major types of chemical bonds. Ionic bonds result from the attraction between oppositely charged ions. Covalent bonds result when atoms share electrons. If the electrons are shared equally, the covalent bond is nonpolar, whereas unequal sharing of electrons results in polar covalent bonds. Polar molecules such as water dissolve ionic compounds such as table salt. Substances that dissolve in liquids are called solutes, and liquids that dissolve solutes are called solvents. A solution forms when a solute is dissolved in a solvent. Polar molecules that dissolve in water are hydrophilic, whereas nonpolar compounds that do not dissolve in water are hydrophobic. Hydrogen bonds are electrical attractions between hydrogen atoms and atoms with negative charges. Hydrogen bonds help large molecules retain their shape.

# Understanding Acids and Bases

Grapefruits and lemons have a sour taste, whereas baking soda tastes bitter. Pure water has no taste at all. These taste differences are attributed, in part, to the level of acidity, ranging from acidic (such as citrus) to neutral (such as water) to alkaline (such as baking soda). To understand acid–base chemistry, it is best to begin with an understanding of water, the medium in which chemical reactions take place.

## The Ionization of Water

As you know, water is a polar compound that consists of 2 hydrogen atoms and 1 oxygen atom held together by polar covalent bonds. Water molecules form hydrogen bonds with other water molecules. Although the covalent bonds holding the atoms in a water molecule together are quite strong, water molecules can also separate, or dissociate, into their charged (ionic) components (Figure 3.13). This **ionization** of water is expressed as $H_2O \longleftrightarrow H^+ + OH^-$. The double-sided arrow ($\longleftrightarrow$) indicates that this chemical reaction is reversible. That is, water molecules dissociate to form hydrogen and hydroxide ions ($H^+$ and $OH^-$, respectively), which can reform back into water molecules ($H_2O$). Technically, this reaction involves two water molecules and forms a hydronium ion ($H_3O^+$), and the ionization reaction is written as $2H_2O \longleftrightarrow H_3O^+ + OH^-$; however, for simplicity we will describe water ionization as forming $H^+$ and $OH^-$.

If you analyzed a sample of pure water, you would find that the number of hydrogen and hydroxide ions is extremely small. That is, in a sample of pure water only a small fraction of the water molecules are in a dissociated, or ionized, state. You would also find that the concentration of hydrogen ions equals the concentration of hydroxide ions. Thus, water is neither acidic nor basic, but neutral. As we explain in the next section, the concentration of hydrogen and hydroxide ions determines whether a solution is acidic or basic.

## FIGURE 3.13 Ionization of Water Molecules

Water molecules can separate, or dissociate, into their charged (ionic) components. This process is called the ionization of water and is reversible.

The double-sided arrow indicates that this chemical reaction is reversible.

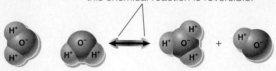

Water molecules    Hydronium ion ($H_3O^+$)    Hydroxide ion ($OH^-$)

## The pH Scale

The ionization of water is the basis for the **pH scale,** which ranges from 0 to 14. Because water is neutral, it has a pH of 7. Fluids that contain a higher concentration of hydrogen ions ($H^+$) than hydroxide ions ($OH^-$) have a pH less than 7. These fluids are said to be **acidic.** Fluids that contain a higher concentration of hydroxide ions ($OH^-$) than hydrogen ions ($H^+$) have a pH greater than 7, and are said be **basic** or **alkaline.** In other words, acidic solutions have a high concentration of hydrogen ions, and basic solutions contain few hydrogen ions, depending on how basic they are.

As a solution becomes more acidic, its pH value decreases. Likewise, the pH value of a solution increases as it becomes more basic. Each consecutive number on the pH scale represents a 10-fold increase or decrease in the concentration of hydrogen ions ($H^+$). Therefore, a fluid with a pH of 3 is 10 times more acidic than a fluid with a pH of 4. Body fluids all have a pH that is maintained within a specific range. For example, the pH of urine typically ranges between 5.5 and 7.5, whereas the pH of blood is usually between 7.3 and 7.5. The pH of various substances is shown in Figure 3.14. The ability to

**ionization** (i – on – i – ZA – tion) The dissociation of a compound into ions.

**pH scale** A scale, ranging from 0 to 14, that signifies the acidity or alkalinity of a solution.

**acidic** Having a pH less than 7.

**basic** or **alkaline** Having a pH greater than 7.

## FIGURE 3.14    The pH Scale

The numbers on the pH scale range from 0 to 14. A substance with a low pH is acidic, a substance with a pH of 7 is neutral, and a substance with a high pH is basic.

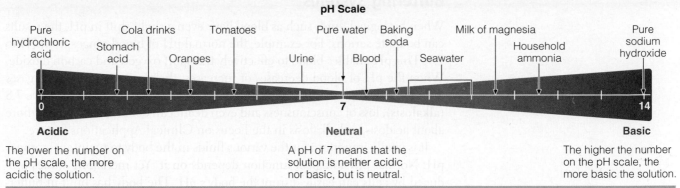

**pH Scale**

Pure hydrochloric acid · Stomach acid · Cola drinks · Oranges · Tomatoes · Urine · Pure water · Blood · Baking soda · Seawater · Milk of magnesia · Household ammonia · Pure sodium hydroxide

0 — 7 — 14

**Acidic** — **Neutral** — **Basic**

The lower the number on the pH scale, the more acidic the solution.

A pH of 7 means that the solution is neither acidic nor basic, but is neutral.

The higher the number on the pH scale, the more basic the solution.

maintain the pH of various body fluids and tissues is important in maintaining optimal function and health.

## Acids and Bases

**Acids** release hydrogen ions ($H^+$) when dissolved in water, whereas **bases** release hydroxide ions ($OH^-$). For example, when hydrochloric acid (HCl) is dissolved in water, hydrogen ($H^+$) and chloride ($Cl^-$) ions are released (Figure 3.15). The increased concentration of hydrogen ions causes the pH of the solution to decrease—or to become more acidic. Conversely, a basic substance such as sodium hydroxide (NaOH) releases sodium ($Na^+$) and hydroxide ($OH^-$) ions when dissolved in water. When sodium hydroxide (NaOH) is dissolved in an acidic solution, the hydroxide ions ($OH^-$) combine with hydrogen ions ($H^+$) present in the solution to form water. As a result, the concentration of hydrogen ions ($H^+$) in the solution decreases, thus increasing the pH. Strong acids, such as hydrochloric acid and sulfuric acid, readily give up hydrogen ions when dissolved in water. Weak acids, such as acetic acid (vinegar), dissociate less readily in water. The same can be said of strong and weak bases. For example, a strong base such as sodium hydroxide, found

## FIGURE 3.15    Acids and Bases

An acid releases hydrogen ions ($H^+$) in solution, whereas a base releases hydroxide ions ($OH^-$) in solution.

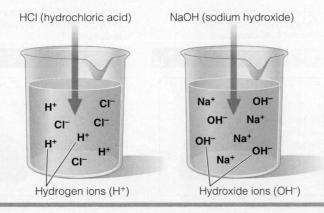

HCl (hydrochloric acid)

NaOH (sodium hydroxide)

Hydrogen ions ($H^+$)

Hydroxide ions ($OH^-$)

**acid**  A substance that releases hydrogen ions ($H^+$) when dissolved.

**base**  A substance that releases hydroxide ions ($OH^-$) when dissolved in water.

in lye, readily dissociates in water. As you will learn later, weak acids play an important role in regulating the pH of the blood.

## Buffering Systems

When biological fluids such as blood have even a slight shift in pH, the results can be quite serious. For example, the normal pH of blood ranges from 7.3 to 7.5. This pH enables blood to effectively transport oxygen and carbon dioxide. When the pH of blood decreases or increases, the ability to carry these gases diminishes. When the pH of the blood goes below 7.0 (**acidosis**) or above 7.8 (**alkalosis**), loss of consciousness and even death can occur. You can read more about acidosis and alkalosis in the Focus on Clinical Applications feature.

It is vitally important for the various fluids in the body to maintain a proper pH. Normal physiological function depends on it. Yet metabolic wastes produced by cells can easily disrupt the body's pH. The body has built-in buffering systems designed to prevent increases and decreases in pH. A **buffer** is a solution that reacts with both acids and bases to maintain a constant pH. As shown in Figure 3.16, buffers can bind hydrogen ions. Removing hydrogen ions from solution increases the pH and decreases the acidity. In this way, buffers act like metabolic "mops" that soak up excess hydrogen ions. Buffers can also donate hydrogen ions. The release of hydrogen ions decreases the pH of a fluid and increases the acidity. These components of buffering systems, found in the blood, kidneys, and lungs, work together to resist changes in pH. Buffers can take corrective actions to prevent shifts in pH by releasing or binding

**acidosis** (a – ci – DO – sis) A condition resulting from the accumulation of acids in body fluids.

**alkalosis** (al – ka – LO – sis) A condition resulting from excess base in body fluids.

**buffer** A substance that releases or binds hydrogen ions in order to resist changes in pH.

## FIGURE 3.16 Buffers

A buffer can act like an acid and like a base.

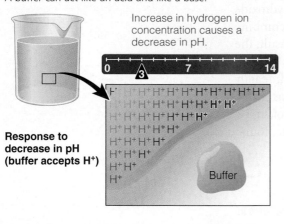

**Response to decrease in pH (buffer accepts H⁺)**

Increase in hydrogen ion concentration causes a decrease in pH.

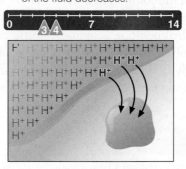

Buffer accepts hydrogen ions. Hydrogen concentration of the fluid decreases.

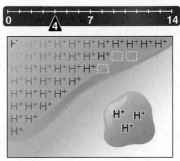

By decreasing the concentration of hydrogen ions, the pH increases.

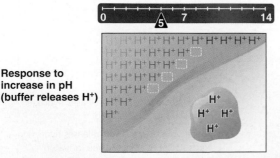

**Response to increase in pH (buffer releases H⁺)**

Decrease in hydrogen ion concentration causes an increase in pH.

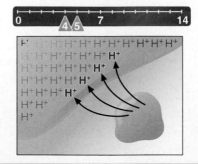

Buffer releases hydrogen ions. Hydrogen concentration of the fluid increases.

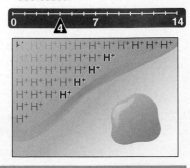

By increasing the concentration of hydrogen ions, the pH decreases.

# *Focus On* CLINICAL APPLICATIONS
## Acidosis and Alkalosis

Acidosis and alkalosis are serious conditions caused by shifts in the pH of body fluids. Acidosis results from the accumulation of acids, whereas alkalosis results from excess base. Both affect the nervous system. For example, mild acidosis causes depression of the nervous system. Symptoms of mild acidosis include headaches, lethargy, loss of appetite, and difficulty breathing. Severe acidosis can cause coma or death. Conversely, alkalosis stimulates the nervous system. Mild alkalosis makes a person feel dizzy and light-headed. However, severe alkalosis can lead to muscle cramps and convulsions.

There are several causes of acidosis. Health conditions that affect breathing, such as emphysema, asthma, or congestive heart failure, can lead to respiratory acidosis. Respiratory acidosis results when the lungs have difficulty removing carbon dioxide from the blood. As a result, the pH of the blood decreases. Even holding your breath can result in mild acidosis, as can sleep medications that slow breathing. Another type of acidosis, called metabolic acidosis, is associated with starvation, chronic alcoholism, and diabetes. These conditions cause the body to rely heavily on stored fat as an energy source. When this occurs, the fat is not completely broken down, resulting in the release of acid compounds into the blood. When the kidneys are not able to excrete sufficient amounts of acid in the urine, the acid accumulates in the blood, disrupting its pH.

Metabolic alkalosis can result from a variety of conditions, including excessive vomiting and overuse of certain drugs such as antacids, diuretics, or laxatives. These conditions cause excessive fluid, sodium, and potassium loss from the body. As a result, the ability of the kidneys to maintain the pH of the blood is impaired, resulting in metabolic alkalosis. This is one reason why people with eating disorders who restrict food intake and/or abuse diuretics or laxatives can develop serious health problems. In contrast to metabolic alkalosis, respiratory alkalosis can result from rapid, deep breathing, called hyperventilation. This can occur when a person feels anxious or panicky. Rapid breathing decreases the level of carbon dioxide in the blood, which in turn causes body fluids to become more alkaline. Respiratory alkalosis can make a person feel dizzy and light-headed. Breathing into a paper bag increases carbon dioxide intake, which in turn restores pH.

hydrogen ions. Collectively, buffering systems maintain pH in a tight range and provide an effective defense against acidosis and alkalosis.

For some people, excess production of stomach acid can lead to indigestion. To relieve the discomfort associated with indigestion, many people take antacids. Nonprescription antacids contain active ingredients that act as bases, such as aluminum salts, magnesium salts, calcium carbonate, and sodium bicarbonate. The base in the antacid reacts with the stomach acid to form water and salt. For example, adding sodium bicarbonate ($NaHCO_3$) to hydrochloric acid (HCl) results in the formation of sodium chloride (NaCl), water ($H_2O$), and the gas carbon dioxide ($CO_2$).

# ESSENTIAL *Concepts*

Small amounts of water molecules dissociate into their ionic components. Water is considered neutral because when ionized, the concentration of hydrogen ions equals the concentration of hydroxide ions. The ionization of water is the basis for the pH scale, which ranges from 0 to 14. Water is neutral and has a pH of 7. Acidic fluids have more hydrogen ions than hydroxide ions and have a pH less than 7. Basic fluids have more hydroxide ions than hydrogen ions and have a pH greater than 7. Each consecutive number on the pH scale represents a 10-fold increase or decrease in the concentration of hydrogen ions. Acids release hydrogen ions ($H^+$) when dissolved in water, whereas bases release hydroxide ions ($OH^-$). The body has built-in buffering systems that prevent changes in pH. A buffer releases or binds hydrogen ions ($H^+$) to maintain a constant pH.

# Macromolecules: The Molecules of Life

So far, we have talked about molecules that are small and simple. However, molecules can also be very large, consisting of thousands of atoms bonded together. Carbohydrates, proteins, and nucleic acids (DNA and RNA) are large molecules, or **macromolecules,** that are vital to the functions of cells. The raw materials used to make macromolecules come from the chemical compounds in foods that we eat. As described in previous chapters, these chemical compounds are called nutrients. Although macromolecules differ in structure and function, they all contain carbon atoms bonded to hydrogen, oxygen, and/or other elements such as nitrogen. Next we discuss macromolecule function, structure, and chemical reactions of importance.

## Macromolecules Have Diverse Functions

Macromolecules are extraordinary molecules that play roles in practically all aspects of life. For example, carbohydrates are important sources of energy (ATP) for the body and are integral parts of cell structure. Proteins are amazingly diverse and have both structural and functional roles. Nucleic acids are the molecules of genetic information and protein synthesis. Together, the nucleic acids deoxyribonucleic acid (DNA) and ribonucleic acid (RNA) direct and carry out protein synthesis. DNA is the genetic material that stores information necessary for protein synthesis, whereas RNA interprets and transmits these instructions to sites within the cell where proteins are made.

## Macromolecules Are Made Up of Subunits

Macromolecules are **polymers,** which are large molecules made up of repeating subunits called **monomers.** Monomers are joined together by covalent bonds. You can think of a polymer as a beaded necklace, where each bead represents a monomer. As shown in Figure 3.17, proteins are polymers of amino acids, and some complex carbohydrates are polymers of glucose. Nucleic acids are polymers of nucleotides. A nucleotide consists of three parts: a nitrogenous base, a phosphate molecule, and a sugar molecule (deoxyribose in DNA and ribose in RNA).

### FIGURE 3.17    Polymers

A polymer is like a beaded necklace, where each bead represents a monomer.

**Protein**

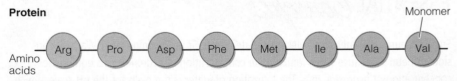

Proteins are polymers of amino acids.

**Carbohydrate**

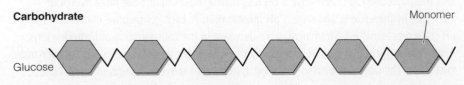

Some carbohydrates are polymers of glucose.

**macromolecules** (MAC – ro – mol – e – cule) Large molecules made by cells such as proteins, that are made up of smaller subunits.

**polymer** (POL – y – mer) Large molecules made up of repeating subunits.

**monomer** (MON – o – mer) Small molecules that join together to form a polymer.

## FIGURE 3.18    Condensation and Hydrolysis

Condensation and hydrolysis are "make-and-break" reactions.

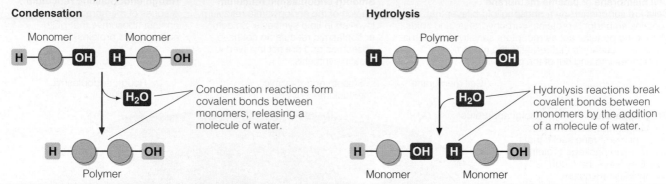

**Condensation**

Monomer        Monomer

Condensation reactions form covalent bonds between monomers, releasing a molecule of water.

Polymer

**Hydrolysis**

Polymer

Hydrolysis reactions break covalent bonds between monomers by the addition of a molecule of water.

Monomer        Monomer

## Condensation and Hydrolysis: Make-and-Break Reactions

Polymers are assembled and disassembled within cells. The chemical reaction that joins monomers together is called **condensation**, whereas **hydrolysis** reactions break polymers apart (Figure 3.18). Water plays an important role in both of these types of chemical reactions. The process of condensation, which makes covalent bonds, results in the formation and release of a water molecule. In contrast, a molecule of water is used to break covalent bonds during hydrolysis reactions. You can think of condensation and hydrolysis as opposite "make-and-break" reactions. These reactions play important roles in digestion and metabolism.

## ESSENTIAL *Concepts*

Carbohydrates, proteins, and nucleic acids (DNA and RNA) are macromolecules made by our cells. The raw materials used to make these molecules come from the nutrients in foods that we eat. Carbohydrates are an important source of energy for the body and are integral parts of cell structure. Proteins are diverse molecules that have both functional and regulatory roles. The nucleic acids—deoxyribonucleic acid (DNA) and ribonucleic acid (RNA)—direct and carry out protein synthesis. Macromolecules are polymers, made up of small subunits, or monomers, joined together by covalent bonds. The chemical reaction that joins monomers together is called condensation, whereas hydrolysis reactions break polymers apart.

# Cells and Their Components

All living organisms consist of cells, the "building blocks" of the body. Cells, which have both structural and functional roles, are surrounded by a protective cell membrane. Within the cell, small membrane-bound structures called organelles carry out specialized functions that are critical for life (Figure 3.19).

## Cell (Plasma) Membranes

Cell membranes (also called plasma membranes) provide a boundary between the **extracellular** (outside the cell) and **intracellular** (within the cell) environments. In addition, cell membranes regulate what gets into and out of cells. The unique structure of cell membranes can be compared to that of a

**condensation** A chemical reaction that results in the formation of water.

**hydrolysis** (hy – DRO – ly – sis) A chemical reaction whereby compounds react with water.

**extracellular** (ex – tra – CEL – lu – lar) Situated outside of a cell or cells.

**intracellular** (in – tra – CEL – lu – lar) Situated within a cell.

## FIGURE 3.19 A Typical Cell

**Cell membrane** or **plasma membrane**
Cells are surrounded by a phospholipid bilayer that contains embedded proteins, carbohydrates, and lipids. Membrane proteins act as receptors sensitive to external stimuli and channels that regulate the movement of substances into and out of the cell.

**Smooth endoplasmic reticulum**
Region of the endoplasmic reticulum involved in lipid synthesis. Smooth endoplasmic reticula do not have ribosomes and are not involved in protein synthesis.

**Rough endoplasmic reticulum**
A series of membrane sacks that contain ribosomes that build and process proteins.

**Golgi apparatus**
The Golgi apparatus is a series of membrane sacks that process and package proteins after they leave the rough endoplasmic reticulum.

**Lysosome**
Contains digestive enzymes that break down proteins, lipids, and nucleic acids. They also remove and recycle waste products.

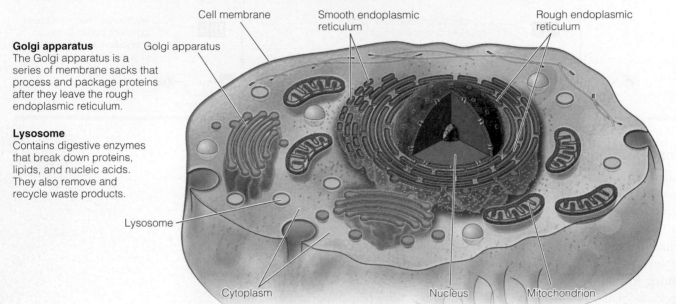

Cell membrane

Smooth endoplasmic reticulum

Rough endoplasmic reticulum

Golgi apparatus

Lysosome

Cytoplasm

Nucleus

Mitochondrion

**Cytoplasm** or **cytosol**
The cytoplasm is the gel-like substance inside cells. Cytoplasm contains cell organelles, protein, electrolytes, and other molecules.

**Nucleus**
The nucleus contains the DNA in the cell. Molecules of DNA provide coded instructions used for protein synthesis.

**Mitochondrion**
Organelles that produce most of the energy (ATP) used by cells.

sandwich. In cell membranes, the "slices of bread" are made up of two phospholipid sheets, called the phospholipid bilayer. Whereas a sandwich may have meat and cheese between the slices of bread, the phospholipid bilayer is embedded with proteins, carbohydrates, and cholesterol. Phospholipids consist of lipids with a phosphate group attached, and are both hydrophilic and hydrophobic (remember, hydrophilic means water-loving, whereas hydrophobic means water-fearing). The phospholipids are aligned so that the hydrophilic heads (phosphate groups) are oriented toward the watery environments both within and outside of the cell. The hydrophobic tails (lipids) point inward, away from the water. This is described in more detail in Chapter 7.

Proteins associated with cell membranes have both structural and functional roles. These include transporting materials into and out of cells, acting as receptors for other molecules, and providing cell-to-cell communication. Cholesterol, a type of lipid, is important for membrane stability and fluidity. Carbohydrates form hairlike projections, which act like antennae and enable cells to recognize and interact with each other. Carbohydrates also help communicate conditions outside of cells across the cell membrane to the intracellular compartment.

## Transport Across Cell Membranes

Cell membranes have many functions, one of which is to control the movement of substances into and out of cells. Cell membranes are considered selectively permeable, allowing some substances to cross them more readily than others. Nutrients and other extracellular substances are transported across cell membranes in a variety of ways, as described next.

## Passive Transport Mechanisms

Cell transport processes that do not require energy are called **passive transport** mechanisms, as shown in Figure 3.20. The three main types of passive transport mechanisms are simple diffusion, facilitated diffusion, and osmosis.

***Simple Diffusion*** **Simple diffusion** enables substances to move through the cell membrane from a region of higher concentration to a region of lower concentration, without any energy expenditure. When this occurs, the substance is said to move passively "down its concentration gradient." Once the concentration of the substance is the same on both sides of the cell membrane, there is no further movement of the substance in either direction. This condition is called **equilibrium.** Small lipids, ions, and charged molecules readily cross cell membranes by way of simple diffusion, as do gases such as oxygen and carbon dioxide. Oxygen, which typically is in high concentration outside the cell, readily diffuses down its concentration gradient into the intracellular environment. Likewise, carbon dioxide, a metabolic waste product, moves across the cell membrane from a higher intracellular concentration to a lower extracellular concentration. This favorable gas exchange provides cells with a plentiful supply of oxygen while eliminating excess carbon dioxide.

***Facilitated Diffusion*** **Facilitated diffusion** also involves the movement of substances down a concentration gradient (high concentration to low concentration) but differs from simple diffusion because it requires the assistance of a membrane-bound transport protein. Transport proteins escort materials across cell membranes. This mechanism of transport occurs at a much faster

**passive transport** A non–energy-requiring mechanism whereby a substance moves from a region of higher concentration to that of a lower concentration.

**simple diffusion** A passive transport mechanism whereby a substance crosses a cell membrane without the assistance of a transport protein.

**equilibrium** The state when substances separated by a membrane are in equal concentration.

**facilitated diffusion** A passive transport mechanism whereby a substance crosses a cell membrane with the assistance of a transport protein.

## FIGURE 3.20 Passive Transport Mechanisms

Passive transport mechanisms—such as simple diffusion, facilitated diffusion, and osmosis—do not require energy (ATP).

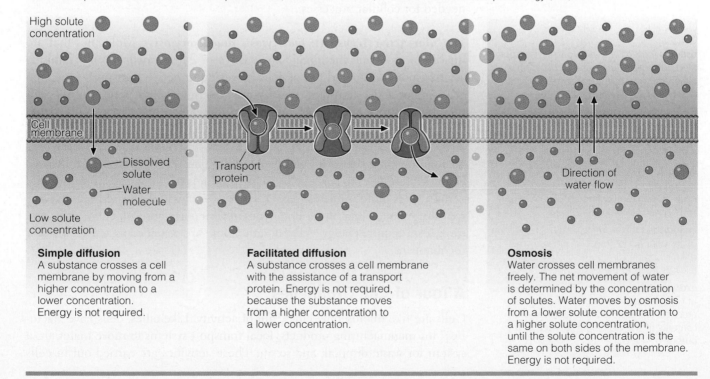

**Simple diffusion**
A substance crosses a cell membrane by moving from a higher concentration to a lower concentration. Energy is not required.

**Facilitated diffusion**
A substance crosses a cell membrane with the assistance of a transport protein. Energy is not required, because the substance moves from a higher concentration to a lower concentration.

**Osmosis**
Water crosses cell membranes freely. The net movement of water is determined by the concentration of solutes. Water moves by osmosis from a lower solute concentration to a higher solute concentration, until the solute concentration is the same on both sides of the membrane. Energy is not required.

rate than simple diffusion, except when all transport proteins are occupied. When this happens, the rate of facilitated diffusion decreases. An increase in the availability of transport proteins increases the rate of facilitated diffusion. Cells can regulate the rate of facilitated diffusion by altering the availability of transport proteins.

*Osmosis*   Unlike many dissolved substances (solutes), water is unique in that it passes freely across cell membranes. However, cells rupture if they contain too much water and collapse if they contain too little. Therefore, the movement of water across cell membranes must be carefully regulated. This is accomplished by concentrating various solutes within and surrounding the cell. A solution with a low solute concentration contains relatively more water than a solution with a high solute concentration. Therefore, water moves from the region of lower solute concentration to that of a higher solute concentration, until equilibrium is reached. This process, called **osmosis,** determines the net movement of water in either direction across the cell membrane. This is described in more detail in Chapter 13.

## Active Transport Mechanisms

Cell transport processes that require energy (ATP) to transport material across cell membranes are called active transport mechanisms. Examples of active transport mechanisms include carrier-mediated active transport and vesicular active transport, as shown in Figure 3.21.

*Carrier-Mediated Active Transport*   Similar to swimming upstream, some molecules must cross cell membranes against the prevailing concentration gradient, moving from a region of lower concentration to that of a higher concentration. In cells, this uphill journey is accomplished by **carrier-mediated active transport.** This transport mechanism requires both energy (ATP) and the assistance of a transport protein. Active transport mechanisms "pump" molecules or ions across cell membranes against their concentration gradients, restoring and maintaining an optimal concentration gradient needed for cellular processes.

*Vesicular Active Transport: Endocytosis and Exocytosis*   Molecules that are unable to cross cell membranes by passive or carrier-mediated mechanisms are transported into and out of cells by **vesicular active transport.** One type of vesicular active transport, **endocytosis,** occurs when a portion of the cell membrane surrounds an extracellular particle, enclosing it in a saclike structure called a vesicle. The contents of the vesicle are then released to the inside of the cell. For example, using endocytosis white blood cells take up bacteria that are then destroyed within the cell. The reverse process, **exocytosis,** enables substances to leave cells. Many products made by cells (for example, certain types of hormones) are packaged in vesicles and released into the surrounding extracellular fluid. Both types of vesicular transport, endocytosis and exocytosis, require energy (ATP) and therefore are classified as active transport mechanisms.

## A Tour of Cell Organelles

Cells are like microscopic cities, full of activity. Like cities, cells have "factories" for manufacturing products, local transport systems to move materials, a system for waste disposal, and so on. These activities are carried out in cells

**osmosis** (os – MO – sis) A passive transport mechanism whereby water moves from a region of lower solute concentration to that of a higher solute concentration.

**carrier-mediated active transport** An energy-requiring mechanism whereby a substance moves from a region of lower concentration to a region of higher concentration.

**vesicular active transport** (ve – SIC – u – lar) An energy-requiring mechanism whereby large molecules move into or out of cells by an enclosed vesicle.

**endocytosis** (en – do – cy – TO – sis) A form of vesicular active transport whereby the cell membrane surrounds substances and releases them to the cytoplasm.

**exocytosis** (ex – o – cy – TO – sis) A form of vesicular active transport whereby cell products are enclosed in a vesicle and the contents of the vesicle are released to the outside of the cell.

# FIGURE 3.21   Active Transport Mechanisms

Active transport mechanisms—such as carrier-mediated active transport and vesicular active transport—require energy (ATP).

**A. Carrier-mediated active transport**
A solute crosses a cell membrane with the assistance of a transport protein. Energy is required, because the substance moves against its concentration gradient, moving from a lower concentration to a higher concentration.

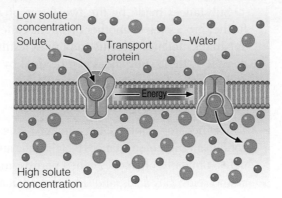

**B. Vesicular active transport**
Part of the cell membrane surrounds solutes. When it closes, the solutes are released outside (exocytosis) or inside (endocytosis) the cell.

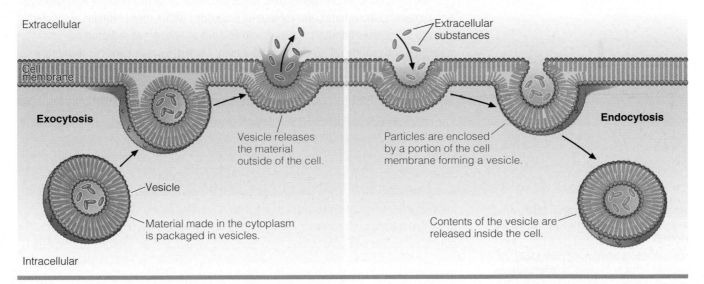

by structures called **organelles,** which are distributed in the gel-like intracellular matrix called the **cytoplasm,** or **cytosol.** Each organelle is responsible for a specific function. Some of the major organelles found in cells and their related functions are discussed next.

## Production, Receiving, and Distribution Organelles

Cellular structures called **mitochondria** serve as power stations, converting energy-yielding nutrients (glucose, fatty acids, and amino acids) into a form of energy that cells can use. Mitochondria contain inner and outer compartments, both of which play a role in generating adenosine triphosphate (ATP), a high-energy molecule used to fuel cellular activity. Another organelle, called the cell **nucleus,** houses the genetic material DNA, which provides the "blueprint" for protein synthesis. Information encoded in the DNA is transported out of the nucleus to organelles called **ribosomes.** Some ribosomes are attached to a network of folds called the **endoplasmic reticulum (ER),** which serves as the "work surface" for protein synthesis. The ribosomes give the ER a bumpy appearance. Therefore, protein-producing ER are referred to as the

**organelle** (or – gan – ELLE)  Within a cell, a structure that has a particular function.

**cytoplasm** (also called **cytosol**)  The gel-like matrix inside cells but outside of cell organelles.

**mitochondria** (mi – to – CHON – dri – a)  Cellular organelles involved in generating energy (ATP).

**nucleus**  A membrane-enclosed organelle that contains the genetic material DNA.

**ribosome** (RI – bo – some)  A cellular organelle involved in protein synthesis.

**endoplasmic reticulum** (EN – do – plas – mic re – TIC – u – lum)  A cell organelle involved in the synthesis and transport of materials within and from cells.

rough endoplasmic reticula (RER). Other endoplasmic reticula are involved with lipid synthesis. These "ribosome-free" ER have a smooth appearance and are therefore referred to as **smooth endoplasmic reticula (SER).**

Substances made by the two forms of endoplasmic reticula are sent for further processing to another organelle called the **Golgi apparatus.** The Golgi apparatus modifies proteins and lipids, resulting in the finished product. Once complete, these substances can be used by the cell or packaged into vesicles and exported from the cell via exocytosis.

### Cleanup Organelles

**Lysosomes** are organelles that function as the waste disposal system in cells. They contain a variety of substances that degrade and recycle worn-out cellular components. Similarly, organelles called **peroxisomes** contain substances that degrade amino acids, fatty acids, and substances that are toxic or harmful to cells.

## ESSENTIAL *Concepts*

Cells are surrounded by cell membranes, which are made up of two phospholipid sheets, with proteins, carbohydrates, and cholesterol embedded within them. Proteins associated with cell membranes transport materials into and out of cells, act as receptors for other molecules, and provide cell-to-cell communication. Cholesterol maintains cell membrane structure and function. Carbohydrates function as antennae that enable cells to interact. Cell membranes control the transport of substances into and out of cells. Transport mechanisms that do not require energy are called passive transport systems, whereas mechanisms that require energy are called active transport systems. Passive transport systems include simple diffusion, facilitated diffusion, and osmosis. Active transport systems include carrier-mediated active transport and vesicular transport. Structures called organelles carry out intracellular activities. Mitochondria serve as power generators, extracting energy from energy-yielding nutrients and converting it to ATP. The cell nucleus contains the genetic material DNA. Information encoded in the DNA is used for protein synthesis. The Golgi apparatus modifies substances made by the cells and prepares the final products for distribution. Lysosomes and peroxisomes help remove worn-out or toxic substances from cells.

**rough endoplasmic reticulum (RER)** An organelle studded with ribosomes that synthesizes proteins.

**smooth endoplasmic reticulum (SER)** An organelle that is involved in lipid synthesis.

**Golgi apparatus** (GOL – gi ap – pa – RA – tus) An organelle that packages macromolecules and other substances in vesicles.

**lysosome** (LY – so – some) An organelle that contains enzymes that degrade molecules.

**peroxisome** (per – OX – i – some) An organelle that contains enzymes that break down amino acids, fatty acids, and toxic substances in cells.

**tissue** (TIS – sue) An aggregation of specialized cell types that are similar in form and function.

**epithelial tissue** (ep – i – THE – li – al) Tissue that forms a protective layer on bodily surfaces and lines internal organs, ducts, and cavities.

**connective tissue** (con – NEC – tive) Tissue that supports, connects, and anchors body structures.

**muscle tissue** Tissue that specializes in movement.

# Tissues, Organs, and Organ Systems

So far, we have discussed how atoms make up molecules, how molecules make up cells, and how cells carry out the basic functions of life. The next level of complexity is the **tissue,** which is a group of cells that carries out specialized functions. In the human body, four types of tissues comprise over 40 organs, which in turn, make up 11 unique organ systems.

## Tissues

The human body contains four general types of tissue: (1) epithelial, (2) connective, (3) muscle, and (4) neural, shown in Figure 3.22. **Epithelial tissue** (or epithelium) provides a protective layer on body surfaces (skin) as well as lining body cavities, ducts, and organs. **Connective tissue** is the "glue" that holds the body together. Tendons, cartilage, and some parts of bones are examples of connective tissue. Blood is also a type of connective tissue, consisting of cells and platelets in a liquid called plasma. **Muscle tissue** specializes in move-

**FIGURE 3.22 Four Basic Types of Tissue**

Epithelial, connective, neural, and muscle tissue make up all the organs in the human body.

**Epithelial tissue**

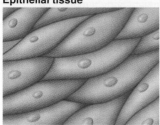

Epithelial tissue covers and lines body surfaces, organs, and cavities.

**Connective tissue**

Connective tissue provides structure to the body by binding and anchoring body parts.

**Neural tissue**

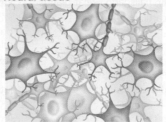

Neural tissue plays a role in communication by receiving and responding to stimuli.

**Muscle tissue**

Muscle tissue contracts and shortens when stimulated, playing an important role in movement.

ment. Movement of skeletal muscle tissue is under voluntary control, whereas movement of smooth and cardiac muscle tissue is not. Smooth muscle tissue is found around many structures and organs such as the esophagus, stomach, and small intestine, whereas cardiac tissue is found exclusively in the heart. The brain, spinal cord, and nerves consist of **neural tissue,** which plays an important communicative role in the body.

## Organs and Organ Systems

**Organs** consist of two or more different types of tissue, functioning collectively to perform a variety of related tasks. For example, the small intestine, which has several digestive functions, contains all four types of tissue. Together, these tissue types enable the small intestine to move (muscle tissue), communicate (neural tissue), and provide a surface (epithelial and connective tissues) for nutrients to be transported into the blood. An outer layer of connective tissue anchors the small intestine within the abdominal cavity.

Typically, several organs associate together as part of an **organ system.** Each organ in an organ system carries out important physiological functions. The human body has 11 organ systems, all of which are pertinent to the study of nutrition. The major organ systems and their basic functions are summarized in Table 3.1. Organ systems maintain a stable internal environment, despite an ever-changing external environment, through communication and feedback systems. The mechanisms by which the body maintains a stable internal environment are varied and complex.

## ESSENTIAL *Concepts*

The human body contains four types of tissue: (1) epithelial tissue, (2) connective tissue, (3) muscle tissue, and (4) neural tissue. Each carries out a specialized function. Epithelial tissue provides a protective layer on body surfaces as well as lining body cavities, ducts, and organs. Connective tissue, such as tendons, cartilage, and parts of bones, hold the body together. The brain, spinal cord, and nerves consist of neural tissue and play an important communicative role in the body. Muscle tissue provides an important role in movement. Organs consist of two or more different types of tissue. Over 40 different organs make up 11 organ systems. Organ systems maintain a stable internal environment through communication and feedback systems.

**neural tissue** Tissue that specializes in communication via neurons.

**organ** A group of tissues that combine to carry out coordinated functions.

**organ system** Organs that work collectively to carry out related functions.

## TABLE 3.1 Organ Systems and Related Major Functions

| Organ System | Major Organs and Structures | Major Function |
|---|---|---|
| Integumentary system | Skin, hair, nails, and sweat glands | Protects against pathogens and helps regulate body temperature. |
| Skeletal system | Bones, cartilage, and joints | Provides support and structure to the body. The bone marrow of some bones produces blood cells. Also provides a storage site for certain minerals. |
| Muscular system | Smooth, cardiac, and skeletal muscle | Assists in voluntary and involuntary body movements. Maintains body position and posture. |
| Nervous system | Brain, spinal cord, nerves, and sensory receptors | Interprets and responds to information. Controls the basic senses, movement, and intellectual functions. |
| Endocrine system | Endocrine glands | Produces and releases of hormones that control physiological functions such as reproduction, hunger, satiety, blood glucose regulation, metabolism, and stress response. |
| Respiratory system | Lungs, nose, mouth, throat, and trachea | Governs gas exchange between the blood and air. Also assists in regulating blood acid–base (pH) balance. |
| Circulatory system | Heart, blood vessels, blood, lymph vessels, lymph nodes, and lymph organs | Transports nutrients, waste products, gases, and hormones. Also plays a role in regulating body temperature. Helps remove foreign substances and plays a role in immunity. |
| Digestive system | Mouth, esophagus, stomach, small intestine, and large intestine | Governs the physical and chemical breakdown of food into a form that can be absorbed into the blood. Eliminates solid wastes. |
| Reproductive system | Gonads and genitals | Carries out reproductive functions and is associated with sexual characteristics, sexual function, and sexual behaviors. |
| Urinary system | Kidneys, bladder, and ureters | Removes metabolic waste products from the blood; governs nutrient resorption, acid–base balance, and regulates water balance. |
| Immune system | White blood cells, lymph vessels, and lymphatic tissue | Provides a defense against foreign bodies such as bacteria and viruses. |

# Homeostasis: A Matter of Balance

The ability of organ systems to work together to carry out common functions requires constant communication. In other words, the right hand must know what the left hand is doing. Through what are called feedback systems, the body is able to maintain a relatively stable internal environment. This process, known as **homeostasis,** is an important concept in physiology and nutrition. Diseases that disrupt homeostasis can have serious health consequences.

## The Role of Nervous and Endocrine Systems in Homeostasis

The two major homeostatic control systems in the body are the nervous and endocrine systems. The nervous system receives and transmits information via electrical impulses between nerve cells, whereas the endocrine system communicates via chemical messengers, called **hormones,** in the blood. The nervous system, which consists of the brain, spinal cord, sensory receptors, and nerves, receives input from a variety of sensors that detect change. For example, some sensory receptors detect temperature changes, whereas others detect stretch or tension. These sensory receptors then transmit signals to integrative control centers, which in turn respond by sending signals to muscles and/or glands (effectors). For instance, when nerves (sensors) are stimulated by the sight and/or smell of food, signals are sent to the brain (control center), which then signals the digestive tract (effector) to prepare for the arrival of food.

Hormones are released from glands or cells in response to various stimuli that include chemical changes in the blood, other hormones, and signals

**homeostasis** (ho – me – o – STA – sis) A state of balance or equilibrium.

**hormone** A chemical substance produced by an endocrine gland that is released into the blood and stimulates a response elsewhere in the body.

from the nervous system. Hormones exert their effects by binding to receptors on specific tissues, which then initiate an appropriate response to the initial stimulus. The endocrine system is slower to respond to these stimuli than is the nervous system. Whereas the endocrine system takes seconds to hours to respond to stimuli, the nervous system can respond immediately. In general, the endocrine system regulates metabolism, growth, and reproduction. Together, the nervous and endocrine systems continuously monitor our internal environment, responding to change and restoring balance. These mechanisms allow complex organisms, such as humans, to adapt in an ever-changing environment.

## Negative and Positive Feedback Systems

The body has two types of feedback systems—negative feedback and positive feedback. **Negative feedback systems** play important roles in homeostasis by opposing changes in the internal environment and by initiating corrective responses that restore balance (Figure 3.23). An example of a negative feedback system is the regulation of blood glucose. In response to a carbohydrate-rich meal, blood glucose levels rise (a change in the internal environment). The pancreas receives this information (the stimulus) and initiates a response—the release of the hormone insulin. Insulin takes corrective action, that is, it facilitates the uptake of glucose out of the blood by cells. The decline in blood glucose (restored internal environment) signals the pancreas to stop

### FIGURE 3.23 Negative Feedback Systems and Homeostasis

A disruption in homeostasis causes the body to respond in such a way as to restore homeostatic conditions. When homeostasis is restored, negative feedback systems halt the continuation of the body's response to the initial stimulus.

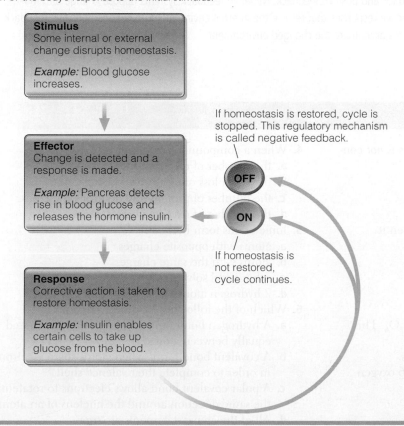

**Stimulus**
Some internal or external change disrupts homeostasis.

*Example:* Blood glucose increases.

If homeostasis is restored, cycle is stopped. This regulatory mechanism is called negative feedback.

**Effector**
Change is detected and a response is made.

*Example:* Pancreas detects rise in blood glucose and releases the hormone insulin.

OFF

ON

If homeostasis is not restored, cycle continues.

**Response**
Corrective action is taken to restore homeostasis.

*Example:* Insulin enables certain cells to take up glucose from the blood.

**negative feedback system** Physiological response that works to restore a system to normal.

releasing insulin, thus completing the negative feedback loop. Conversely, if blood glucose levels decrease too much, the pancreas releases a hormone called glucagon, which restores balance by increasing the level of glucose in the blood. Thus both insulin and glucagon are hormones that are involved in negative feedback systems regulating blood glucose homeostasis.

Whereas negative feedback systems restore balance, **positive feedback systems** accentuate the changed environment. As might be expected, positive feedback loops are rare in the body. However, one example of a positive feedback system occurs during the process of giving birth. During labor, contractions of the uterus cause the cervix to stretch (the stimulus), which in turn triggers the release of a hormone called oxytocin (the response). Under the influence of oxytocin, the uterus (the effector) contracts more forcibly, causing the cervix to stretch even more. This positive feedback system continues until the baby is born. Homeostasis is restored when the stretching of the cervix subsides, diminishing the release of oxytocin, causing the contractions to stop. Once again, the feedback loop is complete.

## ESSENTIAL *Concepts*

The concept of homeostasis encompasses a complex set of important adaptive mechanisms. Homeostasis requires that the body be able to detect change (the stimulus), activate an effector, and carry out a response. The two major homeostatic control systems in the body are the nervous and endocrine systems. The nervous system receives and transmits information via electrical impulses between nerve cells, whereas the endocrine system communicates via chemical messengers, called hormones, in the blood. Together, the nervous and endocrine systems continuously monitor our internal environment, responding to change and restoring balance. The body has two types of feedback systems—negative feedback and positive feedback. Negative feedback systems oppose change and make corrective responses that restore the internal environment. In contrast, positive feedback systems accentuate the changed environment.

**positive feedback system** Physiological response that works to destabilize a system, accentuating an abnormal stimulus.

## Review Questions   Answers are found in Appendix G.

1. Which of the following subatomic particles is *not* contained in the nucleus of an atom?
   a. Protons
   b. Electrons
   c. Neutrons
2. A valence shell of an atom is complete when it contains
   a. an octet of electrons.
   b. an octet of protons.
   c. 4 neutrons and 4 protons.
   d. 4 cations and 4 anions.
3. The molecular formula of glucose is $C_6H_{12}O_6$. This means that glucose
   a. consists of three different types of atoms.
   b. consists of 6 carbon, 12 hydrogen, and 6 oxygen atoms.
   c. fits the definition of a molecule.
   d. all of the above are *true*.

4. When a compound is oxidized,
   a. the number of electrons increases.
   b. there is a loss of electrons.
   c. the number of protons and neutrons increases.
   d. the number of oxygen atoms decreases.
5. Ionic bonds form between
   a. atoms with opposite charges.
   b. atoms with the same charge.
   c. solutes and solvents.
   d. 2 hydrogen atoms.
6. Which of the following is a *true* statement?
   a. A hydrogen bond forms when protons are shared equally between atoms.
   b. A covalent bond forms when atoms share electrons in order to complete their valence shell.
   c. A polar covalent bond allows electrons to rotate in the same direction around the nucleus of an atom.
   d. All of the above statements are *true*.

**7.** A solution with a pH of 2 is considered _____.
   **a.** alkaline
   **b.** neutral
   **c.** acidic
   **d.** ionic

**8.** A compound that prevents increases and decreases in pH is called a(n) _____.
   **a.** solute
   **b.** solvent
   **c.** buffer
   **d.** ion

**9.** A condensation reaction
   **a.** neutralizes acids.
   **b.** is needed to form covalent bonds between monomers.
   **c.** is needed to break covalent bonds in polymers.
   **d.** dissolves ionic compounds.

**10.** Cell membranes primarily consist of two _____ layers.
   **a.** protein
   **b.** carbohydrate
   **c.** phospholipid
   **d.** nucleic acid

**11.** Define the terms *oxidation* and *reduction*. Why does an atom become more negative when it becomes reduced?

**12.** How are ionic and covalent bonds different? How are these chemical bonds formed?

**13.** What is the difference between a nonpolar and polar covalent bond?

**14.** When an ionic compound dissolves in water, it forms a solution. In this example, which is the solvent and which is the solute?

**15.** What is an acid, and what is a base? How do acids and bases react in water?

**16.** Explain how each of the following transport mechanisms enable substances to cross cell membranes.
   **a.** Simple diffusion
   **b.** Facilitated diffusion
   **c.** Osmosis
   **d.** Carrier-mediated active transport
   **e.** Vesicular active transport

**17.** Explain how negative feedback systems help maintain homeostasis.

## Media Links   A variety of study tools for this chapter are available at our website, www.thomsonedu.com/nutrition/mcguire.

Prepare for tests and deepen your understanding of chapter concepts with these online materials:

- Practice tests
- Flashcards
- Glossary
- Student lecture notebook
- Web links
- Animations
- Chapter summaries, learning objectives, and crossword puzzles

## Suggested Readings

**1.** Bettelheim FA, Brown WH, March J. Introduction to general, organic, and biochemistry, 7th ed. Belmont, CA: Wadsworth Thomson Learning; 2004.

**2.** Campbell MK, Farrell SO. Biochemistry, 4th ed. Belmont, CA: Thomson Brooks/Cole; 2003.

**3.** Chang, R. General chemistry: the essential concepts, 4th ed. New York: McGraw-Hill; 2005.

**4.** Freeman S. Biological science. Upper Saddle River, NJ: Prentice Hall; 2002.

**5.** Gropper SS, Smith JL, Groff JL. Advanced nutrition and human metabolism, 4th ed. Belmont, CA: Wadsworth Thomson Learning; 2005.

**6.** Kotz JC, Treichel PM. Chemistry and chemical reactivity, 5th ed. Belmont, CA: Thomson Brooks/Cole; 2003.

**7.** Morris H, Best LR, Miner RL, Richey JM. Introduction to general, organic, and biochemistry, 8th ed. Hoboken, NJ: Wiley; 2005.

**8.** Starr C, Taggart R. *Biology*: The unity and diversity of life, 10th ed. Belmont, CA: Thomson Brooks/Cole; 2004.

Chapter 4

# Nutritional Physiology: Digestion, Absorption, Circulation, and Excretion

F ood is the essence of life. Without nourishment, cells—the basic units of all living organisms—die. To satisfy its nutritional needs, the body extracts nutrients from thousands of complex and diverse foods. The first step in this process takes place in the gastrointestinal tract, where food is physically and chemically broken down into nutrients. Once absorbed, these nutrients circulate through an extensive network of arteries and veins that make up the vascular system. Nutrients taken up by cells undergo a series of chemical transformations. Metabolic waste products produced by cells are then excreted from the body primarily by the kidneys, lungs, and skin. To ensure that these activities take place under optimal conditions, the endocrine and nervous systems maintain a nonstop communication network. This chapter provides a description of the physiological events that take place every time we eat, and should give you an appreciation for the intricate and varied tasks required to nourish our bodies.

# Overview of the Digestive System

The digestive system is made up of the digestive tract and accessory organs. The digestive tract, more commonly known as the **gastrointestinal (GI) tract** or alimentary tract, can be thought of as a hollow tube that runs from the mouth to the anus (Figure 4.1). Organs that make up the gastrointestinal tract include the oral cavity (mouth and pharynx), esophagus, stomach, small intestine, and large intestine. The accessory organs participate in digestion but are not part of the GI tract, and include the salivary glands, pancreas, and biliary

**gastrointestinal (GI) tract** A tubular passage that runs from the mouth to the anus that includes several organs that participate in the process of digestion; also called the digestive tract.

**FIGURE 4.1** Organs of the Digestive System

The digestive system consists of the gastrointestinal tract and the accessory organs.

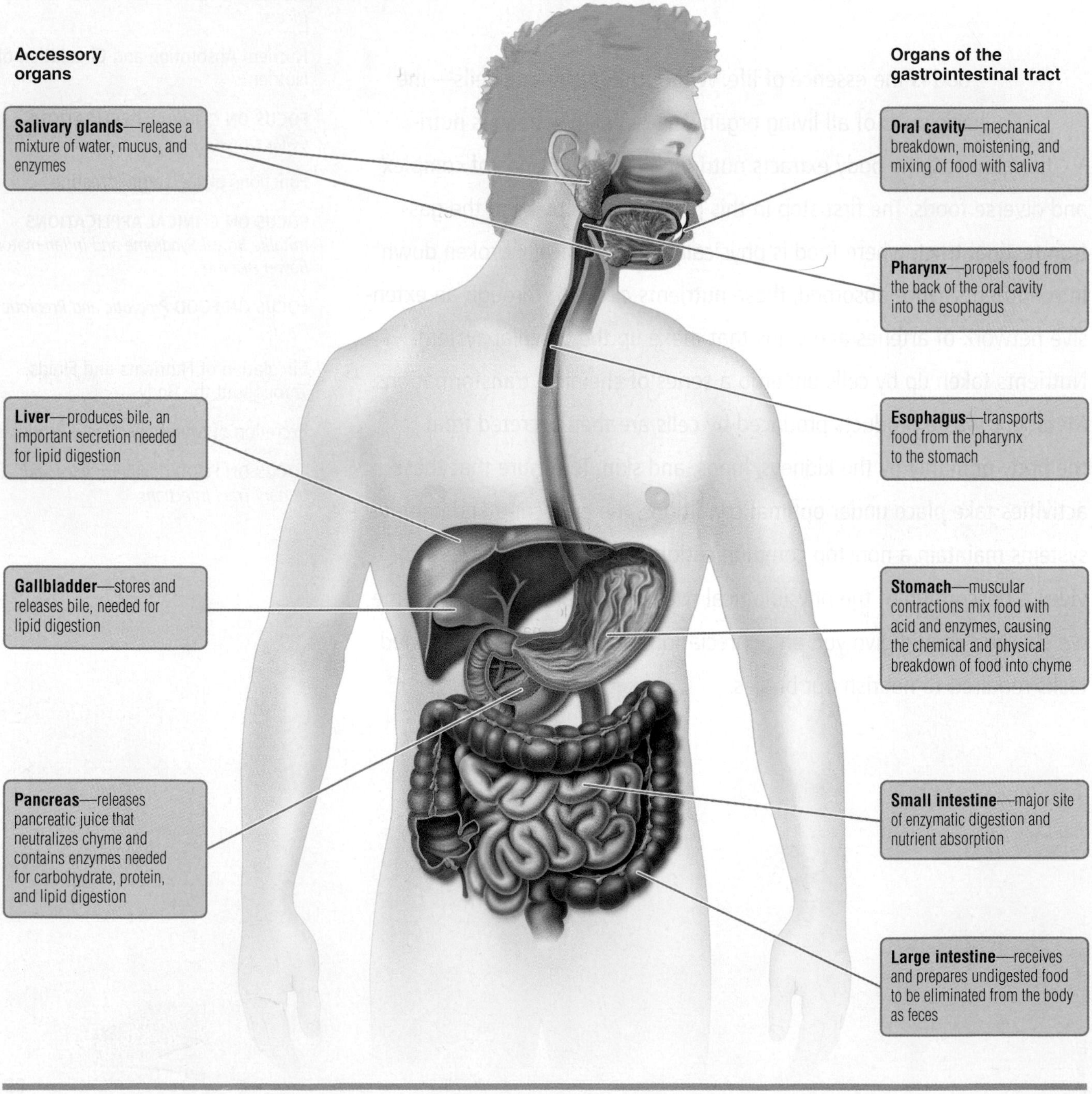

**Accessory organs**

**Salivary glands**—release a mixture of water, mucus, and enzymes

**Liver**—produces bile, an important secretion needed for lipid digestion

**Gallbladder**—stores and releases bile, needed for lipid digestion

**Pancreas**—releases pancreatic juice that neutralizes chyme and contains enzymes needed for carbohydrate, protein, and lipid digestion

**Organs of the gastrointestinal tract**

**Oral cavity**—mechanical breakdown, moistening, and mixing of food with saliva

**Pharynx**—propels food from the back of the oral cavity into the esophagus

**Esophagus**—transports food from the pharynx to the stomach

**Stomach**—muscular contractions mix food with acid and enzymes, causing the chemical and physical breakdown of food into chyme

**Small intestine**—major site of enzymatic digestion and nutrient absorption

**Large intestine**—receives and prepares undigested food to be eliminated from the body as feces

system (liver and gallbladder). The accessory organs release their secretions needed for the process of digestion into ducts, which empty into the **lumen,** the inner cavity that spans the length of the GI tract. Together, the GI tract and accessory organs carry out three important functions: (1) **digestion,** the physical and chemical breakdown of food; (2) **absorption,** the transfer of nutrients from the digestive tract into the blood or lymphatic circulatory systems; and (3) **elimination,** the process whereby solid waste (feces) is formed and expelled from the body.

The amount of time between the consumption of food (ingestion) and its elimination as solid waste is called transit time. It takes approximately 24 to 72 hours for food to pass from mouth to anus. Many factors affect transit time, such as composition of diet, illness, certain medications, physical activity, and emotions. Bands of smooth muscle called **sphincters** act like one-way valves, regulating the flow of the luminal contents from one organ to the next (Figure 4.2). The GI tract has several sphincters, which are often named according to their anatomical locations. For example, the ileocecal sphincter is between the ileum (last segment of the small intestine) and the cecum (first portion of the large intestine). This makes it very convenient to learn the names of the various sphincters.

**CONNECTIONS** Recall that smooth muscle regulates involuntary movement and is found in the lining of hollow organs and blood vessels (Chapter 3, page 83).

## Organization of the GI Tract

As illustrated in Figure 4.3, the digestive tract contains four major tissue layers—the mucosa, submucosa, muscularis, and serosa. Each tissue layer contributes to the overall function of the GI tract by providing secretions, movement, communication, and protection.

### The Mucosa

The innermost lining of the digestive tract, called the **mucosa,** consists mainly of epithelial cells (mucosal cells) and carries out a variety of digestive functions. The mucosa, often called the mucosal lining, produces secretions

**CONNECTIONS** Recall that epithelial tissue provides a protective layer on body surfaces and lines body cavities, ducts, and organs (Chapter 3, page 82).

## FIGURE 4.2  Sphincters

Sphincters are circular bands of muscle, located between organs, that regulate flow of material through the gastrointestinal tract.

The gastroesophageal sphincter (GES) is located at the juncture between the esophagus and stomach.

The GES relaxes long enough to allow food to enter the stomach, and then closes to prevent the gastric juice from reentering the esophagus.

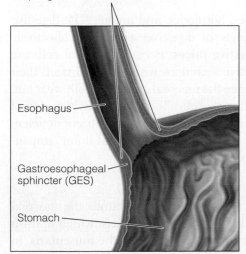

Esophagus

Gastroesophageal sphincter (GES)

Stomach

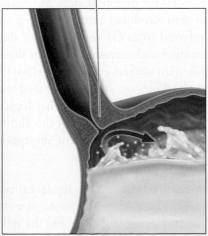

**lumen** (LU – men)  The cavity inside a tubular structure in the body.

**digestion**  The physical and chemical breakdown of food by the digestive system into a form that allows nutrients to be absorbed.

**absorption**  The passage of nutrients through the lining of the GI tract into the blood or lymphatic circulation.

**elimination**  The process whereby solid waste is formed and expelled from the body.

**sphincter**  A muscular band that narrows an opening between organs in the GI tract.

**mucosa** (mu – CO – sa)  The lining of the gastrointestinal tract that is made up of epithelial cells; also called mucosal lining.

## FIGURE 4.3 The Four Layers of Tissues in the Gastrointestinal Tract

The GI tract is made up of four layers: the mucosa, submucosa, muscularis, and serosa.

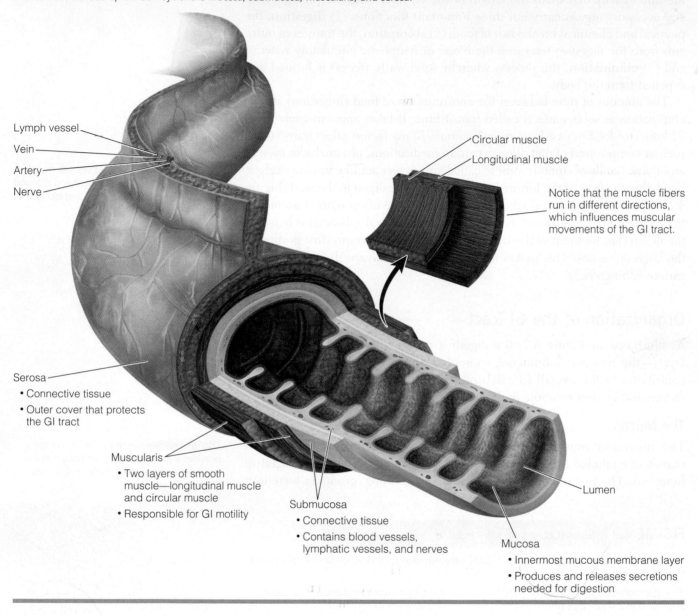

Lymph vessel

Vein

Artery

Nerve

Circular muscle

Longitudinal muscle

Notice that the muscle fibers run in different directions, which influences muscular movements of the GI tract.

Serosa
• Connective tissue
• Outer cover that protects the GI tract

Muscularis
• Two layers of smooth muscle—longitudinal muscle and circular muscle
• Responsible for GI motility

Submucosa
• Connective tissue
• Contains blood vessels, lymphatic vessels, and nerves

Lumen

Mucosa
• Innermost mucous membrane layer
• Produces and releases secretions needed for digestion

---

**GI secretions, digestive juices** Substances released by organs that make up the digestive system that facilitate the process of digestion; also called digestive juices.

**submucosa** (SUB – mu – co – sa) A layer of tissue that lies between the mucosa and muscularis tissue layers.

**muscularis** (mus – cu – LAR – is) The layer of tissue in the gastrointestinal tract that consists of at least two layers of smooth muscle.

CONNECTIONS ▶ Recall that connective tissue connects, supports, and anchors various body parts (Chapter 3, page 82).

needed for digestion such as enzymes, hormones, and mucus. The digestive system produces and releases a variety of digestive secretions collectively referred to as **GI secretions**, or **digestive juices**. Because mucosal cells are continuously exposed to harsh digestive secretions within the GI tract, their life span is a mere two to five days. Once the mucosal epithelial cells wear out, they slough off and are replaced with new cells. To support this high rate of cell turnover, the mucosa has high nutrient requirements. Nutrient deficiencies can profoundly affect the ability to maintain the mucosal lining, impairing digestion and nutrient absorption.

### The Submucosa

A layer of connective tissue called the **submucosa** surrounds the mucosal layer. The submucosa contains a rich supply of blood vessels, which nourish the inner mucosal layer and the next outward layer, called the **muscularis.** In addition to blood vessels, the submucosa contains lymphatic vessels, which are filled with a fluid called **lymph.** Lymph transports fluid away from body tissues and aids in the circulation of fat. The submucosa also contains a network of nerves called the submucosal plexus, which regulates the release of GI secretions from cells making up the mucosal lining.

### The Muscularis

Moving outward from the submucosa, the next layer in the GI tract is the muscularis. The muscularis typically consists of two layers of smooth muscle, which are organized as an outer longitudinal layer and an inner circular layer. Located between these two muscle layers is the myenteric plexus, a network of nerves that control the contraction and relaxation of the muscularis. The contraction and relaxation of the muscularis promotes mixing of the food mass with digestive secretions and keeps food moving through the entire length of the GI tract.

### The Serosa

The final outer layer of connective tissue that encloses the GI tract is the **serosa.** The serosa secretes a fluid that lubricates the digestive organs, preventing them from adhering to one another. In addition, much of the GI tract is anchored within the abdominal cavity by mesentery, a membrane that is continuous with the serosa.

## Gastrointestinal Motility and Secretions

The term **GI motility** refers to the mixing and propulsion of material by muscular contractions in the GI tract. These vigorous movements result from the contraction and relaxation of circular and longitudinal muscle in the muscularis. There are two types of movement in the GI tract—segmentation and peristalsis. **Segmentation** involves mixing movements, whereas **peristalsis** involves propulsive movements. These movements serve different functions, as shown in Figure 4.4. Segmentation occurs when circular muscles in the small intestine move the food mass back and forth. This mixing movement increases the contact between food particles and digestive secretions, giving the intestine an appearance of a chain of sausages. Peristalsis involves rhythmic, wavelike muscle contractions that propel food along the entire length of the GI tract. The contraction of circular muscles behind the food mass causes the longitudinal muscles to shorten. When the longitudinal muscles lengthen, the food is propelled forward. Thus, peristalsis is similar to the motion generated when a crowd at a sporting event does "the wave."

GI secretions (also called digestive juices) are important for digestion and protection of the GI tract, and include water, acid, electrolytes, mucus, salts, enzymes, bicarbonate, and other substances. For example, **mucus** forms a protective coating that lubricates the mucosal lining. **Digestive enzymes** are biological catalysts that facilitate chemical reactions that break down complex food particles. More specifically, digestive enzymes catalyze hydrolysis reactions, which break chemical bonds by adding water. As a result, macromolecules such as starch and protein are broken down into smaller components.

Organs that release digestive secretions include the salivary glands, stomach, pancreas, gallbladder, small intestine, and large intestine. In fact, approximately 7 liters of secretions, most of which is water, are released daily into the lumen of the GI tract. Fortunately, the body has an elaborate recycling system

**lymph** A fluid, found in lymphatic vessels, that is derived from tissue fluids.

**serosa** (se – RO – sa) Connective tissue that encloses the gastrointestinal tract.

**GI motility** Mixing and propulsive movements of the gastrointestinal tract caused by contraction and relaxation of the muscularis.

**segmentation** A muscular movement in the gastrointestinal tract that moves the contents back and forth within a small region.

**peristalsis** (per – i – STAL – sis) Waves of muscular contractions that move materials in the GI tract in a forward direction.

**mucus** (MU – cus) A substance that coats and protects mucous membranes.

**digestive enzymes** Biological catalysts that facilitate chemical reactions that break chemical bonds by the addition of water (hydrolysis), resulting in the breakdown of large molecules into smaller components.

CONNECTIONS ▸ Recall that hydrolysis is the chemical reaction that breaks chemical bonds by adding water (Chapter 3, page 77).

## FIGURE 4.4 Segmentation and Peristalsis

Segmentation helps mix food, whereas peristalsis moves it through the GI tract.

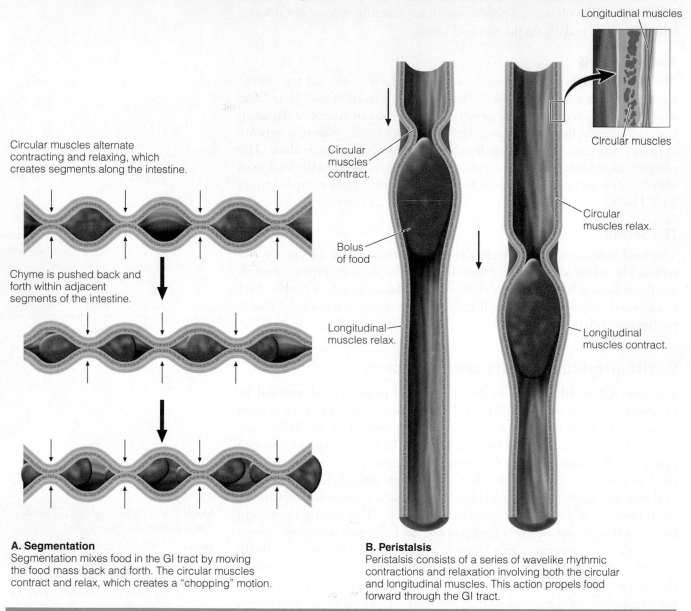

Circular muscles alternate contracting and relaxing, which creates segments along the intestine.

Chyme is pushed back and forth within adjacent segments of the intestine.

Circular muscles contract.

Bolus of food

Longitudinal muscles relax.

Longitudinal muscles

Circular muscles

Circular muscles relax.

Longitudinal muscles contract.

**A. Segmentation**
Segmentation mixes food in the GI tract by moving the food mass back and forth. The circular muscles contract and relax, which creates a "chopping" motion.

**B. Peristalsis**
Peristalsis consists of a series of wavelike rhythmic contractions and relaxation involving both the circular and longitudinal muscles. This action propels food forward through the GI tract.

that enables much of this water to be reclaimed. The major GI secretions and their related functions are summarized in Table 4.1.

## Regulation of Gastrointestinal Motility and Secretions

GI motility and the release of GI secretions are carefully regulated by neural and hormonal signals. These involuntary regulatory activities ensure that complex food particles are physically and chemically broken down and that the food mass moves along the GI tract at the appropriate rate. The GI tract has three regulatory control systems. The enteric nervous system and the central nervous system provide neural control, and the enteric endocrine system provides hormonal control. Note that the word *enteric* always relates to the intestine.

**TABLE 4.1    Summary of Major Secretions Produced and Released by the Gastrointestinal Tract and Accessory Organs, and Their Related Functions**

| Secretion | Source | Function |
|---|---|---|
| Mucus | Mucosal cells of the GI tract | • Protects and lubricates the GI tract |
| Saliva | Salivary glands | • Moistens foods<br>• Helps form the bolus<br>• Facilitates taste<br>• Aids in swallowing<br>• Chemically breaks down nutrients via enzymes |
| Enzymes | Salivary glands, stomach, small intestine, and pancreas | • Chemically break down nutrients |
| Hormones | G cells of gastric pits, and enteric endocrine cells of the small intestine | • Provide communication, and regulate GI motility and release of GI secretions |
| Bile | Made by the liver, stored and released from the gallbladder | • Enables lipid globules to disperse in water |
| Pancreatic juice (contains bicarbonate and enzymes) | Pancreas | • Neutralizes chyme<br>• Provides enzymes needed for the chemical breakdown of nutrients |
| Gastric juice (contains hydrochloric acid, enzymes, water, intrinsic factor) | Exocrine cells of the gastric pits (mucus-secreting cells, parietal cells, and chief cells) | • Provides enzymes needed for the chemical breakdown of food<br>• Hydrochloric acid is needed for forming chyme and activating some enzymes<br>• Intrinsic factor needed for absorption of vitamin $B_{12}$ |

## The Enteric Nervous System

The GI tract has its own local nervous system, the **enteric nervous system.** The enteric nervous system receives information from other nerves called sensory receptors located within the GI tract. There are two kinds of sensory receptors, each monitoring conditions and changes related to digestive activities. **Chemoreceptors** detect changes in the chemical composition of the luminal contents, whereas **mechanoreceptors** detect stretching or distension in the walls of the GI tract. As you might expect, the presence of food in the GI tract can stimulate both chemo- and mechanoreceptors. Information from both kinds of sensory receptors is relayed to the enteric nervous system, which responds by communicating with a variety of muscles and glands. In return, muscles and glands carry out the appropriate response to help digest food, such as an increase in peristalsis or the release of digestive secretions.

## The Central Nervous System

The enteric nervous system controls digestive functions on the local level. However, the GI tract also communicates with the **central nervous system.** The central nervous system consists of the brain and spinal cord, which receive and respond to sensory stimuli from the GI tract. The neural network connecting the central and enteric nervous systems keeps the GI tract and the brain in close communication. This is why sensory and emotional stimuli can affect GI function. For example, the sight, smell, or thought of food stimulates GI motility and secretion. Similarly, emotional factors such as fear, sadness, anger, anxiety, and depression can cause GI distress such as an upset stomach.

## The Enteric Endocrine System

The GI tract consists of many different cell types, some of which are hormone-producing cells referred to collectively as the **enteric endocrine system.** Hormones produced by these cells play an important communica-

**enteric nervous system** (en – TER – ic) Neurons located within the submucosa and muscularis layers of the digestive tract.

**chemoreceptor** (CHE – mo – re – cep – tor) A sensory receptor that responds to a chemical stimulus.

**mechanoreceptor** (mech – A – no – re – cep – tor) A sensory receptor that responds to pressure, stretching, or mechanical stimulus.

**central nervous system** The part of the nervous system made up of the brain and spinal cord.

**enteric endocrine system** (EN – do – crine) Hormones secreted by the mucosal lining of the GI tract that regulate GI motility and secretion.

**CONNECTIONS** Recall that the nervous system receives and transmits information via electrical impulses between nerve cells, whereas the endocrine system communicates via chemical messengers, called hormones, in the blood (Chapter 3, page 83).

tive role in the process of digestion. Enteric hormones, which act as chemical messengers, are released into the blood in response to chemical and physical changes in the GI tract. This information is then communicated to other organs, alerting them to the impending arrival of food. Like neural signals, hormones also influence the rate at which food moves through the GI tract (motility) and the release of GI secretions. In addition to regulating GI motility and secretion, some enteric hormones communicate with appetite centers in the brain—influencing the desire to eat. The four major enteric hormones are gastrin, secretin, cholecystokinin, and gastric inhibitory protein. The specific role of each of these hormones is discussed later in the chapter.

## ESSENTIAL *Concepts*

The digestive system, which is made up of the gastrointestinal (GI) tract and accessory organs, releases a variety of secretions needed for digestion. The three functions of the digestion system are the chemical and physical breakdown of food (digestion), the transfer of nutrients from the digestive tract into the blood or lymphatic circulatory system (absorption), and the elimination of undigested food residue from the body. The GI tract consists of four layers that carry out the functions of GI motility and secretion. Important secretions needed for digestion include water, acid, electrolytes, mucus, salts, bicarbonate, and enzymes. Muscular contractions called segmentation and peristalsis mix and move food in the digestive tract, whereas sphincters regulate the flow of food. GI motility and secretion are regulated by the enteric nervous system, the central nervous system, and the enteric endocrine system.

## Digestion

The intake, or ingestion, of food is the beginning of the long process of digestion. Over the next 24 to 72 hours, food undergoes considerable physical and chemical transformation in the GI tract. To optimize digestion and nutrient absorption, the intensity of GI motility and the release of GI secretions must be synchronized with the arrival of food into the different regions of the GI tract. For this reason, digestion is often subdivided into the cephalic, gastric, and intestinal phases.

The **cephalic phase** of digestion begins even before food enters the mouth. During this phase, the thought, smell, and sight of food stimulate the central nervous system. In response to the anticipation of food, neural signals stimulate GI motility and the release of digestive secretions. This response serves as a "wake-up" call to the digestive system, preparing the GI tract to receive and digest food. The **gastric phase** of digestion begins with the arrival of food in the stomach. During this phase, the release of digestive secretions and GI motility increase even further, preparing the stomach for its role in the digestive process. By the time food reaches the small intestine, it has undergone considerable physical and chemical change. Yet the process of digestion is not complete. In response to hormonal signals from the small intestine, motility and the release of digestive secretions in the stomach decrease. This slows the release of food into the small intestine, signifying that the last phase of digestion, the **intestinal phase,** is underway.

**cephalic phase** (ce – PHAL – ic) The response of the central nervous system to sensory stimuli, such as smell, sight, and taste that occurs before food enters the GI tract, characterized by increased GI motility and release of GI secretions.

**gastric phase** The phase of digestion stimulated by the arrival of food into the stomach characterized by increased GI motility and release of GI secretions.

**intestinal phase** The phase of digestion when chyme enters the small intestine, characterized by both a decrease in gastric motility and secretion of gastric juice.

# The Oral Cavity and the Esophagus

Whereas the cephalic phase prepares the GI tract to receive and digest food, digestion actually begins with chewing, or **mastication.** The forceful grinding action of teeth breaks food into manageable pieces. The presence of food in the mouth stimulates the salivary glands to release **saliva**—as much as a quart of saliva per day. Sour-tasting foods such as lemons stimulate saliva production more than others. In fact, even the thought of lemons can stimulate salivation. Saliva consists of water, mucus, digestive enzymes, and antibacterial agents. As food is broken apart, it mixes with saliva, becoming moist and easier to swallow.

Saliva is also a necessary factor in taste sensation, because food components must first be dissolved before they can be detected by taste buds. Taste buds, located on the tongue, are limited to only four basic tastes: salty, sour, sweet, and bitter. In addition to these basic tastes, a compound found in the seasoning monosodium glutamate (MSG) delivers an additional taste sensation.[1] The Japanese term **umami** is used to describe the meatlike taste produced by MSG. When food is consumed, gustatory (taste) cells and olfactory (smell) cells send neural signals to the brain. Together, these signals enable the brain to distinguish the thousands of different flavors we enjoy in our foods. Olfactory cells in particular have a profound affect on our ability to taste food, accounting for approximately 80% of taste. This is the primary reason why food seems tasteless when we have trouble smelling, such as with a cold.

The tongue, made primarily of muscle, assists in chewing and swallowing. As food mixes with saliva, the tongue manipulates the food mass, pushing it up against the hard, bony palate of the mouth. Figure 4.5 shows that swallowing takes place in two phases. As we prepare to swallow, the tongue directs the soft, moist mass of food, now referred to as a **bolus,** toward the back of the mouth, an area known as the pharynx. The pharynx is the shared space between the mouth and the esophagus that connects the nasal and oral cavities. This phase of swallowing is under voluntary control, but once the bolus reaches the pharynx the involuntary phase of swallowing begins.

During the involuntary phase of swallowing the soft palate rises, blocking the entrance to the nasal cavity. This helps guide the bolus into the correct passageway—the esophagus. The movement of the soft palate pulls the larynx (vocal cords) upward, causing the **epiglottis** (a cartilage flap) to cover the trachea—the airway leading to the lungs. Once the bolus moves past this dangerous intersection, the muscles relax and prepare for the next swallow. Indeed, swallowing is a rather complex physiological event. Disorders that affect skeletal muscles and/or nerves, such as Parkinson's disease and strokes, can affect the ability to swallow. Difficulty in swallowing is called dysphagia. In addition to these types of disorders, anatomical abnormalities affecting the oral and nasal cavities such as a cleft palate and/or cleft lip can also cause dysphagia.

As the bolus moves past the epiglottis, it enters the **esophagus,** a narrow muscular tube that leads to the stomach. The esophagus is lubricated and protected by a thin layer of mucus, which facilitates the passage of food. Peristalsis propels the food toward the stomach, where it encounters the first of several sphincters in the GI tract. The **gastroesophageal sphincter** (also called the lower esophageal sphincter or the cardiac sphincter) forms a juncture between the esophagus and the stomach. To prevent the contents of the stomach from re-entering the esophagus, the gastroesophageal sphincter (GES) remains contracted. However, as the bolus gets closer, nerves signal the GES to relax long enough for the bolus to pass into the stomach. This entire trip, from the pharynx to the stomach, takes 6 to 10 seconds.

**mastication** Chewing and grinding of food by the teeth to prepare for swallowing.

**saliva** A secretion released into the mouth by the salivary glands; moistens food and starts the process of digestion.

**umami** (U – mam – i) A taste, in addition to the four basic taste components, that imparts a savory or meatlike taste.

**bolus** (BO – lus) A soft, rounded mass of chewed food.

**epiglottis** (ep – i – GLOT – tis) A cartilage flap that covers the trachea while swallowing.

**esophagus** (e – SOPH – a – gus) The passageway that begins at the pharynx and ends at the stomach.

**gastroesophageal sphincter (GES)** (gas – tro – e – soph – a – GEAL) A circular muscle that regulates the flow of food between the esophagus and the stomach; also called lower esophageal sphincter or cardiac sphincter.

## FIGURE 4.5  Stages of Swallowing

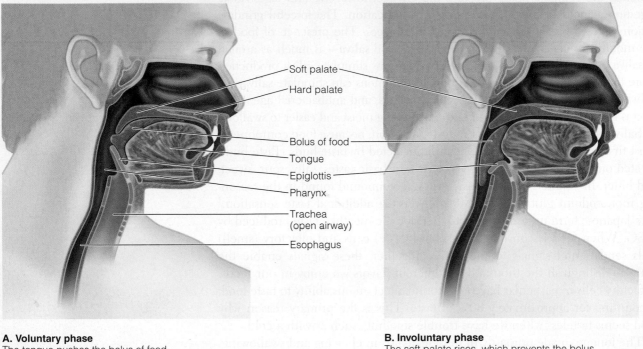

**A. Voluntary phase**
The tongue pushes the bolus of food against the hard palate. Next, the tongue pushes the bolus against the soft palate, which triggers the swallowing response.

**B. Involuntary phase**
The soft palate rises, which prevents the bolus from entering the nasal cavity. The epiglottis covers the trachea, blocking the opening to the lungs. The bolus enters the esophagus and is propelled toward the stomach by peristalsis.

## The Stomach

The stomach is a large, muscular, J-shaped sac divided into three regions: the fundus, the body, and the antrum. These regions are illustrated in Figure 4.6. The fundus is the top portion of the stomach that extends upward, above the esophagus. The middle portion of the stomach is called the body, and the lower portion of the stomach is called the pyloric region or antrum. The **pyloric sphincter,** located at the base of the antrum, regulates the movement of food from the stomach into the **duodenum** (the first portion of the small intestine). Note that the term *gastric* pertains or relates to the stomach.

The stomach is specially equipped to carry out three important functions: (1) temporary storage of food, (2) production of gastric secretions needed for digestion, and (3) mixing of food with gastric secretions. By the time food leaves the stomach, the bolus has been transformed into a semiliquid paste called **chyme.**

### Temporary Storage for Food

The stomach has an amazing capacity to accommodate large amounts of food. When empty, the stomach volume is quite small—approximately one quarter of a cup. As food enters the stomach, its walls expand to increase its capacity to 1 to 2 quarts. The ability to expand to this extent is due to the interior lining of the stomach, which is folded into convoluted pleats called **rugae.** Like an accordion, the rugae unfold and flatten, allowing the stomach to expand as it fills with food. The stretching of the stomach walls trigger mechanoreceptors to signal the brain that the stomach is becoming full. In turn, the brain causes

**pyloric sphincter** (py – LOR – ic SPHINC – ter) A circular muscle that regulates the flow of food between the stomach and the duodenum.

**duodenum** (du – o – DE – num) The first segment of the small intestine.

**chyme** The thick fluid resulting from the mixing of food with gastric secretions in the stomach.

**rugae** (RU – gae) Folds that line the stomach wall.

## FIGURE 4.6   Anatomy of the Stomach and Its Role in Digestion

The stomach is divided into three regions: the fundus, the body, and the antrum.

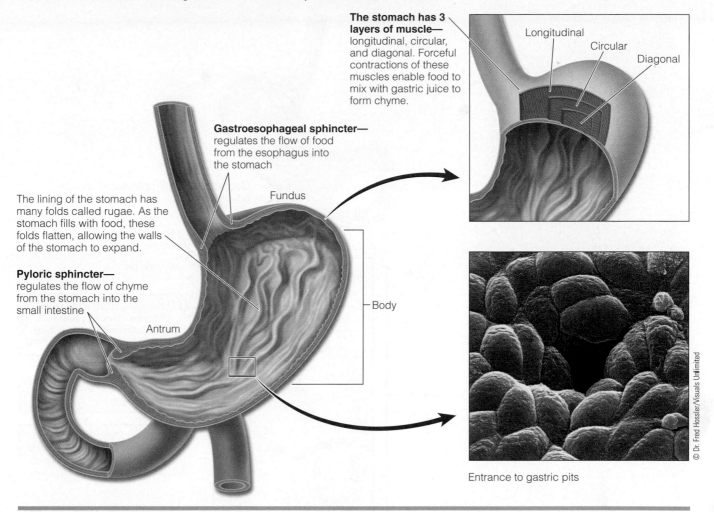

**The stomach has 3 layers of muscle—** longitudinal, circular, and diagonal. Forceful contractions of these muscles enable food to mix with gastric juice to form chyme.

Longitudinal

Circular

Diagonal

**Gastroesophageal sphincter—** regulates the flow of food from the esophagus into the stomach

Fundus

The lining of the stomach has many folds called rugae. As the stomach fills with food, these folds flatten, allowing the walls of the stomach to expand.

**Pyloric sphincter—** regulates the flow of chyme from the stomach into the small intestine

Antrum

Body

Entrance to gastric pits

© Dr. Fred Hossler/Visuals Unlimited

hunger to diminish, causing a person to stop eating. The ability to recognize and respond to these internal physiological cues is an important component of body weight regulation.

## Production of Gastric Secretions

At first glance, the stomach lining appears covered with numerous small holes. When magnified, you can see that these holes penetrate deep into the mucosal layer and form structures called **gastric pits** (Figure 4.7). Gastric pits contain exocrine and endocrine cells that produce and release digestive secretions and hormones. Exocrine cells (such as parietal cells, chief cells, and mucus cells) release their secretions into ducts that empty directly into the stomach. Collectively, these cells release over 2 liters (roughly 2 quarts) of gastric secretions daily, which are needed for the chemical breakdown of food. This juice consists mainly of water, hydrochloric acid, digestive enzymes, mucus, and intrinsic factor—a substance needed for vitamin $B_{12}$ absorption. At the base of the gastric pits are hormone-producing endocrine cells. These cells, called G cells, release their secretion into the blood.

The release of gastric secretions, also called **gastric juice,** is regulated by the hormone **gastrin.** The G cells release gastrin in response to the presence of food

**gastric pits** Invaginations of the mucosal lining of the stomach that contain specialized endocrine and exocrine cells.

**gastric juice** Digestive secretions produced by exocrine cells that make up gastric pits.

**gastrin** (GAS – trin)  A hormone secreted by endocrine cells that stimulates the production and release of gastric juice.

## FIGURE 4.7 Gastric Pits

The mucosal lining of the stomach is made up of exocrine and endocrine cells.

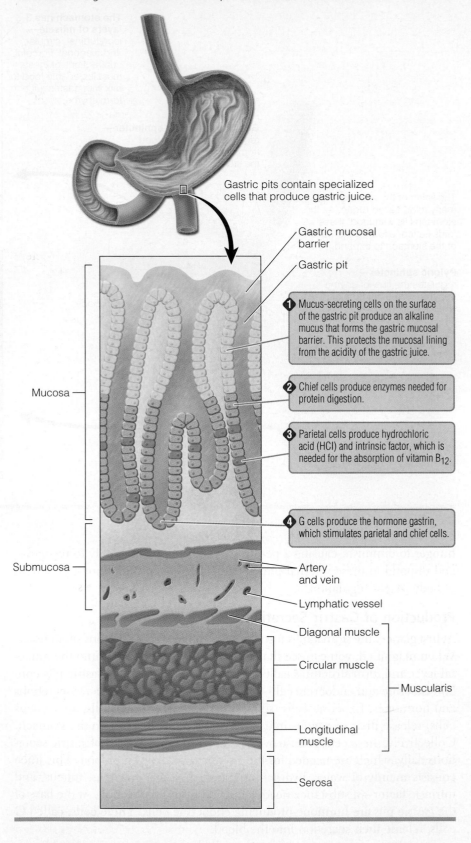

Gastric pits contain specialized cells that produce gastric juice.

Gastric mucosal barrier

Gastric pit

Mucosa

**1** Mucus-secreting cells on the surface of the gastric pit produce an alkaline mucus that forms the gastric mucosal barrier. This protects the mucosal lining from the acidity of the gastric juice.

**2** Chief cells produce enzymes needed for protein digestion.

**3** Parietal cells produce hydrochloric acid (HCl) and intrinsic factor, which is needed for the absorption of vitamin $B_{12}$.

**4** G cells produce the hormone gastrin, which stimulates parietal and chief cells.

Submucosa

Artery and vein

Lymphatic vessel

Diagonal muscle

Circular muscle

Muscularis

Longitudinal muscle

Serosa

in the stomach, and to a lesser extent, alcohol and caffeine. Even when no food is present in the stomach, the cephalic response to sensory stimuli such as the thought, smell, taste, and anticipation of food can trigger the release of gastrin as well. Gastrin stimulates exocrine cells to release gastric juice. Specifically, gastrin stimulates the release of hydrochloric acid (HCl) and intrinsic factor from the **parietal cells** and of pepsinogen from the **chief cells.** HCl, a major component of gastric juice, dissolves food particles, destroys bacteria that may be present in food, and provides an optimal acidic environment (pH 2) for digestive enzymes to function. Pepsinogen, an inactive enzyme, is converted to its active form, pepsin, when exposed to the acidic gastric juice. Once activated, pepsin begins the enzymatic breakdown of protein.

Located near the entrance of the gastric pits are numerous mucus-secreting cells. These exocrine cells release a thin, watery mucus, which forms a protective layer called the gastric mucosal barrier. The gastric mucosal barrier prevents the acidic gastric juice from damaging the stomach lining. Without this layer, the mucosal lining could not withstand its harsh environment, resulting in inflammation and the formation of a gastric ulcer. Similarly, a condition called gastroesophageal reflux disease (GERD) results when the unprotected lining of the esophagus is repeatedly exposed to gastric juice. These conditions are addressed further in the Focus on Clinical Applications feature.

## Gastric Mixing and the Formation of Chyme

Another important function of the stomach is the mixing of food with the gastric juices to form chyme. Unlike the rest of the GI tract, the stomach has three, rather than two, layers of smooth muscle. In addition to circular and longitudinal muscle, the stomach has a layer of diagonal muscle. These powerful muscles generate a forceful churning action, much like the kneading of bread. The strength of these muscle contractions increases under the influence of the hormone gastrin. Within three to five hours of ingestion, the food is thoroughly mixed with the gastric juices. The resulting chyme, now the consistency of a soupy paste, is pushed toward the narrow opening of the pyloric sphincter. With each peristaltic wave, a few milliliters (less than a teaspoon) of chyme squeeze through; the remaining chyme tumbles back and forth, allowing for even more mixing.

## Regulation of Gastric Emptying

The rate of **gastric emptying,** or the rate at which food leaves the stomach, is influenced by several factors, including the volume, consistency, and composition of chyme. For example, large volumes of chyme increase the force and frequency of peristaltic contractions, which in turn increase the rate of gastric emptying. Therefore, large meals leave the stomach at a faster rate compared to small meals. The consistency of food (liquid versus solid) also affects the rate of gastric emptying. Because the opening of the pyloric sphincter is small, only fluids and small particles (<2 mm in diameter) can pass through. Solid foods take more time to liquefy than fluids and therefore remain in the stomach longer. Last, the nutrient composition of chyme also influences gastric emptying. For example, the presence of fat in chyme causes the small intestine to release a hormone called **gastric inhibitory protein (GIP).** This hormone (also called gastric inhibitory peptide) slows the rate of gastric emptying, enabling the small intestine to prepare for the task of fat digestion.

The small intestine itself also influences the rate of gastric emptying. Every day the duodenum receives approximately 10 quarts (roughly 40 cups) of ingested food, drinks, saliva, gastric juice, and so forth. The small intestine can

**parietal cells** (pa – RI – e – tal) Exocrine cells within the gastric mucosa that secrete hydrochloric acid and intrinsic factor.

**chief cells** Exocrine cells in the gastric mucosa that produce the protein-digesting enzyme pepsin.

**gastric emptying** The process by which food leaves the stomach and enters the small intestine.

**gastric inhibitory protein (GIP)** A hormone produced by endocrine cells lining the small intestine that controls GI motility; also called gastric inhibitory peptide.

# *Focus On* CLINICAL APPLICATIONS
## Gastroesophageal Reflux Disease and Peptic Ulcers

Millions of Americans describe the symptoms as "fire" in the belly——that burning sensation often mistaken for indigestion. However, **gastroesophageal reflux disease (GERD)** and **peptic ulcers** are more than just a little indigestion. These terms both refer to conditions of the upper GI tract. In fact, GERD and peptic ulcers are the two most common GI disorders.[2]

GERD is a condition characterized by the movement (reflux) of the stomach contents (chyme) back into the esophagus (Figure 4.8). Because the esophagus does not have a thick protective mucus layer, repeated exposure to acidic chyme can irritate its lining. This causes a burning sensation in mid-chest, the most common symptom of GERD. Because of the location of the pain, people with GERD often complain of what is typically called heartburn. GERD can lead to chronic inflammation of the esophagus, which is considered a risk factor for esophageal cancer if left untreated. A medical procedure called endoscopy can be used to examine the lower esophagus for signs of irritation.

The primary cause of GERD is the relaxation of the gastroesophageal sphincter (GES). Dietary habits that contribute to this include eating large portions of foods or lying down soon after eating. Eating certain foods such as onions, chocolate, mint, high-fat foods, spicy foods, citrus juices, alcohol, caffeinated beverages, and/or carbonated drinks can also cause the GES to relax in some people. Other factors associated with GERD include being overweight, smoking, and wearing tight-fitting clothes. Hormonal changes associated with pregnancy as well as pressure from the enlarged uterus on the GES can also contribute to GERD. Making the appropriate lifestyle changes is an important "first step" in preventing and managing GERD. However, when lifestyle changes are not enough, over-the-counter or prescription medications may be necessary to treat and manage this condition.

A peptic ulcer occurs when gastric juice erodes areas of the mucosal lining in the esophagus (esophageal ulcer), stomach (gastric ulcer), or duodenum (duodenal ulcer). As you can see from Figure 4.9a, an ulcer looks similar to a canker sore. If left untreated, an ulcer can erode through the various tissue layers, sometimes causing severe complications such as bleeding and infection in the abdominal cavity.

Historically, a diet of bland food, milk, and rest was standard treatment for people with ulcers. However, that changed in 1982 when Drs. Barry Marshall and Robin Warren discovered that a small spiral-shaped bacterium called *Helicobacter pylori (H. pylori)* causes most ulcers.[3] The idea that a bacterium could cause ulcers was slow to gain acceptance in the medical community. It seemed highly doubtful that bacteria could survive the acidic environment of the stomach. However, Dr. Marshall was so confident about his discovery that he willingly swallowed a culture of *H. pylori* to prove it could survive the acidic environment of the stomach. Ten days later, Dr. Marshall developed acute gastritis, or inflammation of the gastric lining. The presence of *H. pylori* bacteria was later confirmed by examining a sample of his gastric mucosal lining. Because of Dr. Marshall's scientific conviction, the *H. pylori* bacterium is now widely accepted as the primary cause of ulcers (Figure 4.9b).[4] In 2005, the Nobel Prize in Medicine was awarded to Marshall and Warren for their discovery that peptic ulcers and gastritis could be caused by a bacterium.

*H. pylori* contains several flagella (whiplike tails), which burrow into the thick protective mucus layer, exposing the sensitive underlying stomach layers to acidic gastric juice. *H. pylori* causes even more damage by secreting toxins that irritate the stomach lining. Although most ulcers are caused by *H. pylori*, irritants such as nonsteroidal anti-inflammatory agents (aspirin and ibuprofen) and alcohol can also cause them. The most common symptom associated with an ulcer is a dull, gnawing pain in the stomach that is often relieved by eating. Other symptoms include intermittent pain, weight loss, loss of appetite, and vomiting. Today, most people diagnosed with ulcers are treated with a combination of therapies. These include antibiotics to address the underlying bacterial infection, and acid-blocking medications to help promote healing.

---

**gastroesophageal reflux disease (GERD)** A condition caused by the weakening of the gastroesophageal sphincter, which enables gastric juices to reflux into the esophagus, causing irritation to the mucosal lining.

**peptic ulcer** The presence of irritation and/or erosion of the mucosal lining in the stomach, duodenum, or esophagus.

only process small amounts of chyme at a time. To prevent the small intestine from becoming overwhelmed by too much chyme, neural and hormonal responses inhibit (slow) the rate of gastric emptying. When the small intestine is ready to receive more chyme, the inhibitory response is turned off.

## The Small Intestine

The small intestine is the primary site of chemical digestion and nutrient absorption. This 20-foot-long, narrow tube is about 1 inch in diameter and is uniquely suited to carry out these functions. The small intestine is

## FIGURE 4.8   Gastroesophageal Reflux Disease (GERD)

GERD is caused by dysfunction of the gastroesophageal sphincter (GES).

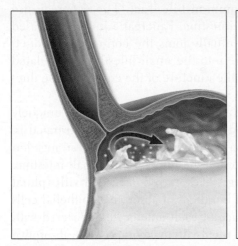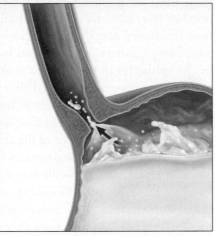

**A.** To prevent food from flowing back into the esophagus, the gastroesophageal sphincter (GES) remains contracted.

**B.** If the GES weakens, the stomach contents flow back into the esophagus. The reflux of stomach acid into the esophagus is called gastroesophageal reflux disease (GERD).

## FIGURE 4.9   Peptic Ulcers

Peptic ulcers are erosions that occur in the mucosal lining of the esophagus, stomach, and/or duodenum.

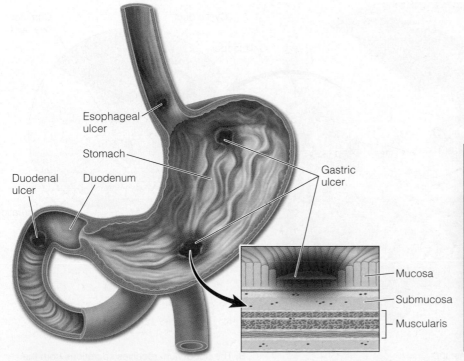

Esophageal ulcer

Stomach

Duodenal ulcer

Duodenum

Gastric ulcer

Mucosa

Submucosa

Muscularis

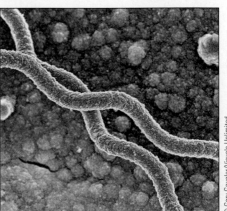

© Gary Gaugler/Visuals Unlimited

**A.** Peptic ulcers occur when the mucosal lining of the esophagus, stomach, and/or duodenum becomes eroded. If left untreated, an ulcer can penetrate through the layers of the GI tract wall.

**B.** Most peptic ulcers are caused by the bacterium *Helicobacter pylori (H. pylori)*.

**jejunum** (je – JU – num) The midsection of the small intestine, located between the duodenum and the ileum.

**ileum** (IL – e – um) The last segment of the small intestine that comes after the jejunum.

**common bile duct** The duct that transports secretions from the liver, pancreas, and gallbladder into the duodenum.

**sphincter of Oddi** The sphincter that regulates the passage of secretions from the common bile duct into the duodenum.

**villi** (plural of *villus*) (VI – li, VI – lus) Small, finger-like projections that cover the inner surface of the small intestine.

**enterocytes** (en – TER – o – cytes) Epithelial cells that make up the surface of each villus.

**microvilli** (MI – cro – vi – li) Hairlike projections on the surface of enterocytes.

**brush border** The absorptive surface of the small intestine made up of thousands of microvilli that line enterocytes.

divided into three regions—the duodenum, the **jejunum**, and the **ileum** (Figure 4.10a). Chyme first passes into the duodenum, the receiving end of the small intestine. In addition to chyme, the duodenum receives secretions from the gallbladder via the **common bile duct.** The pancreas also releases its secretions into the small intestine. Pancreatic juice is released into the pancreatic duct, which eventually joins the common bile duct. The flow of pancreatic juice and bile into the small intestine is regulated by the **sphincter of Oddi,** located at the juncture of the common bile duct and the duodenum (Figure 4.10b).

The lining of the small intestine has a large surface area, making it uniquely suited for the process of digestion and nutrient absorption. As illustrated in Figure 4.11, the inner lining of the small intestine (the mucosa) is arranged in large pleated folds that face inward, toward the lumen of the small intestine. These folds are covered with tiny finger-like projections called **villi** (plural form of *villus*). Each villus consists of hundreds of absorptive epithelial cells called **enterocytes.** The luminal surface of each enterocyte is covered with thousands of minute projections called **microvilli** that comprise the absorptive surface of the small intestine, or what is called the **brush border.** Collectively, these structures create an area that is about 600-fold larger than a simple tube. This is approximately the size of the playing surface on a standard tennis court. The brush border of the small intestine thus provides an enormous surface area where nutrient digestion and absorption takes place. Each villus

## FIGURE 4.10 Overview of Small Intestine and Accessory Organs

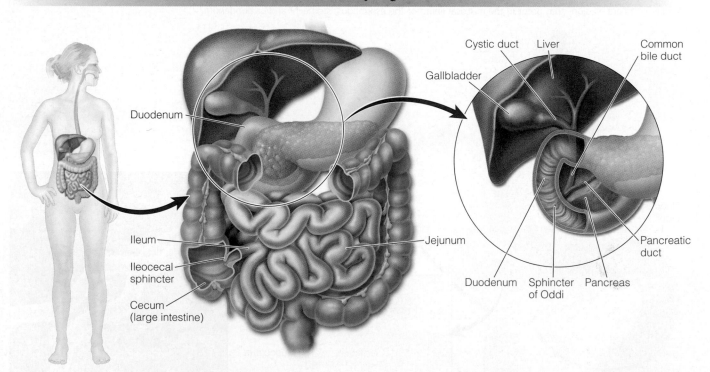

**A.** The small intestine is divided into 3 regions: the duodenum, jejunum, and ileum. The ileocecal sphincter regulates the flow of material from the ileum, the last segment of the small intestine, into the cecum, the first portion of the large intestine.

**B.** The duodenum receives secretions from the gallbladder via the common bile duct. The pancreas releases its secretions into the pancreatic duct, which eventually joins the common bile duct. The sphincter of Oddi regulates the flow of these secretions into the duodenum.

## FIGURE 4.11 Absorptive Surface of the Small Intestine

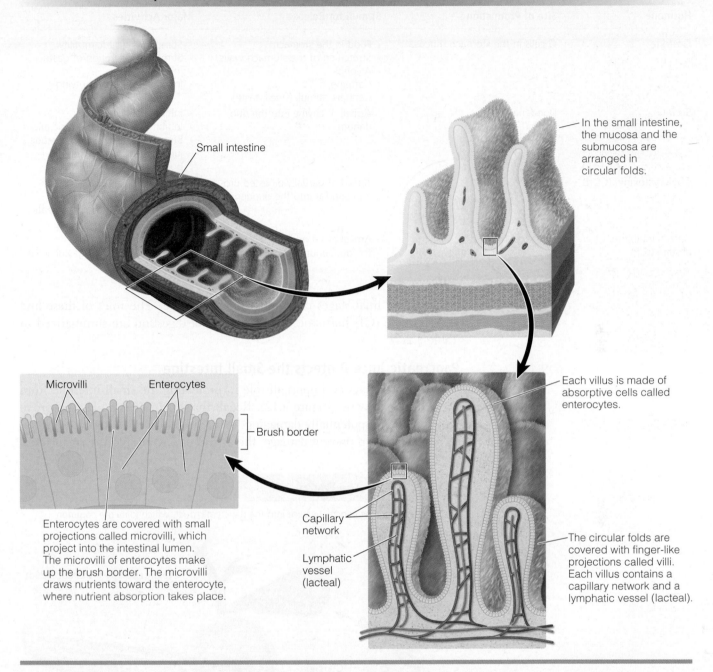

Small intestine

In the small intestine, the mucosa and the submucosa are arranged in circular folds.

Each villus is made of absorptive cells called enterocytes.

The circular folds are covered with finger-like projections called villi. Each villus contains a capillary network and a lymphatic vessel (lacteal).

Microvilli   Enterocytes

Brush border

Enterocytes are covered with small projections called microvilli, which project into the intestinal lumen. The microvilli of enterocytes make up the brush border. The microvilli draws nutrients toward the enterocyte, where nutrient absorption takes place.

Capillary network

Lymphatic vessel (lacteal)

contains a network of capillaries and a lymphatic vessel called a **lacteal,** which together circulate absorbed nutrients away from the small intestine.

### Enteric Hormones Regulate Digestion in the Small Intestine

As in the stomach, the lining of the small intestine contains hormone-producing endocrine cells. These cells release the enteric hormones **secretin, cholecystokinin (CCK),** and gastric inhibitory protein (GIP) in response to conditions within the small intestine. These hormones coordinate the release of secretions from the pancreas and gallbladder, the relaxation of sphincters, and GI motility, all with great precision. The actions of enteric hormones ensure that digestion and nutrient absorption in the small intestine are rapid and efficient. Indeed, within 30 minutes of the arrival of chyme in the small

**lacteal** (LAC – te – al) A lymphatic vessel found in an intestinal villus.

**secretin** (se – CRE – tin) A hormone, secreted by the duodenum, that stimulates the release of sodium bicarbonate and enzymes from the pancreas.

**cholecystokinin (CCK)** (CHO – le – cys – to – KI – nin) A hormone, produced by the duodenum, that stimulates the release of enzymes from the pancreas and stimulates the gallbladder to contract and release bile.

## TABLE 4.2 The Major Enteric (GI) Hormones and Their Related Functions

| Hormone | Site of Production | Stimuli for Release | Major Activities |
|---|---|---|---|
| Gastrin | G cells in the stomach mucosa | • Food in the stomach<br>• Stretching of the stomach walls<br>• Alcohol<br>• Caffeine<br>• Cephalic stimuli (smell, taste) | • Stimulates gastric motility<br>• Stimulates secretion of gastric juice<br>• Increases gastric emptying |
| Secretin | Duodenal mucosa | • Arrival of chyme into the duodenum | • Inhibits gastric motility<br>• Inhibits secretion of gastric juice<br>• Stimulates release of pancreatic juice containing sodium bicarbonate and enzymes |
| Cholecystokinin (CCK) | Duodenal mucosa | • Arrival of partially digested protein and fat into the duodenum | • Stimulates gallbladder to contract and release bile<br>• Stimulates release of pancreatic enzymes |
| Gastric inhibitory protein (GIP) | Duodenal mucosa | • Arrival of fat and glucose into the duodenum | • Inhibits gastric motility<br>• Inhibits secretion of gastric juice |

intestine the final stages of digestion are complete. The roles of these and other enteric (GI) hormones in the process of digestion are summarized in Table 4.2.

## Pancreatic Juice Protects the Small Intestine

The pancreas plays an important role in protecting the small intestine from the acidity of chyme (Figure 4.12). Recall that chyme has a pH of approximately 2 and is potentially damaging to the unprotected lining of the small intestine. To help prevent damage, the arrival of chyme into the small intes-

## FIGURE 4.12 The Pancreas

The pancreas releases pancreatic juice into the pancreatic duct, which joins the common bile duct.

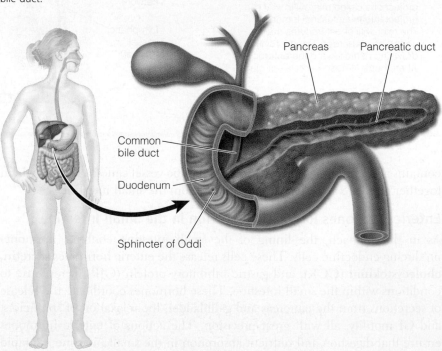

Pancreas    Pancreatic duct

Common bile duct

Duodenum

Sphincter of Oddi

tine stimulates the release of the enteric hormone secretin. Secretin signals the pancreas to release **pancreatic juice,** an alkaline solution consisting of water, sodium bicarbonate, and various enzymes needed for digestion. The sodium bicarbonate in pancreatic juice quickly neutralizes chyme as it enters the duodenum.

### The Role of Bile in Digestion

A substance called **bile** also plays an important role in digestion, especially when fatty foods are consumed (Figure 4.13). Bile, which is made in the liver, is a watery solution that consists primarily of cholesterol, bile acids, and bilirubin—a pigment that gives bile its characteristic yellowish-green color. Once bile is formed, it is transported to the gallbladder, where it is stored. To make it easier to store, the gallbladder concentrates bile by removing as much as 90% of the water.

Fats and other types of lipids are not soluble in the watery environment of the small intestine and are therefore more difficult to digest and absorb. To counteract this, the presence of fat-containing chyme in the small intestine signals the release of the enteric hormone cholecystokinin (CCK) from endocrine cells in the mucosal lining of the duodenum. CCK causes the gallbladder to contract, emptying its contents into the common bile duct. Bile acts like a detergent, dispersing large globules of fat into smaller droplets. Without bile, it would be difficult for enzymes needed for lipid digestion to

**pancreatic juice** Pancreatic secretions that contain bicarbonate and enzymes needed for digestion.

**bile** A fluid, made by the liver and stored and released from the gallbladder, that contains bile salts, cholesterol, water, and bile pigments.

## FIGURE 4.13   The Role of the Gallbladder in Digestion

The gallbladder releases bile into the common bile duct, which empties into the small intestine.

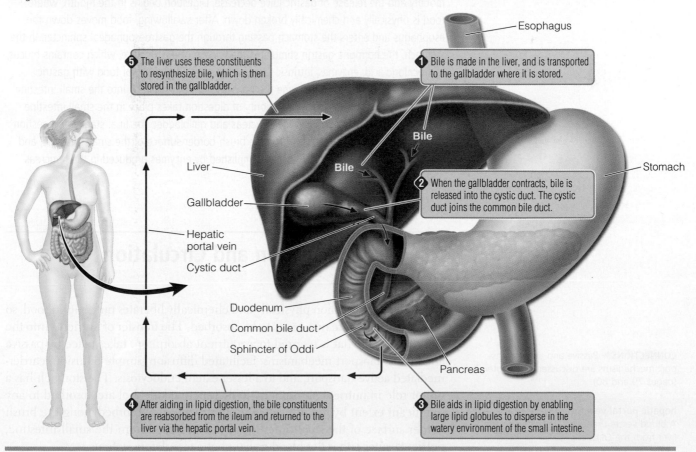

**5** The liver uses these constituents to resynthesize bile, which is then stored in the gallbladder.

**1** Bile is made in the liver, and is transported to the gallbladder where it is stored.

**2** When the gallbladder contracts, bile is released into the cystic duct. The cystic duct joins the common bile duct.

**4** After aiding in lipid digestion, the bile constituents are reabsorbed from the ileum and returned to the liver via the hepatic portal vein.

**3** Bile aids in lipid digestion by enabling large lipid globules to disperse in the watery environment of the small intestine.

Esophagus
Bile
Bile
Stomach
Liver
Bile
Gallbladder
Hepatic portal vein
Cystic duct
Duodenum
Common bile duct
Sphincter of Oddi
Pancreas

make direct contact with the chemical bonds. Once the lipids are absorbed, bile is reabsorbed through the ileum and returned to the liver via the **hepatic portal vein.** This process enables the liver to recycle many of the constituents that make up bile. In fact, only 5% of the bile escapes into the large intestine and is lost in the feces. The role of bile in lipid digestion is presented in more detail in Chapter 7.

### Enzymatic Digestion in the Small Intestine

Both the small intestine and pancreas provide enzymes needed for the final stages of digestion. Pancreatic enzymes are released into the duodenum, via the common bile duct, in response to the hormone CCK. Intestinal enzymes are made in the epithelial cells of the brush border. Both intestinal and pancreatic enzymes facilitate the chemical breakdown of nutrients into their smaller subunits by hydrolysis. Enzymatic digestion takes place in the lumen of the small intestine, along the brush border surface of the enterocytes, and within the enterocytes. The role of each of these enzymes in the process of digestion is discussed in detail in appropriate chapters of this book.

## ESSENTIAL *Concepts*

GI motility and the release of GI secretions are synchronized with the arrival of food into the different regions of the GI tract. For this reason, digestion is often divided into phases. The cephalic phase is initiated by the central nervous system in anticipation of food, whereas the gastric phase begins with the arrival of food in the stomach. During these two phases, GI motility and the release of GI secretions increase. The intestinal phase begins when food moves into the small intestine. During this phase of digestion, the rate of gastric motility and the release of gastric juice decrease. Digestion begins in the mouth, where food is physically and chemically broken down. After swallowing, food moves down the esophagus and enters the stomach passing through the gastroesophageal sphincter. In the stomach, the hormone gastrin stimulates the release of gastric juice, which contains mucus, hydrochloric acid, enzymes, intrinsic factor, and water. The mixing of food with gastric juice turns food into chyme. Chyme then passes from the stomach into the small intestine through the pyloric sphincter. The majority of digestion takes place in the small intestine and is aided by secretions from the pancreas and gallbladder. The final stages of digestion take place in the intestinal lumen, on the brush border surface of the small intestine, and within intestinal enterocytes. This is accomplished by enzymes produced in the pancreas and small intestine.

# Nutrient Absorption and Circulation of Nutrients

The process of digestion physically and chemically liberates nutrients in food, so that the nutrients are now ready to be absorbed. The transfer of nutrients into the enterocytes, or what is referred to as nutrient absorption, takes place by passive and active transport mechanisms: facilitated diffusion, simple diffusion, carrier-mediated active transport, and to a lesser extent endocytosis. The stomach has a minor role in nutrient absorption—only water and alcohol are absorbed to any significant extent by the stomach. Most nutrients are absorbed along the brush border surface of the small intestine. Once absorbed from the small intestine, nutrients enter either the blood or lymphatic circulatory systems.

**CONNECTIONS** Passive and active transport mechanisms are discussed in Chapter 3 (pages 79 and 80).

**hepatic portal vein** (he – PA –tic POR – tal) A blood vessel that circulates blood to the liver from the GI tract.

## Absorption Across the Brush Border

The vast surface area and unique structure of the small intestine makes nutrient absorption very efficient. The sweeping action of the microvilli trap and pull nutrients toward the enterocytes. However, the transfer of nutrients from the lumen of the small intestine and into the enterocyte is really only the first step in nutrient absorption. To enter the blood or lymph, nutrients must also cross the **basolateral membrane,** the cell membrane that faces away from the lumen. Thus, nutrient absorption includes both the entry into, and the exit out of, the enterocyte. The process of nutrient absorption is illustrated in Figure 4.14.

Nearly all the products of carbohydrate, protein, and lipid digestion are absorbed. However, the absorption of some nutrients (such as calcium and iron) is dictated by the physiological needs of the body. For example, the body absorbs only the amount of iron actually required, and excretes the excess in feces. This regulatory step protects the body against nutrient toxicity. The extent to which a nutrient is absorbed is called its **bioavailability.** The bioavailability of a particular nutrient is influenced by physiological conditions, other dietary components, and medication. Disease states that affect the absorptive surface of the small intestine can also affect nutrient bioavailability (malabsorption) and can lead to nutritional deficiencies. An example of such a condition is celiac disease, as discussed in the Focus on Clinical Applications feature.

## Circulation of Nutrients Away from the Gastrointestinal Tract

The intestine and the liver have a unique circulatory arrangement called the **hepatic portal circulation.** Water-soluble nutrients (carbohydrates, proteins, minerals, and some vitamins and lipids) absorbed from the small

## FIGURE 4.14   Nutrient Absorption

Nutrient absorption includes both the entry into and the exit out of the enterocyte.

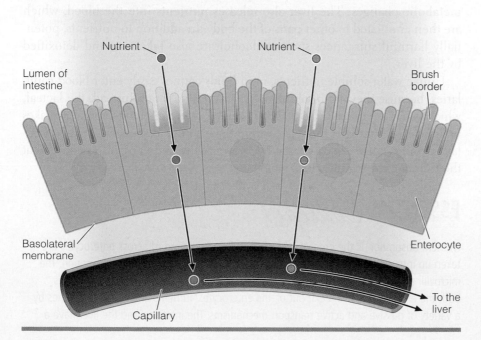

**basolateral membrane**  The cell membrane that faces away from the lumen of the GI tract.

**bioavailability**  The extent to which nutrients are absorbed into the blood or lymphatic system.

**hepatic portal circulation**  A circulatory route that delivers nutrient-rich blood from the small intestine to the liver.

# Focus On CLINICAL APPLICATIONS
## Celiac Disease

Only in the last 50 years have researchers begun to understand **celiac disease** and how to treat it. In 1888, Dr. Samuel Gee recognized that when celiac patients avoided "farinaceous foods" they "suffered" far less. Farinaceous foods are those that contain or consist of starch. Many years later, Dr. Gee's observations were confirmed. Indeed, celiac disease is an inflammatory response to a specific protein called **gluten,** found in a variety of cereal grains (wheat, rye, barley, triticale, and possibly oats).

Researchers now know that when people with celiac disease (also called **gluten-sensitive enteropathy**) consume gluten-containing foods, an immune response results in the production of antibodies.[5] These antibodies attack and damage the intestinal microvilli, causing them to flatten, damaging the absorptive surface area. For this reason, celiac disease is classified as an **autoimmune disease,** a condition that occurs when the immune system attacks and destroys its own healthy cells.

Because of the progressive damage to the absorptive surface of the small intestine, people with celiac disease experience diarrhea, weight loss, and malnutrition from impaired nutrient absorption. Fortunately, this condition can be effectively managed with a gluten-free diet. However, given the numerous food products made with wheat and other cereal grains (such as breads, crackers, cookies, cakes, and pasta) adherence to a gluten-free diet is easier advised than done. In addition to being in wheat products, gluten is often added to many processed foods. People with celiac disease should be vigilant when reading food labels, because gluten may be present in less obvious foods such as meats, soups, hard candies, soy sauce, licorice, jelly beans, malt, and even in some pharmaceutical products. In addition, foods made with modified food starch, hydrolyzed vegetable protein, and binders may contain gluten. For this reason, foods claiming to be wheat free are not always gluten free.

When celiac disease is first suspected, a blood test may be performed to screen for the presence of antibodies made by the immune system in response to gluten. Although antibody testing is important, a definitive diagnosis is made by biopsy, a procedure that requires taking a small piece of tissue from the intestinal lining. Most people with celiac disease can live symptom free by eliminating gluten from their diet. Controlling the condition is important, because the progressive damage caused by gluten increases the risk of developing intestinal cancer. Like many other types of autoimmune diseases, celiac disease runs in families.

**celiac disease** (CE – li – ac) An autoimmune response, to the protein gluten, that damages the absorptive surface of the small intestine; also called gluten-sensitive enteropathy.

**gluten** (GLU – ten) A protein found in cereal grains such as wheat, rye, barley, and possibly oats.

**autoimmune disease** An immune response that results in the destruction of normal body cells.

**thoracic duct** (tho – RAC – ic) The major duct of the lymphatic system; releases lymph into the blood at the subclavian vein.

intestine are circulated directly to the liver via the hepatic portal vein. This arrangement gives the liver first access to the nutrient-rich blood leaving the small intestine. Nutrients taken up by the liver can be stored or can undergo metabolic changes. The liver also releases nutrients into the blood, which are then circulated to other parts of the body. In addition to nutrients, potentially harmful substances such as alcohol are also taken up and detoxified by the liver.

Unlike water-soluble nutrients, most lipids cannot easily enter blood capillaries, because they are insoluble in water and too large. Instead, the lacteal, which is more permeable to these substances than are blood capillaries, takes up lipids and circulates them away from the GI tract in the lymph. This circulatory route initially bypasses the liver, eventually emptying into the blood at the **thoracic duct** in the neck region.

# ESSENTIAL Concepts

Nutrient absorption is the process whereby nutrients transported across enterocytes are taken up into circulation. The small intestine is the major site of nutrient absorption. The microvilli trap and pull nutrients toward the enterocytes. Nutrient absorption includes both the entry into, and the exit out of, the enterocyte. Nutrients cross cell membranes by a variety of passive and active transport mechanisms. The intestine and the liver have a unique circulatory arrangement called the hepatic portal system. Water-soluble nutrients absorbed from the digestive tract are circulated directly to the liver via the hepatic portal vein. Nutrients that are too large or insoluble in water, such as lipids, are taken up through lacteals and circulated by the lymphatic system.

# Functions of the Large Intestine

Approximately 9 to 12 hours after a meal is consumed, the remaining undigested food approaches the last leg of its journey—the large intestine. It will take another 10 to 24 hours before leaving the GI tract entirely. The major functions of the large intestine are (1) absorption of fluids and electrolytes, (2) microbial action, and (3) storage and elimination of solid waste (feces).

## Anatomy of the Large Intestine

The large intestine is divided into four regions: the cecum, colon, rectum, and anal canal (Figure 4.15). The **cecum,** the first portion of the large intestine, is a short, saclike structure with an attached appendage consisting of lymphatic tissue called the **appendix.** On occasion, trapped material can cause the appendix to become inflamed, which can necessitate an appendectomy—the surgical removal of the appendix. The function of the appendix remains unclear, and people do not experience ill effects related to its removal. The **ileocecal sphincter** regulates the intermittent flow of material from the ileum into the cecum.

The **colon,** which makes up most of the large intestine, is shaped like an inverted letter U (∩). The right side of the colon is called the ascending colon, whereas the portion spanning right to left across the abdomen is the

**cecum** (CE – cum) The pouch that marks the first section of the large intestine.

**appendix** A small, finger-like appendage attached to the cecum.

**ileocecal sphincter** (il – e – o – CE – cal) The sphincter that separates the ileum from the cecum and regulates the flow of material between the small and large intestines.

**colon** The portion of the large intestine that carries material from the cecum to the rectum.

## FIGURE 4.15   Overview of the Large Intestine

The large intestine is divided into three regions: the cecum, the colon, and the rectum.

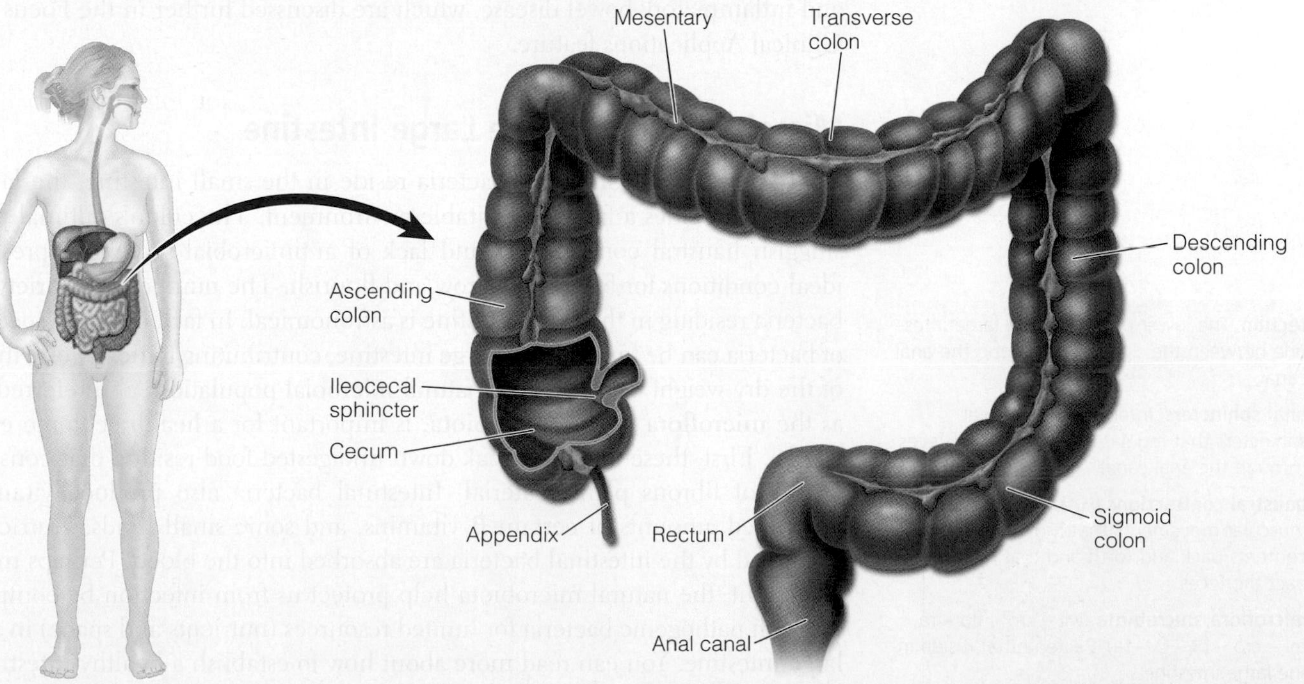

transverse colon. From there, the descending colon continues downward, on the left side of the body. The S-shaped region toward the end of the descending colon is known as the sigmoid colon. Following the sigmoid colon is the **rectum,** which terminates at the anal canal, the segment of the large intestine that leads outside of the body. A thickening of smooth muscle around the anal canal forms the internal and external **anal sphincters,** the latter of which is under voluntary control.

## Absorption of Fluids and Electrolytes in the Large Intestine

**CONNECTIONS** Recall that electrolytes are substances that produce charged particles or ions (such as Na$^+$) when dissolved in fluids (Chapter 3, page 67).

Material entering the large intestine consists mostly of undigested food residue (fibrous material from plants), water, bile, cellular debris, and electrolytes. Peristalsis in the large intestine is slow compared to other regions of the GI tract. Instead, nonpropulsive movements called **haustral contractions** create a mixing motion that exposes the colonic contents to the absorptive (mucosal) lining of the colon. Water and electrolytes are absorbed as material passes through the various regions of the colon. Once extracted, water and electrolytes are circulated away from the GI tract in the blood for use by the body. This is another example of the body's ability to reclaim its important resources.

The consistency of the remaining material, now called feces, reflects the extent of water absorption. For example, diarrhea, characterized by loose, watery fecal matter, can result when material moves too quickly through the colon for sufficient water to be reabsorbed. This is beneficial when it allows the body to eliminate harmful or irritating material quickly. However, prolonged diarrhea can result in excessive loss of fluids and electrolytes from the body, which can lead to serious complications such as dehydration. Conversely, when too much water is removed, the feces can become too hard and dry. This is one cause of constipation, a condition that makes elimination difficult and puts excessive strain on the colonic muscles. Two less common conditions that can cause intestinal discomfort are irritable bowel syndrome and inflammatory bowel disease, which are discussed further in the Focus on Clinical Applications feature.

## Microbial Action in the Large Intestine

**rectum** The lower portion of the large intestine between the sigmoid colon and the anal canal.

**anal sphincters** Internal and external sphincters that regulate the passage of feces through the anal canal.

**haustral contractions** (HAU – stral) Slow muscular movements that move the colonic contents back and forth and that help compact the feces.

**microflora, microbiota** (MI – cro – flo – ra, mi – cro – bi – O – ta) Bacteria that reside in the large intestine.

Although a small number of bacteria reside in the small intestine, the large intestine provides a far more suitable environment. The colon's optimal pH, sluggish haustral contractions, and lack of antimicrobial secretions present ideal conditions for bacteria to grow and flourish. The number and variety of bacteria residing in the large intestine is astronomical. In fact, over 400 species of bacteria can be found in the large intestine, contributing to nearly one third of the dry weight of feces. This natural microbial population, also referred to as the **microflora** or the **microbiota,** is important for a healthy colonic ecosystem. First, these bacteria break down undigested food residue that consists mostly of fibrous plant material. Intestinal bacteria also produce vitamin K, limited amounts of certain B vitamins, and some small lipids. Nutrients produced by the intestinal bacteria are absorbed into the blood. Perhaps most important, the natural microbiota help protect us from infection by competing with pathogenic bacteria for limited resources (nutrients and space) in the large intestine. You can read more about how to establish a healthy intestinal microflora in the Focus on Food feature on page 114.

# *Focus On* CLINICAL APPLICATIONS
## Irritable Bowel Syndrome and Inflammatory Bowel Disease

After we eat and enjoy our food, the GI tract dutifully takes over without us having to give it further thought. That is, unless something goes wrong. Health conditions that affect GI function can seriously impact the ability of the GI tract to digest food and absorb nutrients. This is certainly the case with **inflammatory bowel diseases (IBDs)** such as **ulcerative colitis** and **Crohn's disease.** Another GI disorder is **irritable bowel syndrome (IBS),** a poorly understood condition that affects up to 20% of Americans. These GI disturbances can have serious implications in terms of nutritional health.

Irritable bowel syndrome (IBS) is one of the most common GI disorders. People with IBS experience extreme discomfort such as abdominal cramping, bloating, diarrhea, and/or constipation, but the underlying cause has yet to be determined.[6] Historically, the inability to detect clinical abnormalities led to the belief that IBS was simply a psychological manifestation. Although well-defined signs and symptoms are associated with IBS, the lack of clinical markers can make it difficult to diagnose. In fact, diagnosis of IBS is typically based on ruling out other, better-defined intestinal disorders.

GI disturbances associated with IBS may be related to muscle spasms brought on by disordered GI motility. There is some speculation that IBS sufferers may have a hypersensitive GI tract, which overreacts to stimuli. Although emotional stress is not a cause of IBS, it can be a contributing factor. It is important for people with IBS to identify and avoid foods that trigger IBS episodes and seek out those that bring comfort and relief. IBS is sometimes treated with antispasmodic medication. Fortunately, IBS does not progress to other, more serious illnesses.

Irritable bowel syndrome (IBS) and inflammatory bowel disease (IBD) may sound similar, but they are very different. Inflammatory bowel diseases (IBDs) are a group of disorders including ulcerative colitis and Crohn's disease that are characterized by inflammation of the lining of the gastrointestinal tract.[7] Crohn's disease and ulcerative colitis share many similarities, and often have many of the same symptoms. Crohn's disease tends to affect the lower portion of the small intestine (ileum), although it can occur anywhere along the GI tract. In contrast, ulcerative colitis tends to affect the large intestine. And, unlike ulcerative colitis, Crohn's disease can affect the deeper layers of the intestinal wall, leading to the formation of scar tissue. The accumulation of scar tissue from repeated flareups can cause obstructions, which require surgery.

IBDs are diagnosed by a procedure called colonoscopy that involves the insertion of a small scope into the anus. This scope is threaded through the rectum, allowing the physician to inspect the colon wall for signs of inflammation or ulcers. A biopsy (tissue sample) can also be done at this time. IBDs, which typically develop between 15 and 30 years of age, are classified as autoimmune diseases, although this is somewhat speculative. Many researchers believe that antibodies trigger an inflammatory response in the GI tract, causing IBDs. These antibodies may result from exposure to a virus or bacteria. IBD flareups can cause diarrhea, fatigue, weight loss, abdominal pain, diminished appetite, and, on occasion, rectal bleeding. Because IBDs can increase a person's risk of colon cancer, it is important to have regular medical exams.

Although dietary practices do not cause IBDs, nutritional support is important. Nutrient malabsorption and loss of appetite can cause significant weight loss and a variety of nutritional problems. The right assistance by a qualified team of health care professionals can help prevent further complications associated with IBDs.

## Storage and Elimination of Solid Waste from the Large Intestine

The rectum serves as a holding chamber for feces. Significant accumulation of feces causes the rectum wall to become distended, which in turn stimulates enteric stretch receptors. This ultimately results in muscular contractions and the elimination of fecal waste from the body. The elimination of feces from the body is called **defecation.** Relaxation of the internal anal sphincter allows the feces to move into the anal canal. However, the external anal sphincter, which is under conscious control, enables us to determine whether the time is right for waste elimination. By keeping the external anal sphincter contracted, the person can delay defecation, a skill typically learned by young children around 2 to 3 years of age. Relaxing the external anal sphincter allows feces to be expelled from the body.

**inflammatory bowel diseases (IBDs)** Chronic conditions such as ulcerative colitis and Crohn's disease that cause inflammation of the lower GI tract.

**ulcerative colitis** (co – LI – tis) A type of inflammatory bowel disease (IBD) that causes chronic inflammation of the colon.

**Crohn's disease** A chronic inflammatory condition that usually affects the ileum and/or first portion of the large intestine.

**irritable bowel syndrome (IBS)** A condition that typically affects the lower GI tract, causing abdominal pain, muscle spasms, diarrhea, and/or constipation.

**defecation** The expulsion of feces from the body through the rectum and anal canal.

## *Focus On* FOOD
### Probiotic and Prebiotic Foods

The role of intestinal bacteria in terms of enhanced GI function and disease prevention has only recently become appreciated. There is now considerable evidence that certain types of food contribute to a healthy intestinal microflora.[8] These foods are referred to as probiotic and prebiotic foods. **Probiotic foods** contain live bacterial cultures that increase the population of specific strains of non-pathogenic bacteria in the colon. Dietary supplements are one source of probiotic bacteria, as are some "cultured" dairy products such as yogurt, buttermilk, sour cream, and cottage cheese. For example, most yogurts produced in the United States are made by adding the probiotic bacteria *Lactobacillus acidophilus* to milk. Foods with live bacteria in them are often labeled as such.

Another way to increase colonization of beneficial bacteria is to consume **prebiotic foods.** These foods promote the growth of nonpathogenic bacteria and are found mainly in carbohydrates such as dietary fiber. Because dietary fiber can resist digestion, it passes into the colon and provides a source of nourishment for the microflora. Dietary fiber is found in whole grains, cereals, fruits, vegetables, and legumes, and is discussed in detail in Chapter 5. Together, probiotic and prebiotic foods help maintain a well-colonized microbial population in the large intestine, providing an important defense against pathogenic bacteria. In addition, microflora produce many substances that are likely beneficial to our health.

## ESSENTIAL *Concepts*

The large intestine prepares undigested food residue for excretion as feces by removing water. Bacteria in the large intestine also help form feces by breaking down undigested food residue. Eating prebiotic and probiotic foods helps promote the growth of these bacteria. The rectum serves as a holding chamber for feces until the body eliminates it. If material in the large intestine moves too quickly, diarrhea can result, whereas if the contents move too slowly constipation can result. Other conditions that can affect these functions are irritable bowel syndrome (IBS) and inflammatory bowel diseases (IBDs).

**probiotic foods** (PRO – bi – o – tic) Foods and dietary supplements that contain live bacteria.

**prebiotic foods** (PRE – bi – o – tic) Indigestible foods that stimulate the growth of bacteria that naturally reside in the large intestine.

**plasma** The fluid portion of the blood.

**artery** A blood vessel that carries blood away from the heart.

**vein** A blood vessel that carries blood toward the heart.

**systemic circulation** The division of the cardiovascular system that begins and ends at the heart and delivers blood to all the organs except the lungs.

CONNECTIONS Recall that the cardiovascular system is made up of blood vessels and the heart (Chapter 3, page 84).

# Circulation of Nutrients and Fluids Throughout the Body

Cells require a continuous supply of essential nutrients and oxygen. The cardiovascular system, which consists of the heart and an elaborate vascular network, helps meet these needs. The cardiovascular system consists of two separate loops: (1) the systemic circulation and (2) the pulmonary circulation. As illustrated in Figure 4.16, each of these circulatory routes delivers blood to specific regions within the body. Another major circulatory system, the lymphatic system, plays an important role in circulating lipids, maintaining blood volume, and providing protection against disease.

## The Cardiovascular System: Systemic Circulation

Blood is a connective tissue made up of cells and a fluid called **plasma.** Its function is to transport substances to and from cells. Blood flows through vessels called **arteries** and **veins** in a continuous loop that begins and ends at the heart. This part of the cardiovascular system called the **systemic circulation,**

## FIGURE 4.16   Systemic and Pulmonary Circulation

Blood circulates from the heart to the body and back to the heart via the systemic circulation.
Blood circulates from the heart to the lungs and back to the heart via the pulmonary circulation.

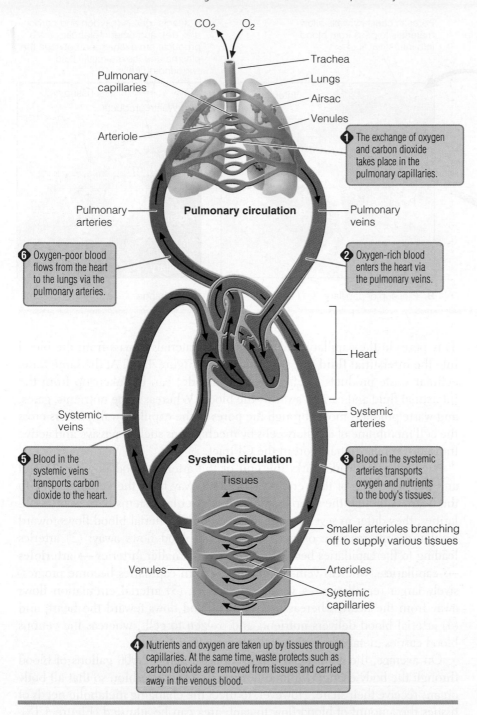

CO₂   O₂

Trachea

Pulmonary capillaries

Lungs

Airsac

Venules

Arteriole

❶ The exchange of oxygen and carbon dioxide takes place in the pulmonary capillaries.

**Pulmonary circulation**

Pulmonary arteries

Pulmonary veins

❻ Oxygen-poor blood flows from the heart to the lungs via the pulmonary arteries.

❷ Oxygen-rich blood enters the heart via the pulmonary veins.

Heart

Systemic veins

Systemic arteries

❺ Blood in the systemic veins transports carbon dioxide to the heart.

**Systemic circulation**

Tissues

❸ Blood in the systemic arteries transports oxygen and nutrients to the body's tissues.

Smaller arterioles branching off to supply various tissues

Venules

Arterioles

Systemic capillaries

❹ Nutrients and oxygen are taken up by tissues through capillaries. At the same time, waste protects such as carbon dioxide are removed from tissues and carried away in the venous blood.

delivers blood to all the body organs with the exception of the lungs. Nutrients and oxygen are delivered to cells through an intricate maze of blood vessels called arteries. Oxygenated blood leaves the heart through the **aorta,** which then branches into arteries. Arteries circulating blood away from the heart divide, subdivide, and eventually form a bed of microscopic blood vessels called **capillaries** around organs and tissues. Capillaries have thin walls and narrow diameters, making them well suited for their primary function—the exchange of materials, nutrients, and gases between the blood and tissues.

**aorta** (a – OR – ta) The main artery that initially carries blood from the heart to all areas of the body except the lungs.

**capillaries** (CAP – il – lar – ies) Blood vessels with thin walls, which allow for the exchange of materials between blood and tissues.

## FIGURE 4.17 Nutrient and Gas Exchange Across the Capillary Wall

The exchange of nutrients, gases (oxygen and carbon dioxide), and other metabolic waste products takes place between the plasma and the interstitial fluid.

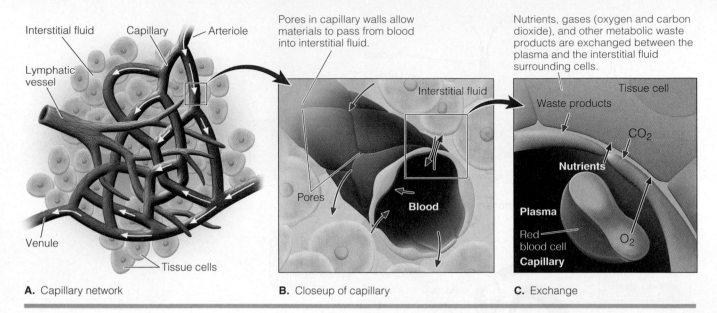

**A.** Capillary network      **B.** Closeup of capillary      **C.** Exchange

Tiny pores in the capillary walls allow most materials to pass from the blood into the **interstitial fluid** that surrounds cells (Figure 4.17). At the same time, cellular waste products (such as carbon dioxide) can be taken up from the interstitial fluid and carried away by the blood. Whereas some nutrients, gases, and waste products move through the pores of the capillary walls, others cross the cell membrane of capillary cells by mechanisms such as passive and active transport, similar to transport in the GI tract.

A capillary network marks the end of the arterial blood flow to the cell and the beginning of the venous blood flow away from the cell and back to the heart. Although the arterial and venous vascular systems have many similarities, they differ in several ways: (1) oxygen-rich arterial blood flows toward capillaries, whereas the oxygen-poor venous blood flows away; (2) arteries leading to the capillaries become progressively smaller (arteries → **arterioles** → capillaries), whereas veins leading away from capillaries become progressively larger (capillaries → **venules** → veins); (3) arterial circulation flows away from the heart, whereas the venous blood flows toward the heart; and (4) arterial blood delivers nutrients and oxygen to cells, whereas the venous blood carries metabolic waste products away from cells.

On average, the adult heart pumps approximately 2,000 gallons of blood through the body each day. Blood flows in an orderly fashion so that all body organs receive their share. However, to meet the changing metabolic needs of tissues the amount of blood flow to each area can be adjusted (Figure 4.18). For example, skeletal muscles receive approximately 15% of the blood flow at rest, increasing to as much as 85% during strenuous exercise. Similarly, increased blood flow to the digestive tract after a large meal aids digestion and nutrient absorption. This is also why a strenuous workout after eating may feel uncomfortable as muscles compete with the GI tract for blood.

Adjustment in blood flow is accomplished, in part, through the contraction and relaxation of the layer of smooth muscles that line arterioles. This change in vascular tone affects the radius of the blood vessels. When these muscles

**interstitial fluid** (in – ter – STI – tial) Fluid that surrounds cells.

**arterioles** Small blood vessels that branch off from arteries.

**venules** Small blood vessels that branch off from veins.

## FIGURE 4.18  Blood Flow (mL/min) at Rest and During Exercise

The cardiovascular system adjusts blood flow based on physiological needs of tissues.

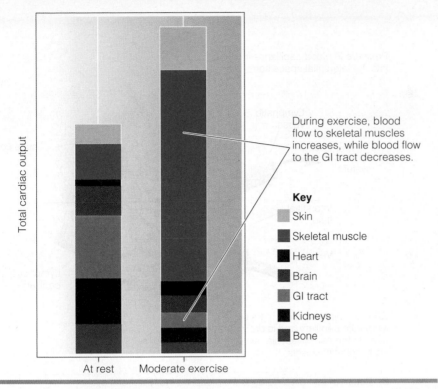

During exercise, blood flow to skeletal muscles increases, while blood flow to the GI tract decreases.

**Key**
- Skin
- Skeletal muscle
- Heart
- Brain
- GI tract
- Kidneys
- Bone

Total cardiac output

At rest        Moderate exercise

constrict, the radius of the blood vessel decreases, resulting in reduced blood flow to the surrounding tissues. Conversely, **vasodilation** associated with relaxation of the arteriole smooth muscle increases blood flow. These changes in vascular tone affect the amount of blood and therefore nutrients received by the various organs.

## The Cardiovascular System: Pulmonary Circulation

Another component of the cardiovascular system is the flow of blood between the heart and lungs. This circuit, referred to as the **pulmonary circulation,** begins with the arrival of partially deoxygenated venous blood to the heart. The **pulmonary arteries** transport blood from the right side of the heart to the lungs, where the exchange of carbon dioxide and oxygen takes place across the pulmonary capillaries. When air is exhaled through the nose and mouth, carbon dioxide is eliminated from the body. Likewise, during inhalation oxygen is taken into the lungs, where it crosses the capillaries and enters the blood. The oxygen-rich blood returns to the heart through the **pulmonary veins** and is pumped out of the heart through the aorta for a return trip through the systemic circulation.

## The Lymphatic System

The **lymphatic system** is another major circulatory system in the body. This system consists of a network of lymph vessels, lymph nodes (small organelles that filter foreign substances out of blood), and lymph organs such as the spleen (Figure 4.19). A clear liquid called lymph circulates through the lymphatic vessels. The lymphatic vessels include a network of lymphatic capillar-

**vasodilation** (VA – so – di – la – tion) Relaxation of the smooth muscles inside a blood vessel; increases the diameter of the vessel.

**pulmonary circulation** (PUL – mo – nar – y) The division of the cardiovascular system that circulates deoxygenated blood from the heart to the lungs, and oxygenated blood from the lungs back to the heart.

**pulmonary arteries** Blood vessels that transport oxygen-poor blood from the right side of the heart to the lungs.

**pulmonary veins** Blood vessels that transport oxygen-rich blood from the lungs to the heart.

**lymphatic system** (lym – PHAT – ic) A circulatory system made up of vessels and lymph that flows from organs and tissues, drains excess fluid from spaces that surround cells, picks up dietary fats from the digestive tract, and plays a role in immune function.

## FIGURE 4.19 Overview of the Lymphatic System

The lymphatic system consists of a network of lymph vessels, lymph nodes, and lymph organs. A clear liquid called lymph circulates through lymphatic vessels.

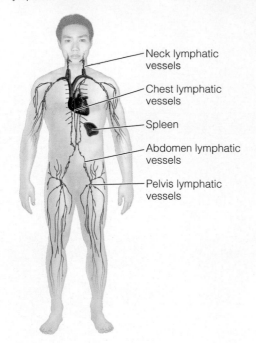

Neck lymphatic vessels

Chest lymphatic vessels

Spleen

Abdomen lymphatic vessels

Pelvis lymphatic vessels

Pressure in blood capillaries forces fluid into the interstitial space surrounding cells.

Lymphatic vessel

To venous system

Interstitial fluid

Venule

Arteriole

Tissue cells

Lymphatic capillary

Excess interstitial fluid enters the lymphatic capillary and is delivered back to the cardiovascular system via lymphatic vessels.

Blood capillary

**A.** Lymphatic system

**B.** Return of interstitial fluids to the blood via the lymphatic system

ies, which become progressively larger, forming **lymphatic ducts.** Lymphatic ducts eventually connect with the cardiovascular system in the lower neck at the thoracic duct. Unlike blood, lymph flows in one direction only, away from tissues. As discussed, when large lipid molecules are absorbed, the lymphatic system transports them away from the GI tract. In addition, the lymphatic system has two other important functions: returning fluids to the blood and protecting against disease (immune function).

### The Return of Fluids to the Blood

Capillaries deliver blood to many organs and tissues of the body. Because of their narrow circumference, the pressure in blood capillaries is very high, forcing plasma out of the capillary wall into the interstitial fluid surrounding cells. The volume of plasma forced out of capillaries is considerable, about 2 to 3 liters a day. Capillary vessels can reabsorb some of this fluid, but the excess must be removed by other means. Left unchecked, excess interstitial fluid would cause swelling and tissue damage. Furthermore, the loss of fluid from the cardiovascular system would decrease blood volume and increase blood viscosity. These conditions can be life threatening. Fortunately, the excess interstitial fluid enters the lymphatic system, where it becomes lymph fluid and is eventually delivered back to the cardiovascular system via lymphatic ducts.

### Protection Against Disease

As the lymph flows through the lymphatic system, it encounters small nodules called lymph nodes. Lymph nodes contain white blood cells that are part of the body's immune system. These cells include **macrophages** and **lymphocytes.**

**lymphatic duct** An enclosed canal that circulates lymph.

**macrophage** (MA – cro – phage) A white blood cell that is part of the body's immune defense.

**lymphocyte** (LYMPH – o – cyte) A type of cell in the immune system that produces antibodies that attack foreign cells.

Macrophages remove and destroy bacteria, cancer cells, and cellular debris from the lymph, whereas lymphocytes monitor and take defensive actions against infectious agents. These actions help prevent disease-causing agents in the lymph from re-entering the blood.

# ESSENTIAL *Concepts*

Cells require a wide variety of nutrients and oxygen. The delivery of these substances is accomplished by the cardiovascular and lymphatic systems. The cardiovascular system consists of the systemic circulation and the pulmonary circulation. These circulatory routes deliver blood to specific regions within the body. The systemic blood circulates between the heart, the body organs, and back again. The pulmonary route circulates blood between the heart and lungs. In addition to circulating fat-soluble nutrients, the lymphatic system plays an important role in maintaining blood volume and protecting against disease.

# Excretion of Metabolic Waste Products

Nutrients taken up by cells undergo considerable metabolic change. That is, nutrients are transformed, broken down, or used to synthesize other materials. These processes result in the formation of a variety of metabolic waste products such as carbon dioxide, water, and urea. Urea is a nitrogen-containing compound, found in urine, that results from the breakdown of amino acids. Because the accumulation of these waste products can be toxic, it is important that they be eliminated from the body. Several body organs assist in this task, each equipped to handle a particular type of waste.

As discussed, the respiratory system removes carbon dioxide. Recall that solid waste, which is not metabolic waste, is eliminated in the feces via the large intestine. The integumentary system, which consists of the skin and its components, also helps eliminate metabolic waste products. For example, sweating eliminates excess water, salts, and to a minor extent urea. In addition to these organ systems, the urinary system is the primary site for excreting metabolic wastes other than carbon dioxide.

## Urinary Excretion of Metabolic Waste Products

The urinary system, which consists of the **kidneys, ureters, urinary bladder,** and **urethra,** plays an important role in removing metabolic wastes from the plasma and their subsequent elimination in the urine. In addition, the urinary system helps maintain water and electrolyte balance by either eliminating or conserving water and other plasma constituents.

Blood flows to the kidneys at a rate of about 1,200 mL per minute. The **nephrons,** which are the functional units of the kidney, perform three functions: (1) filtration, (2) resorption, and (3) secretion. **Filtration** involves removing wastes such as urea as well as excess water, electrolytes, salts, and minerals as the blood flows through the nephrons. **Resorption** reroutes water, amino acids, glucose, and other essential substances and returns them to the blood. This recycling prevents the loss of important materials from the body. Another function of the urinary system is the removal of waste products via secretion. During secretion, the remaining waste products filtered out of the

**kidneys** The organs that filter metabolic waste products from the blood and play a role in maintaining blood volume.

**ureters** (u – RE – ter) Ducts that carry urine from the kidneys to the bladder.

**urinary bladder** The organ that collects urine.

**urethra** (u – RE – thra) The duct that carries urine from the bladder to the outside of the body.

**nephrons** (NEPH – rons) Tubules in the kidneys that filter waste materials from the blood that are later excreted in the urine.

**filtration** The process of selective removal of metabolic waste products from the blood.

**resorption** The return of previously removed materials to the blood.

## FIGURE 4.20 Overview of the Urinary System

The urinary system consists of the kidneys, ureters, urinary bladder, and urethra.

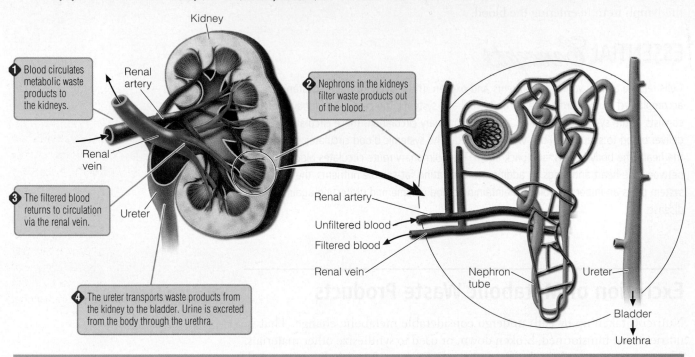

**1** Blood circulates metabolic waste products to the kidneys.

**2** Nephrons in the kidneys filter waste products out of the blood.

**3** The filtered blood returns to circulation via the renal vein.

**4** The ureter transports waste products from the kidney to the bladder. Urine is excreted from the body through the urethra.

Kidney

Renal artery

Renal vein

Ureter

Renal artery

Unfiltered blood

Filtered blood

Renal vein

Nephron tube

Ureter

Bladder

Urethra

blood become urine. Urine leaves the kidneys by ureters, which are the ducts that travel downward and connect with the bladder. As the bladder fills with urine, stretch receptors are activated and the sensation or need to urinate becomes apparent. Urine is eliminated from the body through a tube called the urethra. The role of the urinary system in removing metabolic waste material is illustrated in Figure 4.20.

In addition to water, urine contains a variety of waste products including salts and nitrogenous compounds such as urea. In fact, the composition of urine is so well defined that urine analysis can be used to determine whether certain pathological conditions exist. For example, finding large amounts of glucose in the urine could indicate a condition called diabetes. Normally, urine is sterile, meaning that no microorganisms such as bacteria are present. However, when certain microorganisms gain entry into the urinary system, a urinary tract infection may occur. To determine if a person has a urinary tract infection, urine is analyzed for the presence of microorganisms, which often are bacteria from the digestive tract. You can read more about urinary tract infections in the Focus on Food feature.

## ESSENTIAL *Concepts*

Metabolic waste products such as carbon dioxide, urea, and water are removed by different organs, including the lungs, kidneys, and to a lesser extent, the skin. The urinary system is the primary site for excreting metabolic wastes, other than carbon dioxide. To remove metabolic waste products, the kidneys filter them as the blood flows through the nephrons. Next, water and other essential substances are returned to the blood by a process called resorption. Last, the process of secretion prepares waste products to be eliminated from the body as urine. Urine leaves the kidneys via the ureters, which empty into the bladder. Urine is excreted from the body through the urethra.

# Focus On FOOD

## Cranberry Juice and Urinary Tract Infections

Hype or reality; is there evidence to support the claim that cranberry juice helps prevent urinary tract infections? For years, the evidence was anecdotal, but several intervention studies provide clinical evidence that cranberry juice protects against the recurrence of urinary tract infections (UTIs).[9] UTIs, which are more common in women than men, result from bacteria adhering to the lining of the bladder and urinary tract. Over 11 million prescriptions are written yearly for antibiotics to treat these infections.

The bacteria, *Escherichia coli (E. coli),* is the most common cause of UTIs. Typically, a person with a UTI experiences urgent, frequent, and painful sensations associated with urination. How does drinking cranberry juice help? In a word—proanthocyanidins. It may be a mouthful to pronounce, but proanthocyanidins are phytochemicals found in cranberry juice that may help prevent UTIs. These compounds, which are the pigments responsible for the deep red color of this juice, prevent *E. coli* from adhering to the bladder lining.

A clinical intervention study published in the *British Medical Journal* reported a significant reduction in the incidence of UTIs among women who daily consumed a concentrate consisting mainly of cranberry juice.[10] The rate of UTI recurrence was 16% for women given cranberry juice and 36% for women given a placebo. This and several other studies provide considerable support that cranberry juice helps prevent recurrent UTIs.[11] Although more studies are needed to clarify effective doses, as little as 10 ounces of cranberry juice daily may be enough to help ward off UTIs. It is important to emphasize, however, that cranberry juice is not a treatment for an established UTI. Instead, drinking cranberry juice on a daily basis may offer protection from the recurrence of UTIs and is an example of how a food can offer both medicinal and nutritional benefits.

## Review Questions    Answers are found in Appendix G.

1. Which of the following organs is *not* a part of the digestive tract?
   a. The stomach
   b. The small intestine
   c. The pancreas
   d. The esophagus
2. In the GI tract, which of the following tissue layers provides secretions needed for the process of digestion?
   a. The mucosa
   b. The submucosa
   c. The muscularis
   d. The serosa
3. Segmentation is needed to
   a. propel food through the GI tract.
   b. mix food in the GI tract.
   c. prevent food from moving in the wrong direction in the GI tract.
   d. prevent food from entering the trachea.
4. Which of the following digestive secretions chemically breaks down complex food particles?
   a. Mucus
   b. Enzymes
   c. Hormones
   d. Bile
5. All of the following play a role in regulating GI motility and the release of digestive secretions *except*
   a. enteric hormones.
   b. the enteric nervous system.
   c. bile.
   d. the central nervous system.
6. Which of the following organs releases the hormone gastrin?
   a. The pancreas
   b. The small intestine
   c. The stomach
   d. The gallbladder
7. The _____ regulates the flow of material from the stomach into the small intestine.
   a. epiglottis
   b. ileocecal sphincter
   c. pyloric sphincter
   d. fundus
8. Chyme formation takes place in the
   a. stomach.
   b. small intestine.
   c. large intestine.
   d. pharynx.
9. Which of the following affects the rate of gastric emptying?
   a. The amount of food consumed
   b. The nutrient composition of the food consumed
   c. The consistency of food consumed
   d. All of the above

10. Structures called villi are found in the _____.
    a. mouth
    b. stomach
    c. small intestine
    d. large intestine
    e. all of the above
11. Chyme is neutralized by a secretion released from the _____.
    a. small intestine
    b. gallbladder
    c. stomach
    d. pancreas
12. Haustral contractions occur in the _____.
    a. mouth
    b. stomach
    c. small intestine
    d. large intestine
13. The term used to described bacteria residing in the large intestine is _____.
    a. microflora
    b. chyme
    c. bolus
    d. macrophages
14. Which of the following organ systems filters the blood to remove metabolic waste?
    a. The digestive system
    b. The urinary system
    c. The endocrine system
    d. The nervous system
15. Which of the following vascular structures is the site of nutrient and gas exchange between the plasma and interstitial fluids?
    a. Arteries
    b. Veins
    c. Lymph vessels
    d. Capillaries
16. List and describe the four tissue layers making up the gastrointestinal tract.
17. List the organs that produce and release secretions needed for the process of digestion. For each secretion that you list, describe its role in the process of digestion.
18. Describe the digestive events that take place in the mouth, stomach, and small intestine.
19. Describe the inner surface of the small intestine and how this surface facilitates the process of nutrient absorption.
20. Describe the role of the enteric nervous system, enteric endocrine system and the central nervous system in digestion.

## Media Links

A variety of study tools for this chapter are available at our website www.thomsonedu.com/nutrion/mcguire.

Prepare for tests and deepen your understanding of chapter concepts with these online materials:

- Practice tests
- Flashcards
- Glossary
- Student lecture notebook
- Web links
- Animations
- Chapter summaries, learning objectives, and crossword puzzles

## Notes

1. Kurihara K, Kashiwayanagi M. Physiological studies on umami taste. Journal of Nutrition. 2000;130(4S Suppl): 931S–4S.
2. Spechler SJ. Clinical manifestations and esophageal complications of GERD. American Journal of Medical Sciences. 2003;326(5):279–84.
3. DeCross AJ, Marshall BJ. The role of Helicobacter pylori in acid-peptic disease. American Journal of Medical Sciences. 1993;306(6):381–92.
4. Sanders MK, Peura DA. Helicobacter pylori-associated diseases. Current Gastroenterology Reports. 2002;4:448–54.
5. Nelsen DA. Gluten-sensitive enteropathy (celiac disease): more common than you think. American Family Physician. 2002;66(12):2259–66.
6. Olden KW. Diagnosis of irritable bowel syndrome. Gastroenterology. 2002;122(6):1701–14.
7. Itzkowitz SH, Present DH. Crohn's and Colitis Foundation of America Colon Cancer in IBD Study Group. Consensus conference: Colorectal cancer screening and surveillance in inflammatory bowel disease. Inflammatory Bowel Disease. 2005;11(3):314–21.

8. Rastall RA. Bacteria in the gut: friends and foes and how to alter the balance. Journal of Nutrition. 2004;134 (8 Suppl):2022S–6S.

9. Howell AB. Cranberry proanthocyanidins and the maintenance of urinary tract health. Critical Reviews in Food Science and Nutrition. 2002;42(3 Suppl):273–8.

10. Kontiokari T, Sundqvist K, Nuutinen M, Pokka T, Koskela M, Uhari M. Randomised trial of cranberry-lingonberry juice and Lactobacillus GG drink for the prevention of urinary tract infections in women. British Medical Journal. 2001;30;322(7302):1571.

11. Henig YS, Leahy MM. Cranberry juice and urinary-tract health: science supports folklore. Nutrition. 2000; 16(7–8):684–7.

**Check out the following sources for additional information.**

12. Baum C, Moxon D, Scott M. Gastrointestinal Disease. In: Bowman BA, Russell RM, editors. Present knowledge in nutrition, 8th ed. Washington, DC: ILSI Press; 2001.

13. Bercik P, Verdu EF, Collins SM. Is irritable bowel syndrome a low-grade inflammatory bowel disease? Gastroenterology Clinics of North America. 2005;34(2):235–45.

14. Brown AC, Valiere A. Probiotics and medical nutrition therapy. Nutrition Clinical Care. 2004; 7(2):56–68.

15. Chan FK; Leung WK. Peptic-ulcer disease. Lancet. 2002;360(9337):933–41.

16. Chow J. Probiotics and prebiotics: a brief overview. Journal of Renal Nutrition. 2002;12(2):76–86.

17. Cremonini F, Talley NJ. Irritable bowel syndrome: epidemiology, natural history, health care seeking and emerging risk factors. Gastroenterology Clinics in North America. 2005;34(2):189–204.

18. DeCross AJ, Marshall BJ. The role *of Helicobacter pylori* in acid-peptic disease. American Journal of Medical Sciences. 1993;306(6):381–92.

19. Drossman DA. What does the future hold for irritable bowel syndrome and the functional gastrointestinal disorders? Journal of Clinical Gastroenterology. 2005;39(4 Suppl 3): S251–6.

20. Fasano A. Celiac disease: the past, the present, the future. Pediatrics. 2001;107(4):768–70.

21. Fooks LJ, Gibson GR. Probiotics as modulators of the gut flora. British Journal of Nutrition. 2002;88 (Suppl 1):S39–49.

22. Gropper SS, Smith JL, Groff JL. Advanced nutrition and human metabolism, 4th ed. Belmont, CA: Wadsworth Thomson Learning; 2005.

23. Henig YS, Leahy MM. Cranberry juice and urinary-tract health: science supports folklore. Nutrition. 2000; 16(7–8):684–7.

24. Howell AB. Cranberry proanthocyanidins and the maintenance of urinary tract health. Critical Reviews in Food Science and Nutrition. 2002;42(3 Suppl):273–8.

25. Itzkowitz SH, Present DH. Crohn's and Colitis Foundation of America Colon Cancer in IBD Study Group. Consensus conference: colorectal cancer screening and surveillance in inflammatory bowel disease. Inflammatory Bowel Disease. 2005;11(3):314–21.

26. Guyton AC, Hall JE. Textbook of medical physiology, 10th ed. Philadelphia: W.B. Saunders; 2000.

27. Kuipers EJ, Janssen MJ, de Boer WA. Good bugs and bad bugs: indications and therapies for Helicobacter pylori eradication. Current Opinion in Pharmacology. 2003; 3(5):480–5.

28. Kurihara K, Kashiwayanagi M. Physiological studies on umami taste. Journal of Nutrition. 2000;130(4S Suppl):931S–4S.

29. Kuzminski LN. Cranberry juice and urinary tract infections: is there a beneficial relationship? Nutrition Reviews. 1996;54(11 Pt 2):S87–90.

30. Lacy BE, Lee RD. Irritable bowel syndrome: a syndrome in evolution. Journal of Clinical Gastroenterology. 2005; 39(4 Suppl 3):S230–42.

31. Lea R, Whorwell PJ. The role of food intolerance in irritable bowel syndrome. Gastroenterology Clinics in North America. 2005;34(2):247–55.

32. Longstreth GF. Definition and classification of irritable bowel syndrome: current consensus and controversies. Gastroenterology Clinics in North America. 2005;34(2): 173–87.

33. Lowe FC, Fagelman E. Cranberry juice and urinary tract infections: what is the evidence? Urology. 2001;57(3): 407–13.

34. Marteau P, Seksik P, Jian R. Probiotics and intestinal health effects: a clinical perspective. British Journal of Nutrition. 2002; 88 (Suppl 1)1:S51–7.

35. Marsh MN. The natural history of gluten sensitivity: defining, refining and re-defining. QJM: Quarterly Journal of the Association of Physicians. 1995;88(1):9–13.

36. Nelsen, DA. Gluten-sensitive enteropathy (celiac disease): more common than you think. American Family Physician. 2002;66(12):2259–66.

37. Olden KW. Diagnosis of irritable bowel syndrome. Gastroenterology. 2002;122(6):1701–14.

38. Rastall RA. Bacteria in the gut: friends and foes and how to alter the balance. Journal of Nutrition. 2004;134(8 Suppl):2022S–6S.

39. Rhoades, R., Pflanzer R. Human physiology, 4th ed. Belmont, CA: Brooks/Cole-Thomson Learning; 2003.

40. Saavedra JM, Tschernia A. Human studies with probiotics and prebiotics: clinical implications. British Journal of Nutrition. 2002;87(Suppl 2):S241–6.

41. Sanders MK, Peura DA. Helicobacter pylori-associated diseases. Current Gastroenterology Reports. 2002;4:448–54.

42. Sclafani, A. The sixth taste? Appetite. 2004;43(1):1-3.

43. Sherwood, L. Human physiology—from cells to systems, 5th ed. Belmont, CA: Brooks/Cole-Thomson Learning; 2004.

44. Spechler SJ. Clinical manifestations and esophageal complications of GERD. American Journal Medical Sciences. 2003;326(5):279–84.

45. Sullivan A, Nord CE. Probiotics and gastrointestinal diseases. Journal of Internal Medicine. 2005;257(1):78–92.

46. Talley NJ. Environmental versus genetic risk factors for irritable bowel syndrome: clinical and therapeutic implications. Review of Gastroenterology Disorders. 2005;5(2):82–8.

47. Williamson D, Marsh MN. Celiac disease. Molecular Biotechnology. 2002;22(3):293–9.

48. Wood DW, Block KP. Helicobacter pylori: a review. American Journal of Therapeutics. 1998;5(4):253–61.

## Chapter 5

# Carbohydrates

**C**arbohydrates are a diverse group of compounds produced primarily by plants. The simplest carbohydrates consist of single-sugar units, which can join together to make complex carbohydrates. Carbohydrates serve many important functions such as providing a vital source of energy. In addition, they are components of ribonucleic and deoxyribonucleic acids (RNA and DNA, respectively). Carbohydrates are plentiful in a variety of foods and are important components of a healthy, well-balanced diet. This chapter discusses carbohydrate chemistry, dietary sources of carbohydrates, functions of carbohydrates in the body, and guidelines for carbohydrate intake.

# Classifying Carbohydrates: Simple Carbohydrates

**Carbohydrates** are made up of varying numbers of sugar units and provide approximately 4 kcal/g. A carbohydrate consisting of a single sugar unit is called a **monosaccharide;** a carbohydrate made of two sugar units is a **disaccharide.** Because of their small size, monosaccharides and disaccharides are called **simple carbohydrates** or simple sugars. This section describes the chemistry, classification, and food sources of mono- and disaccharides.

## Monosaccharides

Monosaccharides are single-sugar units that consist of carbon, hydrogen, and oxygen atoms in the ratio of 1:2:1, respectively. For example, glucose, the most abundant carbohydrate in the body, consists of 6 carbon, 12 hydrogen, and 6 oxygen atoms (written $C_6H_{12}O_6$). Monosaccharides are subclassified by the number of carbon atoms in their structure, as indicated by the prefix *tri-*, *tetr-*, *pent-*, *hex-*, and so on. For example, monosaccharides with 3 carbons are triose sugars ($C_3H_6O_3$), whereas those with 5 carbons are pentose sugars. Note that the suffix -*ose* denotes a sugar (such as gluc*ose*). The most abundant monosaccharides found in food are 6-carbon sugars, called hexoses. These include glucose, fructose, and galactose.

Monosaccharides typically exist as cyclic or ring structures, which provide the basis for a numbering system used to describe them. This numbering system assigns each carbon in the ring a number, as illustrated in Figure 5.1. Although glucose, fructose, and galactose have the same molecular formula ($C_6H_{12}O_6$), each has a different arrangement of atoms. Although fructose has a five-sided ring structure, it is still classified as a hexose because it contains

**carbohydrate** Organic compound made up of varying numbers of monosaccharides.

**monosaccharide** (mo – no – SAC – cha – ride) (*mono-*, one; -*saccharide*, sugar) Carbohydrate consisting of a single sugar.

**disaccharide** (di – SAC – cha – ride) (*di-*, two) Carbohydrate consisting of two monosaccharides bonded together.

**simple carbohydrate,** or **simple sugar** Category of carbohydrates consisting of mono- or disaccharides.

## FIGURE 5.1 The Structure of Monosaccharides

Glucose, galactose, and fructose are monosaccharides that contain 6 carbon atoms. Monosaccharides that have 6 carbon atoms are called hexose sugars. All hexose sugars have 6 carbon, 12 hydrogen, and 6 oxygen atoms (1:2:1).

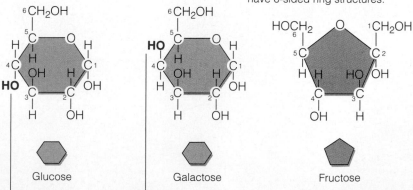

Fructose has a 5-sided ring structure, whereas glucose and galactose have 6-sided ring structures.

Glucose

Galactose

Fructose

Notice that glucose and galactose have a similar chemical structure with the exception of the hydroxyl groups (—OH), which face in opposite directions.

a total of 6 carbons. You may also notice that at first glance glucose and galactose look identical. However, the hydroxyl (−OH) groups on carbon 4 actually face in opposite directions. This dissimilarity may seem insignificant, but even minor structural differences affect digestion, absorption, and physiological function.

## Glucose

The chemical reaction called **photosynthesis** enables chlorophyll-containing plants to combine carbon dioxide and water in the presence of sunlight to produce a sugar called **glucose** (Figure 5.2). The plant then uses glucose to form larger carbohydrates. When animals consume plants, the carbohydrates are broken down (digested), providing a very important source of glucose. In this way, plants and animals are each part of the delicate balance in nature.

Glucose, sometimes called "blood sugar," is the most abundant monosaccharide in the body. The primary function of glucose is to provide a source of energy (ATP) to cells. Importantly, the nervous system (for example, the brain) and the red blood cells generally use glucose as their primary energy source. Glucose is also used to synthesize other compounds in the body. For example, it can be converted to some amino acids (the building blocks of proteins) and fat for long-term energy storage. In addition, the body can store small amounts of glucose as a compound called glycogen.

## Fructose

**Fructose**, also called levulose, is a naturally occurring monosaccharide found primarily in honey, fruits, and vegetables. Of all the monosaccharides, fructose is the sweetest. High-fructose corn syrup, derived from corn, is the major sweetener used in soft drinks, fruit beverages, and a variety of other foods. High-fructose corn syrup, which consists of almost equal amounts of fructose and glucose, is used extensively by food manufacturers, contributing

**photosynthesis** (*photo-*, light; *-synthesis*, product) Process whereby plants trap energy from the sun to produce glucose from carbon dioxide and water.

**glucose** (GLU – kose) A 6-carbon monosaccharide produced in plants by photosynthesis.

**fructose** (FRUC – tose) A 6-carbon monosaccharide found in fruits and vegetables; also called levulose.

## FIGURE 5.2    Photosynthesis

The process of photosynthesis enables plants to capture energy from sunlight and transfer it to chemical bonds of glucose. When we consume plants, the glucose provides us with a source of energy.

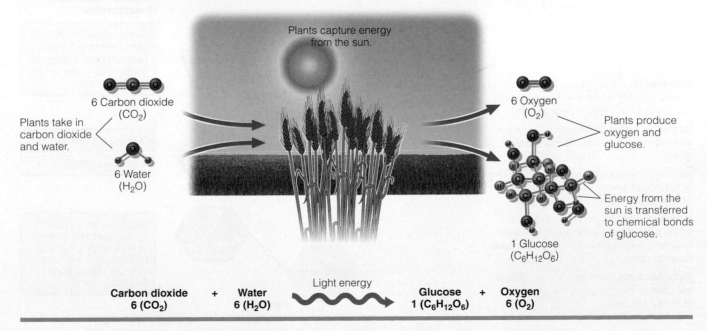

Plants capture energy from the sun.

6 Carbon dioxide ($CO_2$)

Plants take in carbon dioxide and water.

6 Water ($H_2O$)

6 Oxygen ($O_2$)

Plants produce oxygen and glucose.

1 Glucose ($C_6H_{12}O_6$)

Energy from the sun is transferred to chemical bonds of glucose.

| Carbon dioxide | + | Water | Light energy | Glucose | + | Oxygen |
|---|---|---|---|---|---|---|
| 6 ($CO_2$) | | 6 ($H_2O$) | | 1 ($C_6H_{12}O_6$) | | 6 ($O_2$) |

approximately 5% of total energy intake in the United States. Some scientists speculate that excess consumption of beverages containing high-fructose corn syrup has contributed to the increased prevalence of obesity today.[1]

### Galactose

Few foods contain **galactose** in its free state. Rather, it is one of the monosaccharides that make up the disaccharide lactose, the major carbohydrate in milk. Galactose is also found as the carbohydrate component of glycolipids and glycoproteins, which are part of cell membranes. Glycolipids are lipids that contain carbohydrates, whereas glycoproteins are proteins with carbohydrates attached. By far most of the galactose in the body is converted to glucose and used as energy.

## Disaccharides

Disaccharides consist of two monosaccharides bonded together. The most common disaccharides are **lactose, sucrose,** and **maltose.** These disaccharides share one characteristic—at least one of the monosaccharides in the pair is glucose (Figure 5.3). A condensation reaction chemically joins monosaccharides together by a **glycosidic bond.** This occurs when the hydroxyl group ($-OH$) from one monosaccharide interacts with a hydrogen group ($-H$) from another monosaccharide, resulting in the loss of one molecule of water ($H_2O$). Disaccharides have a molecular formula of $C_{12}H_{22}O_{11}$, which reflects the loss of one water molecule in the formation of the glycosidic bond ($2C_6H_{12}O_6 \rightarrow C_{12}H_{24}O_{11} + H_2O$). The formation of a glycosidic bond is shown in Figure 5.4.

Numbers are used to designate which carbon atoms form the glycosidic bond. For example, a glycosidic bond between carbon 1 on the glucose molecule and carbon 4 on the other monosaccharide is referred to as a 1,4 glycosidic bond.

**CONNECTIONS** A condensation reaction occurs when molecules are bonded together with the release of a molecule of water (Chapter 3, page 77).

**galactose** (ga – LAC – tose) A 6-carbon monosaccharide found mainly bonded with glucose to form the milk sugar lactose.

**lactose** Disaccharide consisting of glucose and galactose; produced by mammary glands.

**sucrose** (SU – crose) Disaccharide consisting of glucose and fructose; found primarily in fruits and vegetables.

**maltose** (MAL – tose) Disaccharide consisting of two glucose molecules bonded together; formed during the chemical breakdown of starch.

**glycosidic bond** (gly – co – SI – dic) Chemical bond formed by condensation of two monosaccharides.

**FIGURE 5.3 Disaccharides: Sucrose, Maltose, and Lactose**

Disaccharides consist of two monosaccharides bonded together.

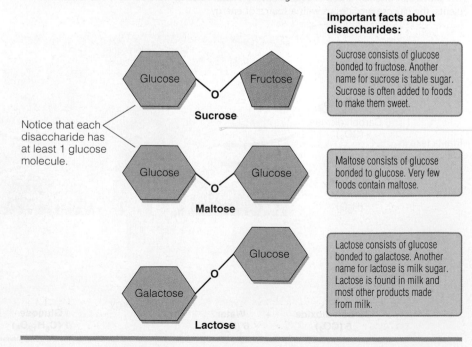

Notice that each disaccharide has at least 1 glucose molecule.

**Important facts about disaccharides:**

Sucrose consists of glucose bonded to fructose. Another name for sucrose is table sugar. Sucrose is often added to foods to make them sweet.

Maltose consists of glucose bonded to glucose. Very few foods contain maltose.

Lactose consists of glucose bonded to galactose. Another name for lactose is milk sugar. Lactose is found in milk and most other products made from milk.

## FIGURE 5.4 Formation of Glycosidic Bonds

A glycosidic bond is formed by a condensation reaction between two monosaccharides. In this example, two glucose molecules join to form water and the disaccharide maltose.

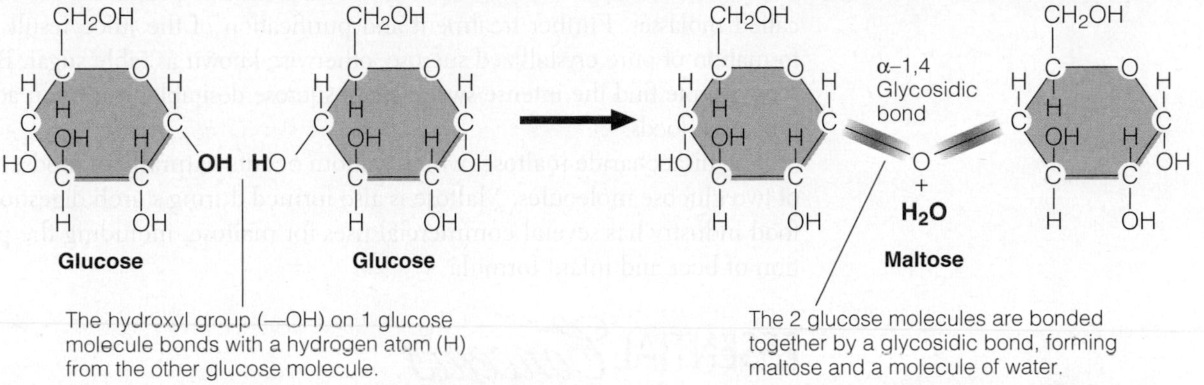

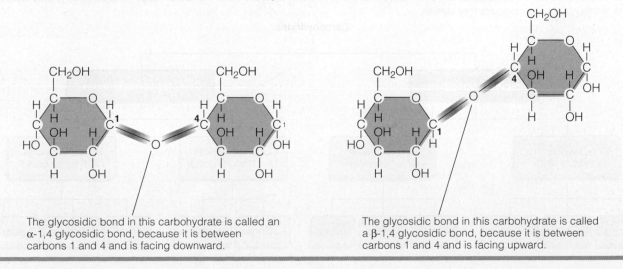

Glucose + Glucose → Maltose

The hydroxyl group (—OH) on 1 glucose molecule bonds with a hydrogen atom (H) from the other glucose molecule.

The 2 glucose molecules are bonded together by a glycosidic bond, forming maltose and a molecule of water.

Furthermore, some glycosidic bonds are called alpha (α) bonds, whereas others are beta (β) bonds (Figure 5.5). Whereas numbers are used to specify the placement of the glycosidic bond, the α and β designations indicate whether the glycosidic bond is facing up or down. This system offers a convenient way to describe important structural features of carbohydrates and becomes particularly important when considering carbohydrate digestion. This is because the type of glycosidic bond determines whether a carbohydrate is digestible or indigestible.

## Lactose

Lactose is a major disaccharide in the diet and is the most abundant carbohydrate in milk. Lactose consists of glucose and galactose and is produced in the mammary glands during lactation. During this period of the life cycle, enzymes in the mammary glands (breasts) convert glucose into galactose, which is then used to synthesize lactose. Thus, milk and milk products (such as yogurt, cheese, and ice cream) contain lactose, although its concentration can vary.

## FIGURE 5.5 The Naming of Glycosidic Bonds

A glycosidic bond joins two monosaccharides. The numbers in a glycosidic bond refer to the carbon atoms that form the bond. If the glycosidic bond is facing down, it is called an alpha (α) glycosidic bond. A glycosidic bond facing up is called a beta (β) glycosidic bond.

The glycosidic bond in this carbohydrate is called an α-1,4 glycosidic bond, because it is between carbons 1 and 4 and is facing downward.

The glycosidic bond in this carbohydrate is called a β-1,4 glycosidic bond, because it is between carbons 1 and 4 and is facing upward.

### Sucrose and Maltose

Sucrose consists of the monosaccharides glucose and fructose. Sucrose is found in many plants, and is especially abundant in sugar cane and sugar beets. When these plants are crushed, the resulting juice is treated to produce a brown liquid called molasses. Further treatment and purification of the juice results in the formation of pure crystallized sucrose, otherwise known as table sugar. Because most people find the intense sweetness of sucrose desirable, it is often added to processed foods.

The disaccharide maltose, which seldom occurs naturally in foods, consists of two glucose molecules. Maltose is also formed during starch digestion. The food industry has several commercial uses for maltose, including the production of beer and infant formula.

## ESSENTIAL *Concepts*

Carbohydrates consisting of a single sugar are called monosaccharides, and those consisting of two sugars are called disaccharides. Together, mono- and disaccharides are considered the simple carbohydrates. Examples of monosaccharides include glucose, fructose, and galactose. Lactose, maltose, and sucrose are disaccharides and consist of at least one glucose molecule. Lactose consists of glucose bonded to galactose, sucrose consists of glucose bonded to fructose, and maltose consists of glucose bonded to glucose.

**oligosaccharide** (o – li – go – SAC – cha – ride) (*oligo-*, few; *-saccharide*, sweet) Carbohydrate made of relatively few (3 to 10) monosaccharides.

**polysaccharide** (po – li – SAC – cha – ride) (*poly-*, many; *-saccharide*, sweet) Complex carbohydrate made of many monosaccharides.

**complex carbohydrates** Category of carbohydrate that includes oligosaccharides and polysaccharides.

## Complex Carbohydrates

In contrast to the simple carbohydrates, which contain 1 or 2 monosaccharides, **oligosaccharides** are made up of 3 to 10 monosaccharides and **polysaccharides** consist of 10 or more monosaccharides. However, most polysaccharides are made up of hundreds of monosaccharides bonded together. Together, the oligosaccharides and polysaccharides are called **complex carbohydrates.** Figure 5.6 provides an overview of carbohydrate classification.

### FIGURE 5.6  Classification of Carbohydrates

Carbohydrates are classified as simple or complex—categories that are further subdivided into mono-, di-, oligo-, and polysaccharides depending on the number of monosaccharides they contain.

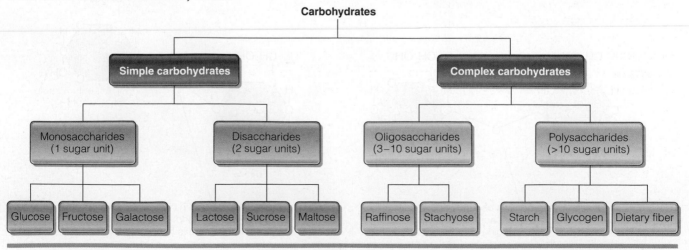

## Oligosaccharides

Oligosaccharides are present in a variety of foods, including dried beans, soybeans, peas, and lentils. Raffinose and stachyose are the two most common oligosaccharides. However, because humans lack the enzymes needed to digest raffinose and stachyose, they pass undigested into the large intestine, where bacteria ferment them. As a result, some people experience abdominal discomfort (bloating and cramps) and flatulence (gas). Commercial products such as Beano® supply the enzymes needed to digest these oligosaccharides, making them more digestible and therefore less available to intestinal bacteria.

In the body, oligosaccharides are components of cell membranes, allowing cells to recognize and interact with one another. Oligosaccharides are also made in the breast, where they are incorporated into human milk. These compounds are part of a complex system that helps protect the infant from disease-causing pathogens, and are one of the many reasons why women are encouraged to breastfeed their infants.

## Polysaccharides

Starch, glycogen, and dietary fiber consist of many monosaccharides bonded together, and are examples of polysaccharides. Plants store glucose as starch, whereas animals store glucose as glycogen. Dietary fiber is not used for energy as are starch and glycogen. Rather, it has a structural role, providing rigidity and strength to cell walls in plants. Polysaccharides have many different shapes and forms, depending on the arrangement of their monosaccharide molecules. Some polysaccharides consist of sugar units linked together in an orderly linear fashion, whereas others are shaped like branches on a tree. More detail is provided next concerning the structures and function of the different types of polysaccharides.

### Starch: The Plant Polysaccharide for Glucose Storage

There are two forms of starch, **amylose** and **amylopectin,** both of which consist entirely of glucose molecules. However, the arrangement of glucose molecules is quite different. As illustrated in Figure 5.7, amylose is a linear (straight) chain of glucose molecules held together by α-1,4 glycosidic bonds, whereas amylopectin consists of glucose molecules bonded together in a highly branched arrangement. Although the linear portion of amylopectin contains α-1,4 glycosidic bonds, α-1,6 glycosidic bonds occur at branch points.

Although plants typically contain a mixture of both forms of starch, amylopectin is more abundant than amylose. Examples of starchy foods include grains (such as corn, rice, and wheat), products made from them (such as pasta and bread), and legumes. Potatoes and winter (hard) squashes are also sources of starch. Starch is sometimes added to food to enhance texture and stability. For example, because cornstarch forms a thick gel when heated it is often used to thicken sauces and gravies. Food scientists have also developed ways to chemically modify starch to improve its functionality. Thus "modified food starch" is an ingredient commonly used in food production.

### Glycogen: The Animal Polysaccharide for Glucose Storage

The body stores small amounts of glucose in the form of **glycogen,** which is found primarily in the liver and in skeletal muscles. Glycogen is a highly branched arrangement of glucose molecules, consisting of both α-1,4 (linear portion) and α-1,6 (branch points) glycosidic bonds (Figure 5.8). A physiological advantage of the numerous branch points in glycogen is that enzymes can

**amylose** (A – my – lose) A type of starch consisting of a linear chain of glucose molecules.

**amylopectin** (a – my – lo – PEK – tin) A type of starch consisting of highly branched arrangement of glucose molecules.

**glycogen** (GLY – co – gen) Polysaccharide consisting of highly branched arrangement of glucose molecules found primarily in liver and skeletal muscle.

## FIGURE 5.7 Structure of Starch

Plants store glucose in the form of starch. The two forms of plant starch, amylose and amylopectin, both consist of glucose molecules bonded together.

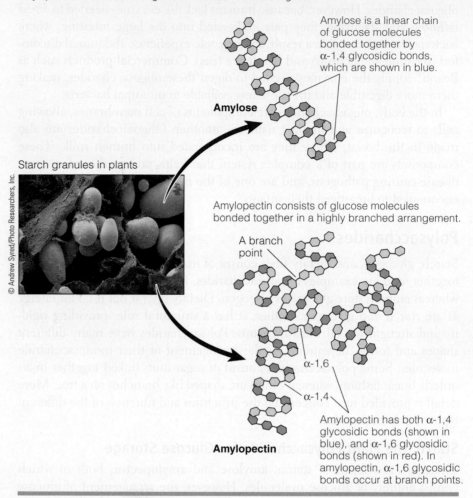

Amylose is a linear chain of glucose molecules bonded together by α-1,4 glycosidic bonds, which are shown in blue.

**Amylose**

Starch granules in plants

© Andrew Syred/Photo Researchers, Inc.

Amylopectin consists of glucose molecules bonded together in a highly branched arrangement.

A branch point

α-1,6

α-1,4

**Amylopectin**

Amylopectin has both α-1,4 glycosidic bonds (shown in blue), and α-1,6 glycosidic bonds (shown in red). In amylopectin, α-1,6 glycosidic bonds occur at branch points.

**CONNECTIONS** Hydrolytic reactions break chemical bonds by the addition of water (Chapter 3, page 77).

hydrolyze multiple glycosidic bonds simultaneously. As a result, glycogen can be rapidly broken down, allowing glucose to be quickly released when needed.

Because most cells use glucose for energy, the body turns to glycogen during periods of limited glucose availability, such as fasting and strenuous exercise. Because liver glycogen is broken down into glucose and released directly into the blood, it plays a very important role in blood glucose regulation. Muscle glycogen serves a different purpose. Unlike the liver, muscle lacks the enzyme needed to release glucose into the blood. Therefore glucose released during glycogen breakdown in muscle is used locally to fuel physical activity. Some athletes try to increase the amount of glycogen stored in the body using a technique called carbohydrate loading. This is described in greater detail in the Focus on Sports Nutrition feature.

## Dietary Fiber

**dietary fiber** Polysaccharide found in plants that is not digested or absorbed in the human small intestine.

The term **dietary fiber** refers to a diverse group of plant polysaccharides that are not digested or absorbed in the human small intestine. Dietary fibers are indigestible because they contain β-glycosidic bonds, which are resistant

## FIGURE 5.8 Structure of Glycogen

Animals store glucose in the form of glycogen, a polysaccharide consisting of many glucose molecules.

Glycogen granules in muscle

Glycogen is a highly branched arrangement of glucose molecules consisting of both α-1,4 glycosidic bonds (shown in blue), and α-1,6 glycosidic bonds (shown in red).

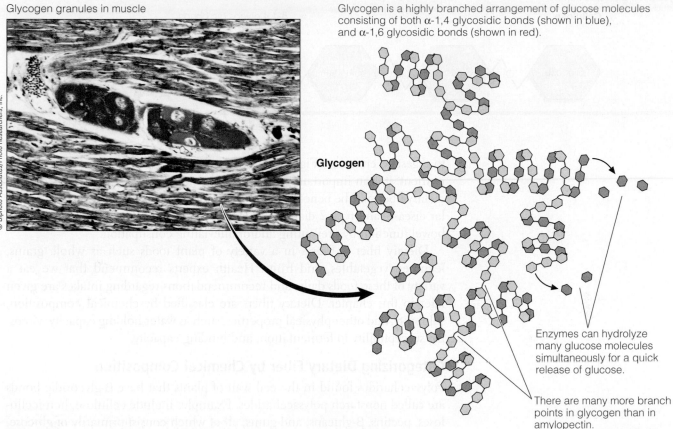

© Biphoto Associates/Photo Researchers, Inc.

Glycogen

Enzymes can hydrolyze many glucose molecules simultaneously for a quick release of glucose.

There are many more branch points in glycogen than in amylopectin.

## *Focus On* SPORTS NUTRITION
### Carbohydrate Loading and Athletic Performance

Carbohydrate loading is a dietary technique used by some athletes to increase the amount of glycogen stored in liver and muscle tissues. When this method is done properly, increased amounts of stored glycogen can delay the onset of fatigue during strenuous physical activities and those that last longer than 60 minutes.[2] Carbohydrate loading typically begins seven to eight days prior to competition and follows a progression of phases.

The first phase of carbohydrate loading (days 1 and 2) involves depleting glycogen stores by high-intensity workouts coupled with reducing carbohydrate intake (5 g of carbohydrate per kilo-

gram of body weight). For example, a person weighing 200 pounds (91 kg) would consume approximately 455 (5 × 91) g of carbohydrate per day during this phase. This is followed by a gradual tapering down in exercise in combination with a moderately high carbohydrate intake of about 7 g per kilogram of body weight (637 g of carbohydrate per day for our 91-kg athlete). The last phase of carbohydrate loading (days 6 and 7) involves a further reduction in physical activity coupled with a high carbohydrate intake of approximately 10 g of carbohydrates per kilogram of body weight. During this phase, the 91-kg athlete would consume approxi-

mately 910 g of carbohydrate daily. This is equivalent to about 40 slices of bread.

For most recreational athletes, extensive carbohydrate loading is not necessary. Nor is it beneficial for athletic competitions that are short in duration (less than 60 minutes). A healthy diet that satisfies both nutrient and energy requirements is more than adequate. However, serious athletes participating in high-intensity or endurance competitions may find that carbohydrate loading gives them a competitive edge. Still, glycogen stores alone do not make an athlete. True athletic ability requires a combination of natural talent, long hours of training, and a healthful balanced diet.

## FIGURE 5.9 Humans Are Unable to Digest Dietary Fiber

When humans consume dietary fiber, it remains intact in the gastrointestinal tract. This is because humans lack digestive enzymes that can hydrolyze β-1,4 glycosidic bonds.

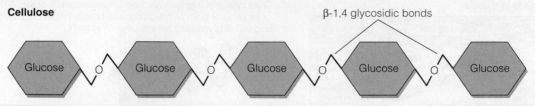

to digestive enzymes (Figure 5.9). Although dietary fiber is not an essential nutrient, it is an important part of a healthful diet. For example, diets high in dietary fiber may be beneficial in preventing and treating obesity, cardiovascular disease, and type 2 diabetes. Furthermore, dietary fiber promotes healthy bowel function by preventing and/or alleviating constipation.

Dietary fiber is found in a variety of plant foods such as whole grains, legumes, vegetables, and fruits. Health experts recommend that we eat a variety of these foods daily, and recommendations regarding intakes are given later in this chapter. Dietary fibers are classified by chemical composition, solubility, and other physical properties, such as water-holding capacity, viscosity, susceptibility to fermentation, and binding capacity.

### Categorizing Dietary Fiber by Chemical Composition

Polysaccharides found in the cell wall of plants that have β-glycosidic bonds are called nonstarch polysaccharides. Examples include cellulose, hemicelluloses, pectins, β-glucans, and gums, all of which consist primarily of glucose. The dietary fiber lignin is quite different from nonstarch polysaccharides and technically is not even a carbohydrate. In fact, lignin is classified as a nonpolysaccharide dietary fiber. Because it is found in the tough and woody portion of plants, it is not consumed in large quantities. The different types of dietary fiber are described briefly in Table 5.1.

### Categorizing Dietary Fiber by Solubility

Dietary fiber can also be classified by solubility. In water, **soluble fiber** dissolves or swells, whereas **insoluble fiber** does not. For example, if you soaked a carrot, which contains mostly insoluble fibers, in water overnight, the carrot would remain intact. If you did the same with oatmeal, which contains mostly soluble fibers, it would become soft and paste-like. Most foods contain both soluble and insoluble fibers. However, foods such as oats, rice, bran, psyllium seeds, soy, and some fruits contain mostly soluble fibers, whereas whole wheat, whole-grain cereals, broccoli, and other vegetables contain primarily insoluble fibers.

The distinction between soluble and insoluble fibers is often used as the basis for specific dietary and health recommendations. For example, soluble dietary fiber is recommended to help lower blood cholesterol levels and, therefore, risk of cardiovascular disease. Likewise, insoluble fiber is recommended to alleviate constipation. However, recent evidence suggests there are many exceptions to these generalizations. For example, both soluble and insoluble fibers help relieve constipation, and not all soluble fibers lower cholesterol.

**soluble fiber** Dietary fiber that is capable of being dissolved in water.

**insoluble fiber** Dietary fiber that is incapable of being dissolved in water.

**TABLE 5.1   Common Dietary Fibers and Food Sources**

| Dietary Fiber | Description | Food Sources |
|---|---|---|
| **Cellulose** | Insoluble fiber consisting of glucose molecules with β-glycosidic bonds; the main structural component of plant cell walls. | Whole-wheat flour<br>Bran<br>Broccoli<br>Cabbage<br>Dried peas/beans<br>Apples<br>Root vegetables |
| **Hemicellulose** | Insoluble fiber consisting of a variety of different monosaccharide molecules (e.g., glucose, arabinose, mannose, and xylose). | Bran<br>Cereals<br>Whole grains |
| **Pectin** | Soluble fiber found in the skin of ripe fruits that consists of a variety of different monosaccharide molecules, and is used commercially to make jams and jellies. | Apples<br>Citrus fruits<br>Strawberries<br>Raspberries |
| **β-Glucan** | A nonstarch polysaccharide composed of branched chains of glucose molecules. | Mushrooms<br>Barley<br>Oats |
| **Gums** | Highly soluble and viscous nonstarch polysaccharide used to thicken foods. | Oatmeal<br>Dried beans<br>Other legumes |
| **Lignin** | Insoluble nonpolysaccharide dietary fiber consisting of numerous alcohol units found within the woody portion of plants. | Berries<br>Wheat |

SOURCE: Institute of Medicine. *Dietary Reference Intakes for Energy, Carbohydrate, Fiber, Fat, Fatty Acids, Cholesterol, Protein, and Amino Acids.* Washington, DC: National Academies Press; 2005.

## Categorizing Dietary Fiber by Other Physical Properties

Dietary fiber can also be categorized by differences in physical properties. For example, some fibers have high water-holding capacities. These fibers form gel-like solutions in water and are viscous. Foods with these properties include oat bran, barley, and legumes. Consuming these types of foods can increase stool weight and fecal volume, helping to prevent and alleviate constipation. In general, large amounts of fecal mass move through the colon more quickly than small amounts. Presumably, this is because large stools stimulate peristaltic contractions in the colon, propelling the material forward. Large, soft stools cause less strain, making elimination easier. Furthermore, studies suggest that consuming fibers with high water-holding capacities helps normalize blood glucose levels and reduce blood cholesterol levels.[3] Water-holding properties may be especially important for people at risk for diabetes and/or cardiovascular disease.

Some fibers are fermented by intestinal bacteria, producing gas and lipids. These types of fibers, called **fermentable fibers,** include pectin, corn, oat bran, carrageen, agar, psyllium, guar gum, and alfalfa. Although gas production (flatulence) by intestinal bacteria may be an annoyance, lipids produced as by-products of this fermentation may serve a useful purpose by nourishing cells that line the colon. In addition, scientists think that consuming fermentable fibers promotes the selective growth of beneficial intestinal bacteria.[4] This is important, because growth of these bacteria helps inhibit the growth of other, disease-causing bacteria.

**fermentable fiber** Fiber that can be fermented by bacteria in the large intestine.

Fibers that are highly resistant to bacterial fermentation may also benefit health. For example, cellulose, which is found in whole grains, wheat bran, broccoli, and brown rice, resists bacterial breakdown and remains largely intact in the gastrointestinal tract. Consuming unfermentable fiber helps increase fecal volume and mass in the same way that consuming fibers with high water-holding capacity does. A mixed high-fiber diet, along with plenty of fluids, provides the best protection from constipation. This is important, because chronic constipation may lead to a more serious condition called **diverticular disease,** which is discussed in the Focus on Diet and Health feature.

Another physical property of some fibers is their ability to bind other compounds. For example, the ability of some fibers to bind cholesterol-containing bile may represent a health benefit. Bile plays an important role in lipid digestion. After being used for this purpose, bile is typically returned to the liver, where it is recycled and reused. Bile that is not returned to the liver is eliminated in the feces. Because some fibers bind bile, less bile is recycled. For

> **diverticular disease,** or **diverticulosis** (di – ver – TI – cu – lar) (di – ver – ti – cu – LO – sis) Condition in the large intestine; characterized by the presence of pouches that form along the intestinal wall.
>
> **diverticulitis** (di – ver – ti – cu – LI – tis) Inflammation of diverticula (pouches) in the lining of the large intestine.

# *Focus On* DIET AND HEALTH
## Diverticular Disease

The production of hard, dry feces, characteristic of a low-fiber diet, not only makes elimination more difficult but may also contribute to a condition called diverticular disease (Figure 5.10). Strain associated with long-term constipation can cause areas along the colon wall to become weak. As a result, protruding pouches, called diverticula, can form. It is not uncommon for the elderly to develop diverticula. By the age of 70 years 50% of adults have diverticulosis, a term used to describe the presence of diverticula. However, only 10 to 20% of those with diverticular disease develop complications such as **diverticulitis.**[5] Thus diverticulitis is the condition in which there is inflammation of the diverticula. Note that the suffix –*itis* refers to inflammation.

Diverticulitis generally occurs when fecal material becomes trapped in the diverticula. Symptoms include cramping, diarrhea, fever, and on occasion bleeding from the rectum. Diverticular disease is most common in older adults with low intakes of fibrous foods, and the incidence is relatively low in populations that consume high-fiber diets. Scientists think that dietary fiber helps prevent the for-

mation of diverticula by increasing fecal mass, thereby reducing strain during bowel movements. Therefore, a diet high in fiber with plenty of fluids may help protect against diverticular disease. This is especially important as people age.

### FIGURE 5.10  Diverticular Disease

Diverticular disease is most common in older adults who consume low intakes of fibrous foods. By the age of 70 years, almost half of all adults have diverticulosis. Diverticulitis is the inflammation of diverticula.

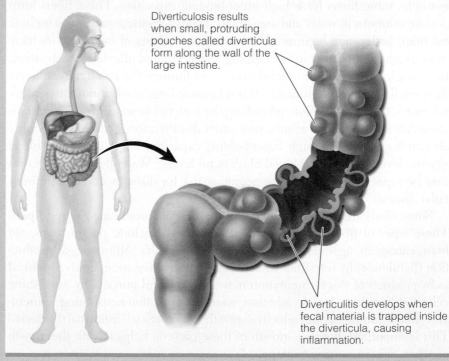

Diverticulosis results when small, protruding pouches called diverticula form along the wall of the large intestine.

Diverticulitis develops when fecal material is trapped inside the diverticula, causing inflammation.

this reason, consuming these fibers is thought to help lower blood cholesterol levels and decrease the risk of cardiovascular disease.[6] Some fibers also bind minerals such as calcium, zinc, iron, and magnesium. However, it is doubtful that consuming dietary fiber in recommended amounts affects mineral status in healthy adults.

## ESSENTIAL *Concepts*

Oligosaccharides and polysaccharides are classified as complex carbohydrates. Oligosaccharides consist of 3 to 10 monosaccharides, and polysaccharides consist of hundreds of monosaccharides. Starch, glycogen, and fiber are important polysaccharides. Amylose and amylopectin are two forms of plant starch. Glucose molecules in amylose are arranged in a linear chain, whereas those in amylopectin are in a highly branched arrangement. Glycogen, which also consists of glucose molecules bonded together in a highly branched arrangement, is found mainly in liver and skeletal muscle. Dietary fibers are a diverse group of plant substances that are not digested or absorbed in the human small intestine. However, some are broken down by bacteria in the large intestine. The physical properties associated with dietary fiber are thought to determine their physiological effects. These include solubility in water, water-holding capacity, viscosity, fermentability, and capacity to bind other substances such as bile and minerals.

# Carbohydrate Digestion, Absorption, and Circulation

During digestion, carbohydrates undergo extensive chemical transformations as they move through the gastrointestinal tract. With the help of digestive enzymes, the glycosidic bonds that hold disaccharides and starch together are broken. The ultimate goal of carbohydrate digestion is the breakdown of large, complex molecules into small, absorbable units—that is, monosaccharides. This process begins in the mouth and requires a series of enzymes produced in various organs, including the salivary glands, pancreas, and small intestine.

## Digestion of Starch

Recall that there are two basic forms of starch—amylose and amylopectin. Although both forms are made entirely of glucose subunits, amylose consists only of α-1,4 glycosidic bonds, whereas amylopectin has both α-1,4 and α-1,6 glycosidic bonds. Even though some structural differences exist between these starches, digestion of amylose and amylopectin requires many of the same enzymes, as illustrated in Figure 5.11.

Starch digestion begins in the mouth when the salivary glands release the enzyme **salivary α-amylase.** Salivary α-amylase hydrolyzes the α-1,4 glycosidic bonds in both amylose and amylopectin. Given the relatively short amount of time that food stays in the mouth, very little starch digestion actually takes place there. Nonetheless, salivary α-amylase can partially hydrolyze starch, resulting in shorter polysaccharide chains of varying lengths called **dextrins.** Once dextrins enter the stomach, the acidic environment stops the enzymatic activity of salivary α-amylase. Dextrins then pass unchanged from the stomach into the small intestine, where they encounter **pancreatic α-amylase,** an enzyme produced by the pancreas and released into the small intestine as

**salivary α-amylase** (A – my – lase) Enzyme released from the salivary glands, that digests starch by hydrolyzing α-1,4 glycosidic bonds.

**dextrin** A partial breakdown product formed during starch digestion consisting of varying numbers of glucose units.

**pancreatic α-amylase** (pan – cre – A – tic al – pha – A – my – lase) Enzyme released from the pancreas that digests starch by hydrolyzing α-1,4 glycosidic bonds.

# FIGURE 5.11 Digestion of Amylose and Amylopectin

Amylose and amylopectin digestion require many of the same enzymes. Through the process of digestion, both amylose and amylopectin are broken down into molecules of glucose. Most starch digestion takes place in the small intestine.

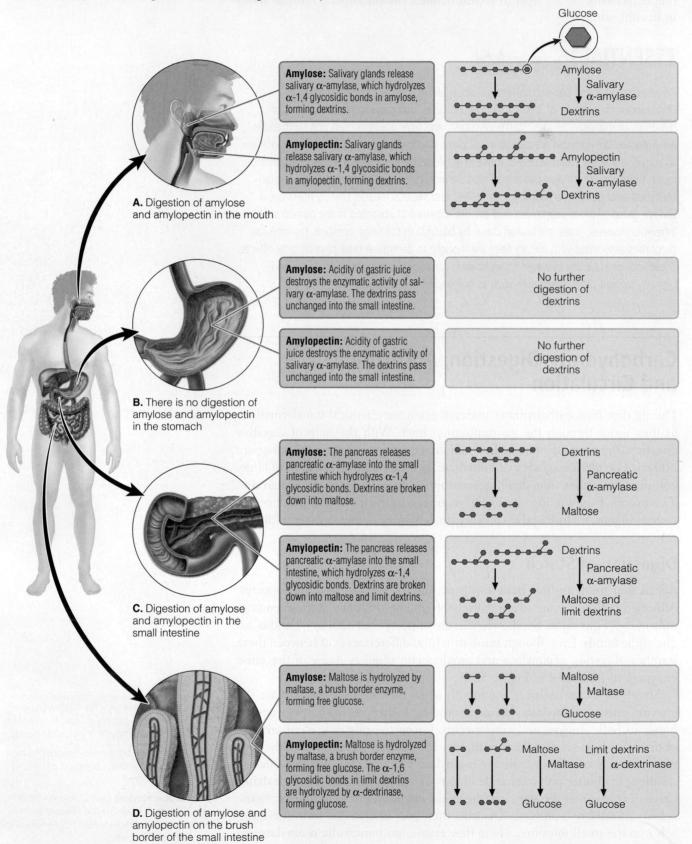

**Amylose:** Salivary glands release salivary α-amylase, which hydrolyzes α-1,4 glycosidic bonds in amylose, forming dextrins.

**Amylopectin:** Salivary glands release salivary α-amylase, which hydrolyzes α-1,4 glycosidic bonds in amylopectin, forming dextrins.

**A.** Digestion of amylose and amylopectin in the mouth

**Amylose:** Acidity of gastric juice destroys the enzymatic activity of salivary α-amylase. The dextrins pass unchanged into the small intestine.

**Amylopectin:** Acidity of gastric juice destroys the enzymatic activity of salivary α-amylase. The dextrins pass unchanged into the small intestine.

**B.** There is no digestion of amylose and amylopectin in the stomach

**Amylose:** The pancreas releases pancreatic α-amylase into the small intestine which hydrolyzes α-1,4 glycosidic bonds. Dextrins are broken down into maltose.

**Amylopectin:** The pancreas releases pancreatic α-amylase into the small intestine, which hydrolyzes α-1,4 glycosidic bonds. Dextrins are broken down into maltose and limit dextrins.

**C.** Digestion of amylose and amylopectin in the small intestine

**Amylose:** Maltose is hydrolyzed by maltase, a brush border enzyme, forming free glucose.

**Amylopectin:** Maltose is hydrolyzed by maltase, a brush border enzyme, forming free glucose. The α-1,6 glycosidic bonds in limit dextrins are hydrolyzed by α-dextrinase, forming glucose.

**D.** Digestion of amylose and amylopectin on the brush border of the small intestine

part of the pancreatic juices. Like salivary α-amylase, pancreatic α-amylase hydrolyzes α-1,4 glycosidic bonds, transforming dextrins into the disaccharide maltose. Last, the enzyme **maltase,** produced in and located on the absorptive surface of the brush border cells, hydrolyzes maltose into glucose molecules.

The combined efforts of salivary α-amylase, pancreatic α-amylase, and maltase complete the chemical transformation of amylose into multiple glucose molecules. However, an additional step, the hydrolysis of α-1,6 glycosidic bonds, is needed to complete the digestion of amylopectin. Because α-amylase only hydrolyzes α-1,4 glycosidic bonds, units called **limit dextrins** form during amylopectin digestion. Limit dextrins contain three to four glucose molecules and have the α-1,6 glycosidic bond that was located at branching points in the original amylopectin molecule. The hydrolysis of the α-1,6 glycosidic bond is accomplished by the enzyme **α-dextrinase,** produced in and located on the absorptive surface of the brush border. Once α-dextrinase hydrolyzes the α-1,6 glycosidic bonds in the limit dextrins, the digestion of amylopectin is complete. Thus amylose and amylopectin digestion results in the production of multiple glucose molecules that are now ready to be absorbed into the enterocytes.

## Digestion of Disaccharides

Digestion of disaccharides (maltose, sucrose, and lactose) takes place entirely in the small intestine (Figure 5.12). Enzymes, collectively called **disaccharidases,** are produced within and located on the absorptive surface of the small intestine cells. The digestion of each disaccharide requires a specific disaccharidase. Specifically, **sucrase** hydrolyzes sucrose into glucose and fructose, maltase hydrolyzes maltose into two glucose molecules, and **lactase** hydrolyzes lactose into glucose and galactose. Once a disaccharide has been

**CONNECTIONS** The absorptive surface of the small intestine cell has minute projections called microvilli, forming the "brush border" (Chapter 4, page 104).

**maltase** (MAL – tase) Enzyme that hydrolyzes maltose into two glucose molecules.

**limit dextrin** (DEX – trins) A partial breakdown product formed during amylopectin digestion that contain three to four glucose molecules and an α-1,6 glycosidic bond.

**α-dextrinase** (DEX – stri – nase) Intestinal enzyme that hydrolyzes α-1,6 glycosidic bonds.

**disaccharidase** (di – SAC – cha – ri – dase) Intestinal enzyme that hydrolyzes glycosidic bonds in disaccharides.

**sucrase** Intestinal enzyme that hydrolyzes sucrose into glucose and fructose.

**lactase** Intestinal enzyme that hydrolyzes lactose into glucose and galactose.

## FIGURE 5.12   Digestion of Disaccharides

Disaccharides are digested along the brush border by intestinal enzymes collectively known as disaccharidases.

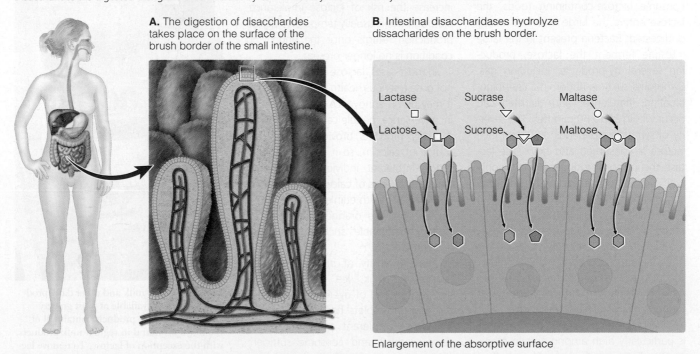

**A.** The digestion of disaccharides takes place on the surface of the brush border of the small intestine.

**B.** Intestinal disaccharidases hydrolyze dissacharides on the brush border.

Enlargement of the absorptive surface

**lactose intolerance** Inability to digest the milk sugar lactose, caused by a lack of the enzyme lactase.

digested into its components, the monosaccharides can be absorbed into the enterocytes.

Although most babies produce enough lactase to digest the high amounts of lactose found in milk, some adults produce very little or none at all of this enzyme, so they have difficulty digesting lactose. This food intolerance, called **lactose intolerance,** is described in the Focus on Diet and Health feature.

## Carbohydrate Absorption and Circulation

**CONNECTIONS** Carrier-mediated active transport requires a transport protein and energy (ATP) to transport material across cell membranes moving from a lower to a higher concentration. Facilitated diffusion requires only a transport protein moving from a higher to a lower concentration (Chapter 3, pages 79, 80).

Once digestion of disaccharides and starch is complete, the resulting mono-saccharides (glucose, galactose, and fructose) are readily taken up by the absorptive cells of the small intestine. In some cases, absorption occurs via energy-requiring, active transport mechanisms, whereas sometimes it occurs via facilitated diffusion. Active transport of monosaccharides requires sodium and is therefore referred to as being "sodium dependent." Monosaccharide absorption is depicted in Figure 5.13. Once absorbed, monosaccharides are circulated away from the small intestine in the blood directly to the liver via the hepatic portal vein.

**CONNECTIONS** The hepatic portal vein is the vascular connection from the small intestine to the liver that circulates absorbed nutrients (Chapter 4, page 109).

# *Focus On* DIET AND HEALTH
## Lactose Intolerance

Some people do not digest significant amounts of the milk sugar lactose. This condition, called lactose intolerance, is caused by insufficient production of the intestinal enzyme lactase. When people with lactose intolerance consume lactose-containing foods, the lactose enters the large intestine mostly undigested. Bacteria present in the large intestine ferment the lactose, producing several by-products, including gas. Symptoms such as abdominal cramping, bloating, flatulence, and diarrhea can occur within 30 to 60 minutes. The severity of symptoms depends on how much lactose is consumed and how little lactase the person produces. Most people with lactose intolerance can consume small amounts of dairy products without experiencing any discomfort. Some dairy products, such as yogurt and cheese, are more easily tolerated because the manufacturing process removes some of the lactose. In addition, bacteria used to make cheese and yogurt convert most of the remaining lactose to lactic acid.

The prevalence of lactose intolerance is particularly high among certain ethnic groups. For example, more than half of

Asian Americans, Native Americans, and African Americans are reportedly lactose intolerant.[7] In comparison, the prevalence of lactose intolerance in those of northern European descent is only 12%. Certain diseases, medications, and surgery that damage the intestinal mucosa can also increase the risk of lactose intolerance, although this is usually temporary; lactase production returns once the underlying condition is no longer present.

In most cases, lactose intolerance does not pose a serious health threat. Although it may be an annoyance and at times an inconvenience, people can easily live with it. Dairy products provide an excellent source of calcium, so it is important for lactose-intolerant individuals to choose alternate sources of calcium such as fortified tofu, fish with edible bones, fortified soymilk, fortified orange juice, and calcium-rich vegetables such as swiss chard and rhubarb.

The availability of products such as lactose-free milk, ice cream, cheese, and yogurt, and of over-the-counter lactase enzyme tablets has made it easier for lactose-intolerant people to enjoy dairy products and consume sufficient calcium. However, if these calcium-

rich alternatives or lactose-reduced foods are not available or acceptable, calcium supplements may be needed.

Lactose-reduced milk and other dairy products are widely available at most grocery stores. These dairy products contain all of the nutrients found in regular milk products, with the exception of lactose. To remove lactose from milk, the enzyme lactase is added.

## FIGURE 5.13    Absorption of Monosaccharides

Glucose and galactose are absorbed into the enterocytes by carrier-dependent, energy-requiring active transport. Fructose is absorbed by facilitated diffusion. Once absorbed, monosaccharides are circulated to the liver via the hepatic portal vein.

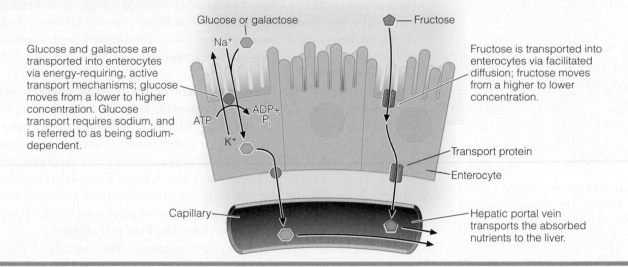

Glucose and galactose are transported into enterocytes via energy-requiring, active transport mechanisms; glucose moves from a lower to higher concentration. Glucose transport requires sodium, and is referred to as being sodium-dependent.

Glucose or galactose

Na⁺

ATP    ADP+ Pᵢ

K⁺

Fructose

Fructose is transported into enterocytes via facilitated diffusion; fructose moves from a higher to lower concentration.

Transport protein

Enterocyte

Capillary

Hepatic portal vein transports the absorbed nutrients to the liver.

## The Glycemic Response

A rise in blood glucose levels can be detected shortly after eating carbohydrate-rich foods. However, not all carbohydrates have the same effect on blood glucose levels. Some foods cause blood glucose levels to rise quickly and remain elevated, whereas others do not. The impact of different foods on the magnitude and duration of the rise in blood glucose after a meal is called the **glycemic response.**

A rating system known as the **glycemic index (GI)** is used to classify foods according to their relative glycemic response (Figure 5.14).[8] Specifically, a food's glycemic index is a numeric value that reflects the glycemic response of a food in comparison to that of a reference food, such as pure glucose or white bread. The glycemic index of the reference food is set at 100, reflecting

## FIGURE 5.14    Glycemic Index (GI) of Food

Foods that are digested slowly have a low glycemic index. These types of foods cause a gradual and more moderate response in blood glucose than do foods with a high glycemic index.

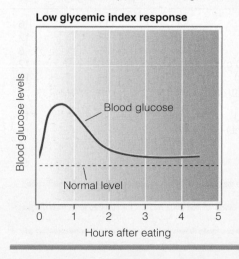

**Low glycemic index response**

Blood glucose levels

Blood glucose

Normal level

Hours after eating

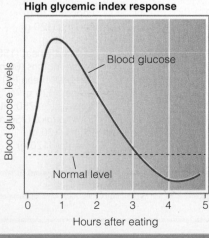

**High glycemic index response**

Blood glucose levels

Blood glucose

Normal level

Hours after eating

**glycemic response** (*glyc-*, sugar) Effect of a food on the extent and/or duration of the rise in blood glucose levels.

**glycemic index (GI)** Measure based on the extent to which a food containing 50 g of carbohydrates increases blood glucose concentrations.

a relatively rapid and large rise in blood glucose following its consumption. Foods with glycemic indices greater than 70 (such as bagels and potatoes) are considered high-glycemic foods, because consuming them elicits a large and/or prolonged rise in blood glucose. Foods with glycemic indices lower than 55 (such as skim milk and kidney beans) are considered low-glycemic index foods and elicit a less dramatic or briefer rise in blood glucose. Glycemic index is of interest, because the consumption of low-glycemic index foods is associated with lower risk of developing certain chronic diseases, such as type 2 diabetes and cardiovascular disease.

Factors that influence the effect of a food on the glycemic response include carbohydrate type, presence of fat or protein, viscosity, processing, ripeness, and food additives. Scientists have long believed that simple carbohydrates had higher glycemic indices than complex carbohydrates. However, this clearly is not the case. In fact, starchy foods containing complex carbohydrates, such as potatoes, refined cereal products, white bread, some whole-grain breads, and white rice, have higher glycemic indices than many simple carbohydrate-rich foods, such as soft drinks. Although simple carbohydrates are rapidly absorbed, their effect on glucose can be appreciably less than that of starch.

The use of the glycemic index has limitations. For example, it does not account for the amount of carbohydrate found in a typical serving size. Rather, it is based on blood glucose response to consuming 50 g of carbohydrate in a given food. For example, to consume 50 g of carbohydrates from carrots a person would need to eat more than a pound, whereas 1 cup of rice (about 8 ounces) provides approximately 42 g of carbohydrates. Therefore, another measure used to assess the effect of a food on blood glucose is the **glycemic load (GL)**. The glycemic load may be a more useful measure, because it takes into account the glycemic index of a food as well as the amount of carbohydrate typically found in a single serving of the food. Glycemic index and glycemic load values of commonly consumed foods are listed in Table 5.2.

**glycemic load (GL)** Measure that assesses the effect of a food on blood glucose that takes into account both the glycemic index of a food as well as its carbohydrate content.

## TABLE 5.2  Glycemic Index (GI) and Glycemic Load (GL) of Selected Foods

| Value | Glycemic Index (GI) | Glycemic Load (GL) |
|---|---|---|
| High | ≥70 | ≥20 |
| Medium | 56–69 | 11–19 |
| Low | ≤55 | ≤10 |

| Food | GI | Carbohydrates per Serving (g) | GL | Food | GI | Carbohydrates per Serving (g) | GL |
|---|---|---|---|---|---|---|---|
| Dates (60 g) | 103 | 40 | 41 | Popcorn (20 g) | 54 | 11 | 6 |
| Baked Potato (150 g) | 85 | 30 | 26 | Banana (120 g) | 48 | 25 | 12 |
| Corn Chex (30 g) | 83 | 30 | 25 | Corn (80 g) | 48 | 16 | 8 |
| Waffles (35 g) | 76 | 13 | 10 | Cracked wheat (150 g) | 48 | 26 | 12 |
| French fries (150 g) | 75 | 29 | 22 | Apple muffins (60 g) | 44 | 29 | 13 |
| Bagel (70 g) | 72 | 35 | 25 | Oat bran bread (30 g) | 44 | 18 | 8 |
| Angel food cake (50 g) | 67 | 29 | 19 | Grapes (120 g) | 43 | 17 | 7 |
| Pancakes (80 g) | 67 | 58 | 39 | Apple juice (250 mL) | 39 | 25 | 10 |
| Raisins (60 g) | 64 | 44 | 28 | Tomato juice (250 mL) | 38 | 9 | 3 |
| Raisin Bran (30 g) | 61 | 19 | 12 | Plums (120 g) | 24 | 14 | 3 |
| White rice (150 g) | 56 | 42 | 24 | Cherries (120 g) | 22 | 12 | 3 |

NOTES: Glycemic load (GL) is the glycemic index (GI) divided by 100 and multiplied by its available carbohydrate content (in grams). GL takes the glycemic index into account but is based on how much carbohydrate is in the food or drink tested. Glycemic load is numerically lower than the glycemic index of a food or drink.

SOURCE: Foster-Powell K, Holt SH, Brand-Miller JC. International table of glycemic index and glycemic load values: 2002. *American Journal of Clinical Nutrition.* 2002;76:5–56. The published tables show GI values based on glucose = 100 or on white bread = 100.

To calculate the glycemic load, divide the glycemic index of a food by 100, times the number of grams of carbohydrate in a serving of food. For example, one serving of carrots (½ cup) has a glycemic index of 16 and 8 g of carbohydrate. Therefore, the glycemic load of carrots is approximately 1 (16 ÷ 100 × 8 g = 1). By comparison, the glycemic load of a 12-ounce soft drink that has a glycemic index of 63 and 39 g carbohydrate is approximately 25 (63 ÷ 100 × 39 g). These values are probably closer to what you may have predicted.

## Uses of Monosaccharides in the Body

After absorption, monosaccharides are circulated directly to the liver via the hepatic portal vein where galactose and fructose are converted into other compounds—most notably glucose. The body has many uses for monosaccharides. For example, some are converted to ribose, a constituent of many vital compounds including ATP, RNA, and DNA. In addition, glucose serves as a major energy source (ATP) for all cells. The liver is ultimately in charge of deciding how much glucose to store as glycogen, how much to convert to fat, and how much to release into the blood. Details related to glucose and energy metabolism are presented in Chapter 8.

## ESSENTIAL *Concepts*

Carbohydrate digestion involves a variety of enzymes produced in the salivary glands, pancreas, and the small intestine. Although some starch digestion begins in the mouth, most occurs in the small intestine via pancreatic α-amylase. Digestion is completed by the enzyme maltase, resulting in glucose. Unlike amylose, the digestion of amylopectin requires the enzyme α-dextrinase to hydrolyze α-1,6 glycosidic bonds. Disaccharides are easily digested by the disaccharidases, which are produced in and found along the brush border of the small intestine. Once carbohydrate digestion is complete, monosaccharides are transported into the absorptive cells of the small intestine and carried to the liver in the blood. After eating carbohydrate-rich foods, blood glucose levels increase, and the magnitude and duration of this rise is referred to as the glycemic response. The glycemic index and glycemic load are rating systems used to assess the glycemic response to specific foods.

# Hormonal Regulation of Blood Glucose and Energy Storage

The concentration of glucose in the blood fluctuates throughout the day. Blood glucose levels are the lowest in the morning after an overnight fast, returning to normal shortly after eating. In general, blood glucose levels rise in response to eating, and decrease between meals. Because cells need glucose 24 hours a day, the pancreatic hormones **insulin** and **glucagon** work vigilantly to maintain blood glucose levels within an acceptable range at all times.

## The Role of the Pancreas

The pancreas, which is partially composed of hormone-secreting cells, plays a major role in glucose homeostasis. These endocrine cells, collectively known as the islets of Langerhans, consist of beta (β) cells that produce the hormone

**insulin** Hormone secreted by the pancreatic β-cells in response to increased blood glucose.

**glucagon** (GLU – ca – gon) Hormone secreted by the pancreatic α-cells in response to decreased blood glucose.

CONNECTIONS ▶ Homeostasis is a state of balance or equilibrium (Chapter 3, page 84).

## FIGURE 5.15 Release of Insulin and Glucagon from the Pancreas

Insulin is made and released by the pancreatic β-cells, and glucagon is made and released by the pancreatic α-cells. Both cell types are found in the islets of Langerhans.

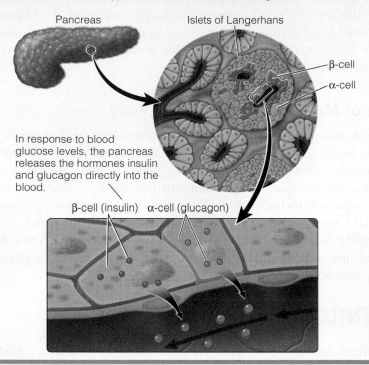

Pancreas

Islets of Langerhans

β-cell

α-cell

In response to blood glucose levels, the pancreas releases the hormones insulin and glucagon directly into the blood.

β-cell (insulin)  α-cell (glucagon)

insulin and alpha (α) cells that produce the hormone glucagon (Figure 5.15). The pancreas has a rich blood supply, allowing these hormones to be released directly into the circulation.

Insulin and glucagon play important roles in blood glucose regulation and energy storage. When blood glucose levels increase, the pancreas responds by releasing more insulin, which in turn lowers blood glucose. This is accomplished by facilitating the uptake of glucose into many kinds of cells. In addition, when meals provide more glucose than we require, insulin favors its storage as glycogen and, indirectly, as fat. The hormonal balance shifts toward glucagon when blood glucose levels decrease. To increase glucose availability, glucagon stimulates the breakdown of glycogen stored in the liver. Note that the release of insulin and glucagon is not an all-or-nothing situation. That is, the *relative* concentrations of these two hormones in the blood determine the shift between energy storage and mobilization (Figure 5.16).

## Response to High Blood Glucose

Glucose enters cells via facilitated diffusion, requiring the presence of carrier proteins known as **glucose transporters** that initially reside within the cell cytoplasm (Figure 5.17). Some glucose transporters require the presence of insulin to transport glucose across the cell membrane, others do not. Most tissues in the body have insulin-requiring glucose transporters, and these tissues are said to be insulin sensitive. Tissues such as the brain and liver can transport glucose across their cell membranes without insulin and are said to be insulin *in*sensitive.

After a person eats carbohydrate-containing foods, blood glucose levels quickly rise, causing the release of insulin from the pancreas. When insulin encounters insulin-sensitive cells, it binds to **insulin receptors** on the surface

**glucose transporter** Intracellular transport proteins that assist in the transport of glucose molecules across cell membranes.

**insulin receptor** Protein found on the surface of certain cell membranes that bind insulin.

## FIGURE 5.16 Hormonal Regulation of Blood Glucose

Insulin is released when blood glucose increases and helps lower the level of glucose in the blood. Glucagon, released when blood glucose is low, helps increase the level of glucose in the blood.

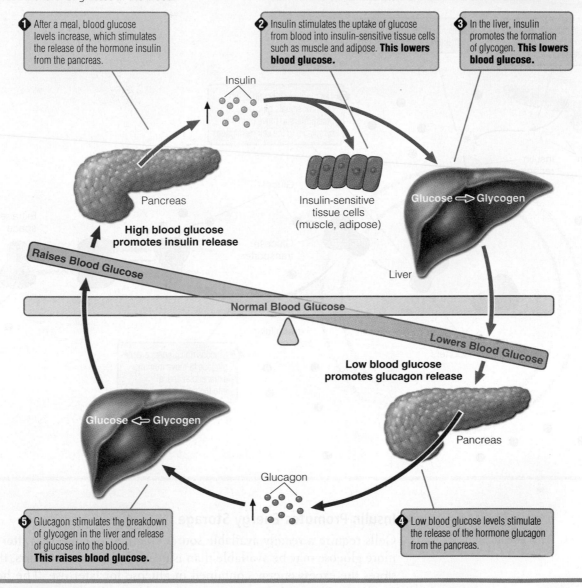

**1** After a meal, blood glucose levels increase, which stimulates the release of the hormone insulin from the pancreas.

**2** Insulin stimulates the uptake of glucose from blood into insulin-sensitive tissue cells such as muscle and adipose. **This lowers blood glucose.**

**3** In the liver, insulin promotes the formation of glycogen. **This lowers blood glucose.**

Insulin

Pancreas

Insulin-sensitive tissue cells (muscle, adipose)

Glucose ⟹ Glycogen

Liver

**High blood glucose promotes insulin release**

Raises Blood Glucose

Normal Blood Glucose

Lowers Blood Glucose

**Low blood glucose promotes glucagon release**

Glucose ⟸ Glycogen

Pancreas

Glucagon

**5** Glucagon stimulates the breakdown of glycogen in the liver and release of glucose into the blood. **This raises blood glucose.**

**4** Low blood glucose levels stimulate the release of the hormone glucagon from the pancreas.

of cell membranes. In response to the binding of insulin to its receptor, glucose transporters relocate from the cytoplasm to the cell membrane, allowing glucose to enter the cell. Skeletal muscle cells, which are usually insulin sensitive, can take up glucose without insulin during exercise. This is because muscle contractions, like insulin, can activate the movement of glucose transporters from the cytoplasm to the cell membrane, allowing for glucose uptake. Muscles return to their insulin-sensitive state within three hours after exercise stops.

Sometimes glucose has difficulty crossing cell membranes, and accumulates in the blood. This condition is called **hyperglycemia.** When this occurs, a person is said to have **impaired glucose regulation,** or in more serious cases, diabetes. There are several types of diabetes, all of which have different underlying causes. Impaired blood glucose regulation and diabetes are discussed in the Nutrition Matters following this chapter.

**hyperglycemia** (hi – per – gly – CE – mi – a) (*hyper-,* excessive) Abnormally high level of glucose in the blood.

**impaired glucose regulation** Condition characterized by elevated levels of glucose in the blood.

## FIGURE 5.17  Role of Insulin in Cellular Uptake of Glucose

Proteins called glucose transporters facilitate the movement of glucose molecules across the cell membrane. Binding of insulin to its receptor causes glucose transporters to relocate from the cytoplasm to the cell membrane.

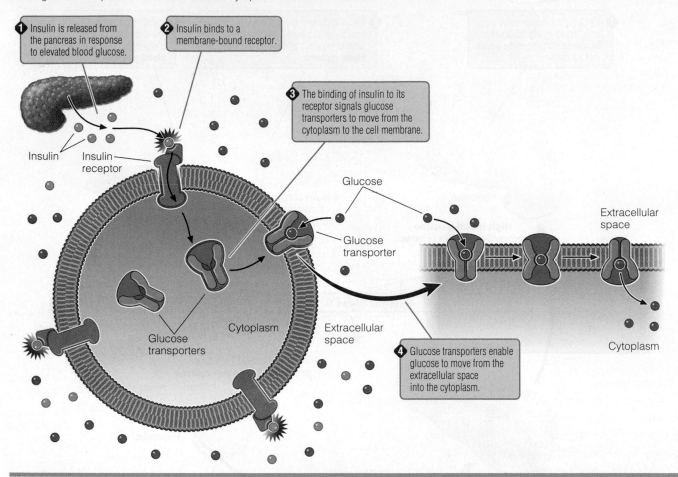

❶ Insulin is released from the pancreas in response to elevated blood glucose.

❷ Insulin binds to a membrane-bound receptor.

❸ The binding of insulin to its receptor signals glucose transporters to move from the cytoplasm to the cell membrane.

Insulin

Insulin receptor

Glucose

Extracellular space

Glucose transporter

Glucose transporters

Cytoplasm

Extracellular space

❹ Glucose transporters enable glucose to move from the extracellular space into the cytoplasm.

Cytoplasm

## Insulin Promotes Energy Storage

Cells require a readily available source of glucose. However, after a meal more glucose may be available than is needed. When this occurs, the body stores the excess energy contained in glucose for later use. The hormone insulin promotes the storage of excess glucose in the form of glycogen and body fat (Figure 5.18). In addition to increases in these energy stores, insulin also stimulates protein synthesis and inhibits the breakdown of muscle.

Before excess energy from glucose can be stored, it must first be converted into glycogen or fat. Insulin controls the synthesis of glycogen, a process called **glycogenesis**, which occurs mainly in the liver and skeletal muscle. There is a limit to how much glycogen can be stored. Once this limit is reached, insulin promotes the uptake of excess glucose by adipose tissue, where the glucose is converted to fatty acids (a type of lipid). Unlike glycogen storage, the body has a seemingly endless capacity to store excess energy from glucose as fatty acids. It is important to note that the conversion of glucose into a fatty acid is irreversible. That is, once glucose is transformed into a fatty acid, the fatty acid cannot be converted back into glucose. This is very different from the reversible metabolic transformation between glucose and glycogen.

**glycogenesis** (gly – co – GE – ne – sis) (*-genesis,* coming into being) Formation of glycogen.

## FIGURE 5.18 Insulin Promotes Energy Storage

The hormone insulin is released from the pancreas in response to rising blood glucose levels. Insulin stimulates glucose transport into cells. Insulin also promotes energy storage, glycogen synthesis, protein synthesis (muscle), and fat synthesis (adipose tissue).

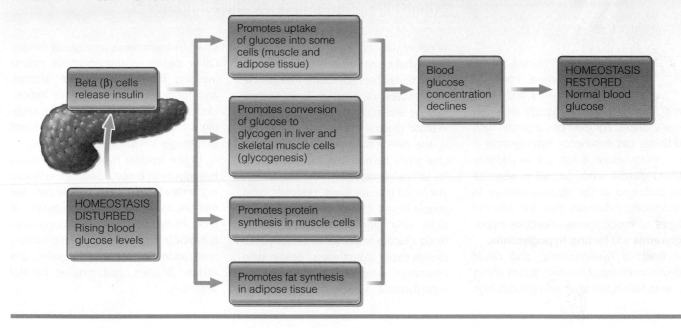

## Response to Low Blood Glucose

Clearly insulin plays a major role in helping the body use glucose after a meal and in storing excess energy from glucose as glycogen and fatty acids. However, the concept of homeostasis requires the body to have another hormone that predominates during periods of low blood glucose. Here the hormone glucagon plays an important role. Several hours after a person eats, blood glucose levels begin to fall. Unless the person eats, the body must begin to use liver glycogen to maintain adequate levels of blood glucose. The hormone glucagon provides the signal for the breakdown of liver glycogen.

Maintaining blood glucose levels is especially important for tissues such as the brain, which relies on glucose as its main energy source. The brain and other components of the nervous system cannot store glucose and therefore depend on circulating blood glucose for energy. Because the brain is very sensitive to low blood glucose levels, even a relatively small decrease in blood glucose, a condition referred to as **hypoglycemia,** can make some people feel nauseated, dizzy, anxious, lethargic, and irritable. This is one reason why some people find it hard to concentrate when they have not eaten for a long time. You can read more about hypoglycemia in the Focus on Diet and Health feature.

### The Breakdown of Glycogen

Glucagon stimulates the breakdown of glycogen in the liver and release of glucose into the blood. This metabolic process, called **glycogenolysis,** literally means the breakdown of glycogen. Liver glycogen can supply glucose for approximately 24 hours before being depleted. Although glycogenolysis occurs in both the liver and skeletal muscle, there are two important differences between these tissue types. First, muscle tissue is not responsive to the hormone glucagon; in other words, it is glucagon insensitive. Rather, muscle glycogenolysis occurs in response to the hormone **epinephrine,** which is produced in the adrenal glands. Another important difference is that muscle lacks the enzyme needed to release

**hypoglycemia** (*hypo-,* under or below) Abnormally low level of glucose in the blood.

**glycogenolysis** (gly – co – ge – NO – ly – sis) (*-lysis,* to break apart) The breakdown of liver and muscle glycogen into glucose.

**epinephrine** (e – pi – NEPH – rine) Hormone released from the adrenal glands in response to stress; helps increase blood glucose levels by promoting glycogenolysis.

## *Focus On* DIET AND HEALTH
### Recognizing and Treating Hypoglycemia

When the level of glucose in the blood is too low—a condition called hypoglycemia—the body does not function properly. Hypoglycemia has many causes. For example, a person with diabetes can experience hypoglycemia if too much insulin is injected. In addition, blood glucose levels can fall in response to prolonged and/or vigorous exercise. In nondiabetic individuals, there are two main types of hypoglycemia—**reactive hypoglycemia** and **fasting hypoglycemia.**

Reactive hypoglycemia, also called postprandial hypoglycemia, occurs within one to four hours after eating foods high in carbohydrates. There is disagreement as to what causes reactive hypoglycemia. Some researchers speculate that people who experience reactive hypoglycemia may be especially sensitive to increases in blood glucose levels after a meal. That is, the pancreas responds to high glucose levels by releasing too much insulin, ultimately resulting in extraordinarily low blood glucose levels. However, most people tested for reactive hypoglycemia show what appears to be a normal blood glucose response to carbohydrate consumption. Nonetheless, people who experience symptoms associated with hypoglycemia are encouraged to eat small, frequent meals throughout the day. Other dietary recommendations include avoiding foods with caffeine, alcohol, and foods with high glycemic indices. Until blood glucose levels stabilize, keeping a food diary can help identify foods that trigger a hypoglycemic response.

Unlike reactive hypoglycemia, fasting hypoglycemia is not associated with eating. Instead, it occurs when the pancreas releases too much insulin regardless of food intake. This form of hypoglycemia is typically caused by pancreatic tumors, medications, hormonal imbalances, and certain illnesses, and requires medical attention.

---

**reactive hypoglycemia** Low blood glucose that occurs after eating carbohydrate-rich foods, caused by the release of too much insulin.

**fasting hypoglycemia** Low blood glucose that occurs when the pancreas releases excess insulin regardless of food intake.

glucose into the blood. Rather, the glucose released from glycogen during muscle glycogenolysis is used by muscle cells themselves and is not made available to other tissues via the blood. Therefore, muscle glycogen does not play a role in blood glucose regulation. The role of glucagon in blood glucose regulation and in the breakdown of energy stores is summarized in Figure 5.19.

## Fight-or-Flight Response

In addition to glugogen, other hormones can increase blood glucose when cells have an immediate need for energy. These hormones include epinephrine and

---

### FIGURE 5.19   Glucagon Promotes the Mobilization of Stored Energy

The hormone glucagon is released from the pancreas in response to declining blood glucose levels. Glucagon promotes the breakdown of liver glycogen, fat (adipose tissue), and protein (muscle), and the synthesis of ketones and glucose from noncarbohydrate sources.

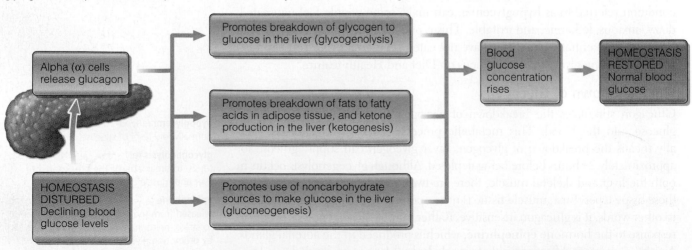

cortisol, produced in the adrenal glands. Unlike glucagon, these hormones are not involved in day-to-day homeostatic regulation of blood glucose. Instead, they are situational and stimulate liver and muscle glycogenolysis in response to danger or stress. This response, sometimes called the fight-or-flight response, ensures glucose availability, especially under extreme circumstances.

## Gluconeogenesis and Ketogenesis

The breakdown of liver glycogen is an effective short-term solution for providing cells with glucose. However, this reserve can be quickly depleted, so the body must find an additional source of glucose. Because fatty acids cannot be converted to glucose, the body must use amino acids from muscle protein to make glucose during these times. The synthesis of glucose from noncarbohydrate sources such as amino acids is called **gluconeogenesis.** The same hormones that stimulate glycogenolysis—glucagon and epinephrine—stimulate gluconeogenesis. Not surprisingly, insulin inhibits gluconeogenesis.

### Ketones Spare Glucose

Although gluconeogenesis increases glucose availability, there are negative consequences to the body. For example, amino acids used for gluconeogenesis come from muscle proteins. To minimize muscle loss associated with gluconeogenesis, the body also uses an alternative energy source (instead of glucose) called **ketones.** Ketones are formed from fatty acids when there is a relative absence of glucose. This process, called **ketogenesis,** occurs in the liver and is stimulated by the hormone glucagon. Once ketones are formed, they are released into the blood and used for energy by the brain, heart, skeletal muscle, and kidney. Thus ketones help spare glucose (and thus muscle loss) by providing the body with an alternative energy source. The importance of ketone use in overall energy metabolism is discussed in Chapter 8.

A condition called **ketosis** occurs when ketone production exceeds the rate of ketone use, resulting in the accumulation of ketones in the blood. This happens when energy intake is very low and/or when the diet provides insufficient amounts of carbohydrate. Ketosis causes a variety of complications, including loss of appetite. In fact, many of the currently popular low-carbohydrate diets promote ketosis as the "fat-burning key." These diets are discussed in more detail in Chapter 9.

## ESSENTIAL *Concepts*

Because glucose must be available to cells at all times, the body has a well-orchestrated homeostatic system to maintain blood glucose concentrations. The pancreatic hormones insulin and glucagon play major roles in this process. Insulin lowers blood glucose levels by (1) enabling glucose to cross cell membranes, (2) stimulating the conversion of glucose to glycogen and fatty acids, and (3) decreasing the use of amino acids for glucose production (gluconeogenesis). When blood glucose levels fall, glucagon stimulates the breakdown of liver glycogen and the subsequent release of glucose into the blood. The hormones epinephrine and cortisol also increase blood glucose levels during periods of stress. Once glycogen is depleted, the body generates glucose via gluconeogenesis. To protect itself from excessive muscle loss, the body decreases its need for glucose by using ketones as an energy source.

**cortisol** (COR – ti – sol) Hormone secreted by the adrenal glands in response to stress that helps increase blood glucose availability via gluconeogenesis and glycogenolysis.

**gluconeogenesis** (glu – co – ne – o – GE – ne – sis) (*neo-,* new; *-genesis,* bringing forth) Synthesis of glucose from noncarbohydrate sources.

**ketone** (KE – tone) Organic compound used as an energy source during starvation, fasting, low-carbohydrate diets, or uncontrolled diabetes.

**ketogenesis** (ke – to – GE – ne – sis) Metabolic pathway that leads to the production of ketones.

**ketosis** (ke – TO – sis) Condition resulting from overproduction of ketone bodies.

**TABLE 5.3 List of FDA-Approved Unqualified Health Claims Concerning Fiber**

- "Diets low in fat and rich in high-fiber foods may reduce the risk of certain cancers."
- "Diets low in fat and rich in soluble fiber may reduce risk of heart disease."
- "Diets low in fat and rich in fruits and vegetables may reduce the risk of certain cancers."
- "Diets low in fat and rich in whole oats and psyllium seed husk can help reduce the risk of heart disease."
- "Diets high in whole-grain foods and other plant foods and low in total fat, saturated fat, and cholesterol may help reduce the risk of heart disease and certain cancers."

# Carbohydrates and Health

There is considerable evidence that dietary carbohydrate choices can influence human health. For example, fiber may be important in preventing and treating diseases such as diabetes and heart disease. In addition, high- and low-carbohydrate diets have both, at one time or another, been advocated as good weight-loss plans. Furthermore, consuming some types of carbohydrates may induce poor dental health. This section briefly discusses some of these issues as they relate to dietary carbohydrates.

## Dietary Fiber and Health

Studies consistently show that dietary fiber is important in preventing disease and maintaining health. Although dietary fiber is not a magic bullet, its credentials are impressive. It has been shown to improve blood glucose regulation, lower blood cholesterol, promote satiety, and contribute to healthy bowel function.[9] In fact, the evidence supporting the health benefits of fiber-rich foods is so substantive that the FDA has approved several fiber-related health claims (Table 5.3).[10] In addition, more than 900 nonnutritive phytochemicals have been identified in fiber-rich, plant-derived foods.

## Carbohydrates and Weight Loss

Diets that are low in fat and protein and high in carbohydrates have long been considered effective in terms of weight loss and weight maintenance. Still, many weight-loss enthusiasts now view carbohydrates as the "fattening" nutrient. Although low-carbohydrate diets appear safe in the short term (up to six months), little, if anything, is known about long-term effects.[11]

Studies comparing weight loss associated with low- versus high-carbohydrate diets show that at six months, greater weight loss is achieved on low-carbohydrate diets.[12] Weight loss associated with low-carbohydrate (high-protein) diets may be caused by a reduced caloric intake rather than by alterations in the macronutrient composition of the diet. That is, limited food choice and increased satiety associated with high-protein, high-fat foods may cause people to eat less and lose weight. There is no evidence that carbohydrate restriction causes the body to burn fat more efficiently. The topic of low-carbohydrate diets and weight loss is more fully addressed in Chapter 9.

## Carbohydrates and Dental Health

Although most carbohydrates can contribute to tooth decay, foods that are high in sugars and stick to teeth are of particular concern. This is because bacteria in the mouth produce enzymes that convert carbohydrates into acids. The combination of food debris, saliva, bacteria, and acid forms a sticky substance called plaque that adheres to the surface of the tooth. The acids in plaque can erode the hard enamel surface of teeth, creating holes, or caries. As the cavities grow, the internal structure of the tooth becomes damaged.

Although tooth decay is caused by a variety of factors, good dental hygiene and good nutrition are particularly important in preventing dental caries. Regular brushing and flossing helps remove food debris, thus decreasing plaque formation. A healthy diet that limits sugary foods and drinks can reduce the incidence of cavities as well.[13] Of particular importance is preventing tooth decay in young children.[14] Infants allowed to fall asleep with bottles filled with milk, formula, juice, or any carbohydrate-containing beverage are at risk for developing baby bottle tooth decay. This is due to the pooling of these carbohydrate-containing beverages in the mouth while the infant sleeps. For this reason, infants should not be put to bed with bottles that contain anything but water.

## ESSENTIAL *Concepts*

Carbohydrate consumption can influence human health in many ways. For example, high-fiber diets are often associated with reduced risk of several chronic diseases, such as diabetes, cardiovascular disease, and obesity. Similarly, some weight-loss "enthusiasts" tout low-carbohydrate diets as promoting weight loss, although the evidence is inconsistent. However, the relationship between sugar consumption and dental health is clear. Nutritionists recommend that children and adults limit foods high in sugars and brush and floss teeth frequently and that infants not be allowed to fall asleep with a bottle of carbohydrate-containing fluid.

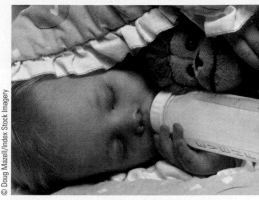

Baby bottle tooth decay can result when infants are put to sleep with bottles containing sugary liquids, including juice and milk.

# Carbohydrate Requirements and Dietary Recommendations

Carbohydrates are needed to provide energy to cells. Some tissues, such as the brain, typically rely on glucose for energy. Furthermore, there is evidence that certain carbohydrates may help prevent chronic diseases. The Dietary Guidelines for Americans and accompanying MyPyramid food guidance system provide recommendations about consuming carbohydrate-rich foods such as whole grains, legumes, fruits, and vegetables. These recommendations—developed to minimize risk for chronic disease and promote optimal health—point out that it is important to pay attention to the amount as well as to the source of dietary carbohydrates.

## Dietary Reference Intakes for Carbohydrates

The Institute of Medicine has issued a variety of recommendations regarding dietary intake of carbohydrates. Dietary Reference Intakes (DRIs) are based mainly on ensuring that the brain has adequate glucose for its energy needs. The Recommended Dietary Allowance (RDA) for carbohydrate is 130 g per day for adults. This value increases during pregnancy and lactation. The Institute of Medicine does not provide any special recommendations about carbohydrate intake for athletes. However, people who are physically active and work out regularly are often advised to consume more carbohydrate to prevent fatigue and replenish glycogen stores. Depending on the person's age and sex, Adequate Intake (AI) values for dietary fiber range from 21 to 38 g per day for adults; DRI values for carbohydrates are provided inside the front cover of this book.

There is no Tolerable Upper Intake Level (UL) for any individual carbohydrate class or for total carbohydrates. Although UL values for dietary fiber have not been established, some people may experience indigestion

**CONNECTIONS** Dietary Reference Intakes (DRIs) are reference values for nutrient intake. When adequate information is available, Recommended Dietary Allowances (RDAs) are established, but when less information is available, Adequate Intake (AI) values are provided. Tolerable Upper Intake Levels (UL) are intake levels that should not be exceeded (Chapter 2, pages 36–37).

**TABLE 5.4    Calculating Total Carbohydrate Requirements**

Intake goals are based on the Acceptable Macronutrient Distribution Ranges (AMDRs).

| Total Energy Requirement (kcal/day) | % kcal from Carbohydrate (AMDR) | Calculation | Total Carbohydrate Requirement (g/day) |
|---|---|---|---|
| 1,500 | 45–65 | 1,500 × 0.45/4<br>1,500 × 0.65/4 | 169–244 |
| 2,000 | 45–65 | 2,000 × 0.45/4<br>2,000 × 0.65/4 | 225–325 |
| 2,500 | 45–65 | 2,500 × 0.45/4<br>2,500 × 0.65/4 | 281–406 |

and other gastrointestinal disturbances when they consume large amounts of fiber. However, the DRI committee recognizes that there is considerable evidence that overconsumption of added sugars is associated with various adverse health effects, such as dental caries and obesity. The Institute of Medicine recommends that added sugars contribute no more than 25% of total calories.

To meet the body's need for energy, the Institute of Medicine suggests an Acceptable Macronutrient Distribution Range (AMDR) of 45 to 65% of total energy from carbohydrates. As there are 4 kcal per gram of carbohydrate, this means a person needing 2,000 kcal daily should consume between 225 and 325 g of carbohydrate (900 and 1,300 kcal, respectively) each day. You can easily determine how much carbohydrate is recommended for you by following the steps outlined in Table 5.4.

## Dietary Guidelines and MyPyramid

The USDA Dietary Guidelines for Americans stress the importance of eating a diet high in fiber and low in added sugars. In response to this, carbohydrates are represented in several food categories of the USDA MyPyramid food guidance system, emphasizing whole grains and cereals as the foundation of a healthy diet. The current recommendation for adults is to consume an equivalent of 5 to 10 ounces of grain-based foods daily, half of which should be whole grain. Examples of whole-grain foods are brown rice, pasta made with whole wheat, and whole-grain bread. A 1-ounce equivalent from the grains group is 1 slice of bread, 1 cup of ready-to-eat cereal, or ½ cup of cooked rice, cooked pasta, or cooked cereal.

In addition to whole grains and cereals, fruits and vegetables make significant contributions to carbohydrate intakes. It is recommended that adults consume approximately 1½ to 2½ cups of fruit and 2 to 4 cups of vegetables daily. To determine the specific number of recommended servings, you can visit the MyPyramid website (http://www.mypyramid.gov).

### Increasing Dietary Fiber Intake

The Dietary Guidelines recommend that adults consume 14 g of dietary fiber per 1,000 kcal. The average daily intake of fiber in the United States is about half this amount.[15] It is helpful to know that foods labeled "high in dietary fiber" contain 20% or more of the Daily Value for fiber per serving. The fiber contents of selected foods are shown in Table 5.5.

CONNECTIONS  The daily value is the recommended intake of a nutrient based on either a 2,000- or 2,500-kcal diet (Chapter 2, page 52).

**TABLE 5.5   The Dietary Fiber Content of Selected Foods**

| Food | Serving Size | Insoluble Fiber (g) | Soluble Fiber (g) | Total Dietary Fiber (g) |
|------|------|------|------|------|
| **Fruits** | | | | |
| Apple | 1 medium | 2.0 | 0.9 | 2.9 |
| Orange | 1 medium | 0.7 | 1.3 | 2.0 |
| Banana | 1 medium | 1.4 | 0.6 | 2.0 |
| **Vegetables** | | | | |
| Broccoli | 1 stalk | 1.4 | 1.3 | 2.7 |
| Carrots | 1 large | 1.6 | 1.3 | 2.9 |
| Tomato | 1 small | 0.7 | 0.1 | 0.8 |
| Potato | 1 medium | 0.8 | 1.0 | 1.8 |
| Corn | ⅔ cup | 1.4 | 0.2 | 1.6 |
| **Grains** | | | | |
| All-Bran® cereal | ½ cup | 7.6 | 1.4 | 9.0 |
| Oat bran | ½ cup | 2.2 | 2.2 | 4.4 |
| Cornflakes® cereal | 1 cup | 0.5 | 0 | 0.5 |
| Rolled oats | ¾ cup | 1.7 | 1.3 | 3.0 |
| Whole-wheat bread | 1 slice | 1.1 | 0.3 | 1.4 |
| White bread | 1 slice | 0.1 | 0.3 | 0.4 |
| Macaroni | 1 cup cooked | 0.3 | 0.5 | 0.8 |
| **Legumes** | | | | |
| Green peas | ⅔ cup | 3.3 | 0.6 | 3.9 |
| Kidney beans | ½ cup | 4.9 | 1.6 | 6.5 |
| Pinto beans | ½ cup | 4.7 | 1.2 | 5.9 |
| Lentils | ⅔ cup | 3.9 | 0.6 | 4.5 |

SOURCE: Anderson JW, Bridges SR. Dietary fiber content of selected foods. *American Journal of Clinical Nutrition.* 1988;47:440–7.

There are many ways to assure adequate fiber intake. Whole grains and cereals, legumes, fruits, and vegetables are probably the best options. It is important to select foods made with whole grains. Nutritionally, white bread, white rice, and white pasta are no match for their whole-grain counterparts. This is because the nutritional value of whole grain is greatest when all three layers of the wheat kernel—bran, germ, and endosperm—are intact (Figure 5.20). Milling removes the bran and germ layers, resulting in finely ground, refined flour. To restore some of the lost nutritive value, food manufacturers fortify their products with a variety of vitamins and minerals. However, many other important nutrients lost during processing are not added back. Clearly, the nutritional tally weighs heavily on the side of whole-grain products.

When reading food labels, it is important to look for the words "whole-grain cereals" and "whole-wheat flour." Foods made with "wheat flour" are not necessarily high in fiber. Try to choose bread and breakfast cereals that have at least 3 to 5 g of dietary fiber per serving. Based on USDA food-labeling guidelines, a food labeled as "high" in fiber must have 5 g or more of fiber per serving. Foods with labels claiming to be "good" sources of fiber have 2.5 to 4.9 g of fiber per serving.

## Decreasing Added Sugars

On food labels the term *sugar* usually refers to both added sugar and sugars that occur naturally. Chemically, added sugars are the same as naturally occurring sugars. However, foods in which naturally occurring sugars predominate, such

**FIGURE 5.20   Anatomy of a Wheat Kernel**

Foods made with whole-wheat flour are more nutritious than foods made with refined wheat flour, because the three layers of the wheat kernel—the bran, the germ, and the endosperm—are intact. Each layer contributes important nutrients needed for good health.

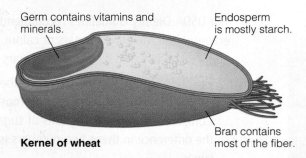

Germ contains vitamins and minerals.

Endosperm is mostly starch.

Bran contains most of the fiber.

**Kernel of wheat**

The processing of grains often removes the germ and bran portions, which contain the majority of the vitamins, minerals, and fiber.

## TABLE 5.6 Terms Used on Food Labels for Added Sweeteners

Brown rice syrup
Brown sugar
Concentrated fruit juice sweetener
Confectioner's sugar
Corn syrup
Dextrose
Fructose
Glucose
Granulated sugar
High-fructose corn syrup
Honey
Invert sugar
Lactose
Levulose
Maltose
Maple sugar
Molasses
Natural sweeteners
Raw sugar
Sucrose
Turbinado sugar
White sugar

as milk, fruits, and vegetables, provide not only energy but also fiber and micronutrients, such as vitamins and minerals. In contrast, foods with large amounts of added sugars, such as soft drinks, cakes, cookies, and candy, often have little nutritional value beyond the calories they contain. Sugars commonly added to food include white sugar, brown sugar, raw sugar, corn syrup, maple syrup, honey, and molasses. These and other terms used for added sugars are listed in Table 5.6. Some people believe that sweeteners such as honey are healthy alternatives to such refined sweeteners as white sugar. Like refined sugars, unprocessed sweeteners also have limited nutritional value other than energy.

In recent years, the consumption of added sugars in the United States has increased, with approximately 33% coming from soft drinks alone.[16] In fact, Americans consume about 20 teaspoons (80 g) of added sugar per day—for a total of 320 kcal. Although this may not sound like much, the calories can add up quickly. To lower the caloric content of foods, food manufacturers often use artificial sweeteners. Additional detail about these products is presented in the Focus on Food feature on the next page.

The USDA Dietary Guidelines suggest that added sugars be limited to no more than 32 g per day. This is equivalent to 1 candy bar or a 16-ounce soft drink. For the average American consuming 2,000 kcal a day, that amounts to approximately 8% of total daily energy intake. The MyPyramid food guidance system also emphasizes the importance of limiting added sugars in our diet. Table 5.7 lists the amount of added sugars in some common foods. Suggestions for reducing your intake of added sugars are provided in the Food Matters feature below.

## Food Matters

### Working Toward the Goal

The USDA Dietary Guidelines recommend that we choose and prepare foods and beverages with little added sugars or caloric sweeteners. The following suggestions can help reduce the amount of added sugars in the diet.

- The Nutrition Facts panel on food labels lists both total carbohydrates and *total* sugars. In highly processed foods, most of the sugar is likely to be *added* sugar. Compare the number of grams of sugar to the number of grams of carbohydrate. If the difference in these two numbers is small, the food is likely to be high in added sugar.
- Limit intake of soft drinks, fruit punch, and other sweet beverages. Water is the best way to quench your thirst. If you do drink these beverages, avoid the large sizes.
- When you want something sweet, choose fresh fruit rather than candy or cookies. Also, when buying canned fruit, choose brands that are packed in natural juices rather than sweetened varieties.
- Avoid presweetened breakfast cereals. Small amounts of sweeteners can be added to oatmeal and other cereals if desired.
- Add fruit to plain yogurt rather than buying sweetened varieties with fruit added.

## TABLE 5.7 Added Sugars in Your Diet

For each food listed, record the average number of servings you consume daily. Total the number of grams of added sugar in your diet.

| Food | Serving Size | Added Sugar (g) (5 g = 1 teaspoon) |
|---|---|---|
| Soft drink | 12 oz. | 36 |
| Milkshake | 10 oz. | 36 |
| Fruit punch | 8 oz. | 25 |
| Chocolate | 1.5 oz. | 21 |
| Sweetened breakfast cereal | 1 cup | 12 |
| Yogurt with fruit | 1 cup | 28 |
| Ice cream/sherbet | 1 cup | 40 |
| Cake with frosting | 1 slice | 24 |
| Cookies | 3 (33 g) | 24 |
| Jam/jelly | 1 tbsp. | 4 |

# *Focus On* FOOD
## Artificial Sweeteners

Nutritive sweeteners are naturally occurring, digestible carbohydrates such as table sugar. However, food manufacturers often use "nonnutritive" artificial sweeteners to sweeten foods, including saccharin, aspartame, acesulfame K, and sugar alcohols. Because most artificial sweeteners are actually sweeter than naturally occurring sweeteners, only small amounts are needed. Choosing which artificial sweetener to use is not as simple as you might think. For example, some lose their sweetness when heated and therefore are not recommended for baking or cooking. Others are not chemically stable and become bitter over time. Thus it is useful to understand some of the characteristics of the most commonly used artificial sweeteners.

Saccharin was one of the first artificial sweeteners to be widely used in the United States. During World War I, the use of saccharin increased because of sugar rationing. Although there was concern that saccharin might cause cancer, recent studies support its safety for human consumption. Saccharin is extremely sweet, very stable, and inexpensive to produce. Commercial products with saccharin include Sweet 'N Low® and Sugar Twin®.

Aspartame, another artificial sweetener, sold by its trade names NutraSweet® and Equal®, consists of two amino acids (phenylalanine and aspartic acid) bonded together. Although aspartame has the same energy content as sucrose (4 kcal per gram), it is almost 200 times as sweet. The food industry uses aspartame in sugar-free beverages. However, it is not heat stable and cannot be used in products that require cooking. Products containing aspartame must be clearly labeled, because of a potential risk to people with a genetic condition called phenylketonuria (PKU). Individuals with PKU cannot metabolize the amino acid phenylalanine, a component of this sweetener. Although the FDA has judged aspartame safe, some people claim it has adverse effects, including headaches, dizziness, nausea, and seizures.

Another artificial sweetener, called acesulfame K, is actually a salt that contains potassium (on the periodic table of elements, K is the symbol for potassium). Acesulfame K has been used extensively in Europe, and was approved in 1998 for use in the United States, where it is sold under the trade name of Sweet One®. Unlike aspartame, this artificial sweetener is heat stable and can be used in a wide variety of commercial products. Acesulfame K has been approved for use in refrigerated and frozen desserts, yogurt, dry dessert mixes, candies, gum, syrups, and alcoholic beverages.

Sucralose (trade name Splenda®), another low-calorie sweetener, was approved by the FDA in 1990. Sucralose is about 600 times sweeter than sucrose. Although derived from sucrose, sucralose is difficult for the body to digest and absorb. Therefore, it provides minimal calories. Because it is water soluble and stable, sucralose is used in a broad range of foods and beverages.

Sugar alcohols, which are actually classified as *nutritive sweeteners,* are neither sugars nor alcohols. Rather, they are "polyols," meaning that the sugar molecule has multiple alcohol groups attached. Sugar alcohols occur naturally in plants, particularly fruits, and have half the sweetness and calories of sucrose (only 2 to 3 kcal per gram, instead of 4 kcal per gram). Sorbitol, mannitol, and xylitol, the most common sugar alcohols, are found in "sugar-free" products such as chewing gums, breath mints, candies, toothpastes, mouthwashes, and cough syrups. One advantage of sugar alcohols is that unlike sucrose, they do not readily promote tooth decay. However, when eaten in excessive amounts sugar alcohols can have a laxative effect, leading to diarrhea. Once again, moderation is a key to good health.

## ESSENTIAL *Concepts*

In general, it is recommended that adults consume 45 to 65% of total energy as carbohydrates. In addition, to decrease the risk for certain diseases, adults are advised to consume 21 to 38 g of fiber daily. As this amount is well above the average U.S. intake, consumption of high-fiber foods such as whole-grain products and fruits and vegetables is encouraged. The term *sugar* is often used to refer to a wide variety of sweet substances, including honey, corn syrup, and brown sugar. Current recommendations suggest limiting added sugar intake to about 32 g/day.

# Carbohydrate Nutrition: Putting Concepts into Action

Although carbohydrate intake in the United States is typically adequate, there are other factors to consider. Some carbohydrate-rich foods provide many essential nutrients in addition to dietary fiber, but others have little nutritional value beyond the energy they provide. It is thus important for people to monitor their food choices to make sure they are getting the right types of carbohydrates in their diet—whole grains, fruits, and vegetables. You can evaluate your own food choices using the dietary assessment and planning techniques discussed in the following section.

## Setting the Stage and Setting Goals

Consider a woman named Malena who is 50 years old and weighs 190 pounds (86 kg). Although she is healthy, her physician has advised her to lose 50 pounds. Malena has tried many different diets, only to regain the lost weight. Frustrated by her previous attempts at weight loss, she decides to try a plan that restricts certain types of carbohydrates. This diet plan encourages people to eliminate foods made with highly processed carbohydrates, such as white flour, pasta, and foods containing added refined sugars. Unlike some carbohydrate-restricting diets, this plan includes plenty of fruits and vegetables. Malena understands that whole grains, fruits, and vegetables are important for overall health. And although she wants to lose weight, Malena knows it is important that her diet provide adequate amounts of essential nutrients.

## Assessing Nutritional Status

After four weeks on her diet, Malena has lost a total of 8 pounds. However, she is somewhat concerned about the nutritional adequacy of her diet. Malena conducts a diet analysis using the three-day food record method. She carefully records everything that she eats and drinks for two weekdays and one weekend day and enters this information into the MyPyramid website (http://www .mypyramid.gov). Using this software, she determines her dietary intake and compares it to the DRI values. In addition, she compares her dietary intake of certain food groups to those recommended in the USDA Dietary Guidelines and MyPyramid food guidance system.

Malena is especially interested to find out how her intake of dietary fiber compares to current recommendations. On average, Malena is consuming about 10 g of dietary fiber daily, which is below her estimated requirement of 25 g. However, she is getting the recommended number of servings of grains (8 ounce-equivalents), fruits (2 cup-equivalents), and vegetables (3 cup-equivalents). Approximately 50% of Malena's energy is coming from carbohydrates, and her diet provides adequate amounts of the essential vitamins and minerals. As the Institute of Medicine's Acceptable Macronutrient Distribution Ranges (AMDR) suggest that 45 to 65% of energy come from carbohydrate, this indicates that her carbohydrate intake is within the desired range. Malena concludes that she should continue on this weight-loss plan but that changes are needed to increase her intake of dietary fiber.

## Setting the Table to Meet the Goals

To accomplish this goal of increasing her intake of dietary fiber, Malena starts to read the nutrition information panel on food labels. To her surprise, many foods made with wheat flour provide very little dietary fiber. Even though her bread is brown and is made from wheat flour, it has only 1 g of fiber per slice. When she compares different varieties of breads, she discovers that those made with "whole-wheat flour" provide significantly more dietary fiber than those made with "wheat flour." This is also the case with breakfast cereals. Only those made with whole grains provided any fiber at all. Thus Malena quickly determines that the best way to increase her intake of dietary fiber, while limiting refined carbohydrates as called for in her weight-loss plan, is to include foods made with whole grains. Malena also discovers that legumes such as dried lentils and beans are excellent sources of dietary fiber. And even though Malena is getting the recommended number of servings of fruits and vegetables, some are higher in fiber than others. For example, a cup of berries provides twice as much dietary fiber as a cup of melon (8 versus 4 g, respectively).

## Comparing the Plan to the Assessment: Did It Succeed?

Several weeks later, Malena evaluates her diet again using the same three-day food record method. She finds that her daily caloric intake has not changed, and that she is still consuming the same amount of carbohydrates. However, her intake of dietary fiber increased by 10 g a day. Malena also finds that her intake of many essential vitamins and minerals increased, while the number of grams from saturated fats decreased. More important, Malena discovers that these dietary changes, along with increased physical activity, are helping her lose weight. Plus, she feels more energetic. Unlike her experience of previous attempts at weight loss, Malena realizes these changes can become a permanent part of her diet.

## Review Questions    Answers are found in Appendix G.

1. Which of the following carbohydrate sequences lists a monosaccharide followed by a disaccharide followed by a polysaccharide?
   a. sucrose        starch        glucose
   b. glucose        lactose       fructose
   c. glucose        maltose       amylose
   d. amylose        sucrose       maltose

2. A monosaccharide that has 6 carbon atoms also has _____ hydrogen atoms and _____ oxygen atoms.
   a. 12        12
   b. 6         6
   c. 12        6
   d. 6         12

3. Which of the following carbohydrates is found in milk?
   a. Glycogen
   b. Pectin
   c. Amylose
   d. Lactose

4. Plants store glucose in the form of _____, whereas animals store glucose in the form of _____.
   a. dietary fiber      amylose
   b. starch             maltose
   c. glycogen           amylopectin
   d. starch             glycogen

5. Which of the following carbohydrates consists of a highly branched arrangement of glucose molecules?
   a. Lactose
   b. Sucrose
   c. Amylose
   d. Glycogen

6. One difference between dietary fiber and amylose is that
   a. amylose is a monosaccharide and dietary fiber is a polysaccharide.
   b. amylose has α-glycosidic bonds, whereas dietary fiber has β-glycosidic bonds.
   c. amylose is not found in plant foods, whereas dietary fiber is abundant in plants.
   d. amylose is a disaccharide, whereas dietary fiber is a monosaccharide.

7. Dextrins are formed during the digestion of _____.
   a. lactose
   b. dietary fiber
   c. amylose and amylopectin
   d. sucrose

8. Foods with a high glycemic index
   a. help to lower the level of glucose in the blood.
   b. cause a rapid and large surge in blood glucose levels.
   c. do not contain glucose.
   d. are not easily digested.

9. The hormone released from the pancreas that helps decrease blood glucose is called _____.
   a. glucagon
   b. epinephrine
   c. insulin
   d. All of the above hormones increase blood glucose levels.

10. Which of the following foods provides a good source of dietary fiber?
    a. Yogurt
    b. Milk
    c. Cheese
    d. Dried beans

11. List the two monosaccharides that make up sucrose, lactose, and maltose. Describe the chemical reaction that takes place when two monosaccharides react to form a disaccharide.

12. Describe the steps associated with the digestion of amylose and amylopectin to form glucose.

13. Explain how insulin lowers the level of glucose in the blood and how glucagon increases the level of glucose in the blood.

14. Why does ketone formation increase when glucose is limited?

## Practice Calculations   Answers are found in Appendix G.

1. The Dietary Reference Intakes recommend that healthy adults consume approximately 45 to 65% of their total energy as carbohydrates. Calculate the minimum amount of carbohydrates (g) a person could consume (based on a total caloric intake of 1,800 calories per day) and still be within the range set by the DRIs.

2. The USDA recommends that added sugars be limited to approximately 10% of our daily total energy intake.

If a person consumes 25 g of sugar daily and that person's total energy intake is 1,500 calories, how close is he or she to meeting this recommendation?

3. The USDA Dietary Guidelines for Americans recommend that we consume 14 g of dietary fiber per 1,000 calories. What are the fiber requirements for a person who consumes 2,500 calories daily?

## Media Links   A variety of study tools for this chapter are available at our website. _____ nsonedu.com/nutrition/mcguire.

Prepare for tests and deepen your understanding of chapter concepts with these online materials:
- Practice tests
- Flashcards
- Glossary
- Student lecture notebook
- Web links
- Animations
- Chapter summaries, learning objectives, and crossword puzzles

# Notes

1. Bray GA, Nielsen SJ, Popkin BM. Consumption of high-fructose corn syrup in beverages may play a role in the epidemic of obesity. American Journal of Clinical Nutrition. 2004;9:537–43.

2. Pizza FX, Flynn MG, Duscha BD, Holden J, Kubitz ER. A carbohydrate loading regimen improves high intensity, short duration exercise performance. International Journal of Sport Nutrition. 1995;5:110–6. Rauch LH, Rodger I, Wilson GR, Belonje JD, Dennis SC, Noakes TD, Hawley JA. The effects of carbohydrate loading on muscle glycogen content and cycling performance. International Journal of Sport Nutrition. 1995;5:25–36.

3. Jenkins DJ, Kendall CW, Axelsen M, Augustin LS, Vuksan V. Viscous and nonviscous fibres, nonabsorbable and low glycaemic index carbohydrates, blood lipids and coronary heart disease. Current Opinion in Lipidology. 2000;11:49–56.

4. Slavin J. Why whole grains are protective: biological mechanisms. Proceedings of the Nutrition Society. 2003;62:129–34.

5. Stollman N, Raskin JB. Diverticular disease of the colon. Lancet. 2004;363(9409):631–9.

6. Kendall CW, Jenkins DJ. A dietary portfolio: maximal reduction of low-density lipoprotein cholesterol with diet. Current Atherosclerosis Reports. 2004;492–8.

7. Vesa TH, Marteau P, Korpela R. Lactose intolerance. Journal of the American College of Nutrition. 2000;19:165S–75S.

8. Foster-Powell K, Holt SH, Brand-Miller JC. International table of glycemic index and glycemic load values: 2002. American Journal of Clinical Nutrition. 2002;76:5–56.

9. Anderson JW. Whole grains protect against atherosclerotic cardiovascular disease. Proceedings of the Nutrition Society. 2003;62:135–42.

10. Marquart L, Wiemer KL, Jones JM, Jacob B. Whole grains health claims in the USA and other efforts to increase whole-grain consumption. Proceedings of the Nutrition Society. 2003;62:151–60.

11. Astrup A, Meinert Larsen T, Harper A. Atkins and other low-carbohydrate diets: hoax or an effective tool for weight loss? Lancet. 2004;364:897–9.

12. Klein S. Clinical trial experience with fat-restricted vs. carbohydrate-restricted weight-loss diets. Obesity Research. 2004;2:141S–4S.

13. Moynihan P, Petersen PE. Diet, nutrition and the prevention of dental diseases. Public Health Nutrition. 2004;7:201–26.

14. Nainar SM, Mohummed S. Role of infant feeding practices on the dental health of children. Clinical Pediatrics. 2004;43:129–33.

15. Lang R, Jebb SA. Who consumes whole grains, and how much? Proceedings of the Nutrition Society. 2003;62:123–7.

16. Guthrie JF, Morton JF. Food sources of added sweeteners in the diets of Americans. Journal of the American Dietetic Association. 2000;100:43–51.

**Check out the following sources for additional information.**

17. Augustin LS, Francesch IS, Jenkins DJA, Kendall CWC, Lavecchia C. Glycemic index in chronic disease: a review. European Journal of Clinical Nutrition. 2002;56:1049–71.

18. Duffy VB, Anderson GH. Position of the American Dietetic Association: use of nutritive and nonnutritive sweeteners. Journal of the American Dietetic Association. 1998; 98:580–7.

19. Flatt JP. Use and storage of carbohydrates. American Journal of Clinical Nutrition. 1995;61:952S–9S.

20. Foster-Powell K, Hold SHA, Brand-Miller JC. International table of glycemic index and glycemic load values. American Journal of Clinical Nutrition. 2002;76:55–6.

21. Gallaher DD, Schneeman B. Dietary fiber. In: Present knowledge in nutrition, 8th ed. Bowman BA, Russell RM, editors. Washington, DC: ILSI Press; 2001.

22. Groff JL, Gropper SS, eds. Advanced nutrition and human metabolism, 4th ed. Belmont, CA: Wadsworth/Thomson; 2005.

23. Hu FB, Willett WC. Optimal diets for prevention of coronary heart disease. JAMA (Journal of the American Medical Association). 2002;288:2569–78.

24. Jones JM, Elam K. Sugars and health: Is there an issue? Journal of the American Dietetic Association. 2003;103:1058–60.

25. Ludwig DS. The glycemic index: physiological mechanisms relating to obesity, diabetes, and cardiovascular disease. JAMA (Journal of the American Medical Association). 2002;287:2414–23.

26. Institute of Medicine. Dietary Reference Intakes for energy, carbohydrate, fiber, fat, fatty acids, cholesterol, protein, and amino acids. Washington, DC: National Academies Press; 2005.

27. Position of the American Dietetic Association: Health implications of dietary fiber. Journal of the American Dietetic Association. 2002;102:993–1000.

28. Mann J. Carbohydrates. In: Present knowledge in nutrition, 8th ed. Bowman BA, Russell RM, editors. Washington, DC: ILSI Press; 2001.

29. McVeagh P, Miller JB. Human milk oligosaccharides: only the breast. Journal of Pediatrics and Child Health. 1977;33:28–286.

30. Slavin JL. Mechanisms for the impact of whole grain foods on cancer risk. Journal of the American College of Nutrition. 2000;19:300S–7S.

31. Southgate DA. Dietary fiber: analysis and food sources. American Journal of Clinical Nutrition. 1978;1:S107–10.

32. U.S. Department of Agriculture, U.S. Department of Health and Human Services. 2005 Dietary guidelines for Americans. Available from: http://www.healthierus.gov/dietaryguidelines/.

33. Weihrauch MR, Diehl V. Artificial sweeteners—do they bear a carcinogenic risk? Annals of Oncology. 2004;15:1460–5.

34. Wildman RC, Medeiros DM. Advanced human nutrition. CRC Series in Modern Nutrition. Washington, DC: CRC Press; 2000.

35. World Health Organization. Carbohydrates in human nutrition: report of a joint FAO/WHO expert consultation, Rome, 1997. FAO Food and Nutrition Paper 66. Rome: Food and Agriculture Organization; 1997.

# Nutrition, Impaired Glucose Regulation, and Diabetes

© Yoav Levy/Phototake

$\mathscr{E}$at a healthy diet, exercise, and watch your weight are all recommendations we have heard before. Yet there is a convincing reason for us to take this advice seriously—in a word, diabetes. In the last 40 years, the incidence of diabetes has dramatically increased. In fact, diabetes has now reached epidemic proportions. According to the Centers for Disease Control and Prevention (CDC), 7% of the U.S. adult population has diabetes, which adds up to approximately 21 million Americans.[1] Of these, 6.2 million are undiagnosed. Furthermore, an additional 41 million Americans have a prediabetic condition characterized by blood glucose levels that are elevated but not high enough to be considered diabetes.[2] These numbers are staggering, but not surprising to many health experts.

Although some types of diabetes are not preventable, many professionals believe the most common form of diabetes, type 2, is largely attributable to lifestyle. Although genetics plays a role, physical inactivity, obesity, and unhealthy dietary practices have paved the way for this modern epidemic. The good news is that diet and other lifestyle practices can play an equally powerful role in managing and preventing type 2 diabetes.[3] Although some aspects of diabetes are not within our control, action is the key to diabetes prevention, treatment, and management.

## What Is Diabetes?

Diabetes was first described over 2,000 years ago as an affliction that caused excessive thirst, weight loss, and honey-sweet urine that attracted ants and flies. By the 16th century, physicians began treating diabetic patients with special diets. To replace the sugar lost in the urine, physicians recommended diets consisting of milk, barley, water, and bread. Physicians later tried diets of meat and fat. Regardless of diet, patients died within a few months of diagnosis. With nothing else to offer, physicians attempted to keep patients alive by restricting their intake of food. Until the discovery of insulin there was little hope for people with diabetes.

## Discovery of Insulin

The discovery of insulin began more than 100 years ago when two German physiologists, Oskar Minkowski and Joseph von Mering, accidentally discovered that in dogs surgical removal of the pancreas caused diabetes. This discovery captured the interest of Frederick Banting, who was certain it would lead to a cure for diabetes. Although several researchers had successfully extracted secretions from the pancreas, none of these compounds reversed diabetes. However, Banting, along with his assistant Charles Best, tried a new approach. Banting reasoned that enzymes produced by the pancreas were destroying other substances released by the pancreas. By first allowing the enzyme-producing pancreatic cells to die, Banting was able to collect the remaining functional pancreatic secretions. When he injected these secretions into diabetic dogs, their symptoms disappeared. Leonard Thompson, a 14-year-old boy who was dying of diabetes, was the first to be injected with Banting and Best's pancreatic extract. Amazingly, the young boy gained weight and seemed

**Billy Leroy, one of the first patients to receive insulin. Pictures were taken of Billy as an infant before diabetes, with diabetes, and after several months of treatment with insulin.**

to recover from the debilitating effects of diabetes. The pancreatic extract was given the name insulin.

Early insulin was obtained from pancreatic extracts of cattle and pigs. Today, human insulin is manufactured by genetic engineering. The gene is inserted into bacterial DNA, enabling the bacteria to produce human insulin. Insulin produced by genetic engineering is purer and more reliable than insulin derived from animals and is less likely to cause allergic reactions.

## Classification of Diabetes

**Diabetes mellitus** is a metabolic disorder characterized by elevated levels of glucose in the blood. The medical term for this is hyperglycemia. People have long known

that there are different types of diabetes, and over the years researchers have classified them in a variety of ways. For example, diabetes was once categorized according to typical age of onset: juvenile-onset diabetes versus adult-onset diabetes. This classification scheme was confusing, because both types of diabetes can occur at any age. Diabetes was later classified on the basis of whether insulin was required as a treatment: insulin-dependent diabetes versus non–insulin-dependent diabetes. The problem with this classification system was that some people with non–insulin-dependent diabetes require insulin to control hyperglycemia. In 1997 the American Diabetes Association developed a new system of diabetes classification based on etiology, or underlying cause.[4] Today diabetes is categorized into four major groups: **type 1 diabetes, type 2 diabetes,** gestational diabetes, and secondary diabetes. These are summarized in Table 1.

> **diabetes mellitus** (di – a – BE – tes MEL – lit – tus) (*diabetes,* to siphon; *mellitus,* sugar) Medical condition characterized by a lack of insulin or impaired insulin utilization that results in elevated blood glucose levels.

## TABLE 1  Classification of Types of Diabetes

| Category | Typical Age of Onset | Underlying Cause | Description |
|---|---|---|---|
| **Type 1 diabetes;** formerly called insulin-dependent diabetes mellitus (IDDM) or juvenile-onset diabetes | Childhood and adolescence; can develop in adults | Lack of insulin production by the pancreatic β-cells | An autoimmune disorder that destroys the insulin-producing β-cells of the pancreas, resulting in little or no insulin production. Person requires insulin delivered via daily injections or external pump. |
| **Type 2 diabetes;** formerly called non–insulin-dependent diabetes mellitus (NIDDM) or adult-onset diabetes mellitus | Middle-aged and older adults; increasing incidence in childhood and adolescence | Insulin resistance | Skeletal muscle and adipose tissue develop insulin resistance, resulting in blood glucose levels that are above normal. Genetics, obesity, and physical inactivity play a major role in insulin resistance. Can often be managed by diet, weight loss, and exercise; may require glucose-lowering medication or insulin injections. |
| Gestational diabetes | Occurs in pregnant women | Insulin resistance | A temporary form of diabetes that develops in 4–7% of pregnant women. Characterized by insulin resistance brought on by hormonal changes that take place during pregnancy. Women who develop gestational diabetes are at increased risk for developing type 2 diabetes. |
| Secondary diabetes | Varies | Varies | Brought on by other diseases, medical conditions, and/or medications. |

## Type 1 Diabetes

Approximately 5 to 10% of all people with diabetes fall into the category of type 1.[5] Although type 1 diabetes can develop at any age, it most often occurs during childhood and early adolescence, typically around 12 to 14 years of age. Because of this, type 1 diabetes was once called juvenile-onset diabetes. Type 1 diabetes was also known as insulin-dependent diabetes mellitus (IDDM), because this form of diabetes requires injections to deliver the insulin that the pancreas is no longer able to produce.

Without insulin, most cells cannot take up glucose, an energy source on which they normally depend (Figure 1). Cells compensate for the lack of glucose by metabolizing fat and protein (muscle) for energy, resulting in rapid weight loss. Using large amounts of fat for energy results in ketone formation. The accumulation of ketones in the blood brings on flulike symptoms such as abdominal pain, nausea, and vomiting. If the situation is not corrected, a life-threatening condition called **ketoaci-**

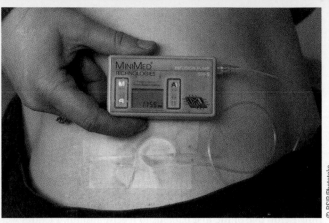

Individuals with type 1 diabetes must either inject insulin or use an insulin pump. Insulin pumps deliver insulin directly under the skin through narrow tubing, and are more accurate than injections. A pump can make diabetes easier to manage.

**dosis** can occur. This is caused by the accumulation of ketones, which can make the blood acidic.

In addition to rapid weight loss, other symptoms associated with type 1 diabetes include extreme thirst, hunger, frequent urination, and fatigue. These symptoms are caused by high levels of blood glucose. For example, the kidneys must work overtime to filter excess glucose out of the blood, resulting in the "spilling" of glucose over into the urine. The kidneys must dilute the glucose in the urine by drawing water out of the blood. As a result, people with diabetes urinate frequently, resulting in dehydration and thirst. Type 1 diabetes also causes a person to feel hungry and weak, because cells are starved for energy. In fact, diabetes is often described as "starvation in the midst of plenty."

Because the pancreas cannot produce insulin, people with type 1 diabetes need multiple daily injections of insulin to control blood glucose. As an alternative, some people use an insulin pump, which looks like a small pager, and is worn on a belt or on the waistband of clothing. Insulin cannot be taken orally because it is a protein and therefore would be destroyed by enzymes in the digestive tract.

***Cause of Type 1 Diabetes***    Type 1 diabetes is an **autoimmune disorder,** meaning that the immune system produces **antibodies** that attack and destroy the insulin-

## FIGURE 1    Type 1 Diabetes

Blood glucose regulation depends on the release of insulin from the pancreas in response to elevated blood glucose. In the case of type 1 diabetes, the pancreas is not able to produce insulin.

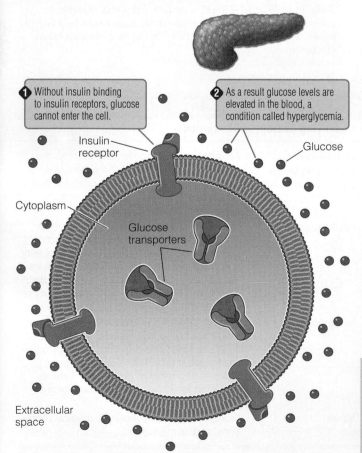

**ketoacidosis** (ke – to – a – ci – DO – sis) Severe metabolic condition resulting from the accumulation of ketones.

**autoimmune disorder** Condition that occurs when the immune system produces antibodies that attack and destroy tissues in the body.

**antibody** Protein produced by the immune system; responds to the presence of foreign proteins in the body.

producing beta cells (β-cells) of the pancreas. Why this occurs is not fully understood. However, as with other autoimmune disorders, genetics plays an important role. Scientists have long recognized that type 1 diabetes seems to run in families and is more common in certain ethnic groups.[6]

Type 1 diabetes is caused by complex interactions among genetics, the immune system, and environmental factors. The destruction of the insulin-producing beta cells of the pancreas begins when something in the environment stimulates the immune system to produce antibodies. Antibodies, which normally protect us from disease by killing foreign invaders, usually "know" which cells to attack and which ones to leave alone. In the case of autoimmune disorders, antibodies mistakenly attack and destroy the body's own cells. These types of antibod-

### FIGURE 2    Cause of Type 1 Diabetes

Type 1 diabetes is caused by an autoimmune disorder in which antibodies destroy the insulin-producing cells of the pancreas.

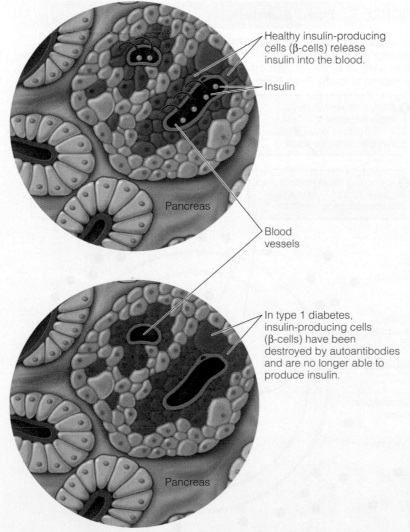

Healthy insulin-producing cells (β-cells) release insulin into the blood.

Insulin

Pancreas

Blood vessels

In type 1 diabetes, insulin-producing cells (β-cells) have been destroyed by autoantibodies and are no longer able to produce insulin.

Pancreas

ies, called **autoantibodies,** are found in people with autoimmune disorders. In type 1 diabetes, autoantibodies incorrectly target the insulin-producing cells of the pancreas and destroy them. Many researchers believe type 1 diabetes is triggered by a viral infection.[7]

The destruction of pancreatic β-cells is a gradual process. Only after the majority of them have been destroyed do symptoms of type 1 diabetes appear (Figure 2). With little or no insulin-producing ability left, the pancreas cannot keep up with the body's need for insulin, resulting in type 1 diabetes.

## Type 2 Diabetes

Type 2 diabetes is by far the most common form of diabetes, with 90 to 95% of people with diabetes falling into this category.[8] In the United States, type 2 diabetes has become so widespread that an estimated one out of four people have or will have this disease, or have a family member with it.[9] Because type 2 diabetes is often managed without insulin treatments, it was once called non–insulin-dependent diabetes mellitus (NIDDM).

Although type 2 diabetes can occur at any age, it most frequently develops in middle-aged and older adults. For this reason, type 2 diabetes was once called adult-onset diabetes. However, the growing number of children and teens diagnosed with type 2 diabetes is alarming. Approximately 800,000 people are diagnosed with type 2 diabetes every year, and 16% of these new cases are in people under 19 years of age.[10] Therefore, type 2 diabetes can no longer be thought of as a condition that affects adults only. The risk of developing type 2 diabetes is influenced in part by genetics.[11] If one identical twin develops type 2 diabetes, there is a strong probability that the other twin will also develop it. Several risk factors are associated with type 2 diabetes, including obesity, lack of physical activity, and ethnicity. These are summarized in Table 2.

Unlike type 1 diabetes, most people with type 2 diabetes have normal or even elevated levels of insulin in their blood. Instead, type 2 diabetes is caused by **insulin resistance,**

**autoantibodies** Antibodies produced by the immune system; attack the body's own cells.

**insulin resistance** Condition characterized by the inability of insulin receptors to respond to the hormone insulin.

## TABLE 2 Risk Factors Associated with Type 2 Diabetes

- Having a parent, brother, or sister with diabetes
- Diagnosis of gestational diabetes or delivery of a baby weighing more than 9 pounds at birth
- Sedentary lifestyle (exercise fewer than three times per week)
- Obesity, especially central obesity
- Being over 45 years of age
- Being of African, Hispanic, Native American, or Pacific Island descent
- Having high blood pressure (≥140/90 mm Hg)
- Having low HDL cholesterol (< 35 mg/dL)
- Having high triglycerides (> 250 mg/dL)

SOURCE: National Institute of Diabetes and Digestive and Kidney Disease, of the National Institutes of Health. Available from: http://diabetes.niddk.nih.gov/dm/pubs/riskfortype2/.

## TABLE 3 Symptoms Associated with Type 1 and Type 2 Diabetes

| Type 1 Diabetes | Type 2 Diabetes |
| --- | --- |
| Frequent urination | Frequent urination |
| Excessive thirst | Excessive thirst |
| Fatigue | Fatigue |
| Unusual weight loss | Frequent infections |
| Extreme hunger | Blurred vision |
| Ketosis | Vaginal itching |
| | Cuts or bruises that are slow to heal |
| | Tingling or numbness in hands or feet |
| | Frequent urinary tract infections |

which means that insulin-sensitive cells do not respond to insulin in a normal way.[12] As a result, these cells do not readily take up glucose even when insulin is present. The inability of insulin to do its job results in the accumulation of glucose in the blood. Although many people with type 2 diabetes can control their blood glucose effectively by diet and exercise, others may require glucose-lowering medications that help make cells more responsive to insulin. Even though the pancreas produces insulin, some people with type 2 diabetes need additional insulin to lower blood glucose. This is why some people with type 2 diabetes require insulin injections.

**A Closer Look at Insulin Resistance** As discussed in Chapter 5, insulin acts like a cellular key that binds to insulin receptors on the surface of some cell membranes. The binding of insulin to its receptor enables glucose to enter these cells. For reasons that are not clear, some people develop resistance to insulin (Figure 3). Insulin resistance results when insulin receptors have difficulty recognizing or binding insulin. To overcome this resistance, the pancreas produces more insulin. In many cases, this response keeps blood glucose levels within a relatively normal range. This is why not everyone with insulin resistance develops type 2 diabetes; however, all people with type 2 diabetes are insulin resistant.

It is not clear why some people with insulin resistance develop type 2 diabetes whereas others do not. Perhaps after many years of working overtime to produce extra insulin, the pancreas becomes worn out and can no longer produce insulin in amounts

needed to overcome insulin resistance. As a result, blood glucose levels rise, and a person begins to experience symptoms associated with type 2 diabetes.

Some of the symptoms associated with type 2 diabetes are similar to those of type 1 diabetes (see Table 3). These

## FIGURE 3 Type 2 Diabetes

Blood glucose regulation depends on the release of insulin from the pancreas in response to elevated blood glucose. Insulin binds to insulin receptors, allowing glucose to be taken up by the cell. Type 2 diabetes is caused by an inability of insulin receptors to respond to insulin. This is called "insulin resistance."

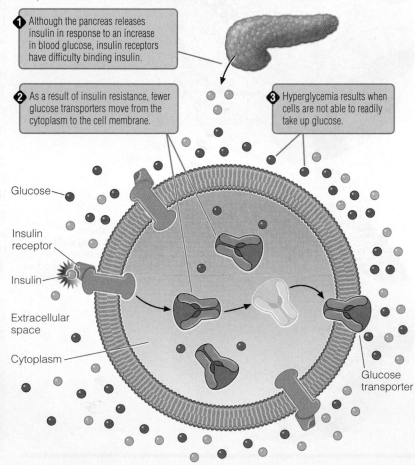

1. Although the pancreas releases insulin in response to an increase in blood glucose, insulin receptors have difficulty binding insulin.

2. As a result of insulin resistance, fewer glucose transporters move from the cytoplasm to the cell membrane.

3. Hyperglycemia results when cells are not able to readily take up glucose.

Glucose

Insulin receptor

Insulin

Extracellular space

Cytoplasm

Glucose transporter

include fatigue, frequent urination, and excessive thirst. Because these symptoms tend to develop gradually, they are often ignored. However, in untreated or uncontrolled type 2 diabetes, chronic hyperglycemia causes additional symptoms, such as blurred vision, frequent urinary tract infections, slow wound healing, and vaginal itching.

***Risk Factors for Type 2 Diabetes***    In addition to genetics, obesity also influences the risk of developing type 2 diabetes. Approximately 80% of people diagnosed with type 2 diabetes are overweight. Overweight adults are significantly more likely to develop type 2 diabetes than are lean individuals.[13] In addition to overall obesity, the distribution of body fat is also a risk factor for type 2 diabetes. Body fat

stores in the abdominal region of the body pose a greater risk than do lower body fat stores.[14] As the prevalence of obesity in the United States continues to climb, so does the prevalence of type 2 diabetes (Figure 4). If these trends continue, researchers estimate that 18 million Americans will have type 2 diabetes by the year 2020.[15]

The mechanism by which obesity causes insulin resistance remains unclear. However, a recent discovery that adipose tissue functions as a hormone-producing endocrine "gland" has generated new insights.[16] In addition to storing fat, adipose tissue secretes a variety of hormones that influence other tissues. Hormone production is greatly influenced by the amount of adipose tissue in the body, which in turn may influence insulin resistance. This may

## FIGURE 4    Obesity and Type 2 Diabetes Trends in U.S. Adults

The prevalence of obesity and type 2 diabetes in the United States is increasing. Note that obesity is defined as being approximately 30 pounds overweight.

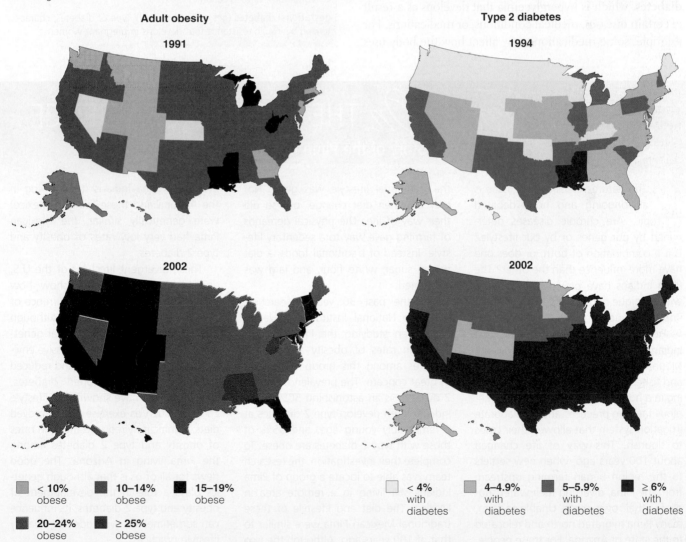

SOURCE: Centers for Disease Control and Prevention, Behavioral Risk Factor Surveillance System (BRFSS).

also explain why weight loss often restores insulin sensitivity. Although there is more to learn about hormones, the link between obesity and insulin resistance is indisputable.

The good news about type 2 diabetes is that in many cases it can be prevented. For people at risk for type 2 diabetes, weight management, regular exercise, and a healthy diet can reduce the risk by as much as 60%.[17] Even small changes can make a big difference. For example, losing as little as 10 to 15 pounds or taking a 30-minute walk daily have been shown to lower blood glucose. We cannot do much about our genetics; however, we can do a lot to prevent weight gain and to stay fit. The story of the Pima Indians, presented in the Focus on the Process of Science feature, provides an example of the way genetics can interact with lifestyle to determine the risk of developing type 2 diabetes.

## Gestational Diabetes and Secondary Diabetes

The majority of people with diabetes have type 1 or type 2. However, a small percentage of people have **secondary diabetes,** which is hyperglycemia that develops as a result of certain diseases, medical conditions, or medications. For example, some medications may affect how the body uses or produces insulin. Once the underlying cause of secondary diabetes is treated, normal blood glucose control is often restored. Another type of diabetes, called **gestational diabetes,** is brought on by pregnancy. During pregnancy, most women experience insulin resistance to some extent. This normal and healthy response to pregnancy helps make glucose more available to the fetus. However, some women develop a more severe form of insulin resistance called gestational diabetes, which occurs in approximately 4 to 7% of all pregnancies.[19] This temporary form of diabetes typically develops after the 24th week of pregnancy, disappearing within 6 weeks after delivery. Gestational diabetes is more common in obese women and in women who have a family history of type 2 diabetes.[20] Risk factors associated with gestational diabetes are listed in Table 4.

The exposure of the fetus to high levels of glucose and insulin lead to increased glucose uptake and the deposit of extra fat in the growing baby. For this reason, infants

**secondary diabetes** Diabetes that results from other diseases, medical conditions, or medication.

**gestational diabetes** (ges – TA – tion – al) Type of diabetes; characterized by insulin resistance that develops in pregnant women.

# Focus On THE PROCESS OF SCIENCE
## The Story of the Pima Indians

The nature-versus-nurture issue is an ongoing and hotly debated topic. Are chronic diseases determined by our genes or by our lifestyle? Is it a combination of both, or does one have more influence than the other? The Pima Indians have provided researchers with a unique opportunity to study these important questions.[18]

For thousands of years the Pima Indians, originally from the Sierra Madre Mountains in Mexico, farmed, hunted, and fished as a way of life. The Pima, living in a hot, dry climate, developed ingenious farming practices and an elaborate irrigation system that allowed their crops to flourish. This way of life changed about 100 years ago, when new settlers to the region began farming upstream from the Pima, diverting their water supply for their own crops. Unable to farm, many Pima migrated north and relocated to the state of Arizona. For these people,

the traditional lifestyle was gone. Not only did their diet change, but so did their way of life. The physical demands of farming gave way to a sedentary lifestyle. Instead of traditional foods, a diet rich in sugar, white flour, and lard was consumed.

For the past 30 years researchers from the National Institutes of Health have been studying the Pima Indians. The high rates of obesity and type 2 diabetes among this group have been of great concern. The prevalence of type 2 diabetes is an astonishing 50%. Pima Indians often develop type 2 diabetes at a remarkably young age, and 95% of those with type 2 diabetes are obese. To complete their investigation, the research team was able to locate a group of Pima Indians still living in a remote area in Mexico. The diet and lifestyle of these traditional Mexican Pima were similar to that of 100 years ago. Although the two

groups of Pima Indians (those living in the Arizona and those living in Mexico) were genetically similar, the Mexican Pima had very low rates of obesity and type 2 diabetes.

These divergent lifestyles of the U.S. and Mexican Pima clearly show how lifestyle can influence the occurrence of obesity and type 2 diabetes. Although both groups of Pima have similar genetics, only those living in a permissive environment of abundant food and reduced physical activity developed diabetes. Further studies have shown that lifestyle changes, such as exercise and improved diets, significantly reduced the high rates of obesity and type 2 diabetes among the Pima living in Arizona. The good news for all of us is that although genetics plays a role in the development of obesity and type 2 diabetes, its influence can sometimes be overridden by healthy lifestyle practices.

## TABLE 4    Risk Factors Associated with Gestational Diabetes

- Obesity
- Central obesity
- Weight gain between pregnancies
- Over 35 years of age
- Being of African, Hispanic, Native American, or Pacific Islander descent
- Family history of type 2 diabetes or gestational diabetes
- Previous delivery of a stillborn or a very large baby (>9 lbs)
- Having a mother who was born with a low birth weight
- Having gestational diabetes with previous pregnancy
- Having high blood pressure prior to pregnancy

SOURCE: U.S. Department of Health and Human Services, National Institutes of Health, and the National Institute of Child Health and Human Development. NIH Pub. No. 00-4818. 2005.

## TABLE 5    Criteria for Metabolic Syndrome

Metabolic syndrome is defined as having three or more of the following variables.

| Variable | Value |
| --- | --- |
| Abdominal obesity | Waist circumference in males > 37 inches and in females > 31.5 inches |
| Fasting triglycerides | ≥150 mg/dL |
| HDL cholesterol | <40 mg/dL in males and < 50 mg/dL in females |
| Blood pressure | ≥130/85 mm Hg |
| Fasting glucose | ≥100 mg/dL |

SOURCE: International Diabetes Federation (IDF). Available from: http://www.idf.org/webdata/docs/MetSyndrome_Final.pdf.

born to women with gestational diabetes tend to be large, making the delivery difficult. Babies born to mothers with poorly controlled gestational diabetes can weigh as much as 10 to 12 pounds. It is important for women who develop gestational diabetes to receive information about diet and blood glucose control.

Pregnant women are routinely tested for gestational diabetes around the 24th week of pregnancy. Women who develop gestational diabetes are at increased risk of for developing type 2 diabetes within the next 15 years. Infants born to women with gestational diabetes may also be at increased risk for developing diabetes, especially if there is a family history of diabetes.[21]

## ESSENTIAL *Concepts*

The four types of diabetes are categorized by the underlying cause. Type 1 diabetes is caused by a lack of insulin production that develops when antibodies mistakenly attack and destroy the insulin-producing cells of the pancreas. Type 2 diabetes is caused by insulin resistance, meaning that insulin receptors do not respond appropriately to insulin. Most people with diabetes have type 2, which is strongly associated with obesity. Insulin resistance also causes gestational diabetes, which can develop during pregnancy. This type of diabetes is temporary and disappears within 6 weeks of delivery. Secondary diabetes results from other diseases or medications. People with type 1 diabetes require daily insulin injections to manage blood glucose. Type 2 diabetes is often controlled by diet, exercise, and weight loss. Both type 1 and type 2 diabetes are caused in part by genetics.

# Metabolic Syndrome

Type 2 diabetes is not the only disease associated with insulin resistance. A metabolic disorder, characterized by elevated fasting blood glucose, abnormal blood lipids,

obesity, and high blood pressure, was first described in 1988.[22] Because of its mysterious nature, it was called Syndrome X. Now referred to as **metabolic syndrome,** it is now known that this is not a single disease but rather a cluster of conditions that appear to be associated with insulin resistance.[23] It is defined by having three or more of the following conditions: abdominal obesity, abnormal blood lipids, high blood pressure, and high fasting blood glucose (Table 5). According to the Centers for Disease Control and Prevention (CDC), approximately 47 million adults in the United States (20%) have metabolic syndrome.[24] People with metabolic syndrome are at increased risk for developing type 2 diabetes and cardiovascular disease. The same factors that are associated with type 2 diabetes — obesity, poor eating habits, and lack of exercise — are also associated with metabolic syndrome. Although type 2 diabetes and metabolic syndrome are in many ways similar, there are some important differences.[25]

## A Closer Look at Metabolic Syndrome

As noted, insulin resistance occurs when insulin-sensitive cells (such as skeletal muscle and adipose tissue) cannot respond appropriately to insulin. This causes glucose in the blood to rise. The pancreas tries to lower blood glucose by releasing large amounts of insulin. The medical term for elevated insulin levels is **hyperinsulinemia.** Often the extra insulin helps maintain blood glucose levels within a somewhat normal range. In some people the pancreas cannot produce high levels of insulin long enough to maintain blood glucose levels in a normal range, ultimately resulting in type 2 diabetes. Most people

**metabolic syndrome** Condition characterized by an abnormal metabolic profile, abdominal body fat, and insulin resistance.

**hyperinsulinemia** (hy – per – in – su – li – NE – mi – a) Condition characterized by high blood insulin.

with insulin resistance do not actually develop type 2 diabetes. Rather, the pancreas continues to release insulin in amounts sufficient to overcome insulin resistance. For these people, blood glucose levels are high but not high enough to be considered type 2 diabetes. This may sound like a good thing, but it comes at a cost.

In addition to problems caused by moderately elevated blood-glucose levels, a high insulin level, or hyperinsulinemia, also has adverse effects on many tissues.[26] For example, hyperinsulinemia stimulates the kidneys to return sodium to the blood rather than excreting it in the urine. Sodium retention can lead to high blood pressure. In response to high insulin levels, the ovaries increase production of the hormone testosterone. This can disrupt menstruation and lead to fertility problems. Also of great concern is the effect of hyperinsulinemia on lipid metabolism in the liver. High insulin levels cause blood lipid levels to increase, which can lead to heart disease.

## ESSENTIAL *Concepts*

Most people with insulin resistance do not develop type 2 diabetes, because the pancreas releases insulin in amounts sufficient to overcome insulin resistance. However, chronically high insulin levels can lead to a condition called metabolic syndrome. This condition is defined by having at least three of the following conditions: abdominal obesity, abnormal blood lipids, high blood pressure, and high fasting blood glucose levels. People with metabolic syndrome are at increased risk for developing type 2 diabetes and cardiovascular disease. Metabolic syndrome can lead to hypertension, fertility problems in women, and disturbances in blood lipid levels.

# Long-Term Complications Associated with Diabetes and Insulin Resistance

Regardless of the type of diabetes, chronic hyperglycemia can lead to serious long-term complications.[27] These complications appear to be caused largely by the effect of excess glucose on blood vessels and nerves (Figure 5). When a person is hyperglycemic, excess glucose in the blood can attach to proteins, causing blood vessels to thicken and lose their elasticity. This vascular damage is one reason why people with diabetes are at increased risk for having heart attacks and strokes. Chronic hyperglycemia can also damage small blood vessels and nerves, leading to blindness, limb amputation, impaired kidney function, and a loss of feeling in feet and hands. Maintaining blood glucose close to normal is the best

## FIGURE 5    Hyperglycemia Damages Blood Vessels and Nerves

Complications associated with diabetes are caused by the effect of glucose on blood vessels and nerves. The damage to large blood vessels increases the risk for heart attacks and strokes.

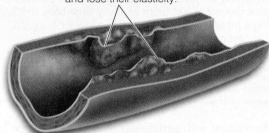

Excess glucose in the blood attaches to proteins, causing blood vessels to thicken and lose their elasticity.

Chronic hyperglycemia damages blood vessles and nerves, leading to blindness, lower leg amputations, and impaired kidney function.

defense in preventing these long-term complications associated with diabetes.

## The Importance of Screening and Early Diagnosis

Type 1 diabetes is typically diagnosed shortly after symptoms appear. This is because symptoms develop rapidly and, because of their severity, are not easily overlooked. This is not the case with type 2 diabetes. Symptoms associated with type 2 diabetes develop gradually and are easily ignored until the disease becomes more advanced. Type 2 diabetes can therefore go undiagnosed for many years. The American Diabetes Association recommends that all adults be screened for type 2 diabetes. Criteria based on blood glucose values used to diagnose diabetes are shown in Table 6.

## ESSENTIAL *Concepts*

Excess glucose in the blood can attach to proteins, causing blood vessels to thicken and lose elasticity, and predisposing people with diabetes to increased risk for heart attacks and strokes. Chronic hyperglycemia also damages small blood vessels and nerves, leading to blindness, limb amputation, impaired kidney function, and loss of feeling in feet and hands. Type 1 diabetes tends to be diagnosed shortly after symptoms develop. However, type 2 diabetes can go unnoticed. Early detection of diabetes and prediabetes, followed by appropriate intervention, can help prevent or delay serious long-term complications.

**TABLE 6    Diagnostic Values for Diabetes**

| Category | Fasting Plasma Glucose[a] | Two-Hour Oral Glucose-Tolerance Test[b] |
|---|---|---|
| Normal | <100 mg/dL | <140 mg/dL |
| Diabetes | ≥126 mg/dL | ≥200 mg/dL |

[a] Person must fast 10 to 12 hours prior to testing.
[b] Person must fast 10 to 12 hours prior to testing. After an initial fasting blood sample is taken, a 50-g glucose beverage is consumed. A second blood sample is taken two hours later.

SOURCE: American Diabetes Association. Standards of medical care in diabetes. *Diabetes Care.* 2004;27:S15–S35.

# Controlling Blood Glucose

For many years, it was unclear if the occurrence of long-term complications associated with diabetes could be delayed or reduced by controlling blood glucose levels as close to normal as possible. However, a large clinical study, called the Diabetes Control and Complications Trial (DCCT), was conducted to answer this important question.[28] This study showed that intensive management of blood glucose levels reduces complications associated with type 1 diabetes. Several years later the United Kingdom Prospective Diabetes Study (UKPDS) revealed that people with type 2 diabetes also benefit from controlling blood glucose within near-normal ranges.[29]

## Components of a Diabetes Care Plan

Controlling blood glucose is the best defense in preventing long-term complications associated with diabetes. The most successful diabetes care plans focus on individual needs, preferences, and cultural practices. There is no such thing as "one-plan-fits-all" when it comes to diabetes care, management, and prevention.

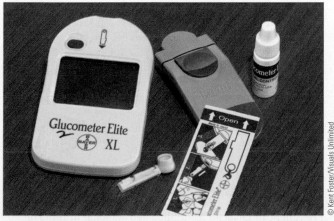

Glucometers are an important part of diabetes management. It is important for people who have diabetes to monitor blood glucose levels daily.

For type 1 diabetes, the focus is balancing insulin injections with a healthy diet and physical activity. For people with type 2 diabetes, the emphasis is on a healthy diet, weight management, and physical activity. Lifestyle changes that promote weight loss are discussed in Chapter 9. In addition to diet and exercise, daily blood glucose monitoring is important for all people with diabetes. Glucose meters are used to monitor blood glucose levels so that adjustments in diet, insulin, or both can be made.

## Diet and Diabetes

Nutrition guidelines developed for people with diabetes can be applied by anyone who wants to eat a healthy diet. These guidelines, developed by the American Diabetes Association (ADA), are summarized in Table 7.[30] Although the current recommendation is that carbohydrate and monounsaturated fats provide 60 to 70% of total energy intake, the ADA recognizes that the percentages of total calories from carbohydrate, protein, and lipid need to be adjusted based on personal needs.

As described in Chapter 5, consuming foods that provide soluble fiber is important for good health. This is particularly important for people with diabetes, because soluble fiber helps lower glucose and cholesterol in the blood. Foods that provide soluble fiber include oats, rice, bran, psyllium seeds, soy, and some fruits.

For years, people with diabetes have been told to avoid foods high in sugar. However, there is no need to totally eliminate sugars from the diet. Still, whole

**TABLE 7    American Diabetes Association Dietary Recommendations for Diabetes**

- Carbohydrates and monounsaturated fat should provide 60 to 70% of total energy intake. Monounsaturated fats are primarily in canola, olive, and peanut oils.
- Soluble dietary fiber helps lower glucose and cholesterol in the blood. Foods with soluble fibers include oats, rice, bran, psyllium seeds, soy, and some fruits.
- There is no need to totally eliminate sugars from the diet, but whole grains, fruits, and vegetables are recommended over high-sugar foods.
- Protein should provide 15 to 20% of the total energy in the diet, unless there is evidence of impaired kidney function.
- There is not sufficient evidence to recommend the use of vitamin and mineral supplements to prevent or treat diabetes. Some studies have shown improvements in blood glucose regulation in response to certain micronutrients such as chromium, potassium, magnesium, and zinc.
- Moderate alcohol consumption, defined as one drink for adult women and two drinks for adult men, poses no significant health concern for people with diabetes.

SOURCE: American Diabetes Association. Nutrition principles and recommendations in diabetes. *Diabetes Care.* 2004;27:S36–S46.

grains, fruits, and vegetables are preferred over high-sugar foods because they are good sources of micro-nutrients, phytochemicals, and dietary fiber. At this time there is insufficient evidence to recommend the use of vitamin and mineral supplements to prevent or treat diabetes. Rather, nutritionists recommend that a well-balanced diet be the source of these important nutrients.

## Physical Activity and Diabetes

Exercise is especially important for people with diabetes. Exercise improves cardiovascular health, facilitates weight loss and weight management, helps lower blood glucose levels, reduces stress, and promotes healthy lipid levels in the blood. As little as 30 minutes of daily exercise can benefit health.

## Diabetes Can Be Demanding

Making dietary and other lifestyle changes is not easy and often requires assistance from health care professionals trained in diabetes care and management. The management of diabetes is demanding, and the emotional side of the task is often overlooked. It is not uncommon for people with diabetes to follow "all the rules" and still have difficulties controlling blood glucose, creating a feeling of frustration. However, people who participate actively as members of their diabetes care team by asking questions and making good choices can avoid complications and achieve positive effects on their health and well-being.

## ESSENTIAL *Concepts*

Managing blood glucose reduces the complications associated with diabetes. Treatment of type 1 diabetes focuses on balancing insulin injections with a healthy diet and physical activity. The treatment focus of type 2 diabetes is on maintaining near-normal blood glucose through a healthy diet, weight management, and physical activity. Many of the nutrition guidelines developed for people with diabetes apply to anyone who wants to eat a healthy diet. As little as 30 minutes of moderate daily exercise can benefit health and lower blood glucose. Making these lifestyle changes can require assistance from health care professionals trained in diabetes care and management.

## Notes

1. Centers for Disease Control and Prevention. National Diabetes Fact Sheet: Total prevalence of diabetes in the United States, 2005. Available from: http://www.cdc.gov/diabetes/pubs/pdf/ndfs_2005.pdf.

2. Centers for Disease Control and Prevention. National Diabetes Fact Sheet: Total prevalence of diabetes in the United States, 2005. Available from: http://www.cdc.gov/diabetes/pubs/pdf/ndfs_2005.pdf.

3. Lindstrom J, Peltonen M, Tuomilehto J. Lifestyle strategies for weight control: experience from the Finnish Diabetes Prevention Study. Proceedings of the Nutrition Society. 2005;64:81–8.

4. Expert Committee on the Diagnosis and Classification of Diabetes Mellitus. Report of the Expert Committee on the diagnosis and classification of diabetes mellitus. Diabetes Care. 1997;20:1183–97.

5. Sperling MA, editor. Type 1 diabetes: etiology and treatment. Totowa, NJ: Human Press; 2003.

6. Eisenbarth GS. Type 1 diabetes: molecular, cellular and clinical immunology. Advances in Experimental Medicine and Biology. 2004;552:306–10. Hirschhorn JN. Genetic epidemiology of type 1 diabetes. Pediatrics and Diabetes. 2003;4:87–100.

7. Knip M. Environmental triggers and determinants of beta-cell autoimmunity and type 1 diabetes. Endocrine and Metabolic Disorders. 2003;4:213–23.

8. Goldstein BJ, editor. Textbook of type 2 diabetes. London: Martin Dunitz; 2003.

9. American Diabetes Association. Position statement from the American Diabetes Association: the prevention or delay of type 2 diabetes. Diabetes Care. 2002;25:742–9.

10. Lewis C. Diabetes: a growing public health concern. FDA Consumer (U.S. Food and Drug Administration). 2002; Jan.–Feb. Available from: http://www.fda.gov/fdac/. features/2002/102_diab.html. Botero D, Wolfsdorf JI. Diabetes mellitus in children and adolescents. Archives of Medical Research. 2005;36:281–90.

11. Malecki MT. Genetics of type 2 diabetes mellitus. Diabetes Research and Clinical Practice. 2005;68:S10–21. Tusie Luna MT. Genes and type 2 diabetes mellitus. Archives of Medical Research. 2005;36:210–22.

12. Leahy JL. Pathogenesis of type 2 diabetes mellitus. Archives of Medical Research. 2005;36:197–209.

13. Hu FB, Manson JE, Stampfer MJ, et al. Diet, lifestyle, and the risk of type 2 diabetes mellitus in women. New England Journal of Medicine. 2001;345:790–7. van Dam RM, Rimm EB, Willett WC, Stampfer MJ, Hu FB. Dietary patterns and risk for type 2 diabetes mellitus in U.S. men. Annals of Internal Medicine. 2002;136:201–9.

14. Solomon CG, Manson JE. Obesity and mortality: a review of the epidemiologic data. American Journal of Clinical Nutrition. 1997;66:1044S–50S.

15. Green A, Christian Hirsch N, Pramming SK. The changing world demography of type 2 diabetes. Diabetes/Metabolism and Research Reviews. 2003;19:3–7.

16. Frayn KN. Obesity and metabolic disease: is adipose tissue the culprit? Proceedings of the Nutrition Society. 2005;64:7–13.

17. American Diabetes Association and National Institute of Diabetes, Digestive and Kidney Disease. Position Statement from the American Diabetes Association: the prevention or delay of type 2 diabetes. Diabetes Care. 2002;25:742–9.

18. Bennett PH. Type 2 diabetes among the Pima Indians of Arizona: an epidemic attributable to environmental change? Nutrition Reviews. 1999;57:S51–4. National Institute of Diabetes and Digestive and Kidney Diseases. The Pima Indians: pathfinders for health. Available from: http://diabetes.niddk.nih.gov/dm/pubs/pima/.

19. American Diabetes Association. Fact sheet on gestational diabetes. Available from: http://www.diabetes.org/gestational-diabetes.jsp.

20. American Diabetes Association. Position statement from the American Diabetes Association: gestational diabetes mellitus. Diabetes Care. 2003;26:S103–5.

21. Stocker CJ, Arch JR, Cawthorne MA. Fetal origins of insulin resistance and obesity. Proceedings of the Nutrition Society. 2005;64:143–51.

22. Reaven GM. The role of insulin resistance in human disease. Diabetes. 1988;37:1495–507.

23. Shaw DI, Hall WL, Williams CM. Metabolic syndrome: what is it and what are the implications? Proceedings of the Nutrition Society. 2005;64:349–57.

24. Centers for Disease Control and Prevention: National Center for Chronic Disease Prevention and Health Promotion. Prevalence among U.S. adults of a metabolic syndrome associated with obesity findings from the Third NHANES Survey. Available from: http://www.cdc.gov/nccdphp/dnpa/obesity/trend/metabolic.htm.

25. Aguilar-Salinas CA, Rojas R, Gomez-Perez FJ, Mehta R, Franco A, Olaiz G, Rull JA. The metabolic syndrome: a concept hard to define. Archives of Medical Research. 2005;36:223–31.

26. Reaven GM. The insulin resistance syndrome: definition and dietary approaches to treatment. Annual Review of Nutrition. 2005;25:391–406.

27. Williams R, Airey M, Killilea T. Epidemiology and public health consequences of diabetes. Current Medical Research and Opinion. 2002;18:S1–S12.

28. Diabetes Control and Complications Trial Research Group. The effect of intensive treatment of diabetes on the development and progression of long-term complications in insulin-dependent diabetes mellitus. New England Journal of Medicine. 1993;329:977–86.

29. UK Prospective Diabetes Study (UKPDS) Group. Intensive blood-glucose control with sulphonylureas or insulin compared with conventional treatment and risk of complications in patients with type 2 diabetes (UKPDS 33). Lancet. 1998;352:837–53. Anderson JW, Kendall CW, Jenkins DJA. Importance of weight management in type 2 diabetes: Review with meta-analysis of clinical studies. Journal of the American College of Nutrition. 2003;22:331–9.

30. American Diabetes Association. Position Statement from the American Diabetes Association: nutrition principles and recommendations in diabetes. Diabetes Care. 2004;27:S36–S46.

**Check out the following sources for additional information.**

31. American Diabetes Association. Complete guide to diabetes. New York: Bantam Dell; 2002.

32. American Diabetes Association. Position statement from the American Diabetes Association: diagnosis and classification of diabetes mellitus. Diabetes Care. 2005;28:S37–S42.

33. American Diabetes Association. Position statement from the American Diabetes Association: nutrition principles and recommendations in diabetes. Diabetes Care. 2004;27:536–46.

34. American Diabetes Association. Position statement from the American Diabetes Association: standards of medical care in diabetes. Diabetes Care. 2005;28:S14–S36.

35. Anonymous. Translation of the diabetes nutrition recommendations for health care institutions: position statement. Journal of the American Dietetic Association. 1997;97:52–53.

36. Bliss M. The discovery of insulin. Edinburgh: Paul Harris Publishing, 1983.

37. National Center for Chronic Disease Prevention and Health Promotion, Diabetes Projects. Available from: http://www.cdc.gov/diabetes/projects/cda2.htm.

38. Costacou T, Mayer-Davis EJ. Nutrition and prevention of type 2 diabetes. Annual Review of Nutrition. 2003;223:147–70.

39. Expert Committee on the Diagnosis and Classification of Diabetes Mellitus. Follow-up report on the diagnosis of diabetes mellitus. Diabetes Care. 2003;26:3160–7.

40. Grundy SM. The optimal ratio of fat-to-carbohydrate in the diet. Annual Review of Nutrition. 1999;19:325–41.

41. Haffner S. Rationale for new American Diabetes Association Guidelines: are national cholesterol education program goals adequate for the patient with diabetes mellitus? American Journal of Cardiology. 2005;96:8E–10E.

42. Ludwig DS. The glycemic index: physiological mechanisms relating to obesity, diabetes, and cardiovascular disease. JAMA (Journal of the American Medical Association). 2002;287:2414–23.

43. MacNeill S et al. Rates and risk factors for recurrence of gestational diabetes. Diabetes Care. 2001;24:659–62.

44. Maki KC, Kurlandsky S. Syndrome X: a tangled web of risk factors for coronary heart disease and diabetes mellitus. Topics in Clinical Nutrition. 2001;16:32–41.

45. Roberts K, Dunn K, Jean SK, Lardinois CK. Syndrome X: medical nutrition therapy. Nutrition Reviews. 2000;58:154–60.

46. Roche HM, Phillips C, Gibney MJ. The metabolic syndrome: the crossroads of diet and genetics. Proceedings of the Nutrition Society. 2005;64:371–7.

47. Rohlfing CL, Wiedmeyer HM, Little RR, England JD, Tennill A, Goldstein DE. Defining the relationship between plasma glucose and $HbA_{1c}$: analysis of glucose profiles and $HbA_{1c}$ in the Diabetes Control and Complications Trial. Diabetes Care. 2002;25:275–8.

48. Stewart KJ. Exercise training and the cardiovascular consequences of type 2 diabetes and hypertension: plausible mechanisms for improving cardiovascular health. JAMA (Journal of the American Medical Association). 2002;288:1622–31.

49. U.S. Department of Agriculture, U.S. Department of Health and Human Services. 2005 Dietary Guidelines for Americans. Available from: http://www.healthierus.gov/dietaryguidelines/.

50. Wootten W, Turner RE. Macrosomia in neonates of mothers with gestational diabetes is associated with body mass index and previous gestational diabetes. Journal of the American Dietetic Association. 2002;102:241–3.

Chapter 6

# Protein

Proteins constitute the most abundant organic substance in the body, making up at least 50% of our dry weight. The body uses proteins from foods to make thousands of proteins in the body that maintain structure, direct and regulate metabolism, allow movement, provide a means of communication, and protect from infection. Although protein is found in a variety of foods, meat, fish, eggs, dairy products, and legumes such as dried beans, lentils, soybeans, and peanuts provide especially good sources.

The term "protein" was derived over 160 years ago from the Greek *proteios,* meaning "of first importance."[1] Proteins define our bodies, making each of us a unique human being. In addition, the new millennium has brought with it unprecedented advances in our understanding of how a person's genetic makeup, otherwise known as the human genome, can influence which proteins each of us makes. In this chapter you can learn what proteins are and how the body uses them to maintain health. You can also learn about the exciting new field of genomics and how it relates to protein nutrition.

Photos © Matthew Farruggio

# Introduction to Proteins

**Proteins** are macromolecules made from amino acids. **Amino acids** are joined together by **peptide bonds** to form hundreds of thousands of proteins. In total, humans need 20 different amino acids. Some proteins are very simple, containing only a few amino acids, whereas others contain thousands. However, most proteins are of intermediate size, having 250 to 300 amino acids. Proteins can be classified based on their number of amino acids: dipeptides have 2 amino acids, tripeptides have 3, and so forth. Proteins with 12 to 20 amino acids are called oligopeptides, and those with over 20 amino acids are called polypeptides.

## Amino Acids: The Building Blocks of Proteins

The numerous proteins in the body are very chemically diverse, because of which amino acids they contain and the ways they are linked together. Although at least 100 amino acids are found in nature, the body uses only about 20 different amino acids to make its proteins. Except for proline, which has a somewhat different, ringlike structure, all amino acids consist of three common components: (1) a central carbon bonded to a hydrogen, (2) an **amino group** $(-NH_2)$ containing nitrogen, and (3) a carboxylic acid group $(-COOH)$. In addition, each amino acid contains a unique side-chain group, called an **R-group.** The R-group makes each amino acid different from the others. Figure 6.1 shows a "generic" amino acid. Note that in the body the amino and carboxylic acid groups almost always exist in "charged" states.

Each amino acid is defined by its R-group, which can be as simple as a single hydrogen atom or as complex as an organic ring structure. Some R-groups also contain sulfur atoms. These subtle differences in the R-groups give each amino acid a unique chemical and physical nature. For example,

**protein** Nitrogen-containing macronutrient made from amino acids.

**amino acid** Nutrient composed of a central carbon bonded to an amino group, carboxylic acid group, and a side-chain group (R-group).

**peptide bond** A chemical bond that joins amino acids.

**amino group** $(-NH_2)$ The nitrogen-containing component of an amino acid.

**R-group** The portion of an amino acid's structure that distinguishes it from other amino acids.

## FIGURE 6.1  The Main Components of an Amino Acid

Amino acids have four main components: a central carbon, an amino group, a carboxylic acid group, and an R-group.

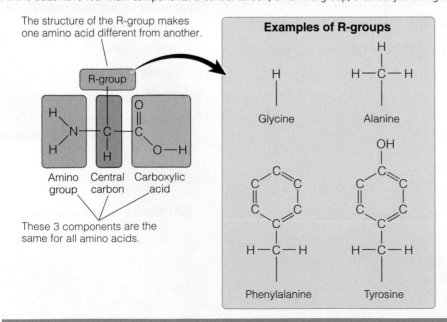
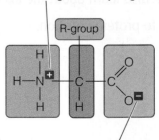

**TABLE 6.1    Essential and Nonessential Amino Acids**

| Essential | Nonessential |
|---|---|
| Histidine | Alanine |
| Isoleucine | Arginine[a] |
| Leucine | Asparagine |
| Lysine | Aspartic acid |
| Methionine | Cysteine[a] |
| Phenylalanine | Glutamic acid |
| Threonine | Glutamine[a] |
| Tryptophan | Glycine[a] |
| Valine | Proline[a] |
|  | Serine |
|  | Tyrosine[a] |

[a]These amino acids are also classified as conditionally essential.

some of the R-groups are negatively charged, some are positively charged, and some have no charge at all. The charges associated with R-groups help determine the final shape and function of the protein.

## Classification of Amino Acids

To make proteins required for the body to function, we need all 20 amino acids. However, the body can make about half of these. Thus, like the other classes of nutrients, amino acids can be categorized as essential, nonessential, or conditionally essential (Table 6.1).*[2] The 9 essential amino acids must be consumed in the diet, because the body cannot make them or cannot make them in required amounts. When necessary, the body makes the 11 nonessential amino acids from the essential amino acids or glucose. A common metabolic process for synthesizing nonessential amino acids, called **transamination,** involves the transfer of an amino group from 1 amino acid to another organic compound to form a different amino acid.

Many factors can influence the essentiality of amino acids. For example, some infants (especially those born early—premature infants) cannot make several of the nonessential amino acids. These amino acids are conditionally essential, because they must be obtained from the diet during this period of the life span. Also, certain diseases can cause a nonessential amino acid to become essential, such as the disease **phenylketonuria (PKU).** You can read more about PKU in Focus on Life Cycle Nutrition: Phenylketonuria (PKU). In all, six of the amino acids the body needs are conditionally essential in specific circumstances.

## Sources and Quality of Amino Acids and Proteins

In general, meat, poultry, fish, eggs, dairy products, legumes, and nuts provide more protein (per gram) than grains, fruits, and vegetables. However, even foods with the same amount of total protein may contain different amounts and varieties of amino acids. Foods of animal origin tend to have larger amounts of certain essential amino acids than do plant-derived foods. An exception to this rule is foods made from soy, which contain all the essential amino acids. Soybeans, like dried beans, lentils, peas, and peanuts, are

**transamination** The transfer of an amino group from one amino acid to another organic compound to form a different amino acid.

**phenylketonuria (PKU)** (phe – nyl – ke – ton – UR – i – a)  An inborn error of metabolism in which phenylalanine cannot be converted to tyrosine.

*Essential amino acids are also called indispensable amino acids, nonessential amino acids are called dispensable amino acids, and conditionally essential amino acids are called conditionally indispensable amino acids.

## *Focus On* LIFE CYCLE NUTRITION
### Phenylketonuria (PKU)

Most healthy adults need to consume the nine essential amino acids from their diets. The other amino acids can be synthesized in the body when needed. However, there are certain conditions when the body cannot synthesize one or more of the nonessential amino acids. When this occurs, these normally nonessential amino acids must be consumed in the diet and are considered conditionally essential. For example, people with a condition called phenylketonuria (PKU) lack the enzyme phenylalanine hydroxylase, which converts phenylalanine to tyrosine, making tyrosine conditionally essential.[3]

PKU is an example of an **inborn error of metabolism** and is due to an alteration in a person's genetic material (DNA).

Inborn errors of metabolism are a group of inherited disorders caused by a deficiency or absence of an enzyme involved in a metabolic pathway. About 1 in every 15,000 babies born in the United States each year is born with PKU. If untreated, PKU results in dangerously high circulating levels of phenylalanine metabolites as well as insufficient levels of tyrosine. Because this can cause serious health problems, including brain damage, most babies in the United States are tested for PKU shortly after birth. This is done with a simple blood test that measures the level of circulating phenylalanine. An infant found to have PKU is fed a special formula containing minimal amounts of phenylalanine and adequate amounts of tyrosine. Although human milk is the

best nutrition for almost all babies, in some conditions such as PKU breastfeeding may not be an option.[4] As the child grows and begins eating regular food, a phenylalanine-restricted diet must be continued. Because phenylalanine is found in almost all proteins, maintaining an appropriate diet can be difficult. However, because the complications of PKU can be completely prevented, early detection and dietary intervention are well worth the effort. Note that the artificial sweetener aspartame is made of phenylalanine and aspartic acid, and should not be consumed by people with PKU. The U.S. Food and Drug Administration (FDA) therefore requires that all products containing aspartame must be labeled with a warning.

Animal-derived foods such as meat, milk, fish, and eggs are excellent sources of complete protein; soy products also supply complete protein.

**inborn error of metabolism** A disease that is caused by the absence of an enzyme needed for metabolism.

**complete protein source** A food that contains all the essential amino acids in relative amounts needed by the body.

**incomplete protein source** A food that lacks or contains very low amounts of one or more essential amino acids.

**limiting amino acid** An essential amino acid in the lowest concentration in an incomplete protein source.

**protein complementation** Combining incomplete protein sources to provide all essential amino acids in relatively adequate amounts.

legumes. Legumes are special in that they are associated with bacteria that can take nitrogen from the air and incorporate it into proteins. Thus legumes tend to be good sources of protein. You can find out the protein content of many commonly consumed foods by consulting the food composition table that accompanies this text.

### Complete and Incomplete Proteins

Foods can be categorized according to their essential amino acid content. Those containing adequate amounts of all essential amino acids are said to be **complete protein sources,** whereas those supplying low amounts of one or more of the essential amino acids are **incomplete protein sources.** The amino acids that are missing or in low amounts are called **limiting amino acids.** For example, corn protein has low amounts of lysine and tryptophan—these are the limiting amino acids. The limiting amino acid in wheat protein is lysine.

### Protein Complementation

When combined, foods with incomplete proteins can provide adequate amounts of all the essential amino acids. This is called **protein complementation,** and is a customary dietary practice around the world, especially in societies that rely heavily on plant sources for protein.[5] Examples of commonly consumed meals representing protein complementation are rice and beans as well as corn and beans. Both rice and corn have several limiting amino acids (for example, lysine) but provide adequate methionine. Beans and other legumes are limited in methionine but provide adequate lysine. In general, protein complementation allows diets containing a variety of plant protein sources to provide all the essential amino acids. Recent advances in our under-

## Focus On FOOD
### Altering Protein Quality of Foods by Genetic Modification

The proteins produced by plants and animals are determined by their genetic makeup. Altering the genetic material of an organism can influence the proteins that it makes. For example, the protein quality of a food can be enhanced by the production of a **genetically modified plant** or **animal** (also called genetically modified organism, or GMO).[6] Genetic modification, also called biotechnology or genetic engineering, involves manipulating the genetic material of an organism to produce an altered protein. For example, a food with one or more limiting amino acids can be fortified with those amino acids by altering its genetic code. The result is the production of a modified food source providing a complete protein to the consumer. An example of a genetically modified organism (GMO) is the Opaque-2 corn plant.[7] To produce this modified corn, scientists altered the corn's genetic code so that it produces proteins with more lysine and tryptophan—the two limiting amino acids in unmodified corn.

Although a discussion of the many issues related to the genetic modification of our food supply is beyond the scope of this book, note that there is controversy about the development and use of GMOs. Some people worry that GMOs may increase risk of allergies, increase antibiotic resistance, and result in the unintended transfer of genetic materials to other organisms. Others have expressed concern that it is unethical to tamper with nature by mixing genes among species. In response, the U.S. Food and Drug Administration (FDA) has developed policies and guidelines that must be followed in developing new foods made with GMOs. For example, genetic engineering may not negatively alter the nutrient quality of the food. Currently, the FDA does not require foods made with GMOs to be so labeled, although those that do *not* contain GMOs can be so labeled.

standing of genetics now enable scientists to alter the amino acid composition of foods. You can read more about this in the Focus on Food feature.

### Protein Quality

The adequacy of food proteins for supplying the body with essential amino acids depends on several factors. These include the total amount of amino acids in food—whether the protein is complete or incomplete in its pattern of amino acids—and the body's ability to digest and absorb them.[8] A protein that provides easily digested and absorbed amino acids is said to have high bioavailability. If a food is a complete protein source and its protein is bioavailable, it is a **high-quality protein source.** In general, animal-derived foods are sources of high-quality protein. In contrast, foods containing incomplete proteins and/or those in which the protein has low bioavailability are **low-quality protein sources.** For example, protein from processed rice is of low quality, because it lacks an essential amino acid (lysine) and has low bioavailability. Thus people who rely solely on rice for their amino acid requirements become lysine deficient unless they consume extremely large amounts of rice.

© Polara Studios, Inc.

Protein complementation involves consumption of a variety of incomplete protein sources so that all essential amino acids are consumed.

## ESSENTIAL Concepts

Proteins are made from amino acids that are linked together with peptide bonds. Molecules with 2 amino acids are called dipeptides, those with 3 are tripeptides, and those with over 20 are polypeptides. Amino acids contain a central carbon, a carboxylic acid group, an amino group, and a side chain, or R-group. The R-group defines each amino acid's chemical nature. The body needs 20 amino acids, 9 of which are essential. However, in certain conditions a nonessential amino acid becomes essential, such as in phenylketonuria (PKU), in which tyrosine becomes conditionally essential. Foods that contain relatively

**genetically modified organism (GMO)** An organism (plant or animal) made by genetic engineering.

**high-quality protein source** A complete protein source with adequate amino-acid bioavailability.

**low-quality protein source** A food that is either an incomplete protein source or one that has low amino acid bioavailability.

high amounts of all the essential amino acids in the appropriate proportions are said to be complete protein sources. Foods lacking or having low amounts of at least 1 of the essential amino acids are said to contain incomplete proteins. In general, animal-derived foods contain complete proteins, whereas plant-derived foods contain incomplete proteins. Combining incomplete proteins can result in consumption of all the essential amino acids; this is called protein complementation.

# Protein Synthesis

The body takes the proteins in foods, breaks them down to amino acids, and then uses the amino acids to synthesize the exact proteins it needs. Like the millions of shapes that snowflakes take as they form in the sky, proteins acquire their distinct structures and shapes as they are made in our cells. Because a protein's shape determines its function, it is important to gain a basic understanding of how and why proteins acquire their many configurations. To understand this, one must first understand the process of protein synthesis. Three basic steps are involved in this process: (1) cell signaling, (2) transcription, and (3) translation. These steps are shown in Figure 6.2.

## FIGURE 6.2    The Steps of Protein Synthesis

Protein synthesis involves three basic steps: cell signaling, transcription, and translation.

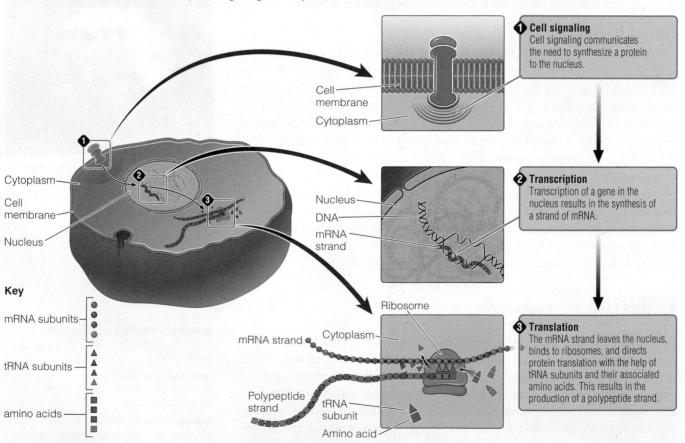

## Step 1: Cell Signaling

In the first step of protein synthesis, the cell receives a signal that tells it to make a protein. This process, called **cell signaling,** is often initiated by proteins or other compounds in the cell membrane. Cell signaling communicates extracellular conditions to the inside of the cell, just as an "indoor/outdoor" thermometer can be used to monitor the outside temperature from inside a house. The "turning on" of protein synthesis by cell signaling is called **up-regulation.** As you will see in later chapters, vitamins often up-regulate genes involved in cell growth and differentiation. Conversely, sometimes cells are instructed to "turn off" the synthesis of a certain protein. This is called **down-regulation.** Nutrients are involved in this process as well. Thus protein synthesis is constantly regulated through cell signaling. The body's ability to orchestrate the up- and down-regulation of genes via cell signaling allows it to produce only the proteins that it needs—and to stop producing proteins that are not needed.

## Step 2: Transcription

The up-regulation of a gene initiates the second step of protein synthesis, **transcription.** Transcription is similar to reading an instruction manual to put together a newly purchased product. In protein synthesis, the instruction manual is in the nucleus of each cell within strands of molecules made from deoxyribonucleic acid, or DNA. Each strand of DNA is called a **chromosome,** and each chromosome is subdivided into thousands of units called **genes.** Each gene provides instructions on how to assemble a particular polypeptide. In other words, a gene tells the cell which amino acids are needed, and which order they must be in to make a specific protein.

For transcription to happen, the DNA code must be communicated to organelles outside the cell nucleus. This is done by a chemical called **messenger ribonucleic acid (mRNA).** To participate in transcription, mRNA relocates from the cytoplasm into the nucleus. Next a series of mRNA subunits bind to the up-regulated gene, forming a strand of mRNA. The newly formed strand of mRNA separates from the DNA, leaves the nucleus, and re-enters the cytoplasm, where it participates in the next step of protein synthesis, called translation.

## Step 3: Translation

Once in the cytoplasm, the mRNA binds to a particle called a **ribosome** where the third step of protein synthesis, called **translation,** occurs. Translation requires another form of RNA, called **transfer ribonucleic acid (tRNA).** The tRNAs carry amino acids to the ribosomes for protein synthesis. For translation to proceed, the ribosome moves along the mRNA strand, "reading its sequence." The sequence of the mRNA, in turn, instructs specific tRNAs to transfer the amino acids they are carrying to the ribosome. One by one, amino acids join together via peptide bonds to form a growing polypeptide chain. When translation is complete, the newly formed protein detaches from the surface of the ribosome. However, this is not the last step in the creation of a new fully functional protein. The final shape of the protein is yet to be determined.

**cell signaling** The first step in protein synthesis in which the cell receives a signal to produce a protein.

**up-regulation** The "turning on" of protein synthesis.

**down-regulation** The "turning off" of protein synthesis.

**transcription** The process by which mRNA is made using DNA as a template.

**chromosome** A strand of DNA in a cell's nucleus.

**gene** (*gen-,* forming) A portion of a chromosome that codes for the primary structure of a polypeptide.

**messenger ribonucleic acid (mRNA)** A form of RNA involved in gene transcription.

**ribosome** A particle associated with the endoplasmic reticulum in the cytoplasm, involved in gene translation.

**translation** The process by which amino acids are linked together via peptide bonds on ribosomes, using mRNA and tRNA.

**transfer ribonucleic acid (tRNA)** A form of RNA in the cytoplasm involved in gene translation.

## ESSENTIAL *Concepts*

The series of steps involving cell signaling, transcription, and translation produce hundreds of thousands of different polypeptides in the body. These steps involve the genetic material DNA as well as mRNA, ribosomes, tRNA, and amino acids. Each gene in a DNA strand contains information about how a protein should be constructed. This information is transcribed into strands of mRNA in the nucleus and then is translated into a polypeptide chain via tRNAs and amino acids in the cytosol.

# Protein Structure

When the genetic material has been transcribed and translated, the polypeptide chain is released from the ribosome, resulting in the production of a protein with a specific sequence of amino acids. However, this is often not the last step in the creation of a functional protein—polypeptide chains can fold and combine to form very complex proteins. Because the final shape and components of a protein must be just right for the protein to work correctly, it is important to consider the many levels of protein structure.

## Primary Structure of Proteins

The number and sequence of amino acids in a single polypeptide chain is called the **primary structure** of a protein (Figure 6.3). The primary structure of a protein is determined by the DNA. Each protein has a unique primary structure, and is therefore a unique molecule.

One way to imagine the enormous number of proteins that can be made from 20 amino acids is to think of the English alphabet, which has a somewhat similar number of letters—26 in all. Think of the number of words that can be made with only 26 letters. Remember that all the words in a dictionary are spelled with these 26 letters. Some words are short, some are long. Some

### FIGURE 6.3    The Primary Structure of a Protein

The primary structure of a protein is the sequence and number of the amino acids in the polypeptide chain.

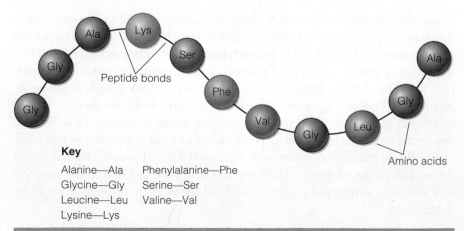

**Key**

| | |
|---|---|
| Alanine—Ala | Phenylalanine—Phe |
| Glycine—Gly | Serine—Ser |
| Leucine—Leu | Valine—Val |
| Lysine—Lys | |

**primary structure** The sequence of amino acids that make up a single polypeptide chain.

words contain just a few letters, others contain many different letters. The same holds true for proteins; some are short, some are long, some contain a handful of amino acids, whereas others may contain all 20 amino acids. The possibilities are seemingly endless.

## The Importance of Primary Structure

The amino acid sequence, or primary structure, is critical to a protein's function, because it determines the protein's basic chemical and physical characteristics. In other words, the primary structure represents the basic identity of the protein. Although some changes in the primary structure of a protein may not affect the function, others can profoundly affect the ability of the protein to do its job. Alterations in the primary structure can be caused by inherited genetic variations. An example of a disease caused by an inherited genetic variation is **sickle cell anemia** (also called sickle cell disease). The complications associated with sickle cell anemia are caused by a small "error" in the DNA resulting in the production of a defective protein.[9] Additional detail on sickle cell anemia is provided in the Focus on Clinical Applications feature.

**sickle cell anemia** A disease in which an alteration in the amino acid sequence of hemoglobin causes red blood cells to become misshapen and decreases the ability of the blood to carry oxygen and carbon dioxide.

**gene therapy** The use of altered genes to enhance health.

# Focus On CLINICAL APPLICATIONS
## Sickle Cell Anemia

Sickle cell anemia is a disease caused by a single error in the amino acid sequence of hemoglobin. More specifically, people with sickle cell anemia have an alteration, or mutation, in the gene that codes for one of the proteins that makes up hemoglobin. Hemoglobin is found in red blood cells, where it carries oxygen from the lungs to the body's tissues and carbon dioxide back to the lungs for removal. The sickle cell mutation causes the incorporation of an incorrect mRNA subunit during transcription. This results in the insertion of an incorrect amino acid (valine) for the correct amino acid (glutamic acid) during translation.

This seemingly small error in the amino acid sequence of the hemoglobin molecule completely alters its shape and has major health implications. Red blood cells with normal hemoglobin are smooth, and glide through blood vessels. Red blood cells with the abnormal hemoglobin become rigid, sticky, and shaped like a sickle. These crescent-shaped red blood cells—sickle cells—can become lodged in small blood vessels,

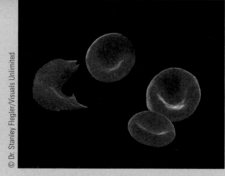

© Dr. Stanley Flegler/Visuals Unlimited

Normal red blood cells are disc-shaped, whereas those of people with sickle cell anemia are crescent-shaped.

block blood flow, and cause pain and serious damage to organs. Signs and symptoms of this disease include anemia (the inability of the blood to carry oxygen); pain in the chest, abdomen, and joints; swollen hands and feet; frequent infections; stunted growth; and vision problems.

Scientists hypothesize that, hundreds of years ago, the genetic alteration responsible for sickle cell anemia protected people from a deadly malaria epidemic that killed tens of thou-

sands of people in parts of Africa, the Mediterranean, the Middle East, and India. People with sickle cell anemia survived the malaria outbreak and were therefore able to reproduce and pass on their genetic code to their offspring. Today, millions of people all over the world have sickle cell anemia, especially those with African, Mediterranean, Middle Eastern, or Indian ancestry. In the United States sickle cell anemia is most common in people of black African heritage and is found in about 1,000 newborn infants each year; 1 in 500 African Americans has the disease.

Although there is currently no cure for sickle cell anemia, the potential technology for scientists to "correct" the defective DNA in the future is promising.[10] This technology, called **gene therapy,** may be able to cure this and other genetically based diseases. In the meantime, complications of this disease are treated with drugs to reduce pain, antibiotics to treat infection, blood transfusions, supplemental oxygen, and bone marrow transplants.

## Secondary Structure of Proteins

As you have learned, each protein has a unique primary structure. However, most proteins have a three-dimensional shape that is more complex than the primary structure's linear chain of amino acids. Because the backbone of the peptide chain is made of amino and carboxylic acid groups with positive and negative charges, the groups attract and repel each other like magnets, folding the protein into organized and predictable patterns. These interactions form weak chemical bonds, called hydrogen bonds, that twist and fold the primary structure, transforming it into a three-dimensional structure. This folding is called the **secondary structure** of a protein. The two most common folding patterns are the **α-helix** and **β-folded sheets**, illustrated in Figure 6.4. Note that Dr. Linus Pauling received the Nobel Prize in chemistry in 1954 for his contributions to understanding the α-helix structure.[11]

## Tertiary Structure of Proteins

The next level of protein complexity is called **tertiary structure,** which is the additional folding of the protein because of interactions between the R-groups. Note that this differs from secondary structure, which is caused by hydrogen bonding between regions of the amino and carboxylic acid groups. Some of

CONNECTIONS ▶ Recall that hydrogen bonds are weak electrical attractions that form between positively charged hydrogen atoms and other atoms with negative charges (Chapter 3, page 70).

**secondary structure** Folding of a protein because of hydrogen bonds that form between elements of the amino acid backbone (not R-groups).

**α-helix** (AL – pha – he – lix) A common configuration that makes up many proteins' secondary structures.

**β-folded sheet** (BE – ta – fol – ded) A common configuration that makes up many proteins' secondary structures.

**tertiary structure** (TER – ti – a- ry) Folding of a polypeptide chain because of interactions among the R-groups of the amino acids.

### FIGURE 6.4    The Secondary Structure of a Protein

Hydrogen bonds fold proteins into α-helix and β-folded sheet patterns, resulting in a secondary structure.

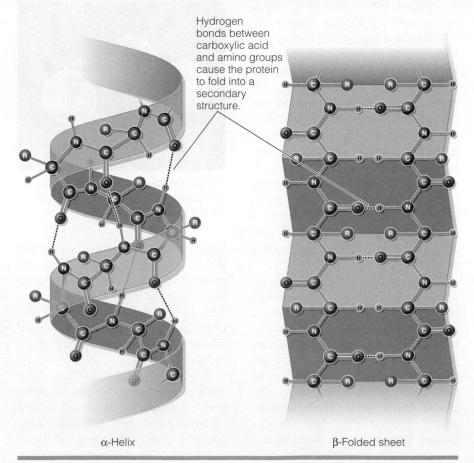

Hydrogen bonds between carboxylic acid and amino groups cause the protein to fold into a secondary structure.

α-Helix                                    β-Folded sheet

the strongest interactions between R-groups occur between amino acids that contain sulfur atoms (for example, cysteine), which react with other sulfur-containing amino acids to form disulfide bonds. These bonds and others are very stable, and anything that disrupts them can disrupt the protein's tertiary structure.

## Quaternary Structure of Proteins

The final level of protein structure is called **quaternary structure.** Quaternary structure consists of several polypeptide chains coming together to form the final protein, as is shown in Figure 6.5. Only proteins made from more than one polypeptide chain can have quaternary structure. In addition, there are often nonprotein components called **prosthetic groups** that must be precisely positioned among the polypeptide chains for the protein to function. Hemoglobin is an example of a protein with quaternary structure and a prosthetic group. Hemoglobin is made from four separate polypeptide chains, each of which combines with an iron-containing prosthetic group called a heme. Heme is the portion of the hemoglobin molecule that actually holds the oxygen and carbon dioxide gases as they are transported in the blood.

## Disruption of Protein Structure: Denaturation and Mercury Poisoning

A protein's final shape determines its ability to carry out its function. However, there are many conditions that can disrupt a protein's shape. One example is **denaturation.** Denaturation occurs when a protein unfolds in unusual ways. Compounds and conditions that cause denaturation are called denaturing agents and include physical agitation, heat, detergents, acids, bases, salts, alcohol, and heavy metals. Once denatured, a protein cannot resume its three-dimensional shape.

### FIGURE 6.5 The Quaternary Structure of Hemoglobin

Hemoglobin is made from four polypeptide chains and four iron-containing prosthetic groups called heme.

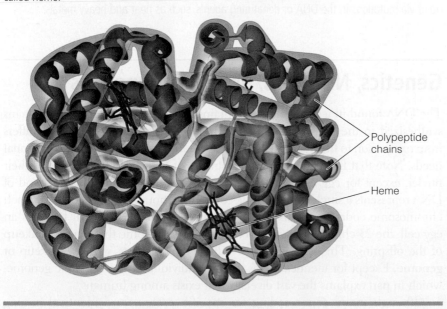

Polypeptide chains

Heme

**quaternary structure** (QUAT – er – nar – y) The combining of polypeptide chains with other polypeptide chains and/or prosthetic groups.

**prosthetic group** (*prostheses*, that which stands in for) A nonprotein component of a protein that is part of the quaternary structure.

**denaturation** The alteration of a protein's three-dimensional structure by heat, acid, enzymes, or agitation.

A familiar example of protein denaturation occurs when an egg white is heated; the proteins unfold, and the egg white changes from a thick, clear liquid to an opaque solid. Another example of a denaturing agent is mercury (a heavy metal), which can disrupt disulfide bonds and thus tertiary structure. High levels of mercury exposure can cause numbness, hearing loss, visual problems, difficulty walking, and severe emotional and cognitive difficulties, and most of these signs and symptoms are due to the denaturing actions of mercury. Interestingly, we now know that a condition termed "mad hatter's disease" that was widespread in the 19th century was due to mercury poisoning. During that period, hat makers used large amounts of mercury-containing compounds to treat felt and fur, and suffered the consequences in that they went "mad."

Because of the denaturing effects of heavy metals such as mercury, people are currently making efforts to decrease the concentrations of these compounds in the environment. For example, some are concerned that certain types of fish contain dangerously high amounts of mercury because of contaminated water. Although most fish are safe to consume, the FDA and the Environmental Protection Agency (EPA) have made the following recommendations for women who might become pregnant, women who are pregnant or lactating, and young children.[12]

- Do not eat shark, swordfish, king mackerel, or tilefish.
- Eat up to 12 ounces (2 average meals) a week of a variety of fish and shellfish that are lower in mercury, such as shrimp, canned light tuna, salmon, pollock, and catfish.
- Check local advisories about the safety of fish caught in local lakes, rivers, and coastal areas. If no advice is available, limit consumption to 6 ounces per week of fish from these waters, and do not consume any other fish during that week.

## ESSENTIAL *Concepts*

The sequence of amino acids making up a protein is called its primary structure. The peptide chain then folds into secondary and tertiary structures depending on the chemical nature of its unique amino acid sequence. Finally, peptide chains can join and combine with other substances such as minerals to form a quaternary structure. Disruption of a protein's shape can occur via mutations in the DNA or denaturing agents, such as heat and heavy metals.

## Genetics, Nutrition, and Nutrigenomics

The DNA found in a cell's nucleus contains the master plan for all the proteins synthesized in the body. Thus it is important to understand how DNA differs from one person to the next and how genetics can influence a person's nutritional needs. Note that human cells have 46 strands of DNA (chromosomes) in their nuclei, except for egg and sperm cells, which have 23 strands. Each strand of DNA represents a single chromosome containing thousands of genes. Thus each chromosome codes for thousands of proteins. When a sperm cell fertilizes an egg cell, the 23 chromosomes in each combine to become the **genetic makeup** of the offspring. This is how infants inherit their parents' genetic makeup or **genome**. Except for identical twins, no two individuals have the same genome, which in part explains the vast diversity that exists among humans.

Although many genes coding for specific proteins are identical among people, we all have modifications in other genes. For example, some people

**genetic makeup (genome)** The particular DNA contained in a person's cells.

have genes that result in the production of proteins needed to make eyes green, whereas others have genes that code for proteins that make eyes brown. These differences in genes result in different physical characteristics, such as eye color, that do not really affect health. However, alterations in other genes can have more important physiological consequences.

## Mutations and Altered Protein Synthesis

Sometimes alterations in our genes result in the formation of proteins with altered function. These modifications are called **mutations.** Some mutations influence metabolism, whereas others cause cells to experience uncontrollable growth, a condition that can lead to cancer. Although not all mutations are inherited, some can be. That is, alterations (mutations) in the DNA of egg and sperm cells can be passed on to the offspring. This is the case for mutations such as those causing PKU and sickle cell anemia, which is why both are considered inherited diseases.

## The Human Genome Project and Nutrigenomics

Until recently, people knew very little about the genes that make up the human genome. However, the Human Genome Project was carried out in the 1990s to describe all the genes within our chromosomes. The Human Genome Project, described in more detail in the Focus on the Process of Science feature, allows scientists to better understand how genetic variations and mutations influence protein synthesis, overall health, and risk for disease. In addition, the Human Genome Project has opened up the new field of nutrition research referred to as **nutrigenomics.**[13] Nutrigenomics is the study of how nutrition and genomics interact to influence health. Nutritionists hope that by understanding a person's nutritional needs, nutritional status, and genome, they will someday be able to optimize health and well-being by precisely matching dietary recommendations with the unique genetic makeup of the individual.

# ESSENTIAL *Concepts*

Most cells in the body contain genetic material, or DNA, in their nuclei. Except for egg and sperm cells, human cells contain 46 strands of DNA—each strand representing a chromosome. At conception, the parents' chromosomes are passed on to the embryo. These chromosomes constitute the offspring's genetic makeup or genome. An alteration in a cell's genome, called a mutation, can influence the cell's ability to produce a functional protein. Some mutations are harmless, perhaps influencing physical characteristics. Other mutations can be dangerous, such as those causing cancer. The study of how nutrition interacts with a person's genome is called nutrigenomics.

# Protein Digestion, Absorption, and Circulation in the Body

The process of digestion disassembles food proteins into amino acids that are then absorbed and circulated to cells where they are reassembled into the exact proteins that we need. In addition, intestinal enzymes, digestive secretions, and degraded cells (all of which are made of proteins) can be broken down and

**mutation** The alteration of a gene.

**nutrigenomics** (nu – tri – gen – O – mics) The science of how genetics and nutrition together influence health.

# *Focus On* THE PROCESS OF SCIENCE
## The Human Genome Project

Scientists, philosophers, and psychologists have long been interested in what makes one person different from the next. Although this question may represent the ultimate mystery, researchers continue to try to understand the physiological and chemical reasons for why we all differ. In the 1800s, Gregor Mendel, a Czechoslovakian monk, discovered that many physical characteristics could be passed on to offspring, thus launching the field of modern genetics. Genetics has progressed significantly since that time. For example, James Watson, Francis Crick and Maurice Wilkins received the Nobel Prize in physiology or medicine in 1962 for their discovery of the chemical structure of DNA. Since then the advances in human genetics have been unprecedented.

In 1990, the U.S. government, in conjunction with an international team of scientists and private industry involvement, launched the Human Genome Project to identify and understand all the genes making up human chromosomes. This information is invaluable for our understanding of genetic diversity and how it influences health and well-being. Our ability to study the human genome and what regulates protein synthesis has far-reaching implications to nutrition and health. At the completion of the Human Genome Project in the year 2000, scientists were surprised to learn that the human genome consists of as few as 30,000 genes. Previous estimates had placed this number between 50,000 and 140,000 genes. When most of the human genome had been published, Francis Collins, director of the Human Genome Project, described the importance of the information: "It's a history book—a narrative of the journey of our species through time. It's a shop manual, with an incredibly detailed blueprint for building every human cell. And it's a transformative textbook of medicine, with insights that will give health care providers immense new powers to treat, prevent and cure disease."[14]

Nutritional scientists can now directly test what they had long thought true: that nutrition interacts with genetics to influence health. We now know that dietary factors can influence whether a gene is turned on or turned off. How a person's genome and, ultimately, protein synthesis are influenced by nutrition has become a new scientific field in itself, called nutrigenomics. In the near future, you may be able to go to a health care provider and receive a specialized list of dietary recommendations based on your genome. For example, if your genes code for poorly functioning amino acid transport proteins, your estimated dietary amino acid requirements may be greater than those of most other people. In other words, you may receive personalized dietary prescriptions. This amazing concept is no longer science fiction, but is clearly the direction in which modern nutritional science is moving.

---

recycled, thus contributing another source of amino acids. The various stages of protein digestion are illustrated in Figure 6.6 and described here.

## Protein Digestion in the Stomach

Chemical digestion of protein begins in the stomach. When specialized cells in the stomach (gastric cells) come in contact with food, a hormone called gastrin is released. Gastrin enters the blood where it triggers the release of hydrochloric acid (HCl), pepsinogen, mucus, and other chemicals from other cells in the stomach. Remember from Chapter 4 that these substances are referred to as the gastric juices.

Hydrochloric acid (HCl) has two major tasks in protein digestion. First, it disrupts the bonds responsible for the protein's secondary, tertiary, and quaternary structure. This process of unfolding, or denaturation, straightens out the complex protein structure so that the peptide bonds can be exposed to digestive enzymes. Second, hydrochloric acid converts **pepsinogen** (also a component of the gastric juices) into its active form, called **pepsin.** Pepsinogen is an example of a **proenzyme,** which is an inactive form of an enzyme. Pepsin, the active enzyme, breaks peptide bonds, and is an example of a **protease.** Proteases are digestive enzymes that break peptide bonds via hydrolysis reactions.

**CONNECTIONS** Recall that hydrochloric acid (HCl) is released from parietal cells, whereas pepsinogen is released from chief cells (Chapter 4, page 101).

**pepsinogen** (pep – SIN – o – gen) The inactive form of pepsin.

**pepsin** (PEP – sin) (*peptein*, to digest) An enzyme needed for protein digestion.

**proenzyme** An inactive precursor of an enzyme.

**protease** An enzyme that cleaves peptide bonds.

**CONNECTIONS** Recall that hydrolysis is a chemical reaction in which chemical bonds are split by the addition of a water molecule (Chapter 3, page 77).

## FIGURE 6.6 Overview of Protein Digestion

Protein digestion occurs in both the stomach and small intestine, via pepsin, pancreatic proteases, and intestinal proteases.

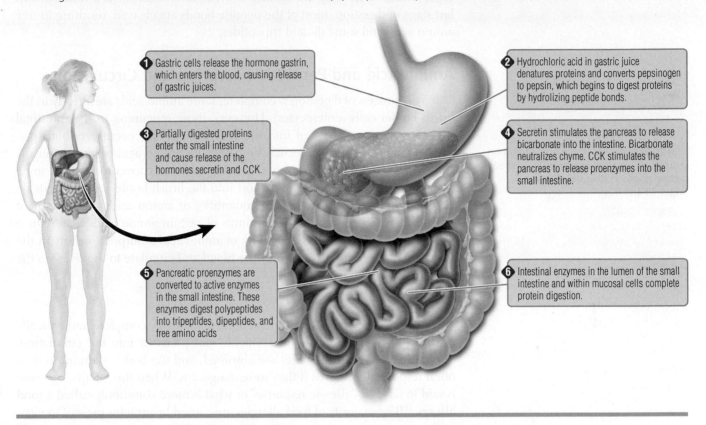

**1** Gastric cells release the hormone gastrin, which enters the blood, causing release of gastric juices.

**2** Hydrochloric acid in gastric juice denatures proteins and converts pepsinogen to pepsin, which begins to digest proteins by hydrolizing peptide bonds.

**3** Partially digested proteins enter the small intestine and cause release of the hormones secretin and CCK.

**4** Secretin stimulates the pancreas to release bicarbonate into the intestine. Bicarbonate neutralizes chyme. CCK stimulates the pancreas to release proenzymes into the small intestine.

**5** Pancreatic proenzymes are converted to active enzymes in the small intestine. These enzymes digest polypeptides into tripeptides, dipeptides, and free amino acids

**6** Intestinal enzymes in the lumen of the small intestine and within mucosal cells complete protein digestion.

As a result of the actions of gastric juices, proteins in foods are partially digested to shorter polypeptides and some free amino acids. The partially digested products of protein digestion are then ready to leave the stomach and enter the small intestine to be digested further.

## Protein Digestion in the Small Intestine

Amino acids and smaller polypeptides move from the stomach to the small intestine where they stimulate the release of the hormone secretin. Secretin signals the pancreas to release bicarbonate into the lumen of the duodenum, neutralizing the acid from the stomach and inactivating pepsin. The presence of amino acids in the small intestine causes specialized cells there to release the hormone cholecystokinin (CCK). Cholecystokinin stimulates the release of **trypsinogen, chymotrypsinogen, proelastase,** and **procarboxypeptidase,** all of which are proenzymes made in the pancreas. These inactive proenzymes are converted in the small intestine to their active protease forms: **trypsin, chymotrypsin, elastase,** and **carboxypeptidase.** Each of these active enzymes hydrolyzes specific peptide bonds. As a result, most proteins are broken down into di- and tripeptides and free amino acids. Note that the stomach and the pancreas do not produce the active protease enzymes. Instead, inactive proenzymes are produced and stored, thus protecting the stomach and pancreatic tissue from the protein-digesting functions of the enzymes produced there.

Next, the di- and tripeptides are broken down to amino acids by a multitude of enzymes produced in the mucosal cells making up the brush border of the

**CONNECTIONS** Recall that cholecystokinin (CCK) is a hormone produced by the duodenum that stimulates the release of pancreatic juice from the pancreas and bile from the gallbladder (Chapter 4, page 105).

**trypsinogen, chymotrypsinogen, proelastase,** and **procarboxypeptidase** Inactive proenzymes produced in the pancreas and released into the small intestine in response to CCK.

**trypsin, chymotrypsin, elastase,** and **carboxypeptidase** Active enzymes involved in protein digestion in the small intestine.

small intestine. In some cases, the enzymes are released into the lumen of the small intestine. However, most di- and tripeptides are transported into the enterocytes, where they are broken down to amino acids. As a result of this last stage of digestion, most of the peptide bonds are cleaved, resulting in free amino acids and some di- and tripeptides.

## Amino Acid and Peptide Absorption and Circulation

Once the process of digestion is complete, some amino acids are already in the brush border cells (enterocytes). However, those remaining in the intestinal lumen must be transported into the enterocytes. This occurs via both passive and active transport mechanisms. Because amino acids with chemically similar R-groups are often transported using the same carrier molecules, some compete with each other for transport into the brush border cell. Experts do not recommend that we take large quantities of amino acids in supplement form, because consuming large amounts of certain amino acids may reduce the bioavailability of others. The bulk of amino acid absorption occurs in the duodenum, where they then enter the blood and circulate to the liver via the hepatic portal system.

## Food Allergies

The breakdown of proteins to amino acids is relatively complete and typically results in the absorption of amino acids (not proteins) into the circulation. Sometimes larger polypeptides are absorbed, and the body's immune system often responds to them as if they were dangerous. When this occurs, a person is said to have an "allergic response" or what is more commonly called a **food allergy**.[15] The majority of food allergies are caused by proteins present in eggs, milk, peanuts, soy, and wheat. Reactions to food allergies can range from mild rashes to severe swelling in the throat and airways, resulting in difficulty with breathing. Researchers estimate that approximately 2% of adults and 5% of infants and young children in the United States suffer from food allergies, and the U.S. Food and Drug Administration (FDA) requires that all foods containing the most common allergens be thus labeled.

## ESSENTIAL *Concepts*

Protein digestion begins in the stomach where the gastric juices denature the protein via hydrochloric acid (HCl). Next, the enzyme pepsin begins the process of selectively hydrolyzing peptide bonds holding the amino acids together. Additional proteases produced in the pancreatic and intestinal cells complete the task of protein digestion. Amino acids and very small peptides are then transported into the cells lining the small intestine by both passive and active transport systems and circulated in the blood to the liver. Sometimes larger polypeptides are absorbed, triggering a food allergy.

## Functions of Amino Acids and Proteins in the Body

The body uses amino acids to synthesize the hundreds of thousands of proteins it needs. These proteins can be classified into general categories related to their function (Table 6.2). Note that these are not the only functions of proteins—

---

**CONNECTIONS** Recall that active transport requires the input of energy (ATP), whereas passive transport does not (Chapter 3, pages 79 and 80).

**food allergy** A condition in which the body's immune system reacts against a ptotein in food.

**TABLE 6.2    The Major Functions of Proteins in the Body**

| Function | Examples |
|---|---|
| Structure | Collagen |
| Catalysis | Enzymes (such as lingual lipase) |
| Movement | Muscle proteins (such as actin and myosin) |
| Transport | Membrane-bound transport proteins, albumin |
| Communication | Hormones (such as insulin), cell-signaling proteins |
| Protection | Skin proteins, fibrinogen, antibodies |
| Regulation of fluid balance | Albumin |
| Regulation of pH | Proteins with charged amino acids (such as hemoglobin) |

simply some of the most important. Furthermore, amino acids themselves serve vital purposes such as glucose synthesis and energy (ATP) production.

## Structure

Proteins provide most of the structural materials in the body. For example, they are important constituents of muscle, skin, bone, hair, and finger nails. An example of a structural protein is collagen, which forms a supporting matrix in bones, teeth, ligaments, and tendons. Proteins are also important structural components of cell membranes and cell organelles. The synthesis of structural proteins such as those in skeletal muscle is especially important during periods of active growth and development such as infancy and adolescence.

## Catalysis: Enzymes

Chemical reactions that occur in the body are all catalyzed by enzymes, all of which are proteins. Enzymes are needed for both catabolic and anabolic reactions. For example, amylase and pepsin catalyze hydrolytic (catabolic) reactions involved in digestion. Without enzymes, none of the body's vital digestive or metabolic reactions would occur. The role of specific enzymes in energy metabolism is discussed further in Chapter 8.

## Movement: Muscle Proteins

Movement results from the contraction and relaxation of skeletal muscles, which are under voluntary control, and of smooth muscles, which assist the function of the internal organs and blood vessels. Special proteins in muscle tissue, called actin and myosin, govern this movement. Nearly half of the body's protein is present in skeletal muscle, and adequate protein intake is required to form and maintain muscle mass throughout life.

## Transport: Membrane-Bound Proteins and Transport Proteins

Transport proteins are responsible for carrying substances into and around the body, as well as across cell membranes. For example, many nutrients require transport proteins to cross the cell membrane of the brush border, and water-insoluble nutrients are often attached to proteins so they can be transported in the blood. Examples of transport proteins include hemoglobin, which transports gases (oxygen and carbon dioxide), and albumin, which transports lipids.

## Communication: Hormones and Cell-Signaling Proteins

Tissues and organs have a variety of ways to communicate with each other, and most of these involve proteins. Although some hormones are not proteins, most of those that we have already discussed are (for example, secretin, gastrin, insulin, and glucagon). The others, called steroid hormones, include many of the reproductive hormones such as estrogen and testosterone. There are also special proteins embedded in cell membranes that communicate information about the extracellular environment to the cytoplasm and the nucleus. For example, some of these proteins are involved in the cell-signaling process that initiates protein synthesis itself. Others regulate cellular metabolism. Together, hormones and cell-signaling proteins make up much of the body's communication system.

## Protection: Skin, Blood Clotting, and Antibodies

Many proteins are involved in protecting the body from physical danger and infection. At the forefront is the skin, which is made mainly of proteins and forms a barrier between the outside world and the internal environment. Note that skin is a multifunctional organ, providing both structure and protection. In addition, proteins can provide other forms of protection. For example, if you cut yourself, blood clots close off this possible entry point for infection. In this way, blood clotting acts as a second level of defense. Blood clotting is carried out by a protein called fibrinogen. However, if a bacterium or other foreign substance does enter the body, the immune system fights back by producing proteins called **antibodies.** Antibodies bind foreign substances so they can be destroyed. Thus, because proteins provide several levels of protection, protein deficiency makes it difficult for the body to prevent and fight certain diseases.

## Regulation of Fluid Balance

Most of the body is made of water. This important fluid is found both inside of cells (intracellular space) and outside of cells (extracellular space). In addition, the extracellular space can be subdivided into that found in blood and lymph vessels (intravascular fluid) and between cells (interstitial fluid). The amount of fluid in these spaces is highly regulated by a variety of means, some of which involve proteins. For example, a protein called **albumin** is present in the blood in relatively high concentrations. In part because of the blood pressure exerted with each beat of the heart and the narrowness of the capillaries, fluid and nutrients in the blood get pushed out of the capillaries into the interstitial space. However, albumin remains in the blood vessels, gradually increasing in concentration as more fluid is lost. When the albumin concentration reaches a certain level in the blood, albumin draws some of the interstitial fluid back into the blood vessel via osmosis. This is illustrated in Figure 6.7. Severe protein deficiency can impair the body's ability to produce adequate albumin, resulting in the accumulation of fluid in the interstitial space, which is called **edema.**

## Regulation of pH

The body must maintain a certain pH or acidity for optimal function. As you learned in Chapter 3, the pH of a fluid is determined by its hydrogen ion ($H^+$) concentration, and the body's fluids are kept in specific pH ranges. For

**CONNECTIONS** Recall that osmosis is a passive transport system in which water moves from a region of low solute concentration to one of high solute concentration (Chapter 3, page 80).

**antibody** A protein, produced by the immune system, that helps fight infection.

**albumin** A protein important in regulating fluid balance between intravascular and interstitial spaces.

**edema** (e – DE – ma) The buildup of fluid in the interstitial spaces.

## FIGURE 6.7 Regulation of Fluid Balance by Albumin

Albumin is a protein in the blood that helps regulate fluid balance between the intravascular space and the interstitial space.

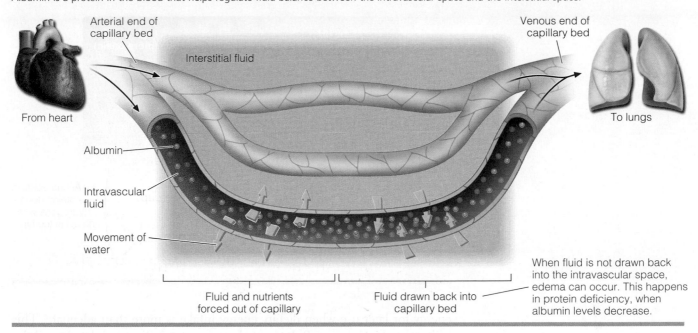

Arterial end of capillary bed

Venous end of capillary bed

Interstitial fluid

From heart

To lungs

Albumin

Intravascular fluid

Movement of water

Fluid and nutrients forced out of capillary

Fluid drawn back into capillary bed

When fluid is not drawn back into the intravascular space, edema can occur. This happens in protein deficiency, when albumin levels decrease.

example, whereas the pH range for the stomach fluid is 1.5 to 3.5, that for the blood is 7.3 to 7.5. One way that the pH of the blood is maintained is through the buffering action of proteins such as hemoglobin. Recall that components of amino acids, including the R-groups, carry charges. Put another way, they can accept and donate charged hydrogen ions easily. When the hydrogen ion concentration in the blood is too high (acidic), proteins can bind excess hydrogen ions. This restores the blood to the proper pH. Conversely, proteins can release hydrogens into the blood when the hydrogen concentration is too low (basic). This is illustrated in Figure 6.8. During periods of severe protein deficiency, the body has difficulty maintaining the correct pH balance.

**CONNECTIONS** Recall that buffers are able to both accept and donate hydrogen ions (H⁺), thus maintaining pH in a specific range (Chapter 3, page 74).

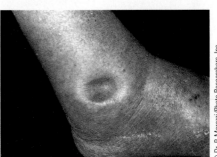

© Dr. P. Marazzi/Photo Researchers, Inc.

**Because protein is needed for fluid balance, severe protein deficiency can cause edema.**

## Additional Roles for Amino Acids

Aside from their roles in protein synthesis, amino acids themselves serve many purposes in the body. Some (such as leucine and glutamine) regulate protein breakdown, others (such as arginine) are involved in cell signaling, and still others (such as tryptophan and glutamate) are converted to chemical messengers called neurotransmitters. Some amino acids can also stimulate or inhibit the activities of enzymes involved in metabolism. Amino acids also provide nitrogen for the synthesis of many important nonprotein, nitrogen-containing compounds such as DNA and RNA. Thus amino acids are needed not only for protein synthesis but for a multitude of other functions as well.

## Amino Acids as a Source of Glucose and Energy (ATP)

Some amino acids can also be used for glucose synthesis and energy (ATP) production as well as energy storage as fats. Together, these processes allow the body to (1) maintain blood glucose at appropriate levels and (2) store excess

### FIGURE 6.8 Proteins Act as Buffers

Proteins accept and donate hydrogen ions (H⁺), helping to maintain pH.

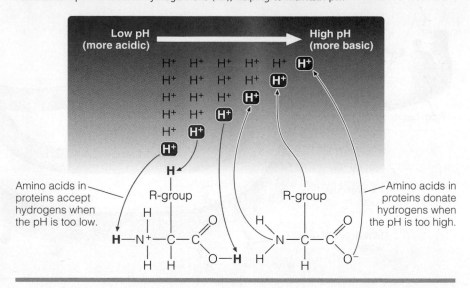

### Use of Amino Acids During Times of Need

energy for later use when dietary energy intake is more than adequate. This is illustrated in Figure 6.9. Integrated energy metabolism as related to all the macronutrients, including protein, is described in more detail in Chapter 8.

### Use of Amino Acids During Times of Need

When the body's available supply of energy (ATP) is low, it first turns to glycogen and fatty acids. However, when glycogen is depleted and fatty acid reserves are low, the body turns to its protein for help. To provide glucose and ATP to cells,

### FIGURE 6.9 Protein and Energy Metabolism

The fate of dietary protein depends on the body's need for glucose and energy (ATP).

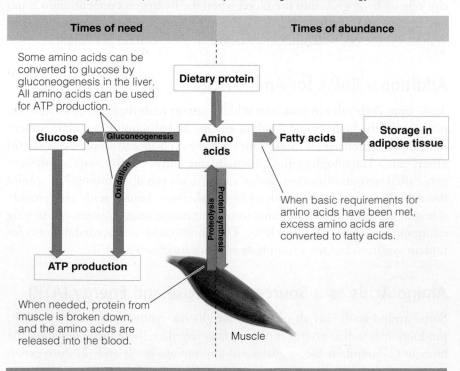

the body can disassemble proteins and convert some amino acids, called **gluco-genic amino acids,** to glucose via gluconeogenesis. In addition, many cells can harvest the energy stored in amino acids by oxidizing them directly. Oxidation of 1 g of protein yields approximately 4 kcal of energy as ATP. Thus, consuming 10 g of protein in the diet is roughly equivalent to consuming 40 kcal.

### Use of Amino Acids in Times of Abundance

During times of glucose and energy excess, the body redirects the flow of amino acids away from gluconeogenesis and ATP producing pathways and instead converts them to lipids. To do this, the nitrogen-containing amino group of each amino acid is removed in the liver via a process called **deamination.** The remaining carbon skeletons are then converted to lipids and stored in adipose tissue. Thus eating extra protein during times of glucose and energy sufficiency contributes to fat stores, not to muscle growth.

## ESSENTIAL *Concepts*

Amino acids are used to synthesize proteins needed for structure, catalysis, movement, transport, communication, and protection. All enzymes are proteins, and muscle is made primarily of the proteins actin and myosin. Many hormones and transport molecules are also made of proteins, as are antibodies important for immunity. Proteins such as albumin regulate fluid balance between intravascular and interstitial spaces, whereas others help maintain appropriate pH. Amino acids, themselves, also have diverse roles such as regulation of metabolic reactions and formation of nitrogen-containing, nonprotein substances such as neurotransmitters, DNA, and RNA. Amino acids also play a key role in energy (ATP) production being converted to glucose and broken down for energy (ATP) during periods of need. During times of energy abundance, amino acids are transformed into fat and stored in adipose tissue.

# Protein Turnover, Urea Excretion, and Nitrogen Balance

Although proteins in the body serve many functions, they eventually wear out. Fortunately, the body is able to recycle and reuse most of the amino acids from these proteins to synthesize new ones. In fact, most of the amino acids used for protein synthesis are obtained from the breakdown, or degradation, of the body's own proteins. The processes by which our proteins are continuously broken down and resynthesized are collectively known as **protein turnover.** Regulation of protein turnover allows the body to adapt to periods of growth and development and to maintain relatively stable amounts of protein during adulthood.

## Protein Turnover

Although we have a dietary requirement for protein, many of the amino acids used for protein synthesis in the body come from recycled proteins. About half of these amino acids come from protein turnover in the liver and intestine, where metabolic activity is exceptionally high.[16] Protein degradation (proteolysis) is catalyzed by special protein-cleaving enzymes and results in the release of amino acids into what is called the body's **labile amino acid pool.** Of course, dietary amino acids also contribute to the body's labile amino acid pool.

**CONNECTIONS** Recall that *gluconeogenesis* refers to the synthesis of glucose from non-carbohydrate sources (Chapter 5, page 149).

**glucogenic amino acid** An amino acid that can be converted to glucose via gluconeogenesis.

**deamination** The removal of an amino group from an amino acid.

**protein turnover** The balance between protein degradation and protein synthesis in the body.

**labile amino acid pool** In the body, amino acids that are immediately available to cells for protein synthesis and other purposes.

Regulation of protein turnover is complex, being controlled by both hormonal and dietary factors. For example, high concentrations of the hormone insulin decrease protein degradation and increase protein synthesis. In contrast, thyroid hormone and cortisol increase protein turnover by stimulating protein degradation and inhibiting protein synthesis. This is important for maintaining sufficient amounts of blood glucose during periods of stress. That is, adequate levels of amino acids must be available for gluconeogenesis when glucose is needed.

## Nitrogen Excretion as Urea

Protein turnover requires proteins to be continually broken down and resynthesized. During times of energy need, amino acids can be removed from the labile amino acid pool and converted to glucose or used as a source of energy. For this to occur, amino acids must first be deaminated, resulting in the release of ammonia ($NH_3$). Because ammonia is toxic to cells, it is quickly converted in the liver to a less toxic substance called **urea**. Urea is released into the blood, filtered out by the kidneys, and excreted in the urine. This is shown in Figure 6.10.

### FIGURE 6.10    Urea Synthesis and Excretion

Urea is synthesized in the liver and excreted by the kidneys in urine.

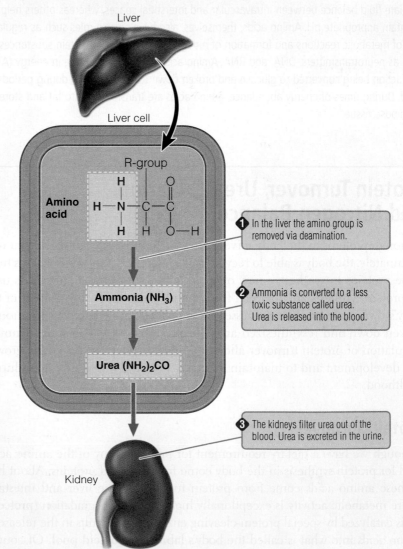

Liver

Liver cell

R-group

Amino acid

**1** In the liver the amino group is removed via deamination.

**Ammonia (NH$_3$)**

**2** Ammonia is converted to a less toxic substance called urea. Urea is released into the blood.

**Urea (NH$_2$)$_2$CO**

**3** The kidneys filter urea out of the blood. Urea is excreted in the urine.

Kidney

**urea** (u – RE – a) A relatively nontoxic, nitrogen-containing compound that is produced from ammonia after deamination.

## Nitrogen Balance

To study protein turnover, scientists often determine whether an individual is losing, gaining, or maintaining protein (or nitrogen). This is done by measuring both protein intake and the amount of nitrogen contained in body secretions such as urine, sweat, and feces.[17] When protein (or nitrogen) loss equals protein intake, a person is in **nitrogen balance. Negative nitrogen balance** occurs when nitrogen loss exceeds intake; this can occur during starvation or illness. When nitrogen intake exceeds loss, as occurs during growth or recovery from illness, the person is in **positive nitrogen balance.**

Growing children are in a state of positive nitrogen balance.

## ESSENTIAL *Concepts*

Protein turnover represents the balance between protein synthesis and degradation. When protein (or nitrogen) intake exceeds loss, the body is in positive nitrogen balance. This typically occurs during periods of growth. When nitrogen loss is greater than intake, the body is in negative nitrogen balance. This occurs during illness and starvation. When amino acids are degraded, they undergo deamination. The resultant ammonia is converted to urea in the liver and excreted in the urine.

# Amino Acid and Protein Requirements and Dietary Recommendations

Dietary protein is required for two major reasons: (1) to supply adequate amounts of the essential amino acids and (2) for the additional nitrogen needed to make the nonessential amino acids and other nonprotein, nitrogen-containing compounds such as DNA. Thus recommendations for dietary amino acids as well as overall protein consumption have been established to assure adequate intake of essential amino acids as well as total nitrogen.

## Dietary Reference Intakes (DRIs) for Amino Acids and Proteins

The Recommended Dietary Allowances (RDAs) for the essential amino acids are provided in Figure 6.11.[18] These values represent relative requirements of the essential amino acids for a given body size. Because there is no compelling evidence that high intake of essential amino acids poses health risks, the Institute of Medicine did not establish Tolerable Upper Intake Levels (ULs) for them.

Dietary Reference Intakes (DRIs) for protein have also been published. However, as you have learned, not all protein sources are created equal. Researchers nonetheless generally agree that most diets in affluent countries such as those consumed in the United States provide a balanced mix of essential amino acids. Therefore, the dietary recommendations for protein intake do not distinguish between people who consume high-quality proteins and those that do not.

The Institute of Medicine has established RDAs for protein, and these are listed inside the front cover of this book. These values are provided

**nitrogen balance** The condition in which protein (nitrogen) intake equals protein (nitrogen) loss by the body.

**negative nitrogen balance** The condition in which protein (nitrogen) intake is less than protein (nitrogen) loss by the body.

**positive nitrogen balance** The condition in which protein (nitrogen) intake is greater than protein (nitrogen) loss by the body.

CONNECTIONS Dietary Reference Intakes (DRIs) are reference values for nutrient intake. Recommended Dietary Allowances (RDAs) are set so that 97% of people have their needs met. Adequate Intake (AI) values are provided when RDAs cannot be established. Tolerable Upper Intake Levels (ULs) indicate intake levels that should not be exceeded (Chapter 2, page 36).

## FIGURE 6.11   The RDAs for the Essential Amino Acids in Adults

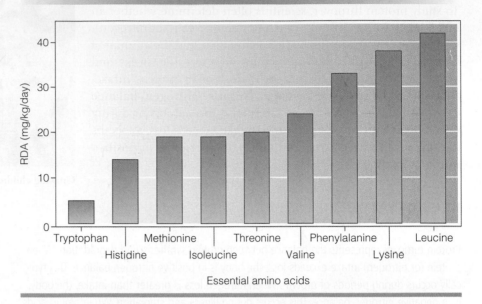

in two ways—the first (grams per day) is for a "typical" person in the particular population group, and the second (grams per kilogram of body weight per day) adjusts for body weight. The Institute recommends that healthy adults consume approximately 0.8 g of protein for each kilogram of body weight. During infancy, the most rapid phase of growth, protein requirements are relatively high when considered on a per body weight basis. Protein requirements are also relatively high during pregnancy and lactation, because additional protein is needed to support growth and milk production.[19] Recommended protein intakes are also somewhat higher for males than females because, in general, males are larger than females and have more muscle mass. These estimates of protein requirements apply only to healthy individuals. People recovering from trauma (such as burns) or illness may require more protein.[20] Because healthy people show little evidence of harmful effects of high protein intake, no UL values are set for this macronutrient.

## Do Athletes Need More Protein?

Although many people believe that individuals who participate in high levels of physical exercise necessarily have higher protein requirements, there is very little evidence to support this. In fact, the DRI committee that established the recommendations for amino acid and protein intake carefully considered this question. They concluded that given the lack of compelling evidence to the contrary, healthy adults undertaking these activities require similar amounts of protein on a body weight basis. Thus an adult athlete can generally estimate his or her protein requirement the same as other adults by using the conversion factor of 0.8 g protein per kilogram of body weight. The special nutritional needs of athletes are described in more detail in the Nutrition Matters following Chapter 8, and the use of protein supplements by athletes is discussed in the Focus on Sports Nutrition feature on the next page.

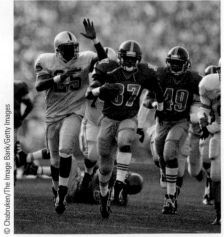

Athletes with increased muscle mass, such as football players and body builders, can estimate their protein requirements using the same equations used by nonathletes. All nonpregnant, nonlactating adults need about 0.8 grams protein for each kilogram of body weight.

© Chabruken/The Image Bank/Getty Images

## Focus On SPORTS NUTRITION
### Protein and Amino Acid Supplements

Optimizing performance is the ultimate goal for most athletes. Indeed, coaches and trainers have long sought training regimens and dietary alterations that increase strength, speed, and agility. Methods thought to increase athletic performance are called ergogenic aids. One example of an ergogenic aid that has gained significant popularity is the use of protein and amino acid supplements. Because protein is the major constituent of muscle, one could argue that increased protein or amino acid intake should result in greater muscle mass and, ultimately, increased strength and performance. This relatively simple idea has motivated the nutritional supplement industry to produce and sell a variety of protein and amino acid powders and drinks.

Although it is tempting to believe that protein or amino acid supplements may increase muscle growth, supporting evidence is weak at best. Keep in mind that before muscle actually grows, the rate of protein synthesis must increase. The synthesis of specific proteins is highly regulated and requires an initial step called cell signaling. In other words, something must initiate the complex processes resulting in DNA transcription and translation. It is now clear that increased dietary protein or amino acid intake does not, by itself, signal these processes—the presence of amino acids in the blood does not mean they will be taken up and synthesized into muscle proteins. In fact, contrary to long-held belief, research now suggests that physical activity, and especially resistance exercise, may actually decrease a

person's dietary protein requirements. This is because physical activity triggers the body to become more efficient in its use of amino acids and proteins, resulting in decreased protein turnover and ultimately in a decreased requirement for dietary protein. Clearly, muscle growth and maintenance is a complex process, and more research is needed before we fully understand how exercise affects the protein requirements of athletes.

So, can protein or amino acid supplements give you that winning edge? Right now, there is little evidence to support this notion.[21] The bottom line: (1) Eat a well-balanced diet providing sufficient energy with an appropriate mix of carbohydrates, fats, and protein; and (2) train long and hard. You can use that money that you saved on supplements to treat yourself to a wonderful meal after the game or race.

## Additional Recommendations for Protein Intake: AMDRs, USDA Dietary Guidelines, and MyPyramid

Several other sets of recommendations are also available for protein intake. For example, the Institute of Medicine's Acceptable Macronutrient Distribution Ranges (AMDRs) recommend that adults consume 10 to 35% of their energy intake as protein. However, this is reasonable only if energy intakes are appropriate. For example, consider an adult with an energy requirement of 2,000 kcal/day. This translates to a protein requirement of about 200 to 700 kcal (50 to 175 g) of protein. However, if this person consumes only 500 kcal, his or her protein requirement does not somehow become 50 to 175 kcal (13 or 44 g). Thus it is important to remember that the AMDR percentages assume that caloric intake is sufficient. This is true for all the macronutrients.

The U.S. Department of Agriculture's (USDA) Dietary Guidelines for Americans and accompanying MyPyramid food guidance system provide additional recommendations concerning intake of high-protein foods such as dairy products, meat, and dried beans (legumes). Aside from supporting the AMDR for protein (10 to 35% of calories from protein), the Guidelines specifically encourage intake of fat-free or low-fat milk, lean meats, and legumes. These food groups represent nutrient-dense, high-protein foods. More specifically, the USDA Food Guide and MyPyramid recommend 2 to 7 ounces of lean meat or legumes and 2 to 3 cups of fat-free or low-fat dairy products daily depending on caloric needs. To determine precisely how many servings are

## Food Matters

*Putting the Guidelines into Perspective*

The USDA Dietary Guidelines recommend that we meet our protein requirements by choosing and preparing lean meat and poultry, including legumes, nuts, and seeds in our meals, and by selecting low-fat or fat-free milk or milk products. This helps assure adequate amino acid intake while minimizing calories, total fat, saturated fat, and cholesterol. The following selection and preparation tips will help you meet your protein requirements.

- Eat low-fat ice cream or frozen yogurt instead of regular varieties.
- Select high-protein foods that are labeled as being "low fat" or "fat free."
- Instead of foods with lower nutrient densities such as chips, choose reasonable amounts of nuts and seeds as snack items.
- When comparing similar foods, choose higher-protein, lower-fat options by comparing Daily Values found on Nutrition Facts labels.
- Add slivered almonds or other nuts to steamed vegetables.
- Trim visible fat from meats and poultry before cooking.
- Instead of frying, broil, grill, roast, poach, or boil meat, poultry, or fish, and drain off any fat that appears during cooking.

recommended for you, visit the MyPyramid website (http://www.mypyramid .gov). Suggestions for including low-fat, high-protein foods in the diet are provided in the Food Matters feature.

## ESSENTIAL *Concepts*

Dietary Reference Intakes (DRIs) have been established for both the essential amino acids and total protein. In general, requirements are greater for males than females and are highest during infancy, pregnancy, and lactation. It is recommended that healthy adults consume approximately 0.8 g of protein for each kilogram of body weight. This value is the same whether a person is physically active or not. There are no Tolerable Upper Intake Levels (ULs) for the amino acids or protein. The Institute of Medicine's Acceptable Macronutrient Distribution Ranges (AMDRs) recommend that adults consume 10 to 35% of energy intake from protein. The USDA Dietary Guidelines for Americans and MyPyramid food guidance system suggest that we meet this goal by choosing a variety of protein sources such as low-fat dairy foods, lean meats, and legumes.

## Vegetarian Diets

People have many different reasons for deciding which foods they will and will not eat. This seems especially true about meat and other animal products. For example, some religious groups avoid some or all meats and animal products. In addition, economic considerations and personal preference can also determine both whether people eat meat and which meats they choose. Consequently, there are many dietary patterns related to which animal products people consume or do not consume. Because animal products tend to provide most high-quality protein in the diet, it is important to consider the effect of animal food consumption (or lack thereof) on issues related to nutritional status.

## Forms of Vegetarianism

The term **vegetarian** (from the Latin *vegetus*, meaning "whole," "sound," "fresh," "lively") was first used in 1847, by the Vegetarian Society of the United Kingdom, to refer to a person who does not eat meat, poultry, and fish or their products, such as milk and eggs. Today most people who consider themselves vegetarians consume dairy products and eggs.[22] These vegetarians are called **lacto-ovo-vegetarians.** Alternatively, **lactovegetarians** include dairy products, but not eggs, in their diets. Vegetarians who avoid all animal products are referred to as **vegans.**

## Dietary Implications of Vegetarian Diets

In general, a well-balanced lacto-ovo- or lactovegetarian diet can easily provide adequate protein, energy, and micronutrients to the body. This is because dairy products and eggs are convenient sources of high-quality protein and many vitamins and minerals. However, because meat is often the primary source of bioavailable iron in most people's diets, eliminating it can result in iron deficiency. Although a well-balanced vegan diet that includes a variety of complementary proteins can supply enough of the essential amino acids, vegans may be at increased risk of being deficient in several micronutrients, including calcium, zinc, iron, and vitamin $B_{12}$.[23] This is especially true during pregnancy, lactation, and periods of growth and development such as infancy and adolescence.[24] In Chapters 10 to 13 you will learn more about the dietary sources and functions of these micronutrients.

## Dietary Recommendations for Vegetarians from the USDA MyPyramid System

The MyPyramid food guidance system specifically recognizes protein, iron, calcium, zinc, and vitamin $B_{12}$ as nutrients for vegetarians to focus on, making specific recommendations as to how to get adequate amounts. In addition, it is recognized that some meat replacements, such as cheese, can be very high in calories, saturated fat, and cholesterol. The following comments and suggestions are offered to help assure optimal health for vegetarians.

- Build meals around protein sources that are naturally low in fat, such as legumes.
- Minimize the amount of high-fat cheese used as meat replacements.
- Consume calcium-fortified, soy-based beverages. These can provide calcium in amounts similar to milk and are usually low in fat and do not contain cholesterol.
- Add vegetarian meat substitutes (such as tofu) to soups and stews to boost protein without adding saturated fat and cholesterol.

**vegetarian** A person who does not consume any or selected foods and beverages made from animal products.

**lacto-ovo-vegetarian** A type of vegetarian who consumes dairy products and eggs in an otherwise plant-based diet.

**lactovegetarian** A type of vegetarian who consumes dairy products (but not eggs) in an otherwise plant-based diet.

**vegan** (VE – gan) A type of vegetarian who consumes no animal products.

- Recognize that most restaurants can accommodate vegetarian modifications to menu items by substituting meatless sauces, omitting meat from stir-fries, and adding vegetables or pastas in place of meat.
- Consider eating out at Asian or Indian restaurants, as they often offer a varied selection of high-protein, vegetarian dishes.

The key to a healthy vegetarian diet, as with any diet, is to enjoy a wide variety of foods. Because no single food provides all the nutrients the body needs, eating a wide variety can help ensure that a vegetarian gets the necessary nutrients and other substances that promote good health.

## ESSENTIAL *Concepts*

There are varying degrees of being vegetarian, from vegans who eat no animal products to lacto-ovo-vegetarians who consume dairy and egg products. Consuming adequate amounts of protein is typically not difficult even for the strictest of vegetarians. However, consuming a vegetarian diet does increase a person's risk for selected micronutrient deficiencies such as calcium, zinc, iron, and vitamin $B_{12}$. Consuming a wide variety of nutrient-dense foods can help vegetarians get enough of these nutrients.

# Protein Deficiency

Although generally not a concern in industrialized countries, protein deficiency is common in regions where the amount and variety of foods is limited. This is especially true for children. Protein deficiency is also seen in adults with some debilitating conditions. Because of the importance of proteins and amino acids for optimal health, protein deficiency can have significant health implications.

## Protein Deficiency in Early Life

Protein deficiency is rare during the first months of life when most infants are consuming most of their energy from human milk or infant formula. However, weaned from these high-quality protein sources to foods that lack adequate protein, infants become at greater risk of protein deficiency. Because protein-deficient diets are almost always also lacking in energy, and vice versa, this condition is called **protein-energy malnutrition (PEM).** Children with PEM are typically deficient in one or more micronutrients as well. Thus PEM is a condition of overall malnutrition, and the World Health Organization estimates that it plays an important role in at least 5 million child deaths each year.[25] The region of the world with the highest PEM rates is Asia, followed by Africa, Latin America, and the Caribbean.

Severe PEM actually encompasses a continuum of malnutrition with two clinical extremes.[26] At one extreme is a condition called **marasmus,** which results from severe, chronic malnutrition. In marasmus, fat and muscle tissue are depleted, and the skin hangs in loose folds, with the bones clearly visible beneath the skin. Children with marasmus tend at first to be alert and ravenously hungry, although with increasing severity they become apathetic and lose their appetites. Clinicians often say that marasmus represents the body's response to long-term, chronic dietary insufficiency. The other extreme form of PEM, called **kwashiorkor,** is often distinguished from marasmus by the presence of edema. The edema typically starts in the legs but may involve the

**protein-energy malnutrition (PEM)** Protein deficiency accompanied by inadequate intake of energy and often of other essential nutrients as well.

**marasmus** (ma – RAS – mus) (*marainein,* to waste away) A form of PEM characterized by extreme wasting of muscle and adipose tissue.

**kwashiorkor** (kwa – she – OR – kor) (Kwa, a language of Ghana; referring to "what happens to the first child when the next is born") A form of PEM often characterized by edema in the extremities (hands, feet).

entire body. Sometimes fluid accumulates in the abdominal cavity, resulting in a condition called **ascites.** Remember that one of the roles that protein plays in the body is regulation of fluid balance. Children with kwashiorkor sometimes have large, distended abdomens. However, ascites can also occur in children with marasmus. Intestinal parasites such as worms sometimes contribute to this abdominal distension. Children with kwashiorkor often are apathetic and have cracked and peeling skin, enlarged fatty livers, and sparse unnaturally blond or red hair. Children with either form of PEM are at great risk of infection.

Although many characteristics of kwashiorkor were once thought simply caused by protein deficiency, this does not appear to be the case.[27] Researchers now believe that many of the signs and symptoms of kwashiorkor are the result of micronutrient deficiencies in combination with infection or other environmental stressors.

## Protein Deficiency in Adults

PEM can also occur in adults. Unlike children, however, adults with PEM rarely experience the signs and symptoms associated with kwashiorkor. Instead, adult PEM generally takes the form of marasmus. There are many causes of PEM in adulthood, including inadequate dietary intake (such as occurs in alcoholics and people with eating disorders), protein malabsorption (such as occurs with many gastrointestinal disorders), excessive and chronic blood loss, cancer, infection, and injury (especially burns).[28]

Adults with PEM can experience extreme muscle loss, because the body's muscles are broken down to provide glucose and energy. In addition, fatty liver and edema are common. Adults with severe PEM experience decreased function of many of the vital physiological systems, including the cardiovascular system, renal system (kidneys), digestive system (gastrointestinal tract and accessory organs), and endocrine and immune systems. Because there are many causes of PEM in adults, treatment is often long and difficult. For example, if the cause is due to infection treatment may involve both dietary intervention and use of antibiotics. If, in contrast, protein deficiency is a result of an eating disorder, psychological counseling becomes a key component of the health care plan.

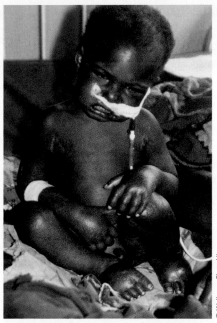

Children with kwashiorkor often have swollen abdomens (ascites), edema in their hands and feet, cracked and peeling skin, and an apathetic nature. These children are at increased risk for infection.

© Wellcome Photo Library

## ESSENTIAL *Concepts*

Although most people consume adequate amounts of protein, protein deficiency remains a significant health concern worldwide. Protein deficiency is almost always accompanied by some degree of energy deficiency. Thus this state of malnutrition is typically referred to as protein energy malnutrition, or PEM. In children, PEM takes two forms: kwashiorkor and marasmus. Most adults with PEM have symptoms associated with marasmus. In almost all cases, treatment of children and adults with PEM is multifaceted, involving a variety of factors, including provision of adequate nutrition.

## Protein Excess: Is There Cause for Concern?

People have long thought that high-protein diets cause adverse health outcomes such as osteoporosis, kidney problems, heart disease, obesity, and cancer. However, more recent clinical studies do not support these claims. This fact is acknowledged by the Institute of Medicine, which carefully considered

**ascites** (a – SCI – tes) Edema that occurs in the abdominal cavity.

CONNECTIONS ▶ Recall that causal relationships are typically determined from intervention studies, whereas epidemiologic studies are powerful in showing relationships or associations within a population (Chapter 1, pages 13–15).

the peer-reviewed literature related to the potential health consequences of high-protein diets in their 2005 DRI publication. This panel of experts concluded that, although epidemiologic studies offer limited evidence that high protein intake has an *association* with some indicators of adverse health, experimental data do not support this contention. In fact, the upper limit of the AMDR for protein intake was developed, not because there was evidence that additional protein might pose a health risk, but solely to complement the recommendations for carbohydrate and fat intakes.

Nonetheless, note that high intakes of protein are often accompanied by high intakes of fat, saturated fat, and cholesterol. Because these dietary components are risk factors for heart disease, it is important to choose a variety of lean and low-fat protein foods, such as those recommended by the MyPyramid food guidance system. In addition, growing evidence shows that very high, chronic intakes of red meat (beef, lamb, and pork) or processed meats (bacon, sausage, hot dogs, ham, and cold cuts) are associated with increased risk for colorectal cancer.[29] However, intakes must be very high for the risk of cancer to increase, and the nature of this relationship is not understood. A more complete discussion of how nutrition influences risk of cancer is provided in the Nutrition Matters following Chapter 11.

## ESSENTIAL *Concepts*

For most people, excess protein intake does not result in health complications. However, be careful to limit intake of fat, saturated fat, and cholesterol, which tend to be found in high-protein foods.

# Protein Nutrition: Putting Concepts into Action

Getting enough protein from the diet is important for many reasons. Although protein intake in the United States is typically adequate, there are still situations when protein may be lacking. However, as with the other nutrients, you can evaluate your diet for its protein content and change your food choices using various dietary assessment and planning techniques.

## Setting the Stage and Setting the Goals

In this hypothetical example, we consider a student named Kristin—a 20-year-old woman who follows a vegan diet and is a member of her university's cross-country track team. Kristin weighs 130 pounds (59 kg). Although she does not appear to have any signs or symptoms of protein deficiency, her coach is concerned that she may not be getting adequate protein. And because Kristin is a competitive athlete and understands the importance of protein nutrition in overall health, she too wants to make sure her diet is appropriate.

## Assessing Nutritional Status

Kristin decides to conduct a diet analysis using the three-day food record method. She carefully records everything that she eats and drinks for three days—Thursday, Friday, and Saturday—and enters this information into the

MyPyramid website (http://www.mypyramid.gov). Using this software, she determines her dietary intake and compares it to the DRI values. She also compares her dietary intake of certain food groups to intakes recommended in the USDA Dietary Guidelines and MyPyramid.

First Kristin compares her protein intake to recommendations for someone of her size and physical activity level and learns that her total protein intake of 40 g is lower than her calculated requirement of 47 g (0.8 g protein per kilogram of body weight). In addition, only 6% of her energy is coming from protein. Because the Institute of Medicine's Acceptable Macronutrient Distribution Ranges (AMDRs) suggest that 10 to 35% of energy comes from protein, this indicates that her protein intake may indeed be marginal. Although her intake of most micronutrients is adequate, her calcium intake is only 71% of the AI. Thus Kristin concludes she should consume more protein and calcium-rich foods.

## Setting the Table to Meet the Goals

To accomplish her goal, Kristin studies a food composition table and explores the MyPyramid website. From these sources, she determines that the easiest way to include both additional calcium and high-quality protein in her diet is to eat foods from the meats and legumes group and drink low-fat dairy products. Although she does not want meat in her diet, she is comfortable with adding legumes and dairy products. So she consumes a serving of low-fat yogurt for breakfast and a glass of skim milk with her lunch. She also asks the kitchen staff in the dining hall to add nonmeat protein sources to selected vegetarian entrees. For example, she asks them to add calcium-fortified tofu to the vegetable stir-fry and lentils to the vegetable soup. In addition, Kristin examines food labels and chooses whole-grain products, especially those fortified with calcium. She is also careful to add garbanzo or kidney beans to her dinner salad and to consume a variety of protein sources to assure adequate intake of all the essential amino acids.

## Comparing the Plan to the Assessment: Did It Succeed?

Several weeks later, Kristin evaluates her diet again. This time, she discovers that her daily protein intake is 55 g—slightly higher than her calculated requirement. Because Kristin has a very lean body from hours of training and competition, she and her coach are not concerned that she is consuming too much protein. In addition, her calcium intake is now adequate. Thus Kristin has succeeded in making positive changes to her diet that support both her health and athletic endeavors.

# ESSENTIAL *Concepts*

Although the availability of high-quality protein foods makes it easy to consume sufficient amounts of this nutrient, at times protein intake may still be inadequate. However, it is relatively simple to assess dietary protein intake using a three-day food record in addition to public-access dietary assessment programs. If protein intake is limited, changes such as consuming low-fat dairy products and lean meats can add important protein to the diet without adding fat.

# Review Questions  Answers are found in Appendix G

1. Which of the following is the basic "building block" of a protein?
   a. Amino acid
   b. Glucose
   c. Phospholipid
   d. DNA
2. Which of the following is *different* among the amino acids?
   a. Central carbon
   b. R-group
   c. Amino group
   d. Carboxylic acid group
3. Which amino acid becomes conditionally essential for people with phenylketonuria (PKU)?
   a. Phenylalanine
   b. Alanine
   c. Tyrosine
   d. Glutamine
4. Creating a genetically modified organism (GMO) can involve which of the following?
   a. Modifying the genetic material of a plant
   b. Modifying the genetic material of an animal
   c. Altering the amino acid or protein content of a food
   d. All of the above
5. Which level of protein structure is due to hydrogen bonding within or between amino acid side chains (R-groups)?
   a. Primary structure
   b. Secondary structure
   c. Tertiary structure
   d. Quaternary structure
6. The chemical digestion of protein occurs in which of these locations?
   a. Mouth
   b. Stomach
   c. Pancreas
   d. Large intestine
7. Cholecystokinin (CCK) is a(n) _____ produced in the _____.

   a. hormone        liver
   b. enzyme         small intestine
   c. hormone        small intestine
   d. enzyme         pancreas
8. Gluconeogenesis is typically stimulated during which of the following periods?
   a. High carbohydrate intake
   b. Low protein intake
   c. High fat intake
   d. Low carbohydrate intake
9. The Institute of Medicine's Acceptable Macronutrient Distribution Ranges (AMDRs) recommend that healthy adults consume how much of their total energy as protein?
   a. 10 to 15%
   b. 15 to 25%
   c. 10 to 35%
   d. 20 to 45%
10. Suboptimal functioning of the _____ may accompany protein deficiency.
   a. immune system
   b. reproductive system
   c. circulatory system
   d. all of the above
11. Describe the basic cause and complications of sickle cell anemia. Your answer should appropriately use the following terms: *genetic code, translation, malaria, iron, DNA, symptoms,* and *protein.*
12. The shape of a protein is very important for its structure. Describe the four levels of protein structure, and explain how mercury can influence the shape of a protein.
13. Protein complementation is important in some regions of the world where people consume limited amounts of animal products. Define *protein complementation,* and describe why it is especially important during periods of rapid growth and development.
14. List four functions of proteins, providing specific examples of each.

# Practice Calculations  Answers are found in Appendix G

1. The Acceptable Macronutrient Distribution Range (AMDR) for protein is 10 to 35% of calories. If your Estimated Energy Requirement (EER) is 2,000 kcal/day, how many calories should come from protein? How many grams of protein would this be? What dietary changes might you make if your protein intake is too high?
2. Using the food composition table that accompanies this text or other sources, calculate the protein content of a lunch containing a hamburger on a white bun, a
small order of french fries, and a large cola soft drink. What percentage of your Recommended Dietary Allowance (RDA) of protein does this represent. What foods might be added to or substituted for the food in this lunch to increase the protein content?
3. Using the MyPyramid food guidance system (http://www.mypyramid.gov), determine your recommended amount of foods from the meats, beans, and dairy groups. Then calculate how much protein you would get if you consumed these amounts of foods.

**Media Links** A variety of study tools for this chapter are available at www.thomsonedu.com/nutrition/mcguire.

Prepare for tests and deepen your understanding of chapter concepts with these online materials:

- Practice tests
- Flashcards
- Glossary
- Student lecture notebook
- Web links
- Animations
- Chapter summaries, learning objectives, and crossword puzzles

# Notes

1. Mulder GJ. Uber die Zusammensetzung einiger thierischen Substanzen. Journal der Praktische Chemie. 1839;16:129–52.

2. Bhutta ZA. Protein. In: Encyclopedia of Human Nutrition. Strain SS, Caballero B, Sadler MJ, editors. New York: Academic Press; 1999. Fuller MF, Garlick PJ. Human amino acid requirements: can the controversy be resolved? Annual Reviews in Nutrition. 1994;14:217–41. Furst P, Stehle P. What are the essential elements needed for the determination of amino acid requirements in humans? Journal of Nutrition. 2004;134:1558S–65S.

3. Erlandsen H, Patch MG, Gamez A, Straub M, Stevens RC. Structural studies on phenylalanine hydroxylase and implications toward understanding and treating phenylketonuria. Pediatrics. 2003;112:1557–65. Hanley WB. Adult phenylketonuria. American Journal of Medicine. 2004;117:590–5.

4. American Academy of Pediatrics. Pediatric Nutrition Handbook. 5th ed. Kleinman RE, editor. Elk Grove Village, IL: American Academy of Pediatrics; 2004.

5. Reeds PJ, Garlick PJ. Protein and amino acid requirements and the composition of complementary foods. Journal of Nutrition. 2003;133:2953S–61S.

6. Fogg-Johnson N, Merolli A. Food biotechnology: nutritional considerations. In: Present knowledge in nutrition, 8th ed. Bowman BA, Russell RM, editors. Washington, DC: ILSI Press; 2001.

7. Huang S, Adams WR, Zhou Q, Malloy KP, Voyles DA, Anthony J, Kriz AL, Luethy MH. Improving nutritional quality of maize proteins by expressing sense and antisense zein genes. Journal of Agricultural and Food Chemistry. 2004;52:1958–64. Zarkadas CG, Hamilton RI, Yu ZR, Choi VK, Khanizadeh S, Rose NGW, Pattison PL. Assessment of the protein quality of 15 new northern adapted cultivars of quality protein maize using amino acid analysis. Journal of Agricultural and Food Chemistry. 2000;48:5351–61.

8. Schaafsma, G. The protein digestibility-corrected amino acid score. Journal of Nutrition. 2000;130:1865S–7S.

9. Buchanan GR, DeBaun MR, Quinn CT, Steinberg MH. Sickle cell disease. Hematology. 2004;35–47. Schnog JB, Duits AJ, Muskiet FAJ, ten Cate H, Rojer RA, Brandjes DPM. Sickle cell disease: a general overview. Journal of Medicine. 2004;62:364–74.

10. Javier C, Milagro FI, Martinez JA. Genetic manipulation in nutrition, metabolism, and obesity research. Nutrition Reviews. 2004;62:321–30. Simopoulos AP. Genetics. Implications for nutrition. Forum of Nutrition. 2003;56: 226–9. Stover PJ. Nutritional genomics. Physiological Genomics. 2004;16:161–5.

11. Hägg, G. The Nobel Prize in chemistry 1954. Presentation speech. Available from: http://nobelprize.org/chemistry/laureates/1954/press.html.

12. U.S. Department of Health and Human Services and U.S. Environmental Protection Agency. What you need to know about mercury in fish and shellfish. 2004 EPA and FDA advice for women who might become pregnant, women who are pregnant, nursing mothers, young children. Available from: http://www.cfsan.fda.gov/~dms/admehg3.html.

13. Kaput, J, Rodrigues RL. Nutritional genomics: the next frontier in the postgenomic era. Physiological Genomics. 2004;16:166–77.

14. National Human Genome Research Institute. The National Human Genome Research Institute: a brief history and timeline. Available from: http://www.genome.gov/10001763.

15. Taylor SL, Hefle SL. Food allergy. In: Present knowledge in nutrition, 8th ed. Bowman BA, Russell RM, editors. Washington, DC: ILSI Press; 2001.

16. Fuller MF, Reeds PJ. Nitrogen cycling in the gut. Annual Review of Nutrition. 1998;18:385–411.

17. Rand WM, Pellet PL, Young VR. Meta-analysis of nitrogen balance studies for estimating protein requirements in healthy adults. American Journal of Clinical Nutrition. 2003;77:109–27. Young VR, Marchini JS, Cortiella, J. Assessment of protein nutritional status. Journal of Nutrition. 1990;120:1496–502.

18. Institute of Medicine. Dietary Reference Intakes for energy, carbohydrate, fiber, fat, fatty acids, cholesterol, protein, and amino acids. Washington, DC: National Academies Press; 2005.

19. Dewey KG. Energy and protein requirements during lactation. Annual Review of Nutrition. 1997;17:19–36.

20. Dickerson RN. Estimating energy and protein requirements of thermally injured patients: art or science? Nutrition. 2002;18:439–42. Gudaviciene D, Rimdeika R, Adamonis

K. Nutrition of burned patients. Medicina. 2004;40:1–8. Wilmore DW. The effect of glutamine supplementation in patients following elective surgery and accidental injury. Journal of Nutrition. 2001;131:2543S–49S.

21. Garlick PJ. Assessment of the safety of glutamine and other amino acids. Journal of Nutrition. 2001,131:2556S–61S. John MS. Popular sports supplements and ergogenic aids. Sports Medicine. 2003;33:921–39. Phillips SM. Protein requirements and supplementation in strength sports. Nutrition. 2004;20:689–95.

22. Haddad EH, Tanzman JS. What do vegetarians in the United States eat? American Journal of Clinical Nutrition. 2003;78:626S–32S.

23. Antony AC. Vegetarianism and vitamin B-12 (cobalamin) deficiency. American Journal of Clinical Nutrition. 2003;78:3–6. Hunt JR. Bioavailability of iron, zinc, and other trace minerals from vegetarian diets. American Journal of Clinical Nutrition. 2003;78:633S–9S. Reijnders L, Soret S. Quantification of the environmental impact of different dietary protein choices. American Journal of Clinical Nutrition. 2003;78:664S–8S.

24. Mangels AR, Messina V. Considerations in planning vegan diets: infants. Journal of the American Dietetic Association. 2001;101:670–7. Messina V, Mangels AR. Considerations in planning vegan diets: children. Journal of the American Dietetic Association. 2001;101:661–9.

25. Pelletier DL, Frongillo EA Jr, Schroeder DG, Habicht JP. The effects of malnutrition on child mortality in developing countries. Bulletin of the World Health Organization. 1995;73:443–8.

26. Jeejeebhoy KN. Protein nutrition in clinical practice. British Medical Bulletin. 1981;37:11–17. Waterlow JC. Classification and definition of protein-calorie malnutrition. British Medical Journal. 1972;3:566–9. Waterlow JC. Metabolic adaptation to low intakes of energy and protein. Annual Reviews of Nutrition. 1986;6:495–526. Waterlow JC. Protein-energy malnutrition. London: Edward Arnold; 1992.

27. Golden M. The development of concepts of malnutrition. Journal of Nutrition. 2002;132:2117S–22S.

28. Hansen RD, Raja C, Allen BJ. Total body protein in chronic diseases and in aging. Annals of the New York Academy of Sciences. 2000;904:345–52.

29. Chao A, Thun MJ, Connell CJ, McCullough ML, Jacobs EJ, Flanders D, Rodriguez C, Sinha R, Calle EE. Meat consumption and risk of colorectal cancer. JAMA (Journal of the American Medical Association). 2005;293:172–82.

**Check out the following sources for additional information.**

30. Gibson RS. Principles of nutritional assessment. New York: Oxford University Press; 2005.

31. Murano PS. Understanding food science and technology. Belmont, CA: Wadsworth; 2003.

32. Pencharz PB, Ball RO. Different approaches to define individual amino acid requirements. Annual Review of Nutrition. 2003;23:101–16.

33. U.S. Department of Health and Human Services, and U.S. Department of Agriculture. Dietary Guidelines for Americans 2005. Washington, DC: U.S. Government Printing Office; 2005. Available from: http:www.healthierus.gov/dietaryguidelines.

34. Young VR, Borgonha S. Nitrogen and amino acid requirements: the Massachusetts Institute of Technology amino acid requirement pattern. Journal of Nutrition. 2000;130:1841S–9S.

35. Young VR. Protein and amino acids. In: Present knowledge in nutrition, 8th ed. Bowman BA, Russell RM, editors. Washington, DC: ILSI Press; 2001.

# Food Safety and Foodborne Illness

© Dynamic Graphics Group/Creatas/Alamy

$\mathscr{P}$erhaps the most beneficial and far-reaching advance in the past century is the ability to buy and prepare abundant amounts of healthful and safe food. This increased availability of food is due to many things, such as increased agriculture productivity, international trade, and the use of preservatives and processing methods that prevent spoilage. As consumers we are living in a time of unprecedented food availability and safety.

However, sometimes food may be less than safe to eat. For example, foods can contain organisms that cause illness. Other contaminants, such as heavy metals, can also be harmful. A **foodborne illness** is a disease caused by ingesting food and is sometimes referred to as "food poisoning." Foodborne disease results in approximately 76 million illnesses and 5,000 deaths annually in the United States.[1] Clearly, making sure that food is safe to eat is an important public health issue.

Most people understand it is important to avoid eating food that is spoiled, unclean, or stored improperly. Have you ever asked yourself why? In other words, what makes some foods risky to eat? In general, these foods are dangerous because they contain illness-causing agents. **Infectious agents** that cause foodborne illness include biological hazards such as bacteria, viruses, molds, fungi, parasites, and possibly prions. **Noninfectious agents** include nonbacterial toxins, other chemicals (such as pesticides and antibiotics), and physical hazards (such as glass and plastic). Understanding how to prevent foodborne illness requires a basic knowledge of how foods come to contain these agents and how the body reacts to them. It is also important to understand how to keep the food we buy and prepare safe.

## Infectious Agents of Foodborne Illness

We are all surrounded by thousands of microscopic organisms (microbes) each day. These microbes make up the world we live in and serve many useful purposes. For example, they reside in the gastrointestinal tract, helping us digest food and preventing some diseases. Others make us sick and are said to be infectious or **pathogenic.** In fact, consuming pathogenic bacteria in foods and beverages is the main cause of foodborne illness in the United States.

A particular organism can have many variations (called species or **serotypes**). Some serotypes are harmless, such as the ones living in our gastrointestinal tracts, whereas others are pathogenic. For example, some of the various serotypes of *Escherichia coli (E. coli)* live harmlessly in the human intestine, whereas others (such as *E. coli* O157:H7) cause foodborne illness.[2] Pathogens can make us sick in many ways. For example, some produce toxins and others invade and destroy intestinal

---

**foodborne illness** A disease caused by ingesting food.

**infectious agent of foodborne illness** A pathogen or compound in food that causes illness and can be passed or transmitted from one infected animal or person to another.

**noninfectious agent of foodborne illness** An inert (nonliving) substance, in food, that causes illness.

**pathogenic** Something that causes disease.

**serotype** A specific strain of a larger class of organism.

cells. Still others are absorbed into the blood, where they cause other complications. Some of the most common infectious agents are summarized in Table 1, and the ways in which they cause illness are described here.

## Some Organisms Make Toxins While in Foods

While they are growing in foods, some organisms produce toxic substances. These toxins are called **preformed toxins,** because they are already present in contaminated food. When we consume these foods, the toxins cause serious and rapid (1 to 6 hours) reactions such as nausea, vomiting, and sometimes neurological damage. For example, the toxin produced by *Staphlococcus aureus* (*S. aureus*) causes 185,000 foodborne illnesses annually in the United States. Foods commonly infected with *S. aureus* include raw or undercooked meat and poultry, cream-filled pastries, and unpasteurized dairy products. Another bacterium that produces preformed toxins, *Clostridium botulinum,* is found mainly in poorly prepared home-canned foods with low acid content, such as green beans and corn.

**preformed toxin** Poisonous substance produced by microbes while they are in a food (prior to ingestion).

## TABLE 1 Infectious Agents of Foodborne Illness, Food Sources, and Symptoms of Infection

| Organism | Common Food Sources | Signs and Symptoms |
|---|---|---|
| **Bacteria** | | |
| *Camplybacter jejuni* | Raw or undercooked poultry, untreated water, unpasteurized milk | Diarrhea (often bloody), abdominal cramping, nausea, vomiting, fever, tiredness; sometimes there are no symptoms |
| *Clostridium botulinum* | Home-canned foods with low acid content, such as asparagus, green beans, beets, and corn; also undercooked baked potatoes | Double vision, drooping eyelids, slurred speech, dry mouth, difficulty swallowing, weak muscles |
| *Clostridium perfringes* | Raw and undercooked meats | Abdominal pain, diarrhea, vomiting |
| *Escherichia coli* O157:H7 | Raw or undercooked hamburgers, salami, alfalfa sprouts, lettuce, unpasteurized milk, apple juice and apple cider; contaminated well water (as might be found in swimming pools) | Nausea, severe abdominal cramps, watery or very bloody diarrhea, fatigue |
| *Listeria monocytogenes* | Raw milk, cheese and vegetables; luncheon meats and frankfurters | Severe headaches, stillbirths |
| *Salmonella* | Raw poultry, eggs, and beef; unwashed fruit and alfalfa sprouts; unpasteurized milk; note that infection frequently occurs after handling pets, particularly reptiles | Diarrhea, fever, abdominal cramps, headaches |
| *Shigella* | Many raw or undercooked foods; pathogen can be carried by flies | Fever, tiredness, watery or bloody diarrhea, nausea, vomiting, abdominal pain |
| *Staphylococcus aureus* | Raw or undercooked meat and poultry, cream-filled pastries, unpasteurized dairy products | Nausea, vomiting, diarrhea, abdominal cramps |
| *Vibrio cholerae* | Contaminated water; raw or undercooked foods, including shellfish | Watery diarrhea and vomiting |
| **Viruses** | | |
| Hepatitis A virus | Mollusks (oysters, clams, mussels, scallops, and cockles) | Jaundice, fatigue, abdominal pain, loss of appetite, nausea, diarrhea, fever |
| Norwalk and Norwalk-like virus | Raw or undercooked shellfish, mishandled foods, contaminated water | Nausea, vomiting, diarrhea, abdominal pain, headache, fever |
| **Parasites** | | |
| *Trichinella* (worm) | Raw or undercooked pork or other meats of carnivorous animals (e.g., bears) | Muscle pain, swollen eyelids, fever |
| *Taenia* (worm) | Raw or undercooked pork or any mishandled food | Worm segments in stools; pain and swelling of muscles, organs, heart, or brain |
| *Giardia intestinalis* (protozoa) | Contaminated water, uncooked foods grown in contaminated water or infected by food handlers | Diarrhea, flatulence, greasy stools, dehydration, loss of appetite, vomiting, jaundice |
| **Molds** | | |
| *Aspergillus flavus* (produces aflatoxin) | Wheat, flour, peanuts, soybeans | Liver damage, perhaps leading to cancer |

Molds can also produce dangerous preformed toxins. An example is **aflatoxin,** which is a poison produced by the *Aspergillus* mold that grows on some agricultural crops such as peanuts. If the crop is not dried properly before storage, *Aspergillus* can continue to grow and produce toxic levels of aflatoxin. Although agricultural practices in the United States minimize contamination of our food with aflatoxin, it remains a significant public health issue in other regions of the world. For example, in 2004 an outbreak of aflatoxin poisoning in Kenya resulted in 317 illnesses and 125 deaths, many of which were children under the age of 5 years.[3]

## Some Organisms Make Enteric (Intestinal) Toxins

Some organisms produce harmful toxins after entering the gastrointestinal tract; these toxins are called **enteric** or **intestinal toxins.** This is different from organisms that make "preformed" toxins, because enteric toxins are produced after ingesting foods. Enteric toxins draw water into the intestinal lumen, resulting in diarrhea. Although the symptoms are variable, people usually begin to feel ill within one to five days after ingesting foods, substantially later than the onset of symptoms from preformed toxins. One of the largest outbreaks of foodborne illness caused by enteric toxins occurred in 1977 when Norwalk virus, an example of this kind of a pathogen, was believed to have caused severe illness in 521 cruise ship passengers. In this outbreak, the food responsible for the illness was never determined. Norwalk virus continues to be a significant challenge for the cruise industry.

In 1999, an outbreak of Norwalk-like virus occurred in 70 long-distance hikers on the Appalachian Trail in the eastern United States. In this case, consumption of contaminated water at a specific general store along the trail was to blame.[4] Another example of a Norwalk-

Foodborne illnesses are of special concern to the travel and hospitality industries.

like virus outbreak occurred in 2000 when at least 333 individuals in several states became ill.[5] This case was caused by contamination of several salads prepared by an infected food handler. These salads were then shipped to 27 states, causing illness in at least 13 of them. Some forms of *E. coli* also produce enteric toxins, and consumption of food and water contaminated with this bacteria can cause what is often called "traveler's diarrhea."[6] Foods and beverages typically contaminated with these forms of *E. coli* include salads, unpeeled fruits, raw or poorly cooked meats and seafood, unpasteurized dairy products, and tap water.

## Some Organisms Invade Intestinal Cells

Some pathogens actually invade the cells of the intestine, seriously irritating the mucosal lining and causing severe abdominal discomfort, diarrhea, and fever. *Salmonella,* which is this kind of pathogen, is one of the most common causes of foodborne illness in the United States (Figure 1). *Salmonella* tends to lurk in raw poultry, eggs, beef, unwashed fruit, alfalfa sprouts, and unpasteurized milk. One of the largest outbreaks of *Salmonella* poisoning occurred in 1984 when about 19,000 attendees of a convention in Chicago ate a contaminated dairy-based dessert.[7] In 2004, 429 cases of *Salmonella* infections followed consumption of raw Roma tomatoes purchased at a delicatessen chain.[8] In 1999, *Salmonella*-infected orange juice caused at least 297 cases of foodborne illness in Washington State and Oregon,[9] and numerous *Salmonella* infections have been reported from consuming raw alfalfa and mung bean sprouts.[10]

> **CONNECTIONS** Recall that the intestinal mucosa is the layer of cells forming the innermost lining of the intestinal tract. These cells produce digestive enzymes and mediate nutrient absorption (Chapter 4, page 91).

Whereas some organisms simply damage the intestinal lining, others cause mucosal cells to die, resulting in bloody diarrhea. An example of a pathogen that invades the intestinal lining is *E. coli* O157:H7, which researchers first identified as a human pathogen in 1982.[11] In 1993, *E. coli* O157:H7 infection resulted in at least 477 cases of foodborne illness in Washington State and California, again caused by consumption of inadequately cooked, contaminated hamburger.[12] Also, *E. coli* O157: H7 infection broke out in California and Arizona when

**aflatoxin** (a – fla – TOX – in) A toxic compound produced by certain molds that grow on peanuts, some grains, and soybeans.

**enteric (intestinal) toxin** A toxic agent produced by an organism after it enters the gastrointestinal tract.

## FIGURE 1    Salmonella

Salmonella causes more cases of foodborne illness than any other pathogen.

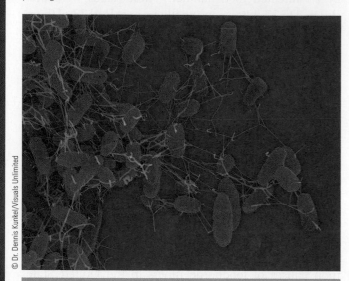

© Dr. Dennis Kunkel/Visuals Unlimited

people ate contaminated alfalfa and clover sprouts in 1998. Unpasteurized milk, apple juice, and apple cider can also harbor this pathogenic organism, as can improperly prepared turkey.[13]

## Parasites: Protozoa and Worms

**Parasites** differ from other pathogens in that they rely on other organisms to survive. Parasites do not offer the host a benefit and are often dangerous. Several types of parasites can cause foodborne illness; the most common belong to the groups of **protozoa** and worms. Protozoa are one-celled organisms, some of which can live as parasites in the intestinal tracts of animals and humans. As part of their reproductive cycle, protozoa living as parasites in humans form **cysts** that are excreted in the feces during a bowel movement. If these cyst-containing feces come in contact with plants or animals, food produced with these products can be contaminated with the protozoa. Thus, consuming foods or beverages contaminated with cyst-containing feces can cause foodborne illness. One such protozoan parasite, *Giardia intestinalis*, causes diarrhea, abdominal discomfort, and cramping; symptoms typically begin one to two weeks after infection. *Giardia intestinalis* can be found in untreated swimming pools and hot tubs, and in rivers, ponds, and streams that have been contaminated with feces of an infected animal or person. For this reason, chlorine (which kills this organism) is typically added to water in public swimming and bathing areas.

Consuming foods that contain worms can also cause foodborne illness. Like protozoa, worms form cysts as part of their life cycles. Once ingested, cysts mature into worms that can cross the intestinal mucosa, travel through the blood, and eventually settle in various locations, including muscles, eyes, and brain. Parasitic invasions can cause a variety of signs and symptoms, including fever, muscle swelling, anemia, and diarrhea. In addition to being colonized by eating contaminated meat, a person can become infected via contact with contaminated feces. For example, *Trichinella* is a roundworm that can invade a variety of animals, including pigs and some fish. Eating undercooked *Trichinella*-contaminated pork and fish can result in the worm entering the body and causing muscle pain, swollen eyelids, and fever. Today, *Trichinella* infection is rare in the United States.

## Prions

Although **prions** are not living organisms, they may pose food safety concerns.[14] Prions are altered proteins created when, for reasons not yet understood, the secondary structure of the normal protein is disrupted, transforming α-helix coils into β-folded sheets. The resulting deformed protein is called a prion. Prions can cause other normal proteins to unravel, setting off a cascade of similar reactions converting hundreds of normal proteins into abnormal prions. When high levels of prions build up, the cell ruptures and releases the prions into the surrounding area, where they destroy other cells. Eventually, this kills an entire group of cells and forms a spongy appearance in the infected tissue. Unfortunately, prions are extremely resistant to breakdown and retain their ability to infect other cells even after extreme heat treatment or exposure to acids. Therefore, if prions are eaten they can be absorbed into the body unaltered and continue to transform normal proteins into prions in their new host.

### Mad Cow Disease

Several diseases in cattle are known to be caused by prions, including **bovine spongiform encephalitis (BSE)** or **mad cow disease**.[15] Mad cow disease is characterized

**parasite** An organism that, during part of its life cycle, must live within or on another organism without benefiting its host.

**protozoa** Very small (single-cell) organisms that are sometimes parasites.

**cyst** A stage of the life cycle of some parasites.

**prion** (derived from "protein" and "infectious") A misshapen protein that causes other proteins to also become distorted, damaging nervous tissue.

**bovine spongiform encephalopathy (BSE; mad cow disease)** A fatal disease in cattle caused by ingesting prions.

by loss of motor control, confusion, paralysis, wasting, and eventually death. An outbreak of BSE in the United Kingdom in the 1980s is thought to have been caused by feeding tissue obtained from infected sheep to cattle. Three cases of BSE have occurred in the United States, one in the state of Washington (2003), one in Texas (2005), and one in Alabama (2006).

A human disease called **Creutzfeldt-Jakob disease,** caused by prions, is very rare and occurs in only one in a million people each year. This disease can be due to either genetic variation or by direct infection of prions from contaminated medical equipment—for example, during surgery. Although no evidence exists that eating prions can cause Creutzfeldt-Jakob disease, during the BSE outbreak in the United Kingdom researchers discovered a new form of Creutzfeldt-Jakob disease, which they called **variant Creutzfeldt-Jakob disease.** Considerable evidence links variant Creutzfeldt-Jakob disease to consumption of BSE-contaminated products.[16]

Although scientists still have much to learn about prions, the World Heath Organization (WHO) prohibits the feeding of nerve tissue from slaughtered cattle to other cattle. This is because prions are found mainly in nerves and brain. Although milk and muscle are safe for humans to consume, experts recommend that humans avoid consuming nerve tissue such as brain. In fact, beginning in 2004 the U.S. Food and Drug Administration (FDA) banned the use of such products in human food, including dietary supplements, and in cosmetics. These measures, along with those established by USDA, provide a uniform national BSE policy and help ensure the safety of human food. Understanding BSE and its potential impact on human health remains of great interest to the health community.

## ESSENTIAL *Concepts*

Infectious foodborne illness can be caused by pathogenic organisms such as bacteria, viruses, parasites, molds, and fungi. Pathogens cause illness in various ways. Some organisms produce toxins while they grow in foods. Consuming these preformed toxins rapidly leads to nausea and vomiting. Other pathogens produce toxins inside the intestinal tract and cause diarrhea. Still other pathogens invade intestinal cells, where they stimulate severe diarrhea, which often contains blood. Some parasites cross the intestinal mucosa, enter the bloodstream, and take up residence in the body. These forms of foodborne illness are considered infectious or contagious because they can be transmitted from person to person. Whether consuming prions causes foodborne illness in humans is not known, although consuming prion-contaminated foods has been *associated* with variant Creutzfeldt-Jakob disease.

# Noninfectious Agents of Foodborne Illness

Consuming foods containing infectious pathogens poses the greatest risk of foodborne illness. However, noninfectious agents in food can also make us sick. These inert (nonliving) compounds include physical contaminants, such as glass and plastic, and other contaminants, such as heavy metals, pesticides, herbicides, antibiotics, hormones, and food additives. Note that Chapters 6, 12, and 13 present additional information regarding heavy metal poisoning.

## Marine Toxins

Consuming certain marine foods such as fish and shellfish (such as clams and oysters) can also cause foodborne illness.[17] This is because these marine animals consume large amounts of algae that sometimes produce poisonous compounds called **marine toxins.** Eating marine toxin–contaminated shellfish or fish, often called **shellfish poisoning,** can result in neurologic symptoms, including tingling, burning, numbness, drowsiness, and difficulty breathing. People with shellfish poisoning may also experience a strange phenomenon called hot–cold inversion, in which they perceive cold drinks as hot and hot drinks as cold. Every year, approximately 30 cases

Some seafood can become contaminated with toxins produced by algae.

**Creutzfeldt-Jakob disease** A fatal disease in humans caused by a genetic mutation or surgical contamination with prions.

**variant Creutzfeldt-Jakob disease** A form of Creutzfeldt-Jakob disease that may be caused by consuming BSE-contaminated foods.

**marine toxin** A poison produced by ocean algae.

**shellfish poisoning** A group of foodborne illnesses caused by consuming shellfish that contain marine toxins.

of shellfish poisoning are reported in the United States, typically in the coastal regions of the Atlantic Northeast and Pacific Northwest.

## Other Agents: Pesticides, Herbicides, Antibiotics, and Hormones

In addition to marine toxins, other noninfectious agents of foodborne illness can make their way into the food chain, including pesticides, herbicides, antibiotics, and hormones. Because exposure to some of these substances can cause illness, many federal and international agencies work together to assure that their presence in foods is negligible or poses no risk to the consumer and the environment. These agencies include the Food and Agriculture Organization (FAO) of the United Nations, the U.S. Environmental Protection Agency (EPA), the FDA, and the U.S. Department of Agriculture (USDA).

For example, in 1972 dichloro-diphenyl-trichloroethane (DDT)—a pesticide once used to kill mosquitoes and increase crop yields—was banned in the United States because it damages wildlife. Similarly, farmers once used diethylstilbestrol (DES) to enhance muscle growth in farm animals. However, the EPA banned its use in 1979 when researchers determined that DES could cause birth defects. More recently, **bovine somatotropin (bST)**, otherwise known as bovine growth hormone, has attracted much public attention. This hormone is used in the dairy industry to increase milk production. Although some people are concerned about the safety of bST, a substantial amount of research suggests that it is safe for both cow and consumer.[18] Thus the FDA allows its use. Even after being approved, the FDA continues to evaluate the safety of compounds such as bST. Clearly, each of these compounds requires intense scrutiny by the many federal agencies that seek to protect the safety and healthfulness of our food supply.

## Food Allergies and Sensitivities

Some compounds in foods can cause illness in small segments of the population. This is because some people are especially sensitive and sometimes even allergic to them. Because the percentage of individuals who have an adverse reaction to these compounds is small, their presence in food is not prohibited. For example, monosodium glutamate (MSG), which is used as a flavor enhancer, has been reported to cause severe headaches, facial flushing, and a generalized burning sensation in some people.[19] Because MSG is used extensively in Asian cuisines, this reaction is often referred to as **Chinese restaurant syndrome.** The presence of MSG must be declared on the label of any food to which it is added. Similarly, **sulfites** are added to some foods (such as wine) to enhance color and prevent spoilage, although consuming them causes breathing difficulties in sulfite-sensitive people. Another potentially harmful substance is aspartame, a commonly used artificial sweetener. Aspartame cannot be metabolized properly by people with phenylketonuria (PKU), and such people should avoid its consumption. Also, some people have allergies to compounds that occur naturally in foods. An example is peanut allergy. Foods that may contain such products must be so labeled.

> **CONNECTIONS** ▶ Recall that aspartame contains phenylalanine, which people with PKU cannot metabolize (Chapter 6, page 175).

## Other Potential Foodborne Toxins

Scientists and public health officials are continually trying to identify compounds in foods that may cause illness. Often these investigations receive attention in the popular press, causing public concern. Generally this is good, because public concern results in scientific scrutiny. A recent example is the investigation of a substance called **acrylamide.** Although acrylamide has probably always been in the food supply, its presence was first reported in 2002.[20] Acrylamide forms in starchy foods exposed to very high temperatures, such as french fries and potato chips. Because very high levels of acrylamide cause cancer in laboratory animals, there is concern that dietary acrylamide may be harmful to humans.[21] However, humans would need to eat an extremely large amount of french fries to consume the cancer-causing dose of acrylamide.[22] Nonetheless, there is much interest in determining whether dietary acrylamide and many other food-related compounds present a health risk to humans, and researchers continue to study these substances.

**bovine somatotropin (bST; bovine growth hormone)** A protein hormone produced by cattle and used in the dairy industry to enhance milk production.

**Chinese restaurant syndrome** Severe headaches and nausea reportedly caused by consuming large amounts of monosodium glutamate, in people who are sensitive to it.

**sulfite** A naturally occurring compound, in some foods, that can be used as a food additive to prevent discoloration and bacterial growth.

**acrylamide** (a – CRYL – a – mide) A compound that is formed in starchy foods (such as potatoes) when heated to high temperatures.

## ESSENTIAL *Concepts*

In addition to infectious agents of foodborne illness, inert (non-infectious) compounds in foods can be dangerous to consume. These include chemical hazards such as marine toxins, pesticides, herbicides, antibiotics, and hormones. In addition, some people are especially sensitive or allergic to selected components of foods, several of which are food additives. Therefore, foods that contain these compounds are often so labeled. Several federal agencies (including the USDA, FDA, and EPA), as well as the international agency FAO, are charged with determining whether the amounts of these compounds in foods present health risks to humans.

# Sources of Foodborne Contaminants and Ways to Decrease Contamination

Many pathogens and inert compounds cause foodborne illness, and it is beyond the scope of this book to describe each one in detail. However, it is useful to have a basic knowledge of the sources of the more common contaminants and how they are transmitted from food to food, also known as **cross-contamination,** and from person to person. In addition, it is important to understand how food manufacturers and consumers alike can keep food safe.

## Sources of Foodborne Pathogens and Toxins

Illness-causing pathogens can theoretically be found in all foods that are grown in nonsterile environments. However, most are incorporated into our foods via exposure to contaminated soil, polluted water, or infected animals. Because the various pathogens tend to prefer specific hosts (or foods), some foods are more common sources of particular pathogens. Some of these, described previously, are listed in Table 1. There is significant overlap among the foods that are likely to harbor illness-causing pathogens or toxins, with raw meat, fish, and dairy foods posing the greatest risk. Fruits and vegetables can also cause foodborne illness if grown in contaminated soil or handled by an infected person.

## Safe Food-Handling Techniques

A person infected with a pathogen can transmit it to almost any food. In this way, a "perfectly clean" food can be made unsafe. Therefore, it is important that food handlers, including those who harvest, process, and prepare foods, avoid coughing or sneezing on foods. In addition, many pathogens can pass through the gastrointestinal tract and be excreted in the feces. Thus fecal contamination can also cause foodborne illness. People who process and prepare food are generally required to wear gloves while handling food and thoroughly wash their hands after using the toilet.

## Food Production, Preservation, and Packaging Techniques

In addition to safe food-handling techniques, the food-processing industry seeks to keep food safe in many other ways. Clearly, the goal is to produce a product that is pathogen free and then to store it in a way that does not allow pathogenic growth. To help reach this goal, the USDA and the FDA have developed guidelines for food processors and handlers. For example, the **Hazard Analysis Critical Control Points (HACCP) system** was developed to help manufacturers identify critical points in their food-processing methods where food contamination is likely. In addition, the USDA requires that meat and poultry be inspected before sale and that instructions for safe handling practices be placed on packages. However, this mandatory inspection does not guarantee that the meat is pathogen free. Although we have no sure way to prevent a food from harboring dangerous pathogens or toxins, many techniques help keep food safe. In general, these remove potential organisms from the product or create an environment in which the organism cannot live.

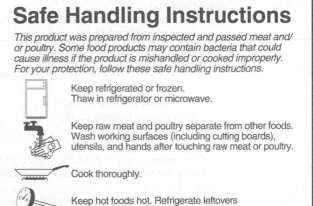

Courtesy FSIS/USDA

**cross-contamination** The transfer of microorganisms from one food to another or from one surface or utensil to another.

**Hazard Analysis Critical Control Points (HACCP) system** A food safety protocol used to decrease contamination of foods during processing.

## Drying, Salting, Smoking, and Fermentation

Meat has long been preserved by salting, smoking, and/or drying to produce ham, sausage, and jerky. People have used these techniques for centuries to preserve food because they remove the water needed for bacterial growth. Another technique called **fermentation** involves adding selected microorganisms to foods and is used in making sauerkraut, yogurt, and wine. During fermentation, growth of nonpathogenic organisms minimizes growth of pathogenic organisms.

## Heat Treatment: Cooking, Canning, and Pasteurization

Because most foodborne pathogens prefer living in an environment between 40 to 140°F, cooling or heating foods below and above this range helps prevent their growth. This temperature region (called the **danger zone**) should be avoided when storing and serving food. Heat can slow the growth of and/or destroy many bacteria, viruses, and parasites. Table 2 provides guidelines for cooking, serving, and reheating foods, and as you can see, to be kept safe different foods require different temperatures.

In addition to cooking, heat can also preserve foods for later consumption. For example, during the canning process foods are packaged (or "canned") in sanitized jars or cans and then heated at high temperatures. Aside from killing pathogens, canning also forms a vacuum within the jar. If you have ever helped someone can foods, you may recall the characteristic "pop" that you hear after the jar has been removed from the boiling water bath. This sound is caused by the formation of a vacuum inside the jar and results in an airtight (anaerobic) seal. The air-free environment helps preserve the food, because most foodborne illness-causing organisms require oxygen to grow. However, it is important to follow basic guidelines to assure that the finished product can be safely stored. Some organisms, such as *Clostridium botulinum*, can survive in low-acid, anaerobic conditions. Making sure that home-canned foods have the proper acidity and are heated sufficiently helps keep them safe. The USDA maintains a set of guidelines for canning that can be easily accessed at http://www.nal.usda.gov/fnic/.

Perhaps one of the most common forms of heat treatment used in food preservation is **pasteurization**. This process was named for the French scientist Louis

**fermentation** Metabolism, by bacteria, that occurs under relatively anaerobic conditions.

**danger zone** The temperature range between 40 and 140°F in which pathogenic organisms grow most readily.

**pasteurization** (pas – ter – i – ZA – tion) A food preservation process that subjects foods to heat to kill bacteria, yeasts, and molds.

---

**TABLE 2    Guidelines for Cooking, Serving, and Reheating Foods to Prevent Foodborne Illness**

Note that internal temperatures should be measured with a thermometer.

| Cooking | Beef and Pork |
|---|---|
| | • Cook **beef roasts** and **steaks** to 145°F for medium rare or to 160°F for medium. |
| | • Cook **ground beef** to at least 160°F. |
| | • Cook **raw sausages** to 160°F. |
| | • Reheat **ready-to-eat sausages** to 165°F. |
| | • Cook **pork roasts, chops, or ground patties** to 160°F for medium or 170°F for well done. |
| | Poultry |
| | • Cook **whole poultry** to 180°F. |
| | • Cook **chicken breasts** to 170°F. |
| | • Cook **stuffing** to 165°F. |
| | Eggs |
| | • Cook **eggs** until the yolks and whites are firm. |
| | • Don't use recipes in which eggs remain raw or only partially cooked. |
| | Fish |
| | • Cook **fish** until it is opaque and flakes easily with a fork. |
| | • For food safety reasons, avoid eating uncooked **oysters** or **shellfish.** People with liver disorders or weakened immune systems are especially at risk for getting sick. |
| Serving | Keep foods at 140°F or higher until served. |
| | Keep foods hot with chafing dishes, crock pots, and warming trays. |
| Reheating | Reheat leftovers to at least 165°F. |
| | When using a microwave oven, make sure food is evenly heated. |

SOURCE: Adapted from the Partnership for Food Safety Education and FightBAC!® (http://www.fightbac.org).

Pasteur, who discovered in the 1860s that spoilage of wine and beer could be inhibited by short periods of intense heating. Indeed, pasteurization is still commonly used today. Foods that are typically pasteurized include milk, juice, spices, ice cream, and cheese. Pasteurization is especially good at killing some forms of *E. coli*, *Salmonella*, *Campylobacter jejuni*, and *Listeria monocytogenes*. A recent outbreak of *Salmonella* poisoning from unpasteurized apple cider prompted laws that require commercially available apple cider to be pasteurized or labeled as being nonpasteurized, and the FDA recommends that homemade cider be heated for 30 minutes at 155°F or 15 seconds at 180°F.

## Cold Treatment: Cooling and Freezing

Foods also spoil less quickly if kept cold. Although this concept seems simple to us today, the successful use of freezing for mass food production was not developed until the early 1900s. At this time, an arctic naturalist named Clarence Birdseye noticed that fish and caribou meat remained surprisingly fresh and tender when exposed to freezing arctic air. On returning home, Birdseye revolutionized food preservation by rapid freezing at extremely low temperatures. The company he founded, Birds Eye Foods™, thrives to this day. To slow or halt the growth of microorganisms, foods should be refrigerated at 40°F or frozen soon after they are prepared. This helps prevent foods from staying in the danger zone for an extended period of time.

## Irradiation

In the 1950s the National Aeronautics and Space Administration (NASA) first used **irradiation** to preserve food for space travel. In 1963 the FDA approved a form of food processing called irradiation, sometimes called "cold pasteurization," as a form of food preservation. Today, irradiation is approved for meat, poultry, shellfish, eggs, and other foods such as fresh fruits, vegetables, and spices. During irradiation, foods are exposed to radiant energy that damages or kills bacteria. It is important to understand that irradiation neither damages nutrients nor makes the foods radioactive—just as the use of x-rays to inspect luggage at an airport does not make luggage radioactive. Irradiation makes foods safer to eat and can dramatically increase their shelf life. For example, compared to nonirradiated strawberries, which have a brief shelf life, irradiated strawberries last for several weeks without spoiling. Irradiation should not be viewed as a substitute for proper food-manufacturing and -handling procedures but is simply an added measure of safety. Irradiated foods must be labeled with the *radura* symbol (Figure 2).

**FIGURE 2    The *Radura* Symbol**

## ESSENTIAL *Concepts*

Most foods have the potential to harbor pathogens or other compounds that cause foodborne illness. Those that tend to pose the greatest risks are raw or undercooked meat, fish, unpasteurized dairy products, and fresh fruits and vegetables. The food-processing industry takes many precautions to help prevent food contamination. These include safe food-handling techniques and methods of packaging and preservation. For example, salting, drying, and smoking remove water from foods, inhibiting pathogenic growth. Fermentation involves adding nonpathogenic organisms that inhibit the growth of dangerous ones. Other methods slow the growth of or even kill pathogens by heat or cold treatment. Irradiation is also used with some foods to damage or kill bacteria, decreasing risk of foodborne illness.

# Food Safety Practices for Consumers

Consumers can do many things to reduce the risk of foodborne illness. First, they should stay aware of recommendations by regulatory groups. For example, in 2001 the FDA warned consumers to avoid eating a certain kind of lollipop because of lead contamination. Similarly, the FDA and the EPA in 2004 announced concern about the mercury content of some forms of fish, such as shark, swordfish, king mackerel, and tilefish. Pregnant women, breastfeeding women, and young children are advised to reduce their consumption of these foods. Both the USDA and FDA maintain websites containing information concerning current food safety recommendations.

**irradiation** A food preservation process that applies radiant energy to foods, to kill bacteria.

Many additional resources are available to you for checking advisories, including the USDA's Meat and Poultry Hotline (1-888-MPHotline) and the FDA's Food Safety Information Hotline (1-888-SAFEFOOD).

Although issues related to foodborne illness change over time, consumers are encouraged to follow some basic rules. The USDA and the Partnership for Food Safety Education have developed a set of food safety guidelines, called **FightBAC!®**, which includes a consumer-friendly icon (Figure 3). The FightBAC!® campaign is a public education program focused on reducing foodborne illness. It specifically addresses avoidance of infectious agents of foodborne illness and emphasizes four main guidelines—*clean, separate, cook,* and *chill.* In addition, the USDA's Dietary Guidelines for Americans recommends that we "keep food safe to eat." A summary of these recommendations is provided next.

## Clean Hands and Cooking Utensils

To remove and/or kill pathogens on hands and cooking utensils, frequent and adequate washing with hot, soapy water is recommended. Cutting boards should be washed after use, and kitchen towels and wash cloths should be washed frequently. To prevent contamination of foods with pathogens found in fecal materials, wash hands after using the bathroom, changing diapers, or handling pets. To be effective, wash hands vigorously with soap for at least 20 seconds and rinse them thoroughly under clean, running warm water. Dry hands using clean paper towels or clean cloth towels. The use of antimicrobial gels can also help assure that hands are pathogen free, although this should not be considered as a substitute for hand washing.

### FIGURE 3    FightBAC!®

The FightBAC!® campaign was created by the Partnership for Food Safety Education, a nonprofit organization, in partnership with the USDA, the FDA, and the CDC. FightBAC!® was developed to reduce the incidence of foodborne illness by educating Americans about safe food handling practices at home and at work.

## Wash Fresh Fruits and Vegetables, But Not Meat

Washing fresh fruits and vegetables provides an important protection from foodborne pathogens and noninfectious agents such as pesticides. To do this effectively, wash produce under clean, cool, running water and, if possible, scrub with a clean brush. Fruits and vegetables should then be dried with a clean paper or cloth towel. Do not wash raw meat, poultry, and fish, because doing so increases the danger of cross-contaminating otherwise noninfected surfaces and foods. In other words, washing contaminated meat and fish can cause pathogens to splatter onto previously clean foods and objects—increasing risk of foodborne illness.

## Separate Foods: Avoid Cross-Contamination

To prevent transferring disease-causing pathogens from food-to-food (cross-contamination), separate raw meat and seafood from other foods in the grocery cart, in the refrigerator, and during food preparation. For example, put raw meats in separate, sealed plastic bags in the grocery cart and use separate cutting boards when preparing fresh produce and raw meat. Also, do not put cooked meat back on the same plate that was used to hold the raw meat unless the plate has been thoroughly washed with hot, soapy water.

## Cook Foods to the Proper Temperature

As previously described, heat kills most dangerous organisms. It can also alter the chemical composition of some preformed toxins, making them less dangerous. As shown in Table 2, the temperatures and cooking times required to kill pathogens depend on the particular food and the organism or toxin. Measure the *internal temperature* of a cooked food, and thoroughly clean meat thermometers between uses. Also, just because a food has been previously cooked does not mean that it is pathogen free. Reheat foods to their appropriate internal temperatures before eating them. This step may be especially important for pregnant women, infants, older adults, and people with impaired immunity such as those with HIV or cancer. For example, compared to other people, pregnant women and children are more likely to become seriously ill when exposed to *Listeria monocytogenes,* an organism sometimes found in lunch meats. A *Listeria*

**FightBAC!®** A public education program developed to reduce foodborne bacterial illness.

*monocytogenes* outbreak in the late 1990s that resulted in at least 6 deaths and two spontaneous abortions has prompted the recent recommendation that lunch meats and frankfurters be heated before eating.[23]

## Keep Foods Cold, and Chill Food Quickly

Because cold temperatures can slow the growth of microorganisms, perishable foods should be refrigerated to 40°F or colder, or frozen, within two hours. Do not marinate or thaw foods at room temperature. Instead, marinate foods in the refrigerator, and thaw foods either in the refrigerator or under cold, running water. Refrigerate foods quickly, separating large amounts of foods into small, shallow containers to allow foods to cool more rapidly. Even properly chilled foods can become sources of foodborne illness, and in many cases it is difficult to determine whether a food is safe to eat after it has been stored. For this reason, use refrigerated leftovers within three to four days of storage, and "If in doubt, throw it out!" Labeling a food storage container with the date it was originally refrigerated can help you determine if food should be thrown away.

## Consumer Tips for Eating Out, Picnics, and Traveling

Making sure food is safe to eat can be especially difficult when eating at restaurants, picnics, and buffets, and when traveling. However, many things can help avoid foodborne illness in these situations. In general, ask yourself whether the four basic concepts of the Fight Bac!® program have been followed:

- Did the people handling and preparing the foods use sanitary practices?
- Were raw foods kept separate from cooked foods? In other words, was cross-contamination avoided?
- Were foods properly cooked and kept out of the "danger zone?"
- Were cold foods kept cold?

Unfortunately, it is common at picnics and social gatherings to neglect these basic rules of food safety, resulting in foodborne illness. Therefore, use good judgment and eat only foods you know to be safe. Better safe than sorry!

Travelers have added concerns related to food safety. In general, when traveling, camping, or hiking, take special care to drink only water that is known to be safe. Because water can harbor dangerous organisms, it is wise to drink bottled water and avoid using ice when traveling in less industrialized regions of the world. In addition, when backpacking or camping in the wilderness, boil, filter, or treat water before drinking. Although fresh produce contaminated with bacteria may not cause illness for local people, it can often cause serious illness (usually diarrhea) in visitors. Thus, when traveling to foreign locations avoid or carefully wash fresh fruits and vegetables that are not peeled. No one can completely prevent foodborne illness. However, when traveling you can greatly reduce it by being vigilant in choosing which foods and beverages to consume—and which to avoid.

## Special Considerations for At-Risk Populations

Although we are all at risk for getting foodborne illness, certain people are more susceptible than others. These include infants, pregnant women, the elderly, and those who are immunologically challenged, such as people with HIV, transplant recipients, those receiving cancer therapy, and people taking immunosuppressive drugs (such as steroids). These people are at increased risk of both *getting* foodborne illness and *having more severe symptoms.*

The USDA Dietary Guidelines for Americans has several specific recommendations for these at-risk populations. For example, they recommend that pregnant women, older adults, and people who are immunocompromised avoid certain lunch meats and frankfurters that have not been reheated. In addition, at-risk individuals should take special care to not eat or drink raw (unpasteurized) milk or any products made from raw milk, raw or partially cooked eggs, raw or undercooked meat and fish, unpasteurized juices, and raw alfalfa sprouts.

## Emerging Issue of Food Biosecurity

Recent events in the United States and abroad such as the September 11, 2001, attacks have raised new concerns about the safety of our food. People quickly realized that terrorists and other malicious people can do serious and widespread harm by contaminating the nation's food supply. **Food biosecurity**—the prevention of terrorist attacks on our food supply—has gained intense national attention, and Congress authorized the "Public Health Security and Bioterrorism Preparedness and Response Act" (also called the **Bioterrorism Act**) in 2002.[24] Since then, additional regulations have been approved to help ensure that foods grown and produced both in the

---

**food biosecurity** Measures aimed at preventing the food supply from falling victim to planned contamination.

**Bioterrorism Act** Federal legislation aimed to ensure the continued safety of the U.S. food supply from intentional harm by terrorists.

United States and abroad cannot be tampered with. Clearly, this focus of deep individual and governmental interest will remain strong in coming years.

## ESSENTIAL *Concepts*

It is important to help ensure that food remains safe to eat. Precautions include proper cleaning, handling, heating, chilling, and storing techniques. Both the FightBAC!® program and the Dietary

Guidelines for Americans make recommendations in this regard. These can be summarized as follows: *Clean, separate, cook,* and *chill.* These food safety recommendations are even more important for people who eat out often, attend picnics frequently, and travel—especially when traveling to foreign lands or in the wilderness. In addition, recent concern that terrorists can cause serious outbreaks of foodborne illness has prompted governmental programs targeted at assuring food biosecurity in the United States.

## Notes

1. Mead PS, Slutsker L, Dietz V, McCaig LF, Bresee JS, Shapiro C, Griffin PM, Tauxe RV. Food-related illness and death in the United States. Emerging Infectious Diseases. 1999;5:607–25.

2. Trabulsi LR, Keller R, Tardelli Gomes TA. Typical and atypical enteropathogenic *Escherichia coli.* Emerging Infectious Diseases. 2002;8:508–13.

3. Centers for Disease Control and Prevention. Outbreak of aflatoxin poisoning—eastern and central provinces, Kenya, January–July 2004. Morbidity and Mortality Weekly Report. 2004;53:790–3.

4. Peipins LA, Highfill KA, Barrett E, Monti MM, Hackler R, Iluang P, Jiang X. A Norwalk-like virus outbreak on the Appalachian Trail. Journal of Environmental Health. 2002;64:18–23.

5. Anderson AD, Garrett VD, Sobel J, Monroe SS, Fankhauser RL, Schwab KJ, Bresee JS, Mead PS, Higgins C, Campana J, Glass RI, Outbreak Investigation Team. Multistate outbreak of Norwalk-like virus gastroenteritis associated with a common caterer. American Journal of Epidemiology. 2001;154:1013–19.

6. Yates J. Traveler's diarrhea. American Family Physician. 2005;71:2095–100.

7. Centers for Disease Control and Prevention. Salmonella isolates from humans in the United States, 1984–1986. Morbidity and Mortality Weekly Report. 1988;37:25–31.

8. Centers for Disease Control and Prevention. Outbreaks of *Salmonella* infections associated with eating Roma tomatoes—United States and Canada, 2004. Morbidity and Mortality Weekly Report. 2005,54:325–8.

9. Centers for Disease Control and Prevention. Outbreak of *Salmonella* serotype Muenchen infections associated with unpasteurized orange juice—United States and Canada, 1999. Morbidity and Mortality Weekly Report. 1999;48:582–5.

10. Gill CJ, Keene WE, Mohle-Boetani JC, Farrar JA, Waller PL, Hahn CG, Cieslak PR. Alfalfa seed decontamination in a *Salmonella* outbreak. Emerging Infectious Diseases. 2003;9:474–9. Taormina PJ, Beuchat LR, Slutsker L. Infections associated with eating seed sprouts: an international concern. Emerging Infectious Diseases. 1999; 5:626–34. Thomas JL, Palumbo MS, Farrar JA, Farver TB, Cliver DO. Industry practices and compliance with U.S. Food and Drug Administration guidelines among California sprout firms. Journal of Food Protection. 2003;66:1253–59.

11. Besser RE. *Escherichia coli* O157:H7 gastroenteritis and the hemolytic uremic syndrome: an emerging infectious disease.

Annual Review of Medicine (Annual Reviews Inc., Palo Alto, CA). 1999;50:355–67.

12. Bell BP, Goldoft M, Griffin PM, Davis MA, Gordon DC, Tarr PI, Bartleson CA, Lewis JH, Barrett TJ, Wells JG. A multistate outbreak of *Escherichia coli* O157:H7—associated bloody diarrhea and hemolytic uremic syndrome from hamburgers. The Washington experience. JAMA (Journal of the American Medical Association).1994; 17:1349–53. Tuttle J, Gomez T, Doyle MP, Wells JG, Zhao T, Tauxe RV, Griffin PM. Lessons from a large outbreak of *Escherichia coli* O157:H7 infections: insights into the infectious dose and method of widespread contamination of hamburger patties. Epidemiology and Infection. 1999;122:185–92.

13. Centers for Disease Control and Prevention. Outbreaks of *Escherichia coli* O157:H7 infection and Cryptosporidiosis associated with drinking unpasteurized apple cider—Connecticut and New York, October 1996. Morbidity and Mortality Weekly Report. 1997;46:4–8. Centers for Disease Control and Prevention. Salmonellosis associated with a Thanksgiving dinner—Nevada, 1995. Morbidity and Mortality Weekly Report. 1996;45:1016-17. Silapalasingam S, Friedman CR, Cohen L, Tauxe R. Fresh produce: a growing cause of outbreaks of foodborne illness in the United States, 1973–1997. Journal of Food Protection. 2004;67:2342–52.

14. Castilla J, Saa P, Hetz C, Soto C. In vitro generation of infectious scrapie prions. Cell. 2005;121:195–206. Wickner RB, Edskes HK, Roberts BT, Baxa U, Pierce MM, Ross ED, Brachmann A. Prions: proteins as genes and infectious entitites. Genes and Development. 2004;18:470–85.

15. Mostl K. Bovine spongiform encephalopathy (BSE): the importance of the food and feed chain. Forum in Nutrition. 2003;56:394–96.

16. Beghi E, Gandolfo C, Ferrarese C, Rizzuto N, Poli G, Tonini MC, Vita G, Leone M, Logroscino G, Granieri E, Salemi G, Savettieri G, Frattola L, Ru G, Mancardi GL, Messina C. Bovine spongiform encephalopathy and Creutzfeldt-Jakob disease: facts and uncertainties underlying the causal link between animal and human diseases. Neurological Sciences. 2004;25:122–9. Belay ED, Schonberger LB. The public health impact of prion diseases. Annual Review of Public Health. 2005;26:191–212. Ironside JW, Head MW. Variant Creutzfeldt-Jakob disease: risk of transmission by blood and blood products. Haemophilia. 2004;4:64–9. Roma AA, Prayson RA. Bovine spongiform

encephalopathy and variant Creutzfeldt-Jakob disease: How safe is eating beef? Cleveland Clinic Journal of Medicine. 2005;72:185–94.

17. Brett MM. Food poisoning associated with biotoxins in fish and shellfish. Current Opinion in Infectious Diseases. 2003;16:461–5.

18. Anonymous. Bovine somatotropin and the safety of cows' milk: National Institutes of Health technology assessment conference statement. Nutrition Reviews. 1991;49:227–32. Bauman DE. Bovine somatotropin and lactation: from basic science to commercial application. Domestic Animal Endocrinology. 1999;17:101–16. Daughaday WH, Barbano DM. Bovine somatotropin supplementation of dairy cows. Is the milk safe? JAMA (Journal of the American Medical Association). 1990;264:1003–05. Etherton TD, Kris-Etherton PM, Mills EW. Recombinant bovine and porcine somatotropin: safety and benefits of these biotechnologies. Journal of the American Dietetic Association. 1993;93:177–80. Juskevich JC, Guyer CG. Bovine growth hormone: human food safety evaluation. Science. 1990;249:875–84.

19. Geha RS, Beiser A, Ren C, Patterson R, Greenberger PA, Grammer LC, Ditto AM, Harris KE, Shaughnessy MA, Yarnold PR, Corren J, Saxon A. Review of alleged reaction to monosodium glutamate and outcome of a multicenter double-blind placebo-controlled study. Journal of Nutrition. 2000;130:1058S–62S. Walker R, Lupien JR. The safety evaluation of monosodium glutamate. Journal of Nutrition. 2000;130:1049S–52S.

20. Stadler RH, Blank E, Varga N, Robert F, Hau J, Guy PA, Robert MC, Riediker S. Food chemistry: acrylamide from Maillard reaction products. Nature. 2002;419:449–50.

21. Bull RJ, Robinson M, Laurie RD, Stoner GD, Greisiger E, Meier RJ, Stober J. Carcinogenic effects of acrylamide in Sencar and A/J mice. Cancer Research. 1984;44:107–11.

22. Anand V. Acrylamide and fried foods: should we freak out? Nutrition Byte. 2003;9(1), Article 7. Available from: http://repositories.cdlib.org/uclabiolchem/nutritionbytes/vol9/iss1/art7. Mucci LA, Dickman PW, Steineck G, Adami H-O, Augustsson K. Dietary acrylamide and cancer of the large bowel, kidney, and bladder: absence of an association in a population-based study in Sweden. British Journal of Cancer. 2003;88:84–9.

23. Centers for Disease Control and Prevention. Multistate outbreak of Listeriosis—United States, 1990. Morbidity and Mortality Weekly Report. 1998;47:1085–86. Centers for Disease Control and Prevention. Update: Multistate outbreak of Listeriosis—United States, 1998–1999. Morbidity and Mortality Weekly Report. 1999;47:1117–8.

24. United States Congress. Public health security and bioterrorism preparedness and response act of 2002. June 12, 2002. Available from: http://www.fda.gov/oc/bioterrorism/bioct.html.

**Check out the following sources for additional information.**

25. Centers for Disease Control and Prevention. Surveillance for foodborne disease outbreaks—1993–1997. Morbidity and Mortality Weekly Review. 2000;49:1–51.

26. Etherton TD, Bauman DE. Biology of somatotropin in growth and lactation of domestic animals. Physiological Reviews. 1998;78:745–61.

27. Mepham TB. Symposium on "The ethics of food production and consumption: the role of food ethics in food policy." Proceedings of the Nutrition Society. 2000;59:609–18.

28. Murano PS. Understanding food science and technology. Belmont, CA: Wadsworth; 2003.

29. National Center for Health Statistics. Health, United States, 2004; with Chartbook on Trends in the Health of Americans. Hyattsville, MD: National Center for Health Statistics; 2004.

30. Neill MA. Foodborne illness and food safety. In: Present knowledge in nutrition, 8th ed. Bowman BA, Russell RM, editors. Washington, DC: ILSI Press; 2001.

31. U.S. Department of Health and Human Services, U.S. Department of Agriculture. Dietary Guidelines for Americans 2005. Washington, DC: U.S. Government Printing Office; 2005. Available from: http://www.healthierus.gov/dietaryguidelines.

32. Woteki CE, Kineman BD. Challenges and approaches to reducing foodborne illness. Annual Review of Nutrition (Annual Reviews Inc., Palo Alto, CA). 2003;23:315–44.

Chapter 7

# Lipids

*L*ipids are required for every physiological system in the body, and are thus essential nutrients. For many people, the thought of fatty foods conjures up images of unhealthy living. We often shop for "fat free" foods and try to avoid fats all together. Food manufacturers have even developed "fat substitutes" to replace the fats normally found in food. However, although diets high in fat can lead to health complications such as obesity and heart disease, getting enough of the right types of fat is just as essential for optimal health. In this chapter, we discuss the variety of fats and oils in foods and how our bodies use them to create substances vital for health. Guidelines for lipid intake are also discussed.

Photos © Matthew Farruggio

221

# Lipids, Fats, and Oils: What's the Difference?

It is important to begin by defining the terms *lipids*, *fats*, and *oils*, because they are often used interchangeably and sometimes incorrectly. Fats and oils are examples of what chemists call **lipids,** which are relatively insoluble in water and relatively more soluble in organic solvents such as fingernail polish remover and paint thinner. Lipids are hydrophobic ("water fearing"), meaning that they do not mix readily with water, and are made mostly of carbon, hydrogen, and oxygen atoms. Some lipids also contain phosphorus and nitrogen atoms. Lipids that are liquid at room temperature are **oils,** and those that are solid at room temperature are **fats.** The major lipid classes include the fatty acids, triglycerides, phospholipids, sterols, and fat-soluble vitamins. Fatty acids, triglycerides, phospholipids, and sterols are discussed in detail in this chapter, and the fat-soluble vitamins are presented in Chapter 11.

> **CONNECTIONS** ▶ Remember that hydrophobic compounds are those that do not readily mix with water, because they are nonpolar (Chapter 3, p. 70).

# Structure and Classification of Fatty Acids

The most abundant lipids are **fatty acids,** which are made entirely of carbon, hydrogen, and oxygen atoms (Figure 7.1). A fatty acid consists of a chain of carbon atoms, which form the fatty acid backbone. One end of the fatty acid backbone contains a carboxylic acid group ($-COOH$), and this is called the **alpha (α) end.** The other end of the fatty acid backbone contains a methyl group ($-CH_3$), and this is called the **omega (ω) end.**

## Number of Carbons (Chain Length)

Fatty acids differ from each other in several ways. First, the numbers of carbon atoms in the backbone of the fatty acid can differ (Figure 7.2), and the number of carbons is called the fatty acid's **chain length.** Another way to think about this is to consider each fatty acid as if it were a chain, with each link in the chain representing a carbon atom. Most naturally occurring fatty acids have an even number of carbons—usually 12 to 22—although some may be as short as 4 or as long as 26 carbons. Fatty acids with fewer than 8 carbons are

**lipid** (*lipos*, fat) Organic substance that is relatively insoluble in water and soluble in organic solvents.

**oil** (*oleum*, olive) A lipid that is liquid at room temperature.

**fat** A lipid that is solid at room temperature.

**fatty acid** A lipid consisting of a chain of carbons with a methyl ($-CH_3$) group on one end and a carboxylic acid group ($-COOH$) on the other.

**alpha (α) end** (*alpha*, the first letter in the Greek alphabet) The end of a fatty acid with the carboxylic acid ($-COOH$) group.

**omega (ω) end** (*omega*, the final letter in the Greek alphabet) The end of a fatty acid with the methyl ($-CH_3$) group.

**chain length** The number of carbons in a fatty acid's backbone.

## FIGURE 7.1 Fatty Acid Structure

All fatty acids have three components: a carboxylic acid or alpha (α) end ($-COOH$), a methyl or omega (ω) end ($-CH_3$), and a fatty acid backbone.

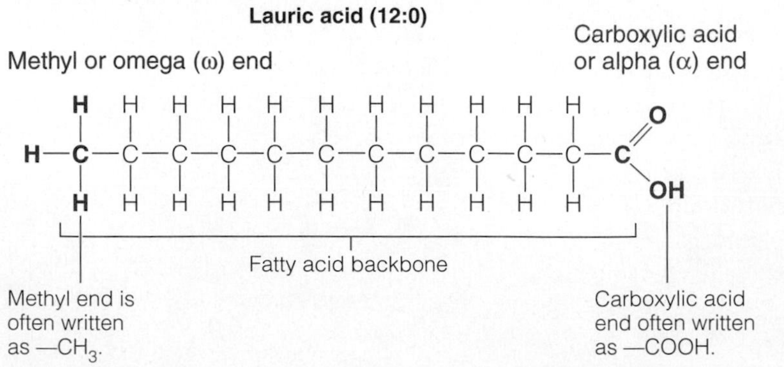

Lauric acid (12:0)

Methyl or omega (ω) end

Carboxylic acid or alpha (α) end

Fatty acid backbone

Methyl end is often written as $-CH_3$.

Carboxylic acid end often written as $-COOH$.

## FIGURE 7.2 Fatty Acids Can Have Different Chain Lengths

Fatty acids can differ in the numbers of carbons making up their backbones.

**Medium-chain fatty acid**

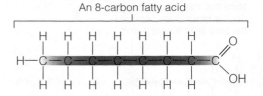

An 8-carbon fatty acid

**Long-chain fatty acid**

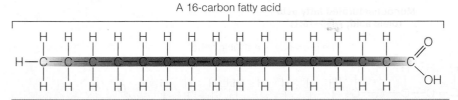

A 16-carbon fatty acid

called short-chain fatty acids; those with 8 to 12 carbons are medium-chain fatty acids; those with more than 12 carbons are long-chain fatty acids.

The chain length of a fatty acid affects its chemical properties and physiological functions. For example, chain length influences the stability of fatty acids. Molecules with high stability tend to have high melting and boiling points. Molecules with low stability lose their structure more easily and tend to have low melting and boiling points. Because long-chain fatty acids tend to be more stable than short-chain fatty acids, they have higher melting and boiling points than short-chain fatty acids. If everything else is equal, long-chain fatty acids tend to be solid (fat) at room temperature, whereas short-chain fatty acids tend to be liquid (oil).

Chain length also affects solubility in water, with short-chain fatty acids being more water soluble than long-chain fatty acids. Because humans are mostly water, it is easier for us to handle water-*soluble* substances such as short-chain fatty acids. Conversely, we must employ complex processes to absorb, transport, and use dietary lipids that are more water *insoluble*, such as long-chain fatty acids.

## Number and Positions of Double Bonds

Another way that fatty acids differ is by the types of chemical bonds between the carbon atoms (Figure 7.3). These carbon–carbon bonds can either be single bonds or double bonds. If a fatty acid contains all single carbon–carbon bonds, it is a **saturated fatty acid (SFA)**; those containing one or more double bonds are **unsaturated fatty acids.** More specifically, fatty acids with one double bond are **monounsaturated fatty acids (MUFAs)**; those with two or more double bonds are **polyunsaturated fatty acids (PUFAs)**. Note that for each carbon–carbon double bond that is present, 2 hydrogen atoms are lost from the fatty acid backbone.

Like chain length, the number of double bonds influences the physical nature of the fatty acid. For example, SFAs have straight, unbending structures that stack neatly together, whereas unsaturated fatty acids can bend and there-

**saturated fatty acid (SFA)** (*saturare,* to fill or satisfy) A fatty acid that contains only carbon–carbon single bonds in its backbone.

**unsaturated fatty acid** A fatty acid that contains at least one carbon–carbon double bond in its backbone.

**monounsaturated fatty acid (MUFA)** A fatty acid that contains one carbon–carbon double bond in its backbone.

**polyunsaturated fatty acid (PUFA)** A fatty acid that contains more than one carbon–carbon double bond in its backbone.

CONNECTIONS Recall that a carbon atom has 4 electrons in its outermost shell, so it can share 4 electrons. Thus the total number of bonds that each carbon atom can have in a fatty acid is 4, with single bonds counting as 1 and double bonds counting as 2 (Chapter 3, p. 69).

## FIGURE 7.3 Saturated and Unsaturated Fatty Acids

Fatty acids can differ by whether or not they have any carbon–carbon double bonds.

**Saturated fatty acid
(stearic acid; 18:0)**

Note that this carbon-oxygen double bond does not make the fatty acid "unsaturated."

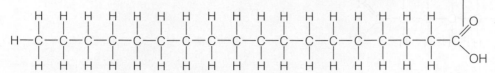

**Monounsaturated fatty acid
(oleic acid; *cis* 9–18:1)**

The presence of a double bond bends the fatty acid backbone.

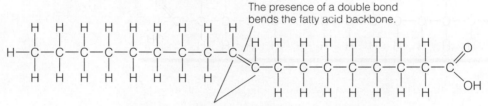

There are two fewer hydrogens for each double bond.

**Polyunsaturated fatty acid
(linoleic acid; *cis* 9, *cis* 12–18:2)**

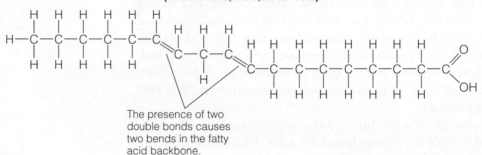

The presence of two double bonds causes two bends in the fatty acid backbone.

fore are more disorganized. As illustrated in Figure 7.3, each carbon atom in a SFA is surrounded (or "saturated") by hydrogen atoms. Being "saturated" with hydrogens inhibits the fatty acid from bending. Because of this rigidity, SFAs are highly organized, making them solid at room temperature. Lipids containing large amounts of SFA (such as butter) tend to be solid fats.

Fatty acids with double bonds have fewer hydrogens and can bend. In fact, whenever there is a carbon–carbon double bond there is a kink or a bend in the fatty acid backbone. These bends cause unsaturated fatty acids to become disorganized. In other words, the bends prevent the fatty acid from becoming densely packed. Imagine the difference between the organization of uncooked spaghetti and elbow macaroni. The straight spaghetti noodles are neatly organized, whereas bent elbow macaroni noodles are disorganized. This is similar to the difference between SFAs (uncooked spaghetti) and PUFAs (elbow macaroni). In general, organized molecules such as SFAs are solid (fats) at room temperature, and disorganized molecules like PUFAs are liquid (oils). MUFAs have chemical characteristics that lie between those of SFAs and

## FIGURE 7.4   *Cis* versus *Trans* Fatty Acids

Unsaturated fatty acids can differ by whether they have *cis* or *trans* carbon–carbon double bonds. *Cis* bonds cause the fatty acid to bend, whereas *trans* bonds do not.

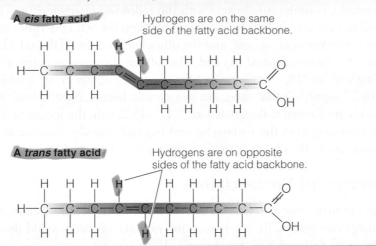

PUFAs, being thick liquids or soft solids at room temperature. You may have noticed that olive oil, which is high in MUFAs, is a thick oil.

### *Cis* versus *Trans* Fatty Acids

Unsaturated fatty acids can differ in the way in which hydrogen atoms are arranged around the carbon–carbon double bonds. Most naturally occurring fatty acids have the hydrogen atoms positioned on the same side of the double bond, resulting in a **cis** **double bond** (Figure 7.4). When the hydrogen atoms are on opposite sides of the double bond, it is called a **trans** **double bond.** Fatty acids containing at least one *trans* double bond are called **trans** **fatty acids,** and have fewer bends in their backbones than their *cis* counterparts. For this reason, *trans* fatty acids are also more likely to be solid (fats) at room temperature.

## ESSENTIAL *Concepts*

There are several kinds of lipids. One type is a fatty acid, which is made of carbon, hydrogen, and oxygen atoms. Fatty acids differ in their chain lengths as well as in number of carbon–carbon double bonds and in types of carbon–carbon double bonds. Saturated fatty acids (SFAs) contain only carbon–carbon single bonds, monounsaturated fatty acids (MUFAs) have one carbon–carbon double bond, and polyunsaturated fatty acids (PUFAs) have two or more carbon–carbon double bonds. Double bonds can be in either the *cis* or *trans* configuration. These differences influence the physical and chemical properties of fatty acids, such as solubility in water and whether they are liquid (oils) or solid (fats) at room temperature.

## Fatty Acid Nomenclature

Many methods are used to describe or label fatty acids. In general, these methods name fatty acids based on the number of carbons, the number and types of double bonds, and the positions of the double bonds. In addition, some fatty acids have "common names." Three methods for naming fatty acids are described here.

**cis double bond** (*cis,* on this side of) A carbon–carbon double bond in which the hydrogen atoms are arranged on the same side of the double bond.

**trans double bond** (*trans,* across) A carbon–carbon double bond in which the hydrogen atoms are arranged on opposite sides of the double bond.

**trans fatty acid** A fatty acid containing at least one *trans* double bond.

## Alpha (α) Nomenclature

The alpha (α) nomenclature system is based on the positions and types of double bonds relative to the carboxylic acid (α) end of the fatty acid. As an example, consider a fatty acid with the following characteristics: 18 carbons and two *cis* double bonds, one being between the 9th and 10th carbons *from the carboxylic acid* (α) *end*, and the other between the 12th and 13th carbons *from the carboxylic acid* (α) *end*. The fatty acid's name is constructed beginning with an "18," signifying that there are 18 carbons. A "2" is added to form "18:2," signifying that there are two double bonds. Next, where the double bonds are located is designated as "9,12"–18:2, with the locations determined by counting from the carboxylic acid (α) end. Finally, because both double bonds are in the *cis* configuration, the name is modified to *cis*9,*cis*12–18:2.

## Omega (ω) Nomenclature

An alternate system for naming a fatty acid is sometimes referred to as the omega (ω) system. In this system, the number of carbons and double bonds are again distinguished (for example, 18:2).

Using the omega nomenclature system fatty acids are categorized into groups based on where the first double bond is located *relative to the methyl* (ω) *end* of the molecule. If the first double bond is between the third and fourth carbons from the ω end, the fatty acid is an **omega-3 (ω-3) fatty acid.** If the first double bond is between the sixth and seventh carbons, it is an **omega-6 (ω-6) fatty acid.** There are also ω-7 and ω-9 fatty acids. For example, oleic acid (a major fatty acid in olive oil and milk) is an ω-9 fatty acid.\* Unlike the α nomenclature, the ω nomenclature does not usually identify whether the double bonds are in the *cis* or *trans* configuration.

## Common Names

Imagine having to describe the chain length, bond position, and bond configuration each time you wanted to mention a specific fatty acid. For this reason, common names are often used. Many common names reflect a prominent food source of the fatty acid. For example, *palmitic* acid (16:0) is found in *palm* oil, and *arachidonic* acid (*cis*5,*cis*8,*cis*11,*cis*14–20:4; from *arachis*, meaning peanuts) is found in peanuts. Some common names of fatty acids are shown in Table 7.1.

## ESSENTIAL *Concepts*

There are several ways to name fatty acids. Alpha (α) nomenclature designates the number of carbon atoms, the number and placement of double bonds, and the type of double bonds relative to the carboxylic acid (α) end. Omega (ω) nomenclature is similar, although the placement of double bonds is determined from the methyl (omega or ω) end. If the first double bond is between the third and fourth carbons from the omega (ω) end, it is an omega-3 (ω-3) fatty acid. If the first double bond is between the sixth and seventh carbons from the omega (ω) end, it is an omega-6 (ω-6) fatty acid. Many fatty acids also have common names.

**omega-3 (ω-3) fatty acid** A fatty acid in which the first double bond is located between the third and fourth carbons from the methyl or omega (ω) end.

**omega-6 (ω-6) fatty acid** A fatty acid in which the first double bond is located between the sixth and seventh carbons from the methyl or omega (ω) end.

---

\*The omega nomenclature system is synonymous with another system called the "*n*" system, in which the "ω" is simply replaced with an "*n*." In other words, an ω-3 fatty acid is sometimes referred to as an *n*-3 fatty acid.

**TABLE 7.1 Names and Food Sources of Some Important Fatty Acids in the Body**

| Alpha (α) Nomenclature | Omega (ω) Family[a] | Common Name | Food Sources |
|---|---|---|---|
| **Saturated Fatty Acids** | | | |
| 12:0 | — | Lauric acid | Coconut and palm oils |
| 14:0 | — | Myristic acid | Coconut and palm oils; most animal and plant fats |
| 16:0 | — | Palmitic acid | Animal and plant fats |
| 18:0 | — | Stearic acid | Animal fats, some plant fats |
| 20:0 | — | Arachidic acid | Peanut oil |
| **Unsaturated Fatty Acids** | | | |
| cis9–16:1 | ω-7 | Palmitoleic acid | Marine animal oils |
| cis9–18:1 | ω-9 | Oleic acid | Plant and animal fats, olive oil |
| cis9,cis12–18:2 | ω-6 | Linoleic acid | Nuts, corn, safflower, soybean, cottonseed, sunflower seeds, and peanut oil |
| cis9,cis12,cis15–18:3 | ω-3 | Linolenic acid (α-linolenic acid) | Canola, soybean, flaxseed, and other seed oils |
| cis5,cis8,cis11,cis14–20:4 | ω-6 | Arachidonic acid | Small amounts in plant and animal oils |
| cis5,cis8,cis11,cis14,cis17–20:5 | ω-3 | Eicosapentaenoic acid (EPA) | Marine algae, fish oils |
| cis4,cis7,cis10,cis13,cis16,cis19–22:6 | ω-3 | Docosahexaenoic acid (DHA) | Animal fats as phospholipid component, fish oils |

[a] The omega (ω) nomenclature only applies to unsaturated fatty acids, because it refers to the position of the first carbon-carbon double bond in relation to the methyl (ω) end of the fatty acid.

SOURCE: Adapted from Gropper SS, Smith JL, Groff JL. *Advanced Nutrition and Human Metabolism*, 4th edition. Thomson/Wadsworth, Belmont, CA, 2005.

# Fatty Acids: Sources, Functions, and Dietary Recommendations

We have two sources of fatty acids: foods and synthesis in the body. In this section, you will learn about several of these fatty acids—including the essential fatty acids—as well as their sources and functions. Dietary recommendations and guidelines are also provided.

## The Essential Fatty Acids: Linoleic and Linolenic Acids

Although there is great diversity in the fatty acids found in foods, only two are essential. These essential fatty acids are **linoleic acid** and **linolenic acid** (also called α-linolenic acid). Linoleic acid has 18 carbons, 2 *cis* double bonds, and is an ω-6 fatty acid. Linolenic acid has 18 carbons, 3 *cis* double bonds, and is an ω-3 fatty acid. Linoleic acid and linolenic acid are essential nutrients, because the body cannot add a double bond in a fatty acid prior to the ninth carbon from the methyl (ω) end. In other words, it cannot synthesize double bonds in the ω-3 and ω-6 positions. Thus dietary linoleic acid and linolenic acid are needed to provide the basic building blocks to make other ω-3 and ω-6 fatty acids. This is done by increasing the number of carbons (elongation) and the number of double bonds (desaturation). For example, the essential fatty acid linoleic acid is used to make **arachidonic acid** (a 20-carbon ω-6 fatty acid) by adding two carbons and two double bonds. These long-chain polyunsaturated fatty acids are important, having many functions in the body.

In addition to linoleic and linolenic acids, other fatty acids may be conditionally essential during infancy. These include arachidonic acid and

**CONNECTIONS** Essential nutrients are needed by the body but cannot be synthesized in sufficient quantities to meet our needs (Chapter 1, p. 3).

**linoleic acid** An essential ω-6 fatty acid.

**linolenic acid** An essential ω-3 fatty acid.

**arachidonic acid** (a – rach- i – DON – ic) A long-chain, polyunsaturated ω-6 fatty acid produced from linoleic acid.

docosahexaenoic acid (DHA; an ω-3 fatty acid). In adults, arachidonic acid and DHA are synthesized from linoleic acid and linolenic acid, respectively. However, babies may not be able to make enough. The conditional essentiality of arachidonic acid and DHA during infancy is described in more detail in the Focus on Life Cycle Nutrition feature.

## Functions of the Essential Fatty Acids

Essential fatty acids, themselves, are required for the proper functioning of all physiological systems. In addition, they can be converted to other fatty acids needed by the body. For example, linoleic acid is elongated and desaturated to form arachidonic acid, whereas linolenic acid is converted to both eicosapentaenoic acid (EPA) and DHA. As previously described, these fatty acids are vitally important for health.

The essential fatty acids can also be converted to other important compounds that are not, themselves, fatty acids. One example is the eicosanoids, which are a group of compounds made from arachidonic acid and EPA. Linoleic acid (an ω-6 fatty acid) is metabolized to ω-6 eicosanoids, whereas linolenic acid (an ω-3 fatty acid) is metabolized to ω-3 eicosanoids. Examples of eicosanoids are the thromboxanes, prostaglandins, and leukotrienes. Eicosanoids have profound, hormone-like effects, assisting and regulating the immune and cardiovascular systems and acting as chemical messengers that direct a variety of additional functions.[1]

The ω-6 and ω-3 series of eicosanoids have somewhat opposing actions. For example, the ω-6 eicosanoids tend to cause inflammation and constriction of blood vessels, whereas the ω-3 eicosanoids stimulate less of an inflammatory response and cause dilation, or opening, of blood vessels. ω-3 eicosanoids also inhibit blood clotting. Both types of eicosanoids are important for health, and the body can shift its balance of them to respond to its needs.

**docosahexaenoic acid (DHA)** (do- cos – a – hex – a – NO- ic) A long-chain, polyunsaturated ω-3 fatty acid produced from linolenic acid.

**eicosapentaenoic acid (EPA)** (ei – co – sa – pen – ta – NO – ic) A long-chain, polyunsaturated ω-3 fatty acid produced from linolenic acid.

**eicosanoids** (ei – COS – a – noids) Biologically active compounds synthesized from arachidonic acid and EPA.

# Focus On LIFE CYCLE NUTRITION
## Optimal Lipid Nutrition During Infancy

Young infants rely solely on either human milk or infant formula for all their nutritional needs. Although manufacturers strive to produce formulas that are similar to human milk, the lipids provided by these two infant foods are sometimes quite dissimilar. For example, human milk has at least 47 different fatty acids in it, most of which are not found in infant formulas.[2] Scientists do not know precisely which of these fatty acids are important for optimal growth and development during this time. However, some of the long-chain PUFAs found in human milk may be conditionally essential nutrients during infancy. These fatty acids, arachidonic acid and docosahexaenoic acid (DHA),

are produced in sufficient amounts from linoleic and linolenic acids in children and adults. Therefore, they are not essential nutrients during these periods of the life cycle. However, infants (especially those born prematurely) have very low stores of arachidonic acid and DHA and may not be able to synthesize them in adequate amounts.[3] These fatty acids are thought to be important for growth, development of the eyes, nervous system, and mental function.[4] Thus many scientists believe that infants must consume adequate amounts of arachidonic acid and DHA during early life to achieve optimal growth and development. Until recently, only breastfed babies received these fatty acids, because infant formu-

las were not fortified with long-chain PUFAs. Because research suggests that fortifying infant formula with long-chain PUFAs may be advantageous, now many companies produce infant formula fortified with these fatty acids.[5]

There is also considerable interest in the importance of other dietary lipids during infancy. An example is cholesterol. Scientists have long known that human milk provides high amounts of cholesterol.[6] However, infant formula generally lacks this compound. Whether early exposure to dietary cholesterol is important for optimal growth and development is unknown but continues to interest the medical and scientific communities.[7]

Our diet can also influence the amount and types of eicosanoids made in the body. For example, Alaskan natives consuming high amounts of ω-3 fatty acids from fish and marine mammals have enhanced physiological responses that are stimulated by ω-3 eicosanoids.[8] In fact, it often takes more time for these people to stop bleeding when they have been cut. For this reason alterations in the balance of ω-3 to ω-6 eicosanoids may influence a person's risk of heart disease and cancer.[9]

### Food Sources of the Essential Fatty Acids and Their Metabolites

In general, linoleic acid is found in high amounts in nuts (such as walnuts), seeds, and certain oils such as those made from soybean, safflower, or corn. Linolenic acid is abundant in some oils such as those made from canola (also called rapeseed) or flaxseed. Because these foods and oils are common in the American diet, getting adequate amounts of linoleic acid and linolenic acid is easy. Longer-chain ω-3 fatty acids such as EPA and DHA are plentiful in fatty fish and seafood; smaller amounts are found in meats and eggs. Arachidonic acid is found in a variety of plant and animal foods.

Many fish are excellent sources of ω-3 fatty acids.

### Dietary Recommendations for the Essential Fatty Acids and Their Metabolites

Dietary Reference Intakes (DRIs) for both essential fatty acids are presented inside the cover of this book. Adequate Intake levels (AIs) for linoleic acid are 17 and 12 g/day for adult males and females, respectively. For linolenic acid, the AIs are 1.6 and 1.1 g/day for adult males and females, respectively. The Acceptable Macronutrient Distribution Ranges (AMDRs) for linoleic and linolenic acids are 5 to 10 and 0.6 to 1.2% of calories, respectively.

The U.S. Department of Agriculture (USDA) Dietary Guidelines for Americans and the American Heart Association suggest that we consume two servings of fish high in ω-3 fatty acids each week as well as other foods rich in linolenic acid such as flaxseed and canola oils. Patients with cardiovascular disease are advised to consume about 1 g of ω-3 fatty acids daily from fish. Considering that one serving of salmon (3 ounces) contains approximately 1.3 g of ω-3 fatty acids, this goal is easy to achieve.

**CONNECTIONS** Chapter 2 discussed the Dietary Reference Intakes (DRIs), which are reference values for nutrient intake that meets the needs of about 97% of the population. When adequate information is available, Recommended Dietary Allowances (RDAs) are established, but when less information is available, Adequate Intake (AI) values are provided. Tolerable Upper Intake Levels (ULs) indicate intake levels that should not be exceeded (Chapter 2, p. 36).

## Essential Fatty Acid Deficiency

If humans do not consume enough of the essential fatty acids, their skin can become irritated and flaky, gastrointestinal problems can occur, and immune function can become impaired. As a result, infections are common and wound healing may be restricted. Children with essential fatty acid deficiencies also exhibit slow growth. Although primary essential fatty acid deficiencies are rare in adults, secondary deficiencies can occur with diseases that disrupt lipid absorption or utilization.

## Saturated and Unsaturated Fatty Acids

Although most saturated and unsaturated fatty acids are not essential in the diet, they serve a variety of functions, which include serving as sources of energy, insulation, and protection. An overview of food sources and dietary recommendations of saturated and unsaturated fatty acids is provided here.

## FIGURE 7.5  Distribution of Fatty Acid Classes in Commonly Consumed Lipids

Dietary lipids contain different relative amounts of saturated fatty acids (SFAs), monounsaturated fatty acids (MUFAs), and polyunsaturated fatty acids (PUFAs).

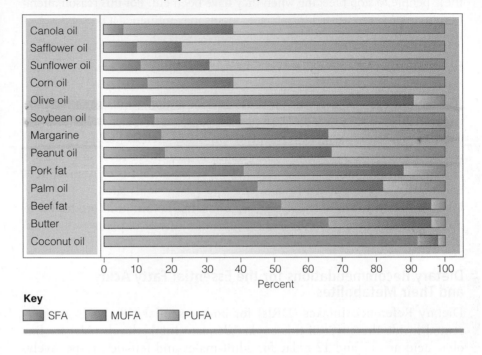

**Key**

■ SFA   ■ MUFA   ■ PUFA

## Food Sources of Saturated and Unsaturated Fatty Acids

In general, animal foods contribute the majority of dietary SFAs, whereas plant-derived foods supply the majority of PUFAs. MUFAs come from both plant and animal foods. However, some tropical oils, such as coconut and palm oils, contain relatively high amounts of SFAs. In addition, many oily fish have high levels of PUFAs. The relative amounts of SFAs, MUFAs, and PUFAs in commonly consumed fats and oils are summarized in Figure 7.5.

## Guidelines for Saturated Fatty Acid (SFA) Consumption

Scientists have long known that SFA intake is positively related to risk for cardiovascular disease. As such, many dietary guidelines are aimed at decreasing SFA intake. For example, although there are no DRIs for SFAs the Institute of Medicine recommends that "intake of SFAs should be minimized while consuming a nutritionally adequate diet." In addition, the Dietary Guidelines suggest that SFA should constitute no more than 10% of total calories. Reading food labels can help you keep track of your SFA intake. You can read more about general suggestions for decreasing SFA intake in the Food Matters feature, and the relationship between SFA and heart disease is discussed in the Nutrition Matters following this chapter.

## *Trans* Fatty Acids

*Trans* fatty acids are found naturally in some foods—especially dairy and beef products. However, most dietary *trans* fatty acids are produced commercially via a process called **partial hydrogenation,** which chemically converts oily lipids (such as corn oil) into solid fats (such as margarine). In addition to decreasing the number of double bonds, some of the *cis* double bonds are converted

**partial hydrogenation** A process by which some carbon–carbon double bonds found in PUFAs are converted to carbon–carbon single bonds, resulting in the production of *trans* fatty acids.

## *Food Matters*

### Working Toward the Goal

The USDA Dietary Guidelines recommend that we limit our intake of saturated fatty acids, *trans* fatty acids, and cholesterol. The following selection and preparation tips will help you reach this goal.

#### Guidelines for Decreasing Intakes of Saturated Fatty Acids, *Trans* Fatty Acids, and Cholesterol

| Saturated Fatty Acids | *Trans* Fatty Acids | Cholesterol |
|---|---|---|
| Consume a diet that provides no more than 35% of total energy from lipids.<br>Limit intake of animal fats (e.g., beef and butter fat).<br>Limit intake of tropical oils, such as coconut oil and palm oil.<br>Decrease intake of commercially prepared cakes, pastries, doughnuts, and french fries.<br>Use oils instead of shortening or animal fats (such as lard). | Use tub or "*trans* fat-free" margarine instead of stick margarine—or use butter.<br>Choose cookies and crackers made with vegetable oil instead of partially hydrogenated oils.<br>Choose "*trans* fat-free" crackers and bakery products when possible. | Limit intake of high-cholesterol foods such as liver, eggs, cheesecake, and custards.<br>Replace high-fat animal products with lower-fat products, such as lean cuts of meat and 2% milk. |

SOURCE: Adapted from the Institute of Medicine. *Dietary Reference Intakes for Energy, Carbohydrate, Fiber, Fat, Fatty Acids, Cholesterol, Protein and Amino Acids.* Washington, DC: National Academies Press; 2005.

to *trans* double bonds. This results in a high amount of *trans* fatty acids. Partial hydrogenation is used in food manufacturing, because partially hydrogenated oils impart desirable food texture and reduce spoilage. Crackers, pastries, bakery products, shortening, and margarine are the main sources of the *trans* fatty acids in our diets.[10] However, the current focus on decreasing *trans* fatty acid intake has resulted in new food-processing methods that decrease or eliminate *trans* fatty acids in many foods.

## Guidelines for *Trans* Fatty Acid Consumption

Many public health agencies suggest that we limit our intake of *trans* fatty acids. This is because studies show that some of them increase the risk for cardiovascular disease.[11] Although there are no absolute guidelines as to how little *trans* fatty acids we should consume, the Institute of Medicine recommends that people "minimize their intakes of *trans* fatty acids while consuming a nutritionally adequate diet." Similarly, the Dietary Guidelines recommend that we obtain less than 1% of calories from *trans* fatty acids. These guidelines apply to commercially produced *trans* fatty acids, not to naturally occurring ones. As of 2006, food manufacturers were required to state the *trans* fatty acid content on their Nutrition Facts panels.

# ESSENTIAL *Concepts*

The two essential fatty acids—linoleic acid and linolenic acid—cannot be synthesized in the human body and are needed to make a variety of other substances, including longer-chain fatty acids and the eicosanoids. The body can make all other fatty acids it needs, except during infancy when arachidonic acid and DHA are likely conditionally essential. Both linoleic and linolenic acids are generally abundant in most diets. In addition, consumption of other nonessential fatty acids appears to influence our health. For example, diets high in SFAs or *trans* fatty acids may increase the risk for cardiovascular disease. SFAs are found mainly in tropical "oils" and animal fats, whereas *trans* fatty acids are in many processed foods. Experts recommend that we limit our intake of SFA and *trans* fatty acids. Conversely, consuming ω-3 fatty acids may decrease risk of cardiovascular disease; these are found mainly in seafood.

# Mono-, Di-, and Triglycerides

Most fatty acids do not exist in their free (unbound) form in foods or in the body. Instead, they are part of larger, more complex molecules called triglycerides or in smaller molecules called diglycerides and monoglycerides. Mono-, di-, and triglycerides play many roles, including serving as sources of energy (ATP), insulation, and protection.

## Structure of Mono-, Di-, and Triglycerides

**Triglycerides** (also called triacylglycerols) consist of a glycerol molecule chemically bonded to three fatty acids via **ester bonds** (Figure 7.6). These fatty acids can be saturated, monounsaturated, polyunsaturated, or a mixture of fatty acid classes. Regardless of which types of fatty acids are attached to the glycerol molecule, all triglycerides are extremely hydrophobic. Whereas a triglyceride has three fatty acids attached to a molecule of glycerol, a **monoglyceride** consists of glycerol with one fatty acid, and a **diglyceride** consists of a glycerol with two fatty acids.

## Functions of Triglycerides

Triglycerides serve many purposes in the body. Perhaps most importantly, they provide the essential fatty acids needed for the body to function. However, triglycerides also supply fatty acids used for energy (ATP) production and storage. Most cells in the body can use fatty acids for energy, and the body stores extra fatty acids for its long-term energy needs. Lipid stored as triglyceride is also important for insulation and protection.

### Use as an Energy Source

Triglycerides provide an important source of energy. For this to happen, they must first be broken down into glycerol and fatty acids. This process, called **lipolysis,** is catalyzed by the enzyme **hormone-sensitive lipase,** whose activity increases when secretion of the pancreatic hormone insulin is low. Lipolysis is also stimulated by exercise and in physiological stress. Compared to the other energy-yielding macronutrients, fatty acids represent the body's richest source of energy. The complete breakdown of 1 g of fatty acids yields approximately

**triglyceride** (also called triacylglycerol) (*tri,* three) A lipid composed of a glycerol bonded to three fatty acids.

**ester bond** The type of chemical bond that holds fatty acids onto triglycerides, phospholipids, and cholesteryl esters.

**monoglyceride** (also called monoacylglycerol) (*monos,* single) A lipid made of a glycerol bonded to a single fatty acid.

**diglyceride** (also called diacylglycerol) (*di,* two) A lipid made of glycerol bonded to two fatty acids.

**lipolysis** The breakdown of triglycerides into fatty acids and glycerol.

**hormone-sensitive lipase** An enzyme that catalyzes the hydrolysis of ester bonds that attach fatty acids to the glycerol molecule; mobilizes fatty acids from adipose tissue.

## FIGURE 7.6    A Triglyceride Molecule

A triglyceride molecule consists of glycerol and three fatty acids.

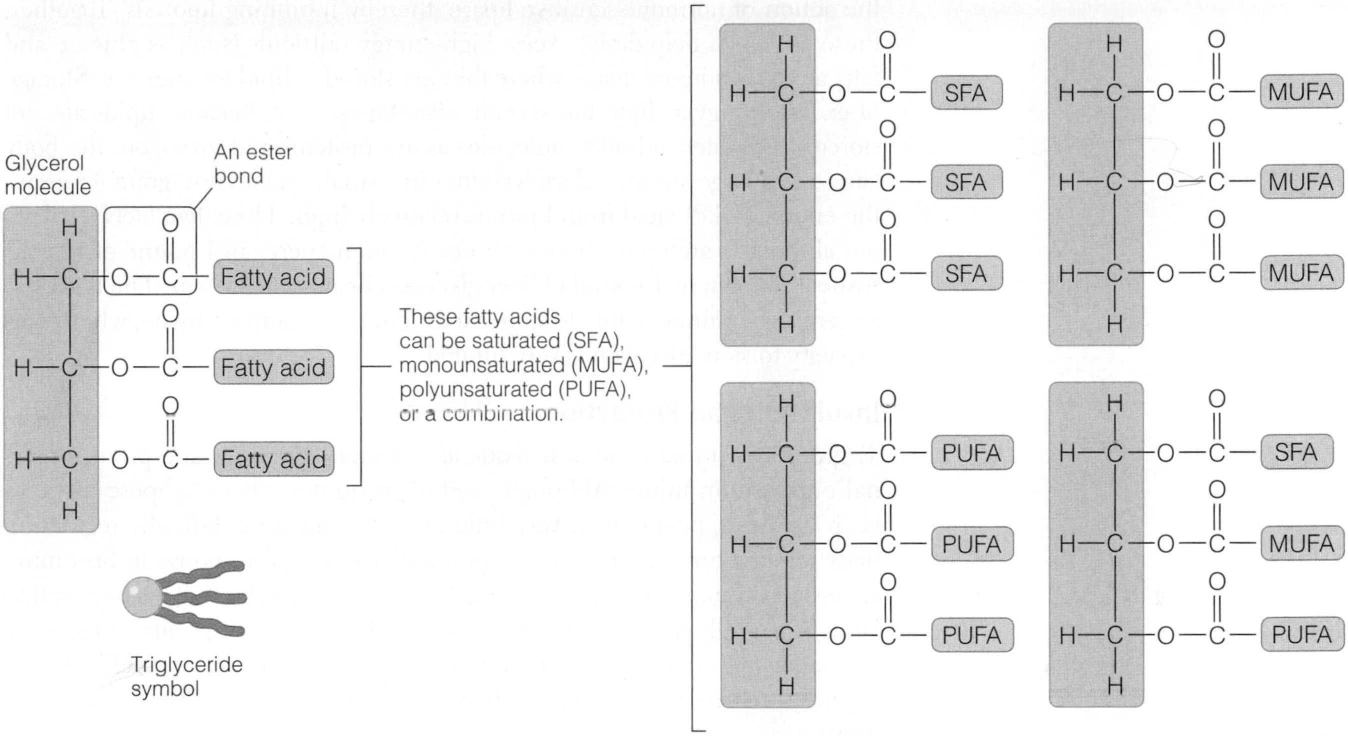

9 kcal of energy, which is more than twice the yield from 1 g of carbohydrate or protein (4 kcal). Therefore, gram for gram, high-fat foods contain more calories than do other foods.

In addition to using fatty acids directly as an energy source, the body can convert them to other energy-yielding compounds called ketones. Recall from Chapter 5 that the production of ketones from fatty acids, called ketogenesis, occurs when the body's supply of glucose becomes limited. Ketogenesis is important, because some tissues such as brain, heart, skeletal muscle, and kidneys can use ketones, in addition to glucose, for energy (ATP). In this way, ketones produced from fatty acids can serve as a major source of energy during times of severe glucose insufficiency (such as starvation) and can spare the body from having to use amino acids to synthesize glucose via gluconeogenesis.

### Use as an Energy Reserve

Fatty acids not required for energy or other functions in the body are stored as triglycerides in adipose tissue and, to a lesser extent, skeletal muscle. Adipose tissue consists of specialized cells called **adipocytes,** which can accumulate large amounts of lipid. Adipose tissue is found in many parts of the body, including beneath the skin (**subcutaneous adipose tissue**) and around the vital organs in the abdomen (**visceral adipose tissue**). Considerable adipose tissue is also associated with many of the body's organs, such as the kidneys and breasts, making it possible for these organs to have ready access to fatty acids for their immediate energy needs.

The pancreatic hormone insulin stimulates the storage of triglycerides during times of energy excess. Insulin causes adipocytes, and to a lesser extent

**CONNECTIONS** Remember that gluconeo-genesis is the synthesis of glucose from non-carbohydrate substances (Chapter 5, p. 149).

**adipocyte** (a – DIP – o – cyte) A specialized cell that makes up adipose tissue.

**subcutaneous adipose tissue** Adipose tissue found directly under the skin.

**visceral adipose tissue** Adipose tissue surrounding the vital organs.

skeletal muscle cells, to take up glucose and fatty acids and convert glucose to fatty acids. In turn, fatty acids are incorporated into triglycerides. The synthesis of fatty acids and triglycerides is called **lipogenesis.** Insulin also inhibits the action of hormone-sensitive lipase, thereby inhibiting lipolysis. Together, these processes help direct excess high-energy nutrients (such as glucose and fatty acids) to adipose tissue, where they are stored as lipid for later use. Storage of excess energy as lipid has several advantages. First, because lipids are not stored with water (a bulky molecule) as are proteins and glycogen, the body can store a large amount of triglycerides in a small space. Also, gram for gram, the energy (ATP) yield from lipids is relatively high. These two facts result in our ability to warehouse about six times as much energy in 1 pound of adipose tissue as we can in 1 pound of liver glycogen or muscle protein. The body has a seemingly infinite ability to store excess energy in adipose tissue, whereas its capacity to store glycogen is very limited.

### Insulation and Protection

Triglycerides stored in adipose tissue also insulate the body and protect internal organs from injury. Although most of us do not rely on adipose tissue to keep us warm, people with very little body fat can have difficulty regulating body temperature. In fact, one common physiological response to becoming excessively thin is to develop very fine hair covering the body. This hair, called lanugo, partially makes up for the absence of subcutaneous adipose tissue by providing a layer of external insulation for the body. The presence of lanugo is common in very lean people, such as those with the eating disorder anorexia nervosa.[12]

## ESSENTIAL *Concepts*

Triglycerides consist of three fatty acids attached to a glycerol molecule. Monoglycerides are made of glycerol and one fatty acid, whereas diglycerides have two fatty acids. Fatty acids found in triglycerides provide an important source of energy in the body. During periods of energy need, fatty acids are mobilized from adipose tissue via hormone-sensitive lipase. When glucose supplies are limited, fatty acids can be converted to ketones, which are used by selected tissues for energy. During times of energy abundance, fatty acids are stored in adipose tissue as triglycerides. There are several types of adipose tissue in the body, including subcutaneous (under the skin) and visceral (within body cavities). Adipose tissue also serves important roles in temperature regulation and protection.

## Phospholipids

The next major lipid class in the body is the **phospholipids,** which are critical for many structures and functions. Because the body can synthesize all that it needs, there is no dietary requirement for this lipid class.

### Structure of Phospholipids

Phospholipids are similar to triglycerides in that they contain a glycerol molecule bonded to fatty acids (Figure 7.7). However, instead of having three fatty acids, a phospholipid has only two fatty acids. Replacing the third fatty acid is a phosphate-containing **polar head group.** There are many different types of

**lipogenesis** (*lipos,* fat; *genus,* birth) The metabolic processes that result in fatty acid and, ultimately, triglyceride synthesis.

**phospholipid** A lipid composed of a glycerol bonded to two fatty acids and a polar head group.

**polar head group** A phosphate-containing charged chemical structure that is a component of a phospholipid.

## FIGURE 7.7 A Phospholipid Molecule

A phospholipid consists of a glycerol, a polar head group, and two fatty acids.

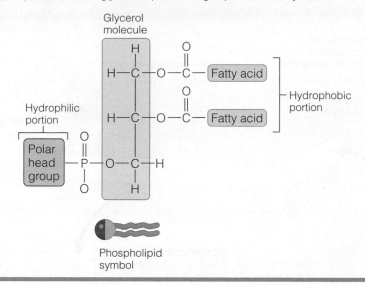

Phospholipid symbol

polar head groups, but the most common are choline, ethanolamine, inositol, and serine.

Phospholipids are **amphipathic**, meaning they contain both polar (hydrophilic) and nonpolar (hydrophobic) portions. The hydrophobic portion consists of the two fatty acids, whereas the hydrophilic portion is the polar head group. This structure allows phospholipids to be major components of cell membranes and play roles in the digestion, absorption, and transport of lipids.

**CONNECTIONS** Polar molecules, such as water, have unequal charge distributions. Polar compounds are hydrophilic (water loving) and dissolve in water (Chapter 3, p. 69).

## Phospholipids Are Components of Cell Membranes

Phospholipids make up the major structural component of all cell membranes (Figure 7.8). More specifically, cell membranes consist of two layers (a bilayer) of phospholipids with the hydrophilic polar head groups pointing to the extra- and intracellular spaces. Remember that these compartments are predominantly water. To function effectively, cell membranes must be able to provide stable barriers between these spaces. If the cell membrane were completely hydrophilic, it would dissolve and not create a barrier. If the cell membrane were completely hydrophobic, there would be no chance of communication between the watery extra- and intracellular compartments. However, the incorporation of amphipathic phospholipids, having both hydrophobic and hydrophilic portions, allows cell membranes to effectively carry out their functions.

Aside from the importance of phospholipids in the overall structure of cell membranes, they also supply fatty acids for cellular metabolism and can act as biologically active compounds. For example, phospholipids activate enzymes important for energy metabolism, blood clotting, and cell turnover.

## Phospholipids Are Carriers of Hydrophobic Compounds

Phospholipids act as carriers of hydrophobic substances in the body. One example is the incorporation of phospholipids on the outer surface of a class of compounds called lipoproteins. Lipoproteins transport lipids in the blood and are described in more detail later in this chapter.

**amphipathic** (*amphi-*, on both sides) Having both nonpolar (noncharged) and polar (charged) portions.

## FIGURE 7.8   A Cell Membrane Consists of a Phospholipid Bilayer

The amphipathic nature of phospholipids allows cell membranes to carry out their functions.

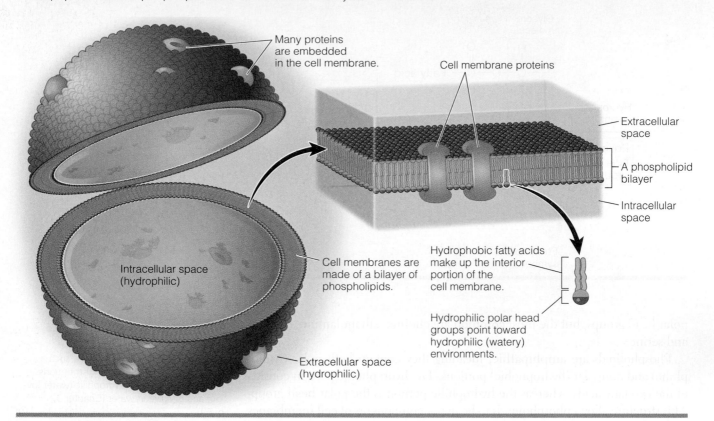

Many proteins are embedded in the cell membrane.

Cell membrane proteins

Extracellular space

A phospholipid bilayer

Intracellular space

Hydrophobic fatty acids make up the interior portion of the cell membrane.

Hydrophilic polar head groups point toward hydrophilic (watery) environments.

Intracellular space (hydrophilic)

Cell membranes are made of a bilayer of phospholipids.

Extracellular space (hydrophilic)

## Sources of Phospholipids

Phospholipids are found naturally in most foods and are also used as food additives. For example, **phosphatidylcholine** (the phospholipid with a choline polar head group) is often added as a stabilizer to foods such as mayonnaise and ice cream. You can check whether a food has phosphatidylcholine by looking for the word *lecithin* (a common name for this compound) on the food label. There are no recommended intakes for phospholipids. However, limited evidence suggests that consuming phosphatidylserine (the phospholipid with a serine polar head group) may enhance mental function in the elderly. This is described in the Focus on Life Cycle Nutrition feature.

## ESSENTIAL *Concepts*

A phospholipid is a lipid consisting of a glycerol, two fatty acids, and a phosphate-containing polar head group. Phospholipids are amphipathic, meaning that one portion is hydrophobic and another is hydrophilic. This property allows the phospholipid to carry out its functions. Fatty acids associated with phospholipids can be used as a source of energy. Phospholipids make up cell membranes and are needed to circulate lipids in the body. Phospholipids are not essential nutrients, and there are no dietary recommendations for them.

**phosphatidylcholine** (also called lecithin) (PHOS – pha – tid – yl – CHO – line)  A phospholipid that contains choline as its polar head group; commonly added to foods as an emulsifying agent.

# Focus On LIFE CYCLE NUTRITION
## Phospholipids and Mental Health in the Elderly

© Jose Luiz Pelaez, Inc./Corbis

*A*s the average life expectancy of the U.S. population increases, society is reminded that the elderly have health concerns that are distinct from those of younger people. For example, risk for many chronic diseases increases with age. In addition, aging sometimes brings loss of some mental abilities. Older people often complain of memory loss or difficulty finding the right words to convey their thoughts. Although annoying and discouraging, these symptoms of aging are considered normal and expected. However, sometimes a more severe condition called dementia develops. Dementia influences a person's ability to think, speak, reason, remember, and move. There are many causes of dementia, including Alzheimer's disease, Parkinson's disease, and stroke.

Dementia typically begins as a mild condition in which the person is unable to remember names or has trouble with tasks such as balancing a checkbook. As dementia worsens, people become increasingly confused, have difficulty recognizing family and friends, and may become withdrawn or aggressive. As we live longer, finding ways to prevent and treat dementia is clearly important. One possible way involves consuming

phospholipid supplements—specifically, phosphatidylserine. Phosphatidylserine is found in relatively high concentration in the brain and is important for neural development and function. Although not found in high concentrations in most foods, it can be produced commercially from soy products. Studies in animals have shown that supplementing the diet with phosphatidylserine can improve brain function.[13] Although very few well-controlled studies have been conducted in humans, some researchers suggest that similar results may be possible.[14] Because

of this, in 2003 the FDA approved the following qualified health claim: "Very limited and preliminary scientific research suggests that phosphatidylserine may reduce the risk of dementia in the elderly. The FDA concludes that there is little scientific evidence supporting this claim." In approving this qualified health claim, the FDA admits that the data do not conclusively support the effect of phosphatidylserine on dementia; however, the link between supplemental phosphatidylserine intake and improved mental health may be promising.

# Sterols and Sterol Esters

**Sterols** also constitute a major lipid class. Sterols are structurally different from other lipids in that they consist of multiring structures (Figure 7.9). A sterol can be attached to a fatty acid via an ester bond, forming a **sterol ester.**

## Cholesterol and Cholesteryl Esters

Of all the sterols, the most abundant and widely recognized is cho*lesterol.* **Cholesterol** is a weakly polar compound. Although some free (unbound) cholesterol is found in the body, most is bonded to a fatty acid. This cholesterol–fatty acid complex is called a **cholesteryl ester** and is an example of a sterol ester. Cholesteryl esters are more hydrophobic than free cholesterol.

**sterol** A type of lipid with a distinctive multiring structure; a common example is cholesterol.

**sterol ester** A chemical compound consisting of a sterol molecule bonded to a fatty acid via an ester linkage.

**cholesterol** (*kholikos*, bile; *stereos*, hard or solid) A sterol found in animal foods and made in the body; required for bile acid and steroid hormone synthesis.

**cholesteryl ester** A sterol ester made of a cholesterol molecule bonded to a fatty acid via an ester linkage.

**FIGURE 7.9 Structures of a Sterol, a Cholesterol, and a Cholesteryl Ester**

**A sterol**

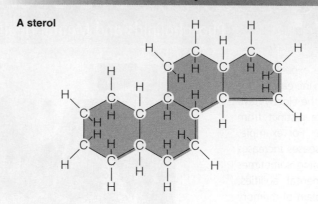

**Cholesterol**

The highlighted areas make this sterol a cholesterol molecule.

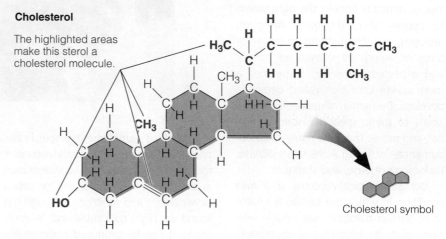

Cholesterol symbol

**A cholesteryl ester**

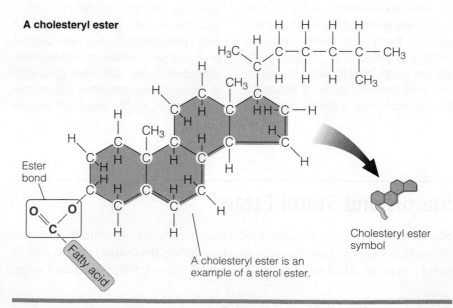

Ester bond

Fatty acid

Cholesteryl ester symbol

A cholesteryl ester is an example of a sterol ester.

**bile acids** Amphipathic substances made from cholesterol in the liver; a component of bile important for lipid digestion and absorption.

Cholesterol is used to synthesize bile acids, which play a critical role in digestion and absorption of lipids. **Bile acids** consist of the basic, multiring structure of cholesterol attached to a very hydrophilic subunit, making bile acids amphipathic, much like phospholipids. Cholesterol and cholesteryl esters are also components of cell membranes, where they help maintain fluidity. If cell membranes were not fluid—or flexible—they would not be able to function. In addition, cholesterol is needed for the synthesis of the steroid hormones important for reproduction (such as estrogen and testosterone), energy metabolism (such as cortisol), calcium homeostasis (such as calcitriol), and electrolyte balance (such as aldosterone). Cholesteryl esters also serve as crucial carrier molecules for fatty acids in the blood. The functions of cholesterol and cholesteryl esters in the human body span practically every physiological system, making them vital molecules for healthy living.

## Sources of Cholesterol in the Body: Synthesis and Diet

Cholesterol is made from glucose and fatty acids in almost every tissue in the body, especially the liver. Many dietary factors influence cholesterol synthesis. For example, eating a low-calorie or low-carbohydrate diet can decrease cholesterol synthesis in some people.[15] However, this is not always the case. Clearly, diet interacts with genetics to influence cholesterol synthesis.[16] Cholesterol synthesis can be lowered by medical means, as is often the case in people at high risk for heart disease. For example, statin drugs, such as Lipitor® (atorvastatin calcium) and Zocor® (simvastatin), decrease blood cholesterol by inhibiting one of the enzymes needed for its synthesis in the liver.

Cholesterol is found only in animal-derived foods, such as shellfish, meat, butter, eggs, and liver (Figure 7.10). Note that the cholesterol content of liver is very high, because it is the body's major site of cholesterol synthesis and storage. However, because cholesterol is a nonessential nutrient, vegan vegetarians are not at risk of cholesterol deficiency.

**CONNECTIONS** Recall that an interaction is when the effect of one factor (such as diet) on another factor (such as cholesterol synthesis) is influenced by a third factor (such as genetics) (Chapter 1, p. 11).

## FIGURE 7.10 Cholesterol Content of Commonly Consumed Foods

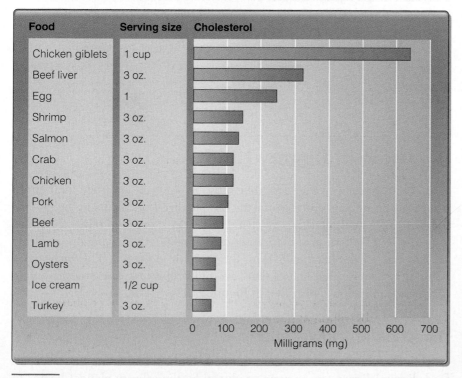

| Food | Serving size | Cholesterol |
|------|--------------|-------------|
| Chicken giblets | 1 cup | |
| Beef liver | 3 oz. | |
| Egg | 1 | |
| Shrimp | 3 oz. | |
| Salmon | 3 oz. | |
| Crab | 3 oz. | |
| Chicken | 3 oz. | |
| Pork | 3 oz. | |
| Beef | 3 oz. | |
| Lamb | 3 oz. | |
| Oysters | 3 oz. | |
| Ice cream | 1/2 cup | |
| Turkey | 3 oz. | |

Milligrams (mg)

SOURCE: USDA National Nutrition Database for Standard Reference, Release 17, 2005.

**steroid hormone** (*stereos,* hard or solid; *hormon,* to urge on) A hormone made from cholesterol.

Only animal-derived foods contain cholesterol.

© Royalty-Free Corbis

## Guidelines for Cholesterol Intake

No DRIs are established for cholesterol. However, the Dietary Guidelines recommend that we consume less than 300 mg of cholesterol daily. This is the amount found in one to two eggs or two servings of beef. In addition, the Institute of Medicine recommends that "cholesterol intake should be minimized while consuming a nutritionally adequate diet."

## Plant Sterols/Stanols

Plants also contain sterol-like compounds called **phytosterols,** which are chemically similar to cholesterol. Phytosterols in foods can be hydrogenated to produce compounds called stanols. An interesting group of phytosterols is found naturally in corn, wheat, rye, and other plants. Similar sterols and stanols are also produced commercially and are marketed under various names, including Benecol®. These products can currently be found in butter substitutes, yogurt drinks, salad dressings, and even dietary supplements.

Studies suggest that consuming products containing these phytosterols or stanols may decrease blood cholesterol concentration, lowering the risk for cardiovascular disease.[17] Therefore the FDA has approved the following unqualified health claim: "Diets low in saturated fat and cholesterol that include two servings of foods that provide a daily total of at least 3.4 grams of plant sterols/stanols in two meals may reduce the risk of heart disease." Because a typical serving of a plant sterol–fortified spread contains about 1.1 gram of the sterol, you would need to consume about three servings daily to reach this goal.

## ESSENTIAL *Concepts*

Sterols are multiring compounds that have many roles in the body. For example, cholesterol is important for lipid digestion and absorption, cell membrane structure, synthesis of steroid hormones, and transport of lipids in the blood. Cholesterol is made in the body and is found in many animal-based foods. It is generally recommended that we limit our consumption of this compound to decrease risk for cardiovascular disease, although the effectiveness of this can depend on genetics. Sterols (but not cholesterol) are also found in some plants. Research suggests that some plant sterols (and stanols) may reduce the risk for heart disease.

## Digestion of Dietary Lipids

Once lipids have been ingested, they must be digested, absorbed, and circulated away from the gastrointestinal tract. Lipid digestion occurs in the mouth, stomach, and small intestine, involving several enzymes and other secretions such as bile. The basic goal of triglyceride digestion is the separation (or cleavage) of most of the fatty acids from the glycerol molecule. The body has a series of enzymes called lipases that accomplish this task, and an overview of triglyceride digestion is provided in Figure 7.11.

## Digestion of Triglycerides

Digestion of triglycerides begins in the mouth. As chewing breaks apart food, **lingual lipase** (an enzyme produced by the salivary glands) begins to hydrolyze fatty acids from glycerol molecules. As food is swallowed, lingual

**phytosterol** (*phuto,* plant) Sterol made by plants.

**lingual lipase** (*lingua,* tongue; *lipos,* fat) An enzyme produced in the salivary glands that hydrolyzes ester bonds between fatty acids and glycerol molecules.

CONNECTIONS▶ Remember that hydrolysis is the breaking of chemical bonds by the addition of water. As a result, larger compounds are broken down into smaller subunits (Chapter 3, p. 77).

## FIGURE 7.11  Overview of Triglyceride Digestion

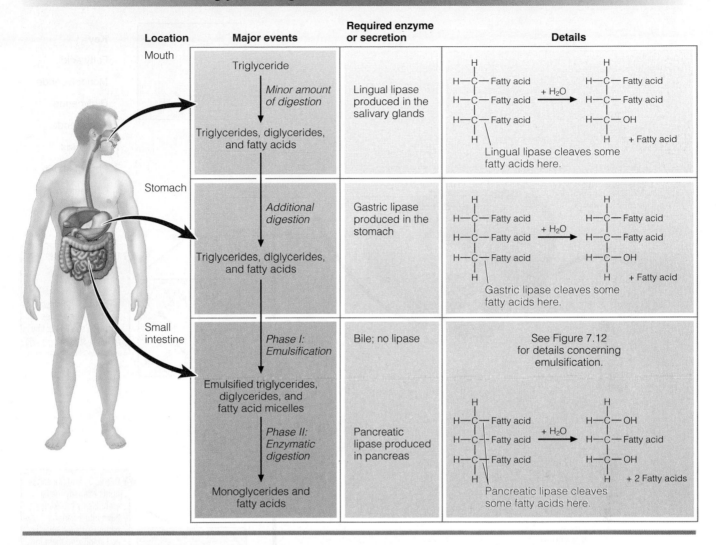

| Location | Major events | Required enzyme or secretion | Details |
|---|---|---|---|
| Mouth | Triglyceride → *Minor amount of digestion* → Triglycerides, diglycerides, and fatty acids | Lingual lipase produced in the salivary glands | Lingual lipase cleaves some fatty acids here. |
| Stomach | *Additional digestion* → Triglycerides, diglycerides, and fatty acids | Gastric lipase produced in the stomach | Gastric lipase cleaves some fatty acids here. |
| Small intestine | *Phase I: Emulsification* → Emulsified triglycerides, diglycerides, and fatty acid micelles → *Phase II: Enzymatic digestion* → Monoglycerides and fatty acids | Bile; no lipase / Pancreatic lipase produced in pancreas | See Figure 7.12 for details concerning emulsification. Pancreatic lipase cleaves some fatty acids here. |

lipase accompanies the bolus into the stomach, where it continues to cleave additional fatty acids.

When food enters the stomach, it stimulates the release of the hormone gastrin from specialized cells lining the gastric pits. Gastrin circulates in the blood and stimulates the release of the enzyme **gastric lipase,** also produced in stomach cells. Gastric lipase is a component of the "gastric juices" and picks up where lingual lipase left off, by continuing to cleave fatty acids from glycerol molecules.

The watery environment of the stomach and small intestine causes lipids to clump together in large lipid globules, making further digestion somewhat challenging. To overcome this, intestinal digestion of triglyceride occurs in two phases and involves several intestinal hormones and secretions from accessory organs.

### Phase 1: Emulsification of Lipids by Bile—Micelle Formation

Phase 1 of intestinal triglyceride digestion does not involve digestive enzymes. In fact, no chemical digestion actually takes place. Instead, large lipid globules are dispersed into smaller lipid droplets in a way that prepares them for further digestion. An overview of this process is provided in Figure 7.12.

**gastric lipase** (*gaster,* belly; *lipos,* fat) Enzyme produced in the stomach that hydrolyzes ester bonds between fatty acids and glycerol molecules.

# FIGURE 7.12 Emulsification of Lipids in the Small Intestine

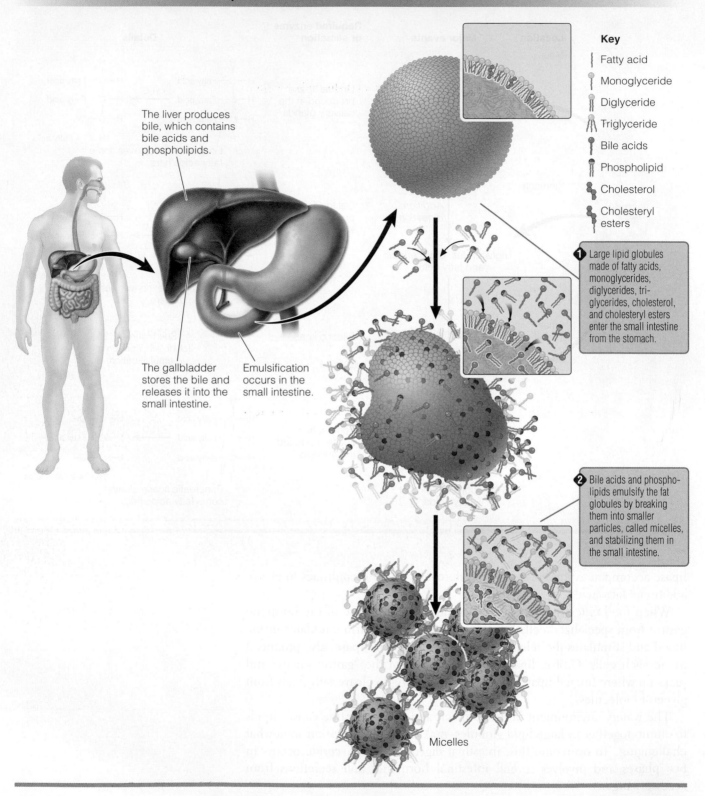

The liver produces bile, which contains bile acids and phospholipids.

The gallbladder stores the bile and releases it into the small intestine.

Emulsification occurs in the small intestine.

**Key**

| Fatty acid
| Monoglyceride
| Diglyceride
| Triglyceride
| Bile acids
| Phospholipid
| Cholesterol
| Cholesteryl esters

**①** Large lipid globules made of fatty acids, monoglycerides, diglycerides, tri-glycerides, cholesterol, and cholesteryl esters enter the small intestine from the stomach.

**②** Bile acids and phospho-lipids emulsify the fat globules by breaking them into smaller particles, called micelles, and stabilizing them in the small intestine.

Micelles

CONNECTIONS Remember that enterocytes are the cells lining the small intestine; micro-villi on enterocytes make up the brush border (Chapter 4, page 104).

The presence of lipids in the small intestine stimulates enterocytes to secrete the hormone cholecystokinin (CCK), which in turn signals the gall-bladder to contract and release bile into the duodenum. Recall that bile is a mixture of bile acids, free cholesterol, and phospholipids made by the liver and stored in the gallbladder until needed. Also recall that bile acids and phospholipids are amphipathic. In the intestine, the hydrophobic portions of these molecules associate with the lipid globules, whereas the hydrophilic portions associate with the surrounding water. These opposing forces pull the lipid globules apart into smaller droplets. Bile acids and phospholipids then stay with the newly formed droplets, preventing them from reforming larger globules.

The process of breaking larger lipid globules into smaller droplets is called **emulsification,** and results in the production of smaller stable lipid particles called **micelles.** Emulsification and micelle formation increase the surface area of the lipid, exposing more ester bonds to the digestive enzymes present in the small intestine. Another way to think about lipid emulsification is to think of what happens when you shake water and cooking oil together. Although the water and oil initially appear to mix, the oil eventually separates from the water. However, if you add a few drops of dish soap and shake the mixture again, the oil becomes suspended as small droplets within the water. This is because soap is an emulsifier, breaking up large lipid droplets into smaller ones and "protecting" them in the watery environment. Although most people never need to think about the importance of bile in lipid digestion, those with gallbladder disease fully understand. This is because gallbladder disease disrupts the process of emulsification. This is described in more detail in the Focus on Clinical Applications feature.

## Phase 2: Digestion of Triglycerides by Pancreatic Lipase

As lipid-containing chyme enters the duodenum, enterocytes are stimulated to release the hormones secretin and CCK which stimulate the **pancreas to release pancreatic juice** that contains the enzyme **pancreatic lipase.** Pancreatic lipase completes triglyceride digestion by hydrolyzing additional fatty acids from glycerol molecules. In general, two of the three fatty acids

**emulsification** The process by which large lipid globules are broken down and stabilized into smaller lipid droplets.

**micelle** A water-soluble, spherical structure formed in the small intestine via emulsification.

**pancreatic lipase** An enzyme produced in the pancreas that hydrolyzes ester bonds between fatty acids and glycerol molecules.

## *Focus On* CLINICAL APPLICATIONS
### Gallbladder Disease and Gallstones

Gallbladder disease is a condition that can result when bile contains an excess amount of cholesterol in relation to its other components. As a result, small amounts of cholesterol separate out of solution. Other factors, such as weak contractility of the gallbladder, are also associated with gallbladder disease. The accumulation of calcium and cellular debris around the cholesterol crystals results in the formation of gallstones, which can range in size from 5 mm to more than 25 mm in diameter. For a point of reference, 1 millimeter is roughly the thickness of a dime. Some people with gallstones have no symptoms, although others experience symptoms of tenderness to extreme pain. Gallstones can also become lodged in the common bile duct, obstructing the flow of bile and pancreatic juice into the intestine.

Surgical removal of the gallbladder is the most common treatment for persistent gallstone-related problems. Initially, some people may have difficulty digesting fat after the gallbladder is removed. However, the common bile duct eventually forms a small pouch, allowing small amounts of bile to be "stored" until needed. Consequently, doctors often recommend that people who have had the gallbladder removed avoid high-fat meals.

Gallbladder disease is more common in women than men, and the risk of developing it increases with age. Other risk factors include obesity, rapid weight loss, and pregnancy. It remains unclear if particular foods or dietary practices influence the formation of gallstones, but some studies report the prevalence to be lower in vegetarians than nonvegetarians.[18] In addition, moderate alcohol consumption, exercise, and taking aspirin appear to have protective roles, decreasing a person's chances of developing gallstones.

are removed from the triglyceride molecule, resulting in the final products of triglyceride digestion: a monoglyceride and two free (unbound) fatty acids.

## Digestion of Phospholipids, Cholesterol, and Cholesteryl Esters

Phospholipids, cholesterol, and cholesteryl esters are also emulsified by the actions of bile in the small intestine. Phospholipids are then digested by an enzyme called **phospholipase A2.** Like pancreatic lipase, phospholipase A2 is produced in the pancreas and released as part of the pancreatic juices in response to the intestinal hormone secretin. The products of digestion of a single phospholipid molecule are one fatty acid and a compound called a **lysophospholipid,** consisting of a glycerol molecule bonded to one fatty acid and a polar head group.

Free cholesterol (not bonded to a fatty acid) does not need to be digested prior to absorption. However, cholesteryl esters must be broken down to their constituents: cholesterol and fatty acids. This is accomplished by another pancreatic enzyme called **bile salt–dependent cholesteryl ester hydrolase.**

## ESSENTIAL *Concepts*

Lipid digestion begins in the mouth where lingual lipase hydrolyzes some fatty acids from glycerol molecules. This process continues in the stomach via gastric lipase. In the small intestine, lipid stimulates the release of the intestinal hormone cholecystokinin (CCK), which causes the gallbladder to release bile. Bile emulsifies large lipid globules into smaller droplets, surrounding them to form micelles. Digestion of triglycerides is complete when pancreatic lipase hydrolyzes additional fatty acids from the glycerol molecules, yielding monoglycerides and free fatty acids. Phospholipids are digested by the pancreatic enzyme phospholipase A2, resulting in the release of fatty acids and lysophospholipids. Although free cholesterol does not undergo digestion, cholesteryl esters are broken down by the pancreatic enzyme bile salt–dependent cholesteryl ester hydrolase into fatty acids and cholesterol molecules.

# Absorption and Circulation of Lipids

The products of lipid digestion are absorbed into the enterocytes and circulated away from the small intestine. This requires special handling, because both the interior of the enterocyte and the circulatory system are hydrophilic in nature. As you will see, the amphipathic properties of bile acids and phospholipids allow both lipid absorption and circulation to proceed.

## Lipid Absorption

Lipid absorption is accomplished in two ways, depending on how water-soluble (hydrophilic) the lipid is. Because they are relatively water soluble, short- and medium-chain fatty acids can be transported into the enterocyte without further assistance. However, more hydrophobic compounds such as long-chain fatty acids, monoglycerides, lysophospholipids, and cholesterol must first be repackaged into a second form of micelle within the intestinal lumen. The formation of these micelles also requires the help of

**phospholipase A2** An enzyme produced in the pancreas that hydrolyzes fatty acids from phospholipids.

**lysophospholipid** A lipid composed of a glycerol bonded to a polar head group and a fatty acid; final product of phospholipid digestion.

**bile salt–dependent cholesteryl ester hydrolase** An enzyme produced in the pancreas that cleaves fatty acids from cholesteryl esters.

bile. Once this type of micelle comes into contact with the brush border surface of the intestine, its contents can be released and transported into the enterocyte.

Most plant sterols/stanols are not readily absorbed. However, that does not mean they cannot influence health. For example, some bind cholesterol in the intestine. Because the resulting sterol/stanol-cholesterol complex is not subsequently absorbed, the cholesterol is eliminated in the feces. This is the mechanism by which products such as Benecol are thought to lower risk for cardiovascular disease.

## Circulation of Lipids Away from the Small Intestine

Because of the inherent difficulties related to the hydrophobic nature of lipids, the process of lipid circulation in the body is more complex than it is for other macronutrients. Short- and medium-chain fatty acids are somewhat hydrophilic and therefore can be circulated away from the small intestine in the blood. However, they are first bound to the protein albumin. Fatty acid–albumin complexes flow in the blood from the small intestine to the liver, where they are either metabolized or packaged for delivery to other cells in the body.

The circulation of larger lipids away from the gastrointestinal tract is more involved. Long-chain fatty acids, monoglycerides, and lysophospholipids must first be resynthesized back into triglycerides and phospholipids in the enterocyte. These large lipids combine with cholesterol and cholesteryl esters and are then incorporated into particles called **chylomicrons** (also called chylomicra), which are released into the lymph for initial circulation. The lymph eventually mixes with the blood via the thoracic duct in the neck region. In this way, chylomicrons gradually enter the bloodstream, where they travel to cells that take up their contents.

Chylomicrons are the largest and least dense members of a class of substances called **lipoproteins**. Lipoproteins are complex globular structures containing varying amounts of triglycerides, phospholipids, cholesteryl esters, free cholesterol, and proteins. Lipoproteins are constructed so that their hydrophilic components (such as the polar head groups of phospholipids and proteins) are situated on the outer surface, and their hydrophobic components (such as triglycerides) are facing inward (see Figure 7.13). Because lipid is less dense than protein, the densities of the lipoproteins depend on their relative amounts (or percentages) of lipids and proteins. Lipoproteins with relatively more lipid than protein have lower densities than those with more protein and less lipid. As you will see, most lipoproteins are named according to their densities. The basic function of lipoproteins is to transport lipids in the hydrophilic (watery) environments of the body.

As a comparison, recall the story of the Trojan horse. This legend tells of a clever war tactic devised by the Greek army to enter the walled city of ancient Troy. To accomplish this goal, the Greeks built a giant wooden horse with a hollow belly. A handful of Greek soldiers climbed into the hollow opening and sealed it up. The giant horse was then left before the gates of Troy, and the gatekeepers, not knowing what the horse contained, brought it into the city. Now consider the lipoprotein, its outer hydrophilic shell representing the Trojan Horse, and its inner hydrophobic (lipid) constituents representing the Greek soldiers. This analogy illustrates the way in which the body can transport hydrophobic lipids in the somewhat

**CONNECTIONS** Lymph is the fluid of the lymphatic system (Chapter 4, p. 117).

**chylomicron** A lipoprotein made in the enterocyte that transports large lipids away from the small intestine in the lymph.

**lipoprotein** A spherical particle made of varying amounts of triglycerides, cholesterol, cholesteryl esters, phospholipids, and proteins.

## FIGURE 7.13   Lipoproteins

The ratio of lipids to proteins determines a lipoprotein's density and its name.

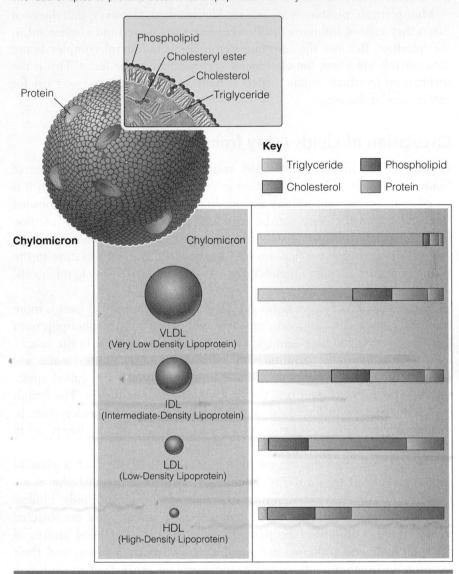

**Key**

Triglyceride    Phospholipid

Cholesterol    Protein

Chylomicron

VLDL
(Very Low Density Lipoprotein)

IDL
(Intermediate-Density Lipoprotein)

LDL
(Low-Density Lipoprotein)

HDL
(High-Density Lipoprotein)

unwelcoming environment of the blood by surrounding them with a hydrophilic shell.

## Functions of Chylomicrons in the Body

Chylomicrons deliver fatty acids to the body via an enzyme called **lipoprotein lipase** produced in many tissues, especially adipose and muscle. After it is produced, this enzyme is relocated out of these cells and into the lumen of the neighboring capillary blood vessels. As blood flows past, lipoprotein lipase "attacks" the chylomicron, releasing fatty acids from triglycerides via hydrolysis. These fatty acids are taken up by surrounding cells. After delivering fatty acids to cells, chylomicrons become more dense and are called **chylomicron remnants.** The remnants are taken up by the liver, where they are broken down and their contents reused or recycled.

**lipoprotein lipase** An enzyme that hydrolyzes the ester bond between a fatty acid and glycerol in a triglyceride molecule.

**chylomicron remnant** The lipoprotein particle that remains after a chylomicron has lost most of its fatty acids.

# ESSENTIAL *Concepts*

Following digestion, lipids must be transported into the enterocytes. For small, water-soluble lipids, this happens easily. However, large, insoluble lipids must first be reassembled into micelles. Lipid circulation away from the small intestine occurs in two ways. Small lipid molecules such as short- and medium-chain fatty acids are circulated directly to the liver in the blood. Large lipids are first packaged into chylomicrons. Chylomicrons are globular particles arranged so that hydrophilic components form an outer shell and hydrophobic components form an inner core. All lipoproteins, including chylomicrons, contain varying amounts of lipids and proteins. Chylomicrons are released into the lymph, where they eventually empty into the blood. Once in the blood, chylomicrons deliver fatty acids to cells via the enzyme lipoprotein lipase.

# Further Delivery of Lipids in the Body

The liver synthesizes and metabolizes lipids as needed. In this way, the liver serves as the central command center for lipid metabolism. The liver also makes lipoproteins required for the delivery of newly synthesized lipids. A summary of their origins and functions is presented in Figure 7.14.

**FIGURE 7.14    The Origins and Major Functions of Lipoproteins**

Both the liver and the small intestine make lipoproteins that circulate lipids in the body.

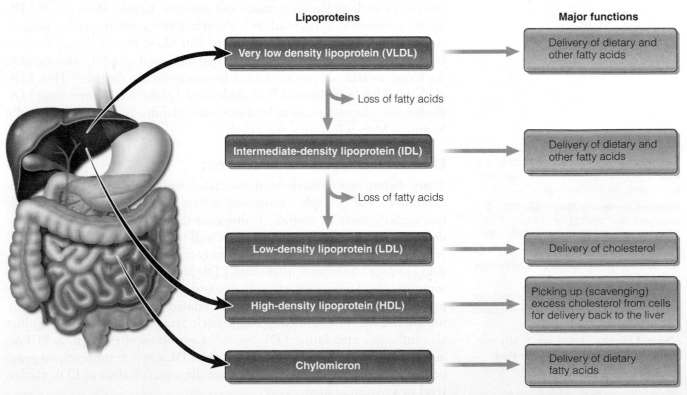

## Very Low Density Lipoproteins (VLDLs)

As you have seen, intestinal cells make one form of lipoprotein—the chylomicron. However, the liver produces additional lipoproteins, each with a special role involving the transport of lipids in the blood. One such lipoprotein is called a **very low density lipoprotein (VLDL)**. VLDLs are similar to chylomicrons, in that they contain triglycerides and cholesteryl esters in their cores surrounded by phospholipids, free cholesterol, and specialized proteins on their surfaces. However, VLDLs have a lower lipid-to-protein ratio than do chylomicrons, making them *more* dense. As with chylomicrons, the primary function of VLDLs is to deliver fatty acids to cells via the actions of lipoprotein lipase. VLDLs deliver fatty acids derived from liver and adipose tissue as well as dietary fatty acids.

## Intermediate-Density Lipoproteins (IDLs) and Low-Density Lipoproteins (LDLs)

As VLDLs lose fatty acids, they become more dense and are called **intermediate-density lipoproteins (IDLs)**. Some IDLs are taken up by the liver, whereas others remain in the circulation, where they continue to lose additional fatty acids. Eventually, IDLs become cholesterol-rich **low-density lipoproteins (LDLs)**. Specialized proteins called **LDL receptors** on cell membranes (especially those of liver, adipose tissue, and muscle) bind LDLs, allowing them to be taken up and broken down by the cell. In this way, cholesterol is delivered to many tissues that use it for structural and metabolic purposes.

LDLs are also taken up and degraded by immune cells, many of which are in the major blood vessels around the heart. The uptake of LDLs into immune cells is necessary, because the cells use their contents to synthesize important substances such as the eicosanoids and immune factors. However, if LDL uptake is excessive, it can result in buildup of a fatty substance called **plaque**, narrowing the blood vessel to the extent that blood flow is slowed or even blocked. Epidemiologic studies suggest that high levels of LDL cholesterol in the blood are related to increased risk for cardiovascular disease.[19] Thus LDL cholesterol has been deemed "bad cholesterol." More detail concerning LDL cholesterol, plaque formation, and cardiovascular disease is presented in the Nutrition Matters following this chapter.

## Effect of Diet on LDL Concentrations

Many dietary factors have been associated with alterations in LDL concentrations. For example, consuming certain SFAs and/or *trans* fatty acids (particularly those in partially hydrogenated oils) can increase LDL concentration via mechanisms that are not well understood.[20] High cholesterol intake is associated with high LDL concentration.[21] However, the relationship between cholesterol intake and LDL production is complex, because some people show no effect of cholesterol intake on LDL levels. This is probably because of genetic variation related to cholesterol absorption, metabolism, and other dietary and lifestyle factors (such as smoking) that also influence circulating LDL levels.[22] Conversely, diets high in PUFAs, ω-3 fatty acids, or dietary fiber can lower LDL cholesterol levels in some people.[23] Clearly, many factors influence the concentration of LDL cholesterol in a person's blood.

---

**very low density lipoprotein (VLDL)** A lipoprotein made by the liver that contains a large amount of triglyceride; its major function is to deliver fatty acids to cells.

**intermediate-density lipoprotein (IDL)** A lipoprotein that results from the loss of fatty acids from a VLDL; IDLs are ultimately converted to LDLs.

**low-density lipoprotein (LDL)** A lipoprotein that delivers cholesterol to cells.

**LDL receptor** Membrane-bound protein that binds LDLs, causing them to be taken up and dismantled.

**plaque** A complex of cholesterol, fatty acids, cells, cellular debris, and calcium that can form inside blood vessels and within vessel walls.

## High-Density Lipoproteins (HDLs)

The liver also makes a series of lipoproteins called **high-density lipoproteins (HDLs)**. Compared to other lipoproteins, HDLs have the lowest lipid-to-protein ratio; thus they have the highest densities. HDLs salvage excess cholesterol from cells, transporting it back to the liver. This transfer of cholesterol from nonhepatic (nonliver) cells back to the liver is called **reverse cholesterol transport,** and represents the major function of HDLs.[24] It is well established that high levels of HDL cholesterol in blood are associated with lower risk of cardiovascular disease.[25] Thus HDL cholesterol is called "good cholesterol."

There are several types of HDL, and not all HDLs are equally as effective in removing excess cholesterol. More specifically, different HDLs have different proteins embedded in them, resulting in somewhat different functions. For example, the presence of particular proteins makes some HDLs less efficient at reverse cholesterol transport. **Tangier disease** is a good example. People with this inherited disease have defective HDLs, because of an altered and dysfunctional protein embedded in their outer shells.[26] Because this negatively affects reverse cholesterol transport, people with Tangier disease are at high risk for developing cardiovascular disease. Learning more about the various HDL proteins and subtypes will undoubtedly lead to a better understanding of how they can alter risk for cardiovascular disease.

> **CONNECTIONS** Inherited diseases are caused by alterations in the DNA and can be passed from parents to offspring (Chapter 6, p. 185).

### Effects of Diet on HDL Concentrations

Although it has long been thought that diets high in carbohydrates offer protection from cardiovascular disease, a considerable amount of research now suggests that these types of diets actually tend to *lower* HDL levels.[27] In other words, very low carbohydrate diets may actually offer some protection from heart disease. In addition, research suggests that high MUFA intake or moderate alcohol consumption can raise HDL levels.[28] Like LDLs, many lifestyle and genetic factors influence HDL concentrations, and researchers continue to study these important interactions.

## ESSENTIAL *Concepts*

Circulating chylomicrons return to the liver as chylomicron remnants. The liver produces other types of lipoproteins including very low density lipoproteins (VLDLs), that deliver fatty acids to cells with the assistance of lipoprotein lipase. The loss of fatty acids from a VLDL results in its conversion to an intermediate-density lipoprotein and ultimately a low-density lipoprotein (LDL). The function of the LDL is to deliver cholesterol to cells. The liver also produces high-density lipoproteins (HDLs) that pick up cholesterol and return it to the liver, a process called reverse cholesterol transport. The concentrations of LDL and HDL in the blood are related to risk for cardiovascular disease, with high levels of LDL and low levels of HDL associated with increased risk. Many dietary and biological factors influence circulating LDL and HDL concentrations and functions.

# Lipids in Health and Disease

After lipids have been digested, absorbed, and circulated they can be used by the body for myriad purposes. As described previously, fatty acids and their metabolites regulate metabolic processes within cells and orchestrate a variety

**high-density lipoprotein (HDL)** A lipoprotein made by the liver that circulates in the blood to collect excess cholesterol from cells.

**reverse cholesterol transport** Process by which HDLs remove cholesterol from nonhepatic (nonliver) tissue for transport back to the liver.

**Tangier disease** A genetic disorder resulting in the production of faulty HDL particles that cannot take up cholesterol from cells.

of physiological responses in the body. Phospholipids are vital components of cell membranes, aid in lipid digestion and absorption, and also contribute fatty acids for cellular use. Cholesterol is incorporated into cell membranes, is a precursor for many hormones, and is involved in lipid digestion and absorption via its role in bile.

Thus lipids are vital for good health. However, sometimes lipids can be associated with poor health. For example, high dietary intake of some lipids is associated with increased risk for obesity, cardiovascular disease, and cancer. These topics are discussed briefly here, although they are also covered in more detail elsewhere in this book.

## Lipids and Obesity

Obesity is defined as the overabundance of body fat, and its causes and consequences are described in detail in Chapter 9. Although obesity is complex, nutrient intake is a major contributor. As fatty acids provide more than twice as many calories per gram than carbohydrates and proteins, fat intake is likely an important piece of the obesity puzzle. Regardless of the causes, obesity is a major public health concern worldwide and is associated with increased risk for many diseases such as cardiovascular disease, type 2 diabetes, and some forms of cancer. Because of an intense interest in decreasing the risk for obesity, experts recommend that we limit our fat consumption. In response to market demand, many food manufacturers produce low-fat and fat-free items in addition to foods that contain fat substitutes. You can read more about fat substitutes in the Focus on Food feature.

## Lipids and Cardiovascular Disease

Like obesity, cardiovascular disease is caused by a complex web of factors, including genetics and diet. Many dietary lipids have been implicated in influencing risk for cardiovascular disease, including total lipids, type of

## *Focus On* FOOD
### The Skinny on Fat Substitutes

Because many people want to consume "low-fat" diets, food manufacturers have developed substances that have desired characteristics of lipids, minus the calories. These "fat substitutes" are diverse in structure, some made from complex carbohydrates, some made from proteins, and others from blends of carbohydrates and fatty acids.

Olestra is an example of a fat substitute made from sucrose (table sugar) bonded with six to eight fatty acids. Olestra can-not be digested by human lipases or colonic bacteria, and therefore provides no usable energy to the body. In fact, it passes through the small intestine relatively intact. In 1996, the FDA approved the use of olestra in savory snacks such as potato chips, cheese puffs, and crackers. As part of the approval, food manufacturers were required to add vitamins A, D, E, and K to olestra-containing foods. This is because olestra interferes with the absorption of the fat-soluble vitamins in the gastrointestinal tract.

There was initial concern that consuming large quantities of olestra would cause intestinal distress such as gas and diarrhea. However, review of many human studies led the FDA to conclude that when consuming reasonable portions of olestra-containing foods, only infrequent and mild gastrointestinal upset is likely to occur. Thus in most situations consuming olestra poses no concern. However, as with any food, those containing olestra or other fat substitutes should be consumed in moderation.

lipid, specific fatty acids, and cholesterol. As noted, high intakes of SFAs, *trans* fatty acids, and cholesterol can increase risk factors for disease in some people, whereas MUFAs may have the opposite effect. However, a person's genetic makeup often interacts with his or her diet to influence health, and this is certainly true for lipids and cardiovascular disease.

Many studies show that consuming diets high in cholesterol increases blood cholesterol and risk for cardiovascular disease. However, some people can eat very high amounts of cholesterol without experiencing this effect. This differential effect may be due to the fact that some people absorb dietary cholesterol better than do others, whereas others may excrete more cholesterol in their feces.[29] Similarly, genetic variation can influence a person's HDL and LDL levels and functioning. For example, genetics can influence the ability of cells to take up LDL particles via mutations in the genes that code for LDL receptor proteins.[30] The Nobel Prize in Physiology or Medicine was awarded in 1985 to Michael Brown and Joseph Goldstein for this discovery.[31] As scientists learn more about how genetics interacts with diet to influence health (the field of nutrigenomics), health professionals will someday be able to "prescribe" the most heart-healthful diet given a person's individual genetic makeup.

## Lipids and Cancer

Although some studies show a link between high-fat diets and risk for cancer, the data are inconclusive. Obesity, however, is a risk factor for several types of cancer, including breast and colorectal cancers.[32] As obesity is often associated with consumption of high-fat diets, dietary lipids may play an indirect role in this disease. You can read more about the relationship between diet and cancer in the Nutrition Matters following Chapter 11.

## ESSENTIAL *Concepts*

Although the mechanisms by which lipids influence health are complex, evidence suggests that dietary fat intake is associated with risk for several chronic diseases. This is especially true for obesity and its related complications, such as cardiovascular disease and type 2 diabetes. These effects are likely due to a combination of factors such as total fat intake, type of fatty acids in the diet, and other dietary lipids such as cholesterol. However, genetic factors as well as other environmental and lifestyle influences interact with lipid intake to influence health and disease.

# Dietary Recommendations for Lipids

Although we have dietary requirements for only the essential fatty acids, we rely on additional fatty acids as important sources of energy. However, consuming too much lipid—or the wrong kind of lipid—can be associated with health problems. Therefore, it is important to maintain a healthy lipid intake throughout life. As with the other macronutrients, there are many recommendations concerning lipid consumption. Some of these are described here.

# Dietary Recommended Intakes (DRIs) and Acceptable Macronutrient Distribution Ranges (AMDRs)

The Institute of Medicine has established no DRIs for total lipid intake, except during infancy when AIs are set at approximately 30 g per day. However, the AMDRs recommend that healthy adults consume 20 to 35% of their energy from lipid. Based on a caloric requirement of 2,000 kcal per day, a person should consume between 400 to 700 kcal of lipid. Considering that 1 g of lipid contains about 9 kcal of energy, the daily lipid intake for most adults should be between 44 and 78 g. This amount is easy to obtain. For example, a typical day's menu for a college student—including three servings of low-fat milk, a bagel with cream cheese, a peanut butter sandwich, spaghetti with meatballs, and a salad with ranch dressing—contains about 56 g of lipid. The lipid content (as well as the SFA and *trans* fatty acid contents) of most packaged foods is easily determined by reading Nutrition Facts panels. This is illustrated in Figure 7.15.

**FIGURE 7.15**  **Using Nutrition Facts Panels to Determine Fat Intake**

You can tell how much total fat, saturated fat, and *trans* fat is in a food by checking its Nutrition Facts panel.

**Nutrition Facts**

Serving Size: 1 cup (244g)

Amount per serving

| Calories 122 | Calories from fat 43 | |
|---|---|---|
| | | **% Daily Value*** |
| **Total fat** 5g | | 7% |
| Saturated fat 3g | | 15% |
| Trans fat 0g | | 0% |
| **Cholesterol** 20mg | | 7% |
| **Sodium** 100mg | | 4% |
| **Total carbohydrates** 11g | | 4% |
| Dietary fiber 0g | | 0% |
| Sugars 12g | | |
| **Protein** 8g | | |
| Vitamin A 9% | Vitamin C 1% | |
| Calcium 29% | Iron 0% | |

*Percent Daily Values are based on a 2,000 calorie diet.

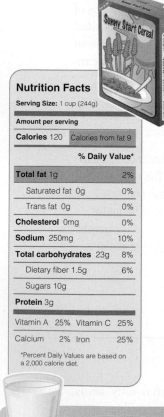

**Nutrition Facts**

Serving Size: 1 cup (244g)

Amount per serving

| Calories 120 | Calories from fat 9 | |
|---|---|---|
| | | **% Daily Value*** |
| **Total fat** 1g | | 2% |
| Saturated fat 0g | | 0% |
| Trans fat 0g | | 0% |
| **Cholesterol** 0mg | | 0% |
| **Sodium** 250mg | | 10% |
| **Total carbohydrates** 23g | | 8% |
| Dietary fiber 1.5g | | 6% |
| Sugars 10g | | |
| **Protein** 3g | | |
| Vitamin A 25% | Vitamin C 25% | |
| Calcium 2% | Iron 25% | |

*Percent Daily Values are based on a 2,000 calorie diet.

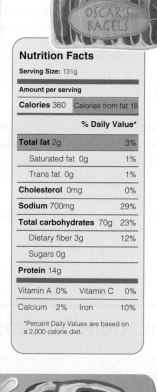

**Nutrition Facts**

Serving Size: 131g

Amount per serving

| Calories 360 | Calories from fat 18 | |
|---|---|---|
| | | **% Daily Value*** |
| **Total fat** 2g | | 3% |
| Saturated fat 0g | | 1% |
| Trans fat 0g | | 1% |
| **Cholesterol** 0mg | | 0% |
| **Sodium** 700mg | | 29% |
| **Total carbohydrates** 70g | | 23% |
| Dietary fiber 3g | | 12% |
| Sugars 0g | | |
| **Protein** 14g | | |
| Vitamin A 0% | Vitamin C 0% | |
| Calcium 2% | Iron 10% | |

*Percent Daily Values are based on a 2,000 calorie diet.

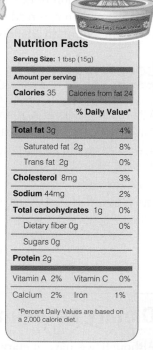

**Nutrition Facts**

Serving Size: 1 tbsp (15g)

Amount per serving

| Calories 35 | Calories from fat 24 | |
|---|---|---|
| | | **% Daily Value*** |
| **Total fat** 3g | | 4% |
| Saturated fat 2g | | 8% |
| Trans fat 2g | | 0% |
| **Cholesterol** 8mg | | 3% |
| **Sodium** 44mg | | 2% |
| **Total carbohydrates** 1g | | 0% |
| Dietary fiber 0g | | 0% |
| Sugars 0g | | |
| **Protein** 2g | | |
| Vitamin A 2% | Vitamin C 0% | |
| Calcium 2% | Iron 1% | |

*Percent Daily Values are based on a 2,000 calorie diet.

**Calories from fat** 94

## Dietary Guidelines for Americans and MyPyramid

The USDA Dietary Guidelines for Americans and the MyPyramid food guidance system also provide recommendations for total lipid intake as well as SFA and *trans* fatty acid intakes. The Dietary Guidelines recommend that we "choose fats wisely for good health," with most fats coming from PUFAs and MUFAs, such as fish, nuts, and vegetable oils. It is also recommended that SFA intake be below 10% of calories, and *trans* fatty acid intake below 1% of calories. These guidelines are only meant to apply to people 2 years of age and older. To help achieve these goals, the MyPyramid food guidance system recommends that we limit our consumption of fats and oils to 3 to 11 teaspoons each day, choose lean meats and poultry, and consume low-fat or fat-free dairy foods. To determine how many servings of these foods groups are appropriate for you, visit the MyPyramid website (http://www.mypyramid.gov).

## ESSENTIAL *Concepts*

Dietary lipid intake should provide the essential fatty acids as well as energy needed by the body. However, lipid intake must be limited so that it does not promote chronic disease. Therefore, it is prudent to consume a wide variety of lipids in moderate amounts. It is recommended that we consume 20 to 35% of our calories from lipid and emphasize foods high in PUFAs and MUFAs. To achieve this goal, the MyPyramid food guidance system recommends limiting consumption of fats and oils and choosing lean meats and poultry as well as low-fat or nonfat dairy products when possible.

# Lipid Nutrition: Putting Concepts into Action

Consuming a nutrient-dense diet helps assure adequate intake of all the essential nutrients while avoiding excess calories that can lead to weight gain. For many people, this is difficult. Indeed, high-fat foods are often the same ones we like to eat the most. Perhaps more than any other nutrient, watching the amounts and the types of fats that we eat is challenging. However, by using dietary assessment tools combined with diet-planning strategies, almost anyone can determine the healthfulness of the lipids in his or her diet and make positive changes to promote health.

## Setting the Stage and Setting the Goals

Consider a 43-year-old man named Chuck who has had his annual physical exam. Much to Chuck's surprise, his doctor finds that he has several risk factors for heart disease, including being overweight and having elevated blood cholesterol levels. Chuck has a family history of cardiovascular disease and is very interested in finding out ways to decrease his risk. Chuck and his doctor decide that Chuck should pay careful attention to his diet—especially his fat intake—in the hope that this will decrease both his weight and his blood cholesterol.

## Assessing Nutritional Status

To assess his diet, Chuck writes down everything that he eats and drinks for three days, including two weekdays and one weekend day, and enters the information into the dietary assessment software on the MyPyramid website.

This program provides him with information concerning total caloric intake and contribution from each macronutrient class. Chuck's energy intake is 3,400 kcal, which is higher than his estimated requirement of 2,900 kcal. And, Chuck learns from his dietary assessment that 50% of the energy (1,450 kcal) in his diet comes from fat. However, the distribution of fatty acid types among SFAs, PUFAs, and MUFAs is within the recommended ranges, as is his cholesterol intake. Given that he is consuming an extra 500 kcal, Chuck and his doctor decide that the primary goal of his new diet plan will be to decrease his fat intake by about this amount.

## Setting the Table to Meet the Goals

This task turns out to be relatively simple—that is, once Chuck begins to pay attention to the foods he chooses. For example, he switches from whole milk, which contains 3.3% fat, to low-fat milk, which contains 1% fat. Because he typically drinks 3 cups of milk each day, this change decreases his caloric intake by about 150 kcal—all from fat. He also checks food packaging for FDA-approved nutrient content claims such as "reduced calories," "light," "fat free," and "low in fat." Compared to regular foods, those carrying these claims are required to be lower in fat, and therefore lower in calories. He also pays attention to foods boasting FDA health claims related to fat intake. Using these tools, as well as information provided on Nutrition Facts panels, Chuck is able to select foods and snacks that are both satisfying and lower in fat and energy.

## Comparing the Plan to the Assessment: Did It Succeed?

Several months later, Chuck completes another three-day diet record and revisits his doctor with good news about his health. He has lost 8 pounds, and is therefore on his way to reaching his target body weight. Chuck believes that his weight loss is due to the relatively simple changes that he has made to his diet, and the results of his second diet-record analysis support this belief. Chuck's doctor also finds that his blood cholesterol level is now within the desirable range. Thus Chuck is succeeding in reducing his risk factors for cardiovascular disease.

## ESSENTIAL *Concepts*

It is relatively simple to assess dietary lipid intake using dietary assessment programs such as that found on the MyPyramid website. This information can be used to determine whether total lipid intake is too high and if fatty acid classes or cholesterol intakes are within recommended ranges. Making changes to dietary lipid intake is made easier by the vast amount of information found on food labels. This includes health claims, nutrient content claims, and Nutrition Facts panels.

## Review Questions   Answers are found in Appendix G.

1. Which of the following generally describes the physical and chemical characteristics of saturated fatty acids?
   a. They tend to be solids at room temperatures.
   b. They tend to be oils at room temperatures.
   c. They typically have low melting points.
   d. They are lower in calories than other fats.

2. A fatty acid with the nomenclature of "*cis*9,*cis*12,*cis*15–18:3" has _____.
   a. 9 carbons
   b. 18 carbons
   c. 9 double bonds
   d. 15 carbons

3. Arachidonic acid and docosahexaenoic acid (DHA) are considered conditionally essential fatty acids for human infants, because
   a. these fatty acids are important for infant health but not adult health.
   b. human milk contains very little of these fatty acids.
   c. babies can synthesize adequate amounts of them from other fatty acids.
   d. babies *cannot* synthesize adequate amounts of them from other fatty acids.

4. Phospholipids are "amphipathic," because they
   a. are polar and easily dissolve in water.
   b. are quite nonpolar and therefore are hydrophobic.
   c. contain both polar and nonpolar regions.
   d. can change from being hydrophobic to being hydrophilic.

5. Digestion of dietary triglycerides occurs in the _____.
   a. mouth
   b. stomach
   c. small intestine
   d. all of the above

6. Cholecystokinin (CCK) is a(n) _____ produced in the _____ in response to _____.
   a. hormone      small intestine      lipids
   b. hormone      liver                lipids
   c. enzyme       small intestine      hunger
   d. enzyme       gallbladder          lipids

7. Which of the following classes of lipoproteins is referred to as "good cholesterol?"
   a. Chylomicron
   b. VLDL
   c. LDL
   d. HDL

8. Short-chain fatty acids are circulated away from the small intestine bound to _____.
   a. cholesterol
   b. albumin
   c. chylomicrons
   d. pancreatic lipase

9. The enzyme lipoprotein lipase is needed to deliver _____ to cells.
   a. phospholipids
   b. cholesterol
   c. chylomicrons
   d. fatty acids

10. The pancreatic hormone insulin causes adipocytes, and to a lesser extent muscle, to
    a. release glucose into the blood.
    b. convert glucose into fatty acids (lipogenesis).
    c. break down triglycerides into fatty acids.
    d. break down glycerol into glucose molecules.

11. Describe the basic structure of a phospholipid. Label the hydrophobic and hydrophilic components. Why is this compound said to be amphipathic? What are phospholipids used for in the body?

12. Provide a description of the following compounds: monoglyceride, diglyceride, and triglyceride. Which of these is most common in the body?

13. What are the current recommendations for lipid intake for adults? Please refer to both the Acceptable Macronutrient Distribution Ranges (AMDRs) and the Dietary Guidelines in your answer.

## Practice Calculations   Answers are found in Appendix G.

1. The Acceptable Macronutrient Distribution Range (AMDR) for linoleic acid is 5 to 10% of calories. If your estimated caloric requirement is 2,000 kcal/day, how much linoleic acid should you consume? What percentage of total calories does this represent? If you find yourself not consuming enough linoleic acid, what foods might you add to your diet?

2. The Acceptable Macronutrient Distribution Range (AMDR) for linolenic acid is 0.6 to 1.2% of calories. If your estimated caloric requirement is 2,000 kcal/day, how much linolenic acid should you consume? What percentage of total calories does this represent? If you find yourself not consuming enough linolenic acid, what foods might you add to your diet?

3. The Acceptable Macronutrient Distribution Range (AMDR) for lipids is approximately 20 to 35% of the energy requirement. If you require 2,750 kcal/day, how many kilocalories should you obtain from lipids? How many *grams* of lipid does this represent?

## Media Links  A variety of study tools for this chapter are available at our website, www.thomsonedu/nutrition/mcguire.

Prepare for tests and deepen your understanding of chapter concepts with these online materials:

- Practice tests
- Flashcards
- Glossary
- Student lecture notebook
- Web links
- Animations
- Chapter summaries, learning objectives, and crossword puzzles

## Notes

1. Calder PC. Polyunsaturated fatty acids and inflammation. Biochemical Society Transactions. 2005; 33:423–7. Shahidi F, Miraliakbari H. Omega-3 fatty acids in health and disease. Part 2: health effects of omega-3 fatty acids in autoimmune diseases, mental health, and gene expression. Journal of Medicinal Food. 2005;8:133–48.

2. Jensen RG. Handbook of Milk Composition. New York: Academic Press; 1995.

3. Innis SM. Essential fatty acid transfer and fetal development. Placenta. 2005;26:S70–5.

4. Alessandri JM, Guisnet P, Vancassel S, Astorg P, Denis I, Langelier B, Aid S, Poumes-Ballihaut C, Champeil-Potokar G, Lavialle M. Polyunsaturated fatty acids in the central nervous system: evolution of concepts and nutritional implications throughout life. Reproduction, Nutrition, and Development. 2004;44:509–38. Heird WC, Lapillone A. The role of essential fatty acids in development. Annual Review of Nutrition. 2005;25:549–71. Innis SM. Polyunsaturated fatty acids in human milk: an essential role in infant development. Advances in Experimental Medicine and Biology. 2004;554:27–43.

5. Auestad N, Halter R, Hall RT, Blatter M, Bogle ML, Burks W, Erickson JR, Fitzgerald KM, Dobson V, Innis SM, Singer LT, Montalto MB, Jacobs JR, Qui Q, Bornstein MH. Growth and development in term infants fed long-chain polyunsaturated fatty acids: A double-masked, randomized, parallel, prospective, multivariate study. Pediatrics. 2001;108:372–81. Auestad N, Montalto MB, Hall RT, Fitzgerald KM, Wheeler RE, Connor WE, Neuringer M, Connor SL, Taylor JA, Hartmann EE. Visual acuity, erythrocyte fatty acid composition, and growth in term infants fed formulas with long chain polyunsaturated fatty acids for one year. Pediatric Research. 1997;41:1–10. Fleith M, Clandinin MT. Dietary PUFA for preterm and term infants: review of clinical studies. Critical Review of Food Science and Nutrition. 2005;45:205–29. Innis SM, Adamkin DH, Hall RT, Kalhan SC, Lair C, Lim M, Stevens DC, Twist PF, Diersen-Schade DA, Harris CL, Merkel KL, Hansen JW. Docosahexaenoic acid and arachidonic acid enhance growth with no adverse effects in preterm infants fed formula. Journal of Pediatrics. 2002;140:547–54.

6. Picciano MF, Guthrie HA, Sheehe DM. The cholesterol content of human milk. A variable constituent among women and within the same woman. Clinical Pediatrics. 1978;17:359–62.

7. Demmers TA, Jones PJ, Wang Y, Krug S, Creutzinger V, Heubi JE. Effects of early cholesterol intake on cholesterol biosynthesis and plasma lipids among infants until 18 months of age. Pediatrics. 2005;115:1594–601. Owen CG, Whincup PH, Odoki K, Gilg JA, Cook DG. Infant feeding and blood cholesterol: a study in adolescents and a systematic review. Pediatrics. 2002;110:597–608.

8. Middaugh JP. Cardiovascular deaths among Alaskan Natives, 1980-1986. American Journal of Public Health. 1990;80:282–5. Oomen CM, Feskens EJM, Rasanen L, Fidanza F, Nissinen AM, Menotti A, Kok FJ, Kromhout D. Fish consumption and coronary heart disease mortality in Finland, Italy, and the Netherlands. American Journal of Epidemiology. 2000;151:999–1006. Vanschoonbeek K, de Maat MP, Heemskerk JW. Fish oil consumption and reduction of arterial disease. Journal of Nutrition. 2003; 133:657–60.

9. Lewis A, Lookinland S, Beckstrand RL, Tiedeman ME. Treatment of hypertriglyceridemia with omega-3 fatty acids: a systematic review. Journal of the American Academy of Nurse Practitioners. 2004;16:384–95. Mori TA, Beilin LJ. Omega-3 fatty acids and inflammation. Current Atherosclerosis Reports. 2004;6:461–7. Shahidi F, Miraliakbari H. Omega-3 (n-3) fatty acids in health and disease. Part 1: cardiovascular disease and cancer. Journal of Medicinal Foods. 2004; 7:387–401. Wijendran V, Hayes KC. Dietary n-6 and n-3 fatty acid balance and cardiovascular health. Annual Review of Nutrition. 2004;24:597–615.

10. Elias SL, Innis SM. Bakery foods are the major dietary source of trans-fatty acids among pregnant women with diets providing 30 percent energy from fat. Journal of the American Dietetic Association. 2002;102:46–51. Steinhart H, Winkler K, Rickert R. Contents and analytical aspects of *trans* and conjugated fatty acids especially in food. Forum in Nutrition. 2003;56:77–79.

11. Aro A, Salminen I. Difference between animal and vegetable trans fatty acids. American Journal of Clinical Nutrition.

1998; 68:918–9. Judd JT, Clevidence BA, Muesing RA, Wittes J, Sunkin ME, Podczasy JJ. Dietary trans fatty acids: Effects on plasma lipids and lipoproteins of healthy men and women. American Journal of Clinical Nutrition. 1994;59:861–8. Mensink RP, Temme HH, Hornstra G. Dietary saturated and trans fatty acids in lipoprotein metabolism. Annals of Medicine. 1994;26:461–4. Stender S, Dyerberg J. Influence of trans fatty acids on health. Annals of Nutrition and Metabolism. 2004;48:61–66.

12. Strumia R. Dermatologic signs in patients with eating disorders. American Journal of Clinical Dermatology. 2005;6:165–73.

13. Satoru S, Yamatoya H, Sakai M, Kataoka A, Furushiro M, Kudo S. Oral administration of soybean lecithin transphosphatidylated phosphatidylserine improved memory impairment in aged rats. Journal of Nutrition. 2001;131:2951–6.

14. U.S. Food and Drug Administration, Center for Food Safety and Applied Nutrition, Office of Nutritional Products, Labeling, and Dietary Supplements. Phosphatidylserine and cognitive dysfunction and dementia (qualified health claim: final decision letter, May 2003). Available from: http://www.cfsan.fda.gov/~dms/ds-ltr36.html.

15. Vidon C, Boucher P, Cachefo A, Peroni O, Diraison F, Beylot M. Effects of isoenergetic high-carbohydrate compared with high-fat diets on human cholesterol synthesis and expression of key regulatory genes of cholesterol metabolism. American Journal of Clinical Nutrition. 2001;73:878–84.

16. Connor WE, Connor SL. Dietary treatment of familial hypercholesterolemia. Arteriosclerosis. 1989;9:91–105.

17. Ostlund RE, Jr. Phytosterols in human nutrition. Annual Review of Nutrition. 2002;22:533–49. Oslund RE, Jr., Racette, SB, Stenson, WF. Inhibition of cholesterol absorption by phytosterol-replete wheat germ compared with phytosterol-depleted wheat germ. American Journal of Clinical Nutrition. 2002;77:1385–9.

18. Leitzmann C. Vegetarian diets: what are the advantages? Forum in Nutrition. 2005;57:147–56.

19. Chilton RJ. Pathophysiology of coronary heart disease: a brief review. Journal of the American Osteopathic Association. 2004;104:S5–S8. Hawkins MA. Markers of increased cardiovascular risk. Obesity Research. 2004;12:107S–14S. Holvoet P. Oxidized LDL and coronary heart disease. Acta Cardiologica. 2004;59:479–84.

20. Judd JT, Clevidence BA, Muesing RA, Wittes J, Sunkin ME, Podczasy JJ. Dietary trans fatty acids: Effects on plasma lipids and lipoproteins of healthy men and women. American Journal of Clinical Nutrition. 1994;59:861–8. Mensink RP, Katan MB. Effect of dietary fatty acids on serum lipids and lipoproteins: a meta-analysis of 27 trials. Arteriosclerosis, Thrombosis, and Vascular Biology. 1992;12:911–9. Mensink RP, Temme HH, Hornstra G. Dietary saturated and trans fatty acids in lipoprotein metabolism. Annals of Medicine. 1994;26:461–4. Stender S, Dyerberg J. Influence of trans fatty acids on health. Annals of Nutrition and Metabolism. 2004;48:61–6.

21. Hopkins PN. Effects of dietary cholesterol on serum cholesterol: A meta-analysis and review. American Journal of Clinical Nutrition. 1992;55:1060–70. Huxley R, Lewington S, Clarke R. Cholesterol, coronary heart disease and stroke: a review of published evidence from observational studies and randomized controlled trials. Seminars in Vascular Medicine. 2002;2:315–23.

22. Berge KE, Tian H, Graf GA, Yu L, Grishin NV, Schultz J, Kwiterovich P, Shan B, Barnes R, Hobbs HH. Accumulation of dietary cholesterol in sitosterolemia caused by mutations in adjacent ABC transporters. Science. 2000;290:1771–5. McMurry MP, Connor WE, Lin DS, Cerqueira MT, Connor SL. The absorption of cholesterol and the sterol balance in the Tarahumara Indians of Mexico fed cholesterol-free and high cholesterol diets. American Journal of Clinical Nutrition. 1985;41:1289–98. Sehayek E, Nath C, Heinemann T, McGee M, Seidman CE, Samuel P, Breslow JL. U-shape relationship between change in dietary cholesterol absorption and plasma lipoprotein responsiveness and evidence for extreme interindividual variation in dietary cholesterol absorption in humans. Journal of Lipid Research. 1998;39:2415–22. Wilson PW. Assessing coronary heart disease risk with traditional and novel risk factors. Clinical Cardiology. 2004;27:7–11.

23. Anderson JW, Randles KM, Kendall CW, Jenkins DJ. Carbohydrate and fiber recommendations for individuals with diabetes: a quantitative assessment and meta-analysis of the evidence. Journal of the American College of Nutrition. 2004;23:5–17. Burr ML, Fehily AM, Gilbert JF, Rogers S, Holliday RM, Sweetnam PM, Elwood PC, Deadman NM. Effects of changes in fat, fish, and fibre intakes on death and myocardial reinfarction: Diet and Reinfarction Trial (DART). Lancet. 1989;2:757–61. Christensen JH, Christensen MS, Dyerberg J, Schmidt EB. Heart rate variability and fatty acid content of blood cell membranes: A dose–response study with n-3 fatty acids. American Journal of Clinical Nutrition. 1999;70:331–7. Vanschoonbeek K, de Maat MP, Heemskerk JW. Fish oil consumption and reduction of arterial disease. Journal of Nutrition. 2003;133:657–60.

24. Lewis GF, Rader DJ. New insights into the regulation of HDL metabolism and reverse cholesterol transport. Circulation Research. 2005;96:1221–32.

25. Gordon T, Castelli WP, Hjotland MC, Kannel WB, Dawber TR. High density lipoprotein as a protective factor against coronary heart disease. The Framingham study. American Journal of Medicine. 1977;62:707–14. Watson AD, Berliner JA, Hama SY, LaDu BN, Faull KF, Fogelman AM, Navab M. Protective effect of high density lipoprotein associated paraoxonase. Inhibition of the biological activity of minimally oxidized low density lipoprotein. Journal of Clinical Investigation. 1995;96:2882–91.

26. Miller M, Zhan M. Genetic determinants of low high-density lipoprotein cholesterol. Current Opinions in Cardiology. 2004;19:380–4. Nofer JR, Remaley AT. Tangier disease: still more questions than answers. Cellular and Molecular Life Sciences. 2005;62:2150–60.

27. Siri PW, Krauss RM. Influence of dietary carbohydrate and fat on LDL and HDL particle distributions. Current Atherosclerosis Reports. 2005;7:455–9. Volek JS, Sharman MJ, Forsythe CE. Modification of lipoproteins by very low-carbohydrate diets. Journal of Nutrition. 2005;135:1339–42.

28. Moorandian AD, Haas MJ, Wong NCW. The effect of select nutrients on serum high density lipoprotein cholesterol and apolipoprotein A-I levels. Endocrine Reviews. 2006;27:2–16

29. Lammert F, Wang DQ. New insights into the genetic regulation of intestinal cholesterol absorption. Gastroenterology; 2005;129:718–34.

30. Brown MS, Goldstein JL. How LDL receptors influence cholesterol and atherosclerosis. Scientific American.

1984;251:52–60. Dedoussis GV, Schmidt H, Genschel J. LDL-receptor mutations in Europe. Human Mutation. 2004;443–59.

**31.** The Nobel Assembly at the Karolinska Institute. The 1985 Nobel Prize in Physiology or Medicine press release. Available from: http://nobelprize.org/medicine/laureates/1985/press.html.

**32.** Al-Serag HB. Obesity and disease of the esophagus and colon. Gastroenterology Clinics of North America. 2005;34:63–82. Key TJ, Schatzkin A, Willett WC, Allen NE, Spencer EA, Travis RC. Diet, nutrition and the prevention of cancer. Public Health Nutrition. 2004;7:187–200. McTiernan A. Obesity and cancer: the risks, science, and potential management strategies. Oncology. 2005;19:871–81.

**Check out the following sources for additional information.**

**33.** Belury MA. Dietary conjugated linoleic acid in health; physiological effects and mechanisms of action. Annual Review of Nutrition (Palo Alto, CA, Annual Reviews Inc.) 2002;22:505–31.

**34.** Berger A, Jones PJH, Abumweis SS. Plant sterols: factors affecting their efficacy and safety as functional food ingredients. Lipids in Health and Disease. 2004;3:5–24.

**35.** Bettelheim FA, Brown WH, March J. Introduction to general, organic, and biochemistry, 7th ed. Belmont, CA: Brooks/Cole, 2004.

**36.** Birch EE, Garfield S, Hoffman DR, Uauy R, Birch DG. A randomized controlled trial of early dietary supply of long-chain polyunsaturated fatty acids and mental development in term infants. Developmental Medicine and Child Neurology. 2000;42:174–81.

**37.** Cunnane SC. Problems with essential fatty acids: time for a new paradigm. Progress in Lipid Research. 2003;42:544–68.

**38.** FAO/WHO (Food and Agricultural Organization/World Health Organization). Sixth World Food and Nutrition Survey. Rome: FAO; 1996.

**39.** German JB, Dillard CJ. Saturated fats: what dietary intake? American Journal of Clinical Nutrition. 2004;80:550–9.

**40.** Goldbourt U, Yaari S, Medalie JH. Factors predictive of long-term coronary heart disease mortality among 10,059 male Israeli civil servants and municipal employees. A 23-year mortality follow-up in the Israeli Ischemic Heart Disease Study. Cardiology. 1993;82:100–21.

**41.** Hajri T, Abumrad NA. Fatty acid transport across membranes: relevance to nutrition and metabolic pathology. Annual Review of Nutrition (Annual Reviews Inc., Palo Alto, CA). 2002;22:383–415.

**42.** Hellerstein MK. Regulation of hepatic de novo lipogenesis in humans. Annual Review of Nutrition (Annual Reviews Inc., Palo Alto, CA). 1996;16:523–57.

**43.** Jones PJ, Papamandjaris AA. Lipids: cellular metabolism. In Present knowledge in nutrition, 8th ed. Bowman BA, Russell RM, editors. Washington, DC: ILSI Press; 2001.

**44.** Keys A, Aravanis C, Blackburn H, Buzina R, Djordevic BS, Dontas AS, Fidanza F, Karvonen JM, Kimura N, Menotti A, Mohacek I, Nedeljkovic S, Puddu V, Punsar S, Taylor HL, van Buchem FSP. Seven countries: a multivariate analysis of death and coronary heart disease. Cambridge, MA: Harvard University Press; 1980.

**45.** Kurzer MS, Xu X. Dietary phytoestrogens. Annual Review of Nutrition (Annual Reviews Inc., Palo Alto, CA) 1997;17:353–81.

**46.** Lichtenstein AH, Deckelbaum RJ. Stanol/sterol ester-containing foods and blood cholesterol levels. A statement for healthcare professionals from the Nutrition Committee of the Council on Nutrition, Physical Activity, and Metabolism of the American Heart Association. Circulation. 2001;103:1177–9.

**47.** Lichtenstein AW, Jones, PJH. Lipids: absorption and transport. In Present knowledge in nutrition, 8th ed. Bowman BA, Russell RM, editors. Washington, DC: ILSI Press; 2001.

**48.** Lu K, Lee M-H, Patel SB. Dietary cholesterol absorption; more than just bile. Trends in Endocrine Metabolism. 2001;12:314–20.

**49.** National Cholesterol Education Program. Third Report of the National Cholesterol Education Program (NCEP) Expert Panel on Detection, Evaluation, and Treatment of High Blood Cholesterol in Adults (Adult Treatment Panel III). NIH Publication No. 01-3670. Bethesda, MD: National Institutes of Health; 2001.

**50.** Institute of Medicine. Dietary Reference Intakes for energy, carbohydrate, fiber, fat, fatty acids, cholesterol, protein, and amino acids. Washington, DC: National Academies Press; 2005.

**51.** U.S. Department of Agriculture Nutrient Data Laboratory, USDA National Nutrient Database for Standard Reference, Release 17. Accessed from: http://www.ars.usda.gov/main/site_main.htm?modecode=12354500.

**52.** U.S. Food and Drug Administration, Center for Food Safety and Applied Nutrition, Office of Nutritional Products, Labeling, and Dietary Supplements. Summary of qualified health claims permitted (September 2003). Accessed from: http://www.cfsan.fda .gov/~dms/qhc-sum.html.

# Nutrition and Cardiovascular Health

© Charles Thatcher/Stone/Getty Images

$\mathcal{A}$s Americans enjoy longer lives, chronic diseases are becoming more common. Of these, heart disease is the leading cause of death in the United States, followed closely by stroke.[1] Heart disease and stroke are forms of **cardiovascular disease,** which describes a variety of diseases of the heart and/or blood vessels. In this section, you will learn basic facts about the physiology of cardiovascular disease and the importance of nutrition in its prevention.

## Cardiovascular Disease

Cardiovascular disease is generally due to a slowing or complete obstruction of blood flow to the heart or other parts of the body, including the brain. Remember that body organs rely on blood vessels to supply oxygen and nutrients to their cells. When these vital compounds are unavailable, the cells begin to die. Decreased blood flow is generally caused by a condition called **atherosclerosis,** which is the narrowing and hardening of the blood vessels that supply blood to the heart muscle (coronary arteries), the brain (cerebral arteries), and other parts of the body as well. Atherosclerosis in coronary arteries can cause **heart disease,** also called coronary heart disease, whereas atherosclerosis in cerebral arteries can cause a **stroke** (Figure 1).

Atherosclerosis can lead to other complications, which can further cause heart disease or stroke. For example, sometimes atherosclerosis causes a blood vessel to become weak and distended, forming an **aneurysm.** An aneurysm is like a bulge on an overinflated inner tube and is dangerous, because it might burst. When an aneurysm in a major blood vessel ruptures, blood pours into the body cavity, causing a rapid drop in blood pressure and depriving other tissues of oxygen and nutrients. Aneurysms can cause both heart disease and stroke, depending on where they occur. In addition, sometimes small pieces of clotted blood, called **blood clots** or thromboses, can become lodged in an artery, reducing or cutting off blood flow to the target tissue. The risk for a blood clot is even greater when the artery has been narrowed by atherosclerosis. Regardless of whether it occurs via atherosclerosis, aneurysm, blood clot, or a combination of factors, cardiovascular disease can be life threatening. This is illustrated in Figure 1.

**cardiovascular disease** (car – di – o – VAS – cu – lar) (*kardia*, heart; *vascellum*, vessel) A disease of the heart or vascular system.

**atherosclerosis** (a – ther – o – scler – O – sis) (*atheroma*, cyst full of pus; *sklerosis*, hardening) The hardening and narrowing of blood vessels caused by buildup of fatty deposits and inflammation in the vessel walls.

**heart disease** (also called coronary heart disease) A condition that occurs when the heart muscle does not receive enough blood.

**stroke** A condition that occurs when a portion of the brain does not receive enough blood.

**aneurysm** (AN – eu – rysm) (*aneurusma*, dilation) The outward bulging of a blood vessel.

**blood clot** (also called thrombosis) A small, insoluble particle made of clotted blood and clotting factors.

## FIGURE 1 Causes of Cardiovascular Disease

Atherosclerosis, blood clots, and aneurysms can all reduce or stop blood flow, causing cardiovascular disease.

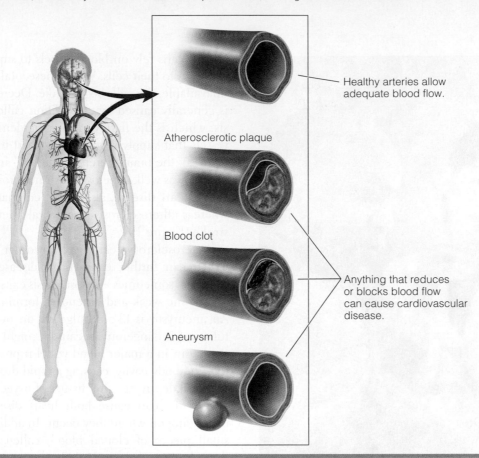

Healthy arteries allow adequate blood flow.

Atherosclerotic plaque

Blood clot

Anything that reduces or blocks blood flow can cause cardiovascular disease.

Aneurysm

## Atherosclerosis

Atherosclerosis is a chronic disease that develops slowly and occurs when fatty deposits, called plaque, form in the lumen or walls of arteries. Plaque contains fatty acids, cholesterol, dead cells, cellular waste products, calcium, and a variety of other compounds. As plaque accumulates within the artery, blood flow is reduced. Sometimes plaque can break off, causing white blood cells to migrate to the injured tissue resulting in an immune response and inflammation of the arterial wall. In addition to hardening and narrowing arteries, atherosclerosis increases the formation of blood clots. Having atherosclerosis is one of the most important risk factors predisposing a person to a subsequent heart attack or stroke.

Although scientists do not know what causes atherosclerosis, emerging research suggests that chronic **inflammation** may play a key role.[2] The term *inflammation* refers to the body's response to a noxious (poisonous) stimulus, injury, or infection. Inflammation causes dilation of blood vessels, movement of white blood cells into the injured tissue, redness, and pain.

One potentially useful biological marker for inflammation is **C-reactive potein (CRP),** which is released into the blood by the liver and smooth muscle as part of the inflammatory response. High circulating concentration of CRP are related to increased risk for cardiovascular disease.[3] Although measuring CRP may someday be standard practice for assessing risk for cardiovascular disease, the Centers for Disease Control and Prevention and the American Heart Association do not currently recommend using it for the general population.[4] The relationships among inflammation, circulating CRP concentrations, and atherosclerosis, however, represent a promising area of research.

**inflammation** A response to cellular injury that is characterized by capillary dilation, white blood cell infiltration, release of immune factors, redness, heat, and pain.

**C-reactive protein (CRP)** A protein produced in the liver and smooth muscles that participates in the immune system's response to a noxious stimulus, injury, or infection; elevated circulating concentrations are related to increased risk for cardiovascular disease.

## Heart Disease

In its less severe state, heart disease causes chest pain and discomfort called **angina pectoris,** or simply angina. Angina occurs when the heart muscle is still receiving some blood but not as much as it needs. Although angina is not usually life threatening, it is an important symptom of increased risk for more serious complications. Heart disease can also lead to a **heart attack,** which occurs when the blood supply to the heart muscle is severely reduced or stopped. The medical term for a heart attack is "myocardial infarction." Because cardiac tissue can survive only for a few minutes without oxygen, heart attacks can permanently damage the heart muscle; if the damage is severe, the results can be fatal.

Warning signs of a heart attack include chest discomfort; pain in one or both arms, back, neck, jaw, or stomach; shortness of breath; cold sweats; nausea; and lightheadedness. It is important that anyone experiencing these symptoms contact emergency medical personnel.

## Stroke

During a stroke, a portion of the brain is deprived of oxygen and critical nutrients. The extent of the resulting brain damage depends on the location, magnitude, duration, and area of the brain affected. Strokes often result in speech impairment or partial paralysis on one side of the body. However, if the stroke affects a large or critical part of the brain it can be life threatening. One important warning sign for stroke is the occurrence of **transient ischemic attacks (TIAs),** sometimes referred to as "ministrokes." TIAs can occur when blood flow to the brain is temporarily disrupted. Warning signs of a TIA or stroke include sudden numbness or weakness (especially on one side of the body), confusion, slurred speech, dizziness, loss of balance, or a severe headache. When a TIA or stroke is suspected, medical professionals should be contacted immediately.

### ESSENTIAL *Concepts*

The constant supply of oxygen and nutrients to the body's tissues is crucial for life. This is especially true for vital tissues such as the heart and brain. However, when blood flow to these organs becomes impaired, conditions such as a heart attack or stroke can occur. Atherosclerosis, aneurysms, and blood clots can cause both heart disease and strokes. Understanding factors that influence these conditions is important, because heart disease is the leading cause of death in the United States, with stroke following close behind.

# Risk Factors and Treatment for Cardiovascular Disease

Biological, lifestyle, and environmental factors are all related to a person's risk for cardiovascular disease. In general, any factor that increases the likelihood that a person will develop atherosclerosis also increases the chance of having a heart attack or stroke. The American Heart Association has categorized these factors as being either major risk factors or contributing risk factors, depending on the availability of supporting evidence. Furthermore, they have subcategorized the risk factors into those that cannot be changed (nonmodifiable or biological risk factors) and those that can be modified, treated, or controlled by lifestyle choices and/or medication (modifiable risk factors).[5]

## Biological and Modifiable Risk Factors

The major nonmodifiable risk factors for cardiovascular disease include age, sex, genetics, race, and prior incidence of stroke or heart attack. Remember that these factors cannot be controlled or treated. However, if you fall into any of these high-risk categories, it is even more important to pay attention to the other risk factors that you may have. These modifiable or lifestyle factors include smoking, high blood pressure, elevated blood lipids, physical inactivity, obesity, diabetes, stress, and alcohol consumption. Although many of these factors are somewhat dependent on biological influences, there are ways we can modify them. Some of these ways, especially those related to nutrition, are briefly described next. The major and contributing risk factors for cardiovascular disease are summarized in Table 1 (see page 262).

### Hypertension

High blood pressure, also called hypertension, increases a person's risk for cardiovascular disease, partly because it puts additional demands on the heart muscle and damages blood vessels. High blood pressure can also cause plaque to break away from arterial walls. Ruptured plaque can act like blood clots, cutting off blood flow.

**angina pectoris** (an – GI – na pec – TOR – is) (*ankhone,* to strangle; *pectos,* breast) Pain in the region of the heart, caused by a portion of the heart muscle receiving inadequate amounts of blood.

**heart attack** (also called myocardial infarction) An often life-threatening condition in which blood flow to some or all of the heart muscle is completely blocked.

**transient ischemic attack (TIA)** A "ministroke" that is caused by a temporary decrease in blood flow to the brain.

## TABLE 1    The Major and Contributing Risk Factors for Cardiovascular Disease

| Factor | Comments |
|---|---|
| **Major Biological Risk Factors** | |
| • Age | Most people who die of heart disease or stroke are over age 65. |
| • Sex | Men are more likely than women to have heart disease or stroke. However, cardiovascular disease is the leading cause of death in both sexes. |
| • Genetics | Children of parents who have had heart disease or stroke are more likely to have these diseases. |
| • Race | Native Americans as well as those of African, Mexican, and Pacific Islander descent are at greatest risk. |
| • Prior stroke or heart attack | People who have had a stroke or heart attack are more likely to have another. |
| **Major Modifiable Risk Factors** | |
| • Smoking | The most important modifiable risk factor for heart disease. |
| • High blood lipids | High blood cholesterol and/or triglycerides increase the likelihood of stroke and heart disease. |
| • High blood pressure | High blood pressure increases risk for heart disease and is the most important risk factor for stroke. |
| • Physical inactivity and obesity | Inactivity, obesity, or both can increase risk of heart disease; at least 30 minutes of activity on most days is recommended. |
| • Diabetes mellitus | Especially if blood glucose regulation is poor, people with diabetes have higher risk of heart disease and stroke. |
| **Contributing Risk Factors** | |
| • Excessive alcohol use | Drinking more than 1–2 drinks (defined as $1\frac{1}{2}$ oz. of spirits, 4 oz. wine, or 12 oz. beer) per day has been linked to increased risk for stroke; however, more moderate alcohol consumption may lower risk of heart disease. |
| • Stress | How an individual responds to stress may influence his or her risk for heart disease. |

SOURCE: Adapted from the American Heart Association. See Note 5, page 266.

Biological, lifestyle, and environmental factors can all contribute to high blood pressure. For example, smoking and obesity are major causes of high blood pressure.[6] Choosing to not smoke and maintaining a healthy body weight can go a long way in preventing high blood pressure and thus cardiovascular disease.

Heavy alcohol consumption can also cause blood pressure to rise.[7] In addition, some people's blood pressures are highly sensitive to sodium, or salt.[8] In other words, high sodium intakes can cause high blood pressure in salt-sensitive individuals. Most people diagnosed with high blood pressure are advised to avoid alcohol consumption and consume relatively low-sodium diets. Consuming a low-sodium diet has become easier as food manufacturers have developed sodium-free and low-sodium food products. Recommendations concerning how to reduce dietary sodium are provided in Table 2.

Other essential dietary minerals such as potassium, calcium, and magnesium are also important for maintaining healthy blood pressure levels. Because higher consumption of these minerals helps lower blood pres-

sure, it is important that people with increased risk for cardiovascular disease consume enough foods that provide these minerals. One way to approach hypertension is the **DASH (Dietary Approaches to Stop Hypertension) diet,** which emphasizes fruits, vegetables, and low-fat dairy products. This diet plan has been shown to lower blood pressure, especially in salt-sensitive people.[9] Although it is not clear whether a single nutrient (such as calcium or potassium) in the DASH diet is responsible for its healthy effects, the diet appears to work by increasing sodium excretion in the urine. Many organizations, such as the American Heart Association and the U.S. Department of Agriculture (USDA) recommend the DASH diet, especially for people predisposed to hypertension. The components of the DASH diet are summarized in Table 3.

## Blood Lipid Levels

A high level of lipids in the blood, or **hyperlipidemia,** is also a risk factor for cardiovascular disease.[10] More specifically, excess triglycerides (**hypertriglyceridemia**) and high levels of cholesterol (**hypercholesterolemia**) are of specific concern. Both forms of hyperlipidemia have genetic and lifestyle components, and are often treated with a combination of medication (to address genetic factors) and lifestyle changes. Of key importance are nutritional modifications.

## TABLE 2    Strategies for Reducing Salt Intake

- Avoid adding salt to foods during cooking or dining.
- Choose fresh, frozen, or canned foods without salt.
- When dining out, ask for foods to be prepared without salt or with half the usual salt.
- Cut down on highly salted snack foods, such as chips.
- Choose low-salt prepared foods when possible.
- Do not drink "sports drinks" unless specifically needed.

**DASH (Dietary Approaches to Stop Hypertension) diet** A dietary pattern emphasizing fruits, vegetables, and low-fat dairy products designed to lower blood pressure.

## TABLE 3  Dietary Approaches to Stop Hypertension (DASH) Diet Basics (based on a 2,000-kcal diet)

| Food Category | Suggested Servings |
|---|---|
| Grains (whole grains preferred) | 7–8 daily |
| Vegetables | 4–5 daily |
| Fruits | 4–5 daily |
| Dairy products (fat-free/low-fat) | 2–3 daily |
| Meat (lean) | 2 or less daily |
| Nuts, seeds, and legumes | 4–5 weekly |
| Fats, oils, and sweets | Use sparingly |

SOURCE: Department of Health and Human Services and U.S. Department of Agriculture. *Facts About the DASH Eating Plan.* Washington, DC: U.S. Government Printing Office, 2003.

## TABLE 4  Reference Values for Blood Lipids

| Blood Lipid Category and Value | Interpretation |
|---|---|
| **Total Cholesterol** | |
| Below 200 mg/dL | Desirable |
| 200–239 mg/dL | Borderline high |
| 240 mg/dL and above | High |
| **Low-Density Lipoprotein (LDL)** | |
| Below 100 mg/dL | Optimal |
| 100–129 mg/dL | Near optimal/above optimal |
| 130–159 mg/dL | Borderline high |
| 160–189 mg/dL | High |
| 190 mg/dL and above | Very high |
| **High-Density Lipoprotein (HDL)** | |
| Below 40 mg/dL | Low |
| 60 mg/dL and above | High |
| **Triglyceride** | |
| Below 150 mg/dL | Normal |
| 150–199 mg/dL | Borderline high |
| 200–499 mg/dL | High |
| 500 mg/dL and above | Very high |

SOURCE: U.S. Department of Health and Human Services. National Cholesterol Education Program. ATPIII Guidelines At-A-Glance Quick Desk Reference, 2001. NIH Publication No. 01-3305. Available from: http://www.nhlbi.nih.gov/guidelines/cholesterol/atglance.pdf.

To better explain how nutrition can influence blood lipids, it is important to briefly review how lipids (particularly cholesterol) are circulated in the blood. Recall that blood lipids are generally packaged as components of lipoproteins, and that cholesterol is circulated mainly by two lipoproteins: low-density lipoproteins (LDLs) and high-density lipoproteins (HDLs). LDLs deliver cholesterol to cells, whereas HDLs pick up excess cholesterol for transport back to the liver. Many studies show that high levels of LDL cholesterol are related to increased risk for plaque formation and cardiovascular disease.[11] Because of this, LDL cholesterol is called "bad cholesterol." Conversely, high levels of HDL cholesterol protect people from cardiovascular disease, and HDL cholesterol is called "good cholesterol."[12]

Medical professionals sometimes use what is called a **cholesterol ratio** to assess a person's risk for cardiovascular disease. The cholesterol ratio is calculated by dividing total cholesterol by HDL cholesterol. As an example, a person with a total cholesterol of 200 mg/dL and an HDL cholesterol of 50 mg/dL would have a cholesterol ratio of 4:1. Doctors recommend that this ratio be below 5:1 and optimally should be 3½:1. Table 4 gives a summary of healthy blood lipid values.

***Dietary Factors Related to Blood Lipid Levels***  Several dietary factors can influence circulating triglyceride and cholesterol levels, although not all people are equally affected. Nonetheless, people with risk factors for cardiovascular disease should consider these dietary factors when planning a healthy diet. For example, in some people high intakes of dietary cholesterol can raise blood cholesterol and triglyceride levels.[13] In addition, consuming high levels of saturated fatty acids and *trans* fatty acids can increase these parameters.[14] Monounsaturated fatty acids (such as in olive oil), polyunsaturated fatty acids (such as in vegetable oils), and omega-3 fatty acids (such as in fish) may help lower blood lipids.[15] Diets high in soluble fiber (such as in

oat or rice bran, oatmeal, legumes, barley, citrus fruits, and strawberries) may help lower LDL cholesterol.[16] Although people have long thought high-carbohydrate diets are heart-healthy, more recent studies show that very low carbohydrate diets may be more beneficial.[17] Clearly, more research will be required to determine the best balance of macronutrients for maintaining a healthy cardiovascular system and which carbohydrates, in particular, are heart-healthy.

In addition to dietary lipid and carbohydrates, many other factors can influence circulating lipid concentrations. Perhaps the most important is being overweight and/or inactive.[18] Even losing small amounts of weight, such as 10 pounds, can lower blood lipid levels. Whereas physical inactivity can also lead to hyperlipidemia, physical activity can raise HDL levels. Doctors recommend that people with hypertriglyceridemia and/or hypercholesterolemia watch their caloric intake, maintain a healthy body weight, and engage in regular physical activity.

**hyperlipidemia** (hy – per – li – pid – EM – i – a) Elevated levels of lipids in the blood.

**hypertriglyceridemia** (hy – per – tri – gly – cer – EM – i – a) Elevated levels of triglycerides in the blood.

**hypercholesterolemia** (hy – per – chol – est – er – ol – EM – i – a) Elevated levels of cholesterol in the blood.

**cholesterol ratio**  The mathematical ratio of total blood cholesterol to high-density lipoprotein (HDL) cholesterol.

Interestingly, consuming alcoholic beverages can be both detrimental and beneficial for blood lipid concentrations. Specifically, any level of alcohol consumption tends to increase triglyceride levels (a negative effect), whereas moderate alcohol consumption (one to two drinks per day) can raise HDL cholesterol levels (a beneficial effect).[19] Doctors recommend that if you drink, do so in moderation. However, they do not recommend that nondrinkers begin to drink to benefit their hearts!

## Obesity

Being overweight or obese is strongly linked to increased risk for cardiovascular disease. However, modifying one's lifestyle to prevent or treat obesity is challenging. Unfortunately, there is no quick fix or magic bullet to help people shed unwanted pounds. However, being physically active and consuming a diet relatively low in fat and high in nutrient-dense foods are important in maintaining a healthy body weight. Information concerning the physiological, psychological, and environmental regulation of body weight is presented in Chapter 9.

## Diabetes

Having diabetes is also a major risk factor for cardiovascular disease. As you have learned, diabetes is a condition in which the body can no longer effectively regulate blood glucose. Although genetics can play a large role in determining who gets diabetes, other factors are also involved. For example, being overweight, having high blood pressure, and having elevated blood lipid levels are risk factors for type 2 diabetes. Keeping these important factors in check can help prevent type 2 diabetes, and, in turn, can reduce the risk for cardiovascular disease. Furthermore, poorly treated diabetes can result in elevated blood glucose levels, which can damage blood vessels and lead to cardiovascular disease. Controlling blood glucose levels with diet, exercise, and medication is important in preventing heart disease and stroke.

## Related Dietary Factors and Patterns

In addition to dietary components known to influence risks of cardiovascular disease, other nutrients may be important as well. For example, the B vitamin folate (folic acid), vitamin $B_6$, and vitamin $B_{12}$ help maintain a healthy cardiovascular system. A deficiency in any of these vitamins can cause a compound called homocysteine to accumulate in the blood. This is important, because raised levels of homocysteine are strongly associated with increased risk for cardiovascular disease.[20]

In addition, compounds—called antioxidants—help rid the body of damaging substances called free radicals. Emerging research suggests that antioxidants provide the body with a natural defense against free-radical damage to LDLs that might contribute to cardiovascular disease.[21] Examples of antioxidant nutrients include vitamins C and E, β-carotene (a precursor of vitamin A), and selenium. However, controlled clinical studies have not conclusively shown that higher consumption of antioxidants decreases risk for disease. Higher vitamin E intake has also been associated with lower circulating concentration of CRP.[22] Further research is needed.

Phytochemicals that may influence heart health include plant sterols and stanols, isoflavones found mainly in soy products, compounds found in red wine and grapes, and a variety of sulfur-containing compounds abundant in garlic, onions, and leeks.[23] However, as with the antioxidants, not enough data are available to confirm whether higher intakes of these compounds reduce the incidence of heart attacks or strokes. In the meantime, consuming a variety of foods—especially fruits and vegetables—is recommended to ensure adequate intake of these dietary factors.[24]

## DASH, Vegetarian, and Mediterranean Diets

In addition to knowing which nutrients are important for heart health, it is also useful to consider which overall types of diets are associated with decreased risk for disease. As such, several dietary patterns have emerged as being associated with lower risk for cardiovascular disease. These include the DASH diet, vegetarian diets, and the **Mediterranean diet.** As previously described, the DASH diet emphasizes fruits, vegetables, and low-fat dairy products. Vegetarian diets are typically those that exclude meat products but include milk, fish, and eggs—in other words, lacto-ovo-vegetarian diets. Mediterranean diets are traditional cuisines consumed by people living in countries that surround the Mediterranean Sea such as Spain, Italy, and Greece. These diets emphasize fruits, vegetables, breads, nuts, and seeds as well as olive oil and wine; meats are consumed in moderation. As you can see, there is much overlap in these diet plans.

Dietary intervention studies show that consumption of the DASH diet decreases risk for cardiovascular disease, and epidemiologic studies show that people consuming either a vegetarian or Mediterranean-type diet have lower rates of heart disease.[25] However, it is difficult to determine which dietary component is protective and whether the reduced risk for disease is due to diet alone. More likely, a variety of confounding factors are involved, such as low rates of obesity and high rates of physical activity.

**Mediterranean diet** A dietary pattern originating from the region surrounding the Mediterranean Sea that includes relatively high intakes of fruits and vegetables, nuts, seeds, fish, and certain oils (such as olive oil) and low intake of meat.

## ESSENTIAL *Concepts*

Many factors are associated with increased risk for cardiovascular disease. Nonmodifiable factors include age, sex, and genetics. Of the modifiable risk factors, smoking is the most dangerous. Whether you have other risk factors, such as high blood pressure, elevated blood lipid levels, and diabetes can be modified by dietary means, such as weight loss, moderation of saturated and *trans* fatty acid intakes, and decreased sodium consumption. In addition, consumption of B vitamins, antioxidants, and a variety of phytochemicals may be important. Several dietary patterns such as the DASH diet, vegetarian diet, and Mediterranean diet are also associated with decreased risk for cardiovascular disease, but which nutrients and other lifestyle factors are involved are not known.

# General Nutrition Guidelines for Healthy Hearts

Clearly, many dietary components are related to risk for cardiovascular disease. In response, agencies such as the American Heart Association, the USDA, and the National Institutes of Health have set forth guidelines regarding dietary intakes that help lower risk for these diseases. These are outlined as follows. The multifaceted and complex relationship between diet and risk for cardiovascular disease is yet another example of how dietary variety, moderation, and balance are essential to health and well-being.

## Energy Intake

The primary goal of recommendations related to energy consumption is to consume only enough calories to maintain a healthy body weight. People who need to lose weight are advised to decrease their energy intake and increase their energy expenditure to meet their body weight goals.

## Lipids and Carbohydrates

There are several recommendations concerning intake of dietary lipids. These include recommendations about total fat intake as well as those for specific groups of fatty acids such as saturated, *trans*, and omega-3 ($\omega$-3) fatty acids. In addition, suggested cholesterol intakes and general recommendations concerning carbohydrate intake have been established. Some of these are described here.

- Total fat intake should be limited to 20 to 35% of total calories.
- Intake of saturated fatty acids should not exceed 10% of total calories. For people who already have some form of heart disease, energy from saturated fatty acids should be less than 7% of total calories. Saturated fatty

acids are found mainly in animal-derived foods.
- Intake of *trans* fatty acids should be less than 1% of calories. *Trans* fatty acids are found mainly in foods that contain partially hydrogenated oils.
- Omega-3 ($\omega$-3) fatty acids should be included in the diet by consuming fish twice each week and a variety of fats and oils, especially those rich in linolenic acid, such as flaxseed, canola, and soybean oils. People with heart disease or elevated blood triglyceride levels should eat fish more frequently.
- Cholesterol consumption should be limited to 300 mg/day. This is the amount found in one to two eggs. People with heart disease should limit their daily cholesterol intake to 200 mg/day, and those with very high levels of blood cholesterol should consider reducing their cholesterol intakes even more.
- Food choices should emphasize complex carbohydrates such as those found in vegetables, fruits, and whole grains rather than refined carbohydrates, such as table sugar. Experts recommend that we consume three or more servings of whole-grain products daily.
- Foods containing soluble fiber, such as oat products, legumes, and citrus fruits, should be consumed on a regular basis.

## Micronutrients

Although less is known about the relationship between micronutrients and cardiovascular disease, many of these compounds are likely important for maintaining good health. Some recommendations concerning intakes of these substances are outlined here.

- Sodium intake should not exceed about 2,300 mg/d—the amount equivalent to 1 teaspoon of salt.
- Calcium intake should at least meet basic dietary requirements, which is 1,000 mg/d for adults. Low-fat dairy products should be emphasized.
- Potassium-rich foods such as legumes, potatoes, seafoods, and some fruits and vegetables (such as bananas) should be consumed regularly.
- A diet providing adequate amounts of folate, vitamin $B_6$, and vitamin $B_{12}$ should be consumed. Eating a variety of both animal- and plant-based foods on a regular basis is critical to meeting this goal.

## ESSENTIAL *Concepts*

A heart-healthy diet is one that emphasizes moderation and variety. In general, experts recommend that we consume a diet low in fat (<35% of calories), saturated fatty acids (<10% of calories), *trans* fatty acids (<1% of calories), cholesterol (<300 mg/d), and salt (the equivalent of no more than 1 teaspoon per day). Fish and other foods rich in omega-3 ($\omega$-3) fatty acids should be consumed on a regular basis as well as whole-grain foods and those containing soluble fiber. Careful attention should also be given to consuming adequate amounts of some vitamins and minerals, such as calcium, potassium, and certain B vitamins. These recommendations are even more important for people who have either had a previous heart attack or stroke or who are at increased risk for such events.

## Notes

1. American Heart Association, American Stroke Association. Heart disease and stroke statistics—2005 update. Available from: http://www.americanheart.org/downloadable/heart/1105390918119HDSStats2005Update.pdf.

2. Kannel WB, Wolf PA, Castelli WP, D'Agostino RB. Fibrinogen and risk of cardiovascular disease: the Framingham Study. Journal of the American Medical Association (JAMA). 1987;258:1183–6. Ross R. Atherosclerosis: an inflammatory disease. New England Journal of Medicine. 1999;340:115–26. Tracy RP. Inflammation in cardiovascular disease. Circulation. 1998;98:2000–2.

3. Koenig W, Sund M, Frohlich M, Fischer HG, Lowel H, Doring A, Hutchison WL, Pepsys MB. C-Reactive protein, a sensitive marker of inflammation, predicts future risk of coronary heart disease in initially healthy middle-aged men: results from the MONICA (Monitoring Trends and Determinants in Cardiovascular Disease) Augsburg Cohort Study, 1984–1992. Circulation. 1999;99:237–42. Ridker PM, Rifai N, Rose L, Buring JE, Cook NR. Comparison of C-reactive protein and low-denisty lipoprotein cholesterol levels in the prediction of first cardiovascular events. New England Journal of Medicine. 2002;347:1557–65. Sakkinen P, Abbott RD, Curb JD, Rodriguez BL, Yano K, Tracy RP. C-reactive protein and myocardial infacrction. Journal of Clinical Epidemiology. 2002;55:445–51.

4. Pearson AP, Mensah GA, Alexander RW, Anderson JL, Cannon RO, Criqui M, Fadl YY, Fortmann SP, Hong Y, Myers GL, Rifai N, Smith SC, Taubert K, Tracy RP, Vinicor. Markers of inflammation and cardiovascular disease. Application to clinical and public health practice. A statement for healthcare professionals from the Centers for Disease Control and Prevention and the American Heart Association. Circulation. 2003;107:499–511.

5. American Heart Association. Risk factors and coronary heart disease. Available from: http://www.americanheart.org/presenter.jhtml?identifier=500. American Heart Association. Stroke risk factors. Available from: http://www.americanheart.org/presenter.jhtml?identifier=54716.

6. Eckel RH. Obesity and heart disease. A statement for health care professionals from the Nutrition Committee, American Heart Association. Circulation. 1997;96:3248–50. Tribble DL, Krauss RM. Atherosclerotic cardiovascular disease. In Present knowledge in nutrition, 8th ed. Bowman BA, Russell RM, editors. Washington, DC: ILSI Press; 2001.

7. Lucas DL, Brown RA, Wassef M, Giles TD. Alcohol and the cardiovascular system research challenges and opportunities. Journal of the American College of Cardiology. 2005;45:1916–24. Miller PM, Anton RF, Egan BM, Basile J, Nguyen SA. Excessive alcohol consumption and hypertension: clinical implications of current research. Journal of Clinical Hypertension. 2005;7:346–51.

8. Kotchen TA, McCarron DA. Dietary electrolytes and blood pressure. A statement for healthcare professionals from the American Heart Association Nutrition Committee. Circulation. 1998;98:613–7. Weinberger MH. Sodium and blood pressure 2003. Current Opinions in Cardiology. 2004;19:353–6.

9. Karanja N, Erlinger TP, Pa-Hwa L, Miller ER, Bray GA. The DASH diet for high blood pressure: from clinical trial to dinner table. Cleveland Clinic Journal of Medicine. 2004;71:745–53. U.S. Department of Health and Human Services, and U.S. Department of Agriculture. Facts About the DASH Eating Plan. Washington, DC: U.S. Government Printing Office; 2003.

10. Hokanson JE, Austin MA. Plasma triglyceride level is a risk factor for cardiovascular disease independent of high-density lipoprotein cholesterol level: a meta-analysis of population-based prospective studies. Journal of Cardiovascular Risk. 1996;3:213–9. Krauss RM. Triglycerides and atherogenic lipoproteins: rationale for lipid management. American Journal of Medicine. 1998;105:S58–S62.

11. Chilton RJ. Pathophysiology of coronary heart disease: a brief review. Journal of the American Osteopathic Association. 2004;104:S5–S8. Hawkins MA. Markers of increased cardiovascular risk. Obesity Research. 2004;12:107S–14S. Holvoet P. Oxidized LDL and coronary heart disease. Acta Cardiologica. 2004;59:479–84.

12. Gordon T, Castelli WP, Hjotland MC, Kannel WB, Dawber TR. High density lipoprotein as a protective factor against coronary heart disease. The Framingham study. American Journal of Medicine. 1977;62:707–14. Watson AD, Berliner JA, Hama SY, LaDu BN, Faull KF, Fogelman AM, Navab M. Protective effect of high density lipoprotein associated paraoxonase. Inhibition of the biological activity of minimally oxidized low density lipoprotein. Journal of Clinical Investigation. 1995;96:2882–91.

13. Mattson FH, Erickson BA, Kligman AM. Effect of dietary cholesterol on serum cholesterol in man. American Journal of Clinical Nutrition. 1972;25:589–94.

14. Judd JT, Clevidence BA, Muesing RA, Wittes J, Sunkin ME, Podczasy JJ. Dietary trans fatty acids: Effects on plasma lipids and lipoproteins of healthy men and women. American Journal of Clinical Nutrition. 1994;59:861–8. Mensink RP, Temme HH, Hornstra G. Dietary saturated and trans fatty acids in lipoprotein metabolism. Annals of Medicine. 1994;26:461–4. Stender S, Dyerberg J. Influence of trans fatty acids on health. Annals of Nutrition and Metabolism. 2004;48:61–6.

15. Moorandian AD, Haas MJ, Wong NCW. The effect of select nutrients on serum high density lipoprotein cholesterol and apolipoprotein A-I levels. Endocrine Reviews. 2006;27:2–16.

16. Pereira MA, O'Reilly E, Augustsson K, Fraser GE, Goldbourt U, Heitmann BL, Hallmans G, Knekt P, Liu S, Pietinen P, Spiegelman D, Stevens J, Virtamo J, Willett WC, Ascherio A. Dietary fiber and risk of coronary heart disease. A pooled analysis of cohort studies. Archives of Internal Medicine. 2004;164:370–6.

17. Siri PW, Krauss RM. Influence of dietary carbohydrate and fat on LDL and HDL particle distributions. Current Atherosclerosis Reports. 2005;7:455–9. Volek JS, Sharman MJ, Forsythe CE. Modification of lipoproteins by very low-carbohydrate diets. Journal of Nutrition. 2005;135:1339–42.

18. Eckel RH. Obesity and heart disease. A statement for health care professionals from the Nutrition Committee. American Heart Association. Circulation. 1997;96:3248–50.

19. Pearson TA. Alcohol and heart disease. Circulation. 1996;94:3023–25. Renaud S, de Logeril M. Wine, alcohol, platelets and the French paradox for coronary heart disease. Lancet. 1992;339:1523–6.

20. Clarke R, Smulders Y, Fowler B, Stehouwer CD. Homocysteine, B-vitamins, and the risk of cardiovascular disease. Seminars in Vascular Medicine. 2005;5:75–6.

21. Knekt P, Ritz J, Pereira MA, O'Reilly EJ, Augustsson K, Fraser GE, Goldbourt U, Heitmann BL, Hallmans G, Liu, S, Pietinen P, Spiegelman D, Stevens J, Virtamo J, Willett WC, Rimm EB, Ascherio A. Antioxidant vitamins and coronary heart disease risk: a pooled analysis of 9 cohorts. American Journal of Clinical Nutrition. 2004;80:1508–20. Lee, IM, Cook NR, Manson JE, Buring JE, Hennekens CH. Betacarotene supplementation and incidence of cancer and cardiovascular disease: the Women's Health Study. Journal of the National Cancer Institute. 1999;91:2102–6. Tribble DL. Antioxidant consumption and risk of coronary heart disease: emphasis on vitamin C, vitamin E and beta-carotene. A statement for health care professionals from the Nutrition Committee, American Heart Association. Circulation. 1999;99:591–5.

22. Liepa GU, Basu H. C-reactive proteins and chronic disease: what role does nutrition play? Nutrition in Clinical Practice. 2003;18:227–33. Singh U, Devaraj S, Jialal I. Vitamin E, oxidative stress, and inflammation. Annual Review of Nutrition. 2005;25:151–74.

23. Castro IA, Barroso LP, Sinnecker P. Functional foods for coronary heart disease risk reduction: a meta-analysis using a multivariate approach. American Journal of Clinical Nutrition. 2005;82:32–40. Gylling H, Miettinen TA. The effect of plant stanol- and sterol-enriched foods on lipid metabolism, serum lipids and coronary heart disease. Annals of Clinical Biochemistry. 2005;42:254–63. Renaud S, de Logeril M. Wine, alcohol, platelets and the French paradox for coronary heart disease. Lancet. 1992;339:1523–6.

24. Law MR, Morris JK. By how much does fruit and vegetable consumption reduce the risk of ischaemic heart disease? European Journal of Clinical Nutrition. 1998;52:549–56.

25. Enas EA, Senthilkumar A, Chennikkara H, Bjurlin MA. Prudent diet and preventive nutrition from pediatrics to geri-atrics: current knowledge and practical recommendations. Indian Heart Journal. 2003;55:310–38. Kok FJ, Kromhout D. Atherosclerosis—epidemiological studies on the health effects of a Mediterranean diet. European Journal of Nutrition. 2004;43:2–5. Leitzmann C. Vegetarian diets: what are the advantages? Forum in Nutrition. 2005;57:147–56.

**Check out the following sources for additional information.**

26. Appel LJ, Sacks FM, Carey VJ, Obarzanek E, Swain JF, Miller ER III, Conlin, PR, Erlinger, TP, Rosner, BA, Laranjo NM, Charleston J, McCarron P, Bishop LM for the OmniHeart Collaborative Research Group. Effects of protein, monounsaturated fat, and carbohydrate intake on blood pressure and serum lipids: results of the OmniHeart randomized trial. JAMA (Journal of the American Medical Association). 2005;294:2455–64.

27. Bloedon LT, Szapary PO. Flaxseed and cardiovascular risk. Nutrition Reviews. 2004;62:18–27.

28. Institute of Medicine. Dietary Reference Intakes for energy, carbohydrate, fiber, fat, fatty acids, cholesterol, protein, and amino acids. Washington, DC: National Academies Press; 2005.

29. Institute of Medicine. Dietary Reference Intakes for thiamin, riboflavin, niacin, vitamin $B_6$, folate, vitamin $B_{12}$, pantothenic acid, biotin, and choline. Washington, DC: National Academy Press; 1998.

30. Institute of Medicine. Dietary Reference Intakes for water, potassium, sodium, chloride, and sulfate. Washington, DC: National Academies Press; 2005.

31. Institute of Medicine. Dietary Reference Intakes for vitamin A, vitamin K, arsenic, boron, chromium, copper, iodine, iron, manganese, molybdenum, nickel, silicon, vanadium, and zinc. Washington, DC: National Academy Press; 2001.

32. Institute of Medicine. Dietary Reference Intakes for vitamin C, vitamin E, selenium, and the carotenoids. Washington, DC: National Academy Press; 2000.

33. Mozaffarian D, Rimm EB, King IB, Lawler RL, McDonald GB, Levy WC. Trans fatty acids and systemic inflammation in heart failure. American Journal of Clinical Nutrition. 2004;80:1521–5.

34. Rimm EB, Stampfer MJ, Ascherio A, Giovannucci E, Colditz GA, Willett WS. Vitamin E consumption and the risk of coronary heart disease in men. New England Journal of Medicine. 1993;328:1450–6.

35. Ross R. Atherosclerosis-an inflammatory disease. New England Journal of Medicine. 1999;340:115–26.

36. Steinberg D, Parthasarathy C, Carew TE, Khoo JC, Witztum JL. Beyond cholesterol. Modifications of low-density lipoprotein that increase its atherogenicity. New England Journal of Medicine. 1989;320:915–24.

37. Upritchard JE, Zeelenberg MJ, Huizinga H, Verschuren PM, Trautwein EA. Modern fat technology: what is the potential for heart health? Proceedings of the Nutrition Society. 2005;64:379–86.

## Chapter 8

# Energy Metabolism

A t any moment in time, thousands of chemical reactions are taking place around us as well as within us. These reactions all have a common purpose, that is, to sustain life. To survive, all living organisms must obtain energy from their surroundings. In humans, that energy comes from the foods we eat. In previous chapters, you learned about the energy-yielding macronutrients—carbohydrates, lipids, and proteins. In this chapter, you will learn how cells actually use these nutrients as an energy source.

**Metabolism** is defined as the chemical reactions that take place in the body. Although a variety of metabolic reactions have been discussed throughout this book, the focus of this chapter is on **energy metabolism,** the chemical reactions that enable cells to obtain and use energy from nutrients. The term *energy metabolism* refers to hundreds of chemical reactions that modify or transform molecules within cells in order to maintain a steady supply of energy (ATP). It is important to understand the basics of how glucose, amino acids, and fatty acids are used to supply energy. It is also important to examine energy metabolism using an integrated approach. This will enable you to see how metabolic pathways are interrelated and how various tissues use different nutrients in response to changes in energy availability. This chapter presents information pertaining to the study of energy metabolism, how metabolism is regulated, and how different macronutrients are used in various states of energy availability.

Photos © Matthew Farruggio

# Understanding Energy Metabolism

The term *energy metabolism* refers to the hundreds of chemical reactions involved in the breakdown, synthesis, and transformation of the energy-yielding nutrients—glucose, amino acids, and fatty acids—to meet the body's need for energy (ATP). Although much of the energy contained in these nutrients is captured as ATP, some of it is lost as heat. Although alcohol can also be used for energy metabolism, this is addressed separately in the Nutrition Matters following Chapter 10. Metabolic pathways work together in a complex way to maintain a steady supply of energy (ATP). These pathways occur simultaneously and are highly coordinated, allowing the body to use different combinations of nutrients in response to various physiological states. This versatility helps ensure that raw materials and energy (ATP) are available to cells at all times.

## Overview of Metabolic Pathways

A **metabolic pathway** consists of a series of interrelated, enzyme-catalyzed chemical reactions. Some metabolic pathways are simple, consisting of 3 to 4 chemical reactions, whereas others are more complex, with 10 to 15 chemical reactions. Chemical reactions transform molecules, sometimes breaking them down and at other times forming new ones. A molecule that enters a chemical reaction is a **substrate,** also called a reactant, and the resulting molecule is a **product.** A chemical reaction can be expressed as an equation whereby the substrate is written on the left of an arrow that points to the product(s); $A + B \rightarrow C + D$. In this example, A and B are substrates and C and D are products. The arrow means "yields." As shown in Figure 8.1, the product of one chemical reaction in a pathway becomes the substrate in the reaction that follows. Products formed before a metabolic pathway reaches completion are called **intermediate products,** whereas the final product(s) in a pathway

**metabolism** (me – TAB – o – lism) Chemical reactions that take place in the body.

**energy metabolism** Chemical reactions that enable cells to store and use energy from nutrients.

**metabolic pathway** A series of interrelated enzyme-catalyzed chemical reactions that take place in cells.

**substrate** (reactant) (SUB – strate) A molecule that enters a chemical reaction.

**product** A molecule produced in a chemical reaction.

**intermediate product** A product formed before a metabolic pathway reaches completion, often serving as a substrate in the next chemical reaction.

## FIGURE 8.1   Metabolic Pathways

A metabolic pathway consists of a series of interrelated, enzyme-catalyzed chemical reactions.

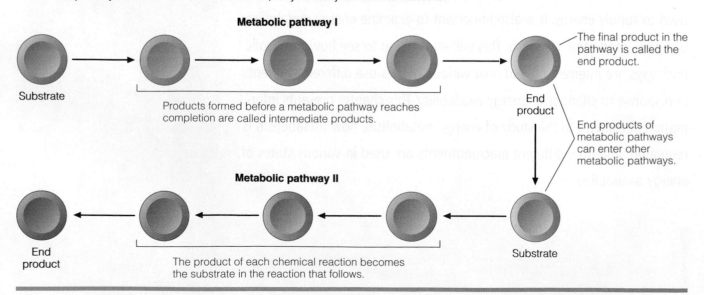

**Metabolic pathway I**

Substrate

Products formed before a metabolic pathway reaches completion are called intermediate products.

The final product in the pathway is called the end product.

End product

End products of metabolic pathways can enter other metabolic pathways.

**Metabolic pathway II**

End product

The product of each chemical reaction becomes the substrate in the reaction that follows.

Substrate

is/are the **end product(s).** It is common for intermediate products and/or end product(s) of one metabolic pathway to enter other metabolic pathways.

## Metabolism Can Be Catabolic or Anabolic

Metabolic pathways can be **catabolic** or **anabolic.** Catabolic pathways release energy through the breakdown of complex molecules into simpler ones, whereas anabolic pathways require energy (ATP) to construct complex molecules from simpler ones. As illustrated in Figure 8.2, the energy made available by catabolic reactions often drives anabolic reactions. Likewise, the products of catabolic pathways provide many of the building blocks needed for anabolic pathways. The breakdown of glycogen into glucose, for example, involves a series of catabolic reactions. Conversely, the synthesis of glycogen from glucose involves a series of anabolic reactions.

The availability of metabolic fuels (glucose, fatty acids, and amino acids) fluctuates throughout the day. After a meal, these energy sources are readily available and tend to exceed the immediate needs of the body. During this time, anabolic pathways favor the storage of excess glucose, amino acids, and fatty acids as glycogen, protein, and triglycerides, respectively. Conversely, catabolic pathways increase fuel availability by breaking down the body's stored energy reserves—glycogen, protein, and triglycerides—into glucose, amino acids, and fatty acids, respectively. In this way, the body shifts from anabolic to catabolic pathways in response to energy availability and need.

**end product** The final product in a metabolic pathway.

**catabolic pathway** (ca – ta – BOL – ic) A series of metabolic reactions that break down complex molecules into simpler ones, often releasing energy in the process.

**anabolic pathway** (an – a – BOL – ic) A series of metabolic reactions that require energy to make complex molecules from simpler ones.

## FIGURE 8.2 Catabolic and Anabolic Metabolism

The energy made available by catabolic pathways provides fuel needed for anabolic pathways.

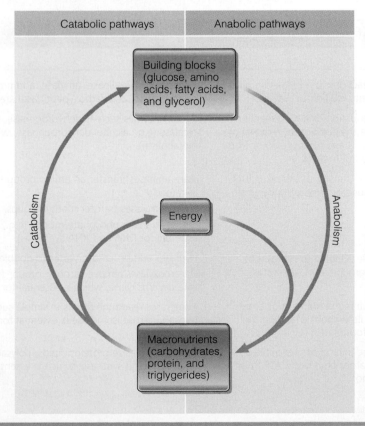

In catabolic pathways, carbohydrates, proteins, and triglycerides are broken down to form glucose, amino acids, fatty acids, and glycerol. Energy is released during these processes.

In anabolic pathways, glucose, amino acids, fatty acids, and glycerol are used to synthesize carbohydrates, proteins, and triglycerides, respectively. Energy is required for these processes.

# Chemical Reactions Require Enzymes

For a chemical reaction to occur, reacting molecules (substrates) must make contact with each other and then be chemically transformed into products. Because these reactions are not likely to occur on their own, the body relies on catalysts to speed them up. In cells, biological catalysts are proteins called enzymes. If it were not for enzymes, metabolic reactions would occur very slowly, if at all. Although enzymes increase the rate at which chemical reactions occur, they themselves do not undergo change.

The names of most enzymes end in *–ase*, such as the names of the digestive enzymes sucrase, maltase, and lactase. There are exceptions. When scientists first discovered that enzymes were proteins, enzymes were often given names ending in *–in*. The names of the enzymes pepsin, trypsin, and chymotrypsin—needed for protein digestion—reflect this older practice. Enzymes can be classified according to the type of chemical reaction they catalyze. For example, enzymes that catalyze hydrolysis reactions are called hydrolases. Similarly, enzymes that transfer atoms from one molecule to another are called transferases. In total, six categories of chemical reactions are catalyzed by enzymes, as summarized in Table 8.1.

Substrates attach to a special surface on the enzyme called the **active site,** forming an **enzyme-substrate complex.** Initially researchers thought that the shape of the active site was complementary to the shape of the substrate—like the pieces of a jigsaw puzzle. However, it is now thought that the shape of the active site changes to fit the shape of the substrate. The active site wraps around the substrate, altering its chemical structure and transforming it into the product. The product is released from the active site, and the enzyme is then free to bind yet another substrate, as illustrated in Figure 8.3.

**CONNECTIONS** Recall that enzymes are biological catalysts that facilitate chemical reactions (Chapter 4, page 93).

**CONNECTIONS** A hydrolysis reaction breaks chemical bonds by the addition of water (Chapter 3, page 77).

**active site** An area on an enzyme that binds substrates in a chemical reaction.

**enzyme-substrate complex** A substrate attached to an enzyme's active site.

## TABLE 8.1 Enzyme Classification

| Enzyme Class | Function | Examples |
|---|---|---|
| **Hydrolases** | Catalyze reactions that break chemical bonds by the addition of water | Enzymes such as lipase, amylase, and protease (e.g., pepsin, trypsin, chymotrypsin, and carboxypeptidase) are essential for digestion. |
| **Oxidoreductases** | Catalyze reduction-oxidation reactions that involve the addition or removal of hydrogen ions and electrons to or from one compound to another | Redox enzymes such as dehydrogenase, oxidase, reductase, and catalase (e.g., alcohol dehydrogenase, which is essential for alcohol metabolism). |
| **Transferases** | Catalyze the transfer of atoms or functional groups from one compound to another | Transaminases transfer an amino group from an amino acid to an α-ketoacid. Transmethylases transfer methyl groups. Kinases transfer phosphate groups (e.g., polymerase, which is essential for DNA and RNA synthesis). |
| **Lyases** | Catalyze the addition or removal of functional groups from substrates | Hydrases add or remove water to double bonds. Decarboxylases remove carboxyl groups from compounds (e.g., pyruvate decarboxylase, which is essential for pyruvate metabolism). |
| **Isomerases** | Catalyze the rearrangement of atoms in a molecule without changing the molecular formula | Isomerases rearrange atoms in simple sugars and fatty acids (e.g., phosphohexose isomerase is essential for the conversion of glucose to fructose). |
| **Ligases or synthetases** | Catalyze the joining of two molecules to form a larger one, using ATP as an energy donor | Ligases synthesize proteins, lipids, polysaccharides, and nucleic acids (e.g., fatty acid synthase, which is essential for fatty acid synthesis). |

SOURCE: Nomenclature Committee of the International Union of Biochemistry and Molecular Biology.

## FIGURE 8.3   Enzymes Are Biological Catalysts

Enzymes increase the rate at which chemical reactions occur.

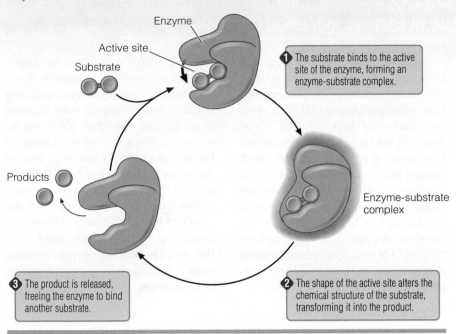

**Enzyme**

**Active site**

**Substrate**

1 The substrate binds to the active site of the enzyme, forming an enzyme-substrate complex.

**Products**

Enzyme-substrate complex

3 The product is released, freeing the enzyme to bind another substrate.

2 The shape of the active site alters the chemical structure of the substrate, transforming it into the product.

Enzymes are said to have specificity, meaning that each enzyme interacts with only certain substrates. Because of this, the body makes thousands of different enzymes to catalyze thousands of different reactions. The ability to measure specific enzymes is the basis for diagnosing certain health conditions such as heart attacks. You can read more about the use of enzymes in medicine in the Focus on Clinical Applications feature.

## Cofactors and Coenzymes Assist Enzymes

Some enzymes require assistance to carry out their catalytic functions. These enzyme "helpers" are called cofactors and coenzymes. **Cofactors** are inorganic substances such as zinc, potassium, iron, and magnesium. For the enzyme to function, cofactors must be attached to its active site. **Coenzymes** are organic molecules derived from vitamins such as niacin and riboflavin. Examples of coenzymes include **nicotinamide adenine dinucleotide (NAD⁺)**, **flavin adenine dinucleotide (FAD)**, and **nicotinamide adenine dinucleotide phosphate (NADP⁺)**. Unlike cofactors, coenzymes are not actually a part of the enzyme structure. Rather, they assist enzymes by accepting and donating hydrogen ions ($H^+$), electrons ($e^-$), and other molecules during chemical reactions. This is one reason why vitamins and minerals, many of which serve as cofactors and coenzymes, are essential components of our diet.

***Coenzymes and Energy Transfer Reactions*** Recall from Chapter 3 that reduction-oxidation (redox) reactions involve the gain and loss of electrons. Specifically, oxidation is the loss of electrons, whereas reduction is the gain of electrons. These reactions often occur simultaneously and are referred to as **coupled reactions.** In other words, when one molecule is being oxidized

**cofactor**  A nonprotein component of an enzyme, often a mineral, needed for its activity.

**coenzyme**  Organic molecule, often derived from vitamins, needed for enzymes to function.

**nicotinamide adenine dinucleotide (NAD⁺)** (nic – o – TIN – a – mide AD – e – nine di – NU – cle – o – tide)  The oxidized form of the coenzyme that is able to accept 2 electrons and 2 hydrogen ions, forming NADH + H⁺.

**flavin adenine dinucleotide (FAD)**  The oxidized form of the coenzyme that is able to accept 2 electrons and 2 hydrogen ions, forming FADH₂.

**nicotinamide adenine dinucleotide phosphate (NADP⁺)**  The oxidized form of the coenzyme that is able to accept 2 electrons and 2 hydrogen ions, forming NADPH + H⁺.

**coupled reactions**  Chemical reactions that take place simultaneously often involving the oxidation of one molecule and the reduction of another.

## *Focus On* CLINICAL APPLICATIONS
### Use of Enzymes in Medical Diagnosis

The body has thousands of enzymes that facilitate thousands of chemical reactions. The concentration of many enzymes in body fluids can be measured and used to diagnose certain diseases and conditions. For example, the amount of specific cardiac enzymes in the blood can be used to diagnose a heart attack.[1] Heart attacks, or myocardial infarctions, can cause heart muscle cells to die, releasing enzymes into the blood. Measuring the level of cardiac enzymes in the blood can help determine the extent of damage that has occurred to the heart.

Other medical tests are based on enzymatic activity as well. For instance, the concentration of the enzyme alanine aminotransferase (ALT) is a sensitive marker of liver function. When liver cells are damaged, alanine aminotransferase is released into the blood, causing its concentration to rise. This is common in people with alcoholism and hepatitis. In addition, certain nutritional conditions can be diagnosed on the basis of circulating enzyme concentrations. For example, blood amylase concentration is low in people with pancreatic disease.

Some analytic tests also take advantage of enzymes as diagnostic tools. For example, blood glucose-monitoring devices used by people with impaired blood glucose regulation are based on enzymatic reactions. When a drop of blood is placed on a test strip, glucose molecules in the blood react with the enzyme glucose oxidase. This causes the test strip to change color, and the extent of the color change indicates the amount of glucose in the blood. Thus the measurement of certain enzymes helps in both diagnosing and monitoring many diseases.

(loss of electrons and hydrogen ions) another is being reduced (gain of electrons and hydrogen ions). Coupled redox reactions allow energy to be transferred from one molecule to another, and require a type of enzyme called a **dehydrogenase.**

Many coenzymes exist in two forms—oxidized (such as $NAD^+$, FAD, and $NADP^+$) and reduced (such as $NADH + H^+$, $FADH_2$, and $NADPH + H^+$). When energy-rich molecules are oxidized, their electrons and hydrogen ions are transferred to $NAD^+$ and FAD. The coenzyme $NAD^+$ can accept 2 electrons (2 $e^-$) and 2 hydrogen ions (2 $H^+$), forming $NADH + H^+$. Although 2 hydrogen ions are transferred to $NAD^+$, only 1 actually attaches. The second hydrogen ion remains in solution. For this reason, the reduced form of $NAD^+$ is written as $NADH + H^+$. Similarly, $FADH_2$ is formed when 2 electrons (2 $e^-$) and 2 hydrogen ions (2 $H^+$) are transferred to FAD. In this case, both hydrogens attach to FAD. The energy carried by these reduced coenzymes ($NADH + H^+$ and $FADH_2$) is used to power the synthesis of the body's most important energy source—adenosine triphosphate (ATP). The oxidation and reduction of coenzymes is illustrated in Figure 8.4.

The reduced form of the coenzyme $NADPH + H^+$ plays an important role in energy-requiring anabolic pathways. Specifically, $NADPH + H^+$ is needed for the synthesis of new compounds in the body such as fatty acids, cholesterol, and five-carbon sugars required for DNA and RNA. $NADPH + H^+$ is transformed to $NADP^+$ when it releases 2 electrons (2 $e^-$) and 2 hydrogen ions (2 $H^+$).

## Regulation of Energy Metabolism

Energy metabolism can be thought of as an elaborate highway system made up of anabolic and catabolic pathways. The pathway or combination of path-

**dehydrogenase** (de – hy – DRO – gen – ase) A type of enzyme that catalyzes the removal and transfer of electrons and hydrogen atoms between organic compounds.

## FIGURE 8.4   The Role of Coenzymes in Energy Metabolism

Energy is transferred to and from energy-yielding nutrients with the assistance of coenzymes.

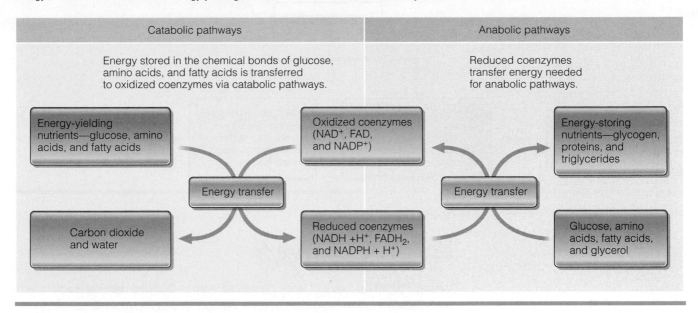

ways used by the body depends primarily on energy (ATP) availability. When ATP levels are more than adequate, the activity of energy-yielding catabolic pathways decreases and the activity of energy-storing anabolic pathways increases. Conversely, low ATP availability increases catabolic activity and decreases anabolic activity. The relative activity of catabolic and anabolic pathways ensures that cells have available energy (ATP) and various substrates at all times.

Hormones are important regulators of energy metabolism, helping the body shift between anabolic and catabolic pathways. Endocrine tissues can detect changes in energy availability, and respond by secreting appropriate hormones. These hormones then suppress or activate key enzymes in metabolic pathways. As a result, metabolic pathways can be "switched" on or off.

The primary hormones involved in the regulation of catabolic and anabolic pathways are insulin, glucagon, cortisol, and epinephrine. As you learned in Chapter 5, insulin is an anabolic hormone that promotes energy storage in the forms of glycogen, triglyceride, and protein. To do this, insulin activates anabolic pathways and inactivates catabolic pathways. When energy availability is limited, the pancreatic hormone glucagon promotes catabolic pathways and inhibits anabolic pathways. In this way, glucagon increases energy availability by mobilizing energy-yielding molecules that have been "stored for a rainy day."

During times of stress and starvation, cortisol and epinephrine—hormones released from the adrenal glands—also play important roles in directing energy metabolism. These hormones stimulate catabolic pathways that help increase fuel availability. For example, cortisol and epinephrine stimulate the breakdown of muscle glycogen to increase glucose availability. The role of hormones in energy metabolism is illustrated in Figure 8.5.

**CONNECTIONS** The hormone insulin is released from the pancreas in response to elevated blood glucose and promotes energy storage. The hormone glucagon is released from the pancreas in response to low blood glucose and promotes the breakdown of liver glycogen and the subsequent release of glucose into the blood (Chapter 5, page 143).

**FIGURE 8.5** Anabolic and Catabolic Pathways Are Regulated by Hormones

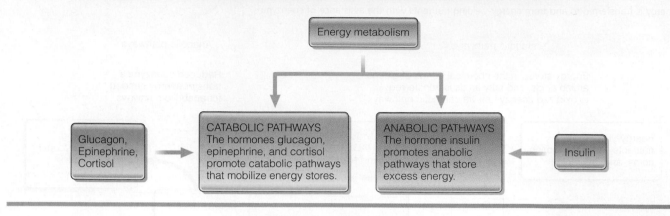

## ESSENTIAL *Concepts*

Metabolic pathways consist of many interrelated, enzyme-catalyzed chemical reactions. These pathways can be categorized as either catabolic or anabolic. Anabolic pathways promote the synthesis of new compounds and energy storage, whereas catabolic pathways promote the mobilization of stored energy and the breakdown of energy-yielding nutrients. Chemical reactions are catalyzed by enzymes, some requiring cofactors or coenzymes to function. A cofactor is an inorganic substance that is part of the enzyme's structure. Several coenzymes are organic molecules that assist enzymes in redox reactions by transferring hydrogen ions ($H^+$) and electrons ($e^-$) to and from molecules. These coenzymes have oxidized ($NAD^+$, FAD, and $NADP^+$) and reduced (NADH + $H^+$, $FADH_2$, and NADPH + $H^+$) forms. Enzymes involved in metabolic pathways are regulated primarily by hormones. The hormone insulin promotes energy storage, whereas the hormone glucagon promotes energy mobilization. The hormones cortisol and epinephrine also promote catabolic pathways, being released in response to stress.

## Cellular Energy and the Role of ATP

Cells are unable to use nutrients directly for energy. That is, the energy stored in the chemical bonds of nutrients must be converted into a form that cells can used—namely adenosine triphosphate (ATP). ATP is a molecule uniquely suited to transfer the energy contained in its chemical bonds to chemical reactions that, to occur, require energy. In fact, ATP provides the energy needed for protein synthesis, muscle contraction, active transport, nerve transmission, and all other energy-requiring reactions that take place in the body. However, the energy contained within certain nutrients cannot be transferred directly to ATP. Rather, that transfer requires intermediate steps that involve the coenzymes $NAD^+$/NADH + $H^+$ and FAD/$FADH_2$.

### ATP as an Energy Source

As illustrated in Figure 8.6, the ATP molecule consists of three basic units: the sugar ribose, a base called adenine, and a chain of three phosphate groups (hence, ATP is a *tri*phosphate). Together, ribose and adenine are referred to as adenosine. Of particular importance is the energy contained in the chemi-

CONNECTIONS Ribose is a 5-carbon sugar and is a constituent RNA (Chapter 5, page 126).

## FIGURE 8.6 Adenosine Triphosphate (ATP)

Adenosine triphosphate (ATP) is a high-energy molecule that transfers its energy to chemical reactions.

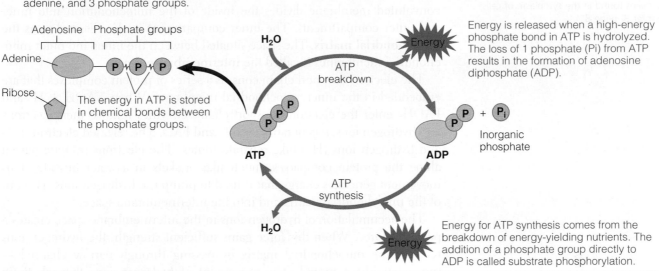

ATP consists of 3 basic units—a sugar called ribose, a base called adenine, and 3 phosphate groups.

The energy in ATP is stored in chemical bonds between the phosphate groups.

Energy is released when a high-energy phosphate bond in ATP is hydrolyzed. The loss of 1 phosphate (Pi) from ATP results in the formation of adenosine diphosphate (ADP).

Energy for ATP synthesis comes from the breakdown of energy-yielding nutrients. The addition of a phosphate group directly to ADP is called substrate phosphorylation.

cal bonds holding the phosphate groups together. These high-energy bonds enable ATP to both store and release energy. When cells need energy, a phosphate group is broken off of ATP, releasing energy, and inorganic phosphate (symbolized as $P_i$). This results in the formation of adenosine *di*phosphate (ADP). The energy released from the breaking of ATP's phosphate bond is used to drive metabolic reactions that require energy.

## ATP Synthesis

Cells use ATP almost as quickly as it is made. Because ATP is not stored to any extent in the body, it is important for cells to be able to synthesize ATP as it is needed. ATP can be synthesized in two ways—substrate phosphorylation and oxidative phosphorylation. **Substrate phosphorylation** occurs when a phosphate group is added directly to ADP. Substrate phosphorylation does not require oxygen and is therefore important when tissues have little oxygen available to them. Although all cells have the capability to carry out substrate phosphorylation, relatively little ATP is synthesized this way. Rather, most ATP is synthesized via **oxidative phosphorylation.**

Recall that the metabolic breakdown of energy-yielding nutrients involves the transfer of electrons ($e^-$) and hydrogen ions ($H^+$) to the oxidized coenzymes $NAD^+$ and FAD. Once reduced, these coenzymes ($NADH + H^+$ and $FADH_2$) must be oxidized back to $NAD^+$ and FAD. This process releases energy and is accomplished by a series of chemical reactions that link the oxidation of $NADH + H^+$ and $FADH_2$ to the synthesis of ATP. The process whereby $NADH + H^+$ and $FADH_2$ are oxidized and ADP is phosphorylated is called oxidative phosphorylation. These reactions are coupled, because the energy needed to phosphorylate ADP is provided by the oxidation of $NADH + H^+$ and $FADH_2$. In this way, the energy originally contained in the reduced coenzymes is transferred to ATP, which is used to fuel the many energy-requiring reactions in the body. We next examine the process whereby ATP is produced via the oxidation of $NADH + H^+$ and $FADH_2$.

**substrate phosphorylation** (SUB – strate phos – pho – ryl – A – tion) The transfer of an inorganic phosphate group to ADP to form ATP.

**oxidative phosphorylation** (ox – i – DA – tive phos – pho – ryl – A – tion) The chemical reactions that link the oxidation of $NADH + H^+$ and $FADH_2$ to the phosphorylation of ADP to form ATP and water.

CONNECTIONS ▶ Recall that mitochondria are organelles found in the cytoplasm of cells (Chapter 3, page 81).

## The Electron Transport Chain

Oxidative phosphorylation accounts for approximately 90% of ATP production and involves a series of linked chemical reactions that make up the **electron transport chain.** The electron transport chain, also called the electron transport system, is located in the inner membrane of the mitochondria. This convoluted membrane divides the inside of the mitochondrion into inner and outer compartments. The inner compartment is also referred to as the **mitochondrial matrix.** The space situated between the inner and outer mitochondrial membranes is called the **intermembrane space.**

The electron transport chain consists a series of protein complexes that are embedded in the inner mitochondrial membrane. When NADH + H$^+$ and FADH$_2$ enter the electron transport chain, enzymes remove their electrons and hydrogen ions, regenerating NAD$^+$ and FAD. The released electrons (e$^-$) and hydrogen ions (H$^+$) take separate routes. The electrons (e$^-$) are passed along the protein complexes, much like buckets in a water brigade. This movement generates energy that is used to pump the hydrogen ions (H$^+$) out of the mitochondrial matrix and into the intermembrane space.

The accumulation of hydrogen ions in the intermembrane space creates a powerful force. When this force gains sufficient strength, the hydrogen ions re-enter the mitochondrial matrix by passing through narrow channels—somewhat like a tunnel. The movement of hydrogen ions through these channels releases energy that is used by the enzyme **ATP synthase** to attach a phosphate group to ADP. Thus ATP is produced.

NADH + H$^+$ and FADH$_2$ enter the electron transport chain at different locations along the protein complexes. This is why the amount of ATP generated by these coenzymes differs. Because NADH + H$^+$ enters "higher" along the chain, the ATP yield is approximately 3 ATP and 2 ATP for NADH + H$^+$ and FADH$_2$, respectively. More current estimates of ATP yield from NADH + H$^+$ and FADH$_2$ are 2.5 and 1.5, respectively. For convenience, ATP yields are often estimated on the basis of 3 ATP per NADH + H$^+$ and 2 ATP per FADH$_2$.

At the completion of the electron transport chain, a group of iron-containing protein complexes called **cytochromes** reunite the electrons (e$^-$) and hydrogen ions (H$^+$) to form hydrogen. The hydrogen molecules then combine with oxygen (O$_2$) to form water (H$_2$O). Because oxygen is essential for these reactions, the electron transport chain is considered an aerobic metabolic process. The process of oxidative phosphorylation via the electron transport chain is illustrated in Figure 8.7.

## ESSENTIAL *Concepts*

Cells rely on the energy contained in the chemical bonds of ATP. Some ATP is generated by substrate phosphorylation, a process that adds a phosphate group (P$_i$) directly to ADP. However, most ATP is synthesized by oxidative phosphorylation, which involves a series of chemical reactions that make up the electron transport chain. When NADH + H$^+$ and FADH$_2$ enter the electron transport chain, their electrons and hydrogen ions are removed. The electrons are passed along protein complexes, and the energy released is used to pump the hydrogen ions (H$^+$) out of the mitochondrial matrix. The movement of hydrogen ions back into the mitochondrial matrix releases energy that is used by the enzyme ATP synthase to attach a phosphate group to ADP, generating ATP. Last, iron-containing protein complexes called cytochromes reunite the electrons (e$^-$) and hydrogen ions (H$^+$), which in turn combine with oxygen to form water.

**electron transport chain** A series of chemical reactions that transfer electrons and hydrogen ions from NADH + H$^+$ and FADH$_2$ along protein complexes in the inner mitochondrial membrane, ultimately producing ATP.

**mitochondrial matrix** The inner compartment of the mitochondrion.

**intermembrane space** The space between the inner and outer mitochondrial membranes.

**ATP synthase** (SYNTH—ase) A mitochondrial enzyme that adds a phosphate to ADP to form ATP.

**cytochromes** (CY – to – chromes) Iron-containing protein complexes that combine electrons, hydrogen ions, and oxygen to form water.

**FIGURE 8.7   Oxidative Phosphorylation and the Electron Transport Chain**

Oxidative phosphorylation involves a series of linked chemical reactions that comprise the electron transport chain, which is located on the surface of inner mitochondrial membrane. The energy needed to phosphorylate ADP is provided by the oxidation of NADH + H⁺ and FADH₂.

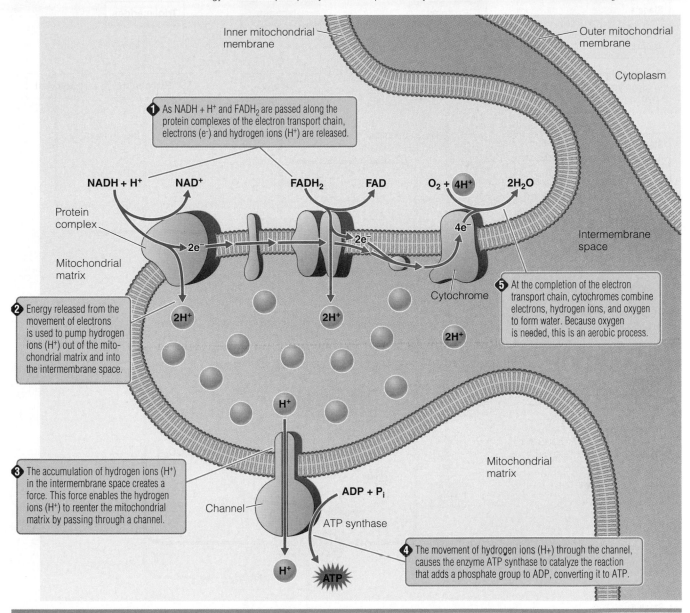

# Catabolism: The Breakdown of Macronutrients for Energy

Energy-yielding nutrients such as glucose, fatty acids, and amino acids store energy in their chemical bonds. For cells to produce ATP, these nutrients must undergo a series of chemical reactions that make up catabolic pathways. These pathways are interrelated in such a way that intermediate products or end products of one pathway often become substrates for other pathways. At first glance, the many catabolic pathways can appear overwhelming. However, they can be simplified by grouping them into four stages, as illustrated in Figure 8.8.

• *Stage 1:* The first stage of catabolism breaks down complex molecules into their fundamental building blocks. That is, glycogen is broken down to glucose,

**FIGURE 8.8**  Stages of Energy Catabolism

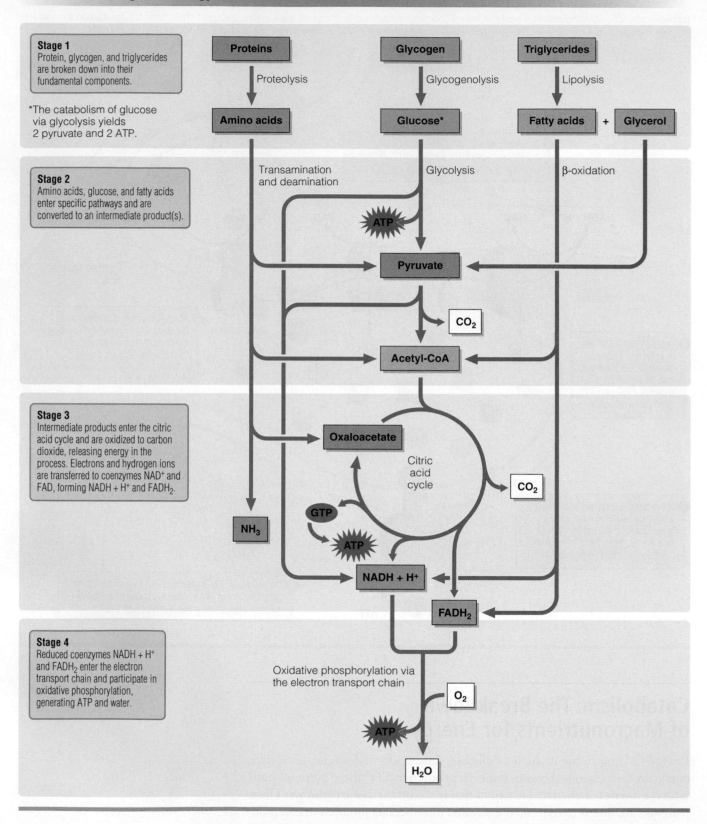

**Stage 1**
Protein, glycogen, and triglycerides are broken down into their fundamental components.

*The catabolism of glucose via glycolysis yields 2 pyruvate and 2 ATP.

**Stage 2**
Amino acids, glucose, and fatty acids enter specific pathways and are converted to an intermediate product(s).

**Stage 3**
Intermediate products enter the citric acid cycle and are oxidized to carbon dioxide, releasing energy in the process. Electrons and hydrogen ions are transferred to coenzymes $NAD^+$ and FAD, forming $NADH + H^+$ and $FADH_2$.

**Stage 4**
Reduced coenzymes $NADH + H^+$ and $FADH_2$ enter the electron transport chain and participate in oxidative phosphorylation, generating ATP and water.

protein to amino acids, and triglycerides to fatty acids and glycerol molecules. These metabolic pathways are glycogenolysis, proteolysis, and lipolysis, respectively.

- *Stage 2:* During the second stage, the basic building blocks enter specific pathways whereby each is converted into an intermediate product that can enter a common pathway called the citric acid cycle, signifying the third stage of catabolism.

- *Stage 3:* The third stage of catabolism begins when intermediate products enter the citric acid cycle and are broken down further to form carbon dioxide, releasing energy in the process. Much of the energy released during Stage 3 of metabolism is transferred to the coenzymes $NAD^+$ and FAD, forming $NADH + H^+$ and $FADH_2$, respectively. In addition, small amounts of ATP are formed via substrate phosphorylation.

- *Stage 4:* The fourth and final stage of catabolism begins when $NADH + H^+$ and $FADH_2$ enter the electron transport chain. It is here where most ATP production occurs, via oxidative phosphorylation.

# Carbohydrates

Glucose is a rich source of energy that can be used by all cells in the body to produce ATP. Most glucose comes from the ingestion, digestion, and absorption of carbohydrate-rich foods. However, when additional glucose is needed, the hormone glucagon stimulates the breakdown of glycogen in the liver and the release of glucose into the blood. This metabolic process, called glycogenolysis, results in the breakdown of glycogen and represents *Stage 1* of carbohydrate catabolism. The hormones epinephrine and cortisol also stimulate glycogenolysis in skeletal muscle.

## Glycolysis

The word *glycolysis* literally means the "splitting of sugar," which is what happens when glucose enters *Stage 2* of carbohydrate metabolism. **Glycolysis** is a series of chemical reactions that ultimately splits the 6-carbon glucose molecule into two 3-carbon subunits called **pyruvate.** These reactions occur in the cytoplasm of cells. Because oxygen is not required for any of these reactions, glycolysis is an anaerobic metabolic pathway.

As you can see from Figure 8.9, glycolysis results in small amounts of energy—two ATPs and two molecules of $NADH + H^+$ per glucose. However, this represents only a small amount of the total energy available in each glucose molecule. This is because the chemical bonds in pyruvate have not yet been broken. What happens next is a "major intersection" in glucose metabolism. That is, the direction taken by pyruvate in *Stage 3* of carbohydrate catabolism is determined by oxygen availability, specifically whether pyruvate encounters anaerobic (oxygen-poor) or aerobic (oxygen-rich) conditions in the cell.

## Anaerobic Conditions: The Cori Cycle

During high-intensity exercise, muscles have increased energy (ATP) requirements. To support high rates of ATP production via the electron transport chain, large amounts of oxygen are needed. However, as exercise intensity increases, it becomes increasingly more difficult for the lungs and cardiovascular system to deliver adequate amounts of oxygen to active muscles. As a result, muscle cells can experience limited oxygen availability, which determines the metabolic fate of pyruvate.

Under relatively anaerobic conditions, pyruvate (the end product of glycolysis) remains in the cytoplasm and is converted to lactate (also called lactic

**glycolysis** The metabolic pathway that splits glucose into two 3-carbon molecules called pyruvate.

**pyruvate** (py – RU – vate) An intermediate product formed during metabolism of carbohydrates and some amino acids.

## FIGURE 8.9 Overview of Glycolysis

Glycolysis is an anaerobic metabolic pathway that consists of a series of chemical reactions that splits a 6-carbon glucose molecule into two 3-carbon molecules called pyruvate. The fate of pyruvate depends on oxygen availability in the cell.

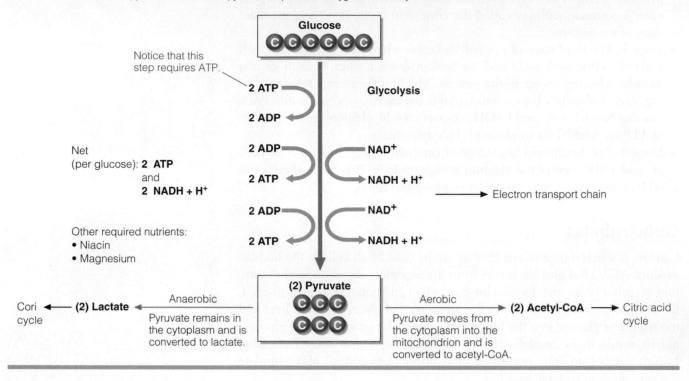

**Glucose**

Notice that this step requires ATP.

2 ATP

2 ADP

**Glycolysis**

2 ADP → NAD$^+$

2 ATP → NADH + H$^+$

Net (per glucose): **2 ATP** and **2 NADH + H$^+$**

2 ADP → NAD$^+$

2 ATP → NADH + H$^+$

→ Electron transport chain

Other required nutrients:
• Niacin
• Magnesium

**(2) Pyruvate**

Cori cycle ← **(2) Lactate** ← Anaerobic
Pyruvate remains in the cytoplasm and is converted to lactate.

Aerobic → **(2) Acetyl-CoA** → Citric acid cycle
Pyruvate moves from the cytoplasm into the mitochondrion and is converted to acetyl-CoA.

---

**CONNECTIONS** Gluconeogenesis is the process whereby glucose is synthesized from noncarbohydrate sources (Chapter 5, page 149).

**citric acid cycle** An amphibolic metabolic pathway that oxidizes acetyl-CoA to yield carbon dioxide, NADH + H$^+$, FADH$_2$, and ATP via substrate phosphorylation.

**Cori cycle** The metabolic pathway that regenerates glucose by circulating lactate from muscle to the liver, where it undergoes gluconeogenesis.

**oxaloacetate** The final product of the citric acid cycle, which becomes the substrate for the first reaction in this pathway.

**citrate** (CIT – rate) The first intermediate product in the citric acid cycle formed when acetyl-CoA joins with oxaloacetate.

acid). Lactate is then released into the blood and taken up by the liver. In the liver, lactate is converted to glucose via gluconeogenesis, which can then undergo glycolysis (Figure 8.10). This sequence of chemical reactions, called the **Cori cycle**, provides a means by which small amounts of ATP can be produced in the relative absence of oxygen. Because very little ATP is actually produced, the Cori cycle cannot sustain physical exertion for very long. As a result, extremely vigorous activity results in muscle fatigue relatively quickly.

### Aerobic Conditions: The Citric Acid Cycle and Oxidative Phosphorylation

Under aerobic conditions, the two molecules of pyruvate (the end products of glycolysis) move from the cytoplasm into the mitochondria. Here pyruvate is chemically transformed into an intermediate product called acetyl-CoA, a 2-carbon unit attached to a compound called coenzyme A (CoA). This irreversible reaction requires several enzymes and vitamin B–derived coenzymes. These B vitamins include thiamin, niacin, pantothenic acid and riboflavin. As shown in Figure 8.11, the formation of acetyl-CoA from pyruvate also results in production of carbon dioxide and NADH + H$^+$. Under aerobic conditions, acetyl-CoA is now ready to enter *Stage 3* of carbohydrate catabolism—the **citric acid cycle.**

The citric acid cycle, also called the tricarboxylic acid (TCA) cycle or the Krebs cycle, is a major pathway used during aerobic conditions. It consists of a series of chemical reactions that take place within mitochondria. These enzyme-catalyzed reactions are often depicted as a circle, because the product of the last reaction of the pathway (**oxaloacetate**) becomes the substrate for **citrate** (the first reaction)—like a chemical carousel. Although the citric acid cycle is primarily a catabolic pathway ultimately generating ATP, it serves

## FIGURE 8.10  The Cori Cycle

The Cori cycle is a metabolic pathway that involves both glycolysis and gluconeogenesis. Under conditions of limited oxygen availability, the Cori cycle provides a means by which small amounts of ATP can continue to be produced in the muscle.

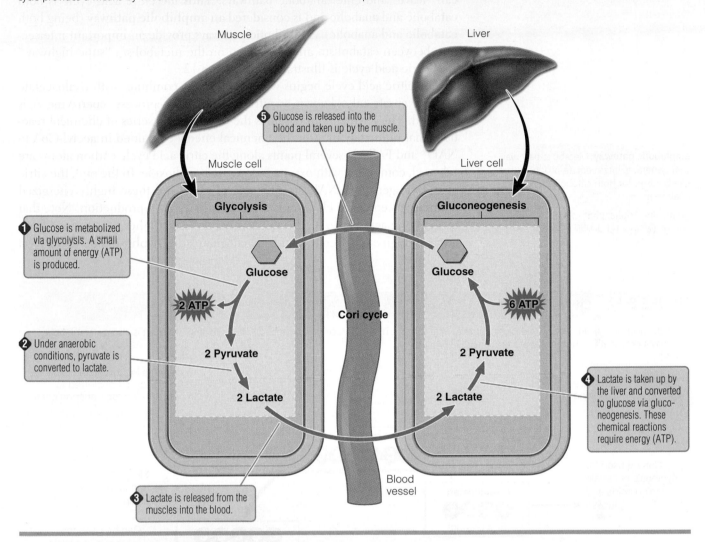

Muscle

Liver

**5** Glucose is released into the blood and taken up by the muscle.

Muscle cell

Liver cell

**Glycolysis**

**Gluconeogenesis**

**1** Glucose is metabolized via glycolysis. A small amount of energy (ATP) is produced.

**Glucose**

**Glucose**

**2 ATP**

**6 ATP**

**Cori cycle**

**2** Under anaerobic conditions, pyruvate is converted to lactate.

**2 Pyruvate**

**2 Pyruvate**

**2 Lactate**

**2 Lactate**

**4** Lactate is taken up by the liver and converted to glucose via gluco-neogenesis. These chemical reactions require energy (ATP).

Blood vessel

**3** Lactate is released from the muscles into the blood.

## FIGURE 8.11  Conversion of Pyruvate to Acetyl-CoA

Under aerobic conditions, pyruvate combines with coenzyme A to from acetyl-CoA.

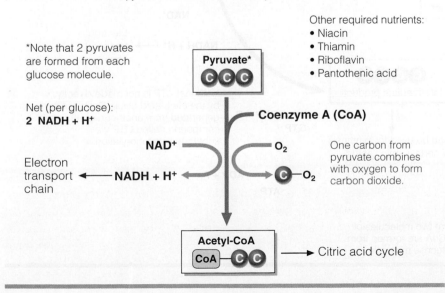

Other required nutrients:
• Niacin
• Thiamin
• Riboflavin
• Pantothenic acid

*Note that 2 pyruvates are formed from each glucose molecule.

**Pyruvate***
C C C

Net (per glucose):
**2 NADH + H⁺**

**Coenzyme A (CoA)**

$NAD^+$

$O_2$

One carbon from pyruvate combines with oxygen to form carbon dioxide.

Electron transport chain ← **NADH + H⁺**

C - $O_2$

**Acetyl-CoA**
CoA - C C → Citric acid cycle

other purposes. For example, intermediate products of the citric acid cycle can "leave" and enter anabolic pathways. Thus the citric acid cycle is both catabolic and anabolic and is considered an **amphibolic pathway** (being both catabolic and anabolic). Amphibolic pathways provide an important intersection between catabolism and anabolism on the metabolism "superhighway." The citric acid cycle is illustrated in Figure 8.12.

The citric acid cycle begins when acetyl-CoA combines with oxaloacetate to form citrate (also known as citric acid). In the process, coenzyme A is released. The formation of citrate is the first step in a series of chemical reactions that ultimately transfer the chemical energy contained in acetyl-CoA to $NAD^+$ and FAD. At several points along the citric acid cycle carbon atoms are released, combining with oxygen to form carbon dioxide. In the end, the citric acid cycle generates $NADH + H^+$ and $FADH_2$, and these highly energized compounds enter the electron transport chain for ATP production. Note that in the citric acid cycle, ATP is not formed directly. Rather, it is formed from another high-energy compound called **guanosine triphosphate (GTP)** via

**amphibolic pathway** Metabolic pathway that generates intermediate products that can be used for both catabolism and anabolism.

**guanosine triphosphate (GTP)** A high-energy compound similar to ATP.

## FIGURE 8.12 Citric Acid Cycle

The citric acid cycle is an aerobic metabolic pathway that forms GTP and reduced coenzymes $NADH + H^+$ and $FADH_2$. It is also an amphibolic pathway, because it has both catabolic and anabolic properties.

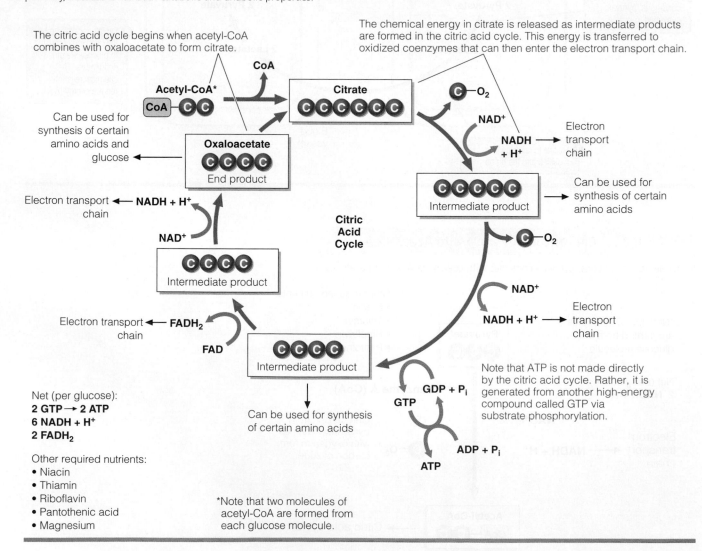

Net (per glucose):
**2 GTP → 2 ATP**
**6 NADH + H⁺**
**2 FADH₂**

Other required nutrients:
• Niacin
• Thiamin
• Riboflavin
• Pantothenic acid
• Magnesium

*Note that two molecules of acetyl-CoA are formed from each glucose molecule.

substrate phosphorylation. In all, the oxidation of two molecules of acetyl-CoA produce 6 NADH + H$^+$, 2 FADH$_2$, and 2 ATPs.

Although small amounts of ATP are produced via substrate phosphorylation, most (90%) of the ATP generated from glucose results from oxidative phosphorylation via the electron transport chain (*Stage 4*). In total, the complete oxidation of one molecule of glucose (via glycolysis, the citric acid cycle, and oxidative phosphorylation) generates up to 38 ATPs, depending on the source of glucose—10 NADH + H$^+$ (30 ATPs via oxidative phosphorylation), 2 FADH$_2$ (4 ATPs via oxidative phosphorylation), and 4 ATPs formed via substrate phosphorylation (Figure 8.13).

## FIGURE 8.13    ATP Formation from Glucose Catabolism

Although some ATP is formed directly via substrate phosphorylation, most ATP is formed via oxidative phosphorylation.

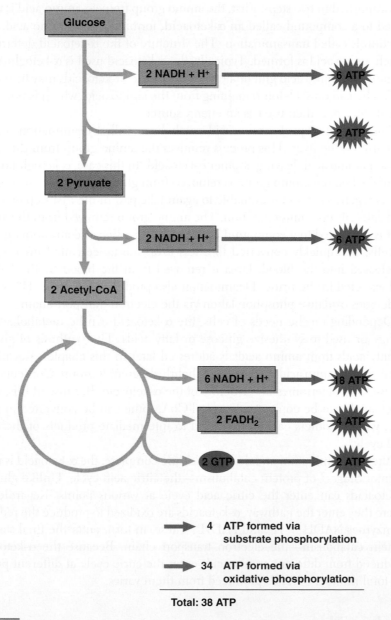

NOTE: Using nonintegral values of 2.5 ATP per NADH + H$^+$ and 1.5 ATP per FADH$_2$, the yield per glucose molecule is estimated to be 30–32 ATP.

## Protein

The major role of amino acids is to serve as the building blocks for proteins and other nitrogen-containing compounds. For this reason, amino acids are not considered a major source of energy, supplying less than 10% of our daily ATP needs. Nonetheless, at times the body must depend on them for energy, particularly during starvation. In addition to dietary amino acids, the body can break down stored protein (primarily skeletal muscle) for a source of amino acids. *Stage 1* of protein catabolism is the process of **proteolysis,** which breaks down protein into amino acids. The liver then takes up amino acids, where they can enter the next stage of protein catabolism.

### Transamination and Deamination

In *Stage 2* of protein catabolism, the nitrogen-containing amino group must first be removed from the central carbon. As illustrated in Figure 8.14, this is accomplished in two steps. First, the amino group from an amino acid is transferred to a compound called an **α-ketoacid,** forming a new amino acid. This reaction is called transamination. The structure of the α-ketoacid determines which amino acid is formed. Typically, the α-ketoacid used is α-ketoglutarate, forming the amino acid glutamate. However, other α-ketoacids may be used as well. The carbon skeleton remaining from the amino acid, which is now itself an α-ketoacid, is then used as an energy source.

The second step of amino acid catabolism, called deamination, occurs primarily in the liver. This process removes the amino group from the newly formed amino acid, leaving another α-ketoacid. In this case, α-ketoglutarate is formed when the amino group is removed from glutamate. Thus α-ketoglutarate is regenerated and is available to again take part in the first step of amino acid catabolism—transamination. The amino group removed from the amino acid is converted to a compound called ammonia. Because ammonia is toxic to cells, it is quickly converted to a less toxic substance called urea, which is released into the blood. Urea is removed from the blood by the kidneys and excreted in the urine. Deamination also produces $NADH + H^+$, which undergoes oxidative phosphorylation via the electron transport chain.

Depending on the needs of cells, the α-ketoacid can be metabolized for energy or used to synthesize glucose or fatty acids. The synthesis of glucose or fatty acids from amino acids is addressed later in this chapter. Recall that there are 20 amino acids, each with a slightly different R-group. Consequently, the R-group determines the structure of the α-ketoacid. Because of this, some α-ketoacids can be converted to acetyl-CoA, some can be converted to pyruvate, whereas others can be converted to intermediate products of the citric acid cycle.

After transamination and deamination are complete, the α-ketoacid is ready to enter *Stage 3* of protein catabolism—the citric acid cycle. Unlike glucose, α-ketoacids can enter the citric acid cycle at various points. Regardless of where they enter the pathway, α-ketoacids are oxidized to produce the reduced coenzymes $NADH + H^+$ and $FADH_2$. These, in turn, enter the final stage of protein catabolism—the electron transport chain. Because the α-ketoacids produced from different amino acids enter the citric cycle at different points, the total number of ATPs produced from them varies.

## Triglycerides

Fatty acids also provide an important source of energy. When the body's readily available energy sources are low, it can break down triglyceride molecules

---

**CONNECTIONS** Transamination is the transfer of an amino group from an amino acid to an α-ketoacid, usually α-ketoglutarate (Chapter 6, page 175).

**CONNECTIONS** Deamination is the removal of an amino group from an amino acid (Chapter 6, page 193).

**proteolysis** The breakdown of proteins into peptides or amino acids.

**α-ketoacid** The structure remaining after the amino group has been removed from an amino acid.

**FIGURE 8.14 Transamination and Deamination**

To use amino acids for a source of energy, the amino group is removed in two steps: transamination and deamination.

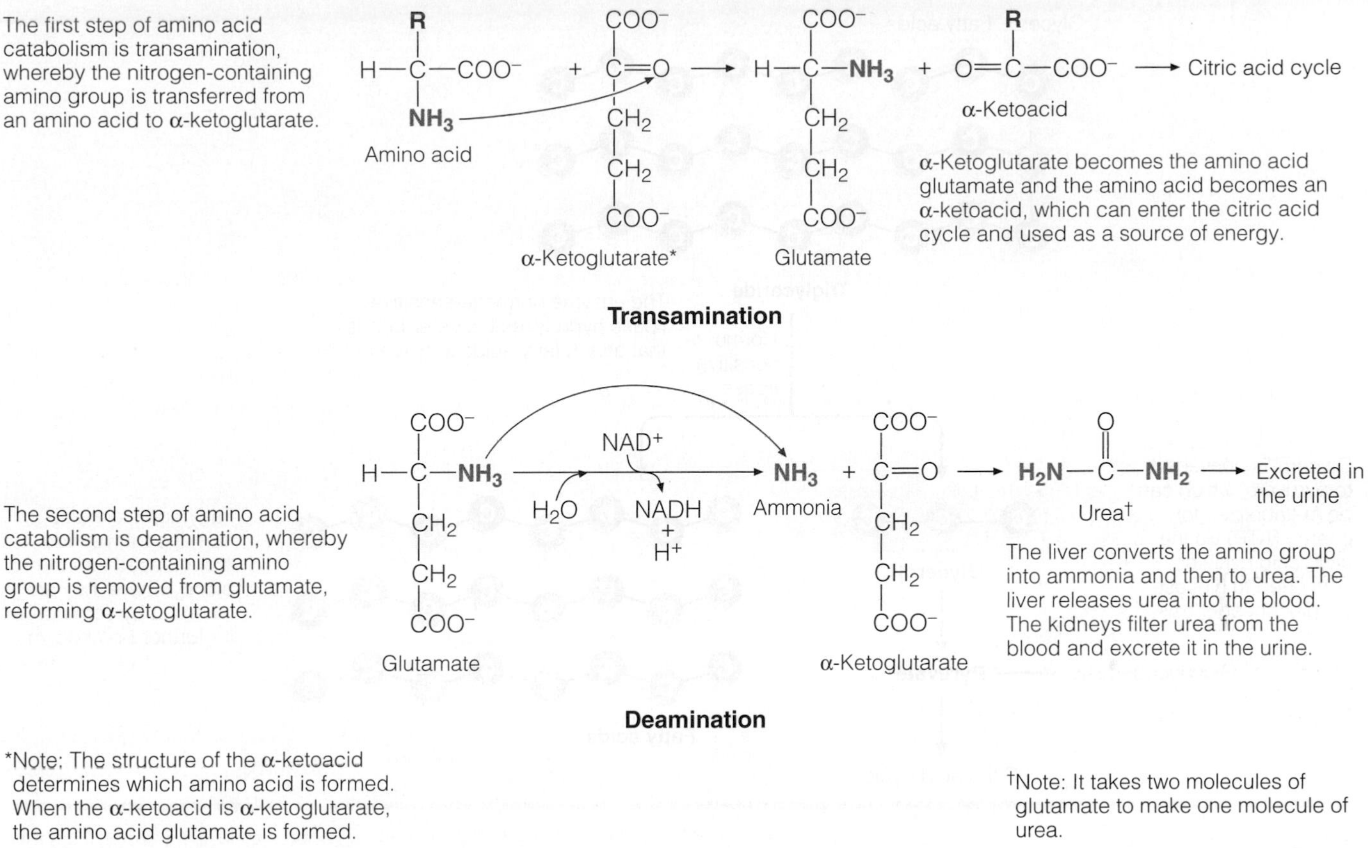

The first step of amino acid catabolism is transamination, whereby the nitrogen-containing amino group is transferred from an amino acid to α-ketoglutarate.
Amino acid
α-Ketoglutarate*
Glutamate
α-Ketoacid
Citric acid cycle
α-Ketoglutarate becomes the amino acid glutamate and the amino acid becomes an α-ketoacid, which can enter the citric acid cycle and used as a source of energy.
Transamination
The second step of amino acid catabolism is deamination, whereby the nitrogen-containing amino group is removed from glutamate, reforming α-ketoglutarate.
NAD+
H2O
NADH + H+
Ammonia
Glutamate
α-Ketoglutarate
Urea†
Excreted in the urine
The liver converts the amino group into ammonia and then to urea. The liver releases urea into the blood. The kidneys filter urea from the blood and excrete it in the urine.
Deamination
*Note: The structure of the α-ketoacid determines which amino acid is formed. When the α-ketoacid is α-ketoglutarate, the amino acid glutamate is formed.
†Note: It takes two molecules of glutamate to make one molecule of urea.


into glycerol and fatty acids, as shown in Figure 8.15. The first stage of lipid catabolism, called lipolysis, is catalyzed by the enzyme hormone-sensitive lipase, whose activity is stimulated by rising levels of the pancreatic hormone glucagon during periods of low glucose availability. Hormone-sensitive lipase is also stimulated by low levels of insulin, exercise, and during times of stress by the adrenal hormones epinephrine and cortisol. The glycerol released during lipolysis can be converted to pyruvate or glucose and therefore used for energy, as well. Once fatty acids have been cleaved from the triglyceride, they are released into the blood, where they are taken up by some cells for further breakdown.

**CONNECTIONS** Lipolysis is the enzymatic breakdown of triglycerides into glycerol and fatty acids by hormone-sensitive lipase (Chapter 7, page 232).

## Beta (β)-Oxidation

*Stage* 2 of lipid catabolism involves the arduous process of disassembling fatty acids into 2-carbon subunits. This takes place in mitochondria by a series of reactions called **β-oxidation,** or beta-oxidation. However, before β-oxidation can take place the fatty acid must be activated in the cytoplasm by the addition of coenzyme A (CoA) to its carboxylic acid end. After the activation step, the fatty acid can be transported across the outer mitochondrial membrane by a molecule called **carnitine.** Athletes have long been hopeful that carnitine supplements might enhance athletic performance by improving the efficiency by which muscle tissue uses fatty acids for energy. You can read more about carnitine and athletic performance in the Focus on Sports Nutrition feature.

**β-oxidation** The series of chemical reactions that breaks down fatty acids to molecules of acetyl-CoA.

**carnitine** A molecule found in muscle and liver cells that transports fatty acids across the mitochondrial membrane.

## FIGURE 8.15 Lipolysis

Triglycerides are broken down to fatty acids and glycerol by the process of lipolysis.

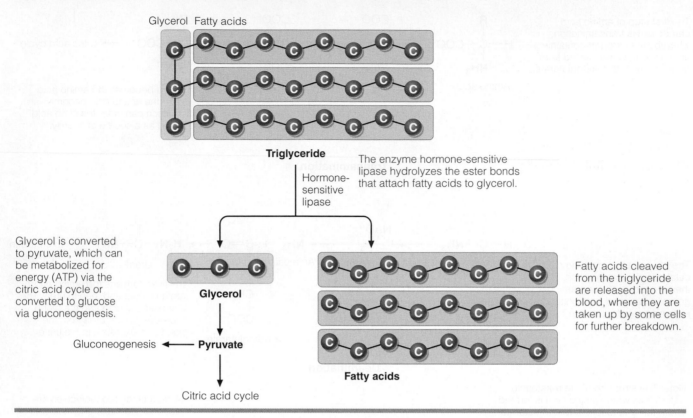

**Triglyceride**

The enzyme hormone-sensitive lipase hydrolyzes the ester bonds that attach fatty acids to glycerol.

Hormone-sensitive lipase

Glycerol is converted to pyruvate, which can be metabolized for energy (ATP) via the citric acid cycle or converted to glucose via gluconeogenesis.

**Glycerol**

Gluconeogenesis ← **Pyruvate**

Citric acid cycle

**Fatty acids**

Fatty acids cleaved from the triglyceride are released into the blood, where they are taken up by some cells for further breakdown.

# *Focus On* SPORTS NUTRITION
## Carnitine Supplementation

Athletes are always seeking ways to be stronger, faster, and more competitive. Special training routines, devices, or dietary modifications used to enhance athletic performance are called ergogenic aids. One such ergogenic aid is carnitine supplements. Carnitine is made in the liver and kidneys and used in the body, especially in skeletal and cardiac muscle. Its purpose is to carry activated fatty acids across the mitochondrial membrane where they are oxidized via β-oxidation and the citric acid cycle for ATP production. Thus carnitine plays a very important role in enabling the body to harvest energy from fatty acids.

During most types of athletic events and training regimens, the body first uses glucose reserves for energy. It then turns to fatty acids to support the energy demands of more long-term activities (for example, marathons). Thus it would seem logical that carnitine supplementation might increase fatty acid oxidation and thus enhance performance, especially for the endurance athlete. Although research is limited, few studies suggest that carnitine supplementation improves speed or endurance.[2] While studies do not show conclusively that athletes benefit from carnitine supplements, neither have negative effects been shown.[3] Until more is known about carnitine, to

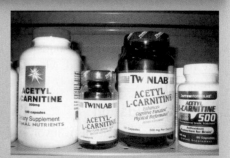

Carnitine is sold over-the-counter and is promoted as an ergogenic aid that improves athletic performance.

boost athletic performance athletes need to rely on adequate training, a well-balanced diet, and mental focus.

β-Oxidation takes place in all cells except brain and red blood cells, which lack the organelles required for this metabolic process. β-Oxidation begins when a series of enzymes systematically cleave off a 2-carbon subunit from the fatty acid's carboxylic acid end, resulting in the synthesis of acetyl-CoA. The remaining (shorter) fatty acid is reactivated by the addition of another CoA and then cleaved again. This process repeats itself until the entire fatty acid has been broken down—2 carbon atoms at a time. Each cleavage releases electrons ($e^-$) and hydrogen ions ($H^+$), which are used to form 1 NADH + $H^+$ and 1 $FADH_2$. Thus β-oxidation of an 18-carbon fatty acid requires eight cleavages, producing 9 molecules of acetyl-CoA, 8 NADH + $H^+$, and 8 $FADH_2$. The reduced coenzymes resulting from β-oxidation are used to generate ATP. The process of β-oxidation is shown in Figure 8.16 on the following page.

*Stage 3* of lipid catabolism begins when molecules of acetyl-CoA resulting from β-oxidation enter the citric acid cycle. A total of 3 NADH + $H^+$, 1 $FADH_2$, and 1 ATP (produced from GTP by substrate phosphorylation) are generated per acetyl-CoA. Recall that a fatty acid with 18 carbon atoms produces 9 molecules of acetyl-CoA. During *Stage 4* of lipid catabolism, the reduced coenzymes produced via β-oxidation and the citric acid cycle enter the electron transport chain, producing large amounts of ATP. Thus an 18-carbon fatty acid generates considerably more ATPs than a single molecule of glucose (148 ATPs versus 38 ATPs, respectively). The ATP yield from an 18-carbon fatty acid is summarized in Table 8.2 (see page 291).

## ESSENTIAL *Concepts*

Liver and muscle cells break down glycogen into glucose by a process called glycogenolysis. Glucose catabolism begins with glycolysis, an anaerobic pathway that converts glucose to pyruvate. Oxygen availability determines if pyruvate is converted to acetyl-CoA or lactate. If oxygen is available, acetyl-CoA is formed and enters the citric acid cycle, resulting in the formation of NADH + $H^+$ and $FADH_2$. These coenzymes can enter the electron transport chain and drive ATP formation via oxidative phosphorylation. Protein is broken down to amino acids by proteolysis. For amino acids to be used as an energy source, the nitrogen-containing amino group is removed via transamination and deamination. The remaining structure (α-ketoacid) can enter the citric acid cycle, ultimately generating ATP via the electron transport chain. Lipid catabolism begins with lipolysis, releasing fatty acids and glycerol. Fatty acids are oxidized via β-oxidation and the citric acid cycle to form NADH + $H^+$ and $FADH_2$, which then enter the electron transport chain to generate additional ATP.

## Anabolism: Energy Storage and Biosynthesis

It is important to recognize that anabolic pathways are not simply catabolic pathways in reverse. Catabolic pathways that break down compounds are distinctly different from anabolic pathways that synthesize compounds. In fact, anabolism tends to take place in the cytoplasm, whereas catabolism takes place primarily in mitochondria. This separation or compartmentalization is important, because it enables both anabolic and catabolic pathways to function simultaneously. The major anabolic and catabolic pathways involved in energy metabolism are reviewed in Table 8.3.

In the previous section you learned that when the body needs energy, cells break down energy-yielding macronutrients to synthesize ATP. When energy intake exceeds requirements however, anabolic pathways predominate over

## FIGURE 8.16 Oxidation of Fatty Acids via β-Oxidation and the Citric Acid Cycle

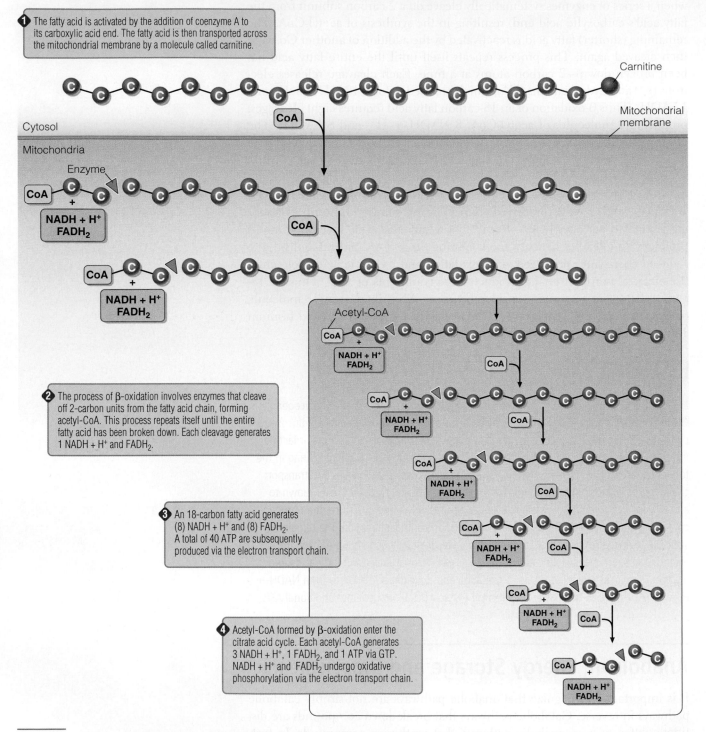

**1** The fatty acid is activated by the addition of coenzyme A to its carboxylic acid end. The fatty acid is then transported across the mitochondrial membrane by a molecule called carnitine.

Carnitine

Cytosol

Mitochondrial membrane

Mitochondria

Enzyme

CoA

NADH + H⁺
FADH₂

CoA

NADH + H⁺
FADH₂

Acetyl-CoA

CoA

NADH + H⁺
FADH₂

CoA

**2** The process of β-oxidation involves enzymes that cleave off 2-carbon units from the fatty acid chain, forming acetyl-CoA. This process repeats itself until the entire fatty acid has been broken down. Each cleavage generates 1 NADH + H⁺ and FADH₂.

**3** An 18-carbon fatty acid generates (8) NADH + H⁺ and (8) FADH₂. A total of 40 ATP are subsequently produced via the electron transport chain.

**4** Acetyl-CoA formed by β-oxidation enter the citrate acid cycle. Each acetyl-CoA generates 3 NADH + H⁺, 1 FADH₂, and 1 ATP via GTP. NADH + H⁺ and FADH₂ undergo oxidative phosphorylation via the electron transport chain.

NOTE: The ATP yield for fatty acid oxidation is based on integral values of 3 ATP per NADH + H⁺ and 2 ATP per FADH₂. Based on nonintegral values, the ATP yield is 2.5 ATP per NADH + H⁺ and 1.5 ATP per FADH₂.

catabolic pathways. During times of energy abundance, anabolic pathways convert energy-yielding nutrients into forms that can be stored—glycogen and triglycerides. However, anabolic pathways serve other functions as well. When the diet does not supply the proper balance or amounts of certain nutrients, anabolic pathways provide alternate routes for their synthesis. Of particular importance is the ability of some cells to synthesize glucose from noncarbo-

## TABLE 8.2 Overview of ATP Production via Fatty Acid Catabolism

From an 18-carbon fatty acid:
  8 cleavages resulting in 8 molecules of NADH + H⁺, 8 FADH₂,[a] and 9 acetyl-CoA

| Pathway | ATP yield |
|---|---|
| **β-oxidation** | |
| By oxidative phosphorylation of 8 NADH + H⁺ | 24 |
| By oxidative phosphorylation of 8 FADH₂ | 16 |
| | 40 |
| **Citric acid cycle (9 acetyl-CoA)** | |
| By substrate phosphorylation via 9 GTP | 9 |
| By oxidative phosphorylation of 9 FADH₂ | 18 |
| By oxidative phosphorylation of 27 NADH + H⁺ | 81 |
| | 108 |

The total yield from the complete catabolism of an 18-carbon fatty acid = 148 (40 + 108) ATPs.

[a]The ATP yield for fatty acid oxidation is based on integral values of 3 ATP per NADH + H⁺ and 2 ATP per FADH₂. Based on nonintegral values of 2.5 ATP per NADH + H⁺ and 1.5 ATP per FADH₂, the ATP yield for an 18-carbon fatty acid is 122 ATP.

hydrate sources when needed via gluconeogenesis. Next we discuss the role of anabolic pathways in energy metabolism.

## Glycogenesis

The anabolic process whereby glycogen is formed in liver and muscle tissue is called glycogenesis, and is stimulated by the hormone insulin. An average person has approximately 70 g of liver glycogen and 200 g of muscle glycogen—enough glucose to provide energy for 8 to 12 hours. Although liver and muscle glycogen stores play different roles, when the body's need

## TABLE 8.3 Summary of Major Energy Metabolism Pathways

| Pathway | Description | Tissue(s) | Anabolic, Catabolic, or Amphibolic |
|---|---|---|---|
| **Glycolysis** | Glucose breakdown forming two molecules of pyruvate; anaerobic energy metabolism | All tissues | Amphibolic |
| **Gluconeogenesis** | Glucose synthesis from noncarbohydrate sources during conditions of fasting and stress | Liver, kidneys | Anabolic |
| **Glycogenolysis** | Breakdown of glycogen for glucose production | Liver, muscle | Catabolic |
| **Glycogenesis** | Formation of glycogen | Liver, muscle | Anabolic |
| **Lipogenesis** | Synthesis of fatty acids and triglycerides | Liver, adipose tissue | Anabolic |
| **Lipolysis** | Breakdown of triglycerides to fatty acids and glycerol | Adipose, muscle | Catabolic |
| **β-Oxidation** | Breakdown of fatty acids to acetyl-CoA | Liver, muscle | Catabolic |
| **Ketogenesis** | Formation of ketones, an alternative energy source, from acetyl-CoA | Liver | Anabolic |
| **Proteolysis** | Breakdown of protein to amino acids | Muscle | Catabolic |
| **Citric acid cycle** | A central metabolic pathway that oxidizes acetyl-CoA to yield carbon dioxide, NADH + H⁺, FADH₂, and GTP | All tissues except red blood cells | Amphibolic |
| **Oxidative phosphorylation** | A coupled process whereby reduced coenzymes NADH + H⁺ and FADH₂ are oxidized to NAD⁺ and FAD and ADP is phosphorylated to ATP | All tissues except red blood cells | — |

for glucose is not being met the breakdown of glycogen occurs in both kinds of tissue.

## Lipogenesis

Because there is a limit to how much glycogen can be stored, insulin also promotes the uptake of excess glucose by adipose tissue where it can be turned into fatty acids and ultimately triglycerides. This process is called lipogenesis. Note that unlike the reversible conversion of glucose to glycogen, the conversion of glucose to a fatty acid is irreversible. That is, once glucose is transformed into a fatty acid, the fatty acid cannot be converted back to glucose.

CONNECTIONS▶ Lipogenesis is the synthesis of lipids (Chapter 7, page 234).

Lipogenesis takes place mainly in liver and adipose tissue, as illustrated in Figure 8.17.[3] First, glucose is broken down via glycolysis into two molecules of pyruvate, which in turn are converted to two molecules of acetyl-CoA. Next, molecules of acetyl-CoA are joined together to make a fatty acid—a process called fatty acid synthesis. After that, three fatty acids are attached to a molecule of glycerol to make a triglyceride. This entire process of lipogenesis is costly in terms of energy, requiring approximately 20 to 25% of the energy originally in the glucose molecule. By comparison, the energy required to transform dietary fatty acids into stored triglyceride is only about 5%. It is easy to see that excess calories from fatty foods are more efficiently stored in adipose tissue than are those from foods containing mostly carbohydrates.

Similarly, amino acids not used for protein synthesis are taken up by the liver, converted to fatty acids, and used to form triglyceride molecules. The energy required to convert amino acids to triglycerides is even greater than that required to convert glucose to triglycerides. This is because the nitrogen-containing amino group must first be removed via transamination and deamination, and the remaining structure is then converted to acetyl-CoA.

## Gluconeogenesis

Although most cells can use glucose and fatty acids for energy, the brain and central nervous system use glucose preferentially, and red blood cells use glucose exclusively. To ensure that these and other tissues have a continual

## FIGURE 8.17 Lipogenesis

Excess glucose and amino acids can be used to synthesize fatty acids, which subsequently join with a molecule of glycerol to make a triglyceride.

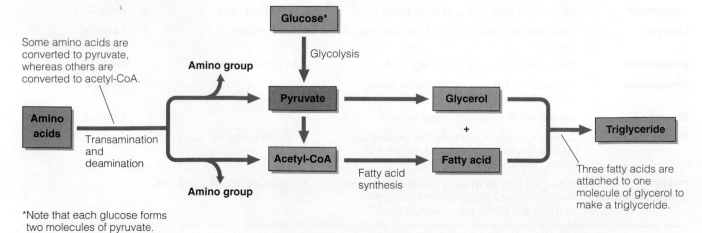

Some amino acids are converted to pyruvate, whereas others are converted to acetyl-CoA.

*Note that each glucose forms two molecules of pyruvate.

supply of glucose, small amounts of glucose are stored as glycogen in the liver and skeletal muscle. However, when glycogen stores are depleted noncarbohydrate molecules are transformed into glucose by various anabolic pathways collectively called gluconeogenesis. Gluconeogenesis occurs primarily within liver cells and, to a lesser extent, kidney cells. During periods of starvation, gluconeogenesis provides an important source of glucose to cells that depend on it as their major or sole source of energy.

Noncarbohydrate sources used for gluconeogenesis include most amino acids, lactate, and glycerol (Figure 8.18). These compounds are first converted to oxaloacetate and then to glucose. The energy needed to fuel these reactions comes from ATP and the oxidation of NADH + H$^+$ to NAD$^+$. Clearly, gluconeogenesis is an energy-intensive process. Gluconeogenesis is stimulated by the hormones glucagon and cortisol. Not surprisingly, insulin inhibits gluconeogenesis.

## Gluconeogenesis from Amino Acids

Amino acids used to generate glucose via gluconeogenesis are referred to as glucogenic amino acids. All but two amino acids (leucine and lysine) can be used for this purpose. There are several routes by which glucogenic amino acids can be converted to glucose, all involving removal of the amino group via transamination and deamination, and conversion of the remaining structure to oxaloacetate. The carbon skeleton of some amino acids is first converted to pyruvate and then to oxaloacetate. Oxaloacetate, which is a 4-carbon compound, loses 1 carbon atom, which combines with oxygen to form carbon dioxide. The resulting 3-carbon compound, phosphoenolpyruvate (PEP), follows many of the steps of glycolysis, only in reverse. It takes two molecules of PEP to form one glucose molecule. In addition, these reactions require many coenzymes that are derived from the B vitamins biotin, riboflavin, niacin, and vitamin B$_6$.

## FIGURE 8.18  Gluconeogenesis

Noncarbohydrate sources used for gluconeogenesis include some amino acids, lactate, and glycerol.

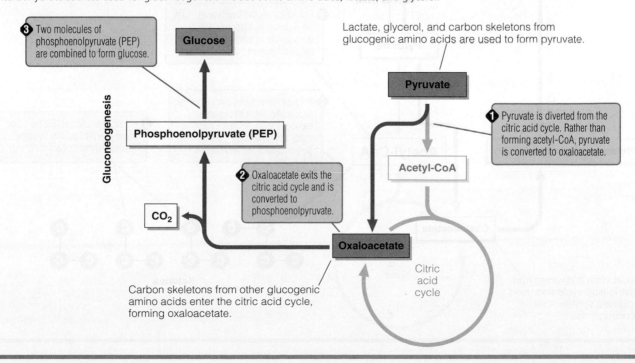

❸ Two molecules of phosphoenolpyruvate (PEP) are combined to form glucose.

**Glucose**

Lactate, glycerol, and carbon skeletons from glucogenic amino acids are used to form pyruvate.

**Pyruvate**

**Phosphoenolpyruvate (PEP)**

❶ Pyruvate is diverted from the citric acid cycle. Rather than forming acetyl-CoA, pyruvate is converted to oxaloacetate.

**Acetyl-CoA**

❷ Oxaloacetate exits the citric acid cycle and is converted to phosphoenolpyruvate.

**CO₂**

**Oxaloacetate**

Citric acid cycle

Gluconeogenesis

Carbon skeletons from other glucogenic amino acids enter the citric acid cycle, forming oxaloacetate.

## Ketogenesis

During starvation and when a person consumes very little carbohydrate, the supply of oxaloacetate can become depleted. This depletion occurs because oxaloacetate is diverted to gluconeogenic (glucose-forming) pathways to produce much-needed glucose for cells that require it for ATP production. With dwindling supplies of oxaloacetate, the citric acid cycle cannot operate at full speed. Therefore, acetyl-CoA formed during β-oxidation is not able to readily enter the citric acid cycle. As a result, the body uses it to produce an alternate form of energy—ketones. This anabolic process, called ketogenesis, provides the body with an important source of energy during periods of low glucose availability.

Ketones can be produced from some amino acids and fatty acids. For example, the structure (α-ketoacids) resulting from the transamination of some amino acids can be converted to acetyl-CoA. These amino acids are referred to as **ketogenic amino acids.** And, as previously described, acetyl-CoA is also produced from β-oxidation of fatty acids. Because high rates of gluconeogenesis can limit the amount of oxaloacetate in a cell, acetyl-CoA from β-oxidation is diverted from the citric acid cycle. Rather, molecules of acetyl-CoA join together to form ketones such as acetoacetate, β-hydroxybutyrate, and acetone. The production of ketones from acetyl-CoA occurs mostly in the liver and is stimulated by the hormone glucagon. In fact, gluconeogenesis and ketogenesis typically occur simultaneously. The process of ketogenesis is illustrated in Figure 8.19.

**CONNECTIONS** Ketones are organic compounds used as an energy source during starvation, fasting, low-carbohydrate diets, or uncontrolled diabetes, and *ketogenesis* refers to the metabolic pathways used to produce them (Chapter 5, page 149).

**ketogenic amino acids** Amino acids that can be used to make ketones.

## FIGURE 8.19  Ketogenesis

Ketones produced from fatty acids and ketogenic amino acids serve as a major source of energy for some tissues during times of glucose insufficiency.

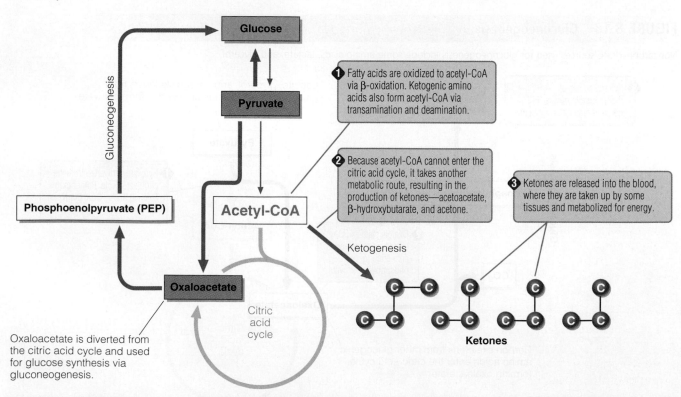

**1** Fatty acids are oxidized to acetyl-CoA via β-oxidation. Ketogenic amino acids also form acetyl-CoA via transamination and deamination.

**2** Because acetyl-CoA cannot enter the citric acid cycle, it takes another metabolic route, resulting in the production of ketones—acetoacetate, β-hydroxybutarate, and acetone.

**3** Ketones are released into the blood, where they are taken up by some tissues and metabolized for energy.

Oxaloacetate is diverted from the citric acid cycle and used for glucose synthesis via gluconeogenesis.

Glucose

Gluconeogenesis

Pyruvate

Phosphoenolpyruvate (PEP)

Acetyl-CoA

Ketogenesis

Oxaloacetate

Citric acid cycle

Ketones

Ketogenesis is important, because some tissues such as muscle, brain, and kidney have special enzymes that allow them to use ketones for ATP production. In fact, ketones produced from fatty acids and ketogenic amino acids can serve as a major source of energy during times of glucose insufficiency (such as starvation and diabetes). In this way, ketones can spare the body from having to use amino acids to synthesize large amounts of glucose via gluconeogenesis.

Although the use of ketones by the body has important survival implications, at times ketone production exceeds ketone use. This can lead to high levels of ketones in the blood, a condition called **ketosis.** Sometimes you can detect that someone is experiencing this condition by the characteristic fruity odor of the person's breath, caused by breathing out acetone. Ketosis can occur during prolonged fasting, consumption of very low carbohydrate diets, strenuous exercise, and some diseases such as type 1 diabetes. Severe ketosis, called **ketoacidosis,** can cause a variety of complications, including lowered blood pH, nausea, coma, and in extreme circumstances, death.

# ESSENTIAL *Concepts*

Anabolic pathways play important roles in storing excess energy and in synthesizing energy-yielding molecules when glucose availability is limited. The hormone insulin stimulates liver and muscle tissues to store excess glucose as glycogen (glycogenesis). Insulin also promotes the conversion of glucose and amino acids to fatty acids and the subsequent production of triglyceride in adipose and liver tissues (lipogenesis). Thus anabolic pathways are important during periods of excess energy availability. During starvation, anabolic pathways are used to synthesize glucose from noncarbohydrate sources (gluconeogenesis). Substances used for gluconeogenesis include glucogenic amino acids, lactate, and glycerol. However, high rates of gluconeogenesis deplete the amount of oxaloacetate. When this occurs, acetyl-CoA cannot participate in the citric acid cycle. Rather, acetyl-CoA is diverted to ketone production. Some amino acids are converted to ketones as well. Tissues such as muscle, brain, and kidney have enzymes that allow them to use ketones for ATP production.

# Shifts in Energy Metabolism in Response to Feeding and Fasting

The availability of metabolic fuels fluctuates throughout the day. After a meal, energy sources are readily available and tend to exceed the immediate needs of cells. During this time, anabolic pathways favor the synthesis and storage of glycogen, protein, and triglycerides. However, after a period of time without eating, the availability of circulating fuels diminishes. At this point, the body is primarily in a catabolic state, relying on ATP generated from the breakdown of the body's stored energy reserves. Thus, energy availability, in part, determines the balance of anabolic and catabolic activity (Figure 8.20). In extreme catabolic states such as prolonged fasting and starvation, the body makes additional metabolic adjustments to maintain physiological functions and to prevent further deterioration.

Which nutrient stores the body uses for energy (ATP) depends on a person's nutritional and physiological state. In fact, energy availability may be the single most influential factor modulating energy metabolism. To better

**ketosis** An accumulation of ketones in body tissues and fluids.

**ketoacidosis** A rise in ketone levels in the blood, causing the pH of the blood to decrease.

## FIGURE 8.20   Overview of Energy Metabolism

The relative activity of anabolic and catabolic pathways is determined by the availability of substrates, intermediate compounds, and energy needs.

Amino acids, glucose, fatty acids, and glycerol can be converted to pyruvate and/or acetyl-CoA. Depending on energy needs, these compounds enter either anabolic or catabolic pathways.

During times of energy excess, acetyl-CoA can be converted to fatty acids, subsequently forming triglycerides.

When energy is needed, acetyl-CoA enters the citric acid cycle and is oxidized to carbon dioxide, liberating energy in the process. Electrons and hydrogen ions are transferred to coenzymes FAD and NAD$^+$, forming FADH$_2$ and NADH + H$^+$.

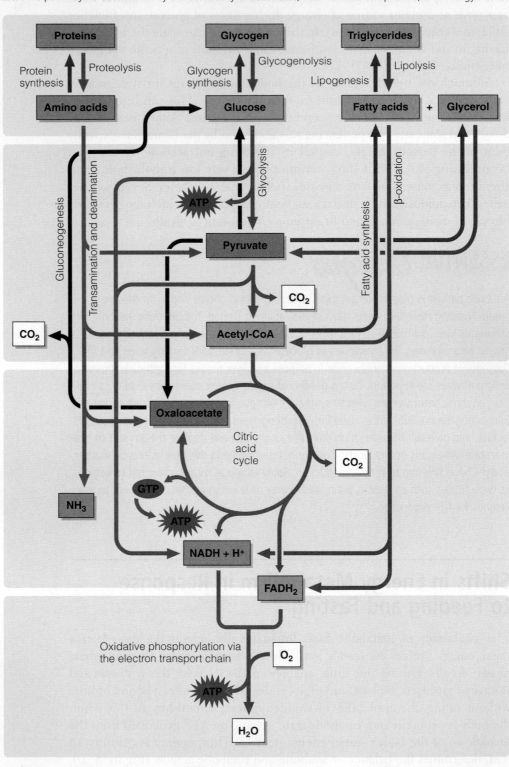

These pathways predominate during times of energy excess. Amino acids, glucose, and fatty acids are converted to protein, glycogen, and triglycerides, respectively.

These pathways predominate during times of energy need, breaking down proteins, glycogen, and triglycerides to amino acids, glucose, and fatty acids, respectively.

These pathways predominate when there is a lack of available glucose. Oxaloacetate can leave the citric acid cycle and is used to synthesize glucose via gluconeogenesis.

describe how energy availability influences energy metabolism, scientists have defined four nutritional states based on the amount of time that has elapsed since the previous meal. These states include the absorptive state, postabsorptive state, acute starvation, and prolonged starvation. As you can see in Figure 8.21, relative fuel availability and hormone levels change in response to how long a person remains fasting. The next section describes alterations in energy metabolism in response to the normal cycles of eating and fasting that most of us experience daily. The adaptive metabolic response to the state of starvation is presented as well.

## The Absorptive State

The first four hours after a meal is referred to as the **absorptive state** or the postprandial period. During this state, absorbed nutrients enter the blood, stimulating the release of insulin and inhibiting the release of glucagon. In turn, energy metabolism in the liver, adipose tissue, and skeletal muscle is affected. Many of the major anabolic pathways operate during the absorptive state, including those involved in the synthesis of proteins, fatty acids, triglycerides, and glycogen. During this time, glucose is the body's major source of energy. Thus, as long as there is sufficient oxygen, most of the ATP produced during the absorptive state is provided by glycolysis, the citric acid cycle, and the electron transport chain.

Throughout the absorptive state, amino acids are used primarily for protein synthesis and little protein degradation (proteolysis) takes place. Once the body's need for protein is met, excess amino acids are converted into fatty acids and ultimately into triglycerides for storage. In addition, insulin stimulates the uptake of fatty acids into adipose tissue, where they are converted to triglycer-

**absorptive state** (postprandial period) The first four hours after a meal.

## FIGURE 8.21 Fuel Availability and Hormonal Changes in Response to Feeding, Fasting, and Acute Starvation

**Absorptive state:**
- The relative concentration of insulin is higher than glucagon, favoring energy storage.
- Blood glucose is elevated.

**Postabsorptive state:**
- Insulin levels decrease, and glucagon increases.
- Blood glucose decreases.
- Liver glycogen broken down for a source of glucose.
- Increase use of fatty acids for energy.

**Acute starvation:**
- Relative concentration of glucagon is higher than insulin.
- Liver glycogen stores are depleted.
- Glucose is supplied mainly by gluconeogenesis.
- Stored triglycerides are broken down with an increase in the use of fatty acids for energy.
- Ketone formation (ketogenesis) increases.

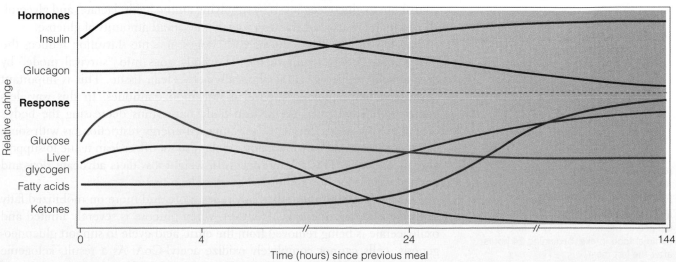

ides for storage. After a meal, elevated insulin accompanied by low glucagon secretion promotes energy storage via anabolic pathways.

## The Postabsorptive State

Because we typically consume at least three distinct meals each day, there are periods between meals when no nutrients are being absorbed, representing what is called the **postabsorptive state.** The postabsorptive state begins about four hours after the last intake of food, when insulin secretion declines and the release of glucagon increases. Throughout this phase, the body's cells rely heavily on energy supplied by the breakdown of stored energy. Blood glucose levels are maintained primarily by liver glycogen via glycogenolysis. In fact, most glucose (75%) used by body tissues during the postabsorptive state comes from liver glycogen.

Most cells continue to produce ATP from glycolysis, the citric acid cycle, and the electron transport chain during the postabsorptive state. However, many cells increase their use of fatty acids as an energy source during this time to ensure that there is enough glucose for red blood cells and the central nervous system. Declining levels of insulin stimulate lipolysis and the subsequent release of fatty acids into the circulation. Some scientists believe that maintaining the human body in the postabsorptive state may have health benefits. This is because chronic, moderate caloric restriction of experimental animals appears to slow the process of aging. Specifically, animals maintained on energy-restricted but nutritionally adequate diets live longer than those on unrestricted diets. You can read more about the possible effect of long-term caloric restriction on longevity in the Focus on Diet and Health feature.

## Acute Starvation

If food is not consumed within 24 hours, the body's ready supply of glucose begins to dwindle. The body then enters the next nutritional state called **acute starvation,** which is defined as the first five days beginning 24 hours after the last meal. Within the first 24 to 36 hours of fasting, glycogen stores are depleted, and the insulin-to-glucagon ratio declines even further. The priority of the body is to supply the nervous system and red blood cells with sufficient glucose. Because of depleted glycogen stores, glucose must now be synthesized from noncarbohydrate substances via gluconeogenesis. Muscles contribute to gluconeogenesis by supplying amino acids, lactate, and glycerol, all of which are used by the liver to generate small amounts of glucose.

The loss of lean tissue during early stages of acute starvation reduces the body's total energy requirements. The body goes into "survival mode" by dramatically reducing its metabolically active lean tissue. This is important because it helps prolong survival if starvation continues. In this way, less energy is needed to maintain lean body mass, thus decreasing the body's overall energy needs. Similarly, self-imposed energy restriction, as with some eating disorders and very low calorie diets, also sacrifices lean tissue to support gluconeogenesis. This is why successful weight-loss diets advocate slow and gradual weight loss to minimize the loss of lean tissue.

During early starvation, the body relies more and more on mobilized fatty acids for a source of energy. However, when glucose is severely limited and oxaloacetate is being removed from the citric acid cycle to support gluconeogenesis, cells cannot completely oxidize acetyl-CoA. As a result, ketogenic pathways in the liver are stimulated. As previously described, the ability of

**postabsorptive state** The period of time (4 to 24 hours after a meal) when no dietary nutrients are being absorbed.

**acute starvation** Early stages of starvation defined as the first five days of fasting or minimal food intake, beginning 24 hours after the last meal.

## *Focus On* DIET AND HEALTH
### Chronic Caloric Restriction and Longevity

Can eating less extend the human life span? Given the results of animal studies, some researchers believe so. There are many theories as to what causes aging, and scientists have long been interested in understanding factors that may slow this process and extend life. Growing evidence suggests that the lifelong or chronic practice of limiting energy intake while consuming adequate amounts of essential nutrients may improve health and longevity.[4] This practice is referred to as "caloric restriction with optimal nutrition" (CRON). Studies suggest that CRON reduces overall metabolic activity, which in turn reduces cell damage. So far, CRON has been studied in a variety of animals, including monkeys and rats. These experiments show that animals on nutrient-dense, energy-restricted diets live longer, have a more youthful appearance, and have fewer age-related health problems compared to their well-fed counterparts. In fact, CRON has been shown to be one of the few effective approaches that extends longevity of experimental animals.

The benefits of CRON on longevity may be the result of reduced overall

*Courtesy of Richard Weindruch, University of WI*

These mice are the same age, 40 months. The two mice on the left, which were fed a low-calorie diet since 12 months of age (early middle age), look younger and are healthier than the normally fed animals on the right.

metabolic activity, which in turn may cause less oxidative damage to cells and tissues. Specifically, oxidative metabolism can cause the formation of unstable molecules called free radicals. Free radicals are destructive and damage cell membranes, DNA, and cell organelles such as mitochondria. Damage caused by free radicals is thought to be one of the leading causes of cellular aging and cancer. In addition to reducing free radical damage, CRON may also prevent damage caused by insulin resistance. Like free radicals, an overabundance of

glucose molecules can damage DNA and proteins, impairing cell function. CRON is thought to improve insulin sensitivity, which in turn reduces cell damage caused by high levels of glucose in the blood.

It is difficult to assess the impact of CRON on longevity in humans. This is because humans generally survive longer than most experimental animals and live in a largely uncontrolled environment. Furthermore, to really examine the effect of CRON on longevity human intervention trials would need to begin early in life, and participants would need to adhere to caloric restriction for many years. Thus it may not be possible to conduct appropriate clinical trials to experimentally assess the effect of CRON on human longevity.

Although living longer through caloric restriction may not appeal to some, there is reason to believe that eating less may also reduce the risk of developing chronic degenerative diseases such as cancer, type 2 diabetes, and cardiovascular disease. Thus, eating fewer calories and more nutrient-dense foods likely has much to offer in terms of health and longevity.

some tissues, especially the brain, to use ketones for ATP production has important survival implications. For example, the ability of the brain to use ketones for energy spares glucose for other tissues. This helps preserve lean body tissue because the body does not have to rely as extensively on gluconeogenesis to support energy requirements. Ketosis resulting from ketogenesis often causes loss of appetite. During times of famine, this disinterest in food can be advantageous. When food becomes available and a person can resume eating, ketogenesis decreases and appetite returns.

## Prolonged Starvation

If food deprivation lasts longer than one week, the body enters a nutritional state called **prolonged starvation.** Maintaining adequate glucose availability to cells and the preservation of lean body mass now take on even higher priority for the body. During this time, only red blood cells continue to rely solely on glucose as a substrate for ATP production. This is because they do not have

**prolonged starvation** Food deprivation lasting longer than one week.

mitochondria, the organelles where fatty acids and ketones are metabolized for energy.

Under conditions of prolonged starvation, the reliance on ketones as a major source of energy is an adaptive response that helps extend life. Thus ketogenesis is further stimulated. However, when stored fat becomes extremely limited, the body has no choice but to break down muscle tissue as its only remaining glucose and energy source. This includes muscle associated with vital body organs such as the heart and the kidneys. Eventually the ability of these organs to carry out their functions is impaired. The body continues to catabolize muscle and vital body organs until death ensues. This may take months, depending on the amount of fat reserves that a person has. As long as adequate water is available, the body typically can maintain minimal functions until food becomes available again. Much of what is known about the physical and psychological affects of prolonged starvation comes from a classic study conducted in 1944 by Ancel Keys. In the Focus on the Process of Science feature you can read more about the physical and psychological effects of prolonged energy restriction.

# Focus On THE PROCESS OF SCIENCE

## Keys Starvation Experiment

During World War II, the U.S. military wanted to know more about how to feed and care for malnourished soldiers returning from war. A study was commissioned to investigate the physical and psychological affects of starvation, called the Keys Starvation Experiment. Over 36 healthy men agreed to participate.[5] Throughout the first three months of the study, measurements were taken to assess the subjects' physical health. In addition, nutrient and caloric requirements were estimated for each participant, and tests were given to assess psychological variables. For the next six months, the men were put on a semistarvation diet, consisting of 1,600 kcal per day. Living together in a dormitory, the men were regularly monitored for physical and psychological changes.

By the end of the six-month semistarvation diet, participants had lost an average of 24% of their body weight, and total energy requirements had decreased substantially. Weight loss was associated with decreases in both adipose and lean tissues. The men became lethargic, lacked endurance, and experienced diminished physical strength. Also striking to the researchers was the extent to which participants' mental health deteriorated. Many of the men developed depression, emotional instability, and what was described as "neurotic tendencies." Participants became apathetic, irritable, moody, and easily distracted. Furthermore, researchers noted deterioration in personal appearance and lack of grooming. Also of interest was the fact that participants talked frequently about food. Food cravings increased, food dislikes disappeared, and participants became possessive about food rations. It became apparent that food restriction had devastating effects, both physically and psychologically.

The next three months of the study focused on refeeding. Participants were free to eat what and how much they wanted. Many men found it difficult to stop eating, even after they reported feeling full—eating 50 to 200% more calories than before the onset of the study.

Three of the 36 study participants, living on a "starvation" diet during tests at the University of Minnesota, are served a measured meal by Mrs. Marietta Anderson, dietitian.

Most participants gained weight quickly, with the added weight mostly deposited as fat, not muscle. As weight was regained, mental state also improved. Although this study was conducted over 50 years ago, the description of the physical and psychological changes that accompany energy restriction and weight regain is no different from what most dieters experience today.

## The Body's Versatile Solution to Meeting Energy Needs

The use of various energy metabolism pathways by the body is affected by what, how much, and how frequently food is consumed. The versatility of these pathways enables cells to respond to a variety of different circumstances to ensure that energy (ATP) needs are always met. Understanding the integrated and coordinated nature of energy metabolism provides important insights as to how nutrients derived from food affect life at the most basic cellular level. However, these countless metabolic reactions can be easily disrupted by imbalances in nutrient and caloric intakes. In the case of potentially life-threatening conditions such as prolonged fasting and starvation, energy metabolism pathways must meet the challenge of securing adequate glucose, while minimizing the loss of muscle tissue. The body's ability to cope with extreme situations such as these demonstrates the extent to which energy metabolism pathways can respond to preserve life.

## ESSENTIAL *Concepts*

Energy metabolism pathways are responsive to intermittent states of feeding and fasting, called the absorptive state, postabsorptive state, acute starvation, and prolonged starvation. Most of the major anabolic pathways operate during the absorptive state, including those promoting protein, triglyceride, and glycogen synthesis. During this time, glucose is the major source of energy for all tissues. The postabsorptive state, which is the period 4 to 24 hours after the last intake of food, relies heavily on energy supplied by the breakdown of stored energy reserves, especially glycogen. As the body enters acute starvation, defined as the first five days after the postabsorptive state, the body begins to rely more on the mobilization of adipose reserves for ATP production, while producing sufficient glucose via gluconeogenesis. The body enters the nutritional state of prolonged starvation when a person is deprived of food for more than one week. During this time, the body must preserve lean tissue from its gluconeogenic fate by increasing the use of ketones as an energy source. This is accomplished via ketogenic pathways. However, prolonged starvation can eventually result in the extensive use of muscle for glucose and, ultimately ATP production. If refeeding is not resumed, the consequences of starvation can be severe.

## Review Questions  Answers are found in Appendix G.

1. Which of the following nutrients is used by cells to generate ATP?
   a. Glucose
   b. Fatty acids
   c. Amino acids
   d. All of the above

2. Which of the following *cannot* provide cells with a source of glucose?
   a. Fatty acids
   b. Glycogen
   c. Certain amino acids
   d. Glycerol derived from triglycerides

3. Which of the following metabolic pathways produces ATP via oxidative phosphorylation?
   a. The citric acid cycle
   b. Lipogenesis
   c. The electron transport chain
   d. Glycolysis

4. Gluconeogenesis
   a. results in the formation of triglycerides.
   b. occurs when liver and muscle glycogen stores are full.
   c. is the process whereby the amine group is removed from amino acids.
   d. occurs when glucose availability is limited.

5. The citric acid cycle begins when _____ combine(s) with oxaloacetate to form citrate.
   a. ketones
   b. pyruvate
   c. acetyl-CoA
   d. water

**6.** The metabolic pathway that splits glucose to form two molecules of pyruvate is called _____.
   **a.** gluconeogenesis
   **b.** glycogenolysis
   **c.** glycolysis
   **d.** the citric acid cycle

**7.** Anabolic pathways are most active during the

   _____.
   **a.** absorptive state
   **b.** postabsorptive state
   **c.** acute starvation
   **d.** exercise

**8.** Which of the following metabolic pathways is not active during the postabsorptive state?
   **a.** Glycogenolysis
   **b.** Lipolysis
   **c.** Lipogenesis

**9.** Which of the following is a *true* statement?
   **a.** Catabolic activity increases when energy (ATP) availability is high.
   **b.** When oxygen is available in cells, pyruvate is converted to acetyl-CoA.

   **c.** Glucogenolysis is the metabolic process whereby fatty acids are converted to glucose.
   **d.** All of the above statements are *true*.

**10.** Ketones production increases
   **a.** after a meal.
   **b.** when glucose is readily available.
   **c.** after a period of prolonged fasting.
   **d.** when glycogen stores are full.

**11.** Describe the metabolic response following a meal and the metabolic response after 24 hours without food.

**12.** Explain how the formation of ketones helps to spare lean body mass during starvation.

**13.** For each of the following nutrients, provide a brief description of the stages of catabolism.
   **a.** Carbohydrates
   **b.** Protein
   **c.** Triglycerides

**14.** Provide a brief description of each metabolic pathway.
   **a.** Glycolysis
   **b.** Citric acid cycle
   **c.** β-oxidation

**15.** Describe two methods used by cells to generate ATP.

## Media Links   A variety of study tools for this chapter are available at our website: www.thomsonedu.com/nutrition/mcguire.

Prepare for tests and deepen your understanding of chapter concepts with these online materials:
- Practice tests
- Flashcards
- Glossary
- Student lecture notebook
- Web links
- Animations
- Chapter summaries, learning objectives, and crossword puzzles

## Notes

1. Malasky BR, Alpert JS. Diagnosis of myocardial injury by biochemical markers: problems and promises. Cardiology in Review. 2002;10:306–17.
2. Karlic H, Lohninger A. Supplementation of L-carnitine in athletes: does it make sense? Nutrition. 2004;20:709–15.
3. Brass EP. Carnitine and sports medicine: use or abuse? Annals of the New York Academy of Sciences. 2004;1033:67–78.
4. Gredilla R, Barja G. Minireview. The role of oxidative stress in relation to caloric restriction and longevity. Endocrinology. 2005;146:3713–7. Piper MD, Mair W, Partridge LJ. Counting the calories: the role of specific nutrients in extension of life span by food restriction. Journals of Gerontology Series A: Biological Sciences and Medical Sciences. 2005;60:549–55. Sinclair DA. Toward a unified theory of caloric restriction and longevity regulation. Mechanisms of Ageing and Development. 2005;126:987–1002.

5. Kalm LM, Semba RD. They starved so that others would be better fed: Remembering Ancel Keys and the Minnesota Experiment. Journal of Nutrition. 2005;135:1347–52. Keys A, Brozek J, Henschel A, Mickelsen O, Taylor HL. The biology of human starvation, Vols I–II. Minneapolis: University of Minnesota Press; 1950.

**Check out the following sources for additional information.**

6. Bettelheim FA, Brown WH, March J. Introduction to general, organic, and biochemistry, 7th ed. Belmont, CA: Wadsworth Thomson Learning; 2004.
7. Gropper SS, Smith JL, Groff JL. Advanced nutrition and human metabolism, 4th ed. Belmont, CA: Wadsworth Thomson Learning; 2005.
8. Garrett RH, Grisham CM. Principles of biochemistry with a human focus, 1st ed. Belmont, CA: Brooks/Cole Thomson Learning; 2002.

9. Morris H, Best LR, Miner RL, Richey JM. Introduction to general, organic, and biochemistry, 8th ed. Hoboken, NJ: Wiley; 2005.

10. Nandi J, Meguid MM, Inui A, Xu Y, Makarenko IG, Tada T, Chen C. Central mechanisms involved with catabolism. Current Opinion in Clinical Nutrition and Metabolic Care. 2002;5:407–18.

11. Pahlavani MA. Influence of caloric restriction on the aging immune system. Journal of Nutrition, Health, and Aging. 2004;8:38–47.

12. Salway JG. Metabolism at a glance, 4th ed. Cambridge, MA: Blackwell Science; 2004.

13. Smith JV, Heilbronn LK, Ravussin E. Energy restriction and aging. Current Opinion in Clinical Nutrition and Metabolic Care. 2004;7:615–22.

14. Starr C, Taggart R. Biology: the unity and diversity of life, 10th ed. Belmont, CA: Thomson Brooks/Cole; 2004.

15. Storlien L, Oakes ND, Kelley DE. Metabolic flexibility. Proceedings of the Nutrition Society. 2004;63:363–8.

# Nutrition, Physical Activity, and Athletic Performance

*P*eople exercise for different reasons, some to stay in shape or to improve overall physical fitness, others because they enjoy the challenge of training and competition. Regardless of the reason, exercise offers many health benefits. Because exercise can be physically demanding, it is important for active people to have a diet that provides adequate nutrients, calories, and fluids. For most recreational athletes, dietary recommendations do not differ much from those suggested for healthy individuals. However, for competitive athletes who often push themselves to the extreme, dietary requirements may be greater. What an athlete eats and drinks before, during, and after exercise can greatly affect performance. The unrelenting stress of training and competition can deplete the body of nutrients and energy needed for cellular processes.

Athletes can optimize athletic performance by understanding the basics of energy metabolism and applying this information to nutritional and training strategies. After all, the foods we eat determine nutrient availability. In this Nutrition Matters, you can learn about the effect of physical activity and training on energy metabolism. The importance of consuming an adequate diet and the benefits of physical activity are also discussed.

## Energy Metabolism During Physical Activity

Throughout the day we are in perpetual motion. We stand, we sit, we walk, we run—all of which require energy. Thus physical activity greatly influences energy metabolism. On average, a person expends approximately 15 to 30% of his or her total daily energy intake on physical activity. This percentage is even higher for people who exercise. At rest the body expends approximately 1 to 1.5 kcal/minute. During physical exertion, energy expenditure can increase to as much as 15 to 36 kcal/minute. Exercise can be very demanding in terms of energy.

Movement requires motor units within skeletal muscle fibers to contract and relax in response to neural signals from the brain. To perform this complex action, individual muscle fibers convert the chemical energy in ATP to mechanical energy. The speed at which this occurs is determined primarily by the availability of fuel sources and the metabolic pathways used by muscle fibers to generate ATP.

### Aerobic and Anaerobic Metabolism

During exercise, ATP is generated from the metabolic breakdown of glucose, fatty acids, and, to a lesser extent, amino acids. The availability of these nutrients is determined by the foods we eat and by the extent to which they are stored in the body. Because the concentration of ATP in muscles is generally low, continuous ATP production in active muscle cells is important. Muscle fibers have multiple metabolic pathways that produce ATP, and these work together to provide energy needed to fuel physical activity.[1] Some of these pathways operate in aerobic conditions (oxygen rich), whereas others are used during anaerobic conditions (oxygen poor). Given the myriad conditions in which athletes compete, both aerobic and anaerobic energy pathways offer certain advantages.

Think about how long it would take your muscles to fatigue while taking a vigorous walk—one hour, perhaps two hours? In contrast, how long would it take your muscles to fatigue during an all-out sprint—1 minute, perhaps 2 minutes? There are several reasons why muscles tire more quickly during high-intensity workouts, but oxygen availability plays an important role. Muscles use both anaerobic and aerobic metabolic pathways to fuel physical activity. However, the intensity and duration of the activity determines the relative contribution of each. For example, short, high-intensity exercise, such as a 100-meter sprint, requires large amounts of energy quickly, compared to endurance sports, such as a marathon, which require a more sustained supply of ATP.

Aerobic metabolic pathways include the citric acid cycle, β-oxidation, and the electron transport chain, which were discussed in detail in Chapter 8. These pathways are complex, consisting of multiple metabolic steps, and therefore generate ATP relatively slowly. However, aerobic metabolic pathways generate a tremendous amount of ATP over an extended period of time. During low-intensity activities, cells can rely mainly on these pathways, because oxygen availability is usually adequate.

CONNECTIONS ➤ Recall that the citric acid cycle is a central metabolic pathway that oxidizes acetyl-CoA to yield carbon dioxide, NADH + H$^+$, FADH$_2$, and GTP. β-Oxidation metabolizes fatty acids, forming acetyl-CoA. The electron transport chain is a series of reactions that use electrons and hydrogen ions from NADH + H$^+$ and FADH$_2$ to generate water, carbon dioxide, and ATP (Chapter 8, pages 282, 284).

Anaerobic pathways, which can generate ATP without relying on oxygen, include the **ATP creatine phosphate (ATP-CP) pathway** and glycolysis. As exercise intensity increases, the rate of ATP production by aerobic means cannot keep pace with energy needs of muscles. Therefore, the body must increasingly rely on anaerobic metabolism for ATP production. Because anaerobic metabolic path-

**FIGURE 1    Three Energy Systems and Relative Contribution to Total Energy Expenditure During Exercise**

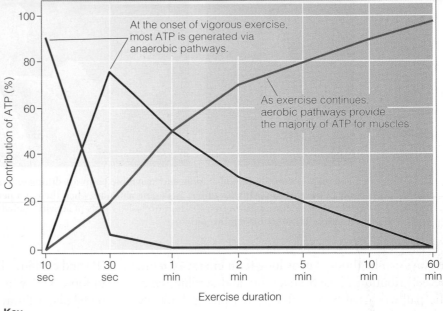

At the onset of vigorous exercise, most ATP is generated via anaerobic pathways.

As exercise continues, aerobic pathways provide the majority of ATP for muscles.

**Key**

— ATP–creatine phosphate pathway          — Glycolysis

— Aerobic pathways (includes citric acid cycle, β-oxidation, and the electron transport chain)

ways are relatively simple compared to aerobic pathways, ATP production is very fast. However, anaerobic pathways cannot maintain a high rate of ATP production for very long. Because of this, the body can rely on anaerobic metabolism only for brief periods of time.

CONNECTIONS ➤ Recall that glycolysis is a series of chemical reactions that split the 6-carbon glucose molecule into two 3-carbon molecules called pyruvate (Chapter 8, page 281).

At the onset of exercise such as jogging or biking, both aerobic and anaerobic pathways are activated to some extent. This is illustrated in Figure 1, which shows the relative contribution of each energy system during exercise. When vigorous exercise begins, muscles primarily use the anaerobic ATP-CP pathway to generate ATP. The ATP-CP pathway can be thought of as an immediate energy system, because it can sustain physical activity for only a short time. The continuation of exercise places a greater demand on glycolysis to generate ATP, which also can be thought of as a short-term energy system. However, if physical activity is to continue, muscles must increase their use of oxygen-requiring (aerobic) pathways (such as the citric acid cycle) for ATP production. These

**ATP creatine phosphate (ATP-CP) pathway** An anaerobic metabolic pathway that uses ADP and creatine phosphate to generate ATP.

The relative contribution of aerobic and anaerobic pathways during exercise depends on the intensity of the exercise. Muscles utilize anaerobic pathways when there is a need to generate ATP rapidly, such as during high intensity short explosive bursts of activity. Activities such as long distance running rely more on aerobic pathways.

pathways can be thought of as long-term energy systems. The contribution of each anaerobic and aerobic metabolic pathway used by muscles to fuel physical activity is examined next.

## The ATP Creatine Phosphate (ATP-CP) Pathway: Immediate Energy System

Only small amounts of ATP are stored in resting muscle, and this is rapidly depleted within the first few seconds of vigorous exercise. Muscles must then quickly replenish this energy source. The ATP creatine phosphate (ATP-CP) pathway is the most simple and rapid means for active muscles to generate ATP. **Creatine phosphate (CP)** is a high-energy, phosphate-containing compound found in muscle tissue. Creatine is synthesized in the liver, kidney, and pancreas from amino acids, and it can also be obtained by eating meat and fish. CP is formed when creatine combines with inorganic phosphate ($P_i$). When ATP supplies diminish, the enzyme **creatine phosphokinase** quickly breaks CP down, forming inorganic phosphate

($P_i$) and creatine. The released $P_i$ is then attached to ADP to form ATP, which can then fuel muscle activity. This metabolic pathway is illustrated in Figure 2.

The ATP-CP pathway is ideally suited to meet the immediate energy demands of intense activities such as power lifting and track or field events that last for only a short period of time. This immediate energy system can sustain this type of physical exertion for only 3 to 15 seconds. High-intensity activities rapidly deplete CP, and resting muscles take several minutes to replenish the supply. Once CP is depleted, muscles must then rely on other metabolic pathways for ATP formation. For this reason, some athletes take creatine supplements as ergogenic aids to enhance athletic endurance.

**creatine phosphate (CP)** (CRE– a – tine) A high-energy compound consisting of creatine and phosphate used to generate ATP.

**creatine phosphokinase** An enzyme that splits creatine phosphate to generate ATP.

## FIGURE 2   ATP Creatine Phosphate (ATP-CP) Pathway

The ATP creatine phosphate (ATP-CP) pathway enables active muscles to generate ATP rapidly.

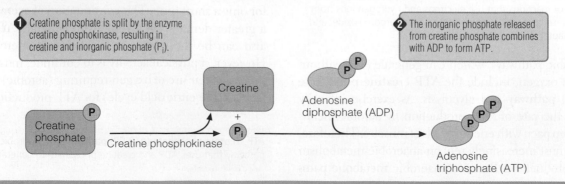

**1** Creatine phosphate is split by the enzyme creatine phosphokinase, resulting in creatine and inorganic phosphate ($P_i$).

**2** The inorganic phosphate released from creatine phosphate combines with ADP to form ATP.

Creatine phosphate

Creatine phosphokinase

Creatine

$P_i$

Adenosine diphosphate (ADP)

Adenosine triphosphate (ATP)

*Creatine Supplements*    Creatine supplements are widely available and are sold as over-the-counter products in most grocery stores and pharmacies. These products are promoted to athletes as a safe and effective ergogenic aid to improve muscle mass, strength, and recovery time. A study conducted in the early 1980s to determine if creatine supplementation could slow the progressive loss of vision in people with a rare eye disease first generated interest in creatine as a performance enhancer. The researchers noted that participants receiving the treatment experienced weight gain and increased strength. Since then, numerous studies on creatine supplementation have been conducted in many different types of athletes.[2] One consistent finding is that creatine supplementation does increase creatine levels in muscle tissue, especially in people who had low creatine levels to begin with. Whether higher muscle creatine translates into improved performance is less clear. Approximately 70% of studies show a beneficial effect of creatine supplementation on athletic performance.[3]

Although large-scale, well-controlled clinical trials are lacking, it appears that creatine supplementation may enhance high-intensity, short-duration athletic performance such as sprinting or power lifting. There is little evidence of a benefit to athletes participating in low-intensity, endurance sports such as long-distance running. In fact, creatine supplements may hinder performance, because they cause weight gain that is mostly attributed to fluid retention.

Although reported side effects associated with short-term use of creatine supplements are minimal, there are insufficient data to determine long-term side effects. A few studies have reported potential concerns regarding the use of creatine on liver and kidney function.[4] However, no adverse effects of creatine supplementation on liver and kidney function in healthy people have been found.[5] Creatine supplementation has not been studied in those under 18 years of age and is not recommended for this age group.

A typical diet that includes meat, fish, and other animal products provides approximately 1 to 2 g of creatine a day. By comparison, athletes taking creatine supplements to enhance performance typically ingest 20 to 30 g per day. However, taking low doses of creatine (3 g/day) appears to have the same effect as higher doses. Although some might argue that creatine supplementation should be banned from sporting events, there is no reliable method to test for its use. Nonetheless, a new rule implemented by the National Collegiate Athletic Association (NCAA) prohibits institutions from providing creatine to athletes.

## Glycolysis: Short-Term Energy System

Because the ATP-CP pathway can only support exercise for a very short period, muscles must be able to obtain

When there is a need to generate ATP rapidly, muscles rely on anaerobic metabolic pathways.

energy from other sources as well. At the onset of exercise, oxygen availability in muscles is somewhat limited, even if the activity is of low intensity. This is caused by a lag in oxygen delivery and uptake by muscle relative to oxygen need. Thus it is difficult for aerobic metabolic pathways to initially meet the energy demands of muscle. However, it is incorrect to think that anaerobic metabolism occurs only under conditions of low oxygen availability. Muscles also use anaerobic pathways when there is a need to generate ATP rapidly. This is certainly the case during high-intensity, short, explosive bursts of activity such as a 100-m swimming event. To generate ATP quickly, muscles rely on the rapid breakdown of glucose via glycolysis (see Figure 3 on the next page).

Unlike aerobic metabolism, which primarily uses glucose and fatty acids to generate ATP, glycolysis (an anaerobic pathway) depends solely on glucose. Glucose from dietary carbohydrates and from the breakdown of glycogen (glycogenolysis) is used for this purpose. The liver and kidneys also make small amounts of glucose from noncarbohydrate precursors (such as some amino acids, glycerol, and lactate) via gluconeogenesis. Muscles rely more on glucose derived from muscle glycogen than on glucose circulating in the blood. Thus muscle glycogen stores can become depleted quickly. By some estimates, the rate of muscle glycogen depletion is 30 to 40 times higher during high-intensity activities such as sprinting, compared to low-intensity activities such as a brisk walk. Depletion of muscle glycogen during anaerobic metabolism can contribute to fatigue, because the rate at which skeletal muscles take up glucose from the blood is not rapid enough to sustain a high rate of physical exertion.

Recall that glycolysis is an anaerobic pathway in which glucose is broken down to two molecules of pyruvate, forming two molecules of ATP in the process. Oxygen availability then determines the fate of pyruvate. If oxygen is available, pyruvate can enter aerobic

## FIGURE 3    Effects of Exercise Duration and Intensity on Glucose Use

As the intensity of exercise increases, muscles rely more on glucose as an energy source.

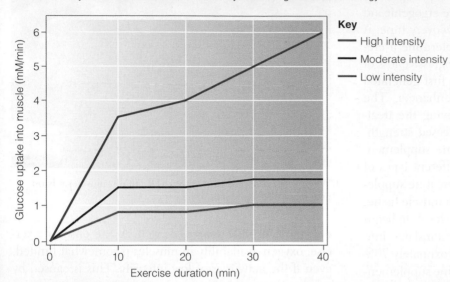

metabolic pathways (such as the citric acid cycle and the electron transport chain) to generate more ATP. If oxygen is limited, however, pyruvate is converted to lactate (also called lactic acid), as is the case during high-intensity workouts. In fact, a sudden burst of all-out exertion lasting 1 to 2 minutes can cause lactate levels to rise significantly (Figure 4). Thus, as exercise intensity increases, lactate formation does as well.

Because very little ATP is actually produced when glucose is converted to pyruvate, anaerobic metabolism

## FIGURE 4    Lactate Production During High-Intensity Exercise

During high-intensity workouts, blood lactate levels rise.

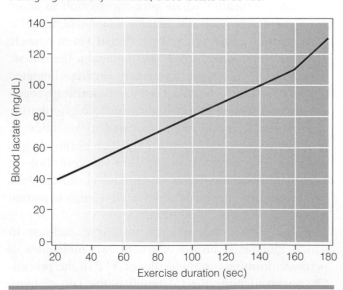

can fuel a high-intensity workout only for a short period of time. Moreover, the accumulation of lactate and other intermediate metabolic products creates an acidic environment (a lower pH), which in turn inhibits the activity of enzymes needed for further glycolysis. Without glycolysis, the muscle can no longer make ATP during anaerobic conditions, contributing to muscle fatigue. Furthermore, a drop in pH causes muscle contractions to weaken, again contributing to fatigue. This change in physiological pH also causes a burning sensation in active muscles, which is relieved when the muscle relaxes.

There is a tendency to view lactate as a metabolic waste product. However, lactate can be recycled into glucose and then "reused" as an energy source. For this to occur, muscles must relax and release lactate into the blood where it is circulated to the liver and converted to glucose via gluconeogenesis. Glucose is released into the blood and then taken up by muscles and other tissues. The release of lactate from the muscles into the blood, and its subsequent conversion to glucose by the liver, is called the Cori cycle. The Cori cycle helps increase glucose availability during periods of intense exercise.

**CONNECTIONS** Recall that the Cori cycle occurs when lactate is released into the blood, taken up by the liver, and converted into glucose via gluconeogenesis (Chapter 8, page 282).

## Aerobic Pathways: Long-Term Energy Systems

After several minutes of light activity (such as walking at a fast pace) to moderate activity (such as jogging and swimming), breathing becomes faster and harder. The heart begins to beat more frequently and more forcefully. These changes in pulmonary and cardiovascular function help increase the delivery of oxygen to muscles. With sufficient oxygen now available, muscle cells are better able to use aerobic metabolic pathways to produce ATP. Aerobic metabolic pathways are important in that substantial amounts of ATP can be generated over an extended period of time—an important advantage over anaerobic pathways.

Aerobic pathways use glucose, fatty acids, and—to a lesser extent—amino acids to generate ATP (Figure 5).[6] These energy sources are catabolized to produce $NADH + H^+$ and $FADH_2$, which in turn are used to generate ATP via the electron transport chain. Although these pathways are relatively slow in terms of the rate

## FIGURE 5    Relative Contributions of Energy Sources During Low-, Moderate-, and High-Intensity Exercise

Aerobic metabolic pathways used during low- to moderate-intensity exercise rely on glucose, fatty acids, and to a lesser extent amino acids, for energy. Anaerobic metabolic pathways used during high-intensity activity rely solely on glucose.

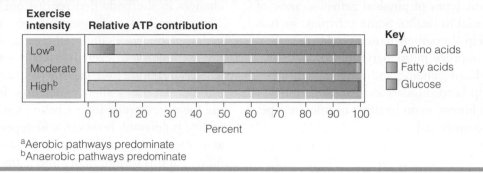

aAerobic pathways predominate
bAnaerobic pathways predominate

of ATP production, the energy yield is rich. Fatty acids used for energy during physical activity are derived primarily from triglycerides stored in adipose tissue. However, intramuscular triglycerides also help fuel activity. Aerobic pathways also use stored muscle glycogen and glucose circulating in the blood.

During prolonged activity, the combined efforts of gluconeogenesis and glycogenolysis help provide glucose to active muscles. However, once glycogen stores are exhausted glycogenolysis can no longer contribute to the glucose pool. Gluconeogenesis alone cannot provide glucose at the rate needed by skeletal muscles, especially those supporting endurance sports such as a marathon. At this point in the race some athletes lose their stamina or "hit the wall." This term is often used by athletes to describe the feeling of profound fatigue that can occur during an athletic event.

Consuming sports drinks, energy bars, or energy gels during exercise can help provide an additional source of glucose. These products provide carbohydrates in a form that is easily digested and rapidly absorbed. Sports drinks are particularly helpful because they not only provide glucose but also help keep athletes hydrated. However, too much fluid can cause cramping, so it is important for athletes to find out what products work best for them.

## ESSENTIAL *Concepts*

During energy metabolism, ATP can be generated in both the presence (aerobic) and relative absence (anaerobic) of oxygen. Aerobic pathways generate large amounts of ATP over an extended period of time, whereas anaerobic pathways generate smaller amounts of ATP quickly. Aerobic metabolic pathways include the citric acid cycle, β-oxidation, and the electron transport chain. Anaerobic metabolic pathways include the ATP creatine phosphate (ATP-CP) pathway and glycolysis. The ATP-CP pathway

serves as an immediate energy system, glycolysis as a short-term energy system, and aerobic pathways as long-term energy systems. The ATP-CP pathway uses creatine phosphate for rapid ATP generation, whereas glycolysis uses glucose. Both immediate and short-term energy systems operate during the first few minutes of exercise and for short bursts of high-intensity activity. The aerobic pathways use glucose, fatty acids, and—to a lesser extent— amino acids for low-intensity, long-duration activities. Although aerobic pathways are relatively slow in regard to the rate of ATP formation, the energy yield is rich.

# Physical Activity and Athletic Training

Physical activity and athletic training provide people with many physiological advantages, and overall improved health. Unfortunately, the majority of adults in the United States are not active enough to promote and maintain good health.[7] Regardless if a person exercises to stay physically fit, is a recreational athlete, or is a competitive athlete, the health-enhancing effects of regular exercise and training on the body are the same— improved muscular strength and endurance.

## Physical Activity and Health

In general, people who are physically active reduce their risk for certain chronic diseases such as high blood pressure, stroke, cardiovascular disease, type 2 diabetes, osteoporosis, and colon cancer. Unfortunately, these benefits are quickly lost once a person becomes inactive. For this reason, it is important to make exercise, along with good nutrition, part of our lifestyle.[8] The Dietary Guidelines for Americans stress the importance of regular physical activity and recommend that children, adolescents, and adults get at least 30 to 60 minutes of moderate physical activity on most days. The overall benefits of exercise on health and fitness are summarized in Table 1.

There are many types of physical activities, most of which are beneficial to health. Some activities, such as weight lifting, help strengthen muscles, whereas others such as jogging improve pulmonary (lung) and cardiovascular fitness. Even bending and stretching exercises such as yoga help keep muscles and joints flexible. To improve physical fitness, a combination of different types of exercise is recommended.

## Athletic Training and Performance

To optimize athletic performance, it is important for athletes to train. **Strength training** promotes muscle growth—or what is called hypertrophy—by challenging muscles with physically demanding, short-duration exercises such as weight training. These types of activities help increase muscle size and strength. Conversely, **endurance training** involves steady, low- to moderate-intensity activities that are longer in duration, such as running and swimming. Endurance training is beneficial to pulmonary and cardiovascular function, which in turn improve aerobic capacity. It also increases the number and size of mitochondria, improving the cell's ability to produce ATP. Some athletic trainers recommend **interval training**, which involves alternating short, fast bursts of intensive exercise with slower, less demanding activity. This type of training helps improve both aerobic and anaerobic capacity. However, it is important for athletes to avoid overtraining because there are potential risks associated with doing too much too soon. Specifically, training beyond the body's abil-

ity to recover can result in injury, muscle soreness, and joint pain.

Training is very important for athletes and can lead to what is called an **adaptation response**, defined as changes in the body that result from regular exercise. These changes help improve athletic conditioning and performance. For example, the expansion of blood vessels in response to exercise increases the body's ability to take in and deliver oxygen to muscles. In addition, muscles develop a higher tolerance for lactate, allowing them to exercise longer before fatigue sets in. Each athlete is different, however, and depending on the type and intensity of training, the adaptation response will differ. For this reason, most athletic trainers recommend that athletes follow a varied training regimen, alternating between workouts that increase both strength and stamina. The physiological response to athletic training and its impact on energy metabolism are discussed next.

### Training Spares Glucose

The ability of muscles to metabolize fatty acids for energy allows the body to use glucose more sparingly. This is particularly important for endurance athletes such as marathon runners, cyclists, and distance swimmers, who depend on stored glycogen for a source of

**strength training** Physical training that increases muscle growth and strength.

**endurance training** Physical training that improves pulmonary and cardiovascular function.

**interval training** Athletic training that alternates between fast bursts of intensive exercise and slower, less demanding activity.

**adaptation response** Physiological changes that result from exercise that help improve physical fitness.

## TABLE 1    Benefits of Exercise on Health and Physical Fitness

| Physiological Change Associated with Exercise | Health Benefit |
| --- | --- |
| Increased blood volume | Improved ability to deliver oxygen to tissues |
| Strengthened heart muscle | More efficient heart functioning |
| Lowered resting heart rate | Less exertion at rest |
| Increased stroke volume | More blood pumped by heart with less effort |
| Improved VO$_2$ max | Improved physical performance and endurance |
| Improved circulation | More oxygen and nutrients received by tissues |
| Increased energy expenditure | Weight loss facilitated and healthy body weight maintained |
| Reduced stress | Improved sleep |
| Improved mental health | Prevents/improves depression |
| Strengthened immune system | Reduced risk of certain types of cancer |
| Improved cardiovascular fitness | Lowered blood pressure |
| Reduced body weight | Reduced strain on joints, reduced risk for many chronic diseases |
| Increased bone density | Reduced risk for osteoporosis |
| Improved blood lipid levels | Decreased risk for heart attacks and strokes |
| Lowered blood glucose levels | Decreased risk for type 2 diabetes and reduced need for medication used to treat type 2 diabetes |

Magnetic resonance imaging (MRI) scans provide cross-section images of muscle shape, bones, and blood vessels. The first two men are physically fit (ages 37 and 60 years, respectively) and exercise regularly. The third man is 35 years of age and leads a sedentary lifestyle.

glucose. Recall that only small amounts of glycogen are stored in the liver and skeletal muscles. Thus the preferential use of fatty acids, of which the supply is virtually unlimited, over glucose is advantageous. Studies have shown that trained athletes use glucose more sparingly and are better able to use fatty acids as an energy source than are untrained individuals.[9] This is because training increases the number and size of mitochondria, the location of aerobic metabolism in muscle.[10] Recall that the use of fatty acids for energy involves β-oxidation and the citric acid cycle—both of which occur in the mitochondria. The increased capacity to use fatty acids as an energy source helps spare glucose, in turn, helping to delay fatigue.

## Training Improves Cardiovascular Function

As the intensity and duration of workouts increase, muscles require significantly more oxygen for ATP production via aerobic pathways. As such, the cardiovascular system must work even harder to increase oxygen delivery to muscle cells. During a hard physical workout, for example, increased heart rate (beats per minute), **ventilation rate** (breaths per minute), and **stroke volume** (the amount of blood pumped per heartbeat) increase the amount of oxygen and nutrients delivered to muscles. Physical training improves cardiac function and blood delivery to muscle tissues. Although at times workouts are so intense and muscle contractions so powerful that the oxygen demand cannot be met, trained athletes with strong cardiovascular function have a clear competitive edge over untrained athletes.

Training increases what is called **maximal oxygen consumption,** also called VO₂ max, which is the maximum ability of the cardiovascular system to deliver oxygen to muscles and the ability of cells to use oxygen to

generate ATP. A trained athlete has a higher VO₂ max than does an untrained individual, which helps delay the onset of fatigue. There are many reasons why training improves VO₂ max. For example, it strengthens the cardiac (heart) muscle, increasing its size and ability to contract forcefully. The heart of a trained athlete is more efficient, thus increasing stroke volume. Because of this, athletes often have lower **resting heart rates** than nonathletes—a sign of physical fitness. In addition, training causes an expansion of capillary vessels supplying blood to muscles—another example of the adaptation response. This aids blood flow and thus delivery of nutrients and oxygen to muscle cells. Training even increases production of red blood cells, which increases the oxygen-carrying capacity of the blood.

## ESSENTIAL *Concepts*

Regardless of activity level, regular exercise is beneficial to health. Because of adaptation responses, athletic performance improves as a result of training. Muscles of trained athletes use glucose more sparingly, and they are better able to use fatty acids as an energy source. Training also increases the number and size of mitochondria, the location of aerobic metabolism. Training increases maximal oxygen consumption (called VO₂ max), and a higher VO₂ max helps delay fatigue. Training strengthens the heart, resulting in more forceful and efficient cardiac function and better nutrient and oxygen delivery to muscles. The expansion of capillary blood vessels in muscles increases blood flow to muscle cells. The increased production of red blood cells in response to training further increases the oxygen-carrying capacity of the blood.

## Dietary Needs of Athletes

It is important for athletes to have the right balance of nutrients, energy, and fluids during training, competition, and recovery. Although no single dietary plan is right for everyone, it is helpful for athletes to be familiar with general guidelines of meal planning and food selection. For the most part, athletes must take in enough water to replace that lost during training and competition, and enough energy to meet the demands of physical activity and prevent weight loss. Although exercise may slightly

**ventilation rate** Rate of breathing.

**stroke volume** The amount of blood pumped out of the heart with each contraction.

**maximal oxygen consumption (VO₂ max)** Maximum volume of oxygen that can be delivered per unit of time.

**resting heart rate** Number of heartbeats per minute measured at rest.

increase micronutrient requirements, these needs are readily met when the person consumes adequate and balanced amounts of food.[11]

## Energy Requirements

It is critical that athletes, especially those who train heavily, have adequate energy intakes. Without enough energy to meet the added demands of physical activity, an athlete can lose muscle and hence strength and endurance. In addition, athletes who consume low-energy diets often lack important micronutrients needed for energy metabolism, hemoglobin synthesis, bone health, protection from oxidative stress, and immune function.

To satisfy energy requirements, athletes must consume enough energy to support normal daily activity as well as that required for exercise. Thus the frequency, duration, and intensity of exercise must be considered in determining energy needs. Estimated energy expenditure associated with various activities is listed in Table 2.

For example, a male who weighs 155 pounds (70 kg) and runs for 1 hour at a 10-minute/mile pace would expend approximately 722 kcal in addition to energy required for normal daily activity. This athlete would require a total of about 3,272 kcal per day—enough energy for normal daily activity (2,550 kcal) and physical activity (722 kcal).

Athletes who are training on a regular basis can have a very high total energy requirement, especially if they are training for long periods of time. In fact, some athletes may require 4,000 to 5,000 kcal per day. Mathematical formulas developed by the Institute of Medicine as part of the Dietary Reference Intakes (DRIs) are often used to estimate total energy requirements, as was discussed in Chapter 2.[12]

To satisfy energy requirements, athletes are encouraged to eat a variety of nutrient-dense foods such as whole grains, cereals, legumes, fruits, vegetables, lean meat, fish, poultry, and low-fat dairy products. In fact, there are few differences in diets recommended for athletes and nonathletes—athletes simply require more of the

## TABLE 2    Energy Expenditure (kcal/hour) for Selected Physical Activities in Relation to Body Weight

| | Body Weight | | | | | | |
| Activity | 50 kg 110 lb. | 57 kg 125 lb. | 64 kg 140 lb. | 70 kg 155 lb. | 77 kg 170 lb. | 84 kg 185 lb. | 91 kg 200 lb. |
|---|---|---|---|---|---|---|---|
| **Aerobic Dance** | | | | | | | |
| Male | 480 | 488 | 506 | 531 | 556 | 582 | 607 |
| Female | 394 | 413 | 433 | 453 | 472 | 492 | 511 |
| **Biking (12 mph)** | | | | | | | |
| Male | 380 | 401 | 422 | 443 | 464 | 486 | 507 |
| Female | 329 | 345 | 361 | 379 | 394 | 410 | 427 |
| **Gardening** | | | | | | | |
| Male | 303 | 320 | 337 | 354 | 371 | 388 | 405 |
| Female | 263 | 276 | 289 | 302 | 315 | 328 | 341 |
| **Golf** | | | | | | | |
| Male | 425 | 448 | 472 | 496 | 519 | 543 | 567 |
| Female | 368 | 386 | 404 | 422 | 441 | 459 | 477 |
| **Running (10 min/mi)** | | | | | | | |
| Male | 619 | 653 | 688 | 722 | 757 | 791 | 826 |
| Female | 536 | 562 | 589 | 615 | 642 | 669 | 695 |
| **Swimming (moderate)** | | | | | | | |
| Male | 364 | 384 | 405 | 425 | 445 | 465 | 486 |
| Female | 315 | 331 | 346 | 362 | 378 | 393 | 409 |
| **Walking (15 min/mi)** | | | | | | | |
| Male | 257 | 271 | 285 | 300 | 314 | 328 | 342 |
| Female | 222 | 233 | 244 | 255 | 266 | 277 | 288 |
| **Weight Lifting** | | | | | | | |
| Male | 340 | 359 | 378 | 397 | 415 | 434 | 453 |
| Female | 294 | 309 | 323 | 338 | 352 | 367 | 382 |

SOURCE: ESHA Research, Salem, Oregon.

## FIGURE 6 Analysis of Energy and Nutrient Requirements of a Male Runner Based on MyPyramid

The daily food record indicates that this athlete is satisfying his need for energy and nutrients by eating a variety of nutrient-dense foods.

**Food record**

| FOODS CONSUMED | AMOUNT CONSUMED |
|---|---|
| Milk (2%) | 16 fluid ounces |
| Cheddar cheese | 4 tbsp |
| Yogurt, fruit, low-fat | 8 oz. |
| Fish, broiled | 7 oz. |
| Bread, cracked wheat | 4 slices |
| Cookie, chocolate chip | 1 each |
| Crackers, wheat | 1/4 cup |
| Pretzels, hard | 1/2 cup |
| Rice, brown prepared with butter | 2 cups |
| Shredded wheat cereal | 2 cups |
| Orange, fresh | 1 each |
| Raisins | 3 tbsp |
| Grapes, fresh | 1 cup |
| Honeydew melon, fresh | 1 cup |
| Broccoli, raw | 1 cup |
| Carrots, raw | 1/2 cup |
| Tomatoes, raw | 1/2 cup |
| Mixed salad greens, raw | 2 cups |
| Butter | 1 tbsp |
| Salad dressing | 1 tbsp |

**Diet analysis**

| NUTRIENT/ ENERGY | ESTIMATED INTAKE | RECOMMENDATION OR ACCEPTABLE RANGE* |
|---|---|---|
| Energy (kcal) | 3,406 | 3,341 |
| Protein (g) | 189 | 56 |
| Carbohydrate (g) | 491 | 130 |
| Total fiber (g) | 45 | 38 |
| Total fat (g) | 86 | 76–133 |
| Saturated fat (g) | 33 | <37.8 |
| Monounsaturated fat (g) | 26 | — |
| Polyunsaturated fat (g) | 19 | — |
| Vitamin C (mg) | 251 | 90 |
| Folate (µg) | 615 | 400 |
| Calcium (mg) | 1,776 | 1,000 |
| Iron (mg) | 11 | 8 |
| Zinc (mg) | 55 | 11 |

*Based on RDA, AI, AMDR

**Pyramid results**

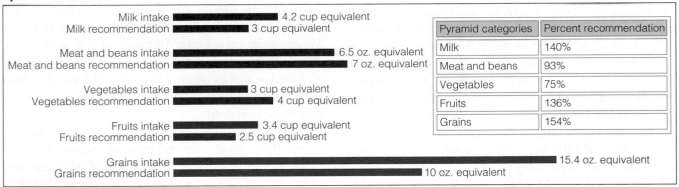

| Pyramid categories | Percent recommendation |
|---|---|
| Milk | 140% |
| Meat and beans | 93% |
| Vegetables | 75% |
| Fruits | 136% |
| Grains | 154% |

Milk intake 4.2 cup equivalent
Milk recommendation 3 cup equivalent

Meat and beans intake 6.5 oz. equivalent
Meat and beans recommendation 7 oz. equivalent

Vegetables intake 3 cup equivalent
Vegetables recommendation 4 cup equivalent

Fruits intake 3.4 cup equivalent
Fruits recommendation 2.5 cup equivalent

Grains intake 15.4 oz. equivalent
Grains recommendation 10 oz. equivalent

same types of foods. To plan a diet that provides adequate energy and nutrients, athletes are encouraged to visit the MyPyramid website, as illustrated in Figure 6.

## Recommendations for Macronutrient Intake

Not only must athletes consider total energy requirements, but the source of these calories is equally important. As previously discussed, which fuels are used during different stages of exercise depend on the intensity and duration of the activity. When the physical activity is brief and intense, muscles rely mostly on glucose for an energy source, because tissues are relatively anaerobic. However, the source of energy shifts in relation to the duration and intensity of the activity. For example, fatty acids can be important energy sources during

exercise, but the extent to which they are used depends on exercise intensity. At high intensities, the need for ATP exceeds ATP availability in the muscle. As a result, muscles increase their use of anaerobic pathways, which generate ATP quickly. Therefore muscles must depend less on fatty acids (which are metabolized by aerobic pathways) and more on glucose (which is metabolized by anaerobic pathways). Amino acids are also used to fuel activity, albeit to a lesser extent than glucose and fatty acids.

In recent years, considerable attention has been given to the "ideal" distribution of total calories in the diet of athletes. Recall that the Dietary Guidelines suggest that 45 to 65% of total calories come from carbohydrates, 20 to 35% from fat, and 10 to 35% from protein. It is important for athletes to take in enough carbohydrates to prevent fatigue and maintain liver and muscle glycogen stores. As discussed in Chapter 5, some athletes try to increase the amount of glycogen stored in muscles, using a technique called carbohydrate loading. The Institute of Medicine (IOM) does not provide any special recommendations concerning carbohydrate intake for athletes as part of the DRIs. However, the American Dietetic Association recommends that athletes consume 6 to 10 g of carbohydrates per kilogram of body weight.[13] Thus, recommended daily carbohydrate intake for a 200-lb (91-kg) athlete is approximately 546 to 910 g. This amount depends somewhat on total daily energy expenditure, sex, and the type of sport.

> **CONNECTIONS** ▶ Recall that the Dietary Reference Intakes (DRIs) are reference values for nutrient intakes (Chapter 2, page 36).

Athletes also require considerable amounts of protein to maintain, build, and repair muscles. There is an ongoing debate about whether athletes require more protein than do nonathletes. Protein recommendations for athletes typically range from 1.2 to 1.7 g of protein per kilogram of body weight per day.[14] However, the Institute of Medicine recently concluded that the estimated protein requirements for healthy adults (0.8 g of protein per kilogram of body weight per day) are adequate to meet increased protein requirements of physically active individuals as well.[15] Until more information suggests otherwise, most experts agree that this amount is likely to be adequate to maintain protein status. For example, consider an athlete who weighs 155 pounds (70 kg) and requires 4,000 kcal per day. If 10 to 20% of total energy intake comes from protein, this person would consume 400 to 800 kcal (100 to 200 g of protein) from protein each day. This is equivalent to 1.4 to 2.8 g of protein per kilogram of body weight, a value well above that thought to be sufficient (0.8 g per kilogram body weight). Thus if the energy intake of athletes is high enough, the amount of protein needed to maintain, build, and repair tissue will likely be adequate.[16]

Similarly, it is important for athletes to consume adequate amounts of fat. This is because fat is an important source of energy, provides essential fatty acids, and is necessary for the absorption of fat-soluble substances in the body. Recommendations regarding the amount and types of dietary fats are the same for athletes and nonathletes. That is, foods rich in monounsaturated (such as olive and canola oils) and polyunsaturated fatty acids (such as vegetable and fish oils) should be emphasized, whereas saturated and *trans* fatty acids should be limited.

## Micronutrient Requirements

As long as an athlete consumes a varied diet that supplies adequate energy, vitamin and mineral needs will likely be met as well. Micronutrients are needed for many functions, including energy metabolism, repair and maintenance of body structures, protection from oxidative damage, and immunity. Supplemental nutrient intake to enhance athletic performance is controversial, and there is little evidence regarding the efficacy of this practice, with the exception of a demonstrated nutrient deficiency.[17]

Athletes who consume foods with low nutrient densities or low-energy foods, as well as athletes who restrict or avoid animal products such as meat, dairy, and eggs, are at increased risk for low intakes of several micronutrients—especially iron, calcium, and zinc. For this reason, it is important for coaches and trainers to be aware of nutritional issues related to athletes who follow strict vegetarian diets.[18]

## Iron

Iron is an important component of hemoglobin, the iron-containing component of red blood cells. Hemoglobin delivers oxygen to cells and picks up carbon dioxide, a waste product of energy metabolism. Iron deficiency can therefore compromise oxygen delivery to tissues. Because oxygen is required for aerobic energy metabolism, poor iron status in athletes can seriously affect training and performance. Vegetarian athletes are at greatest risk for impaired iron status. Although iron is available in cereals, grains, and some vegetables, it is not as readily absorbed as that in meat. Therefore, athletes who don't eat meat should have their blood tested periodically to determine iron status.

Endurance athletes may have higher iron requirements than nonathletes because of increased loss of blood in the feces, iron loss associated with excessive sweating, and rupturing of red blood cells as a result of the repeated striking of the feet against hard surfaces.[19] Because of blood loss associated with menstruation, female athletes are more likely to experience impaired iron status than are males. Unfortunately, very few collegiate athletic programs routinely screen for iron deficiency in female athletes.[20]

In addition to iron deficiency caused by a lack of iron in the diet, some athletes experience a temporary form of anemia referred to as **sports anemia**. Sports anemia often occurs during the onset of training and is caused by a disproportionate increase in plasma volume compared to red blood cells. This **hemodilution** may make the person appear iron deficient when he or she is not. Because

**sports anemia** A physiological response to training caused by a disproportionate increase in plasma volume relative to the number of red blood cells.

**hemodilution** A decrease in the number of red blood cells per volume of plasma caused by plasma volume expansion.

sports anemia is due to a healthy physiological response to training rather than poor nutritional status, there is no need for additional iron in the diet. Unlike true iron deficiency anemia, sports anemia does not appear to hinder athletic performance.

CONNECTIONS ► Recall that plasma is the liquid portion of blood (Chapter 3, page 82).

## Calcium

Although recommended calcium intakes for athletes and nonathletes are the same, adequate calcium intake is especially important for optimal athletic performance. This is because calcium is needed for building and repair of bone tissue. Thus diets that are low in calcium increase an athlete's risk for low bone density, which in turn increases the likelihood of stress fractures. Stress fractures are small cracks that occur in bones, especially of the feet, in response to repeated jarring. Female athletes who have menstrual irregularities are at particularly high risk for bone loss. In fact, female athletes who exercise excessively while restricting their intake of calories are at risk for a syndrome referred to as the female athlete triad, which is addressed further in the Nutrition Matters following Chapter 9. The female athlete triad is characterized by bone loss, disordered eating practices, and amenorrhea—the cessation of menses for three months or more.

Athletes who restrict their intake of dairy products may benefit from foods that are fortified with calcium and vitamin D, which facilitates calcium absorption. These include fortified orange juice, soy products, and breakfast cereals. When dietary intake of calcium-rich foods is lacking, athletes should consider taking a calcium supplement.[21]

## Zinc

Most (70%) of the zinc in the American diet comes from animal products, primarily meat. Thus athletes who restrict their intake of animal products are at increased risk for not getting enough zinc. This mineral plays an important role in energy metabolism and is needed for the maintenance, repair, and growth of muscles, making zinc an especially important mineral for athletes. Vegetarian sources of zinc include legumes, hard cheeses, whole-grain products, wheat germ, fortified cereals, nuts, and soy products.

## Fluid and Electrolyte Requirements

Exercise increases body temperature, and the body dissipates this heat by sweating. Sweat consists mainly of water but also contains substantial amounts of electrolytes such as sodium, chloride, potassium, and to a lesser extent minerals such as iron and calcium. For this reason, extensive sweating can disrupt the balance of fluids and electrolytes in the body. When this occurs, an athlete becomes dehydrated. Dehydration is caused by a lack of fluid intake relative to fluid loss from the body. To meet the body's need for water, most adult men and women need approximately 3.7 and 2.7 liters of water per day, respectively. However, athletes can require substantially more water than this.[22]

Dehydration can reduce blood volume, interfering with the body's cooling system. That is, decreased blood volume reduces blood flow to the skin, which in turn affects the body's ability to dissipate heat. As a result, core body temperature can rise, increasing the risk of heat exhaustion and heat stroke. **Heat exhaustion** can occur when a person loses 5% of his or her body weight due to sweating.[23] When this happens, an athlete often feels ill and experiences muscle spasms, a rapid and weak pulse, low blood pressure, disorientation, and profuse sweating (see Table 3). Heat exhaustion can easily progress into heat stroke if fluid loss continues and the athlete is unable to lower his or her body temperature. **Heat stroke** occurs when 7 to 10% of body weight is lost due to sweating. This very serious condition is characterized by dry skin, confusion, and loss of consciousness. A person experiencing heat stroke requires immediate medical assistance.

Dehydration not only impairs the body's ability to regulate its temperature but can also cause a number of other problems. For example, blood sodium concentration can decrease when an athlete experiences excessive sweating or when large amounts of fluids are consumed without adequate sodium intake. This condition is called hyponatremia. When this occurs, fluids leave the bloodstream and accumulate in surrounding tissues. The accumulation of fluids in lung tissue can impair gas exchange, raising carbon dioxide levels in the blood. Because drinking large amounts of plain water may actually increase risk for hyponatremia, many experts recommend that endurance athletes replenish body fluids by consuming drinks that contain both carbohydrates and electrolytes, such as sports drinks.[24]

## Preventing Dehydration

It is important for athletes to prevent dehydration by consuming adequate amounts of fluids before, during, and after exercise. The National Athletic Trainers' Association (NATA) recommends that athletes consume 2 to 3 cups of fluid two to three hours before exercise.

**heat exhaustion** Rise in body temperature that occurs when the body has difficulty dissipating heat.

**heat stroke** Rise in body temperature due to excess loss of body fluids associated with sweating.

**TABLE 3   Signs and Symptoms of Dehydration**

| Early or Mild Dehydration | Moderate to Severe Dehydration |
|---|---|
| • Flushed face<br>• Extreme thirst<br>• Dry, warm skin<br>• Cannot pass urine or produces reduced amounts (dark, yellow)<br>• Dizziness made worse when standing<br>• Weakness<br>• Cramping in arms and legs<br>• Crying with few or no tears<br>• Difficulty concentrating<br>• Sleepiness<br>• Irritability<br>• Headache<br>• Dry mouth and tongue, with thick saliva | • Low blood pressure<br>• Fainting<br>• Disorientation, confusion, slurred speech<br>• Severe muscle contractions in the arms, legs, stomach, and back<br>• Convulsions<br>• Bloated stomach<br>• Heart failure<br>• Sunken, dry eyes<br>• Skin that has lost its firmness and elasticity<br>• Rapid and deep breathing<br>• Fast, weak pulse |

This ensures that the athlete is fully hydrated while allowing time for excess fluids to be excreted from the body. Athletes training in hot and humid climates are at increased risk of dehydration and therefore may require slightly greater fluid intakes.

Because dehydration can compromise performance, it may be necessary for athletes to consume fluids during exercise. For exercise that lasts less than one hour, plain water is adequate to replenish body fluids. However, if workouts last longer than one hour athletes may benefit from drinking fluids (such as sports drinks) that contain small amounts of carbohydrates and electrolytes.[25] Sports drinks are readily available and well tolerated during exercise. A typical commercial sports beverage contains 10 to 20 g of carbohydrates (40 to 80 kcal) per 8 fluid ounces.

Athletes should be aware that some sports beverages, dubbed as "energy" drinks, contain caffeine. Caffeine is a stimulant and, in moderation, generally has no adverse health effects. Nonetheless, caffeine affects people differently. Whereas some people enjoy the mild stimulatory effect of caffeine, others find that it causes stomach upset,

nervousness, and jitters. As of 2004, the International Olympic Committee (IOC) no longer considers caffeine a prohibited substance.

It is also important that athletes replenish body fluids after exercise. This is best done by consuming 2 to 3 cups of fluids for each pound of weight lost during exercise. Although sweat contains significant amounts of sodium, usually enough sodium is present in foods consumed after exercise to restore electrolyte balance. Therefore, it is neither necessary nor recommended that athletes take sodium (salt) supplements.[26]

## ESSENTIAL *Concepts*

Athletes require enough fluids and energy to meet the demands of physical activity. When energy needs are not met, an athlete can lose muscle and hence lose strength and endurance. Athletes are encouraged to eat an adequate amount of nutrient-dense foods and to emphasize variety in their choices. Carbohydrates are needed to maintain glycogen stores, whereas protein maintains, builds, and repairs muscles. Dietary fat is an important source of energy, providing essential fatty acids and facilitating the use of fat-soluble substances. Low-energy diets often lack micronutrients, which are needed for energy metabolism, hemoglobin synthesis, bone health, protection from oxidative stress, and immune function. Athletes who consume low-energy foods and those who restrict animal products tend to have low intakes of iron, calcium, and zinc. Iron is necessary for oxygen delivery to tissues, and calcium is needed for building and repairing bone tissue. Zinc is needed for energy metabolism and for the maintenance, repair, and growth of muscles. Excess sweating can disrupt fluid and electrolyte balance, impairing body temperature and fluid regulation. Athletes can prevent dehydration by consuming adequate amounts of fluids before, during, and after exercise. To stay fully hydrated, athletes should consume 2 to 3 cups of fluid two to three hours before exercise. After exercise, athletes should replenish body fluids by consuming 2 to 3 cups of fluids for each pound of weight lost.

It is important for athletes to prevent dehydration by consuming adequate amounts of fluids before, during, and after exercise.

# Notes

1. De Feo P, Di Loreto C, Lucidi P, Murdolo G, Parlanti N, De Cicco A, Piccioni F, Santeusanio F. Metabolic response to exercise. Journal of Endocrinological Investigation. 2003;26:851–4.

2. Bizzarini E, DeAngelis L. Is the use of oral creatine supplementation safe? Journal of Sports Medicine and Physical Fitness. 2000;44:411–6. Racette SB. Creatine supplementation and athletic performance. Journal of Orthopedic Sports and Physical Therapy. 2003;33:615–21. Volek JS, Rawson ES. Scientific basis and practical aspects of creatine supplementation for athletes. Nutrition. 2004;20:609–14.

3. Kreider RB. Effects of creatine supplementation on performance and training adaptations. Molecular and Cellular Biochemistry. 2003;244:89–94.

4. Reddy GV, Indira K, Ramanjaneyulu PS, Rao SV. Creatine metabolism of uremic rats. Biochemistry International. 1992;26:343–6.

5. Mayhew DL, Mayhew JL, Ware JS. Effects of long-term creatine supplementation on liver and kidney functions in American college football players. International Journal of Sports Nutrition and Exercise Metabolism. 2002;12:453–60.

6. Hagerman FC. Energy metabolism and fuel utilization. Medicine and Science in Sports and Exercise. 1992;24:S309–14.

7. Macera CA, Ham SA, Yore MM, Jones DA, Ainsworth BE, Kimsey CD, Kohl HW 3rd. Prevalence of physical activity in the United States: Behavioral Risk Factor Surveillance System, 2001. Preventing Chronic Disease. 2005;2:A17–A26.

8. Vuori IM. Health benefits of physical activity with special reference to interaction with diet. Public Health Nutrition. 2001;4:517–28.

9. van Loon LJ, Jeukendrup AE, Saris WH, Wagenmakers AJ. Effect of training status on fuel selection during submaximal exercise with glucose ingestion. Journal of Applied Physiology. 1999;87:1413–20.

10. Jones TE, Baar K, Ojuka E, Chen M, Holloszy JO. Exercise induces an increase in muscle UCP3 as a component of the increase in mitochondrial biogenesis. American Journal of Physiology, Endocrinology, and Metabolism. 2003;284:E96–E101.

11. Position of the American Dietetic Association, Dietitians of Canada, and the American College of Sports Medicine: Nutrition and athletic performance. Journal of the American Dietetic Association. 2000;100:1543–56.

12. Institute of Medicine. Dietary Reference Intakes for energy, carbohydrate, fiber, fat, fatty acids, cholesterol, protein, and amino acids. Washington, DC: National Academies Press; 2005.

13. Position of the American Dietetic Association, Dietitians of Canada, and the American College of Sports Medicine: nutrition and athletic performance. Journal of the American Dietetic Association. 2000;100:1543–56.

14. Tipton KD, Wolfe RR. Protein and amino acids for athletes. Journal of Sports Science. 2004;22:65–79.

15. Lemon PW. Is increased dietary protein necessary or beneficial for individuals with a physically active lifestyle? Nutrition Reviews. 1996;54:S169–S75.

16. Tarnopolsky M. Protein requirements for endurance athletes. Nutrition. 2004;20:662–8.

17. Maughan RJ, King DS, Lea T. Dietary supplements. Journal of Sports Science. 2004;22:95–113.

18. Barr SI, Rideout CA. Nutritional considerations for vegetarian athletes. Nutrition. 2004;20:696–703.

19. Suedekum NA, Dimeff RJ. Iron and the athlete. Current Sports Medicine Reports. 2005;4:199–202.

20. Cowell BS, Rosenbloom CA, Skinner R, Summers SH. Policies on screening female athletes for iron deficiency in NCAA division I-A institutions. International Journal of Sport Nutrition and Exercise Metabolism. 2003;13:277–85.

21. Kunstel K. Calcium requirements for the athlete. Current Sports Medicine Reports. 2005;4:203–6.

22. Sawka MN, Cheuvront SN, Carter R 3rd. Human water needs. Nutrition Reviews. 2005;63:S30–9.

23. Shirreffs SM. The importance of good hydration for work and exercise performance. Nutrition Reviews. 2005;63:S14–S21.

24. Denny S. What are the guidelines for prevention of hyponatremia in individuals training for endurance sports, as well as other physically active adults? Journal of the American Dietetic Association. 2005;105:1323. Hsieh M. Recommendations for treatment of hyponatraemia at endurance events. Sports Medicine. 2004;34:231–8.

25. Convertino VA, Armstrong LE, Coyle EF, Mack GW, Sawka MN, Senay LC Jr, Sherman WM. American College of Sports Medicine position stand. Exercise and fluid replacement. Medicine and Science in Sports and Exercise. 1996;28:1–7.

26. Speedy DB, Thompson JM, Rodgers I, Collins M, Sharwood K, Noakes TD. Oral salt supplementation during ultradistance exercise. Clinical Journal of Sport Medicine. 2002;12:279–84.

# Chapter 9

# Energy Balance and Body Weight Regulation

In Chapter 8 you learned about energy metabolism—the chemical reactions that enable cells to obtain and use energy from nutrients. We also examined how different macronutrients are used in various states of energy availability. This chapter builds on that understanding by taking a closer look at the relationship between energy intake and energy expenditure—how the foods you eat and the activities you engage in affect body weight and body composition. In addition, this chapter also examines how body weight is regulated, the growing prevalence of obesity in the United States, and various approaches to weight loss and weight maintenance.

© Matthew Farruggio; (inset) © USDA

# Energy Balance

Energy balance, the relationship between energy intake and energy expenditure, determines whether the energy in the foods we eat is used for cellular activity or stored in the body for later use. To express the relationship between energy intake and energy expenditure, a simple equation is often used. When energy intake equals energy expenditure, a person is said to be in a state of **energy balance** (energy intake = energy expenditure). When more energy is consumed than the body needs, a person is in **positive energy balance** (energy intake > energy expenditure). Conversely, a person is in **negative energy balance** when energy intake is less than energy expenditure (energy intake < energy expenditure). Under most conditions, change in body weight is a useful indicator of whether a person is in positive or negative energy balance.

## Energy Balance and Body Weight

When energy intake equals energy expenditure, body weight tends to be stable. However, body weight generally increases in response to positive energy balance and decreases during negative energy balance. The relationship between energy balance and body weight is illustrated in Figure 9.1. When a person is in positive energy balance, muscle and adipose tissue—or both—increase. This is desirable during periods of growth such as infancy, childhood, adolescence, athletic training, and pregnancy. However, when increased body weight is primarily associated with increased body fat, positive energy balance is considered unhealthy.

Negative energy balance is a result of decreased energy intake, increased energy expenditure, or both. Under this condition, the mobilization and metabolism of stored energy reserves results in a decrease in body weight. Although some muscle and water are lost during negative energy balance, adipose tissue serves as the body's primary energy reserve. This is because adipose tissue is the body's largest energy depot; 1 pound of adipose tissue, including supporting lean tissue, is equivalent to approximately 3,500 kcal.

## A Closer Look at Adipose Tissue

Adipose tissue is often thought of as a passive reservoir that buffers imbalances between energy intake and expenditure. However, adipose tissue is anything but passive. Instead, it plays an active role in regulating energy balance via the production of hormones. Some experts consider adipose tissue to be the largest endocrine organ in the body.[1]

Adipose tissue is made up of specialized lipid-filled cells called adipocytes, which consist of a central core of triglyceride. The number and size of adipocytes determine the mass of adipose tissue in the body. As an adipocyte fills with triglycerides, its size increases, a process called **hypertrophic growth.** To accommodate large amounts of lipid, the diameter of an adipocyte can increase 20-fold. When existing cells are full, new adipocytes are formed, a process called **hyperplastic growth.**

When a person loses body fat, adipocytes shrink in size; however, the number of adipocytes remains constant. Some scientists believe this may be why formerly obese people, specifically those with a greater number of adipocytes, have difficulty maintaining weight loss. Figure 9.2 illustrates how adipocyte number and size change in response to weight gain and loss.

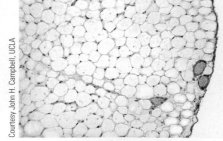

*Courtesy John H. Campbell, UCLA*

**Adipose tissue consists of cells called adipocytes. The accumulation of fat droplets in adipocytes gives adipose tissue a unique look under the microscope.**

**energy balance** A state in which energy intake equals energy expenditure.

**positive energy balance** A state in which energy intake is greater than energy expenditure.

**negative energy balance** A state in which energy intake is less than energy expenditure.

**hypertrophic growth** Growth associated with an increase in cell size.

**hyperplastic growth** Growth associated with an increase in cell number.

## FIGURE 9.1 Energy Balance

Energy balance is the relationship between energy intake and energy expenditure.

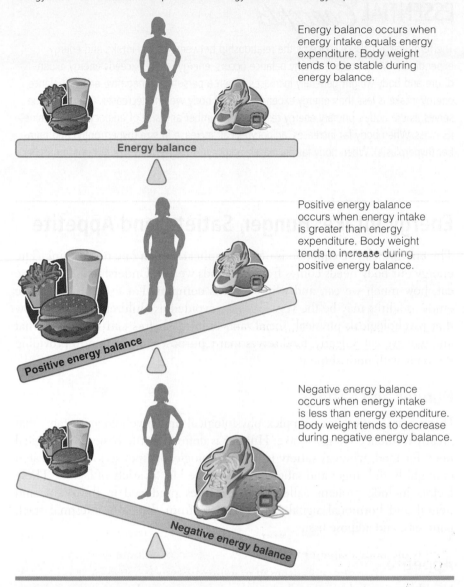

Energy balance occurs when energy intake equals energy expenditure. Body weight tends to be stable during energy balance.

**Energy balance**

Positive energy balance occurs when energy intake is greater than energy expenditure. Body weight tends to increase during positive energy balance.

**Positive energy balance**

Negative energy balance occurs when energy intake is less than energy expenditure. Body weight tends to decrease during negative energy balance.

**Negative energy balance**

## FIGURE 9.2 Adipose Tissue

The amount of adipose tissue a person has depends on the number and size of his or her adipose cells (adipocytes).

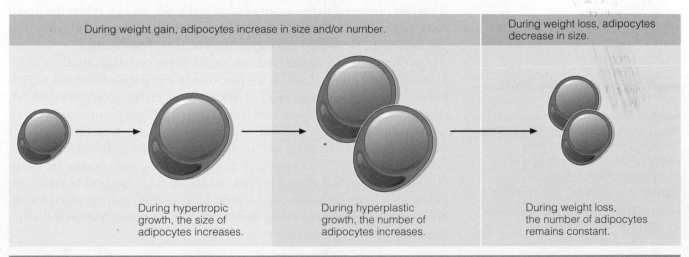

During weight gain, adipocytes increase in size and/or number.

During weight loss, adipocytes decrease in size.

During hypertropic growth, the size of adipocytes increases.

During hyperplastic growth, the number of adipocytes increases.

During weight loss, the number of adipocytes remains constant.

# ESSENTIAL *Concepts*

The term *energy balance* concerns the relationship between energy intake and energy expenditure. When positive energy balance occurs, energy intake exceeds energy expenditure and body weight generally increases. When a person is in negative energy balance, energy intake is less than energy expenditure, and body weight decreases. Adipose tissue serves as the body's primary energy reserve. The number and size of adipocytes determine its mass. When body fat increases, adipocytes can increase in size (hypertrophy) and number (hyperplasia). When body fat decreases, adipocytes decrease in size but not in number.

# Energy Intake: Hunger, Satiety, and Appetite

The energy balance equation is more complicated than one might think. The energy our body needs comes from the foods we eat. Understanding what we eat, how much we eat, and why we eat is complex. For example, foods we enjoy as adults may be the very foods we avoided as children. We also know that psychological, physical, social, and cultural factors can influence what and why we eat. Clearly, food serves many purposes in addition to providing the body with nourishment.

## Hunger and Satiety

Hunger and satiety are complex physiological and psychological states that greatly influence energy intake. **Hunger** is defined as the basic physiological need for food, whereas **satiety** is the physiological response to having eaten enough. Both hunger and satiety are influenced by a variety of factors. These factors include proteins called **neuropeptides** produced in the brain, and neural and hormonal signals that originate from the gastrointestinal tract, pancreas, and adipose tissue.

## Appetite

Healthy eating behaviors include eating in response to hunger cues and stopping in response to satiety cues. However, sometimes we eat for reasons other than hunger, and at times we eat past satiety. The stimuli that override hunger and satiety tend to be more psychological than physiological and are referred to collectively as **appetite**. Appetite, defined as the craving or desire for food, is also an important determinant of what and when we eat. When the desire for a specific food is compelling, it is often referred to as a **food craving**. You can read more about food cravings in the Focus on Food feature.

Appearance, taste, aroma, and emotional states can dramatically affect a person's appetite. In fact, aroma is a particularly strong stimulus in this regard. For example, the pleasing smell of baked bread makes people want to eat, regardless of hunger. Likewise, unpleasant odors can "spoil" our appetite, even when we are hungry. In addition, some people respond to emotions such as fear, depression, excitement, and stress by eating, whereas others respond by not eating at all. In this way, emotions can influence our appetite for foods, overriding our normal cues of hunger and satiety. This can lead to eating disorders such as anorexia nervosa, bulimia, binge eating, and restrained eating. You can read more about eating disorders in the Nutrition Matters following this chapter.

**hunger** The physiological drive to consume food.

**satiety** The state in which hunger is satisfied and a person feels he or she has had enough to eat.

**neuropeptide** A hormone-like protein released by nerve cells.

**appetite** A psychological desire for food.

**food craving** A strong psychological desire for a particular food or foods.

## Focus On FOOD
### Food Cravings

Almost everyone experiences food cravings. A food craving is an irresistible and intense desire for a particular food, and is distinctly different from hunger. For example, hunger can be satisfied by eating almost any food, whereas a food craving is satisfied only when the person consumes the desired foods. The most commonly craved foods tend to be calorie rich, such as cookies, cakes, chips, and chocolate.[2]

Women tend to experience food cravings more frequently than do men.[3] This may, in part, be due to hormonal fluctuations during the menstrual cycle.[4] For example, many women crave certain foods around the time of their menstrual flow. Hormonal changes associated with pregnancy may also cause some women to crave particular foods.[5] In general, food cravings tend to occur at specific times during the day (late afternoon or early evening) and in response to stressful situations, such as when a person feels anxious.[6]

People once thought food cravings were caused by a lack of specific nutrients in the diet. However, this is clearly not the case. For example, a craving for potato chips is not caused by a lack of salt, nor do we crave steak because we need more protein. Yet food cravings are strong, and scientists do not have a clear explanation for their occurrence. However, new information on the role of neurochemicals is unfolding. One explanation is that low serotonin levels in the brain may cause people to crave high-carbohydrate foods.[7] Eating carbohydrates raises serotonin levels, helping people feel calm and relaxed. Food cravings have also been attributed to neuropeptide Y (NPY) and galanin, which are produced in the brain.[8] NPY appears to increase the desire to eat carbohydrate-rich foods, whereas galanin stimulates a preference for high-fat foods.[9]

For most people, food cravings come and go. For others, cravings can take over, and learning how to tame them is important. Most health exerts agree that chronic overconsumption of food is more likely than food cravings, per se, to cause weight gain. In fact, overly restrictive food regimens may actually cause food cravings. Experts believe it is healthier to indulge our cravings within reason, rather than making certain foods "off limits" and/or becoming preoccupied with lingering thoughts and desires for particular foods. Furthermore, getting adequate sleep, regular exercise, practicing relaxation techniques to reduce stress, and eating healthy may help food cravings become less persistent.

## ESSENTIAL Concepts

Energy intake is influenced by both physiological and psychological factors. Hunger is the basic physiological need for food, whereas satiety is the physiological response to having eaten enough. Both are controlled by neuropeptides produced in the brain and by neural and hormonal signals from the gastrointestinal tract, pancreas, and adipose tissue. Psychological factors such as the appearance, taste, and aroma of food can also influence the desire to eat, regardless of hunger or satiety. The desire for food is called appetite, whereas the term food craving refers to the desire for a specific food.

## Energy Expenditure: Basal Metabolism, Physical Activity, and the Thermic Effect of Food

Energy intake is only half of the energy balance equation, the other half being energy expenditure. The body expends energy to maintain physiological functions, for physical activity, and to process food. Collectively, these components of energy use make up **total energy expenditure (TEE)**. As illustrated in Figure 9.3, TEE has three main components: (1) basal metabolism,

**total energy expenditure (TEE)** Total energy expended or used by the body.

**FIGURE 9.3 Components of Total Energy Expenditure (TEE)**

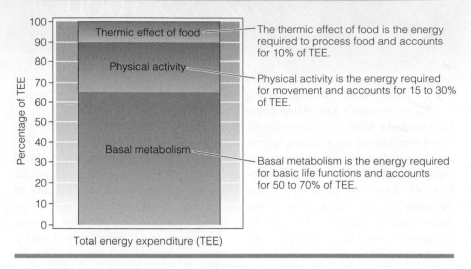

The thermic effect of food is the energy required to process food and accounts for 10% of TEE.

Physical activity is the energy required for movement and accounts for 15 to 30% of TEE.

Basal metabolism is the energy required for basic life functions and accounts for 50 to 70% of TEE.

(2) physical activity, and (3) thermic effect of food. TEE also includes **adaptive thermogenesis** and **nonexercise activity thermogenesis (NEAT)**. Adaptive thermogenesis is a temporary change in energy expenditure that enables the body to adapt to such things as changes in the environment or to physiological conditions such as trauma, starvation, or stress. For example, shivering in response to cold is an example of adaptive thermogenesis. NEAT is energy expended for spontaneous movements such as fidgeting and maintaining posture. The contribution to TEE made by adaptive thermogenesis and NEAT has yet to be determined, but is likely small. The three main components of TEE are discussed next.

## Basal and Resting Metabolism

**Basal metabolism** is the energy expended to sustain basic life functions such as respiration, beating of the heart, nerve function, and muscle tone. **Basal metabolic rate (BMR)**, defined as the amount of energy expended per hour (that is, kcal/hour), accounts for most of TEE — approximately 50 to 70%. **Basal energy expenditure (BEE)** is BMR expressed over a 24-hour period (that is, in kcal/day).

BMR is measured in such a way that energy expenditure associated with the processing of food and physical activity is eliminated. It is typically measured in the morning, after 8 hours of sleep, in a temperature-controlled room, and in a fasting state. Because these stringent conditions make measuring BMR difficult, researchers often measure **resting metabolic rate (RMR)** instead. Measuring RMR is not as difficult as measuring BMR, requiring only a brief resting period and no fasting. Because of this, RMR is approximately 10% higher than BMR. When the RMR is expressed over a 24-hour period, the term **resting energy expenditure (REE)** is used.

The Harris-Benedict equation is a mathematical formula developed almost 100 years ago and still used by clinicians to predict a person's REE. This equation is based on sex, age, height, and weight. It is calculated as follows.

*Males:* REE = 66.5 + (13.8 × weight in kg) + (5 × height in cm) − (6.8 × age in years)

*Females:* REE = 655.1 + (9.6 × weight in kg) + (1.8 × height in cm) − (4.7 × age in years)

**adaptive thermogenesis** Energy expended to adapt to changes in the environment or to physiological conditions.

**nonexercise activity thermogenesis (NEAT)** Energy expended for movement such as fidgeting and maintaining posture.

**basal metabolism** Energy expended to sustain metabolic activities related to basic vital body functions such as respiration, muscle tone, and nerve function.

**basal metabolic rate (BMR)** Energy expended for basal metabolism per hour (expressed as kcal/hour).

**basal energy expenditure (BEE)** Energy expended for basal metabolism over a 24-hour period.

**resting metabolic rate (RMR)** A measure of energy expenditure assessed under less stringent conditions than is BMR.

**resting energy expenditure (REE)** Energy expended for resting metabolism over a 24-hour period.

## TABLE 9.1 Calculating Resting Energy Expenditure (REE)

**Variables**
*Age:* 20 years old
*Sex:* Male
*Weight:* 200 lb. (91 kg)
*Height:* 75 in. (190.5 cm)

**Formula**
REE = 66.5 + (13.8 × weight in kg) + (5 × height in cm) − (6.8 × age in years)

**Calculation**
REE = 66.5 + 13.8 (91) + 5 (190.5) − 6.8 (20)
REE = 66.5 + 1255.8 + 952.5 − 136
REE = 2,138.8 kcal/day

An example of how the Harris-Benedict equation is used to estimate REE is shown in Table 9.1. In this example, you can see that an average-weight, 20-year-old male requires about 2,140 kcal/day to maintain basic body functions. Remember that this value is just an estimate and does not take into account physical fitness or body composition.

## Factors Influencing Basal Metabolic Rate (BMR)

BMR is affected by many factors, including body shape, body composition, age, sex, nutritional status, and genetics.[10] The impact of these and other factors on BMR is summarized in Table 9.2.

In general, tall, thin people tend to have higher BMRs than short, stocky people of the same weight. This is partly because tall, slender people have more surface area, resulting in greater loss of body heat. This is why equations used to estimate energy expenditure take into account both height and weight. Differences in BMR can also be partially explained by body composition. Weight and height being equal, people with high proportions

## TABLE 9.2 Factors Affecting Basal Metabolic Rate (BMR)

| Factor | Effect on BMR |
| --- | --- |
| Age | After physical maturity, BMR decreases with age. |
| Sex | Males have higher BMR than do females of equal size and weight. |
| Growth | BMR is higher during periods of growth. |
| Body weight | BMR increases with increased body weight. |
| Body shape | Tall, thin people have higher BMRs than do short, stocky people of equal weight. |
| Body composition | Because muscle requires more energy to maintain than does adipose tissue, people with more lean tissue have higher BMRs than do people of equal weight with more adipose tissue. |
| Body temperature | Increased body temperature causes a transient increase in BMR. |
| Stress | Stress increases BMR. |
| Thyroid function | Elevated levels of thyroid hormones increase BMR, whereas low levels decrease BMR. |
| Energy restriction | Loss of body tissue associated with fasting and starvation decreases BMR. |
| Pregnancy | BMR increases during pregnancy. |
| Lactation | Milk synthesis increases BMR. |
| Sleep | BMR is lowest while sleeping. |

of lean mass (muscle) tend to have higher BMRs than do people with more fat mass (adipose tissue). This is because, compared to adipose tissue, muscle has greater metabolic activity. Age can also influence BMR such that after 30 years of age, BMR may decrease by about 2 to 5% every 10 years. It is thought this decrease is caused by loss of lean tissue, common with aging.[11] Physical activity can slow the rate of age-related muscle loss and thus can minimize the decline in BMR. Sex differences in BMR are also attributed, in part, to body composition. Because of their smaller size and less lean mass, women tend to have lower BMRs than do men. However, even when adjusting for these differences, average BMR in males remains slightly higher than in females. This is likely due to hormonal differences between men and women.

Other factors affect BMR as well. For example, pound for pound, infants have higher BMRs than do adults. This is because infants are growing. BMR also increases during pregnancy and lactation. Fever and stress can cause a transient increase in BMR, whereas it decreases during sleep. Thyroid function also affects BMR such that thyroid overactivity—called hyperthyroidism—causes BMR to increase. Likewise, thyroid underactivity—hypothyroidism—causes BMR to decrease.

Perhaps the most striking factor that influences BMR is food restriction or dieting.[12] Severe energy restriction over time can decrease BMR by as much as 40%, in part because of the loss of lean body tissue. This energy-sparing response to negative energy balance is undoubtedly beneficial when food is scarce. However, it makes weight loss especially difficult when a person relies solely on energy restriction "dieting" to lose weight.

## Physical Activity

After BMR, energy expended to support physical activity is the second largest component of TEE. The amount of energy required for physical activity is quite variable, accounting for 15 to 30% of TEE.[13] Sedentary people are at the lower end of this estimate, whereas physically active people represent the upper end. In fact, some athletes may require as much as 2,000 to 3,000 extra kcal each day to support physical activity and to maintain a stable body weight.

Many factors affect the amount of energy expended for physical activity. Certainly, rigorous activities such as biking, swimming, and running have higher energy costs than less demanding activities. Body size also affects energy expended for physical activity. Larger people have more body mass to move than smaller people and therefore expend more energy to accomplish the same activity.

## Thermic Effect of Food (TEF)

Another component of total energy expenditure (TEE) is called the **thermic effect of food (TEF)**. About 10% of TEE is attributed to TEF. TEF is the energy expended to digest, absorb, transport, metabolize, and store nutrients following a meal. In other words, TEF is the metabolic cost associated with using the foods we eat.

TEF is influenced by the amount and composition of foods consumed. Because more energy is needed to process large amounts of food, TEF increases as food consumption increases. Some nutrients require more energy to process than others. For example, high-protein foods have the highest TEF, and high-fat foods have the lowest. These differences reflect the metabolic

**thermic effect of food (TEF)** Energy expended for the digestion, absorption, and metabolism of nutrients.

cost associated with processing different nutrients. Because our meals generally supply a mixture of nutrients, TEF is estimated to be about 5 to 10% of total energy intake.[14] For example, after consuming a 500-kcal meal a person typically expends 25 to 50 kcal as TEF. In some ways, TEF is like having a caloric sales tax on energy intake.

## Assessing Total Energy Expenditure (TEE)

Assessing TEE can be difficult because it requires expensive equipment and a high degree of expertise. However, a variety of techniques have been developed to assess TEE, including direct and indirect calorimetry, stable isotopes, and mathematical equations.

### Direct and Indirect Calorimetry

Total energy expenditure (TEE) can be measured by assessing heat loss from the body, because much of the energy (calories) the body uses is eventually lost as body heat. This method, called **direct calorimetry,** requires an airtight chamber surrounded by water-filled tubing within the chamber walls. Heat lost from the body raises the water temperature, which can be measured and used to estimate TEE. Direct calorimetry requires specialized and very expensive equipment, so it is not often used.

An alternative to direct calorimetry is **indirect calorimetry.** Rather than measuring heat loss, indirect calorimetry measures the exchange of respiratory gases—oxygen intake and carbon dioxide output. The use of indirect calorimetry to estimate TEE is based on the assumption that the body uses 1 liter of oxygen to metabolize 4.8 kcal of energy-containing compounds (that is, glucose, amino acids, and fatty acids). Thus, measuring oxygen consumption and carbon dioxide production permits an estimate of TEE.

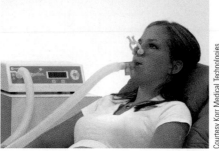

Indirect calorimetry measures the exchange of respiratory gases—oxygen consumption and carbon dioxide production—using portable equipment. These measurements are used to estimate total energy expenditure (TEE), based on the assumption that 1 liter of oxygen is used by the body to metabolize 4.8 kcal of energy.

### Use of Stable Isotopes: Doubly Labeled Water

Another way to estimate TEE involves the use of **stable isotopes**—nonradioactive forms of certain elements. Isotopes have extra neutrons in their nuclei, making them heavier than the more common forms. Because stable isotopes of hydrogen ($^2$H) and oxygen ($^{18}$O) are chemically distinct from normal hydrogen ($^1$H) and oxygen ($^{16}$O), they can be measured in body fluids and expired air.

One technique that uses stable isotopes to estimate TEE is the **doubly labeled water** method. This method requires a person to drink two forms of water that have been labeled with stable isotopes—$^2$H$_2$O and H$_2$$^{18}$O. TEE can be estimated by measuring the elimination of oxygen and hydrogen isotopes from the body as water and carbon dioxide. This technique is now considered the "gold standard" for estimating TEE.

### Mathematical Formulas

Because the equipment and expertise to use calorimetry and stable isotopes are not readily available, relatively simple mathematical formulas can be used to estimate TEE. For example, the Institute of Medicine has developed mathematical formulas as part of the Dietary Reference Intakes (DRIs).[15] These formulas are used to calculate what is called Estimated Energy Requirements (EERs), and take into account TEE and energy needed to support growth. Formulas have been developed to calculate EERs for infants, children, adolescents, pregnant women, lactating women, and adults. These are provided in Appendix B. Note that the EERs are intended to help maintain a healthy

**direct calorimetry** A measurement of energy expenditure obtained by assessing heat loss.

**indirect calorimetry** A measurement of energy expenditure obtained by assessing oxygen consumption and carbon dioxide production.

**stable isotope** A form of an element that contains additional neutrons.

**doubly labeled water** Water that contains stable isotopes of hydrogen and oxygen atoms.

body weight. Once they are determined, adjustments can be made to support weight loss or gain. The following formulas are used to calculate EER for adult males and females. The steps used to calculate EER are shown in Chapter 2.

*Males:* EER = 662 − [9.53 × age (y)] + PA ×
$$[15.91 \times \text{weight (kg)}] + [539.6 \times \text{height (m)}]$$
*Females:* EER = 354 − [6.91 × age (y)] + PA ×
$$[9.36 \times \text{weight (kg)}] + [726 \times \text{height (m)}]$$

## ESSENTIAL *Concepts*

The body expends energy to maintain basal metabolism, for physical activity, and to process food. These components make up a person's total energy expenditure (TEE). Basal metabolism accounts for most of the TEE, and is affected primarily by body size and composition. Energy expended for physical activity is variable, accounting for 15 to 30% of TEE. Thermic effect of food (TEF) accounts for about 10% of the TEE and reflects the energy required to process nutrients after eating. Smaller components of the TEE include adaptive thermogenesis and nonexercise activity thermogenesis (NEAT). Adaptive thermogenesis is the energy expended to adapt to stressful situations. NEAT is the energy expended on such activities as fidgeting. A variety of methods are used to estimate TEE. Direct calorimetry measures the body's heat loss, whereas indirect calorimetry measures the respiratory gases oxygen and carbon dioxide. Other estimates of TEE are based on the metabolism of stable isotopes of hydrogen and oxygen in the body. Mathematical formulas called the Estimated Energy Requirements (EERs) can also be used to estimate TEE.

# Assessing Body Weight and Body Composition

Balancing energy intake with energy expenditure is key to maintaining energy balance, and thus a stable body weight. For many people this is difficult, resulting in unwanted weight gain. But at what point is a person considered overweight? Where do we draw the line between a few extra pounds and a serious health concern? After all, what appears to be a healthy weight for one person may cause health problems for another. To answer these questions, it is important to first understand how body weight and composition are measured and assessed.

## Defining Overweight and Obesity

The terms *overweight* and *obese* are often used interchangeably, although they have very different meanings. **Overweight** refers to excess weight for a given height, whereas **obesity** refers specifically to abundance of body fat, or adiposity. Because simply knowing a person's weight does not provide information about the different components of the body such as muscle or fat, it is possible for muscular people, such as athletes, to be considered overweight but not obese. Conversely, some inactive people may not be considered overweight, yet may still be obese. Nonetheless, most people who are overweight are obese as well, because weight gain in adults is more likely to be associated with increased adipose tissue, rather than muscle. For this reason, body weight is often used as an indirect measure of obesity.

**overweight** Excess weight for a given height.

**obesity** Excess body fat.

## Assessing Body Weight

Because the variation in body weight for a given height is substantial, defining an "ideal body weight" may not be possible. Consequently, recommended body weights are simply reference values, and not necessarily ideal or desirable weights for all people. The reference standards most commonly used to assess body weight are height–weight tables and body mass index.

### Height–Weight Tables

Many years ago, life insurance companies wanted a way to screen applicants to avoid insuring people at high risk for early death. Data from thousands of people who had purchased life insurance between 1935 and 1954 were analyzed to determine the relationship between body weight and longevity.[16] Weights associated with the longest life expectancies were deemed "desirable" or ideal body weights. Based on this information, the Metropolitan Life Insurance Company (MLIC) published tables that listed "ideal weights" for adult males and females. These tables were later revised, and are provided in Appendix F.

### Body Mass Index

Although height–weight tables are often used to assess body weight, body mass index (BMI) is the most widely used measure employed in body weight assessment. BMI is calculated using either of the following formulas.

$$BMI = [weight\ (kilograms)] \div [height\ (meters)^2]$$

$$BMI = [weight\ (pounds)] \div [height\ (inches)^2] \times 703.1$$

Based on these formulas, BMI for a person who weighs 150 lb. (68.2 kg) and is 65 inches (1.65 m) tall is 25 kg/m². People with low BMIs typically have less body fat, whereas those with high BMIs tend to have more body fat. Because each BMI unit represents 6 to 8 pounds for a given height, an increase in just 2 BMI units represents a 12- to 16-pound increase in body weight.

BMI is based on the ratio of weight to height and is therefore a better indicator of obesity than weight alone. Cutoff values for BMI classifications (that is, normal, overweight, and obese) are based on the association between BMI and weight-related morbidity and mortality. Figure 9.4 shows that higher BMI values are strongly associated with increased risk of weight-related health problems and deaths.[17]

Most medical organizations, including the Centers for Disease Control and Prevention, consider an adult with BMI between 25 and 29.9 kg/m² as overweight, and an adult with BMI 30 kg/m² or greater as obese.[18] The BMI range of 18.5 to 24.9 kg/m² is considered healthy for adults, and below 18.5 is classified as underweight. Weight classifications based on BMI are presented in Figure 9.5.

BMI provides a useful way to assess body weight. However, because BMI does not directly measure body fat, at times it may overestimate or underestimate adiposity. For example, an athlete who has a high amount of lean mass (muscle) relative to fat mass may have a BMI that incorrectly classifies that person as overweight.

**CONNECTIONS** Recall that *morbidity* refers to illness and *mortality* refers to death (Chapter 1, page 20).

## Assessing Body Composition

The body can be thought of as having two compartments: (1) the fat compartment and (2) the fat-free (lean) compartment. The fat compartment (adipose

## FIGURE 9.4　Body Mass Index (BMI) and Weight-Related Mortality Rate

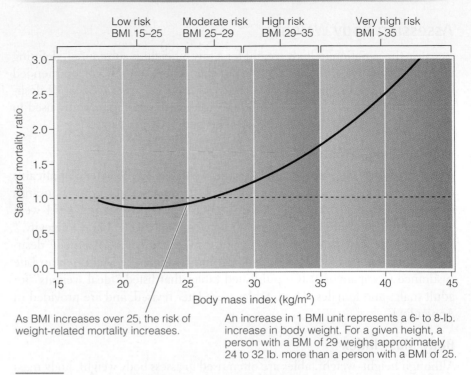

Low risk
BMI 15–25

Moderate risk
BMI 25–29

High risk
BMI 29–35

Very high risk
BMI >35

As BMI increases over 25, the risk of weight-related mortality increases.

An increase in 1 BMI unit represents a 6- to 8-lb. increase in body weight. For a given height, a person with a BMI of 29 weighs approximately 24 to 32 lb. more than a person with a BMI of 25.

SOURCE: Gray DS. Diagnosis and prevalence of obesity. *Medical Clinics of North America.* 1989;73:1–13.

## FIGURE 9.5　Weight Classifications Using Body Mass Index (BMI)

BMI can be determined by locating your weight on the bottom of the chart and your height on the left of the chart. The number located at the intersection of these two values is your BMI.

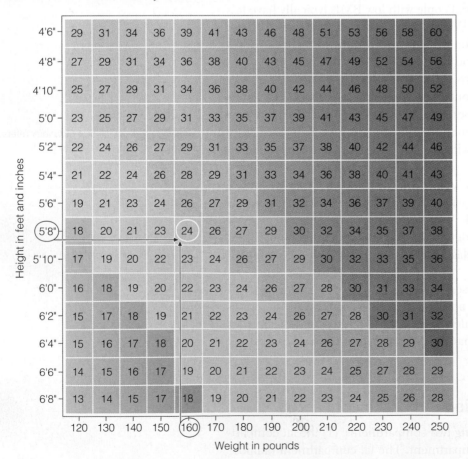

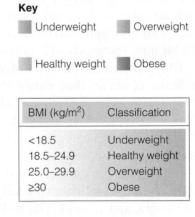

**Key**

Underweight　　Overweight

Healthy weight　　Obese

| BMI (kg/m²) | Classification |
|---|---|
| <18.5 | Underweight |
| 18.5–24.9 | Healthy weight |
| 25.0–29.9 | Overweight |
| ≥30 | Obese |

SOURCE: Centers for Disease Control and Prevention. Overweight and obesity: Defining overweight and obesity. Available from: http://www.cdc.gov/nccdphp/dnpa/obesity/defining.htm.

| TABLE 9.3 | Obesity Classifications Based on Percent Body Fat | |
|---|---|---|
| Criteria | Body Fat (% total body weight) | |
| | Males | Females |
| Normal | 12–20 | 20–30 |
| Borderline obese | 21–25 | 31–33 |
| Obese | >25 | >33 |

SOURCE: Bray G. What is the ideal body weight? *Journal of Nutritional Biochemistry.* 1998;9:489–92.

tissue) consists mostly of stored triglyceride and supporting structures, whereas the fat-free compartment is mostly muscle, water, and bone. How much fat and fat-free mass a person has is determined by many factors, including sex, genetics, physical activity, hormones, and diet.

The amount of fat stored in the body changes throughout the life cycle. For example, nearly 30% of total body weight in a healthy 6-month-old infant is fat, whereas this percentage may be cause for concern in adults. It is recommended that body fat levels be around 12 to 20% and 20 to 30% of total body weight in males and females, respectively (Table 9.3). Body fat over 25% in males and over 33% in females is considered to indicate obesity.

Just as too much body fat can cause health problems, too little body fat can also have harmful effects. Body fat below 5% for males and 12% for females is considered too low. Very low body fat, especially when it is a consequence of disordered eating patterns, can lead to a wide range of health problems.

Obesity increases risk for a variety of health problems, including type 2 diabetes, osteoarthritis, stroke, heart disease, gallstones, and colon cancer (Figure 9.6).[19] Because of this, clinicians use a variety of methods to estimate body fat.[20] Whereas some methods are expensive and quite complex, others are more readily available and easy to use. As you might expect, some methods are more accurate than others. Methods used to estimate body composition are summarized in Figure 9.7 and discussed next.

## Densitometry

Underwater weighing or **hydrostatic weighing** is based on the concept of densitometry, which simply means measuring the density of something. Because adipose tissue is less dense than lean tissue, bodies with more fat are less dense than those with more muscle. Hydrostatic weighing involves measuring a person's weight both in and out of water. The more fat a person has, the less dense he or she is, and the less he or she weighs underwater. To get an accurate estimate of body fat using this method, a person must exhale as much air as possible before being submerged. Also, some people find it difficult to stay motionless while under water. Thus underwater weighing can be challenging.

An instrument called the BOD POD, also based on densitometry, can be used to estimate body fat as well. The BOD POD measures air displacement by having the person sit inside a chamber. Air displaced by the person's body is measured by computerized sensors. Although the BOD POD is easier to use than underwater weighing, its accuracy needs to be more extensively tested.

## Dual-Energy X-Ray Absorptiometry (DEXA)

Dual-energy x-ray absorptiometry (DEXA) is used to estimate total body fat and also provides information about the distribution of fat among different

**hydrostatic weighing** (underwater weighing) Method for estimating body composition that compares weight on land to weight underwater.

**FIGURE 9.6    Health Problems Associated with Obesity**

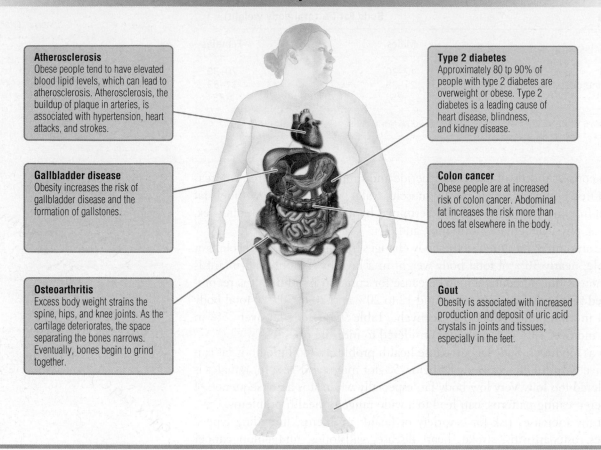

**Atherosclerosis**
Obese people tend to have elevated blood lipid levels, which can lead to atherosclerosis. Atherosclerosis, the buildup of plaque in arteries, is associated with hypertension, heart attacks, and strokes.

**Gallbladder disease**
Obesity increases the risk of gallbladder disease and the formation of gallstones.

**Osteoarthritis**
Excess body weight strains the spine, hips, and knee joints. As the cartilage deteriorates, the space separating the bones narrows. Eventually, bones begin to grind together.

**Type 2 diabetes**
Approximately 80 tp 90% of people with type 2 diabetes are overweight or obese. Type 2 diabetes is a leading cause of heart disease, blindness, and kidney disease.

**Colon cancer**
Obese people are at increased risk of colon cancer. Abdominal fat increases the risk more than does fat elsewhere in the body.

**Gout**
Obesity is associated with increased production and deposit of uric acid crystals in joints and tissues, especially in the feet.

areas of the body. During a DEXA measurement, a person must lie still while a scanning device passes over his or her body. The x-ray beams emitted differentiate between fat and fat-free mass. DEXA, considered the "gold standard" of body composition analysis, is often used by researchers and clinicians.

## Bioelectrical Impedance

**Bioelectrical impedance** is a measure of electrical conductivity—in other words, how easily a weak electric current travels through the body. Electrodes are placed on a person's hand and foot, and a weak electric current is emitted. Lean tissue, which contains a great deal of water, conducts electric currents better than adipose tissue, which has little water associated with it. The accuracy of bioelectrical impedance can be affected by hydration status—estimates of body fat will be incorrectly high if a person is dehydrated. This technique is relatively accurate, simple to use, and is considered an acceptable method for estimating body composition in clinical settings. Bathroom scales with built-in bioelectrical impedance systems are now available for home use, although little is known about their accuracy.

## Skinfold Thickness

The skinfold thickness method has been used for many years to estimate body fat. Adipose tissue directly under the skin is called subcutaneous fat. A measuring device called a **skinfold caliper** is used to measure the thickness of the

**bioelectrical impedance** A method used to assess body composition based on measuring the body's electrical conductivity.

**skinfold caliper** An instrument used to measure the thickness of subcutaneous fat.

## FIGURE 9.7 Methods Used to Estimate Body Composition

Hydrostatic weighing requires the subject to exhale air from the lungs and be submerged in water. It is important to remain motionless while weight underwater is measured.

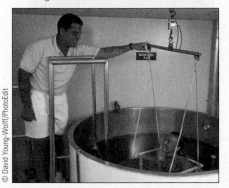

Hydrostatic weighing

The BOD POD uses this sealed chamber to measure air displacement.

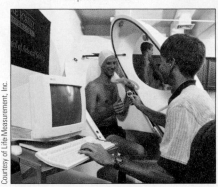

BOD POD

While the person is lying on a table, a scanning device passes over the body. The x-ray beams emitted differentiate between the fat mass, fat-free mass, and skeletal mass. A two-dimentional image of the body is displayed on a computer screen.

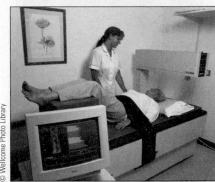

Dual-energy x-ray absorptiometry (DEXA)

Bioelectrical impedance

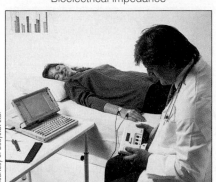

Electrodes are placed on a person's hand and foot, and a weak electrical current is passed through the body. The conductivity of the current is measured, which provides an estimate of the fat mass and fat-free mass.

Skinfold thickness

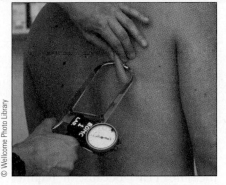

Using a measuring device called a caliper, the heathcare worker measures the thickness of a fold of skin with its underlying layer of fat at precise locations on the body.

Near-infrared interactance

A device that emits light in the infrared spectrum is used to estimate the amounts of fat and fat-free mass.

subcutaneous fat. This is done by pinching a fold of skin (skinfold) with its underlying layer of subcutaneous fat at various locations on the body. Body fat is estimated using mathematical formulas based on the skinfold thickness measures. Although the accuracy of this method depends on a person's ability to use a caliper, it is a fairly accurate and inexpensive way to assess body fat.

### Near-Infrared Interactance

**Near-infrared interactance** is based on the principle that tissues of varying densities reflect and absorb infrared light differently. Infrared light, which has a specific wavelength, is emitted from a probe that is placed on the bicep muscle. The amount of light absorbed and reflected is then measured and used to estimate fat mass and fat-free mass. Near-infrared interactance is often used in health clubs, because it is relatively inexpensive and easy to use. However, most experts agree it does not provide an accurate assessment of body composition.

**near-infrared interactance** A method used to assess body composition, based on the scattering and absorbance of light in the near-infrared spectrum.

**FIGURE 9.8**  Body Fat Distribution Patterns

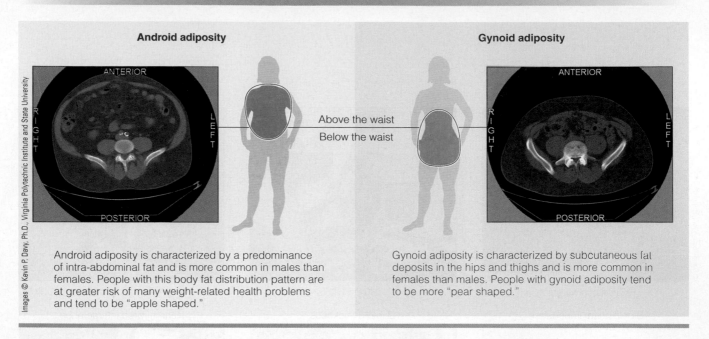

Android adiposity is characterized by a predominance of intra-abdominal fat and is more common in males than females. People with this body fat distribution pattern are at greater risk of many weight-related health problems and tend to be "apple shaped."

Gynoid adiposity is characterized by subcutaneous fat deposits in the hips and thighs and is more common in females than males. People with gynoid adiposity tend to be more "pear shaped."

## Other Methods of Estimating Body Composition

Other methods used to estimate body composition include computed tomography (CT), magnetic resonance imaging (MRI), and isotope dilution. Using stable isotopes to estimate body composition is similar to using isotopes to measure TEE. Because it requires technical equipment, isotope dilution methodology is only used in research. CT scans and MRIs are medical diagnostic tools and are not routinely used for body composition assessment. As shown in Figure 9.8, these state-of-the-art imaging techniques provide amazing views of the body.

## Assessing Body Fat Distribution

It appears that where fat stores are located in the body, in addition to how much fat is stored, can also influence health. A predominance of upper body fat, called **android adiposity,** is characterized by relatively large deposits of **intra-abdominal (visceral) fat;** see Figure 9.8. Android adiposity is more common in males than females. A predominance of lower body fat, called **gynoid adiposity,** is characterized by large amounts of subcutaneous fat in the thighs and hips, and is more common in females than males. A person with android adiposity is often said to be "apple shaped," whereas a person with gynoid adiposity is often referred to as "pear shaped." Compared to those with gynoid adiposity, people with android adiposity are at increased risk for developing cardiovascular disease, type 2 diabetes, and hypertension.[21] Although the reason for this is not clear, it appears that compared to fat stored elsewhere, intra-abdominal fat more easily undergoes lipolysis, readily releasing lipids into the blood.

**android adiposity**  Body fat mostly in the intra-abdominal cavity; also referred to as visceral fat.

**intra-abdominal (visceral) fat**  Body fat associated with internal organs within the abdominal cavity.

**gynoid adiposity**  Subcutaneous body fat in the lower part of the body, typically the thighs and hips.

CONNECTIONS  Recall that lipolysis is the breakdown of triglyceride molecules into glycerol and fatty acids (Chapter 7, page 232).

Body fat distribution patterns are largely determined by biological factors. Daughters tend to be shaped like their mothers, and sons tend to be shaped like their fathers. Reproductive hormones such as estrogen and testosterone also influence where body fat is stored. Testosterone (a male hormone) encourages android adiposity, whereas estrogen (a female hormone) favors gynoid adiposity. As women age, declining estrogen levels can cause body fat to shift toward android adiposity.[22]

## Waist-to-Hip Ratio

A person's **waist-to-hip ratio (WHR)** provides a good indication of body fat distribution pattern—in other words whether a person has gynoid or android adiposity. These measurements are easily done, requiring nothing more than a tape measure. Determining a person's WHR involves measuring the distances (that is, circumference) around the waist and hips. Waist circumference is typically measured at the narrowest area, and hip circumference is measured at the widest area. However, a health professional is likely to use more precise anatomical landmarks. As shown in Figure 9.9, WHR is calculated by dividing waist circumference by hip circumference. In males, a WHR of 0.95 or greater indicates android adiposity, whereas in females the WHR must be 0.85 or greater.[23]

Some researchers believe that waist circumference alone provides a better indication of body fat distribution patterns than does WHR. This is because some ethnic groups have characteristically smaller hips or larger hips. Android adiposity is associated with a waist circumference greater than 40 and 35 inches in males and females, respectively.[24]

**waist-to-hip ratio (WHR)** The ratio of waist circumference to hip circumference used to determine body fat distribution pattern.

---

**FIGURE 9.9    Waist-to-Hip Ratio (WHR): An Indication of Body Fat Distribution**

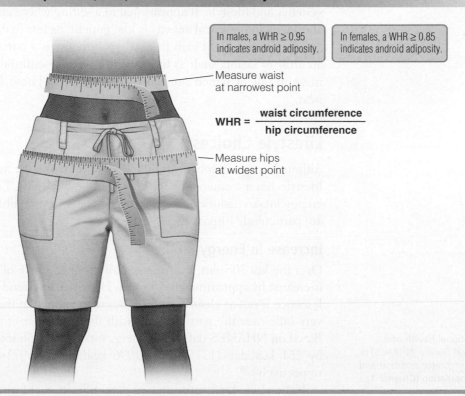

In males, a WHR ≥ 0.95 indicates android adiposity.

In females, a WHR ≥ 0.85 indicates android adiposity.

Measure waist at narrowest point

$$WHR = \frac{waist\ circumference}{hip\ circumference}$$

Measure hips at widest point

## ESSENTIAL *Concepts*

Overweight is excess weight for a given height, whereas obesity is an abundance of body fat in relation to lean tissue. Indices such as height–weight tables and body mass index (BMI) are used to assess body weight. Height–weight tables list "ideal weights" for adults. BMI is based on the ratio of weight to height squared and is considered a good indicator of body fat. Measures of body composition include hydrostatic weighing, dual-energy x-ray absorptiometry (DEXA), bioelectrical impedance, near-infrared interactance, skinfold thickness, computed tomography (CT), magnetic resonance imaging (MRI), and isotope dilution. Body fat centralized within the abdomen is called android adiposity, whereas body fat located in the thighs and hips is gynoid adiposity. Body fat distribution is typically assessed by measuring waist and hip circumferences.

# Obesity: An Outcome of Lifestyle and Genetics

In the late 19th century, only 3% of adults in the United States were overweight. Today, about 65% of adults are overweight or obese; more than 120 million Americans.[25] Furthermore, in the past 20 years the number of overweight children has doubled, and the number of overweight adolescents has tripled. Similar trends are seen around the world.[26] What is causing this global epidemic? To answer this question, we must consider both components of the energy balance equation—energy intake and energy expenditure.

There is general agreement that changes in lifestyle, rather than changes in genetics, are the driving forces behind the obesity epidemic. Nonetheless, like many other aspects of our health, body weight is likely influenced by both genetics and lifestyle. It appears that in a setting where energy-dense foods are abundant and physical activity is low, genetic factors make some people more susceptible to weight gain than others. Obesity is a complex issue caused by an array of factors such as lifestyle choices, sociocultural factors, and genetic makeup, all of which are closely intertwined. These factors are discussed next.

## Lifestyle Choices

Although many factors contribute to the rise in obesity, most experts agree that lifestyle has a tremendous influence on body weight. Factors that influence energy intake (calories consumed) and energy expenditure (physical activity) are particularly important.

### Increase in Energy Intake

Over the last 30 years, the average daily energy intake of American adults has increased by approximately 300 kcal. However, the trend in children and adolescence is not as clear. Surprisingly, energy intake in this group has changed very little over the past 30 years, with the exception of adolescent females.[27] Based on NHANES data, total energy intake for adolescent females increased by 261 kcal/day (1735 versus 1996 kcal/day in 1971–74 and 1999–2000, respectively).[28]

Nationwide food consumption data indicate that for many segments of the population, positive energy balance is due, at least in part, to increased energy

**CONNECTIONS** National Health and Nutrition Examiniation Survey (NHANES) is an ongoing study to monitor nutrition and health in the U.S. population (Chapter 1, page 24).

intake. Contributing factors may include a rise in the number of adults eating away from home, increased portion sizes of foods, increased consumption of energy-dense foods, and changes in snacking habits.[29]

In the United States, energy-dense, inexpensive, and flavorful foods have become readily available and somewhat of a cultural norm. The United States has over 2 million fast-food restaurants, and 37% of adults and 42% of children eat fast food daily.[30] Although consuming fast food is not the sole cause of the obesity epidemic, it may certainly contribute. Today many fast-food restaurants offer healthy food choices such as salads and sandwiches made with lean meats and whole-grain breads. Most fast meals, however, are high in fat, refined starchy carbohydrates, and calories. In addition, a supersized "value" meal can provide more than half the calories required in a day. You can read more about this "supersize" trend in the Focus on Food feature.

## Decrease in Energy Expenditure

While energy intakes may be increasing in some segments of the population, Americans are also becoming less physically active. A reduction in jobs that

## *Focus On* FOOD
### The Cost of "Value" Meals

Americans may be getting more for their money than they realize. A recent trend in food cuisine is big, bigger, and biggest. Over the last 20 years, serving sizes have increased, and so have American waistlines. In fact, some experts believe that the "supersize" phenomenon is largely responsible for the obesity epidemic.[31] For example, in 1950 a meal at McDonald's™—consisting of its original burger, fries, and drink—provided 590 kcal. Today, a meal consisting of a quarter-pound burger with cheese, large serving of fries, and drink, provides 1,550 kcal. Considering that the average American eats at least one meal outside the home every day, these extra calories easily translate into added weight.[32] The desire to get more food for less money is taking a toll on our nation's health.

Studies show that serving size influences food consumption. That is, when larger portions are served, people tend to eat more.[33] For example, study participants ate 39% more M&Ms™ when given a 2-pound bag, as compared to a 1-pound bag.[34] It appears some people depend more on visual cues to judge

A popular "value meal" consisting of a double cheeseburger (760 kcal), medium french fries (360 kcal), and a medium soft drink (230 kcal) provides more than half of a person's daily energy requirements.

© Felicia Martinez/PhotoEdit

how much to eat rather than on physiological cues such as hunger and satiety.

Most fast-food restaurants sell large quantities of food in "value meals" for little additional cost over regular-size menu items. This marketing ploy is very appealing to consumers. You may be wondering how restaurants can afford to sell such large quantities of food for little additional cost. The smallest portion of total retail food operations is the cost of food. Most of the expenses come from labor, advertising, packaging, marketing, rent, and utilities, so restaurants can offer large portions of food for little additional cost. For example, a regular-size (24-ounce) soft drink (250 kcal) typically costs $1. However, a large-size (48-ounce) soft drink (500 kcal) costs $1.50. For retailers, the cost of the additional beverage is minimal and thus still profitable, but the cost to the public in terms of health can be high. Clearly, it is up to the consumer to make wise food choices, whether eating out or at home. With food, more is not always better.

A regular meal consisting of a hamburger (300 kcal), small french fries (230 kcal), and a small soft drink (140 kcal) has half the calories of a "value meal."

© Michael Newman/PhotoEdit

require physical work and an increase in labor-saving devices have made daily life less physically demanding. Not only have modern conveniences reduced energy exertion associated with daily living activities, but in addition almost 60% of adults never engage in any kind of leisure-time physical activity lasting 10 minutes or more per week.[35] Not surprisingly, the vast majority of studies show an inverse relationship between the level of physical activity and risk of being overweight or obese.[36] For most people, walking an extra mile each day would increase energy expenditure by 100 kcal/day.[37] This is equivalent to taking about 2,000 to 2,500 extra steps each day. By making this simple change, a person could expect to lose one pound per month.

Thus physical inactivity has contributed to the growing rates of obesity, especially among children and adolescents. According to the U.S. Department of Health and Human Services, the percentage of students participating in high school physical education classes declined from 42% in 1991 to 25% in 1995.[38] Although this trend has reversed slightly, 33% of high school students do not engage in recommended levels of physical activity, and 10% are completely inactive. The Centers for Disease Control and Prevention report that 62% of adolescence do not participate in any organized physical activity outside of school.

Although less is known about physical activity trends in children than in adults and adolescents, the relationship between media-related sedentary activities (such as watching television) and obesity has been extensively studied. Studies show that overweight children tend to watch more television than children who are not overweight.[39] It is likely that the more time children spend watching television and playing video/computer games, the less likely they are to be physically active. In addition, weight gain may also be due to greater exposure to food advertisements, which in turn may alter food choices.[40] The average child views about 40,000 ads a year on television alone; thus children are exposed to billions of dollars worth of advertising and marketing promotions for candy, cereal, soda, and fast food.

## Sociocultural Factors

Sociocultural factors such as cultural norms, socioeconomic status, marital status, and education can also influence food preferences and dietary practices. Although socioeconomic and cultural factors do not directly cause obesity, associations between these factors and obesity exist (Table 9.4). For example, obesity in the United States is most prevalent in the southern states.[41] This may be due to differences in socioeconomic status, education, and dietary beliefs, as well as to views on the acceptability of being overweight. In addition, obesity rates tend to be higher in certain racial and ethnic groups. For example, the prevalence of obesity among African Americans is almost 30%, compared to 25% in Hispanics and 18% in whites.[42]

## Genetics

CONNECTIONS Recall that a gene is a section of DNA that contains hereditary information needed for cells to make a protein (Chapter 6, page 179).

People have long suspected that genetics plays a role in influencing body weight. Like other inherited physical traits such as height, eye color, hair color, and facial features, risk for obesity is partly inherited as well. Children with obese parents have a high probability that they, too, will become obese. However, in an environment where energy-dense foods are abundant and people are physically inactive, even individuals not genetically predisposed

## TABLE 9.4 Factors Related to the Risk of Obesity

**World-Wide**

- Obesity is more common in developed countries compared to underdeveloped countries.
- Some cultures associate increased body weight with wealth, whereas others associate thinness with wealth.

**In the United States**

- Obesity is more prevalent in African Americans, Hispanics, Native Americans, and Pacific Islanders.
- Obesity is more prevalent in low-income families and in people who work at low-paying jobs.
- Obesity is more prevalent among less educated people.
- Obesity is less prevalent in married people.
- Obesity is more prevalent in people who live in rural settings compared to those who live in urban settings.
- Obesity is more prevalent in people who live in southern states.

SOURCE: Dalton S, editor. *Overweight and Weight Management: The Health Professional's Guide to Understanding and Practice.* Gaithersburg, MD: ASPEN; 1997, 314.

to obesity are likely to gain weight. The good news is that although genetics influences our susceptibility to obesity, lifestyle remains a critical factor.

Although it is difficult to distinguish between genetic and lifestyle influences in the development of obesity, adoption studies help researchers do just that. Studies of genetically identical twins separated at birth suggest that genetics, rather than lifestyle, largely determines body weight.[43] For example, when body weights of children adopted at birth are compared to those of their adoptive parents, there is little similarity.[44] When the children are compared to their biological parents, the similarity is striking. Scientists estimate that at least 50% of our risk for becoming overweight or obese is determined by genetics. There are many theories as to how genetics influences body weight, but perhaps the most widely accepted is the set point theory.

## Set Point Theory

For many years scientists suspected that body weight was, in part, regulated by an internal mechanism. However, evidence for such a system remained elusive. As early as 1953, researchers hypothesized that a complex signaling system regulated body weight by making adjustments in energy intake and energy expenditure.[45] To test this theory, researchers observed weight gain and weight loss cycles in food-restricted mice. The mice appeared to maintain a steady weight by spontaneously adjusting food intake. This phenomenon was called the **set point theory** of body weight regulation.

The set point theory of body weight regulation suggests that factors circulating in the blood communicate to the brain the amount of adipose tissue in the body. When the amount of adipose tissue increases beyond a "set point," a signal causes food intake to decrease and/or energy expenditure to increase, favoring weight loss. Conversely, when the amount of adipose tissue decreases below a "set point" food intake increases and energy expenditure decreases, favoring weight gain. In this way, body weight is restored to its "set point" and remains relatively stable on a long-term basis.

The set point theory was subject to much debate, and without direct evidence, obesity research progressed slowly. That is, until the discovery of an obese mouse, named the *ob/ob* **mouse.** This mouse appeared to have a

**set point theory** A theory that suggests changes in energy intake and energy expenditure help maintain a stable body weight.

***ob/ob* mouse** Obese mouse with a mutation in the gene that produces the hormone leptin.

**CONNECTIONS** Recall that a mutation is an alteration in a gene resulting in the synthesis of an altered protein (Chapter 6, page 185).

**A gene mutation made this mouse obese.**

**db/db mouse** Obese mouse with a mutation in the gene that codes for the leptin receptor.

**ob gene** The gene that codes for the protein leptin.

**leptin** A hormone produced mainly by adipose tissue that helps regulate body weight.

**db gene** The gene that codes for the leptin receptor.

**hypothalamus** An area of the brain that controls many involuntary functions by the release of hormones.

genetic mutation causing it to consume large amounts of food (hyperphagic), be inactive, and therefore gain weight easily. Soon another chance occurrence took place—another obese mouse was identified. Using breeding records, researchers found that this second mouse, called the **db/db mouse**, was not genetically related to the *ob/ob* mouse.[46]

To test for the existence of a circulating factor that regulates body weight, some unusual experiments were designed.[47] Using a technique called parabiosis, an *ob/ob* mouse was surgically joined with a *db/db* mouse so that blood circulated between them. Curiously, the *ob/ob* mouse lost weight, whereas the *db/db* mouse showed no weight change. When an *ob/ob* mouse was joined to a normal mouse, the *ob/ob* mouse lost weight, whereas the normal mouse showed no change. Last, a *db/db* mouse was joined to a normal mouse. Although there was no change in the *db/db* mouse, the normal mouse lost weight. These experiments are illustrated in Figure 9.10.

The search for a circulating signal that regulates body weight continued until 1994 when scientists discovered that the *ob/ob* mice have a defect in the **ob gene**, which codes for a hormone called **leptin**.[48] Leptin is a potent satiety signal, and the lack of leptin in *ob/ob* mice causes them to eat past their "set point." Soon thereafter researchers discovered the **db gene**, which codes for the leptin receptor found primarily in a region of the brain called the **hypothalamus**. Without the leptin receptor, leptin cannot exert its effect. To determine if leptin played a role in regulating body weight, *ob/ob* and *db/db* mice were injected with leptin. When leptin was injected into leptin-deficient *ob/ob* mice, there was significant weight loss. However, there was no change in *db/db* mice in response to leptin injections. This is

---

**FIGURE 9.10 Parabiosis and Weight Loss Experiments Using *ob/ob* and *db/db* Mice**

ob/ob mice lack the satiety signal leptin, whereas db/db mice produce large amounts of leptin but have defective leptin receptors.

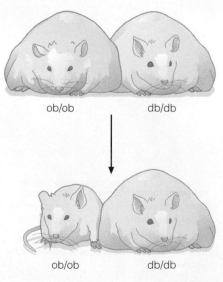

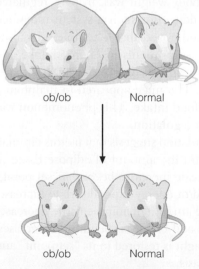

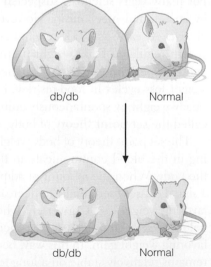

ob/ob    db/db

ob/ob    Normal

db/db    Normal

↓

↓

↓

ob/ob    db/db

ob/ob    Normal

db/db    Normal

When the ob/ob mouse receives leptin from the db/db mouse, the ob/ob mouse loses weight. There is no weight change in the db/db mouse, because it is unable to respond to leptin.

When the ob/ob mouse receives leptin from the normal mouse, the ob/ob mouse loses weight. There is no weight change in the normal mouse.

The normal mouse receives large amounts of leptin from the db/db mouse, causing the normal mouse to lose weight. There is no weight change in the db/db mouse, because it is unable to respond to leptin.

because *db/db* mice have defective leptin receptors, making them unresponsive to this hormone.

People hoped leptin—touted as the antiobesity hormone—would become the miracle cure for obesity. However, the vast majority of obese people produce appropriate or even elevated amounts of leptin. Only a few rare cases of leptin deficiency have been reported in humans.[49]

Although leptin has not proved effective for treating human obesity, its discovery led to important insights into body weight regulation. Many researchers now believe that leptin—rather than protecting against obesity—provides protection from starvation.[50] In times when food is scarce, such as in a famine, leptin may help the body conserve energy, increasing the person's chance of surviving. However, we live in an environment where food is abundant and widely available, making this protective response less advantageous. In fact, James Neel in 1962 proposed the "**thrifty gene hypothesis**" by suggesting that people who are genetically best able to survive famine and starvation are also those most prone to weight gain when food is plentiful.[51]

## ESSENTIAL *Concepts*

Many factors have contributed to the increasing prevalence of obesity in the United States. For example, average energy intake in adults has increased, and flavorful energy-dense foods are widely available and affordable. Furthermore, most Americans do not exercise regularly. Thus the combination of increased energy intake and decreased energy expenditure increases the likelihood of weight gain. Although lifestyle, socioeconomic, and cultural factors influence body weight, genetics greatly affects our susceptibility to obesity as well. Scientists have long suspected that internal signals regulate body weight by adjusting energy intake and expenditure, and developed the "set point" theory to explain this phenomenon. Research later confirmed that the hormone leptin plays a role in regulating body weight. However, most obese people produce adequate amounts of leptin. This suggests that leptin may instead protect against starvation, rather than preventing obesity.

# Regulation of Energy Balance

Significant gains have been made in understanding physiological systems that regulate energy balance and body weight, although much remains unclear. Although body weight is a function of the balance between energy intake and energy expenditure, at times increases or decreases in food intake fail to produce the expected change in body weight. For example, increases in energy intake do not always result in the corresponding increase in body weight. Likewise, a decrease in energy intake does not always cause weight loss. This is because the body can sometimes make adjustments in energy expenditure to coincide with changes in energy intake. For this to occur, the brain receives signals that convey information about food intake and the body's energy reserve. The brain uses this information to coordinate adjustments in energy intake and expenditure to maintain energy balance on both a short- and a long-term basis.

## The Role of the Central Nervous System in Regulating Body Weight

Years ago researchers discovered that hunger and satiety could be controlled in mice by electrically stimulating certain regions within their brains. These areas soon became known as the hunger and satiety centers. However, scien-

**thrifty gene hypothesis** A theory that suggests people who are most likely to survive starvation are those most likely to gain weight when food is abundant.

tists now recognize that energy intake is regulated, in part, by specific neural connections (neurons) to and within the brain, rather than by distinct hunger and satiety centers.[52]

Neural, hormonal, and nutritional signals that originate from the gastrointestinal tract, pancreas, and adipose tissue relay information about energy balance to the brain. For example, signals from the gastrointestinal tract communicate information to the brain about recent food intake, whereas signals from the pancreas and adipose tissue communicate the body's energy reserves. One prime target of these signals is a part of the brain called the **arcuate nucleus** in the hypothalamus. Stimulation of the arcuate nucleus and other areas of the brain causes the release of neuropeptides, hormone-like proteins produced by the nervous system. However, unlike hormones, neuropeptides regulate activities mainly in the central nervous system.

Some neurons release **catabolic neuropeptides,** whereas others release **anabolic neuropeptides.** Catabolic neuropeptides promote weight loss by decreasing energy intake and/or increasing energy expenditure. Anabolic neuropeptides promote weight gain by increasing energy intake and/or decreasing energy expenditure. The relative activities of catabolic and anabolic neuropeptides help determine whether a person is in negative or positive energy balance.[53] Table 9.5 lists catabolic and anabolic neuropeptides thought to be involved in energy balance.

## Short-Term Regulation of Food Intake

Neural and hormonal signals from the GI tract play a role in the short-term regulation of food intake. These signals are triggered before and after food consumption, and elicit meal initiation and termination. Regulation of short-term food intake is influenced by signals that originate from the gastrointestinal (GI) tract. These signals relay information to the brain about the amount of food consumed and the availability of certain nutrients. In response, neuropeptides that influence hunger and satiety are released. Signals from the

**arcuate nucleus** A collection of neurons in the hypothalamus that release anabolic and catabolic neuropeptides, which influence food intake and/or energy expenditure.

**catabolic neuropeptide** A protein released by nerve cells that inhibits hunger and/or stimulates energy expenditure.

**anabolic neuropeptide** A protein released by nerve cells that stimulates hunger and/or decreases energy expenditure.

**TABLE 9.5  Impact of Catabolic and Anabolic Neuropeptides on Energy Balance**

| Anabolic Neuropeptides | Effect on Food Intake | Effect on Energy Expenditure |
|---|---|---|
| Neuropeptide Y (NPY) | ↑ | ↓ |
| Agouti-related peptide (AgRP) | ↑ | ↓ |
| Melanin-concentrating hormone (MCH) | ↑ | — |
| Orexin (ORX) | ↑ | — |

| Catabolic Neuropeptides | Effect on Food Intake | Effect on Energy Expenditure |
|---|---|---|
| α-Melanocyte–stimulating hormone (α-MSH) | ↓ | ↑ |
| Corticotropin-releasing hormone (CRH) | ↓ | ↑ |
| Pro-opiomelanocortin (POMC) | ↓ | ↑ |
| Cocaine- and amphetamine-regulated transcript (CART) | ↓ | ↑ |

SOURCE: Schwartz MW, Woods SC, Porte D, Seeley RJ, Baskin DG. Central nervous system control of food intake. *Nature.* 2000;404:661–71.

## FIGURE 9.11 Regulation of Short-Term Food Intake

Signals originating from the gastrointestinal tract stimulate the brain to regulate short-term food intake by releasing neurotransmitters that influence meal initiation and termination.

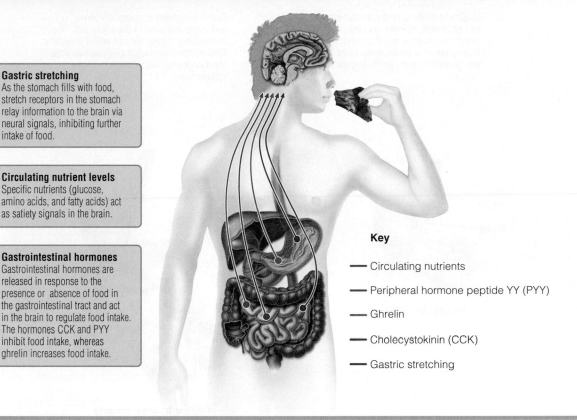

**Gastric stretching**
As the stomach fills with food, stretch receptors in the stomach relay information to the brain via neural signals, inhibiting further intake of food.

**Circulating nutrient levels**
Specific nutrients (glucose, amino acids, and fatty acids) act as satiety signals in the brain.

**Gastrointestinal hormones**
Gastrointestinal hormones are released in response to the presence or absence of food in the gastrointestinal tract and act in the brain to regulate food intake. The hormones CCK and PYY inhibit food intake, whereas ghrelin increases food intake.

**Key**

— Circulating nutrients

— Peripheral hormone peptide YY (PYY)

— Ghrelin

— Cholecystokinin (CCK)

— Gastric stretching

GI tract that play a role in regulating short-term food intake include gastric stretching, circulating nutrient levels, and GI hormones (Figure 9.11).

## Gastric Stretching Signals Satiety

The expansion of the stomach in response to the presence of food, called gastric stretching, provides a powerful satiety signal. When small amounts of food are consumed, the feeling of fullness associated with gastric stretching is barely noticeable. As the volume of food increases stretch receptors in the stomach are stimulated, relaying this information to the brain via neural signals. In response, the brain releases catabolic neuropeptides, causing the sensation of satiety.[54]

Obese people tend to have larger stomachs than do lean people and can accommodate larger volumes of food before gastric stretching triggers satiety. This may be why some severely obese people do not readily feel full. Some severely obese people have weight loss surgery to reduce stomach size. Having a small stomach (gastric) pouch limits how much food can be consumed (Figure 9.12). When the person eats, the pouch quickly fills with food, triggering satiety. You can read more about weight loss surgery in the Focus on Clinical Applications feature.

## Circulating Nutrient Concentrations

The concentration of certain nutrients in the blood also influences hunger and satiety. For example, some neurons in the brain detect changes in blood

## FIGURE 9.12 Types of Weight Loss Surgery

The term *bariatric surgery* refers to medical procedures that promote weight loss. To be considered for weight loss surgery, experts suggest that a person must have a BMI of 40 kg/m² or more or have a BMI of 35 to 39.9 kg/m² with serious medical conditions.

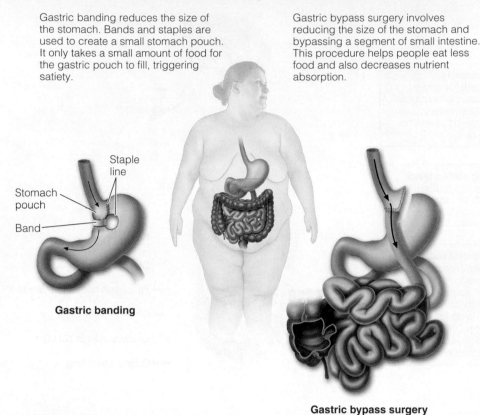

Gastric banding reduces the size of the stomach. Bands and staples are used to create a small stomach pouch. It only takes a small amount of food for the gastric pouch to fill, triggering satiety.

Gastric bypass surgery involves reducing the size of the stomach and bypassing a segment of small intestine. This procedure helps people eat less food and also decreases nutrient absorption.

Staple line

Stomach pouch

Band

**Gastric banding**

**Gastric bypass surgery**

glucose levels. A rise in blood glucose following a meal stimulates satiety, providing a cue to stop eating.[55] Conversely, a decrease in blood glucose causes people to feel hungry and light-headed, cueing them that they need to eat. Perhaps you have experienced this when long periods of time elapse between meals.

Amino acids can also play a role in short-term regulation of food intake. In general, elevated levels of circulating amino acids promote satiety. Of particular interest is the effect of the amino acid tryptophan on food intake. The brain uses tryptophan to synthesize the neurotransmitter serotonin. Some studies suggest that high intakes of foods containing tryptophan, such as turkey and dairy products, may promote both satiety and sleepiness.[56]

In addition to glucose and amino acids, dietary fat also promotes satiety. However, this effect seems weak and is easily overcome by pleasant sensations associated with fat, such as flavor and texture. As a result, consuming high-fat foods tends to lead to excess calorie intake and weight gain, not satiety.[57]

### Gastrointestinal Hormones

The presence of food in the stomach and small intestine can trigger the release of several GI hormones, the majority of which promote satiety (Table 9.6).[58] The exception is a recently discovered hormone, ghrelin, which increases food intake by signaling the release of anabolic neuropeptides—at least in rats. Produced primarily by the stomach, ghrelin is released in response to a lack of food, and circulating concentrations decrease after food is consumed. Ghrelin

# *Focus On* CLINICAL APPLICATIONS
## Bariatric Surgery

**B**ariatrics is the branch of medicine concerned with treating obesity and obesity-related conditions. The term *bariatric surgery* refers to surgical procedures that promote weight loss. There are different types of bariatric surgery, each having potential side effects and risks.[59] However, the health benefits associated with weight loss often outweigh the risks associated with surgery. To even be considered for surgery, experts recommend that a person have a BMI of 40 kg/m$^2$ or more, or have a BMI of 35 to 39.9 kg/m$^2$ with a serious medical condition.[60]

In general, bariatric surgery alters the digestive tract such that only small amounts of foods can be consumed and/or digested. For example, a procedure called **gastric banding** reduces the size of the stomach by creating a small stomach pouch that holds about 2 to 3 tablespoons (about the size of an egg) of food. In time, the pouch stretches and can accommodate slightly larger amounts of food, about 1 cup. Remember that the stomach normally holds 4 to 6 cups of food. Therefore it only takes a small amount of food to signal the feeling of satiety. Individuals with this type of surgery must be careful not to overeat, to eat foods that are soft and moist, and to thoroughly chew food before swallowing. Side effects include heartburn, abdominal pain, and vomiting. Because food intake is severely restricted, care must be taken to ensure that nutritional needs are met, and supplements are typically recommended.

Another type of bariatric surgery, called **gastric bypass,** reduces the size of the stomach and bypasses a segment of the small intestine. To do this, the lower portion of the reduced stomach pouch is connected to the small intestine, bypassing a segment of the small intestine. Not only is there a reduction in food intake, but the intestinal bypass also limits digestion and absorption. The risk of developing nutritional deficiencies is much greater after this combined procedure. Also, many side effects such as nausea and diarrhea are caused by rapid movement of food through the small intestine.

About 20% of people who have weight loss surgery experience complications.[61] Furthermore, depending on an individual's age, sex, and general health, the death rate within one year of surgery is 1 to 5%.[62] Although patients with gastric banding lose weight more slowly than those who have gastric bypass, the banding procedure is somewhat reversible and has a lower mortality rate. Still, one must give serious consideration to risks before undergoing either of these procedures.

The number of weight loss surgeries performed yearly in the United States is on the rise. Most people lose weight immediately following the surgery, and within five years have lost 60% of their excess weight. However, long-term effectiveness depends on a person's willingness to eat a healthy, well-balanced diet and exercise regularly. For many severely obese individuals, weight loss surgery can improve the quality of life, both physically and emotionally.[63]

is thought to cause the sensation of hunger, and there is evidence that overproduction of ghrelin may contribute to obesity.[64] This could, in part, explain why dieters have difficulty maintaining weight loss and why some obese people do not experience the feeling of satiety.

Other GI hormones that regulate food intake include cholecystokinin (CCK) and peripheral hormone peptide YY (PYY). Of these, CCK is the best

**bariatrics** The branch of medicine concerned with the treatment of obesity.

**gastric banding** A band around the upper portion of the stomach, dividing it into a small upper pouch and a larger lower pouch.

**gastric bypass** A surgical procedure that reduces the size of the stomach and bypasses a segment of the small intestine so that fewer nutrients are absorbed.

## TABLE 9.6  Gastrointestinal Hormones and Regulation of Food Intake

| Gastrointestinal Hormone | Stimulus for Release | Site of Production | Effect on Food Intake |
|---|---|---|---|
| Cholecystokinin (CCK) | Protein and fatty acids | Small intestine | ↓ |
| Glucagon-like peptide 1 (GLP-1) | Nutritional signals and neural/hormonal signals from the gut | Small and large intestine | ↓ |
| Ghrelin | Fasting | Stomach | ↑ |
| Enterostatin | Fatty acids | Stomach and small intestine | ↓ |
| Peripheral hormone peptide YY (PYY) | Released in proportion to energy content of meals | Gastrointestinal tract | ↓ |

SOURCE: Marx J. Cellular warriors at the battle of the bulge. *Science.* 2003;299:846–9.

understood. CCK is released from intestinal cells (enterocytes), particularly in response to dietary fat and protein, signaling the brain to decrease food intake by mechanisms that are not clear. Similarly, PYY is also released from the intestinal lining following a meal, inhibiting the release of anabolic neuropeptides, thus decreasing food intake.

## Long-Term Regulators of Energy Balance

In addition to short-term regulation of food intake, the body also regulates energy intake on a long-term basis. These more chronic regulators of food intake serve a different purpose—body weight regulation. Long-term regulatory signals communicate the body's energy reserves to the brain, which in turn releases neuropeptides that influence energy intake and/or energy expenditure.[65] If this long-term system functions effectively, body weight remains somewhat stable over time.

The mechanisms that regulate long-term energy balance are complex and not well understood. However, the hormones leptin, and to a lesser extent, insulin, appear important (Figure 9.13). It seems that leptin and insulin also influence the brain's response to short-term satiety signals released from the GI tract. Thus short-term and long-term regulatory systems are highly coordinated, and together influence energy balance.

---

**FIGURE 9.13** **Leptin and Insulin: Long-Term Regulators of Body Weight**

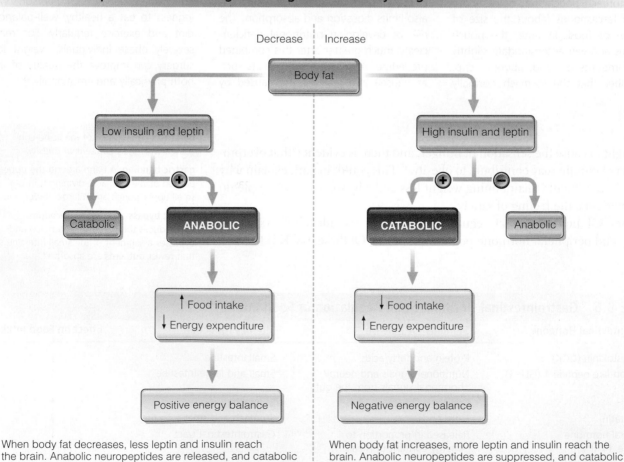

When body fat decreases, less leptin and insulin reach the brain. Anabolic neuropeptides are released, and catabolic neuropeptides are suppressed. This condition favors increased food intake, decreased energy expenditure, and weight gain.

When body fat increases, more leptin and insulin reach the brain. Anabolic neuropeptides are suppressed, and catabolic neuropeptides are released. This condition favors decreased food intake, increased energy expenditure, and weight loss.

## The Roles of Leptin and Insulin

The hormone leptin is produced primarily by adipose tissue, and when body fat increases circulating leptin concentration increases as well. Likewise, when body fat decreases leptin production decreases. Along with leptin, the hormone insulin is also important in communicating adiposity to the brain.[66] In fact, many obese people have high levels of both leptin and insulin.

Increases and decreases in adipose tissue—the body's primary energy reserve—are communicated to the brain, in part, by fluctuating levels of leptin and insulin. Together, these adiposity signals are part of a negative feedback loop that helps maintain a relatively stable body weight. As you may recall, the acute rise in insulin after a meal acts as an anabolic hormone, favoring energy storage in peripheral tissues. Unlike insulin, acute changes in leptin are not easily detected after meals. However, changes in baseline levels of these hormones appear to play a different role: body weight regulation. Specifically, increases in adiposity cause leptin and insulin concentrations to rise, which in turn signals the brain to increase its release of catabolic neuropeptides. Catabolic neuropeptides help the body to resist further weight gain by decreasing energy intake and increasing energy expenditure. Conversely, decreased adiposity causes blood concentrations of leptin and insulin to decrease. When this occurs, the brain increases its release of anabolic neuropeptides. The release of these neuropeptides stimulates energy intake and decreases energy expenditure, thus protecting the body against further weight loss. In this way, leptin and insulin are part of a more long-term homeostatic system that helps prevent large shifts in body weight. Some researchers believe that defects in the leptin/insulin signaling system may lead to impaired body weight regulation in some people.[67] Clearly, there is more to learn about this regulatory system. However, the role of leptin and insulin in body weight homeostasis provides a better understanding of how the body protects itself during times of food scarcity and how defects in this system may promote obesity during times of food surplus.

## ESSENTIAL *Concepts*

The brain receives information about energy intake and the body's energy reserve via nerves, circulating nutrients, and hormones. This complex signaling system strives to maintain energy balance by making adjustments in energy intake and/or energy expenditure. The brain releases catabolic and anabolic neuropeptides that promote weight loss and weight gain, respectively. Signals from the GI tract play a role in regulation of short-term food intake and include gastric stretching and GI hormones. Circulating concentrations of glucose, fatty acids, and amino acids also influence hunger and satiety. The majority of GI hormones inhibit food intake with the exception of ghrelin, which stimulates hunger. The hormones leptin and insulin play a role in long-term regulation of energy balance by suppressing the release of anabolic neuropeptides in the brain. When leptin and insulin concentrations decrease, anabolic neuropeptides are released, resulting in overall increased energy intake and decreased energy expenditure.

# Approaches to Weight Loss

Although the weight loss industry would like us to think otherwise, the truth is clear: There is no quick and easy way to lose weight. Approximately one third of adults in the United States are on a weight loss diet, and there are plenty to choose from. Certainly the current obesity epidemic cannot be attributed to a lack of weight loss advice.

Most popular weight-loss plans have specific rules for people to follow. For instance, some recommend eating foods in the right combination, whereas others require foods be chosen according to blood type. However, the real issue is not just what it takes to lose weight, but what it takes to keep it off. Rather than succumbing to "quick fix" approaches to weight loss, it is important to consider what weight loss experts have to say.

## Eating to Promote Overall Health and Well-Being

Although there are many reasons why people want to lose weight, the most important is to improve health. The reality is that achieving and maintaining weight loss requires making lasting lifestyle changes, including what we choose to eat and how much physical activity we engage in. Most health experts suggest that people focus less on weight loss and more on eating healthy and on overall fitness.[68] Misguided efforts of weight loss at any cost have unfortunately presented food as the enemy, rather than as a means to good health. Most people who successfully lose weight and keep it off do so by eating a balanced diet of nutrient-dense foods and by maintaining a moderately high level of

### Working Toward the Goal

The USDA Dietary Guidelines for Americans recommend that we maintain a healthy body weight by balancing caloric intake with energy expenditure. The following tips will help you maintain a healthy body weight and to prevent gradual weight gain over time.

- Consume foods that are nutrient dense while limiting foods that are high in saturated fats, *trans* fats, refined carbohydrates, and added sugars.
- Select foods that add dietary fiber, such as fruits, vegetables, and whole grains.
- Follow a balanced eating plan that emphasizes variety, moderation, and balance.
- Select low-fat dairy products and lean meats whenever possible.
- Pay attention to internal cues of hunger and satiety.
- Make food portions smaller than normal. If still hungry, eat a second portion.
- Avoid distractions while eating, such as watching television, that can lead to over-consumption of food.
- Eat regularly and avoid skipping meals that can lead to extreme hunger.
- Avoid eating food directly from containers or packages. Pay attention to food portions by serving food on a plate or in a bowl.
- Many restaurants serve generous food portions—enough for two people. Try sharing a meal or if you can't find someone to share with, ask for a carryout bag and take the rest of the meal home.
- Increase your physical activity by using the stairs instead of elevators, walking instead of driving, and parking so as to walk farther.
- Join a fitness center or take a fitness class.
- Take up a hobby, such as gardening, that is both rewarding and physically active.
- Make exercise interesting by changing your routine or by doing it with a friend.

physical activity.[69] Key recommendations for weight management based on the Dietary Guidelines for Americans are provided in the Food Matters.

A healthy weight loss and weight maintenance program consists of three components: (1) setting reasonable goals, (2) choosing nutritious foods in moderation, and (3) increasing energy expenditure by daily physical activity. These are described here.

## Setting Reasonable Goals

Setting reasonable and attainable goals is an important component of any successful weight-loss program. A realistic weight-loss goal is to reduce current weight by 5 to 10%. Studies show that even this modest reduction in weight can improve overall health.[70] When it comes to weight loss, slow and steady is the way to go, and weight loss should not exceed 1 to 2 pounds a week. This can be achieved by decreasing energy intake by 200 kcal each day. Over time, these small changes can result in significant changes in body weight. Rather than making dramatic dietary changes, small changes such as reducing portion sizes and cutting back on energy-dense snack foods make a big difference in overall energy intake.

Once body weight stabilizes and the new lower weight is maintained for a few months, a decision can be made whether additional weight loss is needed. Some people benefit from joining weight management programs such as Weight Watchers™, or Take Off Pounds Sensibly™. These types of programs provide long-term support and motivate people to maintain a healthy diet and lifestyle.

## Choosing Nutritious Foods

Weight loss plans that drastically reduce calories and offer limited food choices leave people feeling hungry and dissatisfied. Thus weight loss diets that encourage people to eat foods that are healthy and appealing tend to have greater success. Contrary to popular belief, to lose weight it is not necessary to avoid foods that contain fat. The Dietary Guidelines for Americans now recommends that we choose our fats as carefully as we choose our carbohydrates. In general, it is best to limit intake of foods containing *trans* fatty acids and saturated fatty acids. Foods containing relatively more polyunsaturated and monounsaturated fatty acids tend to be healthier.

A common misconception is that dairy products and meat are high-fat foods that should be avoided when trying to lose weight. Again, the key to good nutrition is moderation and choosing wisely. For example, switching from whole to reduced-fat milk is one way to lower caloric intake without losing out on the many vitamins and minerals in dairy products. Likewise, how meat is prepared and the types of meat consumed can greatly affect the amount of calories consumed. Lean meats prepared by broiling or grilling are both nutritious and satisfying.

Reducing energy intake is best achieved by cutting back on energy-dense foods that have little nutritional value such as soft drinks, potato chips, cookies, and cakes. Aside from their lower energy densities, foods such as whole grains, legumes, nuts, fruits, and vegetables offer many health benefits such as micronutrients and fiber. In addition, these foods tend to have greater volume compared to more energy-dense foods, thus helping people feel full. In the Focus on Food feature, you can read more about how eating high-volume, low-energy dense foods can help people feel full and satisfied.

Healthy eating also requires people to pay attention to hunger and satiety cues. Rather, the amount of food served or packaged often determines how much we eat. That is, visual cues, rather than internal cues, have a greater

# *Focus On* FOOD
## Eating More and Weighing Less

People trying to lose weight tend to have the most success when they select foods that are both nutritious and satisfying. Too often, weight loss plans emphasize food restriction, causing a state of perpetual hunger. As a result, people typically revert back to former eating practices. However, new studies show that eating more can sometimes help people weigh less.[71] This is because eating high-volume, low–energy-dense foods can help people feel full longer than when they eat low-volume, high–energy-dense foods.[72] In other words, it is possible for people to consume larger volumes of foods while eating fewer calories.

High-volume, low–energy-dense foods are those with high water and fiber contents such as fruits and vegetables. Recall that as the stomach fills with food, the walls stretch. As a result, neural signals are sent to the brain, triggering the release of catabolic neuropeptides. These neuropeptides cause the sensation of satiety. This is why eating foods with a high volume can help people feel full and satisfied. For example, two cups of grapes have the same number of calories (~100 kcal) as ¼ cup of raisins. However, because the volume of the grapes is greater than raisins, a person is more likely to feel fuller longer after eating the grapes.

The energy density of food reflects the relationship between energy (kcal) and weight (grams). High–energy-dense foods are those that provide 4 to 9 kcal per gram, whereas low–energy-dense foods have less than 1.5 kcal per gram.[73] High–energy-dense foods typically have a

© Scott Bauer/ARS/USDA

**Foods that are not energy dense, such as these grapes, have a high water and fiber content and are low in fat.**

low-moisture or high-fat content. These include cookies, cakes, crackers, cheesecake, and butter. Examples of foods with medium energy density (1.5 to 4 kcal/g) include eggs, dried fruits, breads, and cheese. Low–energy-dense foods have a high water and/or fiber content and are low in fat, such as most fruits and vegetables, broth-based soups, low-fat cheese, and certain types of meat such as roasted turkey.

Studies show that people feel fuller and tend to consume fewer calories when fed low–energy-dense foods.[74] For example, people who eat salads or broth-based soups prior to a high–energy-dense main course consume fewer calories overall. Similarly, the effect of volume on satiety was demonstrated by feeding people milkshakes with the

same number of calories but with different volumes (300 mL, 450 mL, and 600 mL). Participants consumed 12% fewer calories when fed 600-mL milkshakes than when they consumed those with less volume. These and other studies show that meal volume provides a stronger satiety signal than the total calories in the food.[75] Most importantly, people consuming low–energy-dense foods lose more weight over time than those following a fat-reduction weight-loss plan.

Here are some suggestions to increase your intake of low–energy-dense foods.

- Eat a low–energy-dense salad before a main course. Lower the energy density of your salad by adding plenty of greens, vegetables, and low-fat salad dressing.
- As an alternative to salad, have a broth-based soup containing vegetables, high-fiber grains (such as barley), and legumes (such as split peas or lentils).
- Reduce energy density by cutting back on the amount of high-fat meat in recipes and adding more vegetables.
- Avoid beverages that have high sugar or fat contents. Instead, add a small amount of fruit juice to carbonated water.
- Read the Nutrition Facts panel on food labels, and compare the number of calories in relation to the number of grams. Foods that have fewer calories than grams are low–energy-dense foods, whereas foods that have more than twice the number of calories than grams are high–energy-dense foods.

influence on the quantity of food consumed. For instance, some commercially made muffins are extremely large, containing as many calories as 8 slices of bread. Therefore, learning to choose reasonable portions of food is a critical component to successful weight management. In fact, reducing portion sizes by as little as 10 to 15% could reduce our daily energy intake by as much as 300 kcal. One way to limit serving size is to consider sharing large and supersized meals the next time you eat at a restaurant.

## Physical Activity

In addition to eating less, physical activity also helps tip the energy balance equation toward negative energy balance. Walking 1 mile a day, which takes most people about 15 to 20 minutes, uses about 100 kcal. This adds up to 700 kcal per week, if done daily. Small amounts of physical activity can make a big difference in terms of weight loss and weight maintenance. Even without weight loss, individuals who exercise show improved physical fitness. Regardless of weight, lack of exercise may prove the greatest health hazard of all.

It is important to realize that people can be physically fit while being overweight. Normal blood pressure, healthy blood glucose regulation, and healthy levels of blood lipids are important indicators of physically fitness. Studies show that obese individuals who are physically fit have fewer health problems than do average-weight individuals who are unfit.[76] Exercise also promotes a positive self-image and helps people take charge of their lives. Although people often resist the idea of starting an exercise program, they rarely regret it once they begin.

Physical activity is an effective strategy in preventing unhealthy weight gain in normal, overweight, and obese individuals. An expert panel assembled by the International Association for the Study of Obesity concluded that 45 to 60 minutes of daily exercise can help prevent normal-weight individuals from becoming overweight, overweight people from becoming obese, and obesity from worsening in already obese individuals.[77] People who were formerly obese, however, may require more exercise—60 to 90 minutes each day—to maintain a lower body weight. This is because a person requires 8 fewer kcal/day for every pound of body weight lost. For example, when a person loses 10 pounds, the energy requirement decreases by 80 kcal/day. Thus, to maintain a 10-pounds weight loss, a person must further reduce energy intake or increase energy expenditure.

Not all obese people are physically unfit. It is the combination of being both overweight and sedentary that increases a person's risk of developing weight-related health problems.

## ESSENTIAL *Concepts*

Stringent dieting alone rarely results in long-term weight loss. Instead, maintaining weight loss requires lasting lifestyle changes, which include eating healthy foods and regular exercise. A healthy weight-loss and weight-maintenance program sets reasonable goals, encourages intake of low–energy-dense, nutrient-dense foods, and promotes regular physical activity. A realistic weight-loss goal is to decrease body weight by 5 to 10%, not exceeding 1 to 2 pounds each week. Choosing reasonable serving sizes is also important for weight loss and weight management. In addition, exercise helps prevent weight gain and improves overall health.

## Does Macronutrient Distribution Matter?

Experts have long had varying opinions as to how weight loss can best be achieved. Today, one of the biggest controversies is the role of dietary carbohydrate versus dietary fat in promoting weight loss and weight gain.[78] Weight loss diets that are low in fat and high in carbohydrates have long been considered the most effective, in terms of weight loss and weight maintenance. In fact, the Dietary Guidelines for Americans advocates low-fat food choices with an emphasis on whole grains, fruits, and vegetables. Similarly, the Acceptable Macronutrient Distribution Range (AMDR) suggests that we consume 45 to

**TABLE 9.7 Caloric Distribution of High-Carbohydrate and Low-Carbohydrate Diets**

| Nutrient | AMDR[a] | High-Carbohydrate Diet[b] | Low-Carbohydrate Diet[c] |
|---|---|---|---|
| | | % of total calories | |
| Fat | 20–35 | 10–15 | 55–65 |
| Carbohydrate | 45–65 | 65–75 | 5–20 |
| Protein | 10–35 | 10–25 | 20–40 |

[a] Acceptable Macronutrient Distribution Ranges. Institute of Medicine. *Dietary Reference Intakes for Energy, Carbohydrate, Fiber, Fat, Fatty Acids, Cholesterol, Protein, and Amino Acids.* Washington, DC: National Academies Press; 2005.
[b] High-carbohydrate, low-fat weight loss diet advocated by Dr. Dean Ornish. In: Ornish D. *Eat More Weigh Less: Dr. Dean Ornish's Life Choice Diet for Losing Weight Safely While Eating Abundant.* New York: HarperCollins; 1993.
[c] Low-carbohydrate, high-fat weight loss diet advocated by Dr. Robert Atkins. In: Atkins RC. *Dr. Atkins' New Diet Revolution, Revised.* National Book Network; 2003.

65% of energy from carbohydrate, 10 to 35% from protein, and 20 to 35% from fat. However, Robert Atkins, one of the first pioneers of the low-carbohydrate diet, turned the nutritional world upside down in 1972 when he proposed that too much carbohydrate, rather than too much fat, may actually cause people to gain weight. The AMDR caloric distribution is compared to that of high- and low-carbohydrate weight loss diets in Table 9.7. We next examine the scientific evidence concerning the effects of macronutrient distribution on weight loss and overall health.

## High-Carbohydrate, Low-Fat Weight Loss Diets

Dean Ornish was one of the first to claim that diets low in fat and high in carbohydrates promote weight loss and have an overall beneficial effect on health. To maintain a low intake of fat (10 to 15% of total calories), Ornish advises dieters to avoid meat, dairy, oils, and olives; low-fat meat and dairy products can be eaten in moderation. With an emphasis on fruits, vegetables, and whole grains, this weight reduction and maintenance plan provides about 65 to 75% of total calories from carbohydrates, with protein making up the difference.

There are several reasons why advocates of low-fat diets believe such diets help prevent obesity. First, gram for gram, fat has twice as many calories as carbohydrate and protein. Therefore it is reasonable to assume that consuming less fat may lead to lower energy intake, which in turn results in weight loss. Fat can also make food more flavorful, contributing to overconsumption. Last, excess calories from fat are more readily stored by the body compared to those from carbohydrate or protein. This is due to the energy cost associated with converting excess glucose and amino acids into fatty acids prior to storage. Some experts claim that low-fat diets benefit overall health as well, by lowering total and LDL cholesterol concentrations, increasing HDL cholesterol concentrations, and improving blood glucose regulation.[79] However, these benefits may be due to weight loss in general rather than to reduced dietary fat.

**CONNECTIONS** Recall that elevated LDL cholesterol and low HDL cholesterol are risk factors for cardiovascular disease (Chapter 7, pages 248–249).

### Trends in Dietary Fat Intake

Long-standing dietary advice aimed at helping people lose weight has consistently focused on reducing dietary fat. Although total energy intake has increased, the percentage of total calories from fat has declined from 45% in the 1960s to approximately 33% today. According to the USDA Center for Nutrition Policy and Promotion, this reduction adds up to an average of 10 to 20 fewer grams of fat per day.[80] However, decreased fat intake has not resulted in a decreased prevalence of obesity. In fact, obesity rates have increased under the low-fat regime.[81] Although it is not clear which dietary factors contributed to this trend, some researchers believe a lack of emphasis on healthy carbohydrate food choices may be reason for concern.

Although nutritionists hoped Americans would replace fatty foods in their diet with more nutritious grains, fruits, and vegetables, they have not done so. The availability of low-fat snack products has made it possible for people to eat snack foods minus the fat. Typically, the ingredient that replaces fat in these products is refined carbohydrate (such as white flour and sugar), and people mistakenly think that low-fat snack foods are healthy snack alternatives. Overconsuming fat-free snack foods may, in part, contribute to weight gain, because many have the same amount of calories as the original product. Some researchers believe that eating foods high in refined carbohydrate, especially those low in fat, make us hungrier, and therefore can make us heavier.[82]

The theory that a low-fat diet is our best defense against weight gain is not without debate, and many health experts believe there is now enough solid evidence to lift the "ban" on dietary fat.[83] On the opposite end of the weight-loss diet spectrum are low-carbohydrate, high-fat diets, which are discussed next.

## Low-Carbohydrate, High-Fat Weight Loss Diets

Health claims made by advocates of low-carbohydrate, high-fat diets include weight loss without hunger and improved cardiovascular health.[84] These experts contend that people are more likely to gain weight from excess carbohydrates as opposed to excess fats or proteins. This premise is based on the fact that high-carbohydrate foods cause insulin levels to rise, which in turn, may cause weight gain. Thus eliminating starch and refined sugars from the diet helps people lose weight.

**CONNECTIONS** Recall that insulin is a hormone needed for blood glucose regulation (Chapter 5, page 143).

There are many low-carbohydrate, high-fat weight loss diets, each differing in terms of the types of foods allowed. Although some exclude nearly all carbohydrates, others, such as the Zone™ and South Beach™ diets, take a more moderate approach by allowing fruits, vegetables, and whole grains. Indeed, low-carbohydrate diets have become enormously popular, with 13% of Americans reporting that are on some type of low-carbohydrate diet to lose weight.[85] Low-carbohydrate diets appear safe in the short term, but little is known about their long-term effects.[86]

### Do Low-Carbohydrate Diets Work?

The lure of a diet plan that does not require counting calories or limiting portion sizes is appealing to many people. However, do low-carbohydrate diets really help one lose weight? When a person restricts intake of carbohydrates, the body breaks down glycogen to provide glucose. Approximately 500 g (about 1 pound) of glycogen is stored in the liver and skeletal muscle, combined. Glycogen is bound to molecules of water, approximately 3 g of water per gram of glycogen. Thus the combined weight of stored glycogen and associated water is about 4 pounds. Consequently, when glycogen is broken down, its associated water is eliminated from the body. Because of this, the initial weight loss associated with low-carbohydrate diets is largely attributed to water loss.

CONNECTIONS ▶ Recall that ketosis is a condition resulting from the accumulation of ketones in the blood (Chapter 5, page 149).

Once glycogen stores are depleted, the body begins to use triglycerides for energy. However, limited glucose availability allows fatty acids to be only partially broken down. As a result, ketone formation increases, a sign that the body is using stored fat as a major source of energy. Limiting carbohydrate intake causes ketosis, and for this reason these diets are called **ketogenic diets.** Ketosis often results in a loss of appetite, further promoting weight loss.

Studies comparing weight loss associated with low-carbohydrate diets and with low-fat diets show that at 6 months, greater weight loss is achieved on low-carbohydrate diets. However, differences disappear by 12 months.[87] Still, weight loss associated with low-carbohydrate diets may not necessarily be caused by alterations in the macronutrient composition of the diet, but rather by a reduction in caloric intake.[88] That is, limited food choice and decreased appetite associated with ketosis may cause people to eat less.

There is no consistent evidence that carbohydrate restriction causes the body to burn energy more efficiently. Therefore, it is not true that a person can maintain a high intake of calories and still lose weight while on a low-carbohydrate diet. In other words, there is no compelling evidence that low-carbohydrate diets are more effective than other types of diets in helping people lose weight.[89]

## Nutritional Adequacy of Low-Carbohydrate Diets

The amount of carbohydrates allowed on most low-carbohydrate diet plans varies from 20 to 60 g per day—approximately 5 to 20% of total energy. By comparison, current recommendations suggest that 45 to 65% of total calories come from carbohydrates (225 to 325 g/day for a diet of 2,000 kcal). Low-carbohydrate diets have been criticized as containing too much total fat, saturated fat, and cholesterol. Current recommendations regarding optimal levels of fat in the diet suggest that 20 to 35% of total calories come from fat. By comparison, most low-carbohydrate diets provide 55 to 65% calories from fat. In addition, not all low-carbohydrate diets distinguish between "healthy" and "unhealthy" fats. Likewise, there is some concern that the amount of protein in some low-carbohydrate diets exceeds recommended amounts. Many low-carbohydrate diets provide 20 to 40% of total calories from protein. It is currently recommended that we consume 10 to 35% of calories from protein. Therefore, not all low-carbohydrate diets are actually high in protein.

Perhaps one of the biggest concerns regarding low-carbohydrate diets is the restriction of healthy, high-carbohydrate foods such as fruits, vegetables, and whole grains—a valid criticism.[90] Low-carbohydrate diets may lack essential micronutrients, dietary fiber, and beneficial phytochemicals. Although most low-carbohydrate diet plans recommend that people take dietary supplements, this cannot replace the many other nutrients (such as fiber and phytochemicals) in these restricted foods.

## Other Health Concerns Regarding Low-Carbohydrate Diets

Because some cells (such as red blood cells) require glucose, consuming a very low-carbohydrate diet causes the body to break down protein (muscle), so that the resulting amino acids can be used for glucose synthesis. Thus concerns have been raised that low-carbohydrate diets may cause loss of lean tissue. However, studies suggest that weight loss associated with low-carbohydrate diets is largely attributed to a loss of body fat, with little loss of lean tissue.[91] This may be because protein intakes associated with some low-carbohydrate diets are high enough to prevent the loss of muscle.

**ketogenic diets** Diets that stimulate ketone production.

Concerns regarding the effect of low-carbohydrate diets on bone health and kidney function have also been expressed. However, limited research shows no harmful effects of low-carbohydrate/high-protein diets on bone health.[92] Similarly, some professionals worry that low-carbohydrate diets may impair kidney function, because high protein intakes and ketones are thought to overburden the kidneys.[93] Although this has not been adequately studied, low-carbohydrate diets are not advised for people with impaired kidney function.

Low-carbohydrate diets also do not appear to have an adverse affect on cardiovascular health, at least in the short term.[94] Rather, several studies suggest that low-carbohydrate diets result in favorable changes in blood lipid levels, glycemic control, and blood pressure, despite increased intakes of fat.[95] However, these improvements are likely due to a reduction in body weight rather than to a direct effect of the macronutrient composition of the diet.

## ESSENTIAL *Concepts*

Some experts recommend low-carbohydrate diets for weight loss, whereas others recommend low-fat diets. Proponents of low-fat diets believe that less fat in the diet leads to consumption of fewer calories and to greater weight loss. However, reductions in fat intake have accompanied increased prevalence of obesity in the United States. In addition, compared to low-fat diets greater short-term weight loss is achieved on low-carbohydrate diets, but long-term differences have not been found. Advocates of low-carbohydrate diets believe that high carbohydrate intake can cause insulin levels to rise, leading to weight gain. In addition, limiting carbohydrates causes the body to go into a ketotic state, which results in decreased appetite. Low-carbohydrate diets have been criticized for being too high in fat and protein; however, not all low-carbohydrate diets are high in protein. Concerns regarding the lack of fruits, vegetables, and whole grains have been also expressed. A positive aspect of weight loss associated with low-carbohydrate diets is the loss of body fat without substantial loss of lean tissue.

## Review Questions    Answers are found in Appendix C.

1. A person in positive energy balance is likely to experience
   a. an increase in body weight.
   b. a decrease in body weight.
   c. a decrease in BMR.
   d. memory loss.
2. Most total energy expenditure (TEE) is associated with
   a. basal metabolism.
   b. physical activity.
   c. thermic effect of food.
   d. adaptive thermogenesis.
3. Body mass index (BMI) is based on the
   a. ratio of weight to height.
   b. ratio of waist to hip circumference.
   c. amount of fat deposited directly under the skin.
   d. self-reported weight and height of people who buy life insurance.
4. Intra-abdominal fat deposits are closely associated with a body fat distribution pattern called
   a. gynoid adiposity.
   b. android adiposity.

   c. subcutaneous adiposity.
   d. lower body fat adiposity.
5. A person's waist-to-hip ratio provides an indication of
   a. body fat distribution.
   b. the amount of lean tissue in the body (muscle mass).
   c. ideal body weight.
   d. relative body weight.
6. An anabolic neuropeptide
   a. stimulates hunger.
   b. stimulates satiety.
   c. causes a decrease in body weight.
   d. is produced by adipose tissue.
7. Which of the following is a *true* statement?
   a. Most obese people do not produce the hormone leptin.
   b. Leptin is primarily produced by adipose tissue.
   c. Leptin stimulates hunger when injected into leptin deficient mice.
   d. All of the above statements are *true*.

8. The release of the gastric hormone _____ increases in response to hunger.
   a. cholecystokinin (CCK)
   b. leptin
   c. ghrelin
   d. insulin
9. Which of the following body composition measurement methods uses a device called calipers?
   a. Bioelectrical impedance (BEI)
   b. Skinfold measures
   c. Dual-energy x-ray absorptiometry (DEXA)
   d. Near-infrared interactance (NIR)
10. When a person loses body fat,
    a. the number of adipocytes decrease.
    b. adipocytes undergo a process called hypertrophy.
    c. adipocytes undergo a process called hyperplasia.
    d. the size of adipocytes decreases.
11. List and describe three methods used to assess body composition.
12. Describe the set point theory of body weight regulation.
13. Describe the role of the hormones leptin and insulin in terms of long-term body weight regulation.
14. Describe the physiological rationale of low-carbohydrate diets in terms of promoting weight loss.
15. List and describe three signals that originate in the gastrointestinal tract and that trigger satiety.
16. List and explain the three components of total energy expenditure (TEE).

## Practice Calculations  Answers are found in Appendix G.

1. Calculate body mass index (kg/m²) for a male who is 5 feet, 10 inches tall and weighs 175 pounds. How would you assess this person's weight?

2. Calculate waist-to-hip ratio of a male with a waist circumference of 36 inches and hip circumference of 34 inches. How would you assess this person in terms of body fat distribution and associated health risk?

## Media Links  A variety of study tools for this chapter are available at our website, www.thomsonedu.com/nutrition/mcguire.

Prepare for tests and deepen your understanding of chapter concepts with these online materials:

- Practice tests
- Flashcards
- Glossary
- Student lecture notebook
- Web links
- Animations
- Chapter summaries, learning objectives, and crossword puzzles

## Notes

1. Prins JB. Adipose tissue as an endocrine organ. Best Practice and Research Clinical Endocrinology and Metabolism. 2002;16:639–51.
2. Yanovski S. Sugar and fat: cravings and aversions. Journal of Nutrition. 2003;133:835S–7S.
3. Lafay L, Thomas F, Mennen L, Charles MA, Eschwege E, Borys JM, Basdevant A. Gender differences in the relation between food cravings and mood in an adult community: Results from the Fleurbaix Laventie Ville Sante study. International Journal of Eating Disorders. 2001;29:195–204.
4. Dye L, Blundell JE. Menstrual cycle and appetite control: implications for weight regulation. Human Reproduction. 1997;12:1142–51.
5. Bayley TM, Dye L, Jones S, DeBono M, Hill AJ. Food cravings and aversions during pregnancy: relationships with nausea and vomiting. Appetite. 2002;38:45–51.
6. Dye L, Warner P, Bancroft JJ. Food craving during the menstrual cycle and its relationship to stress, happiness of relationship and depression; a preliminary enquiry. Journal of Affective Disorders. 1995;34:157–64.
7. Wurtman JJ. Carbohydrate cravings: a disorder of food intake and mood. Clinical Neuropharmacology. 1988;1:S139–45.
8. le Roux CW, Bloom SR. Peptide YY, appetite and food intake. Proceedings of the Nutrition Society. 2005;64:213–6.
9. Leibowitz SF. Regulation and effects of hypothalamic galanin: relation to dietary fat, alcohol ingestion, circulating lipids and energy homeostasis. Neuropeptides. 2005;39:327–32.

10. Hulbert AJ, Else PL. Basal metabolic rate: history, composition, regulation, and usefulness. Physiological and Biochemical Zoology. 2004;77:869–76. Institute of Medicine. Dietary Reference Intakes for energy, carbohydrate, fiber, fat, fatty acids, cholesterol, protein, and amino acids. Washington, DC: National Academies Press; 2005.

11. Henry CJ. Mechanisms of changes in basal metabolism during ageing. European Journal of Clinical Nutrition. 2000;54: S77–91.

12. Luke A, Schoeller DA. Basal metabolic rate, fat-free mass, and body cell mass during energy restriction. Metabolism. 1992;41:450–6.

13. Brooks GA, Butte NF, Rand WM, Flatt JP, Caballero B. Chronicle of the Institute of Medicine physical activity recommendation: how a physical activity recommendation came to be among dietary recommendations. American Journal of Clinical Nutrition. 2004;79:921S–30S. Institute of Medicine. Dietary Reference Intakes for energy, carbohydrate, fiber, fat, fatty acids, cholesterol, protein, and amino acids. Washington, DC: National Academies Press; 2005.

14. Institute of Medicine. Dietary Reference Intakes for energy, carbohydrate, fiber, fat, fatty acids, cholesterol, protein, and amino acids. Washington, DC: National Academies Press; 2005. Flatt JP. The biochemistry of energy expenditure. In: Recent advances in obesity research II. Bray GA, editor. London: Newman Publishing; 1978. Nair KS, Halliday D, Garrow JS. Thermic response to isoenergetic protein, carbohydrate or fat meals in lean and obese subjects. Clinical Science. 1983;65:307–12.

15. Institute of Medicine. Dietary Reference Intakes for energy, carbohydrate, fiber, fat, fatty acids, cholesterol, protein, and amino acids. Washington, DC: National Academies Press; 2005.

16. Metropolitan Life Insurance Company. New weight standards for men and women. Statistical Bulletin (Metropolitan Life Insurance Company). 1959;40:1–4. Metropolitan Life Insurance Company. 1983 Metropolitan height and weight tables. Statistical Bulletin (Metropolitan Life Insurance Company). 1983;64:1–19. Weigley ES. Average? Ideal? Desirable? A brief review of height-weight tables in the United States. Journal of the American Dietetic Association. 1984;84:417–23.

17. Aronne LJ. Classification of obesity and assessment of obesity-related health risks. Obesity Research. 2002;10:105S–15S.

18. Centers for Disease Control and Prevention. U.S. Department of Health and Human Services. Overweight and obesity: Defining overweight and obesity. Available from: http://www.cdc.gov/nccdphp/dnpa/obesity/defining.htm.

19. Katzmarzyk PT, Janssen I, Ardern CI. Physical inactivity, excess adiposity and premature mortality. Obesity Reviews. 2003;4:257–90.

20. Wagner DR, Heyward VH. Techniques of body composition assessment: a review of laboratory and field methods. Research Quarterly for Exercise and Sports. 1999;70:135–49.

21. Lafontan M, Berlan M. Do regional differences in adipocyte biology provide new pathophysiological insights? Trends in Pharmacological Sciences. 2003;24:276–83.

22. Toth MJ, Tchernof A, Sites CK, Poehlman ET. Menopause-related changes in body fat distribution. Annals of the New York Academy of Sciences. 2000;904:502–6.

23. National Institutes of Health. Clinical guidelines on the identification, evaluation, and treatment of overweight and obesity in adults. National Institutes of Health, National Heart, Lung, and Blood Institute, Obesity Education Initiative. Available from: http://www.nhlbi.nih.gov/guidelines/obesity/practgde.htm.

24. National Institutes of Health, National Heart, Lung, and Blood Institute. Clinical guidelines on the identification, evaluation, and treatment of overweight and obesity in adults—the evidence report. Obesity Research. 1998;6:51S–209S.

25. Baskin ML, Ard J, Franklin F, Allison DB. Prevalence of obesity in the United States. Obesity Reviews. 2005;6:5–7.

26. Popkin BM, Gordon-Larsen P. The nutrition transition: worldwide obesity dynamics and their determinants. International Journal of Obesity and Related Metabolic Disorders. 2004;3:S2–9.

27. Briefel RR, Johnson CL. Secular trends in dietary intake in the United States. Annual Review of Nutrition. 2004;24: 401–31.

28. Enns CW, Mickle SJ, Goldman JD. Trends in food and nutrient intakes by children in the United States. Family Economics and Nutrition Review. 2002;14(2):56–68.

29. Enns CW, Mickle SJ, Goldman JD. Trends in food and nutrient intakes by adolescents in the United States. Family Economics and Nutrition Review. 2003;15(2):15–27.

30. Briefel RR, Johnson CL. Secular trends in dietary intake in the United States. Annual Review of Nutrition. 2004;24:401–31.

31. Prentice AM, Jebb SA. Fast foods, energy density and obesity: a possible mechanistic link. Obesity Reviews. 2003;4:187–94.

32. Kant AK, Graubard BI. Eating out in America, 1987–2000: trends and nutritional correlates. Preventive Medicine. 2004;38:243–9.

33. Ello-Martin JA, Ledikwe JH, Rolls BJ. The influence of food portion size and energy density on energy intake: implications for weight management. American Journal of Clinical Nutrition. 2005;82:236S–41S.

34. Rolls BJ, Roe LS, Kral TVE, Meengs JS, Wall DE. Increasing the portion size of a packaged snack increases energy intake in men and women. Appetite. 2004;42:63–9.

35. Brownson RC, Boehmer TK, Luke DA. Declining rates of physical activity in the United States: what are the contributors? Annual Review of Public Health. 2005;26:421–43. Centers for Disease Control and Prevention. Prevalence of physical activity, including lifestyle activities among adults—United States, 2000–2001. Morbidity and Mortality Weekly Report. 2003;52:764–9.

36. Wareham NJ, van Sluijs EM, Ekelund U. Physical activity and obesity prevention: a review of the current evidence. Proceedings of the Nutrition Society. 2005;64:229–47.

37. Hill JO, Wyatt HR, Reed GW, Peters JC. Obesity and the environment: where do we go from here? Science. 2003;299:853–5.

38. DHHS. 1996. Physical activity and health: a report of the surgeon general. Atlanta, GA: United States Department of Health and Human Services, Centers for Disease Control and Prevention, National Center for Chronic Disease Prevention and Health Promotion.; Centers for Disease Control and Prevention. Physical activity levels among children aged 9–13 years—United States 2002. Morbidity and Mortality Weekly Report. 2003;52:785–8.

39. Caroli M, Argentieri L, Cardone M, Masi A. Role of television in childhood obesity prevention. International Journal of Obesity and Related Metabolic Disorders. 2004; 28: S4–108.

40. Kaiser Family Foundation. The role of media in childhood obesity. 2004. Publication number 7030. Available from: http://www.kff.org/entmedia/7030.cfm.

41. Mokdad AH, Bowman AB, Ford ES, Vinicor F, Marks JS, Koplan JP. The continuing epidemic of obesity and diabetes in the United States. JAMA (Journal of the American Medical Association). 2001;286:1197–2000.

42. Grantmakers in Health. Weighing in on obesity: America's growing health epidemic. Issue Brief No. 11. Washington, DC: Grantmakers in Health; 2001.

43. Stunkard AJ, Harris JR, Pedersen NL, McClearn GE. The body-mass index of twins who have been reared apart. New England Journal of Medicine. 1990;322:1483–7.

44. Sorensen TI, Holst C, Stunkard AJ. Adoption study of environmental modifications of the genetic influences on obesity. International Journal of Obesity and Related Metabolic Disorders. 1998;22:73–81.

45. Kennedy AG. The role of the fat depot in the hypothalamic control of food intake in the rat. Proceedings of the Royal Society of London. 1953;140:578–92.

46. Ingalls A, Dickie M, Snell GD. Obese, a new mutation I the house mouse. Journal of Heredity. 1950;41:317–8.

47. Coleman D, Hummel KP. Effects of parabiosis of normal with genetically diabetic mice. American Journal of Physiology. 1969;217:1298–304. Coleman DL. Effects of parabiosis of obese with diabetes and normal mice. Diabetologia. 1973;9:294–8.

48. Zhang Y, Proenca R, Maffei M, Leopold L, Friedman JM. Positional cloning of the mouse obese gene and its human homologue. Nature. 1994;372:125–32.

49. Farooqi IS, O'Rahilly S. Monogenic obesity in humans. Annual Review of Medicine. 2005;56:443–58.

50. Jequier E. Leptin signaling, adiposity, and energy balance. Annals of the New York Academy of Sciences. 2002;967: 379–88.

51. Chakravarthy MV, Booth FW. Eating, exercise, and "thrifty" genotypes: connecting the dots toward an evolutionary understanding of modern chronic diseases. Journal of Applied Physiology. 2004;96:3–10.

52. Schwartz MW, Woods SC, Porte D, Seeley RJ, Baskin DG. Central nervous system control of food intake. Nature. 2000;404:661–71.

53. Leibowitz SF, Wortley KE. Hypothalamic control of energy balance: different peptides, different functions. Peptides. 2004;25:473–504. Strader AD, Woods SC. Gastrointestinal hormones and food intake. Gastroenterology. 2005;128:175–91.

54. de Graaf C, Blom WA, Smeets PA, Stafleu A, Hendriks HF. Biomarkers of satiation and satiety. American Journal of Clinical Nutrition. 2004;79:946–61.

55. Heini AF, Kirk KA, Lara-Castro C, Weinsier RL. Relationship between hunger-satiety feelings and various metabolic parameters in women with obesity during controlled weight loss. Obesity Research. 1998;6:225–30.

56. Wurtman RJ, Wurtman JJ. Brain serotonin, carbohydrate-craving, obesity and depression. Obesity Research. 1995;3:477S–80S.

57. Blundell JE, MacDiarmid JI. Fat as a risk factor for overconsumption: satiation, satiety, and patterns of eating. Journal of the American Dietetic Association. 1997;97:S63–9.

58. Orr J, Davy B. Dietary influences on peripheral hormones regulating energy intake: potential applications for weight management. Journal of the American Dietetic Association. 2005;105:1115–24.

59. Hydock CM. A brief overview of bariatric surgical procedures currently being used to treat the obese patient. Critical Care Nursing Quarterly. 2005;28:217–26.

60. Buchwald H. Consensus Conference Panel. Bariatric surgery for morbid obesity: health implications for patients, health professionals, and third-party payers. Journal of American College of Surgeons. 2005;200:593–604.

61. Ali MR, Fuller WD, Choi MP, Wolfe BM. Bariatric surgical outcomes. Surgical Clinics of North America. 2005;85:835–52.

62. Flum DR, Salem L, Broeckel-Elrod JA, Patchen-Dellinger E, Cheadle A, Chan L. Early mortality among Medicare beneficiaries undergoing bariatric surgical procedures. JAMA (Journal of the American Medical Association). 2005;294:1903–8.

63. Coelho JC, Campos AC. Surgical treatment of morbid obesity. Current Opinion in Clinical Nutrition and Metabolic Care. 2001;4:201–6.

64. Inui A, Asakawa A, Bowers CY, Mantovani G, Laviano A, Meguid MM, Fujimiya M. Ghrelin, appetite, and gastric motility: the emerging role of the stomach as an endocrine organ. Federation of American Societies for Experimental Biology Journal. 2004;18:439–56.

65. Marx J. Cellular warriors at the battle of the bulge. Science. 2003;299:846–9.

66. Kieffer TJ, Habener JF. The adipoinsular axis: effects of leptin and pancreatic β-cells. American Journal of Physiology, Endocrinology, and Metabolism. 2000;278: E1–E14.

67. Popovic V, Duntas LH. Brain somatic cross-talk: ghrelin, leptin and ultimate challengers of obesity. Nutrition and Neuroscience. 2005;8:1–5; Cancello R, Tounian A, Poitou Ch, Clement K. Adiposity signals, genetic and body weight regulation in humans. Diabetes Metabolism. 2004;30:215–27.

68. Volek JS, Vanheest JL, Forsythe CE. Diet and exercise for weight loss: a review of current issues. Sports Medicine. 2005;35:1–9.

69. Wing RR, Phelan S. Long-term weight loss maintenance. American Journal of Clinical Nutrition. 2005;82:222S–5S.

70. National Institutes of Health, and National Heart, Lung and Blood Institute. Clinical guidelines on the identification, evaluation and treatment of overweight and obesity in adults—the evidence report. National Institutes of Health Publication Number 00–4084. Bethesda, MD: National Institutes of Health; October 2000.

71. Bell EA, Castellanos VH, Pelkman CL, Thorwart ML, Rolls BJ. Energy density of foods affects energy intake in normal-weight women. American Journal of Clinical Nutrition. 1998;67:412–20. Yao M, Roberts SB. Dietary energy density and weight regulation. Nutrition Reviews. 2001;59:247–58.

72. Ello-Martin JA, Ledikwe JH, Rolls BJ. The influence of food portion size and energy density on energy intake: implications for weight management. American Journal of Clinical Nutrition. 2005;82:236S–41S.

73. National Center for Chronic Disease Prevention and Health Promotion Division of Nutrition and Physical Activity, Department of Health and Human Services, Centers for Disease Control and Prevention. Can eating fruits and vegetables help people to manage their weight? Research to Practice Series, No.1. Available from: http://www.cdc.gov/nccdphp/dnpa/nutrition/pdf/rtp_practitioner_10_07.pdf.

74. Rolls BJ, Bell EA, Waugh BA. Increasing the volume of a food by incorporating air affects satiety in men. American Journal of Clinical Nutrition. 2000;72:361–8.

75. Rock CL, Thomson C, Caan BJ, Flatt SW, Newman V, Ritenbaugh C, Marshall JR, Hollenbach KA, Stefanick ML, Pierce JP. Reduction in fat intake is not associated with weight loss in most women after breast cancer diagnosis: evidence from a randomized controlled trial. Cancer. 2001;9:25–34.

76. Barlow CE, Kohl HW 3rd, Gibbons LW, Blair SN. Physical fitness, mortality and obesity. International Journal of Obesity and Related Metabolic Disorders. 1995;19:S41–4.

77. International Association for the Study of Obesity. Diet, nutrition and the prevention of chronic diseases. Report of the joint WHO/FAO expert consultation. WHO Technical Report Series, No. 916 (TRS 916). 2003.

78. Westman EC, Yancy WS Jr, Vernon MC. Is a low-carb, low-fat diet optimal? Archives of Internal Medicine. 2005;165:1071–2.

79. Lovejoy JC, Bray GA, LeFevre M, Smith SR, Most MM, Denkins YM, Volaufova J, Rood JC, Eldrige AL, Peters JC. Consumption of a controlled low-fat diet containing olestra for 9 months improves health risk factors in conjunction with weight loss in obese men: The Olé Study. International Journal of Obesity and Related Metabolic Disorders. 2003;27:1242–9.

80. Center for Nutrition Policy and Promotion and the U.S. Department of Agriculture. Nutrition Insights. Is fat consumption really increasing? Insight 5 April 1998. Available from: http://www.cnpp.usda.gov/insght5a.PDF.

81. Willett WC. Dietary fat and body fat: Is there a relationship? Journal of Nutritional Biochemistry. 1998;9:522–4. Willett WC. Is dietary fat a major determinant of body fat? American Journal of Clinical Nutrition. 1998;67:556S–625S.

82. Bray GA, Paeratakul S, Popkin BM. Dietary fat and obesity: a review of animal, clinical and epidemiological studies. Physiology and Behavior. 2004;83:549–55.

83. Taubes G. The soft science of dietary fat. Science. 2001;291:2536–45.

84. Foster GD, Wyatt HR, Hill JO, McGuckin BG, Brill C, Selma B, Szapary PO, Rader DJ, Edman JS, Klein S. A randomized trial of a low-carbohydrate diet for obesity. New England Journal of Medicine. 2003;248:2082–90.

85. Opinion Dynamics Corporation. Measuring the low-carb revolution. Available from: http://www.opiniondynamics.com/lowcarb.html.

86. Bravata DM, Sanders L, Huang J, Krumholz HM, Olkin I, Gardner CD, Bravata DM. Efficacy and safety of low-carbohydrate diets: a systematic review. JAMA (Journal of the American Medical Association). 2003;289:1838–49.

87. Klein S. Clinical trial experience with fat-restricted vs. carbohydrate-restricted weight-loss diets. Obesity Research. 2004;12:141S–4S.

88. Bravata DM, Sanders L, Huang J, Krumholz HM, Olkin I, Gardner CD, Bravata DM. Efficacy and safety of low-carbohydrate diets: a systematic review. JAMA (Journal of the American Medical Association). 2003;289:1838–49.

89. Freedman, MR, King J, Kennedy E. Popular diets: A scientific review. Obesity Research. 2001;9:1S–40S.

90. Schwenke DC. Insulin resistance, low-fat diets, and low-carbohydrate diets: time to test new menus. Current Opinion in Lipidology. 2005;16:55–60.

91. Astrup A, Meinert Larsen T, Harper A. Atkins and other low-carbohydrate diets: hoax or an effective tool for weight loss? Lancet. 2004;364:897–9.

92. Farnsworth E, Luscombe ND, Noakes M, Wittert G, Argyiou E, Clifton PM. Effect of a high-protein, energy-restricted diet on body composition, glycemic control, and lipid concentrations in overweight and hyperinsulinemic men and women. American Journal of Clinical Nutrition. 2003;78:31–9.

93. Martin WF, Armstrong LE, Rodriguez NR. Dietary protein intake and renal function. Nutrition and Metabolism. 2005;2:25.

94. Acheson KJ. Carbohydrate and weight control: where do we stand? Current Opinion in Clinical Nutrition and Metabolic Care. 2004;7:485–92.

95. Aude YW, Agatston AS, Lopez-Jimenez F, Lieberman EH, Almon M, Hansen M, Rojas G, Lamas GA, Hennekens CH. The national cholesterol education program diet vs. a diet lower in carbohydrates and higher in protein and monounsaturated fat: a randomized trial. Archives of Internal Medicine. 2004;164:2141–6.

**Check out the following sources for additional information.**

96. Acheson KJ. Carbohydrate and weight control: where do we stand? Current Opinion in Clinical Nutrition and Metabolic Care. 2004;7:485–92.

97. Ahima RS, Flier JS. Leptin. Annual Review of Physiology. 2000;62:413–37.

98. Arch JR. Central regulation of energy balance: inputs, outputs and leptin resistance. Proceedings of the Nutrition Society. 2005;64:39–46.

99. Barnard ND, Scialli AR, Turner-McGrievy G, Lanou AJ, Glass J. The effects of a low-fat, plant-based dietary intervention on body weight, metabolism, and insulin sensitivity. American Journal of Medicine. 2005;118:991–7.

100. Baskin ML, Ard J, Franklin F, Allison DB. Prevalence of obesity in the United States. Obesity Reviews. 2005;6:5–7.

101. Bray GA, Bouchard C, James WPT. Definitions and proposed current classification of obesity. In: Handbook of obesity. Bray GA, Bouchard C, James WPT, editors. New York: Marcel Dekker. 1998, 31–40.

102. Elia M, Ward LC. New techniques in nutritional assessment: body composition methods. Proceedings of the Nutrition Society. 1999;58:33–8.

103. Foreyt JP. Need for lifestyle intervention: how to begin. American Journal of Cardiology. 2005;96:11E–4E.

104. Freedman MR, King J, Kennedy E. Popular diets: A scientific review. Obesity Research. 2001;9:1S–40S.

105. Grundy SM. The optimal ratio of fat-to-carbohydrate in the diet. Annual Review of Nutrition. 1999;19:325–41.

106. Hill JO, Wyatt HR. Role of physical activity in preventing and treating obesity. Journal of Applied Physiology. 2005;99:765–70.

107. Korner J, Aronne LJ. The emerging science of body weight regulation and its impact on obesity treatment. Journal of Clinical Investigation. 2003;111:565–70.

108. Leibowitz SF, Wortley KE. Hypothalamic control of energy balance: different peptides, different functions. Peptides. 2004;25:473–504.

109. Lev-Ran A. Human obesity: an evolutionary approach to understanding our bulging waistline. Diabetes/Metabolism Research and Reviews. 2001;17:347–62.

110. Moore MS. Interactions between physical activity and diet in the regulation of body weight. Proceedings of the Nutrition Society. 2000;59:193–98.

111. National Institute of Diabetes and Digestive and Kidney Diseases, Bethesda, MD. Overweight, obesity, and health risk. National Task Force on the Prevention and Treatment of Obesity. Archives of Internal Medicine. 2000;160:898–904.

112. National Task Force on the Prevention and Treatment of Obesity. Archives of Internal Medicine. 2000;160:898–904.

113. Nickols-Richardson SM, Coleman MD, Volpe JJ, Hosig KW. Perceived hunger is lower and weight loss is greater in overweight premenopausal women consuming a low-carbohydrate/high-protein vs. high-carbohydrate/low-fat diet. Journal of the American Dietetic Association. 2005;105:1433–7.

114. Cummings S, Parham ES, Strain GW. Position of the American Dietetic Association: weight management. Journal of the American Dietetic Association. 2002;102:1145–55.

115. Saris WHM. Sugars, energy metabolism, and body weight control. American Journal of Clinical Nutrition. 2003;78:850S–7S.

116. Schwenke DC. Insulin resistance, low-fat diets, and low-carbohydrate diets: time to test new menus. Current Opinion in Lipidology. 2005;16:55–60.

117. Takeda S, Elefteriou F, Karsenty G. Common endocrine control of body weight, reproduction, and bone mass. Annual Review of Nutrition. 2003;23:403–11.

118. Tremblay A, Buemann B. Exercise-training, macronutrient balance and body weight control. International Journal of Obesity. 1995;19:79–86.

119. Vermeulen A, Goemaere S, Kaufman JM. Testosterone, body composition and aging. Journal of Endocrinological Investigation. 1999;22:110–6.

120. Volek JS, Westman EC. Very-low-carbohydrate weight-loss diets revisited. Cleveland Clinic Journal of Medicine. 2002;69:849–62.

121. Wagner DR, Heyward VH. Techniques of body composition assessment: a review of laboratory and field methods. Research Quarterly for Exercise and Sport. 1999;70:135–49.

122. Wee CC, McCarthy EP, Davis RB, Phillips RS. Physician counseling about exercise. JAMA (Journal of the American Medical Association). 1999;282:1583–8.

123. Willett WC. Dietary fat plays a major role in obesity: no. Obesity Reviews. 2002;3:59–68.

124. Wing RR, Phelan S. Long-term weight loss maintenance. American Journal of Clinical Nutrition. 2005;82:222S–5S.

# Eating Disorders

© Photodisc/Getty Images

*H*ealthy eating is characterized by the ability to eat almost anything, at any time, and in moderate amounts. It also entails eating without fear, guilt, and shame. For some people, however, preoccupation with food, dieting, exercise, and weight loss reaches obsessive proportions and goes beyond what is considered normal. These troublesome disturbances in eating behaviors are signs of disordered eating. If disordered eating patterns continue, they can eventually progress into an eating disorder.

In the past 25 years, the number of people diagnosed with eating disorders has increased. Not surprisingly, clinicians are very interested in learning more about eating disorders and how they can be treated effectively. In this Nutrition Matters you will learn about disordered eating behaviors and the different types, causes, and complexities of eating disorders.

## Overview of Eating Disorders

**Disordered eating** behaviors include a wide variety of non-normative eating patterns such as irregular eating, consistent undereating, and/or consistent overeating.[1] These types of behaviors are common, often occurring in response to stress, illness, or dissatisfaction with personal appearance. However, in some people disordered eating can progress into a full-blown eating disorder such as anorexia nervosa, bulimia nervosa, or binge eating disorder. **Eating disorders** are characterized by extreme disturbances in eating behaviors that can be both physically and psychologically harmful. People with eating disorders often feel isolated, and their relationships with family and friends become strained.

There is more to understanding what causes eating disorders than just factors related to food intake. That is, eating disorders are complex behaviors that arise from physical, psychological, and social issues. Yet only in the last 35 years have people outside the medical profession become widely aware of such problems and their serious health consequences.

The American Psychiatric Association classifies eating disorders into distinct categories: anorexia nervosa (AN), bulimia nervosa (BN), eating disorders not otherwise specified (EDNOS), and binge-eating disorder (BED).[2] There is a great deal of overlap in behaviors associated with the various eating disorders, so it is often difficult to categorize people (Table 1). The American Psychiatric Association criteria used to diagnose eating disorders are summarized in Table 2.

## Anorexia Nervosa

**Anorexia nervosa (AN)** is characterized by an irrational fear of gaining weight or becoming obese. Often there is a disconnect between actual and perceived body weight

---

**disordered eating** Eating pattern that includes irregular eating, consistent undereating, and/or consistent overeating.

**eating disorder** Extreme disturbance in eating behaviors that can result in serious medical conditions, psychological consequences, and dangerous weight loss.

**anorexia nervosa (AN)** An eating disorder characterized by an irrational fear of gaining weight or becoming obese.

## TABLE 1 Behaviors Associated with Eating Disorders

| Eating Disorder Classification | Food Restriction | Purging[a] | Bingeing | Excessive Exercise |
|---|---|---|---|---|
| Anorexia nervosa, restricting type | X | | | x |
| Anorexia nervosa, binge eating/purging type | X | x | X | X |
| Bulimia nervosa, purging type | x | X | X | x |
| Bulimia nervosa, nonpurging type | x | | X | X |
| Binge-eating disorder | | | X | |
| Restrained eating | X | | X | |

NOTE: Large "X" indicates primary behavior associated with a specific eating disorder, whereas a small "x" indicates that behavior occurs to a lesser extent.
[a]Refers to self-induced vomiting and/or the misuse of laxatives, diuretics, or enemas.

## TABLE 2 Classification and Diagnostic Criteria for Eating Disorders

### Anorexia Nervosa

A. Refusal to maintain body weight at or above a minimally normal weight for age and height.
B. Intense fear of gaining weight or becoming fat, even though underweight.
C. Disturbance in body weight or shape is experienced, or denial of the seriousness of the current low body weight.
D. In postmenarchal females, the absence of at least three consecutive menstrual cycles.

**Two types:**
- *Restricting type:* The person has not regularly engaged in binge eating or purging behavior.
- *Binge-eating/purging type:* The person has regularly engaged in binge eating or purging behavior.

### Bulimia Nervosa

A. Recurrent episodes of binge eating. An episode of binge eating is characterized by both of the following: (1) Eating, in a discrete period of time, an amount of food that is definitely larger than most people would eat during a similar period of time and under similar circumstances, and (2) A sense of lack of control over eating during the episode.
B. Recurrent compensatory behavior to prevent weight gain, such as self-induced vomiting; misuse of laxatives, diuretics, enemas, or other medications; fasting or excessive exercise.
C. The binge eating and inappropriate compensatory behaviors both occur, on average, at least twice a week for 3 months.
D. Self-evaluation is unduly influenced by body shape and weight.
E. The disturbance does not occur exclusively during episodes of anorexia nervosa.

**Two types:**
- *Purging type:* The person has regularly engaged in self-induced vomiting or the misuse of laxatives, diuretics, or enemas.
- *Nonpurging type:* The person has used inappropriate compensatory behaviors, such as fasting or excessive exercise, but has not regularly engaged in self-induced vomiting or the misuse of laxatives, diuretics, or enemas.

### Eating Disorders Not Otherwise Specified (EDNOS)

Includes disorders of eating that do not meet the criteria for any specific eating disorder.
A. The criteria for anorexia nervosa are met except the individual has regular menses.
B. The criteria for anorexia nervosa are met except that despite significant weight loss, the individual's current weight is in the normal range.
C. The criteria for bulimia nervosa are met except that the binge eating and inappropriate compensatory mechanisms occur at a frequency of less than twice a week or for a duration of less than 3 months.
D. The regular use of inappropriate compensatory behavior by an individual of normal body weight after eating small amounts of food.
E. Repeatedly chewing and spitting out, but not swallowing, large amounts of food.
F. Binge eating disorder: recurrent episodes of binge eating in the absence of the regular use of inappropriate compensatory behaviors characteristic of bulimia nervosa.

### Binge-Eating Disorder

These behaviors must occur two times per week over the course of 6 months with no compensatory behaviors.
A. Eating much more rapidly than normal.
B. Eating until feeling uncomfortably full.
C. Eating large amounts of food when not feeling physically hungry.
D. Eating alone because of being embarrassed by how much one is eating.
E. Feeling disgusted with oneself, depressed, or very guilty after overeating.

SOURCE: *Diagnostic and Statistical Manual of Mental Disorders,* 4th ed. (DSM-IV-TR). Washington, DC: American Psychiatric Association Press; 1994.

and shape, so that people with AN believe they are "fat" even though they may be dangerously thin. When people with AN look in the mirror, they tend to be very critical of their body shape and size. Because of this distorted self-perception, weight loss becomes an obsession. In fact, people with AN tend to view their self-worth in terms of weight and body shape. Because they perceive self-starvation as an accomplishment rather than a problem, there is little motivation to change. Although issues associated with AN often center around food, the causes are much more complex. AN and other eating disorders stem mainly from psychological issues, and the denial of food and the relentless pursuit of thinness become ways of coping with emotions, conflict, and stress.[3]

The American Psychiatric Association recognizes two types of AN: **restricting type** and **binge-eating/purging type** (Figure 1). People with restricting type AN maintain a low body weight by food restriction and/or excessive exercise, whereas people with binge-eating/purging type AN engage in periods of bingeing and purging (that is, self-induced vomiting, and/or abuse of laxatives, diuretics, exercise, or enemas) in addition to food restriction.

Typically, people with AN limit their food intake as well as the variety of foods consumed. For example,

**People with a distorted body image perceive themselves to be fat even if they are very thin.**

© David Young-Wolff/PhotoEdit

**anorexia nervosa, restricting type** An eating disorder characterized by food restriction.

**anorexia nervosa, binge-eating/purging type** An eating disorder characterized by food restriction as well as bingeing and/or purging.

## FIGURE 1    Types of Anorexia Nervosa

Anorexia nervosa is an eating disorder characterized by self-starvation and a relentless pursuit of weight loss.

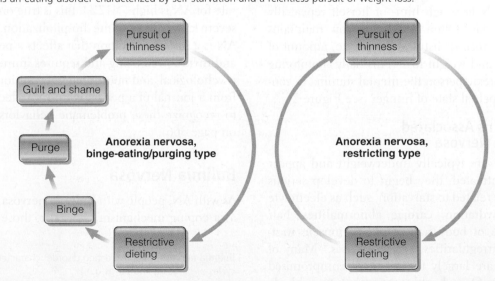

**TABLE 3   Signs, Symptoms, and Health Consequences Associated with Anorexia Nervosa**

| Signs and Symptoms | Health Consequences |
| --- | --- |
| • Rigid dieting resulting in dramatic weight loss<br>• At or below 15% of an ideal body weight<br>• Diets or restricts food even though he or she is not overweight<br>• Complains of being fat when he or she is thin<br>• Preoccupied with food, calories, nutrition, and/or cooking<br>• Denial of hunger<br>• Exercises obsessively<br>• Weighs self frequently<br>• Complains about feeling bloated when eating<br>• Complains of feeling cold<br>• Irregular or cessation of menstruation<br>• Food rituals such as excessive chewing or cutting food into small pieces<br>• Restricts amount and types of foods consumed<br>• Maintains rigid schedule and routine<br>• Tendency toward perfectionism | • Dry skin, dry hair, thinning hair, and/or hair loss<br>• Fainting, fatigue, and overall weakness<br>• Intolerance to cold<br>• Formation of fine hair on the body (lanugo)<br>• Significant loss of fat and lean body mass<br>• Reproductive problems<br>• Loss of bone mass<br>• Electrolyte imbalance<br>• Irregular heartbeat<br>• Bruises easily<br>• Injuries such as stress fractures<br>• Impaired iron status<br>• Impaired immune status<br>• Slow heart rate and low blood pressure |

foods are often categorized as either "safe" or "unsafe" depending on whether they consider the food to cause weight gain or not. Safe foods are those that are not likely to cause weight gain, whereas unsafe foods are perceived as being fattening. Foods often perceived as unsafe, and typically the first to be eliminated, are meat, high-sugar foods, and fatty foods. Eventually the diet becomes so restricted that nutritional needs are no longer met. In addition to limiting food intake, individuals with AN often have food rituals such as chewing food but not swallowing, overchewing, obsessive cooking for others, and cutting food into unusually small pieces. These behaviors are often a display of self-restraint and control over the urge to eat. These and other traits often associated with AN are summarized in Table 3.

In addition to unusual food restrictions and rituals, many individuals with AN exercise regularly and excessively for fear of gaining weight. It is not uncommon for someone with AN to weigh him- or herself repeatedly throughout the day. Often such a person maintains meticulous daily records listing food intake, amount of time exercising, and weight loss. This daily monitoring can give the anorexic person the mental stamina to continue in this perpetual state of hunger (see Figure 2).

## Health Concerns Associated with Anorexia Nervosa

People with AN are typically underweight and appear gaunt. If left untreated, they begin to develop serious health problems related to starvation, such as electrolyte imbalance, dehydration, cardiac abnormalities, hair loss, appearance of body hair (lanugo), muscle wasting, menstrual irregularities, and bone loss.[4] Many of these problems are largely the result of compromised nutritional status. Once healthy eating habits and body weight are restored, most of these medical concerns are reversed.

In females, the loss of body fat leads to a decline in the reproductive hormone estrogen. This can cause endocrine abnormalities, resulting in disrupted menstruation and reproductive function.[5] This response undoubtedly evolved as a means to protect women from pregnancy during periods of low food availability. However, estrogen also plays an important role in bone health by increasing calcium uptake into bone. Thus the longer a woman goes without menstruating—a condition called amenorrhea—the greater the loss of bone. Although estrogen levels return to normal when body weight is restored, bones probably do not completely recover. For this reason, women with eating disorders who have experienced amenorrhea remain at increased risk for bone disease.

Although researchers have reported that the death rate for AN is high (>12%), this is true only for the most severe cases that require hospitalization.[6] Nonetheless, AN is a serious condition that affects a person's physical and mental well-being and requires immediate medical, psychological, and nutritional intervention. A page taken from a journal of a person with AN reflects the inability to recognize these problematic behaviors (see Figure 3 on page 366).

## Bulimia Nervosa

As with AN, people with **bulimia nervosa (BN)** use food as a coping mechanism. However, those with BN turn

---

**bulimia nervosa (BN)**   An eating disorder characterized by repeated cycles of bingeing and purging.

**FIGURE 2    Food, Weight, and Exercise Record of a Person with Anorexia Nervosa**

Some people with eating disorders keep meticulous records regarding exercise and diet.

| MONDAY | TUESDAY |
|---|---|
| Morning | Morning |
| weight - 87 lb. | weight - 87 lb. |
| 5-mile run (42 minutes) | 5-mile run (45 minutes) |
| 100 situps | 120 situps |
| | |
| Breakfast | Breakfast |
| weight-86 lb. | weight-85 lb. |
| rice cake | rice cake |
| 10 grapes | 1 slice melon |
| | |
| Afternoon | Afternoon |
| Lunch | Lunch |
| weight-86 lb. | weight-85 lb. |
| diet pop | diet pop |
| 1/2 grapefruit | 10 grapes |
| rice cake | rice cake |
| | |
| Dinner | Dinner |
| weight-86 lb. | weight-85 lb. |
| 1/2 cup rice | 1/2 slice toast |
| diet pop | diet pop |
| celery and nonfat dressing | carrots and nonfat dressing |
| | |
| Before bed | Before bed |
| 2-mile run (20 minutes) | 2-mile run (18 minutes) |
| 100 situps | 150 situps |
| weight-85 lb. | weight-85 lb. |

to, rather than away from, food during times of stress and emotional conflict. Bulimia nervosa is characterized by repeated cycles of "bingeing and purging." **Bingeing** is best characterized by the compulsive consumption of large amounts of food, followed by **purging** behaviors.[7] Foods often consumed during a binge include cakes, cookies, and ice cream, which can easily add up to thousands of calories. However, not all people with BN purge, and for this reason the American Psychiatric Association recognizes nonpurging as a specific type of BN. Nonpurging individuals with BN exercise or fast after a binge, rather than vomit, or use laxatives, diuretics, or enemas as compensatory behavior. These and other characteristics commonly observed in people with BN are presented in Table 4.

People with BN outnumber people with AN by about two to one. Because bingeing and purging are often carried out in secrecy, family and friends may be unaware there is a problem. In fact, binges are often planned in advance when nobody is likely to be around. Although bulimics may consume several thousand calories during a binge, purging behaviors keep most people with BN

in energy balance. However, even purging immediately after bingeing cannot totally prevent nutrient absorption, and some weight gain is likely. Thus people with BN tend to be within or slightly above their acceptable weight range, and there may be no outward change in physical appearance.

Some individuals with BN experience regret and feel out of control after bingeing. This can often lead to depression, which increases the likelihood of future binge–purge cycles (see Figure 4 on page 367). However, some bulimic people may alternate between cycles of bingeing and purging, and periods of food restriction. Alternatively, some bulimic people try to compensate for their bingeing behavior with excessive exercise. In fact, exercise is a recognized compensatory behavior used to offset excess calories from bingeing.

**bingeing** Uncontrolled consumption of large quantities of food in a relatively short period of time.

**purging** Self-induced vomiting and/or misuse of laxatives, diuretics, and/or enemas.

**FIGURE 3    Thoughts and Behaviors Associated with Anorexia Nervosa**

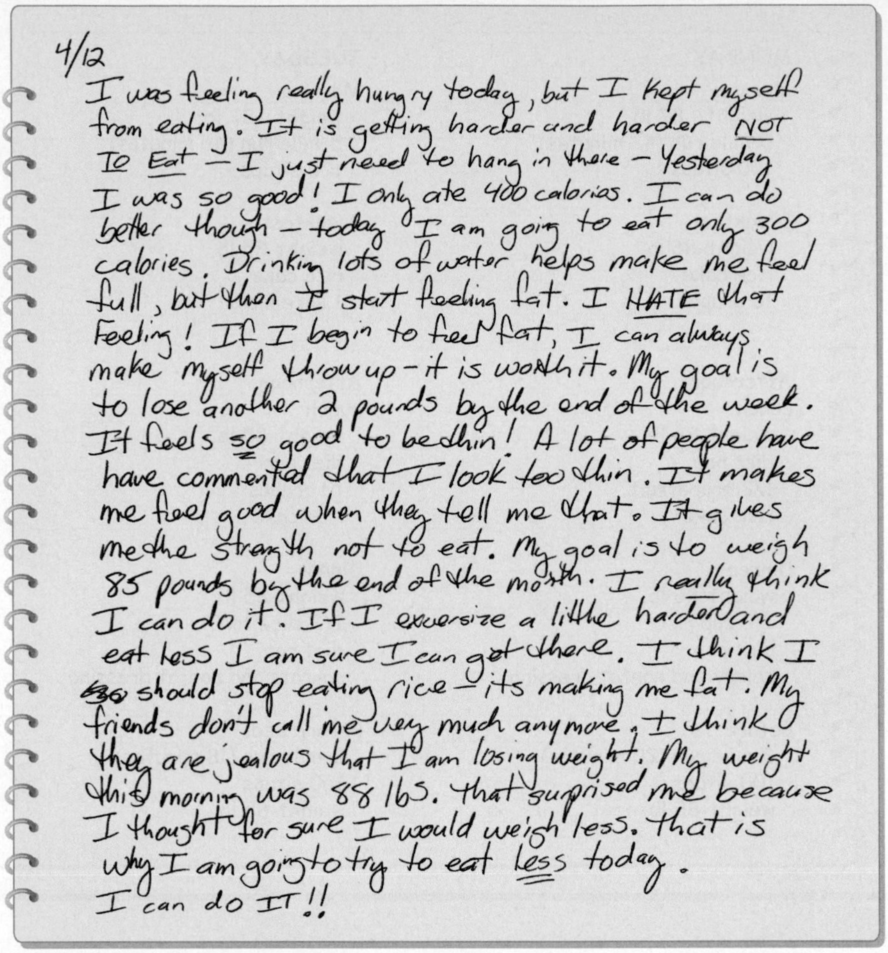

> 4/12
>
> I was feeling really hungry today, but I kept myself from eating. It is getting harder and harder NOT To Eat — I just need to hang in there — Yesterday I was so good! I only ate 400 calories. I can do better though — today I am going to eat only 300 calories. Drinking lots of water helps make me feel full, but then I start feeling fat. I HATE that Feeling! If I begin to feel fat, I can always make myself throw up — it is worth it. My goal is to lose another 2 pounds by the end of the week. It feels so good to be thin! A lot of people have have commented that I look too thin. It makes me feel good when they tell me that. It gives me the strength not to eat. My goal is to weigh 85 pounds by the end of the month. I really think I can do it. If I exercise a little harder and eat less I am sure I can get there. I think I should stop eating rice — its making me fat. My friends don't call me very much any more. I think they are jealous that I am losing weight. My weight this morning was 88 lbs. that surprised me because I thought for sure I would weigh less. that is why I am going to try to eat less today.
> I can do IT!!

Whereas individuals with AN feel a sense of satisfaction associated with dieting and weight loss, many bulimic people feel guilty and depressed when they binge and purge. In addition, people with BN tend to be impulsive, and are prone to other unhealthy behaviors such as substance abuse, self-mutilation, and suicidal tendencies.[8]

**TABLE 4    Signs, Symptoms, and Health Consequences of Bulimia Nervosa**

| Signs and Symptoms | Health Consequences |
| --- | --- |
| • Responds to emotional stress by overeating<br>• Feels guilt or shame after eating<br>• Obsessive concerns about weight<br>• Repeated attempts at food restriction and dieting<br>• Frequent use of bathroom during and after meals<br>• Feels out of control<br>• Moodiness and depression<br>• Fluctuations in body weight (+/− 10 lb.)<br>• Swollen or puffy face<br>• Teeth may appear eroded<br>• Odor of vomit on breath or in bathroom<br>• Frequent purchase of laxatives<br>• Fear of not being able to stop eating voluntarily<br>• Disappearance of food | • Erosion of tooth enamel and tooth decay from exposure to stomach acid<br>• Electrolyte imbalances that can lead to irregular heart function and possible sudden cardiac arrest<br>• Inflammation of the salivary glands<br>• Irritation and inflammation of the esophagus that could lead to hemorrhaging and bleeding during vomiting<br>• Irregular bowel function<br>• Dehydration |

**FIGURE 4**  Thoughts and Behaviors Associated with Bulimia Nervosa

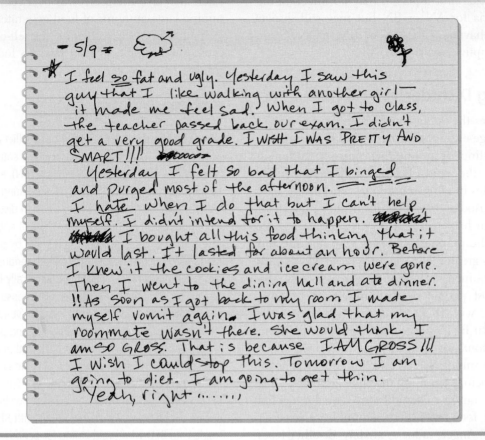

— 5/9 =

# I feel so fat and ugly. Yesterday I saw this guy that I like walking with another girl — it made me feel sad. When I got to class, the teacher passed back our exam. I didn't get a very good grade. I WISH I WAS PRETTY AND SMART!!!

Yesterday I felt so bad that I binged and purged most of the afternoon. = = =
I hate when I do that but I can't help myself. I didn't intend for it to happen.
I bought all this food thinking that it would last. It lasted for about an hour. Before I knew it the cookies and ice cream were gone. Then I went to the dining hall and ate dinner. !! As soon as I got back to my room I made myself vomit again. I was glad that my roommate wasn't there. She would think I am so GROSS. That is because I AM GROSS!!! I wish I could stop this. Tomorrow I am going to diet. I am going to get thin.
Yeah, right........

For these reasons, people with BN are more likely to seek treatment than are those with AN. Help for people with BN often comes after getting caught in the act of bingeing and purging.

### Health Concerns Associated with Bulimia Nervosa

Over time, repeated cycles of bingeing and vomiting damage the body. Much of this damage is caused when the delicate lining of the esophagus and mouth becomes irritated by frequent exposure to stomach acid. The acidity of gastric juice can also damage dental enamel, causing teeth to decay. Dentists and dental hygienists are often the first to notice these signs. Furthermore, frequent vomiting and/or overuse of laxatives and diuretics can cause dehydration and electrolyte imbalance. This can lead to irregular heart function and can even result in sudden cardiac arrest.

### Eating Disorders Not Otherwise Specified (EDNOS)

Eating disorders not otherwise specified (EDNOS) is a general category of eating disorders that do not meet precise diagnostic criteria associated with AN or BN. In

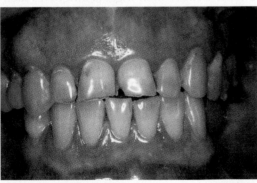

© Wellcome Photo Library

Repeated vomiting can damage the tooth enamel. As a result, teeth appear mottled. This can occur in people with bulimia nervosa or in individuals with other eating disorders, who frequently vomit.

other words, a person may display many of the behaviors associated with AN but the person has normal weight or (if a woman) still menstruates. Similarly, a person may binge and purge once a week, whereas the diagnostic

**eating disorders not otherwise specified (EDNOS)** A category of eating disorders that includes some, but not all, of the diagnostic criteria for anorexia nervosa and/or bulimia nervosa.

criteria for BN states a person must binge and purge at least twice a week. Thus a person may not meet all the diagnostic criteria for AN or BN but may still display many of the behaviors. Like AN or BN, EDNOS are serious and disruptive in a person's life.

## Binge Eating Disorder

Only recently has the American Psychiatric Association recognized **binge-eating disorder (BED)** as an eating disorder distinct from bulimia nervosa. Although BED requires further study, it is so far characterized by recurring episodes of consuming large amounts of food within a short period of time. The diagnostic criterion for BED is having at least two binges per week for at least 6 months, and the number of people who fall into this category is much greater than the number of people with AN and BN combined.

Although most people overeat from time to time, binges associated with BED are distinctly different. For many people with BED, binges provide an escape from stress and emotional pain. In other words, food has a psychological numbing effect and induces a state of emotional well-being. Some studies indicate that people with BED were raised in families affected by alcohol abuse.[9] However, food rather than alcohol becomes their "drug of choice." Anger, sadness, anxiety, or other types of emotional distress often triggers binges. In fact, many people with BED struggle with clinical depression. It is not clear if depression triggers BED or if BED causes depression.[10]

Similar to BN, binges typically take place in private and are often accompanied by feelings of shame. However, binges associated with BED are not followed by purging behaviors that rid the body of the ingested food. A person can easily consume thousands of calories during a binge episode. As such, those with BED tend to be in positive energy balance and significantly overweight or obese. Consequently, BED puts people at increased risk for weight-related health problems such as type 2 diabetes, gallstones, and cardiovascular disease.

> CONNECTIONS ▸ Recall that positive energy balance results when energy intake exceeds energy expenditure (Chapter 9, page 320).

## Restrained Eating

Some individuals with BED can suppress their desire for food and avoid eating for long periods of time between binges; these people are called restrained eaters.[11] **Restrained eaters** limit their food intake to lose weight. However, after an extended period of eating very little food restrained eaters find themselves feeling out of control and respond by bingeing. This cycle of fasting

and bingeing can be difficult to stop. Many restrained eaters perceive themselves as overweight, and overconsuming large amounts of food generates further feelings of inadequacy. These feelings of self-contempt in turn cause such people to turn back to food for emotional comfort.

## ESSENTIAL *Concepts*

Disordered eating patterns result from conflict and stress, and can lead to eating disorders. The main types of eating disorders are anorexia nervosa (AN), bulimia nervosa (BN), eating disorders not otherwise specified (EDNOS), and binge-eating disorder (BED). AN is characterized by a fear of weight gain, a distorted body image, and food restriction. There are two types of AN, one characterized by restriction of food intake and the other by purging in addition to food restriction. If left untreated, serious health problems related to starvation can develop. BN typically involves cycles of bingeing and purging. Bingeing is the consumption of large amounts of food, and purging can involve vomiting, use of laxatives, and/or excessive exercise. Bingeing and purging are done in secrecy, and people with BN tend to be within or slightly above their acceptable weight range. Some people with BN do not purge and are said to have nonpurging-type BN. Binge eating disorder (BED) is characterized by having at least two binges per week for at least 6 months. However, individuals with BED do not purge. People with BED who avoid eating for periods of time between binges are referred to as restrained eaters.

## Causes of Eating Disorders

Although eating disorders in the general population are somewhat rare, they are relatively common among adolescent girls and young women. The American Psychological Association estimates that approximately 8 million females in the United States battle eating disorders. The overall incidence of AN is approximately 8 per 100,000 people per year, and the incidence of BN is somewhat higher—12 per 100,000 people per year. Although most females with eating disorders are adolescents or young adults, a pattern of eating disorders among adult women has also begun to emerge.[12] This trend may be a response to a youth-oriented society that intensifies insecurities associated with aging. As in younger women,

**binge eating disorder (BED)** Recurring episodes of consuming large amounts of food within a short period of time not followed by purging.

**restrained eaters** People who experience cycles of fasting followed by bingeing.

in midlife a person's body image concerns may also pre-cipitate eating disorders.

> **CONNECTIONS** Recall that incidence is the number of new cases in a population during a specified period of time (Chapter 1, page 21).

Although the extent of eating disorders in females is fully recognized, far less is known about males. Researchers have estimated that males account for approximately 10% of people with AN and 15% of people with BN.[13] Because the small number of males with eating disorders makes this group difficult to study, it is not clear what factors lead to eating disorders in males. However, possible risk factors include a history of obesity, participation in a sport that emphasizes thinness, and a heightened emphasis on physical appearance.

Although scientists have offered many theories as to what causes eating disorders, there is no simple answer. Undoubtedly, people develop eating disorders for a variety of reasons. Why some people are more vulnerable than others is not clear. However, several factors are likely, including sociocultural characteristics, family dynamics, personality traits, and biological (genetic) factors.[14]

## Sociocultural Factors

Eating disorders are more prevalent in some cultures than others. Cultures where food is abundant and slim-ness is valued are the ones most conducive to fostering eating disorders. The media, which often portray unre-alistic physiques and glamorize unnaturally thin bod-ies, are criticized for invoking a sense of inadequacy in

young, impressionable people.[15] Even the body shapes of popular dolls are not realistic (Figure 5).

Celebrity role models can give also teenagers an unrealistic standard of thinness, causing some to engage in unhealthy eating practices to achieve this thin and glamorous appearance.[16] Media images suggest that the perfect body is tall, lean, and has well-defined muscles. In fact, female fashion models, beauty pageant contes-tants, and actresses have become increasingly thinner over the years. Even the body weights of Miss America beauty contestants have decreased. When the pageant first began in the 1920s, BMI averaged 20 to 25 kg/m$^2$. Today, nearly all the participants have BMIs below what is considered healthy. Furthermore, some contestants have BMIs consistent with the diagnostic criteria for AN.[17]

Societal pressure to be thin appears more intense for females than males. However, it is not at all uncom-mon for males to also have issues related to body image. Rather than displaying expectations of being thin, the media portrays the ideal male body as strong, devel-oped, and muscular. **Muscle dysmorphia** is a recently recognized syndrome that primarily affects males and is characterized by a pathological preoccupation with increasing muscularity.[18] Individuals with muscle dys-morphia think their muscles are too small and may work out excessively to obtain the "perfect" physique. Unlike AN, people with muscle dysmorphia believe the bigger the better. Signs include intensive workouts that last for

> **muscle dysmorphia** Pathological preoccupation with increasing mus-cularity.

## FIGURE 5    Unrealistic Body Shapes of Popular Dolls

Female dolls are unrealistically tall and slender, and male dolls have unrealistically broad shoulders and muscular physiques. These calculations show the changes needed for a young, healthy adult female and male to attain the proportions of Barbie® and Ken®.

| | If Barbie® were a real female | Female | If Ken® were a real male | Male |
|---|---|---|---|---|
| Height (ft., in.) | 7'2" | 5'2" | 7'8" | 6'0" |
| Chest (in.) | 40 | 35 | 50 | 40 |
| Waist (in.) | 22 | 28 | 43 | 33 |
| Neck circumference (in.) | 12 | 12.2 | 23.4 | 15.5 |
| Neck length (in.) | 6.2 | 3 | 6.4 | 5.5 |

© Robert Brenner/PhotoEdit

SOURCE: Data from Brownell KD, Napolitano MA. Distorting reality for children: body size proportions of Barbie and Ken dolls. *International Journal of Eating Disorders.* 1995;18:295–298.

© John Lamb/The Image Bank/Getty Images

Muscle dysmorphia has been described as the opposite of anorexia nervosa and is characterized by a pathological preoccupation with increasing muscularity.

© Kathy Willens/AP Photos

Models that are portrayed as glamorous tend to have unhealthy body weights, giving some women unrealistic standards of thinness.

hours a day, workouts that interfere with social life, paying excessive attention to diet, and unusual food rituals and eating practices. Reportedly, people with muscle dysmorphia may also engage in health-threatening practices such as use of anabolic steroids to gain sufficient muscle mass.[19]

Dissatisfaction with body weight and shape is thought to be an essential precursor to the development of eating disorders. Therefore, it is important for children to have a healthy body image and to understand that bodies come in all shapes and sizes. Most importantly, older children must be prepared for the physical and emotional changes associated with puberty.

In addition to the media, peers may also be a contributing factor in the development of eating disorders. This is because attitudes and behaviors about the importance of slimness and appearance are often learned from peers. Furthermore, body dissatisfaction and dieting in adolescent females often stem from the desire to gain acceptance among peer groups.[20] Thus, some girls may feel that to be liked and belong, they must be thin.

Although eating disorders often begin with a desire to lose weight, relatively few people on weight loss diets develop them. Furthermore, not everyone living in an affluent society that stigmatizes obesity and advocates slenderness develops an eating disorder. So these factors alone are not sufficient to cause eating disorders. Rather, the social and cultural environment in industrialized countries appears to foster the development of eating disorders in individuals who are vulnerable in some other way.

## Family Dynamics

Parents influence many aspects of their children's lives. Thus it is not surprising that family dynamics can play a role in the development and perpetuation of eating disorders. Although no one particular family type is most conducive to eating disorders, researchers have found certain distinguishing characteristics. These include enmeshment, overprotectiveness, rigidity, conflict avoidance, abusiveness, chaotic family dynamics, and having a mother with an eating disorder.[21]

**Enmeshment** is a term used to describe family members who are overly involved with one another and have little autonomy. Enmeshed families have no clear boundaries and discourage independence and individualism. Children growing up in such families often feel tremendous pressure to please parents and meet expectations. Rather than doing things for themselves, they strive to please others. Growing up in an environment with little open communication, rigidity, and conflict avoidance may also lay the foundation for the emergence of eating disorders. Under these circumstances, food may become the only component remaining in the child's life over which he or she can exert control.

Unlike the closeness of enmeshed families, chaotic family dynamics have also been linked with eating disorders. Chaotic families, or what is also referred to as **disengaged families,** lack cohesiveness and there is little parental involvement. The roles of family members are loosely defined, and children often have a sense of abandonment. For example, parents may be depressed, alcoholic, or "emotionally" absent. A child growing up

**enmeshment** Family interaction whereby family members are overly involved with one another and have little autonomy.

**disengaged families** Family interaction characterized by a lack of cohesiveness and little parental involvement.

in this type of home may later develop eating disorders as a way to fill an emotional emptiness, gain attention, or suppress emotional conflict.

Mothers with eating disorders or those with body dissatisfaction are more likely to negatively influence eating behaviors in their children.[22] The inability to demonstrate a healthy relationship with food and model healthy eating to children is a serious concern. Furthermore, mothers with eating disorders are more likely to encourage their daughters to lose weight and to criticize their appearance. As a result, children of women with eating disorders are at increased risk for developing eating disorders later in life.

## Personality Traits and Emotional Triggers

Scientists have long thought that certain personality traits make some people more prone than others to eating disorders. Some of these traits include low self-esteem, lack of self-confidence, obsessiveness, and feelings of helplessness, anxiety, and/or depression.[23] In people who are emotionally vulnerable, sometimes it takes only a small incident to trigger the development of an eating disorder. For example, many individuals with eating disorders recall having been teased or ridiculed about their weight and/or appearance as a child.[24] Other emotional triggers include puberty, family dysfunction, starting a new school, or something as devastating as rape or incest. As a result, disordered eating behaviors begin to spiral out of control. Mistakenly, individuals with eating disorders associate achieving a certain body weight or shape with emotional fulfillment, happiness, and approval from others.

Individuals with eating disorders are often described as being **food preoccupied.** That is, they spend an inordinate amount of time thinking about food. Being obsessive is a common trait among people with eating disorders, who may spend as much as three hours per day just thinking about food. Similarly, individuals with eating disorders also have a tendency to be perfectionists.[25] Such people have difficulty dealing with shortcomings in themselves and in others. Thus an imperfect body is not easily tolerated.

## Biological (Genetic) Factors

Seeking to better understand eating disorders, researchers are investigating biological influences on their development. Because certain personality traits and eating behaviors are, in part, determined by the nervous and endocrine systems, it makes sense that brain chemicals may play a role in the development of eating disorders. Then again, because eating disorders can disrupt normal signals of hunger and satiety, which in turn may affect neuroendocrine regulation, it may not be possible to identify predisposing factors. For example, studies show that individuals with eating disorders are often clinically depressed. However, it is difficult to determine if clinical depression leads to eating disorders or visa versa. Nonetheless, because medication used to treat clinical depression is often effective in the treatment of certain eating disorders, depression may be a contributing factor.

> **CONNECTIONS** Recall that the nervous and endocrine systems regulate a variety of functions via neural and hormonal controls (Chapter 3, page 84).

Recently, scientists have become interested in identifying genes that might influence susceptibility to eating disorders. For example, studies of identical and fraternal twins provide evidence that eating disorders may, in part, be inherited. Although these types of studies cannot completely differentiate the contribution of genes versus environment, some research suggests that the contribution of genetics is significantly greater than that of the environment, although how this might occur is unclear.[26]

## ESSENTIAL *Concepts*

Although the causes of eating disorders are unclear, a variety of risk factors have been identified. For example, eating disorders are more prevalent in cultures where food is abundant and slimness is valued. Similarly, the media may contribute to body weight and shape dissatisfaction by portraying slimness as the ideal physique. The importance of slimness and appearance is also learned from peers. This peer pressure may also contribute to the occurrence of eating disorders. Dysfunctional family dynamics are also associated with eating disorders. In addition, mothers with eating disorders or those with body dissatisfaction are more likely to negatively influence eating behaviors in their children. Personality traits associated with eating disorders include low self-esteem, lack of self-confidence, obsessiveness, and feelings of helplessness, anxiety, and/or depression. Genetic factors may also predispose some individuals to eating disorders.

## Eating Disorders, Athletics, and the Female Athlete Triad

There are more competitive female athletes today than ever before. The number of female athletes with eating disorders has increased as well. Some studies indicate

**food preoccupied** Spending an inordinate amount of time thinking about food.

that the prevalence of eating disorders among female student-athletes and nonathletes do not differ. However, other studies suggest otherwise.[27] At any rate, losing weight to enhance athletic performance can have serious health consequences. Thus it is of utmost importance for coaches and trainers to recognize early warning signs and symptoms associated with eating disorders in athletes.

## Are Athletes at Increased Risk for Eating Disorders?

The incidence of disordered eating and eating disorders among collegiate athletes is estimated to be somewhere between 15 and 60%.[28] Disagreement and inconsistent estimates may, in part, be due to the reluctance of athletes to admit that such a problem exists. In addition, some athletes may exhibit disordered eating behaviors, yet do not satisfy all the criteria needed for diagnosis. Regardless of the exact number, athletes (especially females) are considered a group "at risk" for developing eating disorders.

For some athletes, physical performance is not only determined by motor abilities, strength, and coordination, but also by body weight. For this reason, some athletes such as ski jumpers, cyclists, rock climbers, and long-distance runners deliberately try to achieve a low body weight to gain a competitive advantage. In addition, sports that demand a thin physical appearance are likely to have more athletes with eating disorders than are activities where greater size is thought an advantage. This may, in part, be due to the fact that judges often consider size and appearance when rating performance. Athletes who participate in dancing, figure skating, synchronized swimming, gymnastics, and diving perceive that body size can affect how judges rate their performance.

Because athletes are invariably competitive people and may equate their self-worth with athletic success, they may be more willing to engage in risky weight-loss

© Robin Sachs/PhotoEdit

Sports that demand a thin physical appearance are likely to have more athletes with eating disorders than do activities where large size is thought an advantage.

practices. In addition, coaches and trainers themselves often believe that excess weight can hinder performance. According to the National Collegiate Athletic Association (NCAA), sports with the highest number of female athletes with eating disorders are cross-country, gymnastics, swimming, and track and field.[29] Sports with the highest number of male athletes with eating disorders are wrestling and cross-country.

## The Female Athlete Triad

Athletes with eating disorders are at extremely high risk for developing medical complications. This is largely because the rigor of athletic training alone is very stressful on the body. As such, it must be adequately nourished to meet these physical demands. Serious health problems can arise when an athlete is restricting food intake, bingeing, and/or purging. For example, female athletes are at increased risk for developing a syndrome known as the **"female athlete triad."** The female athlete triad is a combination of interrelated conditions including disordered eating, amenorrhea, and osteopenia—a condition characterized by a loss of calcium from bones (Figure 6).[30]

Disordered eating can lead to very low levels of body fat, which can lead to amenorrhea. When this occurs, estrogen levels greatly diminish. Without adequate estrogen, bones become dangerously weak, leading to osteopenia. In more severe cases, the entire matrix of the bone begins to deteriorate, a condition called osteoporosis. A study of female distance skiers found that 10% had osteoporosis and nearly 50% had osteopenia.[31] Some female athletes feel unburdened when menstruation stops and therefore are unlikely to report it. However, these athletes often begin to experience frequent injuries such as stress fractures, which can draw attention to the fact that they may have an eating disorder. It is important for parents, coaches, and health care providers to be aware of the spectrum of disordered eating and eating disorders among athletes.

## ESSENTIAL *Concepts*

Athletes, especially females, are considered "at risk" for developing eating disorders, which can have serious health consequences. Sports that value and reward a thin physical appearance are likely to have more athletes with eating disorders than do activities where size is not as important. Also, because athletes are often competitive, they tend to be more willing to engage in risky

**female athlete triad** A combination of interrelated conditions: disordered eating, amenorrhea, and osteopenia.

## FIGURE 6    Female Athlete Triad

The female athlete triad is a combination of interrelated conditions: an eating disorder, amenorrhea, and osteopenia.

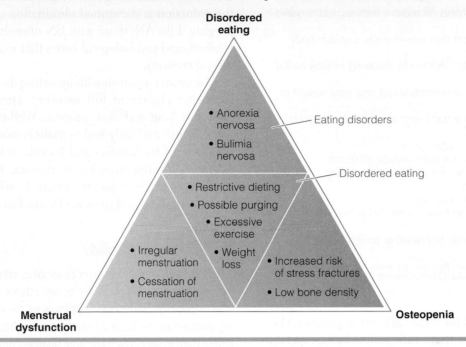

weight-loss practices in order to succeed. Athletes with eating disorders are at high risk for developing medical complications. Female athletes are also at increased risk for developing the "female athlete triad": disordered eating, amenorrhea, and osteoporosis. This can lead to frequent injuries such as stress fractures.

# Prevention and Treatment

People with eating disorders are at risk for serious medical and/or emotional problems and need treatment from qualified health professionals. Typically this includes mental health specialists who can help address and treat underlying psychological issues. An important treatment goal is for individuals with eating disorders to learn how to enjoy food without fear and guilt and to rely on the physiological cues of hunger and satiety to regulate food intake. For this to become possible, specialists help people recognize and appreciate their self-worth. Many treatment options are available for people who have eating disorders.[32] However, the first step is to recognize that there is a problem and to seek help.

## Prevention of Eating Disorders

Programs and educational curricula designed to increase awareness and prevent eating disorders in young girls produce varied results. Although it is important to reach school-age children before eating disorders begin, too many programs simply focus on deterring dangerous eating-disorder behaviors rather than on encouraging healthy attitudes toward eating, dieting, and body image. Furthermore, to prevent eating disorders from developing later educational strategies must focus on issues related to self-esteem. To reduce the risk of eating disorders, especially among the most vulnerable, programs must encourage positive attitudes and behaviors. Suggestions for promoting a healthy body image among children and adolescents are presented in Table 5.

## Treatment Strategies

People with eating disorders often do not recognize or cannot admit there is a problem. Concerns expressed by friends and families are often ignored or dismissed. Friends and family members feel confused and frustrated by their inability to help—a completely normal reaction to a very difficult situation. However, it is important to remember that even though people with eating disorders may resist getting help, family and friends play important supportive roles. Experts recommend that family and friends not focus on the eating disorder per se, because this may make the person feel more defensive. Expressing concerns regarding the person's unhappiness and encouraging him or her to seek help may be of greatest value.

**TABLE 5    Promoting a Healthy Body Image Among Children and Adolescents**

- Encourage children to focus on positive features, not negative ones.
- Help children understand that everyone has a unique body size and shape.
- Be a good role model for children by modeling healthy eating behaviors.
- Resist making negative comments about your own weight or body shape.
- Focus on nonappearance traits such as generosity, kindness, and a friendly laugh.
- Do not criticize a child's appearance.
- Do not associate self-worth with physical attributes.
- Prepare a child in advance for puberty by discussing physical and emotional changes.
- Enjoy meals together as a family.
- Encourage discussions on how the media can negatively affect body image.
- Avoid using food as a way to reward or punish behavior.

SOURCE: Story M, Holt K, Sofka D, editors. *Bright Futures in Practice: Nutrition.* Arlington, VA: National Center for Education in Maternal and Child Health; 2000, Appendix I.

It is important that the medical team or program chosen to treat a person with an eating disorder has specific training and expertise in this area. Treatment goals for people with AN include restoration of lost weight, resolving psychological issues such as low self-esteem and distorted body image, and long-term recovery. Some people with AN are severely malnourished and may require care in an inpatient facility.[33] These types of facilities are staffed with professionals who can provide care in all aspects of recovery, including medical needs, nutritional problems, and psychological issues. This is best achieved by a team of medical professionals, including physicians, nurses, social workers, mental health therapists, and dietitians. Treatment goals for people with BN include the reduction and eventual elimination of bingeing and purging. Like AN, those with BN often have nutritional, medical, and psychological issues that must be addressed during recovery.

The sooner a person with an eating disorder gets help, the better chance of full recovery. However, recovery can be a long and slow process. Well-meaning advice such as "just eat" only makes matters worse. However, it is important for families and friends to know that most people with eating disorders do recover. Treatment often involves counseling for the entire family, which helps everyone to heal and to move forward in life.

## ESSENTIAL *Concepts*

To prevent eating disorders from developing, educational strategies must focus on issues related to self-esteem and encourage healthy behaviors, rather than just focus on the dangers of eating disorders per se. Once an eating disorder has developed, it is important to seek treatment from qualified professionals. This includes mental health specialists, medical professionals, and registered dietitians. Treatment goals for people with AN include restoration of lost weight, resolving psychological issues such as low self-esteem and distorted body image, and long-term recovery. Treatment goals for people with BN include the reduction and eventual elimination of bingeing and purging. The sooner a person with an eating disorder gets help, the better the chance of full recovery.

## Notes

1. Paxton SJ. Body dissatisfaction and disordered eating. Journal of Psychosomatic Research. 2002;53:961–2.
2. American Psychiatric Association. Diagnostic and statistical manual of mental disorders, 4th ed. (DSM-IV). Washington, DC: American Psychiatric Association; 1994.
3. Keel PK, Klump KL, Miller KB, McGue M, Iacono WG. Shared transmission of eating disorders and anxiety disorders. International Journal of Eating Disorders. 2005;38:99–105.
4. Katzman DK. Medical complications in adolescents with anorexia nervosa: a review of the literature. International Journal of Eating Disorders. 2005;37:S52–9.
5. Wolfe BE. Reproductive health in women with eating disorders. Journal of Obstetrics and Gynecology in Neonatal Nursing. 2005;34:255–63.
6. Birmingham CL, Su J, Hlynsky JA, Goldner EM, Gao M. The mortality rate from anorexia nervosa. International Journal of Eating Disorders. 2005;38:143–6.
7. Hay P, Bacaltchuk J. Bulimia nervosa. Clinical Evidence. 2004;12:1326–47.
8. Dawe S, Loxton NJ. The role of impulsivity in the development of substance use and eating disorders. Neuroscience Biobehavorial Reviews. 2004;28:343–51.
9. Dansky BS, Brewerton TD, Kilpatrick DG. Comorbidity of bulimia nervosa and alcohol use disorders: results from the National Women's Study. International Journal of Eating Disorders. 2000;27:180–90.
10. Vanderlinden J, DalleGrave R, Fernandez F, Vandereycken W, Pieters G, Noorduin C. Which factors do provoke binge eating? An exploratory study in eating disorder patients. Eating and Weight Disorders. 2004;9:300–5.
11. Masheb RM, Grilo CM. On the relation of attempting to lose weight, restraint, and binge eating in outpatients with binge eating disorder. Obesity Research. 2000;8:638–45.
12. Clarke LH. Older women's perceptions of ideal body weights: the tensions between health and appearance motivations for weight loss. Ageing and Society. 2002;22:751–3.
13. Sharp CW, Clark SA, Dunan JR, Blackwood DH, Shapiro CM. Clinical presentation of anorexia nervosa in males:

24 new cases. International Journal of Eating Disorders. 1994;15:125–34.

**14.** Becker AE, Keel P, Anderson-Fye EP, Thomas JJ. Genes and/or jeans? Genetic and socio-cultural contributions to risk for eating disorders. Journal of Addictive Disorders. 2004;23:81–103.

**15.** Bulik CM. Exploring the gene-environment nexus in eating disorders. Journal of Psychiatry Neuroscience. 2005;30:335–9.

**16.** Brown JD, Witherspoon EM. The mass media and American adolescents' health. Journal of Adolescent Health. 2002;31:153–70.

**17.** Rubinstein S, Caballero B. Is Miss America an undernourished role model? JAMA (Journal of the American Medical Association). 2000;283:1569.

**18.** Chung B. Muscle dysmorphia: A critical review of the proposed criteria. Perspectives in Biology and Medicine. 2001;44:565–74.

**19.** Wroblewska AM. Androgenic-anabolic steroids and body dysmorphia in young men. Journal of Psychosomatic Research. 1997;42:225–34.

**20.** Neumark-Sztainer D, Falkner N, Story M, Perry C, Hannan PJ, Mulert S. Weight-teasing among adolescents: correlations with weight status and disordered eating behaviors. International Journal of Eating Disorders. 2002;26:123–31. Mathieu J. Disordered eating across the life span. Journal of the American Dietetic Association. 2004;104:1208–10.

**21.** Botta RA, Dumlao R. How do conflict and communication patterns between fathers and daughters contribute to or offset eating disorders? Health Communication. 2002;14:199–219. Coulthard H, Blissett J, Harris G. The relationship between parental eating problems and children's feeding behavior: a selective review of the literature. Eating Behavior. 2004;5:103–15.

**22.** Mazzeo SE, Zucker NL, Gerke CK, Mitchell KS, Bulik CM. Parenting concerns of women with histories of eating disorders. International Journal of Eating Disorders. 2005;37:S77–S9.

**23.** Cassin SE, von Ranson KM. Personality and eating disorders: A decade in review. Clinical Psychology Review. 2005;25:895–916.

**24.** Keery H, Boutelle K, van den Berg P, Thompson JK. The impact of appearance-related teasing by family members. Journal of Adolescent Health. 2005;37:120–7.

**25.** Bulik CM, Tozzi F, Anderson C, Mazzeo SE, Aggen S, Sullivan PF. The relation between eating disorders and components of perfectionism. American Journal of Psychiatry. 2003;160:366–8.

**26.** Bulik CM, Reba L, Siega-Riz AM, Reichborn-Kjennerud T. Anorexia nervosa: definition, epidemiology, and cycle of risk. International Journal of Eating Disorders. 2005;37 Suppl:S2–S9.

**27.** Reinking MF, Alexander LE. Prevalence of disordered-eating behaviors in undergraduate female collegiate athletes and nonathletes. Journal of Athletic Training. 2005;40:47–51.

**28.** Sudi K, Ottl K, Payerl D, Baumgartl P, Tauschmann K, Muller W. Anorexia athletica. Nutrition. 2004;20:657–61.

**29.** Johnson C, Powers PS, Dick R. Athletes and eating disorders: The National Collegiate Athletic Association study. International Journal of Eating Disorders. 1999;26:179–88.

**30.** Waldrop J. Early identification and interventions for female athlete triad. Journal of Pediatric Health Care. 2005;19:213–20.

**31.** Pettersson U, Alfredson H, Nordstrom P, Henriksson-Larsen K, Lorentzon R. Bone mass in female cross-country skiers: relationship between muscle strength and different BMD sites. Calciferous Tissue International. 2000;67:199–206.

**32.** Kotler LA, Boudreau GS, Devlin MJ. Emerging psychotherapies for eating disorders. Journal of Psychiatric Practice. 2003;9:431–41.

**33.** Lock J, Agras WS, Bryson S, Kraemer HC. A comparison of short- and long-term family therapy for adolescent anorexia nervosa. Journal of the American Academy of Child and Adolescent Psychiatry. 2005;44:632–9.

**Check out the following sources for additional information.**

**34.** Donaldson ML. The female athlete triad. A growing health concern. Orthopedic Nursing. 2003;22:322–24.

**35.** Anderson DA, Lundgren JD, Shapiro JR, Paulosky CA. Assessment of eating disorders: review and recommendations for clinical use. Behavior Modification. 2004;28:763–82.

**36.** Engel SG, Johnson C, Powers PS, Crosby RD, Wonderlich SA, Wittrock DA, Mitchell JE. Predictors of disordered eating in a sample of elite Division I college athletes. Eating Behavior. 2003;4:333–43.

**37.** Kaye WH, Frank GK, Bailer UF, Henry SE, Meltzer CC, Price JC, Mathis CA, Wagner A. Serotonin alterations in anorexia and bulimia nervosa: new insights from imaging studies. Physiology and Behavior. 2005;85:73–81.

**38.** Mitchell JE, Cook-Myers T, Wonderlich SA. Diagnostic criteria for anorexia nervosa: looking ahead to DSM-V. International Journal of Eating Disorders. 2005;37:S95–S7.

**39.** Silber TJ. Anorexia nervosa among children and adolescents. Advances in Pediatrics. 2005;52:49–76.

**40.** Waldrop J. Early identification and interventions for female athlete triad. Journal of Pediatric Health Care. 2005;19:213–20.

**41.** Wonderlich SA, Lilenfeld LR, Riso LP, Engel S, Mitchell JE. Personality and anorexia nervosa. International Journal of Eating Disorders. 2005;37:S68–S71.

## Chapter 10

# Water-Soluble Vitamins

So far, you have learned how the body uses macronutrients for a multitude of functions. However, carbohydrates, proteins, and lipids represent only some of the essential nutrients needed for life. We also need vitamins and minerals, also known as micronutrients, and understanding these compounds can help you choose foods to optimize health.

We begin with the vitamins. Recall that vitamins are organic molecules in foods, required in very small amounts to maintain the body's basic functions. The 13 essential vitamins can be classified as water soluble or fat soluble depending on their chemical solubility in water. Vitamins participate in a variety of functions, from regulating energy metabolism to repairing cell damage.

This chapter describes the water-soluble vitamins, including their structures, dietary sources, regulation, major functions, and dietary recommendations. Deficiency and toxicity signs and symptoms are also discussed. In addition, you will learn how to prepare and store foods to optimize their vitamin contents, as well as basic information about when it is appropriate—or inappropriate—to get vitamins from supplements.

## Chapter Outline

Photos © Matthew Farruggio

# The Basics of Water-Soluble Vitamins

As the name implies, water-soluble vitamins dissolve in water. There are nine essential water-soluble vitamins: eight B vitamins and vitamin C. Some people consider choline and carnitine to be conditionally essential, water-soluble nutrients, and perhaps vitamins. Table 10.1 lists the names, structures, basic functions, and some dietary sources of these compounds.

*(Text continues on page 381.)*

**TABLE 10.1  Functions, Deficiency Diseases, Toxicity Symptoms, and Sources for the Essential Water-Soluble Vitamins**

| Vitamin Structure | Major Functions | Deficiency | Toxicity | Excellent Food Sources |
|---|---|---|---|---|
| **Thiamin** | • Coenzyme (TPP)<br>• Energy metabolism<br>• Synthesis of DNA, RNA, and NADPH+H$^+$<br>• Nerve function | Beriberi | — | Pork<br>Whole grains<br>Legumes<br>Fortified and enriched foods<br>Tuna<br>Soy Milk |
| **Riboflavin (B$_2$)** | • Coenzyme (FAD$^+$ and FMN)<br>• Energy metabolism (redox)<br>• Metabolism of folate, vitamin A, niacin, vitamin B$_6$, and vitamin K<br>• Neurotransmitter metabolism | Ariboflavinosis | — | Liver<br>Mushrooms<br>Dairy products<br>Fortified and enriched foods<br>Tomatoes<br>Spinach |
| **Niacin (B$_3$)** | • Coenzyme (NAD$^+$ and NADP$^+$)<br>• Energy metabolism (redox)<br>• Protein synthesis<br>• Glucose homeostasis<br>• Cholesterol metabolism<br>• DNA repair | Pellagra | Skin inflammation and flushing | Liver<br>Fish<br>Meat (including chicken)<br>Fortified and enriched foods<br>Tomatoes<br>Mushrooms |

Phosphates added here to form TPP and TTP

Modified to form FMN and FAD$^+$

Replaced with —NH$_2$ to become nicotinamide

**TABLE 10.1** Functions, Deficiency Diseases, Toxicity Symptoms, and Sources for the Essential Water-Soluble Vitamins (*continued*)

| Vitamin Structure | Major Functions | Deficiency | Toxicity | Excellent Food Sources |
|---|---|---|---|---|
| **Pantothenic acid (B$_5$)** 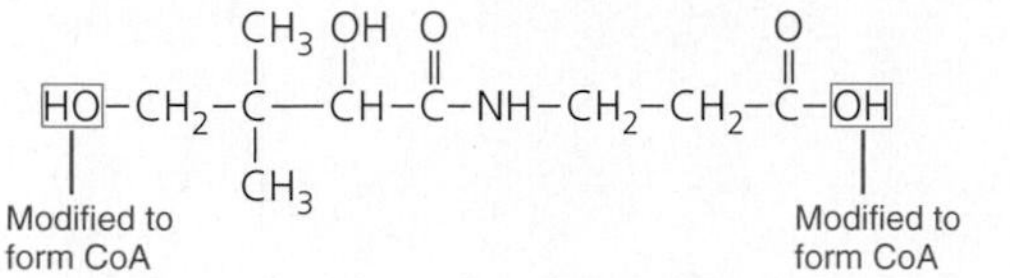 | • Coenzyme (CoA)<br>• Energy metabolism<br>• Heme synthesis<br>• Cholesterol, fatty acid, steroid, and phospholipid synthesis | Burning feet syndrome | — | Liver<br>Mushrooms<br>Sunflower seeds<br>Fortified foods<br>Yogurt<br>Turkey |
| **Vitamin B$_6$** 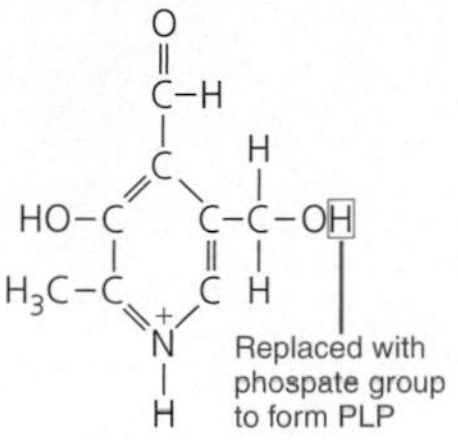 | • Coenzyme (PLP)<br>• Amino acid metabolism (transamination)<br>• Neurotransmitter and hemoglobin synthesis<br>• Glycogenesis<br>• Regulation of steroid hormone function | Microcytic hypochromic anemia | Neurological problems | Fish<br>Chickpeas<br>Liver<br>Fortified foods<br>Potatoes<br>Bananas |
| **Biotin** 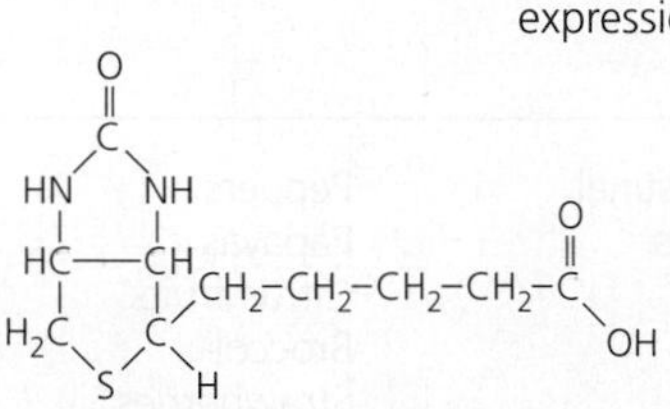 | • Coenzyme<br>• Energy metabolism (carboxylation)<br>• Regulation of gene expression | Depression, loss of muscle control, and skin irritations | — | Peanuts<br>Almonds<br>Mushrooms<br>Egg yolk<br>Tomatoes<br>Avocados |
| **Folate** 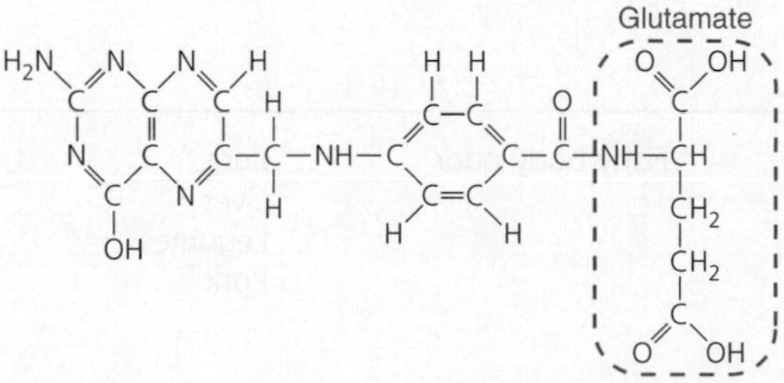 | • Coenzyme (THF)<br>• Single-carbon transfers<br>• Amino acid metabolism<br>• DNA and RNA synthesis | Macrocytic anemia | — | Organ meats<br>Legumes<br>Okra<br>Leafy vegetables<br>Fortified and enriched foods |

Pantothenic acid structure:

CH$_3$ OH O ... O
HO–CH$_2$–C—CH–C–NH–CH$_2$–CH$_2$–C–OH
CH$_3$

Modified to form CoA (left HO–)
Modified to form CoA (right –OH)

Vitamin B$_6$ structure:
Replaced with phosphate group to form PLP

Folate structure:
Glutamate

*continued*

**TABLE 10.1** Functions, Deficiency Diseases, Toxicity Symptoms, and Sources for the Essential Water-Soluble Vitamins (*continued*)

| Vitamin Structure | Major Functions | Deficiency | Toxicity | Excellent Food Sources |
|---|---|---|---|---|
| **Vitamin B$_{12}$** | • Coenzyme<br>• Homocysteine metabolism<br>• Energy metabolism | Macrocytic anemia | — | Mollusks<br>Liver<br>Salmon<br>Meat<br>Fortified foods<br>Cottage cheese |
| **Vitamin C** | • Antioxidant<br>• "Recharging" enzymes<br>• Collagen synthesis<br>• Tyrosine, neurotransmitter, and hormone synthesis<br>• Protection from free radicals | Scurvy | Gastrointestinal problems | Peppers<br>Papayas<br>Citrus fruits<br>Broccoli<br>Strawberries<br>Brussels sprouts |
| **Choline** | • Phospholipid synthesis<br>• Neurotransmitter synthesis | Liver damage | Fishy body odor | Eggs<br>Liver<br>Legumes<br>Pork |

**TABLE 10.2   Common Characteristics of the Water-Soluble Vitamins**

Many of the water-soluble vitamins have similar properties. Although there are exceptions, some principles can be applied to these compounds.

| | General Characteristics |
|---|---|
| **Vitamins in Foods** | • Bound to proteins that must be cleaved prior to absorption<br>• Easily destroyed during cooking |
| **Digestion** | • Digested mostly in small intestine |
| **Absorption** | • Absorbed mostly in small intestine, but also in stomach<br>• Absorbed via simple diffusion when intake is high and active transport when intake is low<br>• Bioavailability is influenced by many factors |
| **Circulation** | • Transported via blood to liver |
| **Functions** | • Many coenzyme and other roles, especially in energy metabolism |
| **Toxicity Effects** | • Minimal, although some exist |

Although there are exceptions, the water-soluble vitamins tend to be digested, absorbed, and transported similarly. For example, many water-soluble vitamins are bound to proteins that enzymes must remove before absorption. The water-soluble vitamins are absorbed mostly in the small intestine, and to a lesser extent, the stomach. The extent to which vitamins are absorbed and used in the body—in other words, bioavailability—is influenced by many factors, including nutritional status, other nutrients and substances in foods, medications, age, and illness. Except for choline, which circulates in the lymph, the water-soluble vitamins are circulated to the liver in the blood. And because the body does not store large quantities of most water-soluble vitamins, they generally do not have toxic effects when consumed in large amounts. Table 10.2 gives an overview of the digestion, absorption, circulation, and functions of the water-soluble vitamins.

The functions of the water-soluble vitamins are diverse. For example, whereas most of the B vitamins serve as coenzymes, many also play noncoenzyme roles in the body.

## Naming the Vitamins

Vitamins were initially sorted into two groups based on their solubility in lipids and water: vitamin A (fat soluble) and vitamin B (water soluble).[1] Later researchers discovered that there were several vitamins in each of these groups and vitamins were given different letter designations (such as vitamins C, D, E). Furthermore, what researchers initially thought was a single B vitamin turned out to be several vitamins. Each was given a number (such as vitamins $B_1$, vitamin $B_2$, vitamin $B_3$). In addition, the B vitamins are often referred to as the **B-complex vitamins,** because they tend to be physically associated with each other. Many vitamins were also given "common names," such as *thiamin* and *riboflavin*, and most have chemical names; for example, vitamin C is *ascorbic acid*.

### Memorizing B Vitamin Names and Numbers

Memorizing details can make a topic difficult to master. For example, remembering which common names go with which B vitamins can be challenging.

**CONNECTIONS** Essential nutrients are substances needed for health but either not synthesized by the body or synthesized in inadequate amounts. Consequently, essential nutrients must be obtained from the diet (Chapter 1, page 3).

**B-complex vitamins** A term used to describe all the B vitamins.

**TABLE 10.3    Naming B Vitamins**

Note that folate does not have a corresponding number.

| Vitamin | Common Name | Mnemonic |
|---------|-------------|----------|
| $B_1$ | Thiamin | *The* |
| $B_2$ | Riboflavin | *Romans* |
| $B_3$ | Niacin | *never* |
| $B_5$ | Pantothenic acid | *painted* |
| $B_6$ | Pyridoxine | *pyramids* |
| $B_7$ | Biotin | *before* |
| $B_{12}$ | Cobalamin | *college.* |

However, there are some "tricks of the trade" for learning basic facts such as these. One of these "tricks" is to come up with a phrase or sentence that helps you remember a certain sequence of words. Table 10.3 shows an example of such a phrase that will help you remember the B vitamins, which is a nutrition fundamental.

## Fortification and Enrichment of Foods

Although water-soluble vitamins are found naturally in foods, sometimes they are added during processing. In other words, foods can be **fortified** with these nutrients. For example, the B vitamin thiamin is often added to milled rice and other processed cereal products (such as flour). This is important, because the outer coating and the inner portion of grains contain most of the thiamin, and these parts are removed during processing (milling) of the grain into "white rice." Adding back vitamins via fortification helps make many grain products more nutritious.

A special form of fortification is **enrichment.** Food products fortified with a specified set of nutrients to certain levels suggested by the U.S. Food and Drug Administration (FDA) can be labeled as being "enriched."[2] Enrichment often makes the nutritional content of the food similar to what it was before processing. At other times, an enriched food is *more* nutrient dense than its unprocessed counterpart. Federal enrichment standards—in other words, recommendations concerning which nutrients to add and how much to add—were first established in 1938 and have since been revised several times. Currently, the nutrients that must be added to a food for it to be labeled "enriched" are four B vitamins (thiamin, niacin, riboflavin, and folate) and the mineral iron. Only select foods can be labeled as enriched. These are rice, flour, breads and rolls, farina (such as cream of wheat), pasta, corn meal, and corn grits. Looking at food packaging and Nutrition Facts panels (Figure 10.1) makes it relatively easy to check whether a food is enriched, contains enriched cereal products, or has been fortified in other ways.

## Protecting Water-Soluble Vitamins in Foods

Water-soluble vitamins can be easily destroyed during cooking and storage. However, many techniques can protect these compounds. For example, not

**fortified food** A food to which nutrients have been added.

**enrichment** The fortification of a select group of foods (rice, flour, bread or rolls, farina, pasta, corn meal, and corn grits) with FDA-specified levels of thiamin, niacin, riboflavin, folate, and iron.

**FIGURE 10.1   Using Food Labels to Determine If a Product Is Enriched or Fortified**

Although only certain foods (such as pasta and corn meal) can be labeled as being "enriched," other foods can contain enriched ingredients or be fortified to other levels.

**Nutrition Facts**

Serving size   3/4 cup (56g)
Servings per container  6

**Amount per serving**

| | |
|---|---|
| Calories | 210 |
| Calories from Fat | 10 |

**% Daily Value**

| | | |
|---|---|---|
| **Total fat** 1g | | 2% |
| Saturated fat 0g | | 0% |
| **Cholesterol** 0mg | | 0% |
| **Sodium** 0mg | | 9% |
| **Total carbohydrates** 41g | | 10% |
| Dietary fiber 2g | | 7% |
| **Protein** 7g | | |

| | |
|---|---|
| Vitamin A 0% | Vitamin C 0% |
| Calcium 0% | Iron 10% |
| Thiamin 30% | Riboflavin 15% |
| Niacin 20% | Folate 25% |

**INGREDIENTS:** Durum semolina, niacin, ferrous sulfate (iron), thiamine mononitrate, riboflavin, folic acid, contains wheat ingredients

This is an "enriched product" because it is fortified with niacin, riboflavin, folate, thiamin, and iron at levels suggested by the FDA.

**Nutrition Facts**

Serving size   55 pieces
Servings per container  ~12

**Amount per serving**

| | |
|---|---|
| Calories from Fat | 0 |

**% Daily Value**

| | | |
|---|---|---|
| **Total fat** 6g | | 9% |
| Saturated fat 1.5g | | |
| Trans fat 0g | | |
| Polyunsaturated fat 1.5g | | |
| Monounsaturated fat 2.5g | | |
| **Cholesterol** Less than 5mg | | |
| **Sodium** 250mg | | 1% |
| **Total carbohydrates** 19g | | 6% |
| Dietary fiber 1g | | 3% |
| Sugars 0g | | |
| **Protein** 3g | | |

| | |
|---|---|
| Vitamin A 0% | Calcium 2% |
| Vitamin C 0% | Iron 6% |

**INGREDIENTS:** Unbleached enriched wheat flour [flour, niacin, reduced iron, thiamin mononitrate (vitamin B1), riboflavin (vitamin B2), folic acid] cheddar cheese [(pasteurized milk, cheese culture, salt, enzymes), water salt], vegetable oils (canola, sunflower and/or soybean), contains 2 percent or less of:salt, yeast, sugar, yeast extract, leavening (baking soda, monocalcium phosphate, ammonium bicarbonate), spices, annatto (color) and onion powder.

This is not an "enriched product" but is made with enriched flour.

**Nutrition Facts**

Serving size   1 cup (55g)
Servings per container  ~12

**Amount per serving**

| | |
|---|---|
| Calories | 120 |
| Calories from Fat | 0 |

**% Daily Value**

| | | |
|---|---|---|
| **Total fat** 0.5g | | 1% |
| Polyunsaturated fat 0g | | |
| Monounsaturated fat 0g | | 0% |
| **Cholesterol** 0mg | | 0% |
| **Sodium** 240mg | | 10% |
| **Potassium** 350mg | | 10% |
| **Total carbohydrates** 41g | | 14% |
| Dietary fiber 5g | | 20% |
| Sugars 20g | | |
| Other carbohydrate 16g | | |
| **Protein** 4g | | |
| Vitamin A | | 10% |
| Vitamin C | | 0% |
| Calcium | | 100% |
| Iron | | 100% |
| Thiamin | | 100% |
| Riboflavin | | 100% |
| Niacin | | 100% |
| Folic acid | | 100% |
| Vitamin B$_{12}$ | | 100% |
| Pantothenic Acid | | 100% |
| Phosphorus | | 10% |
| Magnesium | | 10% |
| Zinc | | 100% |
| Copper | | 6% |

**INGREDIENTS:** Wheat bran with other parts of wheat, raisins, sugar, calcium carbonate, corn syrup, brown sugar syrup, salt, lactose, zinc and iron (mineral nutrients), vitamin E, (tocopheryl acetate),trisodium phosphate, a B vitamin (niacinamide), vitamin C (sodium ascorbate), a B vitamin (calcium pantothenate), vitamin B6 (pyridoxine hydrocloride), vitamin B$_2$ (riboflavin), vitamin B$_1$ (thiamin mononitrate), annatto extract color, a B vitamin (folic acid), vitamin A (palmitate), vitamin B$_{12}$, vitamin D. Freshness preserved by BHT.

This cereal is fortified with many vitamins and minerals, at levels related to nutrient requirements (DVs).

## Focus On FOOD
### Keeping Foods Nutritious

Choosing a balance and variety of foods and beverages can provide all the nutrients needed to maintain optimal health. Unfortunately, no matter how careful you are in choosing foods, some preparation, cooking, and storing methods can destroy some nutrients—especially water-soluble vitamins. For example, many can be lost or destroyed by exposure to water, air, heat, or light. They are also affected by acidity (pH). Fortunately, proper preparation and storage can prevent excessive nutrient loss.

The following recommendations will help prevent vitamin loss from foods.[3]

- *Water.* Because water dissolves water-soluble vitamins, avoid excessive soaking of foods and cook vegetables in pieces that are as large as possible. Also, do not wash enriched rice before cooking. Milled (white) rice is enriched by spraying with vitamins and minerals, and when you wash it, the enrichment is washed off.
- *Heat.* As heating destroys most of the water-soluble vitamins, do not overcook fruits and vegetables. Instead, cook or steam until just tender.
- *Light.* Light can destroy several of the B vitamins. For example, if milk is exposed to light, considerable destruc-

tion of riboflavin can occur. It is best to store foods in opaque containers or wrap them in paper or foil.
- *pH.* High pH can destroy some vitamins such as thiamin and vitamin C. Do not add baking soda to green vegetables to retain color during cooking or to legumes to decrease cooking time.
- *Air.* Vitamin C and many of the B vitamins are destroyed by exposure to air (oxygen). Thus, store properly in airtight containers and wraps, and cook vegetables as soon as possible after cutting.

overcooking vegetables can help prevent the destruction of several vitamins. In the Focus on Food feature, you can read more about how to prepare and store foods to decrease vitamin loss.

## ESSENTIAL *Concepts*

Water-soluble vitamins tend to be absorbed in the small intestine and circulated to the liver in the blood. The absorption and use, or bioavailability, of many vitamins is influenced by a variety of factors, and although there are exceptions, water-soluble vitamins tend not to be stored in the body. Water-soluble vitamins play important functions as coenzymes, but many also have noncoenzyme roles. Water-soluble vitamins are often added to foods (fortification), and when added in certain amounts these foods can be labeled as "enriched." The water-soluble vitamins in foods are easily destroyed or lost during cooking and storage. However, decreasing cooking times and protecting foods during storage can help prevent this loss.

## Thiamin (Vitamin B₁)

**thiamin** (vitamin B$_1$) An essential water-soluble vitamin involved in energy metabolism; synthesis of DNA, RNA, and NADPH+H$^+$; and nerve function.

**thiamin pyrophosphate (TPP)** The coenzyme form of thiamin that has two phosphate groups.

**thiamin triphosphate (TTP)** A form of thiamin with three phosphate groups.

Vitamin B$_1$ was given the name **thiamin** because it contains a *thiol* (sulfur) group and an *amine* (nitrogen) group. In the body, two phosphate groups are combined with thiamin to form the coenzyme **thiamin pyrophosphate (TPP).*** Thiamin is also present as **thiamin triphosphate (TTP),** which has three phosphate groups.

---

*Thiamin pyrophosphate (TPP) is also called thiamin diphosphate (TDP).

## Dietary Sources and Regulation of Thiamin

Thiamin is found in a variety of foods (Figure 10.2). Good sources include pork, peas, fish (such as tuna), legumes (such as black beans), whole-grain foods, enriched cereal products, and soy milk. As you will see, whole-grain foods tend to be good sources of water-soluble vitamins. You can read more about including these foods in the Food Matters feature on page 388. Thiamin is sensitive to heat and easily destroyed during cooking. Shorter cooking times and/or lower temperatures can decrease this, although it is important to cook meat until done.

## Bioavailability and Regulation of Thiamin in the Body

Bioavailability (percent absorption) of thiamin increases when intake is low, because the cells of the small intestine actively transport thiamin from the intestinal lumen into the absorptive epithelial cell. However, when thiamin intake is high, its absorption proceeds more slowly, by simple diffusion, which does not require energy.

Several dietary factors can destroy thiamin or interfere with its absorption, and are found in raw fish, coffee, tea, berries, brussels sprouts, and cabbage. These "antithiamin factors" work in several ways, one of which is by inactivating thiamin via oxidation. Consuming foods high in vitamin C can help counteract this, because vitamin C prevents thiamin from being oxidized. In other words, vitamin C acts as an antioxidant. Alcohol consumption can also decrease the bioavailability of thiamin. Once absorbed, thiamin is circulated away from the small intestine via the blood to the liver. Excess thiamin is not stored, being metabolized and excreted in the urine.

Pork products are especially good sources of thiamin.

© Riccardo Marcialis/StockFood Munich/ Stockfood America

CONNECTIONS Recall that simple diffusion does not require energy (ATP) and moves a substance from a region of higher concentration to one of lower concentration. Active transport requires energy (ATP) and moves a substance from a low to a high concentration (Chapter 3, pages 79–80).

---

**FIGURE 10.2  Good Sources of Thiamin**

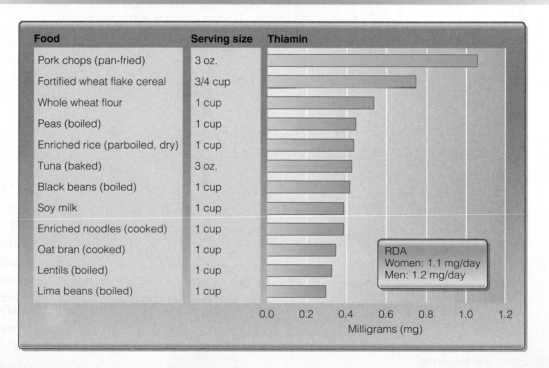

| Food | Serving size | Thiamin |
|------|-------------|---------|
| Pork chops (pan-fried) | 3 oz. | |
| Fortified wheat flake cereal | 3/4 cup | |
| Whole wheat flour | 1 cup | |
| Peas (boiled) | 1 cup | |
| Enriched rice (parboiled, dry) | 1 cup | |
| Tuna (baked) | 3 oz. | |
| Black beans (boiled) | 1 cup | |
| Soy milk | 1 cup | |
| Enriched noodles (cooked) | 1 cup | |
| Oat bran (cooked) | 1 cup | |
| Lentils (boiled) | 1 cup | |
| Lima beans (boiled) | 1 cup | |

RDA
Women: 1.1 mg/day
Men: 1.2 mg/day

Milligrams (mg): 0.0  0.2  0.4  0.6  0.8  1.0  1.2

SOURCE: Data from USDA Nutrient Database, Release 16-1.

**CONNECTIONS** Coenzymes are organic compounds that are required for the actions of some enzymes (Chapter 8, page 273).

The citric acid cycle is an energy metabolism pathway that breaks down intermediate metabolites, generating ATP, NADH + H⁺, FADH₂, and carbon dioxide (Chapter 8, page 282).

## Functions of Thiamin

Thiamin has many functions, including both coenzyme and noncoenzyme roles. Although thiamin, per se, is not an energy-yielding nutrient, its coenzyme form TPP is needed for ATP production. This is illustrated in Figure 10.3. In each of these reactions, TPP removes a molecule of carbon dioxide ($CO_2$) from a substrate. Specifically, TPP is needed for converting pyruvate to acetyl coenzyme A (acetyl-CoA). TPP is also needed for one of the chemical reactions in the citric acid cycle and is involved in reactions allowing some amino acids to enter the citric acid cycle. In this way, thiamin is an integral component of the body's ATP-producing energy metabolism pathways.

TPP is also needed for the synthesis of deoxyribonucleic acid (DNA) and ribonucleic acid (RNA). Remember that DNA is the genetic material that, together with RNA, directs protein synthesis. TPP is also required for the synthesis of nicotinamide adenine dinucleotide phospate ($NADPH+H^+$), which is needed for triglyceride synthesis.

TTP also plays noncoenzyme roles in the body. For example, it is needed for nerve function. Although the mechanisms are not well understood, TTP appears to be involved in either neurotransmitter production or the generation of nerve impulses. Scientists continue to study the many functions of thiamin in the body.

## Thiamin Deficiency: Beriberi

As early as 2600 B.C., Chinese physicians described the life-threatening condition of thiamin deficiency called **beriberi**. Today, although this disease is uncommon in the United States, beriberi is still prevalent in regions of the

---

### FIGURE 10.3    B Vitamins and Energy Metabolism

The B vitamins are important for many aspects of energy metabolism.

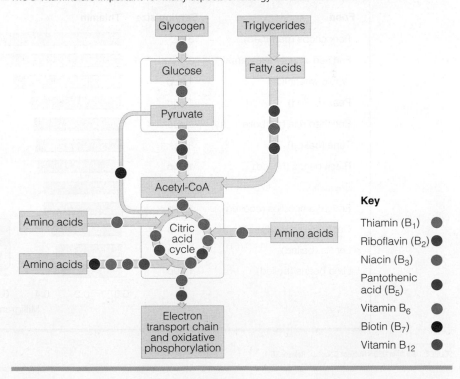

Key

Thiamin (B₁)
Riboflavin (B₂)
Niacin (B₃)
Pantothenic acid (B₅)
Vitamin B₆
Biotin (B₇)
Vitamin B₁₂

**beriberi** (BER – i – BER – i) A disease that results from thiamin deficiency.

world that rely heavily on unfortified, milled rice as a major source of energy. The term *beriberi* is thought to be derived from the Indonesian word meaning "I can't. I can't"—said in response to weakness and impaired functions brought on by thiamin deficiency.

There are four forms of beriberi. **Dry beriberi** is found mostly in adults and is characterized by severe muscle wasting, leg cramps, tenderness, and decreased feeling in the feet and toes. **Wet beriberi** involves severe edema (swelling) in the arms and legs as well as enlargement of the heart and respiratory problems, often resulting in heart failure. It is not clear why some people get dry beriberi whereas others get wet beriberi. **Infantile beriberi** occurs in babies breastfed by thiamin-deficient mothers. Infantile beriberi can cause cardiac arrest and therefore is very serious.

**Cerebral beriberi** (also called Wernicke-Korsakoff syndrome) is typically associated with alcoholism, because alcohol decreases the body's ability to absorb thiamin. Furthermore, thiamin intake by alcoholics can be low, and to compound matters alcoholism-induced liver damage decreases the liver's ability to convert thiamin to TPP. Also, a genetic component predisposes some people to cerebral beriberi. This form of thiamin deficiency is characterized by paralysis of the eye muscles, involuntary eye movement, poor muscle coordination, confusion, and short-term memory loss.

## Recommended Intakes of Thiamin

The Recommended Dietary Allowances (RDAs) for thiamin for adult males and females are 1.2 and 1.1 milligrams (mg), respectively. Because thiamin toxicity is almost unheard of, there is no Tolerable Upper Intake Level (UL) for this vitamin. A complete list of the Dietary Reference Intakes (DRIs) for all vitamins is provided inside the front cover of this book.

**CONNECTIONS** Remember that Dietary Reference Intakes (DRIs) are reference values for nutrient intake. Typically recommended Dietary Allowances (RDAs) are established, but when less information is available Adequate Intake (AI) values are provided. Tolerable Upper Intake Levels (ULs) are intakes that should not be exceeded (Chapter 2, page 36).

## ESSENTIAL *Concepts*

There are three forms of thiamin in the body: free thiamin, thiamin pyrophosphate (TPP), and thiamin triphosphate (TTP). TPP functions as a coenzyme, catalyzing reactions that enable the body to use glucose, amino acids, and fatty acids for energy. Thiamin is also involved in the synthesis of DNA, RNA, and NADP⁺. TTP also has noncoenzyme roles important for nerve function. Good sources of thiamin include pork, peas, whole grains, fish, enriched cereal products, and other fortified foods. Thiamin deficiency causes beriberi, of which there are four forms. High thiamin consumption has no known toxic effects.

**dry beriberi** A form of thiamin deficiency characterized by muscle loss and leg cramps.

**wet beriberi** A form of thiamin deficiency characterized by severe edema.

**infantile beriberi** A form of thiamin deficiency that occurs in infants breastfed by thiamin-deficient mothers.

**cerebral beriberi** (Wernicke-Korsakoff syndrome) A form of thiamin deficiency characterized by poor muscle control and paralysis of the eye muscles.

**riboflavin** (vitamin B₂) An essential water-soluble vitamin involved in energy metabolism, the synthesis of a variety of vitamins, nerve function, and protection of lipids.

**flavin mononucleotide (FMN)** (FLA – vin) A coenzyme form of riboflavin.

**flavin adenine dinucleotide (FAD)** A coenzyme form of riboflavin.

# Riboflavin (Vitamin B₂)

**Riboflavin** (vitamin B₂) consists of a multiring structure attached to the simple sugar ribose. Riboflavin in the body is typically found as one of its coenzymes—**flavin mononucleotide (FMN)** and **flavin adenine dinucleotide (FAD)**—and is important for energy metabolism.

## Dietary Sources and Regulation of Riboflavin

Riboflavin is found in a variety of foods (Figure 10.4), including liver, meat, dairy products, enriched cereal products, and other fortified foods. Most fruits and vegetables provide only marginal amounts, although whole-grain

**CONNECTIONS** Recall that ribose is a 5-carbon monosaccharide that is also a constituent of RNA (Chapter 3, page 76).

# Food Matters

### Working Toward the Goal

The USDA Dietary Guidelines for Americans recommends that we consume three or more servings of whole-grain products per day, with the rest of the grains coming from enriched products. The following selection and preparation tips will help you meet your goal.

- Choose foods that list one of the following whole-grain ingredients *first* on its ingredient list: brown rice, bulgur, graham flour, oatmeal, whole-grain corn, whole oats, whole rye, whole wheat, or wild rice.
- Color is not an indication of a whole grain. Bread can be brown because of molasses or other added ingredients. Read the ingredient list to see if a food is a whole-grain product.
- Substitute a whole-grain product for a refined product—such as whole-wheat bread instead of white bread or whole-wheat pasta instead of plain pasta.
- Substitute whole wheat or oat flour for up to half of the flour in pancake, waffle, muffin, or other flour-based recipes.
- Choose whole-grain chips and snacks, such as baked tortilla chips or popcorn, instead of potato chips or cheese puffs.
- Foods labeled as "multigrain," "stone-ground," "100% wheat," "cracked wheat," "seven-grain," or "bran" are usually not whole-grain products. Foods are usually labeled as "whole grain" if they are.
- Buy bread providing 3 to 5 g of fiber per serving.
- When whole-grain products are not available, choose foods made with enriched flour or other cereal product instead of plain flour.

The B vitamin riboflavin is abundant in meat and dairy products.

foods contribute some riboflavin to the diet. Although riboflavin is relatively stable during cooking, it is easily destroyed by exposure to light. For this reason, milk is packaged in cardboard or cloudy plastic containers, and it is recommended that food be stored in opaque containers or covered with paper or foil.

Riboflavin in food can be in its free form, bound to protein, or as a coenzyme (FMN or FAD). However, only free riboflavin can be absorbed. As such, stomach acid and intestinal enzymes convert all forms of riboflavin to its free (unbound) form prior to absorption in the small intestine. As with thiamin, absorption of riboflavin occurs via simple diffusion when intake is high and active transport when intake is low. In this way, bioavailability increases at low intakes. Riboflavin in animal foods is somewhat more bioavailable than from plant sources, and alcohol can inhibit its absorption. Upon absorption, riboflavin is circulated in the blood to the liver. Riboflavin is not readily stored in the body, and excess is excreted in the urine.

**FIGURE 10.4   Good Sources of Riboflavin**

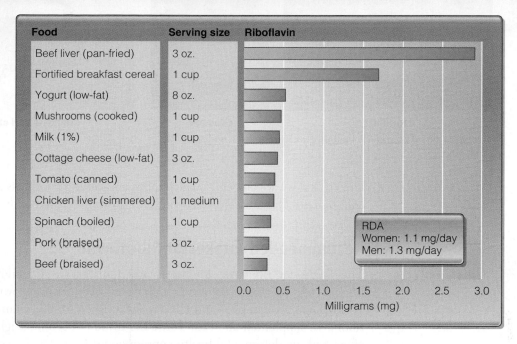

| Food | Serving size | Riboflavin |
|---|---|---|
| Beef liver (pan-fried) | 3 oz. | |
| Fortified breakfast cereal | 1 cup | |
| Yogurt (low-fat) | 8 oz. | |
| Mushrooms (cooked) | 1 cup | |
| Milk (1%) | 1 cup | |
| Cottage cheese (low-fat) | 3 oz. | |
| Tomato (canned) | 1 cup | |
| Chicken liver (simmered) | 1 medium | |
| Spinach (boiled) | 1 cup | |
| Pork (braised) | 3 oz. | |
| Beef (braised) | 3 oz. | |

RDA
Women: 1.1 mg/day
Men: 1.3 mg/day

Milligrams (mg)

SOURCE: Data from USDA Nutrient Database, Release 16-1.

## Functions of Riboflavin

Riboflavin is a coenzyme for several chemical reactions involved in energy metabolism, many of which are reduction-oxidation (redox) reactions. For example, FAD participates in the citric acid cycle, being reduced to FADH$_2$ (Figure 10.3). FADH$_2$ is reoxidized in the electron transport chain, resulting in the formation of ATP, water, and carbon dioxide. FADH$_2$ is also needed for the breakdown, or oxidation, of fatty acids into acetyl-CoA molecules—a process called β-oxidation—which allows the body to use fatty acids to synthesize ATP.

FAD is also required for the synthesis of other compounds. For example, it is needed to convert vitamin A and folate (a B vitamin) to their active forms, convert tryptophan (an amino acid) to niacin (a B vitamin), and form vitamins B$_6$ and K. Riboflavin (FAD) is also needed for the metabolism of some important neurotransmitters (such as dopamine) and is involved in several important reactions that protect biological membranes from oxidative damage. As FMN, riboflavin is needed for activating vitamin B$_6$. In summary, riboflavin is important in energy metabolism, the biosynthesis or activation of several compounds, including some vitamins, and protection of cell membranes from oxidative damage. Thus its deficiency results in numerous complications.

**CONNECTIONS** Oxidation is the loss of electrons, whereas reduction is the gain of electrons (Chapter 3, page 67).

## Riboflavin Deficiency

Riboflavin deficiency, technically called **ariboflavinosis**, is typically associated with broader nutrient deficiencies, and its signs and symptoms are multifaceted. These include weakness, sores on the outside and corners of the lips (**cheilosis**), inflammation of the mouth (**stomatitis**), enlarged and inflamed tongue (**glossitis**), anemia, and confusion. Riboflavin deficiency is rare in the

**ariboflavinosis** (a – ribo – flav – i – NO – sis) A disease caused by riboflavin deficiency.

**cheilosis** (chei – LO – sis) Sores occurring on the outsides and corners of the lips.

**stomatitis** (stom – a – TI – tis) Swollen mouth.

**glossitis** (gloss – I – tis) Inflamed tongue.

**Normal**

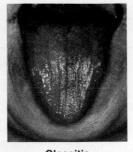
**Glossitis**

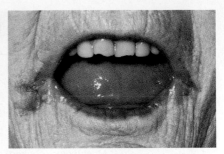
**Cheilosis and stomatitis**

Symptoms of ariboflavinosis include glossitis, cheilosis, and stomatitis.

United States but can occur in alcoholics consuming poor diets and in people with diseases that interfere with riboflavin utilization, such as thyroid disease.

## Recommended Intakes of Riboflavin

The RDAs for riboflavin for males and females are 1.3 and 1.1 mg/day, respectively. Because even at very high doses there are no known toxic effects of riboflavin consumption, no ULs are established for this vitamin. However, riboflavin supplementation can cause the urine to become bright yellow—an effect with no detrimental health consequences.[4]

## ESSENTIAL *Concepts*

There are three forms of riboflavin: free riboflavin, flavin mononucleotide (FMN), and flavin adenine dinucleotide (FAD). Riboflavin functions as a coenzyme in a variety of reduction-oxidation reactions in the body, enabling it to use glucose, amino acids, and fatty acids for energy. Riboflavin is also needed for the activation or synthesis of vitamin A, folate, niacin, vitamins $B_6$ and K, and some neurotransmitters. Good sources of riboflavin include liver, meat, dairy products, whole-grain products, and enriched cereals. Severe riboflavin deficiency causes ariboflavinosis.

## Niacin (Vitamin $B_3$)

**Niacin** (vitamin $B_3$) takes two forms—nicotinic acid and nicotinamide. The body uses both forms to make the coenzymes nicotinamide adenine dinucleotide ($NAD^+$) and nicotinamide adenine dinucleotide phosphate ($NADP^+$). Recall from Chapter 8 that $NAD^+$ and $NADP^+$ are involved in numerous reactions in the body, many of which are required for energy metabolism.

## Dietary Sources of Niacin

We get niacin by consuming dietary niacin and by converting the essential amino acid tryptophan to niacin. About 1 mg of niacin can be made from 60 mg of tryptophan. Both niacin and tryptophan are considered "dietary sources" of niacin, and the unit of measure called the **niacin equivalent (NE)** refers to the combined amounts of niacin and tryptophan in foods.

The niacin contents—or NE—of some foods are provided in Figure 10.5. Niacin or tryptophan is found in a variety of foods such as liver, chicken, fish, tomatoes, beef, and mushrooms. Whole-grain foods, enriched cereal products,

**niacin** (vitamin $B_3$) (NI – a – cin) An essential water-soluble vitamin involved in energy metabolism, electron transport chain, synthesis of fatty acids and proteins, metabolism of vitamin C and folate, glucose homeostasis, and cholesterol metabolism.

**niacin equivalent (NE)** A unit of measure that describes the niacin content and/or tryptophan in food.

**FIGURE 10.5 Good Sources of Niacin**

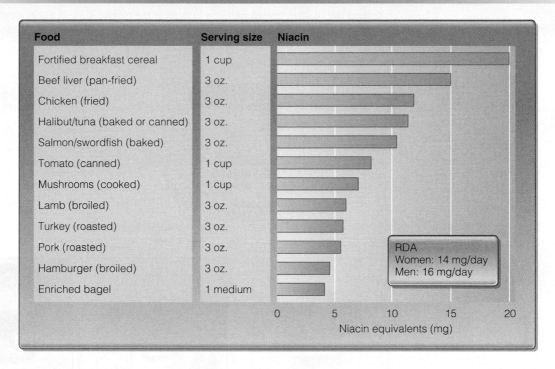

| Food | Serving size | Niacin |
|------|-------------|--------|
| Fortified breakfast cereal | 1 cup | |
| Beef liver (pan-fried) | 3 oz. | |
| Chicken (fried) | 3 oz. | |
| Halibut/tuna (baked or canned) | 3 oz. | |
| Salmon/swordfish (baked) | 3 oz. | |
| Tomato (canned) | 1 cup | |
| Mushrooms (cooked) | 1 cup | |
| Lamb (broiled) | 3 oz. | |
| Turkey (roasted) | 3 oz. | |
| Pork (roasted) | 3 oz. | |
| Hamburger (broiled) | 3 oz. | |
| Enriched bagel | 1 medium | |

RDA
Women: 14 mg/day
Men: 16 mg/day

Niacin equivalents (mg)

SOURCE: Data from USDA Nutrient Database, Release 16-1.

and other fortified foods are also important sources of niacin. Niacin is quite stable and not easily destroyed by cooking or exposure to light.

## Bioavailability of Niacin

Niacin in animal products is more bioavailable than that in grain products. The niacin in grains such as corn and wheat is bound to proteins, and these niacin-protein complexes are difficult to absorb. However, treating grain products with alkaline (basic) substances such as lime water or baking soda can cleave the protein from the niacin, increasing its bioavailability. The traditional technique of soaking corn and corn meal in lime water—a historically common practice in preparing tortillas in Mexico—is an example. Note that lime water—derived from limestone—is not the same as lime juice. As shown in Figure 10.6, lime is still added to some forms of corn meal, making foods (such as tamales) made with these products good sources of niacin. However, because alkaline conditions can destroy other vitamins in foods it is not recommended that we routinely increase the pH of our foods to enhance niacin absorption.

**Meat is an excellent source of niacin.**

## Regulation of Niacin in the Body

Although small amounts of niacin can be absorbed in the stomach, most absorption occurs in the small intestine. Absorption of niacin is by simple diffusion when intake is high and by active transport when intake is low. Niacin is then circulated to the liver, where most of it is attached to transport proteins or converted to NAD⁺ or NADP⁺, the coenzyme forms. When needed, the liver converts the amino acid tryptophan to niacin as well. This reaction requires the assistance of riboflavin and vitamin B₆.

**FIGURE 10.6   Increasing Niacin Bioavailability with Lime**

Ground corn is often treated with lime to increase the bioavailability of niacin, which makes foods made with these products good sources of this vitamin.

Lime is added to increase the bioavailability of niacin found naturally in corn.

**Ingredients:** Specially ground and dehydrated whole kernel corn and lime. No preservatives added.

Tamales

Corn tortillas

Atole

## Functions of Niacin

Niacin plays many roles as the coenzymes $NAD^+$ and $NADP^+$, which work together with over 200 enzymes, catalyzing many redox reactions involved in energy metabolism (Figure 10.3). For example, $NAD^+$ is reduced to $NADH + H^+$ during the oxidation (breakdown) of fatty acids as well as in several reactions of glycolysis and the citric acid cycle. $NADH + H^+$ then participates in the electron transport chain to produce ATP. In addition, $NADP^+$ can be reduced to $NADPH + H^+$—which is needed for synthesizing many compounds, including fatty acids, cholesterol, steroid hormones, and DNA. $NADPH + H^+$ is also required for metabolizing vitamin C and folate.

Niacin has additional functions unrelated to its role as a coenzyme. For example, it is important for maintaining, replicating, and repairing DNA, and may play a role in protein synthesis, glucose homeostasis, and cholesterol metabolism. Consuming large amounts of niacin (2 to 4 g/day) has been shown to lower low-density lipoprotein (LDL) cholesterol and increase high-density lipoprotein (HDL) cholesterol in some people.[5] However, little is understood about the mechanisms by which these effects occur, so consuming large amounts of supplemental niacin for this purpose is not recommended unless monitored by a physician.

CONNECTIONS Remember that low levels of LDL and high levels of HDL are associated with lower risk of cardiovascular disease (Chapter 7, page 248).

## Niacin Deficiency: Pellagra

Niacin deficiency was originally given the name *mal del sol*, which in Italian means "illness of the sun." This is because niacin deficiency causes a variety of skin problems, including severe skin irritation, which is made worse by exposure to sunlight. Later, niacin deficiency was named **pellagra,** Italian for "rough skin."

The symptoms of pellagra are often referred to as the four D's: dermatitis, dementia, diarrhea, and death. For example, pellagra results in dermatitis—a condition characterized by rough, red skin that eventually thickens and turns dark. Pellagra also results in a series of neurological problems (that is, dementia) including depression, anxiety, irritability, and inability to concentrate. The associated gastrointestinal disturbances cause loss of appetite, diarrhea, and a characteristically red and swollen tongue. If not treated, severe pellagra can cause death.

Although once common, pellagra is now typically seen only in conjunction with poverty, general malnutrition, or chronic alcoholism. Diets providing limited amounts of both niacin and tryptophan can put a person at an especially high risk. For this reason, pellagra was once endemic in some portions of the United States; you can read more about this in the Focus on the Process of Science feature. Genetics and some medications can also cause niacin deficiency by inhibiting the synthesis of niacin from tryptophan. For example, people with Hartnup disease have a genetic abnormality that impairs tryptophan absorption. Therefore people with this disease cannot rely on tryptophan to provide niacin to the body.[6]

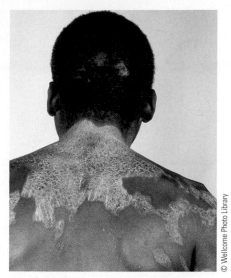

As shown here, pellagra can cause serious dermatitis.

## Niacin Toxicity

Unlike other water-soluble vitamins, large doses of nicotinic acid (1 g/day) can have harmful effects. These include skin inflammation and itchiness, flushing, heartburn, nausea, increased plasma glucose, and liver damage. Thus, consuming large doses of supplemental nicotinic acid is not recommended.

## Recommended Intakes of Niacin

The RDAs for men and women are 16 and 14 mg of niacin (or niacin equivalents, NE) each day, respectively. Because large doses of nicotinic acid have toxic effects, a UL for niacin has also been established; experts recommend that consumption of niacin from supplements and fortified foods not exceed 35 mg/day.

**pellagra** (pell – A – gra)  A disease caused by niacin and/or tryptophan deficiency.

# *Focus On* THE PROCESS OF SCIENCE
## Tryptophan, Niacin, and Pellagra

Pellagra was once prevalent in communities where corn provided the primary source of protein to the diet. Because corn contains low amounts of both niacin and its precursor tryptophan, consuming a corn-based diet increases the risk of pellagra. As such, pellagra was found in epidemic proportions the southern United States at the turn of the 20th century. About 10,000 people died annually from this disease. Pellagra was so common that public health officials believed it was caused by an infectious agent, such as a bacterium.

In 1914, Dr. Joseph Goldberger was asked by the U.S. Public Health Service to travel throughout the southern states and observe pellagra patients. Interestingly, even with exceptionally high exposure to pellagra, he never contracted it. Goldberger concluded that pellagra was not due to infectious disease but instead was related to diet. He confirmed this hypothesis by conducting dietary studies in orphanages. These studies were followed by a dietary intervention study in a Mississippi prison, in which pellagra was induced experimentally by dietary

manipulation. Goldberger demonstrated that a poor diet could cause pellagra and supplementation with a variety of protein sources could prevent and treat this disease.

With time, scientists learned that pellagra was actually due to niacin deficiency. With the advent of government programs that promoted the enrichment of cereal products with the B vitamins (including niacin) and better availability of a variety of foods, it is now rare to find a case of pellagra in the United States.

Pantothenic acid is found in a diverse group of foods including mushrooms.

© Smith-StockFood Munich/Stockfood America

# Pantothenic Acid (Vitamin B₅)

**Pantothenic acid** (vitamin B₅) is a nitrogen-containing vitamin named for the Greek word *pantos*, meaning "everywhere." This is because pantothenic acid is found in almost every plant and animal tissue. Pantothenic acid functions as a component of coenzyme A (CoA) in a variety of metabolic reactions.

## Dietary Sources of Pantothenic Acid

Pantothenic acid is found in many foods, including mushrooms, organ meats (such as liver), and sunflower seeds (Figure 10.7). Dairy products, turkey, fish, and coffee are also good sources. High temperatures can destroy pantothenic acid in foods.

## Regulation and Functions of Pantothenic Acid in the Body

Bioavailability of pantothenic acid increases via active transport when pantothenic acid intake is low. Once absorbed, pantothenic acid is circulated to the liver in the blood. Although pantothenic acid itself is not stored in the body, concentrations of its coenzyme form (CoA) are especially high in liver, kidney, heart, adrenal glands, and brain.

The primary function of pantothenic acid (as CoA) is in the metabolism of glucose, amino acids, and fatty acids for energy (ATP) production via glycolysis and the citric acid cycle (Figure 10.3). Recall from Chapter 8 that one of the pivotal steps in energy metabolism involves converting pyruvate to acetyl-CoA. This reaction requires pantothenic acid. The ability to produce acetyl-CoA is essential for the body to metabolize all the energy-yielding nutrients for ATP production. Pantothenic acid is also required for synthesizing many other critical compounds in the body, including heme (a portion of hemoglobin), cholesterol, bile salts, phospholipids, fatty acids, and steroid hormones.

**pantothenic acid** (vitamin B₅) (pan – to – THE – nic) A water-soluble vitamin involved in energy metabolism, hemoglobin synthesis, and phospholipid synthesis.

## FIGURE 10.7  Good Sources of Pantothenic Acid

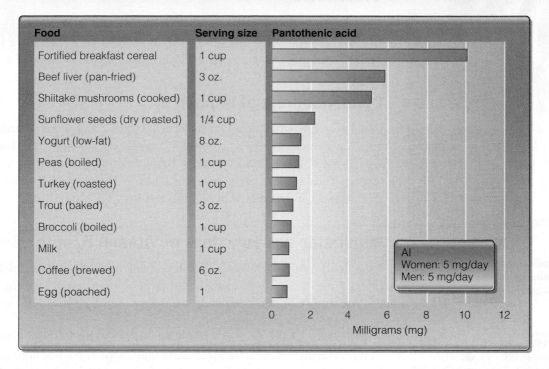

| Food | Serving size | Pantothenic acid |
|------|--------------|------------------|
| Fortified breakfast cereal | 1 cup | |
| Beef liver (pan-fried) | 3 oz. | |
| Shiitake mushrooms (cooked) | 1 cup | |
| Sunflower seeds (dry roasted) | 1/4 cup | |
| Yogurt (low-fat) | 8 oz. | |
| Peas (boiled) | 1 cup | |
| Turkey (roasted) | 1 cup | |
| Trout (baked) | 3 oz. | |
| Broccoli (boiled) | 1 cup | |
| Milk | 1 cup | |
| Coffee (brewed) | 6 oz. | |
| Egg (poached) | 1 | |

AI
Women: 5 mg/day
Men: 5 mg/day

Milligrams (mg)

SOURCE: Data from USDA Nutrient Database, Release 16-1.

## Pantothenic Acid Deficiency and Toxicity

Because it is found in almost all foods, pantothenic acid deficiency is rare. Nonetheless, a condition called "burning feet syndrome" is thought to be due to severe pantothenic acid deficiency. Burning feet syndrome causes a tingling in the feet and legs as well as fatigue, weakness, and nausea. Pantothenic acid toxicity has not been reported, but very high intakes of this vitamin have been associated with nausea and diarrhea.

## Recommended Intakes of Pantothenic Acid

Although there is not enough information to establish RDAs for pantothenic acid, an AI of 5 mg/day has been set for adults. Because no evidence exists of toxicity, no ULs are set for this vitamin.

## ESSENTIAL *Concepts*

Pantothenic acid is a component of coenzyme A (CoA), which is needed to make acetyl-CoA. Acetyl-CoA is required for energy metabolism and ATP production. It is also required for synthesizing heme, cholesterol, bile salts, fatty acids, phospholipids, and steroid hormones. Good food sources of pantothenic acid include fortified cereals, mushrooms, organ meats (such as liver), and sunflower seeds, although heat can destroy it. Severe pantothenic acid deficiency causes burning feet syndrome, characterized by tingling feet, weakness, and gastrointestinal distress. High doses of the vitamin have been reported to cause nausea.

Hummus, a traditional Middle-Eastern dip, is made from chick peas (garbanzo beans) and is a good source of vitamin B$_6$.

**vitamin B$_6$** A water-soluble vitamin involved in the metabolism of proteins and amino acids; the synthesis of neurotransmitters and hemoglobin; glycogenolysis; and regulation of steroid hormone function.

**pyridoxal phosphate (PLP)** (pyr – i – DOX – al) The coenzyme form of vitamin B$_6$.

# Vitamin B$_6$

There are three forms of **vitamin B$_6$**—pyridoxine, pyridoxal, and pyridoxamine—all made of a modified, nitrogen-containing ring structure. All three forms have similar biological activities in the body, involving over 100 chemical reactions.

## Dietary Sources of Vitamin B$_6$

Chickpeas (garbanzo beans), fish, liver, potatoes, and chicken are especially good sources of vitamin B$_6$, as are fortified breakfast cereals and bakery products. A list of foods rich in vitamin B$_6$ is provided in Figure 10.8. Note that vitamin B$_6$ is not one of the vitamins added to "enriched" products. Vitamin B$_6$ is somewhat unstable, and heating and freezing can destroy it.

## Regulation and Functions of Vitamin B$_6$ in the Body

Vitamin B$_6$ is readily absorbed in the small intestine and circulated in the blood to the liver. The liver then adds a phosphate group, forming **pyridoxal phosphate (PLP)**, the coenzyme form of vitamin B$_6$. PLP is efficiently stored in muscle, and to a lesser extent, in the liver.

PLP acts as a coenzyme in over 100 chemical reactions related to the metabolism of proteins and amino acids via transamination. These reactions are required for synthesizing nonessential amino acids from essential amino acids. Recall that only 9 essential amino acids must be obtained from foods, whereas 20 amino acids are needed for life. Without vitamin B$_6$, all 20 amino acids would be essential. Vitamin B$_6$ is also needed for producing nonprotein substances, such as the neurotransmitters serotonin and dopamine as well as heme. For example, vita-

---

**FIGURE 10.8  Good Sources of Vitamin B$_6$**

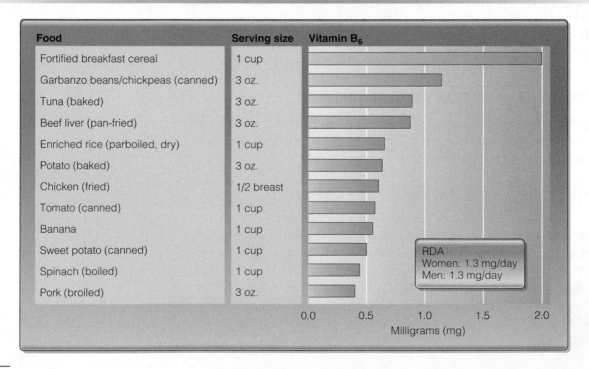

| Food | Serving size | Vitamin B$_6$ |
|------|--------------|---------------|
| Fortified breakfast cereal | 1 cup | |
| Garbanzo beans/chickpeas (canned) | 3 oz. | |
| Tuna (baked) | 3 oz. | |
| Beef liver (pan-fried) | 3 oz. | |
| Enriched rice (parboiled, dry) | 1 cup | |
| Potato (baked) | 3 oz. | |
| Chicken (fried) | 1/2 breast | |
| Tomato (canned) | 1 cup | |
| Banana | 1 cup | |
| Sweet potato (canned) | 1 cup | |
| Spinach (boiled) | 1 cup | |
| Pork (broiled) | 3 oz. | |

RDA
Women: 1.3 mg/day
Men: 1.3 mg/day

0.0    0.5    1.0    1.5    2.0
Milligrams (mg)

SOURCE: Data from USDA Nutrient Database, Release 16-1.

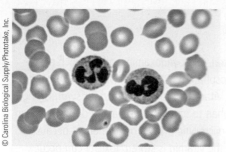

**Normal**

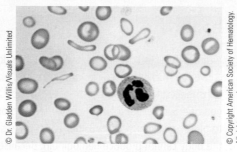

**Microcytic hypochromic anemia**

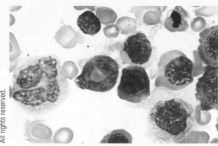

**Macrocytic anemia**

Red blood cells from healthy people can be distinguished from those of people with microcytic hypochromic anemia or macrocytic anemia.

min B$_6$ is needed to convert tryptophan to niacin, to break down glycogen to its glucose subunits, and to regulate some of the steroid hormones. It is also needed to prepare some amino acids to enter the citric acid cycle.

**CONNECTIONS** Recall that transamination involves the transfer of an amino group from one amino acid to an α-ketoacid, forming a different amino acid (Chapter 8, page 286).

## Vitamin B$_6$ Deficiency: Microcytic Hypochromic Anemia

Vitamin B$_6$ deficiency results in inadequate heme production, and thus lower concentrations of hemoglobin in red blood cells. This condition, called **microcytic hypochromic anemia,** decreases oxygen availability in tissues and impairs the ability to produce ATP via aerobic metabolism. As shown in the photo, the resulting red blood cells are small (microcytic) and light in color (hypochromic). Vitamin B$_6$ deficiency also causes cheilosis, glossitis, stomatitis, and fatigue. These symptoms are very similar to those of riboflavin deficiency.

Vitamin B$_6$ deficiency is rare when food is abundant. However, during the 1950s the damaging effect of high temperatures on vitamin B$_6$ was not well understood, and vitamin B$_6$ added to infant formula was destroyed during processing. Thus many formula-fed infants developed vitamin B$_6$ deficiency. This unfortunate "epidemic" of vitamin B$_6$ deficiency caused serious complications such as seizures and convulsions.

**CONNECTIONS** Because oxygen is the final electron acceptor in the electron transport chain, aerobic ATP production is dependent on oxygen availability and, thus, on hemoglobin (Chapter 8, page 278).

## Vitamin B$_6$ Toxicity

Because excess vitamin B$_6$ is stored, vitamin B$_6$ toxicity occurs more frequently than for most water-soluble vitamins. However, toxicity does not result from consuming naturally occurring vitamin B$_6$ but rather from supplement use. Vitamin B$_6$ toxicity causes severe neurological problems, including difficulty walking and numbness in the feet and hands.

Limited studies suggest that very high dosages of vitamin B$_6$ (1,000 mg/day) may relieve the symptoms of premenstrual syndrome (PMS) and carpal tunnel syndrome.[7] Other data suggest that relatively large doses of vitamin B$_6$ may decrease the risk of heart disease.[8] However, most studies do not support a helpful effect of vitamin B$_6$ supplements on these conditions. Consequently, because toxicity is likely at these doses, administration of large doses of vitamin B$_6$ is not generally recommended.

## Recommended Intakes of Vitamin B$_6$

RDAs for vitamin B$_6$ in adults vary from 1.3 to 1.7 mg/day. To prevent the neurological problems associated with vitamin B$_6$ toxicity, a UL has been established at 100 mg/day for adults. Note that supplements containing 500 mg of vitamin B$_6$ are widely available, making it relatively easy to consume amounts above those recommended.

**microcytic hypochromic anemia** (mic – ro – CYT – ic hy – po – CHRO – mic a – NE – mi – a) A condition in which red blood cells are small and light in color due to inadequate hemoglobin synthesis; can be due to vitamin B$_6$ deficiency.

## ESSENTIAL *Concepts*

Vitamin $B_6$ takes three forms: pyridoxine, pyridoxal, and pyridoxamine, all of which are converted to their coenzyme form, pyridoxal phosphate (PLP). PLP is involved in many reactions related to the metabolism of amino acids. Vitamin $B_6$ is also needed for synthesizing neurotransmitters and heme, converting tryptophan to niacin, and breaking down glycogen to glucose. Severe vitamin $B_6$ deficiency causes microcytic hypochromic anemia, because the body cannot produce hemoglobin. Because the vitamin is stored, toxicity can occur, resulting in neurological problems. Good sources of vitamin $B_6$ include chickpeas (garbanzo beans), fish, liver, and potatoes, as well as fortified breakfast cereals and bakery products.

# Biotin (Vitamin $B_7$)

**Biotin** (vitamin $B_7$) is a sulfur-containing molecule with two connected ring structures and a side chain. Humans obtain biotin from both the diet and via biotin-producing bacteria ("biota") in the large intestine. Biotin is required for gluconeogenesis and the citric acid cycle, being important for ATP production.

## Dietary Sources and Bioavailability of Biotin

As shown in Figure 10.9, sources of biotin include peanuts, tree nuts, mushrooms, eggs, and tomatoes. Although biotin frequently is bound to proteins that are cleaved during digestion, its bioavailability can be greatly reduced when it is consumed with foods containing the protein **avidin**. Avidin is present in very large quantities in egg whites and binds tightly to biotin in the

**biotin** (vitamin $B_7$) (BI – o – tin) A water-soluble vitamin involved in energy metabolism and regulation of gene expression.

**avidin** (AV – i – din) A protein present in egg whites that binds biotin, making it unavailable for absorption.

---

**FIGURE 10.9    Good Sources of Biotin**

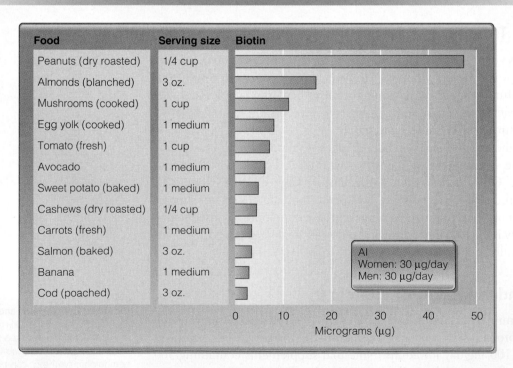

SOURCE: Data from Hands E. *Food Finder Vitamin and Mineral Source Guide,* 3rd ed. Salem, OR: ESHA Research; 1995.

intestinal tract, making it difficult to absorb. Because heat destroys avidin, eating raw eggs can decrease biotin bioavailability, whereas cooked eggs do not have this effect. Consuming alcohol can also decrease biotin absorption, and extreme heat can destroy biotin in foods.

## Regulation and Functions of Biotin in the Body

Enzymes in the small intestine cleave most of the protein-biotin complexes found in foods, releasing free (unbound) biotin, which is then absorbed. Biotin is also produced by bacteria in the large intestine, where it can be absorbed. However, because bacteria do not make enough biotin to meet our needs, biotin is an essential nutrient. Biotin circulates to the liver in the blood, and small amounts are stored in muscle, liver, and brain.

Biotin acts as a coenzyme for several enzymes, all of which catalyze **carboxylation** reactions. In other words, each biotin-requiring enzyme causes a bicarbonate subunit (HCO$_3$) to be added to a molecule. In general, these enzymes are involved in energy metabolism pathways (Figure 10.3). For example, a biotin-requiring enzyme converts pyruvate to oxaloacetate, a key step in gluconeogenesis. Biotin is also a coenzyme for the reactions that allow the body to use some amino acids in the citric acid cycle, for the synthesis of fatty acids, and for the breakdown of the amino acid leucine. In addition to biotin's role as a coenzyme, it has noncoenzyme functions related to gene expression, especially that influencing cell growth and development.

## Biotin Deficiency

Although biotin deficiency is uncommon, it occurs in small portions of the population, such as people who routinely consume large quantities of raw egg whites (containing avidin). However, in theory it would take daily consumption of at least 12 raw egg whites for a prolonged period of time to cause biotin deficiency. Biotin deficiency can also be caused by conditions impairing intestinal absorption (such as inflammatory bowel disease) and in some genetic disorders. Signs and symptoms of biotin deficiency are poorly understood, but include depression, hallucinations, skin irritations, infections, hair loss, poor muscle control, seizures, and developmental delays in infants.

## Recommended Intakes of Biotin

Although there is insufficient information for the development of RDAs for biotin, an AI level for adults of 30 micrograms ($\mu$g) per day has been set. Because very high biotin intake has no known detrimental effects, no ULs are established for this vitamin.

## ESSENTIAL *Concepts*

Biotin acts as a coenzyme for enzymes catalyzing carboxylation reactions. These enzymes allow the body to use glucose, amino acids, and fatty acids to produce ATP. Biotin is also needed for the synthesis of fatty acids, the breakdown of the amino acid leucine, and cell growth and development. Good sources of biotin include nuts, eggs, mushrooms, and tomatoes. Avidin, found in raw egg whites, can decrease biotin bioavailability, and biotin in foods can be destroyed by extreme heat. Biotin deficiency causes a variety of neurological problems and can be severe, especially in infants. There are no known toxic effects of biotin.

**carboxylation reaction** A metabolic reaction in which a bicarbonate subunit (HCO$_3$) is added to a molecule.

Folate is found in a wide variety of foods, such as legumes, okra, spinach, asparagus, liver, and oranges.

**folate** (FO – late) (also called folacin) A water-soluble vitamin involved in single-carbon transfer reactions; needed for amino acid metabolism and DNA synthesis.

**folic acid** The form of folate commonly used in vitamin supplements and food fortification.

# Folate (Folic Acid)

**Folate,** also called folacin, consists of three parts: (1) a nitrogen-containing, double ring structure, (2) a nitrogen-containing single ring structure, and (3) a glutamic acid (also called glutamate). Folate typically has additional glutamic acids attached to it. The interconversion of these "polyglutamate" forms of folate is important for the functions of folate. **Folic acid,** which is its most oxidized and stable form, is rarely found in foods but is used in vitamin supplements and food fortification.

## Dietary Sources of Folate

The term *folate*, from Italian, means foliage, reflecting its abundance in leafy plants. Examples of foods that contain folate are provided in Figure 10.10, and good sources include organ meats, legumes (such as lentils and pinto beans), okra, spinach, and many green leafy vegetables. Since 1998, all enriched cereal products in the United States are fortified with folate, making these foods very good sources of this vitamin as well. Other products, such as orange juice, can now be found fortified with folate. Because heat, light, and oxygen can destroy folate, cooked foods often have less folate than raw foods.

## Bioavailability and Regulation of Folate in the Body

The bioavailability of folate varies, depending on the form of folate present. Genetic factors and drugs can also influence folate absorption. In general, absorption of folic acid from supplements and fortified foods is higher than

**FIGURE 10.10   Good Sources of Folate**

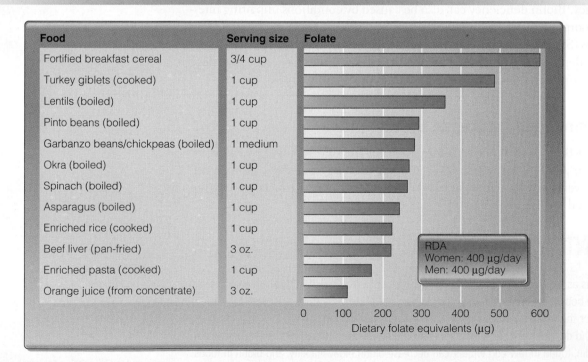

| Food | Serving size | Folate |
| --- | --- | --- |
| Fortified breakfast cereal | 3/4 cup | |
| Turkey giblets (cooked) | 1 cup | |
| Lentils (boiled) | 1 cup | |
| Pinto beans (boiled) | 1 cup | |
| Garbanzo beans/chickpeas (boiled) | 1 medium | |
| Okra (boiled) | 1 cup | |
| Spinach (boiled) | 1 cup | |
| Asparagus (boiled) | 1 cup | |
| Enriched rice (cooked) | 1 cup | |
| Beef liver (pan-fried) | 3 oz. | |
| Enriched pasta (cooked) | 1 cup | |
| Orange juice (from concentrate) | 3 oz. | |

RDA
Women: 400 µg/day
Men: 400 µg/day

Dietary folate equivalents (µg): 0   100   200   300   400   500   600

SOURCE: Data from USDA Nutrient Database, Release 16-1.

folate naturally found in foods. Because folate absorption is so variable, requirements are expressed as **dietary folate equivalents (DFE)**, which take this into account.

Dietary folate typically contains multiple glutamate units that must be cleaved prior to absorption. A variety of foods (such as cabbage) contain compounds that inhibit this process, thus decreasing folate bioavailability. Once taken up by the intestinal cell, folate is converted to **tetrahydrofolate (THF)** by the addition of 4 hydrogen atoms. Finally, a methyl group ($-CH_3$) is added, resulting in the production of **5-methyltetrahydrofolate (5-methyl THF)**, which is released into the blood and circulated to the liver. A limited amount of folate is stored in the liver.

## Function of Folate: Single-Carbon Transfers

Folate, in its active form of THF, acts as a coenzyme for many reactions, all involving the transfer of single-carbon groups (such as $-CH_3$). These reactions shift carbons from one molecule to another to form the many organic substances required for life.

For example, folate is involved in the metabolism and interconversion of amino acids. In these reactions, carbons are moved one at a time from one compound to another to produce what the body needs. An example of folate's single-carbon transfer role is the conversion of **homocysteine** to the amino acid methionine (Figure 10.11). In this reaction, 5-methyl THF transfers a methyl group ($-CH_3$) to homocysteine, making THF and methionine. This important reaction provides the body with the amino acid methionine as well as THF—the active form of folate. However, this reaction does not happen by itself. Instead, it occurs in synchrony with another reaction involving a vitamin $B_{12}$-requiring enzyme. This is an example of a coupled reaction, meaning that the two reactions occur simultaneously. In other words, one reaction cannot happen without the other. More specifically, the methyl group ($-CH_3$) is first transferred from 5-methyl THF to vitamin $B_{12}$. Then vitamin $B_{12}$ transfers the

## FIGURE 10.11    Folate and Vitamin $B_{12}$ Are Involved in Single-Carbon Transfers

The conversion of homocysteine to methionine requires folate and vitamin $B_{12}$. These are coupled reactions, because one cannot happen without the other.

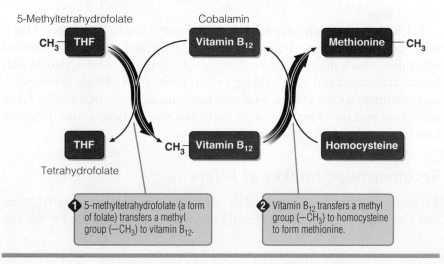

**dietary folate equivalent (DFE)** A unit of measure used to describe the amount of bio-available folate in a food or supplement.

**tetrahydrofolate (THF)** The active form of folate.

**5-methyltetrahydrofolate (5-methyl THF)** An inactive form of folate.

**homocysteine** (ho – mo – CYS – teine) A compound that is converted to methionine in a folate and vitamin $B_{12}$-requiring, coupled reaction.

methyl group to homocysteine, resulting in the synthesis of methionine. So the production of methionine from homocysteine requires both folate and vitamin $B_{12}$. A deficiency in either of these B vitamins results in a buildup of homocysteine in the body. Because high levels of homocysteine are associated with increased risk of heart disease, this may have serious consequences.[9] The relationship between homocysteine and heart disease is described in more detail in the Nutrition Matters following Chapter 7. High concentrations of homocysteine are also related to memory loss and difficulty with abstract thought, and some researchers hypothesize that impaired folate metabolism may be associated with Alzheimer's disease.[10]

THF is also involved in single-carbon transfer reactions required to make **purines** and **pyrimidines**—the molecules that make up DNA and RNA. Because DNA must be synthesized each time a new cell is made, folate is essential for the growth, maintenance, and repair of all tissues in the body. This is especially true during periods of rapid growth and development, like fetal growth, and in cells with very short life spans, such as those lining the gastrointestinal tract. Folate is also required for the normal growth and development of nerve tissue in the fetus, and increased folate intake during the reproductive period has been shown to decrease the risk of **neural tube defects** in newborns. This is described in more detail in the Focus on Life Cycle Nutrition feature.

## Folate Deficiency: Macrocytic Anemia

Mild folate deficiency results in a variety of symptoms, including fatigue, weakness, irregular heart function, and headaches. Severe folate deficiency causes a condition called **macrocytic anemia.**\* Because of folate's importance in DNA synthesis and cell maturation, severe folate deficiency causes many cells, including red blood cells, to remain in an immature state. Immature red blood cells are large (macrocytic) and contain organelles not typically found in mature red blood cells. For example, mature red blood cells do not have nuclei, whereas immature red blood cells do. Thus red blood cells from folate-deficient people are easily recognizable, because they are large and contain nuclei. Note that deficiencies of other nutrients, such as vitamin $B_{12}$, can also cause this form of anemia. In fact, it is sometimes difficult to determine whether macrocytic anemia is due to folate or to vitamin $B_{12}$ deficiency. However, red blood cells from people with macrocytic anemia or microcytic hypochromic anemia appear very different, as shown in the photo of red blood cells on page 397.

Folate deficiency was once relatively common in the United States. This is no longer the case, partly because enriched cereal products are now fortified with folate. Folate deficiency now occurs most often in alcoholics, people with intestinal diseases, and people taking certain medications. Folate deficiency is quite common in the elderly, who take many medications that inhibit folate absorption and use. Finally, genetic variations among people also influence how folate is absorbed and metabolized.

## Recommended Intakes of Folate

Recommendations concerning folate intake have received substantial attention since its relationship with neural tube defects was determined in the late

**purine** (PUR – ine) A compound that makes up DNA and RNA.

**pyrimidine** (pyr - i – MID – ine) A compound that makes up DNA and RNA.

**neural tube defect** A malformation in which the neural tissue does not form properly during fetal development.

**macrocytic anemia** A condition in which red blood cells are large, caused by inability of the cell to mature and divide appropriately; can be due to folate deficiency and/or vitamin $B_{12}$ deficiency.

**spina bifida** A form of neural tube defect in which the spine does not properly form.

---

\*Macrocytic anemia is also called megaloblastic anemia.

# *Focus On* LIFE CYCLE NUTRITION
## Folate, Neural Tube Defects, and Spina Bifida

Perhaps one of the most vital organ systems in the body is the nervous system, which includes the spinal cord and brain. Thus the proper growth and development of the nervous system during fetal life is especially critical to life. Studies show that folate is involved in the very early development of the spinal cord and brain, also called the neural tube. In early fetal development, the neural tube is actually not a tube but a flat sheet of nerve tissue. Consider what would happen if you cut a garden hose lengthwise and pressed it flat in your hand. This "open" form would represent the early stage of neural tube development. During fetal growth, the flat sheet of neural tissue "closes," forming a tube, much as would happen if you let go of the cut garden hose.

Neural tube defects result when closure of the neural tissue or "tube" is incomplete. An example of a neural tube defect is **spina bifida,** a failure of the spine to close properly during the first months of fetal life (Figure 10.12).[11] The National Center for Health Statistics estimates that approximately 1 out of 2,000 children is born each year with some form of neural tube defect, with 20 out of every 100,000 babies having spina bifida. Although some forms of neural tube defects do not cause problems, others are more severe.

Human dietary intervention studies show that maternal folate supplementation decreases the risk of neural tube defects in some women.[12] Why folate is required for normal neural tube development probably involves many folate-containing enzymes, including those required for DNA synthesis, and certain genetic characteristics in the mother or child appear to predispose some fetuses to neural tube defects.[13] For example, some studies suggest that transfer of folate from the mother to the fetus is impaired in pregnant women with certain genetic traits. Future research on nutrigenomics will undoubtedly shed more light on the complexities that modulate how folate intake and genetics interact in this regard. Until then, it is important for all women capable of becoming pregnant to consume plenty of folate-rich foods. Furthermore, the Institute of Medicine recommends taking folic acid supplements and/or consuming folic acid-fortified foods in addition to a varied diet.

## FIGURE 10.12  Neural Tube Defects and Spina Bifida

Spina bifida is a form of neural tube defect affecting the spine.

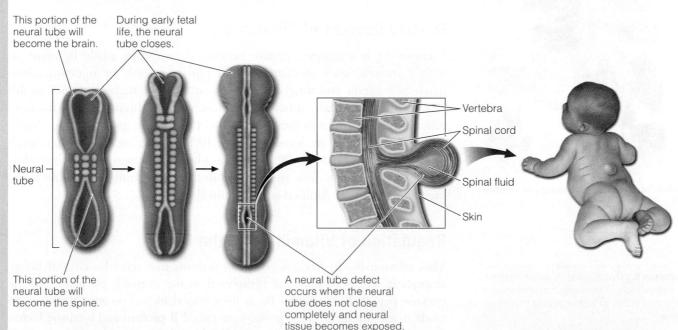

This portion of the neural tube will become the brain.

During early fetal life, the neural tube closes.

Neural tube

This portion of the neural tube will become the spine.

A neural tube defect occurs when the neural tube does not close completely and neural tissue becomes exposed.

Vertebra

Spinal cord

Spinal fluid

Skin

1980s. It is now recommended that adults consume 400 μg of folate (DFE) daily. This increases to 600 μg/day for pregnant women. Women capable of and/or planning to become pregnant are encouraged to consume 400 μg of folate (folic acid) from supplements, fortified foods, or both, in addition to consuming food folate from a varied diet. Because high intakes of folate may make it difficult to detect vitamin $B_{12}$ deficiency, a UL for folate has been set at 1,000 μg/day from fortified foods or supplements. There is no evidence that high intake of naturally occurring folate poses any risk.

## ESSENTIAL *Concepts*

The active form of folate in the body is tetrahydrofolate acid (THF), which is involved in single-carbon transfer reactions. Many of these are needed for amino acid metabolism. For example, 5-methyl THF is converted to THF during the conversion of homocysteine to methionine. This reaction also requires vitamin $B_{12}$. Folate is also required for DNA synthesis and therefore growth, maintenance, and repair of all tissues. Good sources of folate include liver, legumes, mushrooms, and green leafy vegetables. Enriched cereal products and other fortified foods are also good sources. Folate deficiency causes macrocytic anemia because of the inability of red blood cells to produce DNA and divide properly. Folate deficiency also increases the risk for some women of giving birth to children with neural tube defects.

# Vitamin $B_{12}$ (Cobalamin)

**Vitamin $B_{12}$**, also called cobalamin, was the last of the B vitamins to be discovered. Vitamin $B_{12}$ is a complex molecule and gets its name from the fact that it contains the trace element cobalt (Co) and several nitrogen (N) atoms.

## Dietary Sources of Vitamin $B_{12}$

Vitamin $B_{12}$ is a unique vitamin, because it cannot be made by plants or higher animals (such as mammals and birds) but only by microorganisms (such as bacteria and fungi). As such, many of our dietary sources of this vitamin do not actually produce it themselves but obtain vitamin $B_{12}$ from microorganisms either in their environment or in their gastrointestinal tracts. Some of these foods are listed in Figure 10.13. Foods containing high levels of vitamin $B_{12}$ include shellfish (such as clams and crabs), meat (including poultry), organ meats, fish, and dairy products. In addition, many ready-to-eat breakfast cereals are fortified with vitamin $B_{12}$.

## Regulation of Vitamin $B_{12}$ in the Body

Most vitamin $B_{12}$ in foods is bound to proteins that must be cleaved before absorption. These proteins are hydrolyzed in the stomach via acids and the enzyme pepsin. Free vitamin $B_{12}$ is then bound to two proteins that are also made in the stomach. These proteins are called **R protein** and **intrinsic factor.** R protein is thought to protect vitamin $B_{12}$ from destruction. In the intestine, R protein is released whereas intrinsic factor remains bound to vitamin $B_{12}$. This vitamin $B_{12}$–intrinsic-factor complex is then transported into the absorptive cell. Thus an inability to produce intrinsic factor can result in severe vitamin $B_{12}$ deficiency. Early studies relating intrinsic factor, vitamin $B_{12}$ intake, and vitamin $B_{12}$

Shellfish are excellent sources of vitamin $B_{12}$.

© Eising Food Photography/Stockfood America

**vitamin $B_{12}$ (cobalamin)** A water-soluble vitamin involved in energy metabolism and the conversion of homocysteine to methionine.

**R protein** A protein produced in the stomach that binds to vitamin $B_{12}$.

**intrinsic factor** A protein produced by exocrine cells in the gastric pits of the stomach, needed for vitamin $B_{12}$ absorption.

**FIGURE 10.13  Good Sources of Vitamin B$_{12}$**

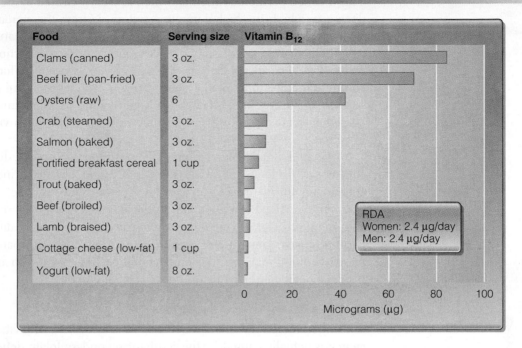

| Food | Serving size | Vitamin B$_{12}$ |
| --- | --- | --- |
| Clams (canned) | 3 oz. | |
| Beef liver (pan-fried) | 3 oz. | |
| Oysters (raw) | 6 | |
| Crab (steamed) | 3 oz. | |
| Salmon (baked) | 3 oz. | |
| Fortified breakfast cereal | 1 cup | |
| Trout (baked) | 3 oz. | |
| Beef (broiled) | 3 oz. | |
| Lamb (braised) | 3 oz. | |
| Cottage cheese (low-fat) | 1 cup | |
| Yogurt (low-fat) | 8 oz. | |

RDA
Women: 2.4 µg/day
Men: 2.4 µg/day

Micrograms (µg)

SOURCE: Data from USDA Nutrient Database, Release 16-1.

deficiency led to a coveted Nobel Prize in physiology or medicine being awarded to Drs. George Whipple, George Minot, and William Murphy in 1934.

Once absorbed, vitamin B$_{12}$ is released from intrinsic factor and bound to another protein called **transcobalamin,** which circulates the vitamin to the liver in the blood. Although most B vitamins are not stored in the body, this is certainly not the case for vitamin B$_{12}$. In fact, the liver stores several years worth of vitamin B$_{12}$.

## Functions of Vitamin B$_{12}$

Vitamin B$_{12}$ participates as a coenzyme in only two reactions. One reaction catalyzes the production of succinyl CoA, an intermediate in the citric acid cycle. This reaction ultimately allows the body to use some amino acids and fatty acids for energy (ATP) production. The other reaction catalyzes the conversion of homocysteine to the amino acid methionine. This was described previously in the section on folate and was shown in Figure 10.11. Recall that during the conversion of homocysteine to methionine, the inactive form of folate (5-methyl THF) is converted to its active form (THF). Without adequate vitamin B$_{12}$, homocysteine levels build up in the blood, and folate becomes "trapped" as its inactive 5-methyl THF form. Thus folate deficiency symptoms appear. In this way, vitamin B$_{12}$ deficiency can cause secondary folate deficiency.

## Vitamin B$_{12}$ Deficiency and Pernicious Anemia

Vitamin B$_{12}$ deficiency can result from inadequate dietary intake or absorption of vitamin B$_{12}$. Primary vitamin B$_{12}$ deficiency is sometimes seen in vegans or infants being breastfed by vitamin B$_{12}$-deficient mothers. Secondary vitamin

**transcobalamin** The protein that transports vitamin B$_{12}$ in the blood.

$B_{12}$ deficiency can occur when gastric cells stop producing intrinsic factor. This is called **pernicious anemia,** which is usually caused by an autoimmune disease in which the body's immune system destroys the stomach cells that produce intrinsic factor.[14] Pernicious anemia can also be caused by other conditions such as aging, genetic defects, gastrointestinal infections, surgeries, and some medications. Because intrinsic factor is essential for vitamin $B_{12}$ absorption, pernicious anemia can be present even when seemingly adequate vitamin $B_{12}$ is consumed. People with pernicious anemia cannot readily be treated with oral vitamin $B_{12}$ supplements, but must be given vitamin $B_{12}$ by injection.

Vitamin $B_{12}$ deficiency is especially common in the elderly population, affecting as many as 15%. This is due to a variety of factors, including inadequate vitamin $B_{12}$ intake, decreased synthesis of intrinsic factor (pernicious anemia), decreased acid secretion in the stomach, and malabsorption. Signs and symptoms typically include macrocytic anemia, fatigue, difficulty sleeping, numbness, memory loss, and severe neurological disturbances. Because these symptoms may resemble other conditions associated with aging, vitamin $B_{12}$ deficiency may be easily overlooked in the elderly.

## Folate, Vitamin $B_{12}$, and Anemia

It is important to understand that the anemia associated with vitamin $B_{12}$ deficiency is actually caused by the resultant secondary folate deficiency. More specifically, vitamin $B_{12}$ deficiency causes a deficiency in THF even when adequate folate is consumed. This is because folate gets "trapped" as 5-methyl THF when vitamin $B_{12}$ is not available. Clinicians often say that folate deficiency can "mask" vitamin $B_{12}$ deficiency, because large doses of folate seem to alleviate some of its symptoms (such as the anemia). However, other complications of vitamin $B_{12}$ deficiency are not alleviated with high doses of folate. Because the other symptoms involve severe neurologic damage and can be fatal, misdiagnosing vitamin $B_{12}$ deficiency for folate deficiency is dangerous.

## Recommended Intakes for Vitamin $B_{12}$

The RDA for vitamin $B_{12}$ is 2.4 μg/day for adults. Although these recommendations do not change for older people, it is recommended that those over 50 years of age be especially careful to choose vitamin $B_{12}$-fortified foods or consume vitamin $B_{12}$ supplements if necessary. Vegans who do not eat any animal products should take a supplement or eat foods that have been fortified with vitamin $B_{12}$. Although vitamin $B_{12}$ is actively stored in the liver, no ULs are established for this vitamin.

## ESSENTIAL *Concepts*

Vitamin $B_{12}$, also called cobalamin, is required as a coenzyme in two important reactions. The first allows some amino acids and fatty acids to enter the citric acid cycle. The second reaction catalyzes the conversion of homocysteine to methionine and regenerates the active form of folate (THF). Good sources of vitamin $B_{12}$ include shellfish, liver, fish, meat, fortified breakfast cereals, and bakery products. Vitamin $B_{12}$ deficiency can be caused by inadequate vitamin $B_{12}$ intake or lack of intrinsic factor, the latter being called pernicious anemia. Vitamin $B_{12}$ deficiency is characterized by macrocytic anemia because it causes secondary folate deficiency. Vitamin $B_{12}$ deficiency can also lead to severe neurological complications. There are no known toxic effects of high vitamin $B_{12}$ intake.

**pernicious anemia** A condition caused by vitamin $B_{12}$ deficiency due to lack of intrinsic factor.

# Vitamin C (Ascorbic Acid)

**Vitamin C** appears to play a role in almost every physiological system. For example, it is important for the immune, cardiovascular, neurological, and endocrine systems. This relatively simple compound can be made from glucose in all plants and most animals, but not in humans. In fact, primates (including humans), fruit bats, and guinea pigs are some of the few animals for which vitamin C is an essential nutrient. Vitamin C is technically referred to as **ascorbic acid.**

**Brightly colored fruits and vegetables are often excellent sources of vitamin C.**

## Dietary Sources of Vitamin C

As shown in Figure 10.14, vitamin C is found in many fruits and vegetables, such as citrus fruits, peppers, papayas, broccoli, strawberries, and peas. The bioavailability of vitamin C is generally high, although it is easily destroyed by heat, oxygen, and high pH. Thus, freshly peeled and/or prepared fruits and vegetables tend to provide more vitamin C than cooked, processed, and/or stored ones.

## Regulation of Vitamin C in the Body

Absorption of vitamin C occurs mainly in the small intestine via active transport using glucose transport proteins. This is because vitamin C is structurally similar to glucose. However, at very high intakes vitamin C is also absorbed by simple diffusion in both the stomach and small intestine. Vitamin C then circulates to the liver in the blood. As in the intestine, the uptake of vitamin C into the body's cells relies, in part, on glucose transporters. Excess vitamin C is not stored, being rapidly metabolized and excreted in the urine.

**CONNECTIONS** Remember from Chapter 5 that glucose transporters are proteins that escort glucose molecules across cell membranes (Chapter 5, page 144).

**vitamin C (ascorbic acid)** A water-soluble vitamin that has antioxidant functions in the body.

**FIGURE 10.14    Good Sources of Vitamin C**

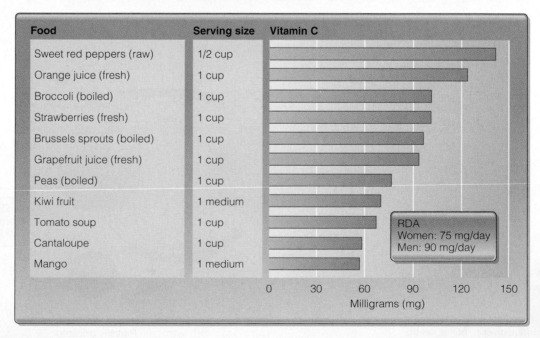

| Food | Serving size | Vitamin C |
|------|--------------|-----------|
| Sweet red peppers (raw) | 1/2 cup | |
| Orange juice (fresh) | 1 cup | |
| Broccoli (boiled) | 1 cup | |
| Strawberries (fresh) | 1 cup | |
| Brussels sprouts (boiled) | 1 cup | |
| Grapefruit juice (fresh) | 1 cup | |
| Peas (boiled) | 1 cup | |
| Kiwi fruit | 1 medium | |
| Tomato soup | 1 cup | |
| Cantaloupe | 1 cup | |
| Mango | 1 medium | |

RDA
Women: 75 mg/day
Men: 90 mg/day

0    30    60    90    120    150
Milligrams (mg)

SOURCE: Data from USDA Nutrient Database, Release 16-1.

## Functions of Vitamin C

Unlike the B vitamins, vitamin C is not a coenzyme. Instead, it acts as an **antioxidant.** Recall from Chapter 3 that atoms contain negatively charged electrons, and an abundance of electrons in an atom or molecule can result in an overall negative charge. Such an atom or molecule is therefore said to be "reduced." When electrons are removed, the atom or molecule has a more positive charge and is said to be "oxidized." Because vitamin C can easily accept and donate electrons, it is involved in a variety of redox reactions.

### The "Recharging" of Enzymes

One way in which vitamin C functions as an antioxidant is in "recharging" enzymes. Recall that enzymes are not used up or destroyed while catalyzing reactions. However, sometimes a portion of the enzyme needs to be restored or regenerated between reactions. This is especially true when the enzyme contains an element, such as copper or iron, that gets oxidized during the reaction.

An enzyme that needs restoration is like a flashlight that requires batteries. The flashlight may work for years, as long as the battery is replaced or recharged periodically. Using this analogy, you can think of an enzyme as a flashlight, copper or iron atoms as batteries, and vitamin C as a battery recharger. During many chemical reactions, copper or iron atoms are oxidized (made more positively charged) and then must be reduced (or "recharged") for the enzyme to catalyze another reaction. This recharging via reduction is the function of vitamin C and is illustrated in Figure 10.15.

An example of how vitamin C "recharges" an enzyme is the role of vitamin C in collagen production. **Collagen** is an important protein found in con-

### FIGURE 10.15 Vitamin C "Recharges" Enzymes

Vitamin C reduces iron- and copper-containing enzymes back to their original forms so they can catalyze more reactions.

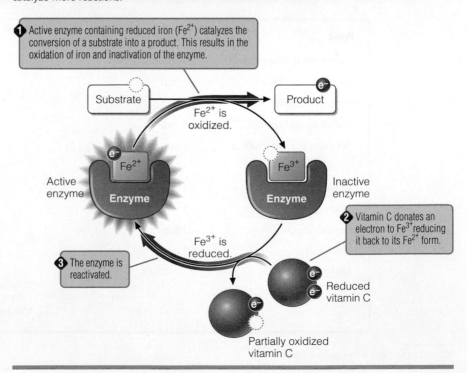

① Active enzyme containing reduced iron ($Fe^{2+}$) catalyzes the conversion of a substrate into a product. This results in the oxidation of iron and inactivation of the enzyme.

Substrate → Product

$Fe^{2+}$ is oxidized.

Active enzyme — $Fe^{2+}$ — Enzyme

$Fe^{3+}$ — Enzyme — Inactive enzyme

② Vitamin C donates an electron to $Fe^{3+}$ reducing it back to its $Fe^{2+}$ form.

$Fe^{3+}$ is reduced.

③ The enzyme is reactivated.

Reduced vitamin C

Partially oxidized vitamin C

**antioxidant** A compound that readily gives up electrons (and hydrogen ions) to other substances.

**collagen** The main protein found in connective tissue, including skin, bones, teeth, cartilage, and tendons.

nective tissue such as skin, muscle, and cartilage. For collagen to function properly, three strands of protein must twist together in just the right way—a process requiring a copper-containing enzyme to be oxidized. Vitamin C is needed to reduce this copper-containing enzyme between reactions so that it can continue to form collagen. Because of this, vitamin C is required for the health of all connective tissue (such as skin, tendons, and gums).

Vitamin C is also involved in the synthesis of carnitine by "recharging" two iron-containing enzymes needed for its synthesis. Recall from Chapter 8 that carnitine allows the body to use fatty acids for energy production. In addition, tyrosine (a nonessential amino acid) is made in the body from phenylalanine (an essential amino acid) via an iron-containing enzyme, and the neurotransmitters norepinephrine and serotonin are synthesized via enzymes that are "recharged" by vitamin C. Enzymes involved in the synthesis of several hormones, such as cholecystokinin (CCK) and gastrin, also require the reducing actions of vitamin C.

### Enhancing Iron, Copper, and Chromium Bioavailability

The reducing power of vitamin C also plays an important role in the gastrointestinal tract. Several minerals, such as iron, copper, and chromium, are better absorbed in their reduced states. As such, consuming vitamin C with these essential minerals can increase their bioavailability by reducing them in the gastrointestinal tract. For example, drinking orange juice with iron-fortified cereal helps increase the amount of iron that is absorbed.

### Vitamin C and Protection from Free Radical Damage

During normal cellular metabolism, many charged compounds are produced. Exposure of the body to toxic substances such as smog, cigarette smoke, and ozone, as well as some drugs and intense sunlight can also result in the production of charged compounds. These charged substances, called **free radicals,** have unpaired electrons in their outer shells, making them unstable and reactive. Most substances try to have an even number of electrons and will take electrons from other molecules to do so. In other words, free radicals readily oxidize other molecules. Oxidation by free radicals harms the body in several ways. First, it can break and damage DNA, potentially causing mutations in genes and possibly cancer. Free radicals can also oxidize fatty acids found in cell membranes, causing them to become weak and break down. Free radicals can also damage proteins.

Fortunately, the body is equipped to both destroy free radicals and repair the damage they cause. Vitamin C is one such "antioxidant system." For example, by donating a hydrogen atom and its electron vitamin C reduces the dangerous hydroxyl free radical ($OH^-$) to water ($H_2O$). Researchers think these antioxidant systems provide protection from the damaging consequences of free radicals.[15] This is illustrated in Figure 10.16. As you will see in Chapters 11 and 12, vitamin E as well as some phytochemicals and several of the essential minerals also have antioxidant functions.

### Vitamin C, the Common Cold, and Other Diseases

There is considerable interest in whether vitamin C may prevent or cure many diseases, including the common cold, cancer, heart disease, and cataracts. In most cases, studies suggest that increased consumption of fruits and vegetables containing vitamin C is associated with decreased risk for these diseases. Yet controlled clinical intervention studies often have not supported a protective effect of vitamin C.[16] Growing evidence, however, shows that large doses of

**CONNECTIONS** The quaternary structure of a protein forms when several smaller polypeptides come together to form a larger protein (Chapter 6, page 183).

**CONNECTIONS** The outermost shell of an atom is called its valence shell, and most atoms strive to have 8 electrons in this shell (Chapter 3, page 63).

**free radical** A reactive molecule with 1 or more unpaired electrons; free radicals are destructive to cell membranes, DNA, and proteins.

**FIGURE 10.16** Vitamin C and Free Radicals

Vitamin C stabilizes free radicals by donating electrons to them. Free radicals are dangerous to cells, because they seek to fill their outer shells by oxidizing other substances such as DNA, proteins, and lipids.

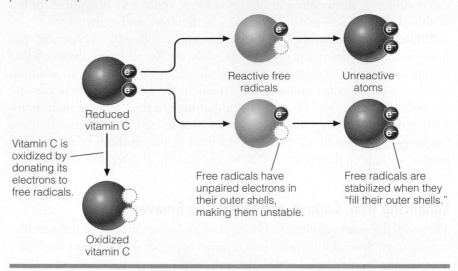

Reduced vitamin C

Vitamin C is oxidized by donating its electrons to free radicals.

Oxidized vitamin C

Reactive free radicals

Unreactive atoms

Free radicals have unpaired electrons in their outer shells, making them unstable.

Free radicals are stabilized when they "fill their outer shells."

One of the signs of scurvy (vitamin C deficiency) is the presence of very small red spots on the skin caused by internal bleeding.

© Dr. P. Marazzi/Science Photo Library/ Photo Researchers, Inc.

vitamin C can benefit the immune system.[17] Because the immune system is directly involved in many of the these diseases, it is possible that vitamin C indirectly affects disease via regulation of immune function. Vitamin C most likely works with other factors, such as vitamin E, which also have antioxidant functions.

## Vitamin C Deficiency and Toxicity

Vitamin C deficiency causes a sometimes deadly condition called **scurvy**. In the 18th century, a British medical doctor named James Lind conducted what was perhaps the first controlled nutrition intervention experiment when he determined that consuming citrus fruits would prevent this disease. Lind studied 12 sailors suffering from scurvy by treating them with lemons, limes, cider, nutmeg, seawater, or vinegar. He found that consuming lemons or limes, but not the other treatments, prevented and cured scurvy. After this discovery, British sailors could often be seen eating limes, which is why people came to call them "Limeys."

Scurvy results in a multitude of signs and symptoms, including bleeding gums, skin irritations, bruising, and poor wound healing, many of which are due to inadequate collagen production. Although it used to be very common, the increased availability of fruits and vegetables has made this disease rare. However, scurvy is still seen in developing countries, alcoholics, and appears to be somewhat common in people with diabetes. Although most people can consume very large doses (2 to 4 g/day) of supplemental vitamin C without experiencing harmful effects, this can cause nausea, diarrhea, cramping, and kidney stones in other people.

## Recommended Intakes for Vitamin C

The RDAs for vitamin C in men and women are 90 and 75 mg/day, respectively. Because of the increased risk for free radical damage from cigarette

**scurvy** A condition caused by vitamin C deficiency; symptoms include bleeding gums, bruising, poor wound healing, and skin irritations.

smoke, smokers are advised to increase their vitamin C intake by an additional 35 mg/day. Whether this recommendation also applies to people exposed to secondhand smoke is not clear. To avoid possible gastrointestinal distress, the UL for vitamin C intake from supplements has been set at 2 g/day (2,000 mg/day).

## ESSENTIAL *Concepts*

Vitamin C provides important antioxidant functions in the body. Many of these reactions involve the regeneration of reduced mineral components of enzymes, modulating the synthesis of important compounds such as collagen, carnitine, and neurotransmitters. Other vitamin C-related reactions stabilize free radicals and repair damage caused by free radical oxidation. Vitamin C is found in many foods, including a variety of fruits and vegetables, and is well absorbed. Vitamin C deficiency causes scurvy, which is characterized by bleeding gums and poor wound healing. Although high consumption of vitamin C may have beneficial effects on the immune system, some people have reported unpleasant effects of large doses of vitamin C intake from supplements.

## Choline: A "New" Essential Vitamin?

Because the body can typically make choline in sufficient amounts, scientists long thought that people did not need to consume it from the diet. However, more recent work shows that otherwise healthy men develop liver damage when fed a choline-free diet.[18] In addition, studies show that humans fed intravenous solutions that do not contain choline develop fatty livers and liver damage.[19] Thus the Institute of Medicine designated **choline** as an essential nutrient—at least in some situations.[20] Researchers do not know whether it is an essential nutrient for all people, but some populations may be at increased risk for choline deficiency. Thus choline may act more as a conditionally essential nutrient than an essential nutrient. Although choline is often considered a vitamin, this classification is not universally accepted. Clearly, we still have much to learn about the way choline functions.

### Dietary Sources of Choline

Choline is abundant in many plant and animal foods, particularly eggs, liver, legumes, and pork. In addition, because the choline-containing compound called lecithin (also called phosphatidylcholine) is often added to foods as an emulsifier, we consume relatively large amounts of choline from products such as mayonnaise and salad dressings. Remember that lecithin is a phospholipid containing a glycerol backbone, two fatty acids, and a choline-containing polar group.

**CONNECTIONS** Phospholipids are amphipathic compounds, are important components of cell membranes and lipoproteins, and help emulsify lipids in the small intestine (Chapter 7, page 234).

### Regulation and Functions of Choline in the Body

Free (unbound) choline found in food is absorbed in the small intestine and circulates to the liver via the blood. However, to be absorbed the choline component of lecithin is first cleaved from the glycerol backbone by pancreatic enzymes. Choline is then taken up by intestinal cells and reconstituted into lecithin molecules, which are released into the lymph. A large portion of

**choline** (CHO – line) A water-soluble compound used by the body to synthesize acetylcholine (a neurotransmitter) and a variety of phospholipids needed for cell membrane structure; considered a conditionally essential nutrient.

the lecithin that is ultimately delivered to the liver becomes a component of lipoproteins (such as low-density lipoprotein or LDL). In most healthy adults, choline is readily made in the body from the essential amino acid methionine, with the assistance of vitamin $B_{12}$ and folate.

Choline is needed for synthesizing a variety of phospholipids (such as lecithin) as well as acetylcholine (a neurotransmitter), and is an important component of cell membranes and lipoproteins. Choline is also needed for muscle control. Other compounds requiring choline for their synthesis play structural roles in the body and act as single-carbon donors in metabolic reactions. The potential functions of choline are still being discovered.

## Choline Deficiency, Toxicity, and Recommendations

Very little is known about the symptoms of choline deficiency in humans. However, there is limited evidence that choline deficiency can cause liver damage in adult men. Choline deficiency is thought to cause a buildup of lipids in the liver because of the liver's inability to produce very low density lipoproteins (VLDLs), which are needed to export triglycerides. Very high choline intake from foods can cause a fishy body odor, excess perspiration, salivation, low blood glucose, and liver damage. Although these symptoms may not be desirable, they are generally not fatal.

Currently, there is not enough information to establish RDAs for choline. However, AIs for men and women are 550 and 425 mg/day, respectively. The UL for choline is 3.5 g/day (3,500 mg/day).

## ESSENTIAL *Concepts*

Choline is a water-soluble compound that has been recently deemed an essential nutrient—at least in men. Choline is needed for synthesizing several phospholipids, including lecithin, and is required for producing acetylcholine, a neurotransmitter. Choline is also needed for muscle control and a variety of metabolic reactions. The functions of choline are still being studied. Choline is found in many foods but is especially high in eggs, liver, legumes, and pork. It is also added to foods as lecithin. Choline deficiency may cause liver damage in some people. Very high intakes of choline cause an unpleasant and characteristic fishy body odor.

# Carnitine

Carnitine is not an essential nutrient for adults, although it is considered conditionally essential for infants. All mammals can synthesize carnitine, although the ability to do this may be inadequate in newborns, especially those born prematurely. As with choline, researchers disagree on how to classify carnitine. The best sources of carnitine in the diet are animal sources, such as meat and milk. Human milk provides relatively high amounts of carnitine, as do many infant formulas.[21] The high amount of naturally occurring carnitine in human milk supports the hypothesis that this compound is conditionally essential during infancy.

## Regulation and Functions of Carnitine in the Body

Carnitine is efficiently absorbed in the small intestine and is also produced in the body from the amino acids methionine and lysine. This requires the assistance of iron, niacin, vitamin $B_6$, and vitamin C. Carnitine is needed for fatty acids to cross membranes, such as the mitochondrial membrane. Therefore, carnitine is essential for the body's use of fatty acids for energy (ATP) production. Although carnitine supplements are sold as aids to help build muscle and increase athletic performance, there is limited scientific evidence to support this claim.

## Carnitine Deficiency, Toxicity, and Recommendations

Carnitine deficiency is rare but it can occur in some genetic conditions, resulting in muscle weakness, hypoglycemia, and heart irregularities. Carnitine deficiency is rare even in vegans who consume very little dietary carnitine.[22] This supports the concept that carnitine is not an essential nutrient for adults. No toxic effects of high doses of carnitine are known. Because researchers do not consider carnitine an essential nutrient and high intakes of carnitine do not seem detrimental, no DRIs are established for this substance.

## ESSENTIAL *Concepts*

Although carnitine is not an essential nutrient for adults, many scientists believe it is conditionally essential for the newborn infant. Carnitine is important for fatty acid transport across biological membranes and is therefore needed to obtain energy from lipids. Carnitine is found most abundantly in meat and milk products. There are no dietary recommendations for carnitine at the present.

## Summary and Use of Supplements

You now know that the water-soluble vitamins serve hundreds of purposes. Although it is not possible to list all the functions of each vitamin, it is useful to group some of these roles into functional categories. For example, most water-soluble vitamins participate as coenzymes in energy metabolism pathways, several are involved in DNA synthesis, and others serve antioxidant roles. A general classification system of the major roles of the water-soluble vitamins is summarized in Table 10.4.

### TABLE 10.4 General Functions of the Water-Soluble Vitamins and Choline

| | Vitamins | | | | | | | | | |
|---|---|---|---|---|---|---|---|---|---|---|
| | Thiamin (B$_1$) | Riboflavin (B$_2$) | Niacin (B$_3$) | Pantothenic Acid (B$_5$) | Vitamin B$_6$ | Biotin (B$_7$) | Folate | Vitamin B$_{12}$ | Vitamin C | Choline[b] |
| **Functions** | | | | | | | | | | |
| Coenzyme | X | X | X | X | X | X | X | X | | |
| Energy metabolism[a] | X | X | X | X | X | X | | X | | |
| Antioxidant function | | | | | | | | | X | |
| Interconversion or activation of nutrients | | X | X | | X | | X | X | X | |
| Blood health | | X | | X | X | | X | X | | |
| DNA or RNA synthesis | X | | X | | | | X | | | |
| Nerve/muscle function | X | X | | | X | | X | | X | X |

[a]Note that these vitamins are not, themselves, energy-yielding nutrients but are involved in energy metabolism via their coenzyme roles.
[b]The roles that choline plays in the body are still being investigated.

Because vitamins serve so many purposes, their consumption is vitally important for health. Several federal agencies, such as the U.S. Department of Agriculture (USDA) and the FDA, have developed guidelines and policies that help consumers choose foods wisely. For example, the MyPyramid food guidance system helps people determine how many servings of each food group are needed to obtain these essential nutrients. To customize these recommendations for yourself, visit the MyPyramid website (http://www .mypyramid.gov). Although whole-grain products, fruits, and vegetables tend to be excellent sources of water-soluble vitamins, remember that vitamins are in all the food groups, as shown in Table 10.5. Thus dietary balance remains a key component of good health—even when it comes to the water-soluble vitamins.

## Vitamin Supplements: Making Wise Choices

At times we may not get enough of an essential vitamin from foods, and taking a vitamin supplement may be beneficial. But how do you know when, and how can you determine which type of supplement to consume? Although answering these questions may seem overwhelming, knowing some basic information about dietary supplements can help you make educated decisions about whether supplements are right for you.

First, it is important to realize that the term **dietary supplement** describes a broad group of products. Dietary supplements can contain one or more of the following ingredients: vitamins, minerals, amino acids, herbs, or other plant-derived substances (botanicals), and a multitude of other compounds derived from foods. Thus only some dietary supplements contain vitamins.

Unlike drugs, dietary supplements are not approved by the FDA for safety and effectiveness. Instead, it is the responsibility of dietary supplement manufacturers and distributors to ensure that their products are safe and that their label claims are accurate and truthful. However, once a product enters the marketplace, FDA has the authority to take action against any dietary supplement that presents a risk of illness or injury.

**dietary supplements** Products intended to supplement the diet that contain vitamins, minerals, amino acids, herbs or other plant-derived substances, and/or a multitude of other food-derived compounds.

## TABLE 10.5  Major Food Sources of the Water-Soluble Vitamins and Choline

| | Types of Food | | | | | | | | | |
|---|---|---|---|---|---|---|---|---|---|---|
| | Grains | Vegetables | Fruits | Dairy | Meat and/ or Eggs | Organ Meats | Seafood | Nuts and Seeds | Legumes | Mushrooms |
| **Water-Soluble Vitamins** | | | | | | | | | | |
| Thiamin (B$_1$)[a] | X | | | | X | | X | | X | |
| Riboflavin (B$_2$)[a] | X | X | | X | X | X | | | | X |
| Niacin (B$_3$) [a] | X | | | | X | X | X | | | X |
| Pantothenic Acid (B$_5$) | X | X | | X | X | X | X | X | X | X |
| Vitamin B$_6$ | X | X | X | | X | X | X | | X | |
| Biotin (B$_7$) | | X | | | X | | X | X | X | X |
| Folate[a] | X | X | X | | | X | | | X | |
| Vitamin B$_{12}$ | | | | X | X | X | X | | | |
| Vitamin C | | X | X | | | | | | X | |
| Choline | | | | | X | X | X | X | | |

[a]Added to enriched cereal products.

## Sorting Fact from Fiction

Although it may be tempting to simply believe what you read and hear about vitamin supplements, this is generally unwise. As you learned in Chapter 1, proving that a nutrient influences health requires proper use of scientific method. Because it is sometimes difficult to determine whether claims regarding vitamin supplements and health are accurate, you should consult a reliable source for expert advice. An excellent source of up-to-date information is the Office of Dietary Supplements (ODS), which is part of the National Institutes of Health. This office was established in 1994 to strengthen knowledge and understanding of dietary supplements by evaluating scientific information and educating the public. The ODS maintains an excellent, user-friendly website (http://ods.od.nih.gov) that contains information concerning most dietary supplements. The ODS recommends the following tips for buying and using any dietary supplement.

- *Safety first.* Some supplement ingredients can be toxic—especially in high doses. Do not hesitate to check with a health professional before taking any dietary supplement.
- *Think twice about chasing the latest headline.* Sound health advice is generally based on research over time, not a single study touted by the media. Be wary of results claiming a "quick fix."
- *Learn to spot false claims.* Remember, if something sounds too good to be true, it probably is.
- *More may not be better.* Some products can be harmful when consumed in high amounts, for a long time. Do not assume that more is better—it might be toxic.
- *The term "natural" does not always mean safe.* The term "natural" simply means something is not synthetic or human-made. Do not assume that this term ensures wholesomeness or safety.

## When Should You Consider Taking a Supplement?

There are no hard-and-fast rules about when you should take a dietary supplement. However, if someone has difficulty consuming a good variety and balance of foods in adequate amounts, taking a dietary supplement may help ensure that the person gets appropriate amounts of the essential nutrients. This is likely in several situations, as listed here.

- When food availability and/or variety is limited by time limitations or cooking constraints (such as college life)
- In situations when a person does not consume certain foods
- During periods of rapid growth and development
- When economic situations are difficult
- When a low-calorie diet is being consumed for weight loss
- In certain health conditions that increase nutrient requirements

Certainly, taking a supplement containing reasonable amounts of the essential vitamins and minerals can cause no harm and may be advantageous in these situations. However, supplements should never replace prescribed medications or the variety of foods important to a healthful diet. Also, be careful that daily intakes of vitamins and minerals do not exceed UL values. Clinicians advise that we keep a record of all dietary supplements that we take and periodically share this information with our health care providers. This is especially important when medications are taken on a regular basis, as is often the case in the elderly.

# ESSENTIAL *Concepts*

We can use many tools and guidelines to choose foods rich in the water-soluble vitamins. For example, the Dietary Guidelines for Americans and MyPyramid were specifically devised to help us consume adequate amounts of all the micronutrients. However, at times getting adequate amounts of the water-soluble vitamins may be difficult, and consuming dietary supplements may be prudent. Clinicians recommend that we be careful to not exceed UL levels and that we keep a record of dietary supplements that we are taking.

## Review Questions   Answers are found in Appendix G.

1. Which of the following vitamins is readily stored in the body?
   a. Thiamin
   b. Riboflavin
   c. Vitamin $B_{12}$
   d. Vitamin C

2. Beriberi is caused by a deficiency in which of the following vitamins?
   a. Riboflavin
   b. Thiamin
   c. Vitamin $B_6$
   d. Folate

3. To help preserve _____, milk is stored in cloudy cartons or paper containers.
   a. niacin
   b. biotin
   c. folate
   d. riboflavin

4. Which vitamin can be made in the body from the amino acid tryptophan?
   a. Vitamin C
   b. Choline
   c. Niacin
   d. Pantothenic acid

5. Which of these molecules requires pantothenic acid for its synthesis?
   a. $NAD^+$
   b. Coenzyme A (CoA)
   c. $FADH_2$
   d. Collagen

6. Microcytic anemia can be caused by a deficiency in which of the following vitamins?
   a. Vitamin $B_6$
   b. Folate

   c. Vitamin $B_{12}$
   d. Thiamin

7. Folate is needed for which of the following types of reactions in the body?
   a. Carboxylation reactions
   b. Decarboxylation reactions
   c. Reduction-oxidation reactions
   d. Single-carbon transfer reactions

8. Which of these vitamins requires the presence of cobalt?
   a. Folate
   b. Vitamin $B_{12}$
   c. Vitamin $B_6$
   d. Vitamin C

9. Which of these vitamin-related conditions can be caused by an autoimmune disease?
   a. Scurvy
   b. Pellagra
   c. Ariboflavinosis
   d. Pernicious anemia

10. The main function of vitamin C in the body is
    a. coenzyme for energy metabolism reactions.
    b. antioxidant functions.
    c. lipid synthesis.
    d. regulation of cell growth.

11. Name at least two good dietary sources of each essential water-soluble vitamin.

12. Folate deficiency is sometimes said to "mask" vitamin $B_{12}$ deficiency. What is meant by this?

13. Describe how an antioxidant functions, and provide two examples of how vitamin C functions in this capacity.

## Practice Calculations   Answers are found in Appendix G.

1. Using the food composition table that accompanies this text or other sources, calculate the folate content of a lunch containing a hamburger on a white bun, a small order of french fries, and a large cola soft drink. What percentage of your folate RDA does this repre-

   sent? What foods might be added to or substituted for the food in this lunch to increase the folate content?

2. Using the MyPyramid food guidance system (http://www.mypyramid.gov), determine the amount of foods from the fruits and vegetables groups that are recom-

mended for you. Then, taking into account fruits and vegetables that you prefer, calculate the amount of vitamin C you would get if you consumed these amounts of foods. What percentage of your RDA does this represent?

## Media Links
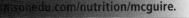
A variety of study tools for this chapter are available at our website, www.thomsonedu.com/nutrition/mcguire.

Prepare for tests and deepen your understanding of chapter concepts with these online materials:

- Practice tests
- Flashcards
- Glossary
- Student lecture notebook
- Web links
- Animations
- Chapter summaries, learning objectives, and crossword puzzles

## Notes

1. Carpenter KJ. A short history of nutritional science: Part 1 (1785–1885). Journal of Nutrition. 2003;133:638–45. Carpenter KJ. A short history of nutritional science: Part 2 (1885–1912). Journal of Nutrition. 2003;133:975–84. Carpenter KJ. A short history of nutritional science: Part 3 (1945–1985). Journal of Nutrition. 2003;133:3023–32. Carpenter KJ. A short history of nutritional science: Part 4 (1945–1985). Journal of Nutrition. 2003;133:3331–42.

2. Institute of Medicine. Dietary Reference Intakes: guiding principles for nutrition labeling and fortification. Washington, DC: National Academy Press; 1998. Park YK, McDowell MA, Hanson EA, Yetley E. History of cereal-grain product fortification in the United States. Nutrition Today. 2001;36:124–37.

3. Adapted from U.S. Department of Agriculture. Building blocks for fun and healthy meals. 2000. Available from: http://www.fns.usda.gov/tn/Resources/buildingblocks.html.

4. West DW, Owen EC. The urinary excretion of metabolites of riboflavin in man. British Journal of Nutrition. 1963;23:889–98.

5. Birjmohun RS, Hutten BA, Kastelein JJP, Stroes ESG. Increasing HDL cholesterol with extended-release nicotinic acid: from promise to practice. Netherlands Journal of Medicine. 2004;62:229–33. Ganji SH, Kamanna VS, Kashyap ML. Niacin and cholesterol: role in cardiovascular disease. Journal of Nutritional Biochemistry. 2003;14:298–305. Krauss RM. Lipids and lipoproteins in patients with type 2 diabetes. Diabetes Care. 2004;27:1496–504.

6. Potter SJ, Lu A, Wilcken B, Green K, Rasko JEJ. Hartnup disorder: polymorphisms identified in the neutral amino acid transporter SLC1A5. Journal of Inherited Metabolic Disease. 2002; 25:437–48. Seow HF, Broer S, Broer A, Bailey CG, Potter SJ, Cavanaugh JA, Rasko JEJ. Hartnup disorder is caused by mutations in the gene encoding the neutral amino acid transporter SLC6A19. Nature Genetics. 2004;36:1003–07.

7. Bendich A. The potential for dietary supplements to reduce premenstrual syndrome (PMS) symptoms. Journal of the American College of Nutrition. 2000;19:3–12. Wyatt KM, Dimmock PW, Jones PW, Shaughn O'Brien PM. Efficacy of vitamin B-6 in the treatment of premenstrual syndrome: a systematic review. British Medical Journal. 1999;318:1375–82.

8. Fairfield KM, Fletcher RH. Vitamins for chronic disease prevention in adults: scientific review. JAMA (Journal of the American Medical Association). 2002;287:3116–26. Schnyder G, Roffi M, Pin R, Flammer Y, Lange H, Eberli FR, Meier B, Turi ZG, Hess OM. Decreased rate of coronary restenosis after lowering of plasma homocysteine levels. New England Journal of Medicine. 2001;29:345:1593–60. Schnyder G, Roffi M, Flammer Y, Pin R, Hess OM. Effect of homocysteine-lowering therapy with folic acid, vitamin B$_{12}$, and vitamin B$_6$ on clinical outcome after percutaneous coronary intervention: the Swiss Heart study: a randomized controlled trial. JAMA (Journal of the American Medical Association). 2002;28:973–4179.

9. Scott JM. Homocysteine and cardiovascular risk. American Journal of Clinical Nutrition. 2000;72:333–4.

10. Shae TB, Lyons-Weiler J, Rogers E. Homocysteine, folate deprivation and Alzheimer neuropathology. Journal of Alzheimer's Disease. 2002;4:261–7.

11. Mitchell LE, Adzick NS, Melchionne J, Pasquariello PS, Sutton LN, Whitehead AS. Spina bifida. Lancet. 2004;364:1885–95.

12. Czeizel AE, Dudas I. Prevention of the first occurrence of neural tube defects by periconceptional vitamin supplementation. New England Journal of Medicine. 1992;327:32–5. Daly S, Mills JL, Molloy AM, Conley M, Lee YJ, Kirke PN, Weir DG, Scott JM. Minimum effective dose of folic acid for food fortification to prevent neural-tube defects. Lancet. 1997;350:1666–9. MRC Vitamin Study Research Group. Prevention of neural tube defects: results of the Medical Research Council Vitamin Study. Lancet. 1991;338:131–7. Moyers S, Bailey LB. Fetal malformation and folate metabolism: review of recent evidence. Nutrition Reviews. 2001;7:215–24.

13. Esfahani S, Cogger EA, Caudill MA. Heterogeneity in the prevalence of methylenetetrahydrofolate reductase gene polymorphisms in women of different ethnic groups. Journal of the American Dietetic Association. 2003;103:200–1. Moyers S, Bailey LB. Fetal malformations and folate metabolism: review of recent evidence. Nutrition Reviews. 59:215–24.

14. Toh BH, Alderuccio F. Pernicious anaemia. Autoimmunity. 2004;37:357–61.

15. Bowen DJ, Beresford SAA. Dietary interventions to prevent disease. Annual Review of Public Health. 2002;23:255–86. Kris-Etherton PM, Hecker KD, Bonanome A, Coval SM, Binkowski AM, Hilpert KF, Griel AE, Etherton TD. Bioactive compounds in foods: their role in the prevention of cardiovascular disease and cancer. American Journal of Medicine. 2002;113:71S–88S.

16. Fairfield KM, Fletcher RH. Vitamins for chronic disease prevention in adults. JAMA (Journal of the American Medical Association). 2002;287:3116–26. Jacob RA, Aiello GM, Stephensen CB, Blumberg JB, Milbury PE, Wallock LM, Ames BN. Moderate antioxidant supplementation has no effect on biomarkers of oxidant damage in healthy men with low fruit and vegetable intakes. Journal of Nutrition. 2003;133:740–3. Padayatty SJ, Katz A, Wang Y, Eck P, Kwon O, Lee J-H, Chen S, Corpe C, Dutta A, Dutta SK, Levine M. Vitamin C as an antioxidant: evaluation of its role in disease prevention. Journal of the American College of Nutrition. 2003;22:18–35.

17. Bhaskaram P. Micronutrient malnutrition, infection, and immunity: an overview. Nutrition Reviews. 2002;60:S60–45.

18. Zeisel SH, daCosta KD, Franklin PD, Alexander EA, Lamont JT, Sheard NF, Beiser A. Choline, an essential nutrient for humans. FASEB Journal. 1991;5:2093–98.

19. Buchman AL, Dubin M, Jenden D, Moukarzel A, Roch MH, Rice K, Gornbein J, Ament ME, Eckhert CD. Lecithin increases plasma free choline and decreases hepatic steatosis in long-term total parenteral nutrition patients. Gastroenterology. 1992;102:1363–70.

20. Institute of Medicine. Dietary Reference Intakes for thiamin, riboflavin, niacin, vitamin $B_6$, folate, vitamin $B_{12}$, pantothenic acid, biotin, and choline. Washington, DC: National Academy Press; 1998.

21. Ferreira I. Quantification of non-protein nitrogen components of infant formulae and follow-up milks: comparison with cows' and human milk. British Journal of Nutrition. 2003;90:127–33.

22. Lombard KA, Olson AL, Nelson SE, Rebouche CJ. Carnitine status of lactoovovegetarians and strict vegetarian adults and children. American Journal of Clinical Nutrition. 1989;50:301–6.

**Check out the following sources for additional information.**

23. Bailey LB, Moyers S, Gregory JF. Folate. In: Present knowledge in nutrition, 8th ed. Bowman BA, Russell RM, editors. Washington, DC: ILSI Press; 2001.

24. Bates CJ. Thiamin. In: Present knowledge in nutrition, 8th ed. Bowman BA, Russell RM, editors. Washington, DC: ILSI Press; 2001.

25. Chawla RK, Wolf DC, Kutner MH, Bonkovsky HL. Choline may be an essential nutrient in malnourished patients with cirrhosis. Gastroenterology. 1989;97:1514–20.

26. Garrow TA. Choline and carnitine. In: Present knowledge in nutrition, 8th ed. Bowman BA, Russell RM, editors. Washington, DC: ILSI Press; 2001.

27. Gibson RS. Principles of nutritional assessment, 2nd ed. New York: Oxford University Press; 2005.

28. Gropper SS, Smith JL, Groff JL. Advanced nutrition and human metabolism, 4th ed. Belmont, CA: Wadsworth; 2005.

29. Jacob RA. Niacin. In: Present knowledge in nutrition, 8th ed. Bowman BA, Russell RM, editors. Washington, DC: ILSI Press; 2001.

30. Johnston CS. Vitamin C. In: Present knowledge in nutrition, 8th ed. Bowman BA, Russell RM, editors. Washington, DC: ILSI Press; 2001.

31. McCormick DB. Vitamin B-6. In: Present knowledge in nutrition, 8th ed. Bowman BA, Russell RM, editors. Washington, DC: ILSI Press; 2001.

32. The Merck Manual of Diagnosis and Therapy, 17th ed. Beers MH, Berkow R, editors. Merck and Co. Inc., 1999–2005. Available from: http://www.merck.com/mrkshared/mmanual/home.jsp.

33. Miller JW, Rogers LM, Rucker RB. Pantothenic acid. In: Present knowledge in nutrition, 8th ed. Bowman BA, Russell RM, editors. Washington, DC: ILSI Press; 2001.

34. Rivlin RS. Riboflavin. In: Present knowledge in nutrition, 8th ed. Bowman BA, Russell RM, editors. Washington, DC: ILSI Press; 2001.

35. Stabler SP. Vitamin $B_{12}$. In: Present knowledge in nutrition, 8th ed. Bowman BA, Russell RM, editors. Washington, DC: ILSI Press; 2001.

36. U.S. Department of Health and Human Services, and U.S. Department of Agriculture. Dietary Guidelines for Americans 2005. Washington, DC: U.S. Government Printing Office; 2005. Available from http:www.healthierus.gov/dietaryguidelines.

37. U.S. Department of Agriculture. National Nutrient Database for Standard Reference Release 18. Available from http://www.nal.usda.gov/fnic/foodcomp/search.

38. Zeisel SH. Choline: needed for normal development of memory. Journal of the American College of Nutrition. 2000;19:528S–31S.

39. Zeisel SH, Blusztajn JK. Choline and human nutrition. Annual Review of Nutrition. 1994;14:269–96.

40. Zempleni J. Biotin. In: Present knowledge in nutrition, 8th ed. Bowman BA, Russell RM, editors. Washington, DC: ILSI Press; 2001.

# Alcohol and Health

© Photodisc/Getty Images

© A. Inden/zefa/Corbis

The ritual toast is one of the oldest celebratory traditions.

*T*he raising of a celebratory glass of alcohol is one of the oldest traditions. Yet this ancient beverage with its rich history of pageantry and ritual has also brought misery and suffering. Although most people who drink do so without harm to themselves or others, long-time abusers know all too well that alcohol can lead to psychological and physical dependency. Today, millions of people worldwide seek help in their effort to abstain from alcohol. There is no easy explanation as to why some people can control their drinking whereas others cannot. Although genetics may predispose some people to alcoholism, cultural factors play a major role as well.

Scientists have long debated whether alcohol is good or bad for health. Studies show that moderate alcohol consumption can reduce the risk of heart disease in middle-aged and older adults, and may even provide some protection against type 2 diabetes and gallstones.[1] However, when it comes to alcohol more is clearly not better. In excess, alcohol alters judgment, can lead to dependency, and can damage the liver, pancreas, heart, and brain. Heavy drinking also increases the risk for accidents and some types of cancer, and can seriously harm an unborn child.

Alcohol has both beneficial and harmful effects depending on the amount, frequency, and circumstances in which it is consumed. Consumed responsibly, alcohol poses little threat physically, socially, or psychologically, and may even be beneficial. However, not all drinkers are responsible, and estimated annual costs associated with alcohol abuse in the United States are more than $180 billion. For this reason, the Dietary Guidelines for Americans clearly state that people who choose to drink alcoholic beverages should do so sensibly and in moderation.[2] In this Nutrition Matters, you will learn how the body metabolizes alcohol as well as about the many effects that alcohol has on health.

## Alcohol and Alcohol Absorption

If you have ever been around someone who has had too much to drink, you know alcohol is a drug with mind-altering effects. Indeed, most people act and behave differently when under the influence of alcohol. Alcohol is a rather simple molecule—but has profound effects on the body.

# What Is Alcohol?

**Alcohol** is a broad term for a class of organic compounds that have common properties. For example, all alcohols have a general chemical formula (an —OH group bonded to a carbon atom; C—OH), are quite volatile, and tend to be soluble in water. There are many different types of alcohol, and most are not safe to drink. For example, methanol, which is used to make antifreeze, can be lethal if consumed. The form of alcohol found in alcoholic beverages is a molecule called **ethanol,** which has a chemical formula of $C_2H_5OH$. Although alcohol is not considered a nutrient, it does provide 7 kcal per gram. The caloric contents of selected alcoholic beverages are presented in Table 1.

A process called **fermentation,** discovered thousands of years ago, is still used today to produce ethanol. Fermentation occurs when single-cell microorganisms called yeast metabolize sugar present in fruits and grains. This involves the metabolic pathway glycolysis. Under anaerobic conditions, yeast converts glucose to pyruvate, which in turn is converted to ethanol. In the process, carbon dioxide is released (Figure 1). Whereas the coenzyme $NAD^+$ is converted to $NADH + H^+$ during glycolysis, fermentation converts $NADH + H^+$ to $NAD^+$. Once the alcohol content reaches 11 to 14%, fermentation stops naturally.

> **CONNECTIONS** Recall that glycolysis is an anaerobic catabolic pathway that breaks down glucose to two molecules of pyruvate (Chapter 8, page 281).

The alcohol content of some alcohol-containing beverages can be increased by a process called **distillation.**

During distillation, fermented beverages are heated, causing the alcohol to become a vaporous gas. The alcohol vapors are collected and cooled until they are liquid again. This pure alcohol concentrate is used to produce distilled alcoholic beverages such as gin, vodka, and whiskey. Distilled alcoholic beverages are also called "hard liquors," and their alcohol content is labeled as **proof.** A beverage's proof is twice its alcohol content. For example, distilled liquors labeled as 80 proof contain 40% alcohol.

## Alcohol Absorption

When alcohol is consumed, it requires no digestion and is readily absorbed by simple diffusion into the blood. Although some alcohol is absorbed from the stomach, most alcohol absorption (80%) takes place in the small intestine. The rate of alcohol absorption is influenced by several factors. For example, alcohol is absorbed more quickly when a person drinks on an empty stomach. This is because the presence of food in the stomach dilutes the alcohol and delays gastric emptying. As a result, alco-

**alcohol** An organic compound containing one or more hydroxyl (–OH) groups attached to carbon atoms.

**ethanol** (ETH – a – nol) An alcohol produced by the chemical breakdown of sugar by yeast.

**fermentation** The process whereby yeast chemically breaks down sugar to produce ethanol and carbon dioxide.

**distillation** A process used to make a concentrated alcohol beverage by condensing and collecting alcohol vapors.

**proof** A measure of the alcohol content of distilled liquor.

## TABLE 1   Alcohol and Energy Content of Selected Alcoholic Beverages

| Beverages | Serving Size (fl oz.) | Energy (kcal)[a] | Alcohol (g) |
|---|---|---|---|
| Light beer | 12 | 103 | 11 |
| Beer | 12 | 139 | 13 |
| White wine | 5.0 | 98 | 14 |
| Red wine | 5.0 | 98 | 14 |
| Distilled beverages (gin, rum, vodka, whiskey) | | | |
| 80 proof | 1.5 | 97 | 14 |
| 86 proof | 1.5 | 105 | 15 |
| 90 proof | 1.5 | 110 | 16 |
| 94 proof | 1.5 | 116 | 17 |
| 100 proof | 1.5 | 124 | 18 |
| Crème de menthe | 1.5 | 186 | 15 |
| Daiquiri cocktail | 4 | 224 | 14 |
| Whiskey sour cocktail | 4.5 | 226 | 19 |
| Piña colada cocktail | 4.5 | 245 | 14 |

[a] Note that some alcoholic beverages such as beer contain energy-yielding nutrients other than alcohol.

SOURCE: USDA National Nutrient Database for Standard Reference. Available from: http://www.nal.usda.gov/fnic/foodcomp/search/.

## FIGURE 1    Fermentation

During fermentation, yeast metabolize sugar, producing alcohol.

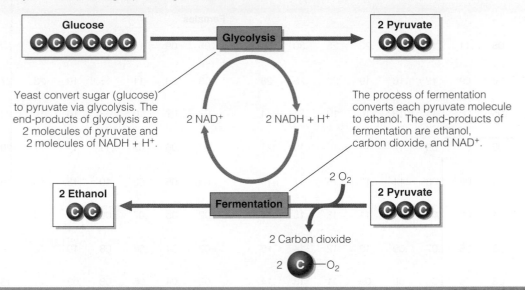

hol cannot reach the small intestine as quickly. The type of food consumed does not appear to have a measurable effect on the rate of alcohol absorption, although the concentration of alcohol in the beverage can. Alcohol is most rapidly absorbed when the concentration is less than 30% (60 proof). Drinks with a high alcohol concentration delay gastric emptying, which slows absorption.

> **CONNECTIONS** Recall that simple diffusion is a type of passive transport whereby molecules move across cell membranes from an area of higher concentration to an area of lower concentration (Chapter 3, page 79).

## Alcohol Circulation and Blood Alcohol Concentration

Once absorbed, alcohol enters the bloodstream and circulates throughout the body. The term **blood alcohol concentration (BAC)** refers to the amount of alcohol in the blood. For example, a person with a BAC of 0.10 has one tenth of a gram of alcohol per deciliter of blood. Within 20 minutes of consuming one standard drink (12 ounces of beer, 5 ounces of wine, or 1.5 ounces of 80-proof distilled liquor), BAC begins to rise, and peaks within 30 to 45 minutes. As more alcohol is consumed, it accumulates in the blood, because the rate of alcohol metabolism is slower than the rate of its absorption. For this reason, people who drink should do so slowly and in moderation. The accumulation of alcohol in the blood leads to intoxication, and in most states a BAC of 0.08 g/dL is the legal limit for driving (Figure 2).

Alcohol is unusual in that it is both water soluble and lipid soluble. As a lipid-soluble molecule, alcohol read-

ily crosses cell membranes. As a water-soluble molecule, alcohol becomes distributed in the water-filled environments inside (intracellular space) and outside (extracellular space) of cells. Because there is great variation in the water content of tissues in the body, some tissues take up alcohol more readily than others. For example, lean tissue (muscle) has more water associated with it than adipose tissue. As a result, a person's body composition can influence BAC. For example, if two people of similar body weight were to ingest the same amount of alcohol, the leaner individual would have a lower BAC. The reason for this is that alcohol diffuses from the blood and into the lean tissue until equilibrium is reached. In contrast, very little alcohol is taken up by adipose tissue, resulting in more alcohol remaining in the blood. Similarly, body size can also influence BAC. An individual with a large body has more blood and body fluids, diluting the concentration of alcohol. Therefore, the BAC of a large person is often lower than that of a small person after drinking the same amount of alcohol.

## Effects of Alcohol on the Central Nervous System

Alcohol is classified as a central nervous system depressant, because it sedates brain activities. This is surprising to many people, because consuming small amounts of alcohol often makes people feel euphoric. These pleasant feel-

**blood alcohol concentration (BAC)** A unit of measurement that describes the level of alcohol in the blood.

## FIGURE 2   Blood Alcohol Concentration (BAC)

Estimates of blood alcohol concentration (BAC) based on alcohol consumption, body weight (pounds), and sex.

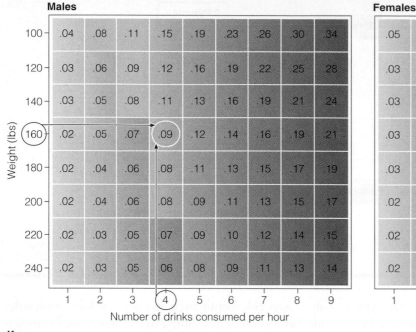

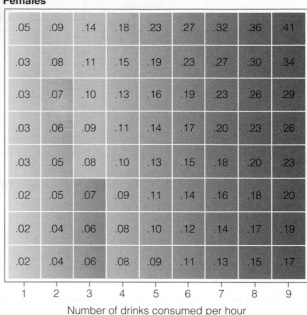

**Key**

◼ Driving skills impaired    ◼ Legally intoxicated    1 drink = 1.5 ounces of 80-proof hard liquor; 5 ounces of wine; 12 ounces of beer.

---

ings arise when alcohol selectively depresses parts of the brain that normally "censor" thoughts and behaviors. The inhibitory effect of alcohol on these portions of the brain is called **disinhibition.** In other words, alcohol can cause a temporary loss of inhibition, making people feel relaxed and more outgoing. However, disinhibition also impairs judgment and reasoning. As BAC increases, areas of the brain that control speech, vision, and voluntary muscular movement become depressed as well. If drinking continues, the person will likely lose consciousness. Alcohol can also reach lethal levels. Table 2 summarizes the brain's response to different levels of alcohol in the blood.

## ESSENTIAL *Concepts*

The type of alcohol in beer, wine, and distilled liquor is called ethanol, and is produced by a process called fermentation. During fermentation, yeast produces ethanol by metabolizing sugar. To make a more concentrated alcohol, fermented beverages can be distilled, a process used to make hard liquors such as gin and vodka. Alcohol requires no digestion and is absorbed into the blood mainly from the small intestine, although some is absorbed from the stomach. When alcohol is consumed without food, the rate of alcohol absorption from the stomach increases.

Once absorbed, alcohol circulates in the blood, and its con-

centration is expressed as blood alcohol concentration (BAC). A high BAC leads to intoxication and impairment. Alcohol is water soluble, and body tissues with a high water content (such as muscle) take up alcohol more readily than those with a low water content (such as adipose tissue). However, because alcohol is also lipid soluble, it readily crosses cell membranes. Alcohol is a central nervous system depressant and acts as a sedative in the brain. At low concentrations in the blood, alcohol causes disinhibition. At higher concentrations, speech, voluntary movements, and coordination become impaired; loss of consciousness is possible. At very high levels, alcohol can be lethal.

---

## Alcohol Metabolism

Although small amounts of unmetabolized alcohol are eliminated from the body by the lungs (expired air), skin (sweat), and kidneys (urine), the liver chemically breaks down most alcohol. However, there is a limit to how much alcohol the liver can metabolize at any given time. The average person metabolizes 0.5 oz. of pure

**disinhibition** (dis – in – hi – BI – tion) A loss of inhibition.

## TABLE 2  Stages of Alcohol Intoxication

| BAC (g/dL of blood) | Stage | Observable Effects |
|---|---|---|
| 0.01–0.03 | Subclinical | • Behavior nearly normal by ordinary observation |
| 0.03–0.09 | Euphoria | • Sociability; talkative, increased self-confidence<br>• Decreased inhibitions; diminution of attention, altered judgment<br>• Beginning of sensory-motor impairment; loss of efficiency in finer performance tests |
| 0.09–0.25 | Excitement | • Emotional instability; loss of critical judgment<br>• Impairment of perception; memory, and comprehension<br>• Decreased sensory response; increased reaction time<br>• Reduced visual acuity; peripheral vision and glare recovery<br>• Reduced sensory-motor coordination; impaired balance<br>• Drowsiness |
| 0.25–0.30 | Confusion | • Disorientation; mental confusion, and dizziness<br>• Exaggerated emotional states<br>• Disturbances of vision and of perception of color, form, motion, and dimensions<br>• Increased pain threshold<br>• Decreased muscular coordination; staggering gait; slurred speech<br>• Apathy, lethargy |
| 0.30–0.40 | Stupor | • General inertia; approaching loss of motor functions<br>• Markedly decreased response to stimuli<br>• Lack of muscular coordination; inability to stand or walk<br>• Vomiting; incontinence<br>• Impaired consciousness; sleep or stupor |
| 0.40–0.45 | Coma | • Complete unconsciousness<br>• Depressed or abolished reflexes<br>• Subnormal body temperature<br>• Impairment of circulation and respiration<br>• Possible death |
| 0.45 + | Death | • Death from respiratory arrest |

SOURCE: Mason MF, Dubowski KM. Alcohol, traffic, and chemical testing in the United States: A résumé and some remaining problems. *Clinical Chemistry.* 1974;20:126–40.

---

alcohol per hour. For example, it takes about one hour to metabolize the alcohol in a 12-oz. can of beer. Two major pathways are used to metabolize alcohol: the alcohol dehydrogenase (ADH) pathway and the microsomal ethanol-oxidizing system (MEOS).

## The Alcohol Dehydrogenase (ADH) Pathway

During light to moderate drinking, most alcohol (78%) is metabolized by a two-step enzyme system called the **alcohol dehydrogenase (ADH) pathway.** The first step in this metabolic pathway requires the enzyme **alcohol dehydrogenase (ADH),** which is found primarily in the cytoplasm of liver cells. However, ADH is also produced by gastric cells, and for this reason a small amount of alcohol is actually metabolized in the stomach before it ever reaches the blood. As shown in Figure 3, ADH converts alcohol to acetaldehyde, generating NADH + H⁺. Because acetaldehyde is a toxic molecule, it must be quickly metabolized. If acetaldehyde accumulates, some will pass from the liver into the blood, causing the

unpleasant side effects (headache, nausea, and vomiting) associated with heavy drinking—what is commonly referred to as a hangover.

> **CONNECTIONS** Recall that NAD⁺ is a coenzyme that accepts electrons and hydrogen ions in energy metabolism pathways (Chapter 8, page 273).

The next step in the ADH pathway is catalyzed by the enzyme **acetaldehyde dehydrogenase (ALDH),** which converts acetaldehyde to acetic acid. This reaction requires NAD⁺ for the transfer of hydrogen, and forms NADH + H⁺. Acetic acid combines with a molecule of coenzyme A to form acetyl-CoA. This step requires the enzyme acetyl-CoA synthase. Acetyl-CoA enters the

**alcohol dehydrogenase pathway** (de – hy – DRO – gen – ase) The primary metabolic pathway that chemically breaks down alcohol in the liver.

**alcohol dehydrogenase (ADH)** An enzyme found mostly in the liver that metabolizes ethanol to acetaldehyde.

**acetaldehyde dehydrogenase (ALDH)** (ac – et – AL – de – hyde) An enzyme that converts acetaldehyde to acetic acid.

## FIGURE 3 Alcohol Metabolism

The alcohol dehydrogenase (ADH) pathway and the microsomal ethanol-oxidizing system (MEOS) are used to metabolize alcohol.

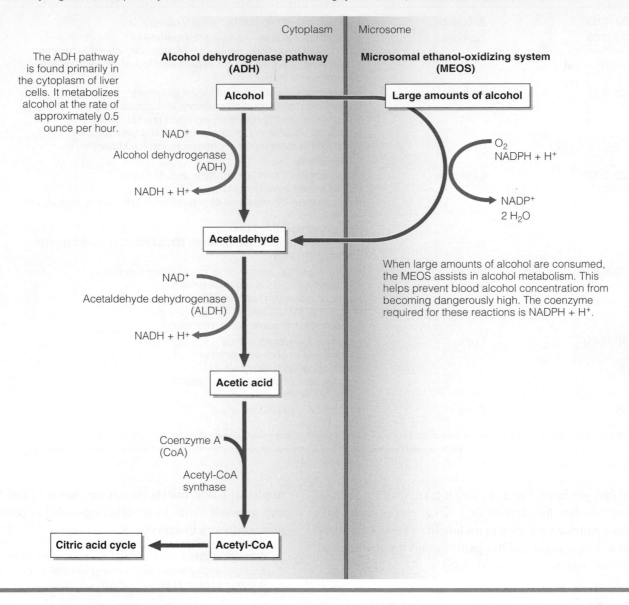

The ADH pathway is found primarily in the cytoplasm of liver cells. It metabolizes alcohol at the rate of approximately 0.5 ounce per hour.

When large amounts of alcohol are consumed, the MEOS assists in alcohol metabolism. This helps prevent blood alcohol concentration from becoming dangerously high. The coenzyme required for these reactions is NADPH + H⁺.

citric acid cycle and is metabolized further. The metabolism of one molecule of alcohol to acetic acid yields approximately six ATP. The metabolism of acetyl-CoA via the citric acid cycle yields additional energy (ATP).

**CONNECTIONS** Recall that acetyl-CoA is a 2-carbon intermediate compound that combines with oxaloacetate to enter the citric acid cycle, ultimately producing ATP (Chapter 8, page 284).

### Genetics and Alcohol Metabolism

Studies have found genetic alterations that appear to affect the enzymatic activities of ADH and ALDH, causing some people to have difficulty metabolizing alcohol.[3] For example, a high percentage of Asians have a less functional form of ALDH. When

these people consume alcohol, acetaldehyde levels increase quickly, causing dilation of blood vessels, headaches, and facial flushing. Differences in alcohol-metabolizing enzymes have also been found between men and women. For example, women tend to have lower ADH activity in gastric cells than men and may have a lower tolerance for alcohol. This may also be why women are more likely to develop alcohol-related health problems.[4]

### The Microsomal Ethanol-Oxidizing System (MEOS)

When large amounts of alcohol are consumed, some of the excess alcohol can be metabolized by an alternate pathway called the **microsomal ethanol-oxidizing system (MEOS)**. This helps prevent alcohol from reaching dangerously high levels in the blood. The components of this accessory pathway are up-regulated in response to frequent intoxication. This is why some heavy drinkers develop a **tolerance** to alcohol. People with high alcohol tolerances metabolize alcohol rapidly, allowing them to drink large amounts before feeling intoxicated. MEOS is also used to metabolize other drugs besides alcohol. For this reason, heavy drinkers often have a **cross-tolerance** to other substances as well. In other words, when tolerance develops to alcohol, the person also develops a tolerance to certain other drugs. However, the need to metabolize alcohol can outcompete the need to metabolize other drugs. Consequently, when a person with cross-tolerance drinks heavily, drug concentrations can reach dangerously high levels in the blood. This is why taking certain drugs in combination with alcohol can be dangerous.

> **CONNECTIONS** Recall that a gene can be turned on or up-regulated when the conditions warrant greater production of the protein for which it codes (Chapter 6, page 179).

Whereas the ADH pathway takes place in the fluid-filled cytosol of the cell, the MEOS takes place inside **microsomes.** Microsomes are small, spherical vesicles embedded within the endoplasmic reticulum. The MEOS uses oxygen and the coenzyme NADPH + H$^+$ to convert alcohol to acetaldehyde. Note that the ADH pathway yields NADH + H$^+$, whereas the MEOS actually uses energy. In addition, much of the energy released from alcohol when metabolized by the MEOS is lost as heat rather than used to supply the body with energy. This may, in part, explain why heavy drinkers often experience unexpected weight loss despite high-energy intake.

> **CONNECTIONS** Recall that the endoplasmic reticulum is an organelle in the cytoplasm that synthesizes and transports molecules within cells (Chapter 3, page 81).

## The Impact of Alcohol Metabolism on the Liver

Frequent and chronic alcohol consumption can interfere with normal liver function. This is, in part, caused by the two main by-products of alcohol metabolism, acetaldehyde and NADH + H$^+$. The accumulation of acetaldehyde damages the liver, impairing its function. Furthermore, alcohol metabolism raises NADH + H$^+$ levels and lowers NAD$^+$ levels in the liver. As a result, the activity of numerous NAD-linked enzymes becomes limited. You may recall that the end product of glycolysis is pyruvate and that NAD$^+$ is needed to convert pyruvate to acetyl-CoA. With limited NAD$^+$, pyruvate is instead converted to lactate (also called lactic acid), as shown in Figure 4 on the next page. High levels of lactate in the blood can interfere with the ability to excrete uric acid in the urine. Uric acid is a nitrogen-containing waste product formed during purine metabolism. Rising levels of uric acid in the blood can exacerbate a condition called **gout,** which is caused by the excess production and deposit of uric acid crystals in joints. People with gout are advised not to drink alcoholic beverages.

## Effects of Alcohol on Lipid Metabolism in the Liver

Alcohol metabolism also causes lipids to accumulate in the liver. This lipid accumulation occurs for several reasons, as listed here.

- Decreased fatty acid breakdown (β-oxidation)
- Increased uptake of fatty acids from the blood
- Increased fatty acid synthesis and triglyceride formation (lipogenesis)
- Decreased transport of triglycerides from the liver into the blood

High concentration of NADH + H$^+$ is an important cause of lipid accumulation in the liver. The amount of acetyl-CoA entering the citric acid cycle decreases when NADH + H$^+$ concentrations are high. Instead, acetyl-CoA is used to synthesize fatty acids, which then form triglycerides. The lipogenic effects of chronic and heavy alcohol consumption cause triglycerides to accumulate in the liver, a condition referred to as **fatty liver.**

## Effects of Alcohol on Vitamin A Metabolism in the Liver

Another consequence of heavy drinking involves the effects of alcohol on vitamin A metabolism in the liver. Vitamin A plays an important role in vision, being especially important at night. One form of vitamin A is retinol. Like ethanol, retinol is an alcohol, and ADH is involved in its metabolism. Specifically, ADH converts retinol (the alcohol form of vitamin A) into another form of vitamin A called retinal (the aldehyde form of

---

**microsomal ethanol-oxidizing system (MEOS)** A pathway used to metabolize alcohol when it is present in high amounts.

**tolerance** Response to high and repeated drug exposure that results in a reduced effect.

**cross-tolerance** Tolerance to one drug that causes tolerance to other similar drugs.

**microsomes** Cell organelles associated with endoplasmic reticula.

**gout** A condition caused by the accumulation of uric acid in the joints.

**fatty liver** A condition caused by excess alcohol consumption characterized by the accumulation of triglycerides in the liver.

**FIGURE 4    Chronic Alcohol Intake Interferes with Conversion of Pyruvate to Acetyl-CoA**

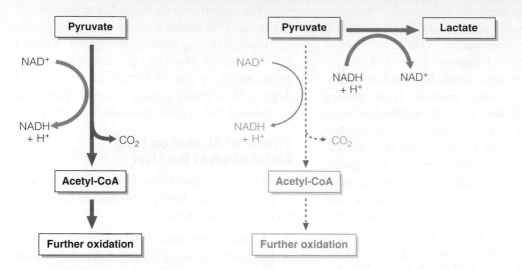

Normally, pyruvate is converted to acetyl-CoA with the aid of NAD$^+$.

Heavy alcohol consumption depletes NAD$^+$ levels and inhibits the conversion of pyruvate to acetyl-CoA in the liver. The liver converts pyruvate to lactate (lactic acid) rather than to acetyl-CoA, and NAD$^+$ is regenerated.

vitamin A), which is needed for night vision. When a person drinks heavily, alcohol is given preferential access to ADH. Thus less retinal is made by the body, and this can lead to vision problems, especially at night.

### ESSENTIAL *Concepts*

The liver metabolizes most of the alcohol that is consumed. The two major pathways used to metabolize alcohol are the alcohol dehydrogenase (ADH) pathway and the microsomal ethanol-oxidizing system (MEOS). When alcohol is consumed in moderation, the ADH pathway is used. During heavy drinking, the MEOS is used as well. The enzymes associated with the MEOS increase in response to heavy drinking, which is why some people develop an alcohol tolerance. Because the MEOS metabolizes others drugs as well, heavy drinkers can develop a cross-tolerance to them. The ADH pathway uses the enzymes alcohol dehydrogenase (ADH) and acetaldehyde dehydrogenase (ALDH). ADH converts alcohol to acetaldehyde, generating NADH + H$^+$. Next, ALDH converts acetaldehyde to acetic acid, also forming NADH + H$^+$. Genetic alterations in these enzymes can affect a person's ability to metabolize alcohol. The MEOS also converts alcohol to acetaldehyde, although different enzymes are used. The accumulation of acetaldehyde and NADH + H$^+$ can lead to serious health problems. For example, acetaldehyde alters liver structure and function, whereas an imbalance between NAD$^+$ and NADH + H$^+$ affects other NAD-linked enzymes. Heavy drinking can also increase the synthesis of lactate, interfere with vitamin A metabolism, and cause triglycerides to accumulate in the liver.

## Health Benefits of Moderate Alcohol Consumption

Although the detrimental effects of heavy alcohol consumption on health have long been known, research now suggests that alcohol consumption in small amounts may be beneficial. For example, adults who drink alcohol in moderation tend to live longer than nondrinkers and those who drink heavily.[5] In moderation, alcohol consumption appears to be associated with decreased risk for cardiovascular disease, gallstones, age-related memory loss, and even type 2 diabetes. Consumed in excess, alcohol can adversely affect virtually every organ system in the body. Thus it is important to consider how much alcohol is too much. The Dietary Guidelines for Americans state that those who choose to drink alcoholic beverages should do so sensibly and in moderation—defined as the consumption of up to one drink per day for women and up to two drinks per day for men. In the United States, a standard drink is equivalent to 12 fluid ounces of beer, 5 fluid ounces of wine, or 1.5 fluid ounces of distilled liquor, and each has approximately 12 to 14 g of alcohol.

It is possible that people who drink in moderation may differ from nondrinkers and heavy drinkers in ways that independently influence health. For example, moderate drinkers are more likely than nondrinkers and heavy drinkers to be better educated, wealthier, and physically active, and to have regular sleep habits and healthier weights.[6] Nonetheless, even when these confounding variables are accounted for, the evidence supporting the relationship between moderate intakes of alcohol and health remains compelling. Only a clinical intervention trial could determine directly the health risks and benefits associated with moderate alcohol intakes. However, this type of study would require randomly assigning hundreds of healthy adults to either a treatment or control group. Participants in the treatment group would be required to consume a determined number of alcoholic drinks per day, whereas participants in the control group would consume placebo drinks made to taste and smell like alcohol. Such a study would be difficult to carry out, and unlikely to be conducted. Clearly, health professionals will need to rely on less controlled studies to assess the effects of alcohol on health.

## Moderate Alcohol Consumption and Cardiovascular Health

The relationship between moderate alcohol consumption and reduced risk of cardiovascular disease has been confirmed by hundreds of epidemiological studies. Given these studies researchers have estimated that adults who consume an average of one to two alcoholic drinks daily have a 25 to 40% lower risk of cardiovascular disease (CVD) than adults who do not consume alcohol. This is true for both men and women.

You may wonder how alcohol lowers a person's risk of CVD. First, recall that CVD results from the formation and accumulation of plaque in the lining of arteries. Furthermore, high-density lipoproteins (HDLs) help protect against heart disease. Studies show that moderate daily intake of alcohol increases HDL-cholesterol levels and therefore may offer some protection from CVD.[7] There is also evidence that alcohol decreases levels of a protein (fibrinogen) that promotes blood clot formation, and increases levels of an enzyme that dissolves blood clots. Blood clots can block the flow of blood in arteries and are common causes of heart attacks and strokes. Last, because chronic inflammation is also associated with CVD, alcohol's anti-inflammatory effect may offer moderate drinkers protection.

**CONNECTIONS** Recall that HDLs are lipoproteins that help scavenge excess cholesterol from cells (Chapter 7, page 249).

Moderate alcohol intake provides little, if any, health benefits among young adults, unlike among middle-aged adults. In fact, alcohol consumption in the young adult group is associated with increased risk of alcohol-related injury and death. Studies show that many adults with alcohol-related problems started drinking at a young age.[8] The frequency of risky drinking behavior as young adults is associated with hazardous drinking patterns later in life.

## Moderate Alcohol Consumption and Other Health Benefits

In addition to protecting cardiovascular health, moderate alcohol intake may help guard against other age-related chronic diseases such as type 2 diabetes, gallstones, and dementia. Although evidence is limited to epidemiological studies, moderate drinkers tend to be more sensitive to insulin. However, consuming higher amounts of alcohol is associated with increased risk for type 2 diabetes.[9] The mechanism by which alcohol modulates insulin sensitivity remains unclear. Alcohol consumption may also lower the risk of gallstone formation. Regardless of the amount and frequency of consumption, the incidence of symptomatic gallstone disease tends to be lower among drinkers than nondrinkers.[10] Moderate drinking may also reduce the risk of age-related memory loss, or what is called dementia.[11] It is important to remember that these possible benefits of moderate alcohol consumption require further study.

**CONNECTIONS** Gallstones form in the gallbladder when cholesterol separates from bile (Chapter 7, page 243).

### ESSENTIAL *Concepts*

When consumed in moderation, alcohol is associated with decreased risk for cardiovascular disease, gallstones, age-related memory loss, and even type 2 diabetes. Adults who consume an average of one to two alcoholic drinks daily have lower risk of cardiovascular disease than do adults who do not consume alcohol. This effect may be due to the ability of alcohol to raise HDL-cholesterol levels and decrease blood clot formation. Moderate alcohol consumption may also have anti-inflammatory effects. However, moderate alcohol intakes provide little, if any, health benefits among young adults.

## Health Risks Associated with Heavy Alcohol Consumption

Although numerous studies suggest that moderate alcohol intakes may provide health benefits for middle-aged adults, when consumed in excess alcohol is clearly hazardous to health. Over time, heavy drinking can lead to impaired nutritional status, liver damage, certain cancers, heart problems, and pancreatitis (Figure 5). High-risk drinking also takes a heavy toll on families, friends, and society.

## Nutritional Status

Alcoholic beverages have calories but very few essential nutrients. Therefore even light to moderate drinking can increase energy intake, which can result in weight gain. Furthermore, heavy drinkers often show many other signs of malnutrition. Alcoholics tend to eat very poor diets, and essential nutrient deficiencies, or primary malnutrition, can become an important consideration. When alcohol accounts for more than 30% of total energy, micronutrient intakes are likely to be inadequate.

Even with an adequate diet, nutritional status is often compromised in heavy drinkers. Alcohol can interfere with digestion, absorption, utilization, and excretion of various nutrients, leading to secondary malnutrition. For example, heavy use of alcohol can interfere with the absorption and metabolism of the B vitamin thiamin, which in turn damages nerves in the brain. As a result, long-term alcohol abuse can cause alcohol-induced dementia, or what is called Wernicke-Korsakoff syndrome. This syndrome results when areas of the brain involved with short-term memory are damaged. When this happens, people often compensate by making up information, a behavior called confabulation.

## FIGURE 5    Effects of Habitual Alcohol Abuse on the Body

Alcohol interferes with digestion, absorption, use, and excretion of nutrients. Chronic alcohol intake can lead to malnutrition, cancer, and heart disease.

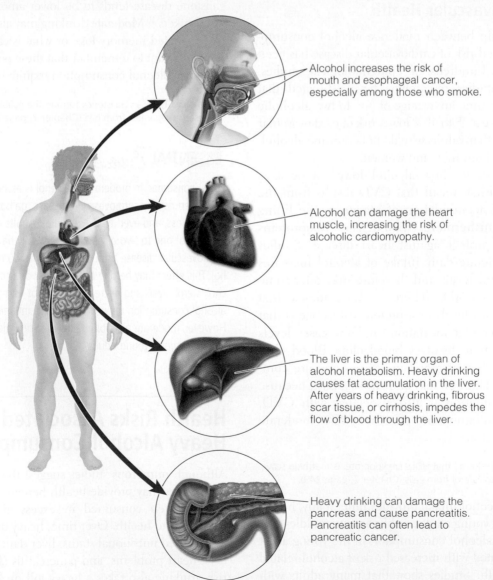

Alcohol increases the risk of mouth and esophageal cancer, especially among those who smoke.

Alcohol can damage the heart muscle, increasing the risk of alcoholic cardiomyopathy.

The liver is the primary organ of alcohol metabolism. Heavy drinking causes fat accumulation in the liver. After years of heavy drinking, fibrous scar tissue, or cirrhosis, impedes the flow of blood through the liver.

Heavy drinking can damage the pancreas and cause pancreatitis. Pancreatitis can often lead to pancreatic cancer.

**CONNECTIONS** Secondary malnutrition is not directly caused by inadequate nutrient intake, but rather by some other factor (Chapter 2, page 30).

**CONNECTIONS** Recall that Wernicke-Korsakoff syndrome is related to a thiamin deficiency and is characterized by impairment of short-term memory, poor muscle coordination, and confusion (Chapter 10, page 387).

Many studies show that heavy drinking leads to decreased nutrient availability and impaired nutritional status via both primary and secondary malnutrition. However, the negative impact of high alcohol intake on nutritional status appears more common among people of low socioeconomic status than among those of high socioeconomic status. This is likely because overall nutritional quality of diets eaten by less economically advantaged people is generally lower than diets eaten by those of greater means. The impact of alcohol consumption on nutritional status for selected nutrients is summarized in Table 3.

## Liver Disease

Long-term alcohol consumption can alter liver function and lead to liver disease. For example, approximately 20% of heavy drinkers develop fatty livers, which can be reversed if the person stops drinking. In most cases no clinical signs are associated with having a fatty liver, but over time fatty liver can lead to more serious conditions such as alcoholic hepatitis and cirrhosis. A person with **alcoholic hepatitis** typically has an enlarged and swollen liver. This is caused by the accumulation of fat in the liver, which obstructs blood flow, which in turn deprives liver cells of oxygen and nutrients. When this occurs, the liver cells can die. After years of heavy drinking, fibrous scar tissue can form in the liver, causing what is called **cirrhosis.** However, alcoholic hepatitis does not progress to cirrhosis in all heavy drinkers, and it is not clear why it develops in some people but not in others. Researchers have estimated that 10 to 15% of heavy drinkers develop cirrhosis of the liver.[13]

## Cancer Risk

Habitual alcohol consumption increases the risk of developing certain types of cancer, especially those of the mouth, esophagus, colon, liver, and breast. In fact, almost 50% of cancers of the mouth and esophagus are associated with heavy drinking.[14] However, it is not clear how alcohol increases the risk of cancer. Carcinogens are compounds that initiate the formation of cancer. Surprisingly, there is no evidence that alcohol per se is carcinogenic. Rather, alcohol may act as a co-carcinogen by enhancing the carcinogenicity of other cancer-

Normal liver, fatty liver, and cirrhosis

© Arthur Glauberman/Photo Researchers, Inc.

causing chemicals. For example, it is well known that heavy drinkers who smoke are at particularly high risk for cancers of the mouth, esophagus, and trachea. It is possible that alcohol interacts with cigarette smoke to make it more dangerous.

It has also been hypothesized that poor nutritional status associated with heavy drinking may hasten the development of certain types of cancer. For example, alcohol interferes with folate availability, which in turn may compromise cell division and DNA repair. Similarly, reduced levels of iron, zinc, vitamin E, certain B vitamins, and vitamin A have been experimentally linked to certain types of cancer. Because adequate intake of many of these nutrients may offer protection against cancer, overall dietary inadequacy from chronic alcohol intake may weaken a person's natural defense mechanisms.

## Cardiovascular Disease

Although the cardiovascular system appears to benefit from moderate alcohol intake, heavy alcohol consumption has harmful effects. Consequences associated with long-term, heavy drinking include hypertension, stroke, irregular heart rhythms, cardiomyopathy, and sudden cardiac death. Although the mechanisms are not completely understood, the relationship between heavy drinking and hypertension is well established. Alterations in liver function may, in part, contribute to a

**alcoholic hepatitis** (*hepatic,* pertaining to the liver; *-itis,* inflammation) Inflammation of the liver caused by chronic alcohol abuse.

**cirrhosis** (cir – RHO – sis) The formation of scar tissue in the liver caused by chronic alcohol abuse.

## TABLE 3    Impact of Alcohol on Nutritional Status

| Nutrient | Impact of Alcohol | Related Nutritional Problems |
|---|---|---|
| **Water-Soluble Vitamins** | | |
| Thiamin | • Impaired absorption<br>• Increased urinary loss<br>• Altered metabolism<br>• Reduced storage | • Paralysis of eye muscles<br>• Degeneration of nerves with loss of sensation in lower extremities<br>• Loss of balance<br>• Abnormal gait<br>• Memory loss<br>• Psychosis |
| Vitamin $B_6$ | • Decreased conversion to the active vitamin form<br>• Increased urinary loss<br>• Displacement of vitamin from its binding protein | • Increased risk of anemia<br>• Impaired metabolic reactions involving amino acids |
| Vitamin $B_{12}$ | • Impaired absorption | • Because of large body stores, vitamin $B_{12}$ deficiency not common in alcoholics |
| Folic Acid (folate) | • Decreased absorption<br>• Increased urinary loss | • Increased risk of macrocytic anemia<br>• Malabsorption of nutrients and diarrhea |
| Riboflavin | • Impaired absorption<br>• Decreased conversion to the active form | • Deficiency not typical in isolation, but occurs in conjunction with other B vitamins |
| **Fat-Soluble Vitamins** | | |
| Vitamin A | • Alcoholic liver disease associated with low levels of vitamin A in the plasma<br>• Decreased synthesis of retinol binding protein needed to circulate vitamin A<br>• Decreased conversion of retinol to retinal<br>• Reduced vitamin A absorption with pancreatic disease | • Impaired vision<br>• Impaired ability to see in dim light (night blindness) |
| Vitamin D | • Alcoholic liver disease impairs ability to convert vitamin D to its active form<br>• Pancreatic impairment leads to malabsorption | • Increased susceptibility to bone fractures and osteoporosis |
| Vitamin E | • Pancreatic impairment leads to malabsorption | • Nerve damage<br>• Tunnel vision<br>• Fragility of cell membranes |
| Vitamin K | • Pancreatic impairment leads to malabsorption<br>• Alcoholic liver disease impairs ability to synthesize vitamin K-dependent factors needed for blood clot formation | • Bruising and prolonged bleeding |
| **Minerals** | | |
| Magnesium | • Increased urinary loss | • Muscle rigidity, cramps, and twitching<br>• Irregular heart rhythm<br>• Possible link to hallucinations |
| Iron | • Alcoholic liver disease linked to increased iron storage<br>• Increased iron absorption<br>• Increased gastrointestinal bleeding leads to iron loss | • Both iron deficiency and overload are possible |
| Zinc | • Decreased absorption<br>• Increased urinary loss<br>• Altered zinc utilization | • Altered taste<br>• Loss of appetite<br>• Impaired wound healing<br>• Difficulty seeing at night |

rise in blood pressure. When heavy drinkers refrain from alcohol, blood pressure often falls.

Alcohol can also trigger disturbances in heart rhythms, a condition called **cardiac arrhythmia**. Cardiac arrhythmia is sometimes referred to as "holiday heart syndrome," because it is associated with **binge drinking**, which often occurs on weekends and around holidays.[15] Binge drink-

ing is defined as consumption of five or more drinks on a single occasion for a man or four or more drinks for a

**cardiac arrhythmia** (ar – RHYTH – mi – a) (*cardiac,* pertaining to the heart) Irregular heartbeat caused by high intakes of alcohol.

**binge drinking** Consumption of five or more drinks in males and four or more drinks in females with the intent to become intoxicated.

woman. Although the cause is not clear, cardiac arrhythmia may be a response to rising levels of acetaldehyde. Alcohol-induced cardiac arrhythmia can lead to sudden cardiac death.

One of the most serious cardiovascular consequences associated with heavy drinking is **alcoholic cardiomyopathy.** This is caused by the direct toxic effect of alcohol on heart muscle cells, possibly from apoptosis.[16] Apoptosis, or what can be thought of as "programmed cell death," may occur when a cell becomes damaged by alcohol and/or acetaldehyde. Over time, this toxicity causes the heart muscle to weaken, making it difficult for the heart to contract forcibly. Blood flow to vital organs such as the lungs, liver, kidneys, and brain is thereby reduced. If drinking continues, alcoholic cardiomyopathy can ultimately result in heart failure. For reasons that are not clear, women are more susceptible to alcohol-induced cardiomyopathy than are males.

## Impaired Pancreatic Function

**Pancreatitis** is a painful condition characterized by inflammation of the pancreas. Chronic consumption of high intakes of alcohol (seven drinks/day for more than 5 years) increases the risk of developing this condition.[17] In fact, 70 to 90% of all cases of chronic pancreatitis are associated with heavy drinking. Inflammation makes it difficult for the pancreas to release pancreatic juice, which supplies the enzymes needed for digestion. The inability to completely digest food impairs nutrient absorption. You may also remember that, in addition to the release of digestive enzymes, the pancreas also releases the hormones insulin and glucagon needed for blood glucose regulation. Consequently, people with pancreatitis often have poorly regulated blood glucose levels.

## ESSENTIAL *Concepts*

Consumed in excess, alcohol is hazardous to health. Aside from increasing energy intake, which leads to weight gain, habitual drinking can adversely affect nutritional status by causing both primary and secondary malnutrition. Long-term alcohol consumption can alter liver function, leading to alcoholic hepatitis. This condition results when fat accumulates in the liver and obstructs blood flow. When liver cells are deprived of oxygen and nutrients, the formation of fibrous scar tissue leads to a condition called cirrhosis. Alcohol consumption can also increase the risk of developing cancers of the mouth, esophagus, colon, liver, and breast. Heavy alcohol consumption can damage heart tissue, causing a condition known as alcoholic cardiomyopathy.

## Other Problems Related to Alcohol Abuse

Although the majority of people who drink alcohol do so responsibly and without negative consequences, the line between alcohol use and abuse is often blurry. When alcohol is the cause of significant problems in a person's life, drinking is problematic. Habitual alcohol abuse affects virtually every aspect of a person's life, including family relationships, performance at work, and school. Arrests related to drinking while under the influence of alcohol can lead to legal problems. Yet because alcohol use can lead to both physical and emotional dependence, many alcohol abusers are not able to stop drinking.

Although popular culture has created a stereotypic view of alcoholics, in reality alcoholism affects men and women, young and old, professionals and nonprofessionals, and both the rich and poor. In fact, most health professionals recognize alcoholism as a widespread, complex disease that can be treated but not cured. One of the oldest and most reputable organizations for the treatment of alcoholism is **Alcoholics Anonymous® (AA®).** This organization promotes a 12-step program that provides fellowship and support for individuals who wish to stop drinking. Currently, more than 2 million people consider themselves lifelong members of AA. The primary purpose of AA is to help alcoholics achieve sobriety and stay sober. Other organizations such as **Al-Anon®** and **Alateen®** provide support to family members of alcoholics.

## Alcohol and College Students

Numerous studies have documented the extent to which alcohol use affects college students and campuses nationwide.[18] Alcohol abuse on many college campuses is rampant and associated with a wide range of negative repercussions such as vandalism, violence, acquaintance rape, unprotected sex, and death. To give some perspective of the magnitude of alcohol-related problems,

**alcoholic cardiomyopathy** (car – di – o – my – O – path – y) Condition that results when the heart muscle weakens in response to heavy alcohol consumption.

**pancreatitis** (pan – cre – a – TI – tis) Inflammation of the pancreas.

**Alcoholics Anonymous (AA)** An organization dedicated to helping people achieve and maintain sobriety.

**Al-Anon** An organization dedicated to helping people cope with alcoholic family members and friends.

**Alateen** An organization dedicated to helping children cope with an alcoholic parent.

**FIGURE 6    Percentage of Fatal Traffic Accidents Involving Alcohol (2001)**

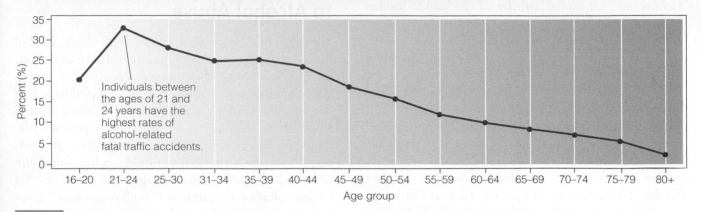

Individuals between the ages of 21 and 24 years have the highest rates of alcohol-related fatal traffic accidents.

SOURCE: Yi H, Williams GD, Dufour MC. *Surveillance Report #65: Trends in Alcohol-Related Fatal Traffic Crashes, United States, 1977–2001.* Bethesda, MD: NIAAA, Division of Biometry and Epidemiology, Alcohol Epidemiologic Data System; August 2003.

data from the National Institute on Alcohol Abuse and Alcoholism indicate that 1,400 college students between the ages of 18 and 24 die each year from alcohol-related injuries, including motor vehicle crashes.[19] Individuals between the ages of 21 and 24 have the highest percentages of alcohol-related fatal traffic accidents (Figure 6).

Although one out of five college students report that they abstain from drinking alcohol, one out of four describe themselves as binge drinkers (4 to 5 drinks/occasion). According to a recent study of over 100 college campuses, binge drinkers are more likely than non-binge drinkers to miss class, have lower academic rankings, be in trouble with campus law enforcement, and drive while under the influence of alcohol.[20]

In an effort to reduce alcohol abuse on college campuses, new comprehensive programs are being developed and implemented. Clearly, traditional educational approaches that distribute pamphlets and conduct one-day alcohol awareness programs are not enough. Colleges and universities must work together with communities to create a culture that discourages high-risk drinking.[21] For example, increasing recreational activities on college campuses that do not involve drinking offer students alternatives to bars and clubs. There are also initiatives to curb practices that encourage drinking among college students.[22] These include such things as discouraging alcohol discounting, two-for-one drink specials, inexpensive beer pitcher sales, and other types of "happy hour" promotions. Alcohol abuse among college students is a problem that affects the entire campus community.

## Summary and Recommendations

Both health benefits and risks are associated with alcohol consumption by adults. Although most Americans who drink alcohol do so safely and responsibly, this is certainly not the case for all those who drink. This is especially true for young people. In the United States alone, several thousand people die each year from alcohol-related problems. For this reason, one of the goals identified in *Healthy People 2010* is to reduce alcohol abuse in order to protect the health, safety, and quality of life for all, especially children.[23] In 1998, the number of alcohol-related deaths was 5.9 per 100,000 people per year, and the target goal is to reduce this number to 4 deaths per 100,000 people by the year 2010.

The Dietary Guidelines for Americans also address the harmful effects of alcohol when consumed in excess.[24] For this reason, the guidelines clearly state that adults who consume alcohol should do so in moderation. Again, moderation is defined as the consumption of up to one drink per day for women and up to two drinks per day for men. Importantly some people should avoid alcohol completely, including children, adolescents, women of childbearing age who may become pregnant, pregnant and lactating women, individuals who cannot restrict alcohol intake, those taking medications that can interact with alcohol, and individuals with specific medical conditions. It is also important for people to abstain from drinking when driving, operating machinery, or taking part in activities that require attention, skill, and/or coordination. In Chapter 14 you can read more about the negative consequences of alcohol consumption during pregnancy.

## ESSENTIAL *Concepts*

The majority of people who drink alcohol do so responsibly and without negative consequences. However, when alcohol is the cause of significant problems in a person's life, drinking is problematic. The organization Alcoholics Anonymous (AA) helps people achieve sobriety and stay sober. Other organizations such as Al-Anon and Alateen provide support to family members of alcoholics. Because alcohol abuse on many college campuses is rampant and associated with a wide range of negative repercussions, many college campuses are developing programs to help curb high-risk drinking behaviors among students. Nonetheless, thousands of Americans die annually from alcohol-related problems, and one of the goals for *Healthy People 2010* is to reduce the number of alcohol-related deaths from 5.9 per 100,000 people/year to 4 per 100,000 by the year 2010. The harmful effects of excess alcohol are also addressed in the Dietary Guidelines for Americans. In addition to promoting drinking only in moderation (1 to 2 drinks/day), the guidelines recognize that some people should completely avoid alcohol consumption.

# Notes

1. Gunzerath L, Faden V, Zakhari S, Warren K. National Institute on Alcohol Abuse and Alcoholism report on moderate drinking. Alcoholism, Clinical and Experimental Research. 2004;28:829–47.

2. U.S. Department of Health and Human Services, and U.S. Department of Agriculture. Dietary Guidelines for Americans 2005. Washington, DC: U.S. Government Printing Office; 2005. Available from: http://www.healthierus.gov/dietaryguidelines.

3. Whitfield JB. Alcohol and gene interactions. Clinical Chemistry and Laboratory Medicine. 2005;43:480–7.

4. Nolen-Hoeksema S. Gender differences in risk factors and consequences for alcohol use and problems. Clinical Psychology Review. 2004;24:981–1010.

5. Ellison RC. Balancing the risks and benefits of moderate drinking. Annals of the New York Academy of Sciences. 2002;957:1–6.

6. Naimi TS, Brown DW, Brewer RD, Giles WH, Mensah G, Serdula MK, Mokdad AH, Hungerford DW, Lando J, Naimi S, Stroup DF. The state sets the rate: the relationship among state-specific college binge drinking, state binge drinking rates, and selected state alcohol control policies. American Journal of Preventive Medicine. 2005;28:369–73.

7. Rimm EB, Williams P, Fosher K, Criqui M, Stampfer MJ. Moderate alcohol intake and lower risk of coronary heart disease: meta-analysis of effects on lipids and haemostatic factors. British Medical Journal. 1999;319:1523–8; Pai JK, Hankinson SE, Thadhani R, Rifai N, Pischon T, and Rimm EB. Moderate alcohol consumption and lower levels of inflammatory markers in U.S. men and women. Atherosclerosis. 2006;186:113–20.

8. Hill EM, Chow K. Life-history theory and risky drinking. Addiction. 2002;97:401–13.

9. Carlsson S, Hammar N, Grill V. Alcohol consumption and type 2 diabetes. Meta-analysis of epidemiological studies indicates a U-shaped relationship. Diabetologia. 2005;48:1051–4.

10. Leitzmann MF, Giovannucci EL, Stampfer MJ, Spiegelman D, Colditz GA, Willett WC, Rimm EB. Prospective study of alcohol consumption patterns in relation to symptomatic gallstone disease in men. Alcoholism, Clinical and Experimental Research. 1999;23:835–41.

11. Espeland MA, Gu L, Masaki KH, Langer RD, Coker LH, Stefanick ML, Ockene J, Rapp SR. Association between reported alcohol intake and cognition: results from the Women's Health Initiative Memory Study. American Journal of Epidemiology. 2005;161:228–38.

12. Salaspuro M. Nutrient intake and nutritional status in alcoholics. Alcohol and Alcoholism. 1993;28:85–8.

13. Mann RE, Smart RG, Govoni R. The epidemiology of alcoholic liver disease. Alcohol Research and Health. 2003;27:209–19.

14. Blot WJ. Alcohol and cancer. Cancer Research. 1992;52:2119S–23S.

15. Menz V, Grimm W, Hoffmann J, Maisch B. Alcohol and rhythm disturbance: the holiday heart syndrome. Herz. 1996;21:227–31.

16. Piano MR. Alcoholic cardiomyopathy: incidence, clinical characteristics, and pathophysiology. Chest. 2002;121:1638–50.

17. Hanck C, Whitcomb DC. Alcoholic pancreatitis. Gastroenterology Clinics of North America. 2004;33:751–65.

18. Hingson R, Heeren T, Winter M, Wechsler H. Magnitude of alcohol-related mortality and morbidity among U.S. college students ages 18-24: changes from 1998 to 2001. Annual Review of Public Health. 2005;26:259–79.

19. Yi H, Williams GD, Dufour MC. Surveillance Report #65: Trends in alcohol-related traffic crashes, United States, 1977–2001. Bethesda, MD: NIAAA, Division of Biometry and Epidemiology, Alcohol Epidemiologic Data System, August 2003.

20. Lange JE, Clapp JD, Turrisi R, Reavy R, Jaccard J, Johnson MB, Voas RB, Larimer M. College binge drinking: what is it? Who does it? Alcoholism, Clinical and Experimental Research. 2002;26:723–30.

21. Dejong W. Finding common ground for effective campus-based prevention. Psychology of Addictive Behaviors. 2001;15:292–6.

22. Kuo M, Wechsler H, Greenberg P, Lee H. The marketing of alcohol to college students: the role of low prices and special promotions. American Journal of Preventive Medicine. 2003;25:204–11.

23. Office of Disease Prevention and Health Promotion, U.S. Department of Health and Human Services. Healthy People 2010. Available from: http://www.healthypeople.gov/Publications/.

24. U.S. Department of Health and Human Services and U.S. Department of Agriculture. Dietary Guidelines for Americans 2005. Washington, DC: U.S. Government Printing Office, 2005. Available from: http://www.healthierus.gov/dietaryguidelines.

# Chapter 11

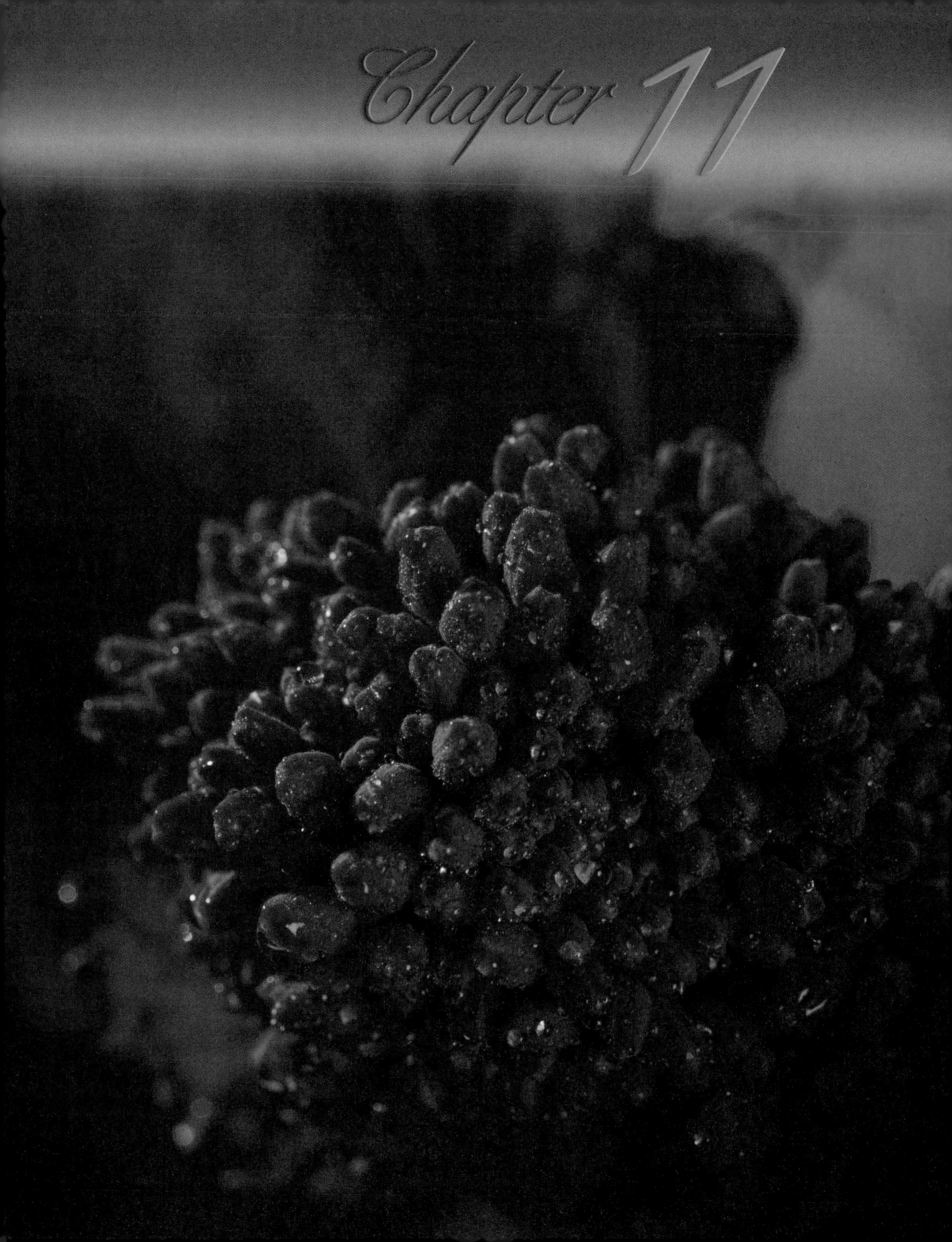

# Fat-Soluble Vitamins

What would you think of an advertisement promoting something that could boost your immune system, fight cancer and heart disease, improve your vision, and keep your bones strong? You might think this is something worth buying, or you could assume that these claims are simply too good to be true. However, all these health benefits can be attributed to the fat-soluble vitamins. Fat-soluble vitamins are organic substances that dissolve in lipids and are vital for health. Because the body needs fat-soluble vitamins in very small quantities, they are considered micronutrients.

There are four essential fat-soluble vitamins—vitamins A, D, E, and K. In this chapter, you will learn about each vitamin's chemical structure, food sources, and functions in the body, as well as information concerning deficiencies, toxicities, and dietary recommendations. Whereas water-soluble vitamins primarily function as coenzymes in energy metabolism, fat-soluble vitamins tend to be key players in cell growth, maturation, and gene regulation.

# The Basics of Fat-Soluble Vitamins

Some general characteristics common among the fat-soluble vitamins are listed in Table 11.1. Fat-soluble vitamins typically are absorbed in the small intestine. This requires the presence of other lipids as well as the action of bile. Fat-soluble vitamins are circulated away from the small intestine in the lymph via chylomicrons, and are eventually circulated in the blood either as components of lipoproteins (such as very low density lipoproteins, VLDLs) or bound to transport proteins. Like water-soluble vitamins, each fat-soluble vitamin has several forms, some of which are more biologically active than others. Because most of the fat-soluble vitamins are stored in the body, consuming large amounts of them (especially in supplement form) can result in toxicities, sometimes with serious consequences. Although vitamin K acts as a coenzyme, most fat-soluble vitamins are involved in other processes such as regulation of gene expression, cell maturation, and stabilization of free radicals. Table 11.2 gives an overview of the functions and sources of the fat-soluble vitamins.

## Naming of the Fat-Soluble Vitamins

Recall that researchers initially thought there were only two vitamins, and they were referred to as *vitamine* A (fat-soluble) and *vitamine* B (water-soluble). However, as with *vitamine* B, there turned out to be several kinds of *vitamine* A—each being assigned a different letter as its individual importance was discovered (vitamin A, vitamin D, and so forth). Because each fat-soluble vitamin has several forms, numbers are sometimes used to distinguish them from each other (such as vitamin $K_1$, vitamin $K_2$). In addition, most have been given names that reflect their chemical composition or function. For example, the active form of vitamin D is called calcitriol, partly because of its role in calcium homeostasis.

CONNECTIONS Recall that bile is a fluid produced in the liver and stored in the gallbladder and aids in lipid digestion (Chapter 4, page 107).

### TABLE 11.1 General Characteristics of the Fat-Soluble Vitamins

| | General Characteristics |
|---|---|
| **Food Sources** | • Typically found in fatty portions of foods<br>• Except for vitamin D, easily destroyed by heat and/or light |
| **Digestion** | • Very little needed |
| **Absorption** | • Occurs mostly in small intestine<br>• Requires incorporation into micelles and the actions of bile<br>• Once transported into the intestinal cell, vitamins are packaged with other lipids into chylomicrons |
| **Circulation Away from Gastrointestinal Tract** | • Via lymph |
| **Functions** | • Gene regulation and various other functions |
| **Toxicity** | • Except for vitamin K, toxicities are dangerous and sometimes fatal |

**TABLE 11.2** Functions, Sources, Deficiency Diseases, and Toxicity Symptoms for the Essential Fat-Soluble Vitamins

| Vitamin | Major Functions | Deficiency | Toxicity | Excellent Food Sources |
|---|---|---|---|---|
| Vitamin A | • Growth<br>• Reproduction<br>• Vision<br>• Cell differentiation<br>• Immune function<br>• Bone health | • Vitamin A deficiency disorder (VADD)<br>• Night blindness<br>• Xerophthalmia<br>• Hyperkeratosis | • Hypervitaminos A<br>• Hypercarotenemia<br>• Blurred vision<br>• Birth defects<br>• Liver damage<br>• Osteoporosis | Liver<br>Pumpkin<br>Sweet potato<br>Carrot |
| Vitamin D | • Calcium homeostasis<br>• Bone health<br>• Cell differentiation | • Rickets<br>• Osteomalacia<br>• Osteoporosis | • Hypercalcemia | Fish<br>Shiitake mushrooms<br>Fortified milk<br>Fortified cereals |
| Vitamin E | • Antioxidant<br>• Cell membranes<br>• Eye health<br>• Heart health | • Neuromuscular problems<br>• Hemolytic anemia | • Hemorrhage | Tomatoes<br>Nuts, seeds<br>Spinach<br>Fortified cereals |
| Vitamin K | • Coenzyme (carboxylation)<br>• Blood clotting<br>• Bone health<br>• Tooth health | • Vitamin K deficiency bleeding (VKDB) | • No known effects | Kale<br>Spinach<br>Broccoli<br>Brussels sprouts |

# Vitamin A and the Carotenoids

In 1916, Elmer V. McCullum and Marguerite Davis at the University of Wisconsin and Thomas Osborne and Lafayette Mendel at Yale University first identified vitamin A as a substance in food important for health in experimental animals. The term *vitamin A* is now used to refer to a series of compounds called **retinoids,** which include retinol, retinoic acid, and retinal. The structures of the retinoids are provided in Figure 11.1. The retinoids are often referred to as **preformed vitamin A.** Retinol is the most potent form of vitamin A and can be synthesized in the body from retinal. Retinal can also be converted to retinoic acid, but retinoic acid cannot be converted to any other retinoid.

Some foods contain preformed vitamin A, whereas others contain compounds that have structures somewhat similar to the retinoids. These compounds, called **carotenoids,** can be subdivided into two categories—those that can be converted to vitamin A and those that cannot. Carotenoids that can be converted to vitamin A are called **provitamin A carotenoids. β-Carotene** is one of the most common provitamin A carotenoids in foods, and the body can split it to form two molecules of retinal. Carotenoids that cannot be converted to vitamin A are called **nonprovitamin A carotenoids.** These include lycopene, astaxanthin, zeaxanthin, and lutein. Nonprovitamin A carotenoids are not classified as vitamins. Rather, they are called phytochemicals, and foods containing them are considered functional foods.

## Dietary Sources of Vitamin A and the Provitamin A Carotenoids

The unit of measure used to describe the vitamin A content of foods is the **retinol activity equivalent (RAE).** The RAE allows food containing preformed vitamin A to be easily compared to food containing provitamin A carotenoids,

**retinoid (preformed vitamin A)** (RE – ti – noid) A term used to describe all forms of vitamin A.

**carotenoids** (car – O – te – noids) Brightly colored compounds found in some foods; structures similar to that of vitamin A.

**provitamin A carotenoid** A carotenoid that can be converted to vitamin A.

**β-carotene** A provitamin A carotenoid.

**nonprovitamin A carotenoid** A carotenoid that cannot be converted to vitamin A.

**retinol activity equivalent (RAE)** A unit of measure used to describe the combined amount of preformed vitamin A and provitamin A carotenoids in foods.

## FIGURE 11.1 Vitamin A

Vitamin A has three forms: retinol, retinal, and retinoic acid. Retinol and retinal can be converted to the other forms; retinoic acid cannot.

**Retinol**

**Retinal**

**Retinoic acid**

CONNECTIONS Recall that tryptophan can be converted to niacin in the body (Chapter 10, page 390).

© MaXx Images/Stockfood America

**Peppers are excellent sources of provitamin A carotenoids.**

such as β-carotene. In this way, the RAE is similar to the niacin equivalent (NE)—a measure of a food's "potential" niacin content. Approximately 12 μg of β-carotene or 1 μg of retinol equals 1 RAE.

In general, preformed vitamin A is found in animal foods such as liver and other organ meats, fatty fish, and dairy products. In fact, a single serving of beef liver provides more than seven times the daily requirement for vitamin A. Whole-fat dairy products such as whole milk, cheese, and butter are good sources of vitamin A, whereas reduced-fat products are not unless they are fortified. Many foods such as margarine and breakfast cereals are fortified with vitamin A, making these good sources as well.

Whereas animal foods provide preformed vitamin A, plants tend to provide most of the provitamin A carotenoids to the diet. Yellow, orange, and red fruits and vegetables such as cantaloupe, carrots, and peppers are especially good sources of the carotenoids. Leafy greens are also good sources. Carotenoids can also be found in brightly colored animal foods, such as egg yolks and shrimp. Carotenoids tend to be yellowish-red in color, and their presence in foods makes them brightly colored as well. In fact, carotenoids are sometimes added to foods as coloring agents. For example, some cheese is orange because a colorant called annatto is added. Annatto is produced from the carotenoid-rich, bright red berries of the achiote plant. It is also used to color foods in Mexican, Puerto Rican, and Indian cooking.

Vitamin A and many of the provitamin A carotenoids are quickly destroyed by extreme cold or hot temperatures as well as exposure to light or oxygen. Thus they are easily lost during processing, cooking, freezing, and storage. However, β-carotene appears to be an exception, because processing and heating seem to increase its bioavailability. A list of foods containing vitamin A or provitamin A carotenoids is provided in Figure 11.2.

**FIGURE 11.2**   **Good Sources of Preformed Vitamin A or Provitamin A Carotenoids**

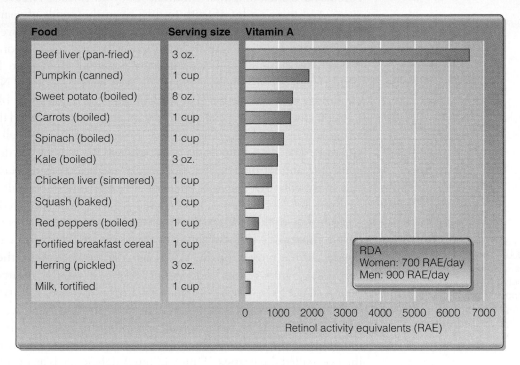

| Food | Serving size | Vitamin A |
|---|---|---|
| Beef liver (pan-fried) | 3 oz. | |
| Pumpkin (canned) | 1 cup | |
| Sweet potato (boiled) | 8 oz. | |
| Carrots (boiled) | 1 cup | |
| Spinach (boiled) | 1 cup | |
| Kale (boiled) | 3 oz. | |
| Chicken liver (simmered) | 1 cup | |
| Squash (baked) | 1 cup | |
| Red peppers (boiled) | 1 cup | |
| Fortified breakfast cereal | 1 cup | |
| Herring (pickled) | 3 oz. | |
| Milk, fortified | 1 cup | |

RDA
Women: 700 RAE/day
Men: 900 RAE/day

Retinol activity equivalents (RAE)
0 1000 2000 3000 4000 5000 6000 7000

SOURCE: Data from USDA Nutrient Database, Release 16-1.

## Regulation of Vitamin A and the Carotenoids in the Body

Vitamin A and the carotenoids are absorbed in the small intestine. This process requires the presence of dietary lipid, involving the formation of micelles with the assistance of bile. A low-fat diet can decrease vitamin A and carotenoid absorption (and thus their bioavailability), because the body is unable to form micelles.

Recall that on their entrance into the enterocyte, lipids are incorporated into chylomicrons. As such, fat-soluble vitamins join with other dietary lipids as they move from the gastrointestinal tract into the circulation. A lack of dietary lipid can inhibit chylomicron formation, in turn inhibiting the bioavailability of vitamin A. Vitamin A and the carotenoids contained in the chylomicrons then enter the lymph, where they ultimately enter the blood at the thoracic duct. Once in the blood, vitamin A and the carotenoids are delivered to many tissues and organs including adipose tissue, muscle, and the eyes. What is not taken up is transported to the liver as part of the chylomicron remnant.

Once taken up by the liver, carotenoids are packaged into VLDLs for further transport. However, vitamin A is not. Instead, vitamin A is attached to transport proteins for delivery to cells. The transport proteins **retinol-binding protein** and **transthyretin** work together in this regard. Vitamin A not taken up by tissues is stored in the liver.

## Functions of Vitamin A

Vitamin A has many roles, including aiding vision, growth, reproduction. In addition, it is needed for maintaining a healthy immune system and building strong bones.

Achiote paste, which gets its vibrant color from carotenoids, is often used in Puerto Rican, Mexican, and Indian cuisines to add color.

**CONNECTIONS** Remember that chylomicrons are lipoproteins made in the intestinal epithelial cell (enterocyte) (Chapter 7, page 245).

**retinol-binding protein** and **transthyretin**
Proteins that carry retinol in the blood.

## Vision

Parents have often told their children, "Eat your carrots. They're good for your eyes." And there's a lot of truth to this. Even thousands of years ago ancient Egyptian physicians prescribed liver to cure poor vision. However, not until the 20th century did scientists begin to understand the relationship between consuming vitamin A-rich food and eye function. In the late 1960s George Wald, Haldan Hartline, and Ragnar Granit were awarded a Nobel Prize in physiology or medicine for their work on vitamin A and night blindness.

Light entering the eye encounters the inner back lining, called the **retina.** The retina consists of a layer of nerve tissue as well as millions of cells called **cones** and **rods.** The cones help us see color, whereas the rods are needed to see black and white and for night vision. Both cones and rods require vitamin A to work effectively, although scientists understand its role in the rod cells better. Figure 11.3 provides an overview of how vitamin A (as retinal) is involved in vision.

To function, the rods must contain thousands of molecules of a substance called **rhodopsin.** Rhodopsin is made of *cis*-retinal (a form of vitamin A) bonded to the protein **opsin.** When light strikes rhodopsin, the *cis*-retinal is converted to *trans*-retinal and separates from the opsin. This, in turn, causes a neural signal to be sent to the brain via the optic nerve. In this way, the light is "seen" by the brain, which can then form an image that we recognize. In addition to its role as a component of rhodopsin, vitamin A is important for maintaining the health of the outermost layer of tissue covering the front of the eye, called the **cornea.** Thus vitamin A deficiency is associated with a variety of vision problems, one of which is difficulty seeing in the dark.

***Night Blindness*** Without vitamin A, vision is impaired in low light situations. This is because *trans*-retinal must be reconverted to *cis*-retinal and then recombined with opsin to re-form rhodopsin each time light reaches the back of the eye. However, *trans*-retinal does not recycle to *cis*-retinal with 100% efficiency. Instead, some *trans*-retinal is metabolized to retinoic acid, which cannot be used to form rhodopsin. Extra retinal must therefore always be available for vision to remain optimal. If there is not enough retinal to re-form rhodopsin, night vision becomes especially difficult—a condition called **night blindness.**

Lack of vitamin A can also delay vision restoration after exposure to bright light. Some delay is normal, and you have probably experienced this when someone has taken your picture using a flash. This is because exposure to a flash of bright light results in the "bleaching" of a spot on the retina, caused by the dissociation of rhodopsin. In people with insufficient vitamin A to re-form rhodopsin, this bleached spot is "seen" for an extended period of time. People with vitamin A deficiency have an especially difficult time driving in the dark, because each time they pass a bright object (such as an oncoming car) the re-formation of rhodopsin is delayed. Thus, for a few seconds, the person cannot see another image, being temporarily blinded by light.

## Cell Differentiation, Growth, and Reproduction

Vitamin A is also needed for normal cell differentiation as well as growth and reproduction. For example, retinoic acid is needed for immature epithelial cells to become mature, functioning cells.[1] This process, called **cell differentiation,** is illustrated in Figure 11.4. The body has many different kinds of cells, each with its own structure and function. Vitamin A helps some young undifferentiated cells become *epithelial* cells instead of some other type of cell (such as a liver cell or adipose cell). For cell differentiation to occur, vitamin A is taken up into the immature cell where it moves into the nucleus and binds to specific genes on the strands of deoxyribonucleic acid (DNA). The

**CONNECTIONS** Recall that *cis* refers to a carbon–carbon double bond with hydrogens positioned on the same side, and *trans* refers to a carbon–carbon double bond with the hydrogens on opposite sides of the double bonds (Chapter 7, page 225).

**retina** (RE – ti – na) The inner lining of the back of the eye.

**cones and rods** Cells in the retina that are needed for vision.

**rhodopsin** (rhod – OPS – in) A compound, in the retina, that consists of the protein opsin and the vitamin *cis*-retinal; needed for night vision.

**opsin** (OPS – in) The protein component of rhodopsin.

**cornea** (COR – ne – a) The outermost layer of tissue covering the front of the eye.

**night blindness** A condition characterized by impaired ability to see in the dark.

**cell differentiation** The process in which an immature cell becomes a specific type of mature cell.

## FIGURE 11.3  Vitamin A and Vision

Vitamin A (retinal) is essential for vision, especially in darkness, when the function of rod cells is most important.

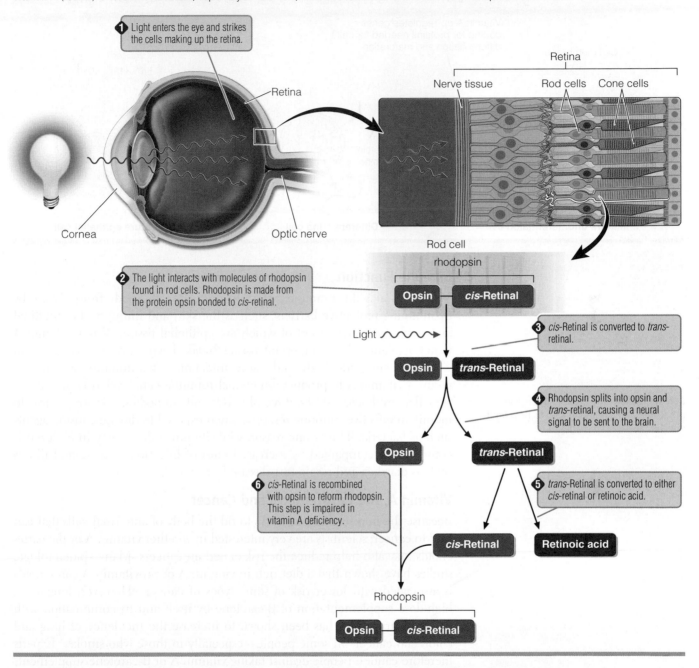

**1** Light enters the eye and strikes the cells making up the retina.

Retina

Cornea

Optic nerve

Retina

Nerve tissue    Rod cells    Cone cells

Rod cell

rhodopsin

**2** The light interacts with molecules of rhodopsin found in rod cells. Rhodopsin is made from the protein opsin bonded to *cis*-retinal.

Opsin    *cis*-Retinal

Light

**3** *cis*-Retinal is converted to *trans*-retinal.

Opsin    *trans*-Retinal

**4** Rhodopsin splits into opsin and *trans*-retinal, causing a neural signal to be sent to the brain.

Opsin    *trans*-Retinal

**6** *cis*-Retinal is recombined with opsin to reform rhodopsin. This step is impaired in vitamin A deficiency.

**5** *trans*-Retinal is converted to either *cis*-retinal or retinoic acid.

*cis*-Retinal    Retinoic acid

Rhodopsin

Opsin    *cis*-Retinal

binding of vitamin A to a gene initiates protein synthesis via cell signaling. The proteins that are produced when vitamin A binds to genes help the young cell differentiate and mature into a functional epithelial cell. Because older epithelial cells are constantly being sloughed off, we need a steady supply of vitamin A so that new cells can take their place.

Vitamin A is also important for growth and reproduction. The mechanisms by which vitamin A influences growth are numerous, including regulation of cell differentiation. For example, during embryonic growth vitamin A directs the differentiation and maturation of various cell types that give rise to specific tissues and organs. Because of this function, vitamin A is needed for successful reproduction.

**CONNECTIONS** Remember that epithelial cells make up our skin and form the lining of our organs and blood vessels (Chapter 3, page 82).

Protein synthesis is initiated when specific genes are "turned on." This step is followed by gene transcription and then gene translation (Chapter 6, page 179).

## FIGURE 11.4 Vitamin A Is Important for Cell Differentiation and Maturation

Vitamin A is involved in cell signaling, stimulating synthesis of proteins needed for cell differentiation and maturation.

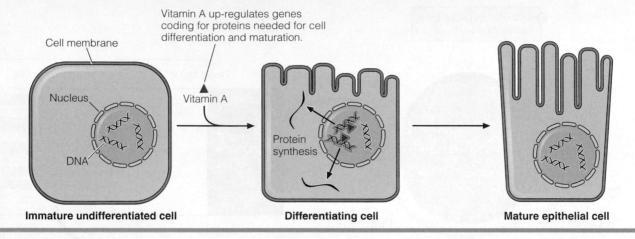

Vitamin A up-regulates genes coding for proteins needed for cell differentiation and maturation.

Cell membrane

Nucleus

Vitamin A

DNA

Protein synthesis

**Immature undifferentiated cell**

**Differentiating cell**

**Mature epithelial cell**

## Immune Function

Vitamin A aids the immune system in protecting the body from disease by maintaining protective barriers, such as the skin and lining of the intestinal and respiratory tracts, most of which are epithelial tissue.[2] When vitamin A is not available, these important tissues break down, allowing bacteria and viruses to enter the body and cause infection. The immune system also requires vitamin A to produce functional immune cells, called lymphocytes, as well as antibodies. Without lymphocytes and antibodies, it is not possible to mount an effective immune response when exposed to disease-causing agents such as bacteria. This is one reason why vitamin A deficiency in children is commonly accompanied by increased rates of infectious disease and illness such as diarrhea and respiratory disease.[3]

**CONNECTIONS** Antibodies are proteins produced by the immune cells to help fight infection (Chapter 6, page 190).

## Vitamin A, the Carotenoids, and Cancer

Because the immune system works to rid the body of abnormal cells that can lead to cancer, scientists are very interested in whether vitamin A or the carotenoids may also help reduce the risk of certain cancers. Many epidemiologic studies have shown that a diet rich in vitamin A or provitamin A carotenoids is associated with lower risk of some types of cancer.[4] However, long-term, high-dose supplementation of β-carotene by itself and in combination with other dietary factors has been shown to increase the incidence of lung and colorectal cancers in some people—especially in those who smoke.[5] Experts therefore caution people *against* taking vitamin A or β-carotene supplements to reduce the risk of cancer.

## Bone Health

Vitamin A is also required for healthy bones. Bone tissue is continually being broken down by cells called osteoclasts and reformed by cells called osteoblasts. Vitamin A appears to be required for a healthy balance of these processes—favoring bone formation over bone breakdown.

Conversely, a growing amount of research suggests that very high vitamin A intake may lead to greater risk of bone fractures in older people. This effect seems to be true for preformed vitamin A but not for the provitamin A carotenoids.[6] Although the mechanism by which this occurs is unclear, some evidence suggests that high intake of vitamin A interferes with calcium absorption.[7] Clearly, more research is needed to understand how vitamin A functions in the body to influence bone health—and, perhaps, disease.

## Functions of the Carotenoids

Scientists once thought the health-promoting effects of the carotenoids were due solely to their conversion to vitamin A. However, they no longer think this is the case. Rather, several carotenoids (such as lycopene) are directly associated with decreased risk for cancer—at least in animal models.[8] Others appear beneficial in decreasing the risk of age-related eye disease.[9] Researchers now think these effects are due to the potent antioxidant functions of some carotenoids in the body.[10] As you learned in Chapter 10, antioxidants protect DNA, proteins, and lipids from the harmful effects of free radicals. Consuming fruits and vegetables that contain antioxidant carotenoids is associated with decreased risk for many diseases such as heart disease and cancer that are thought to be caused, in part, by free radical damage. Many studies suggest that the carotenoids exert their effects by influencing the immune system and that this involves a combination of antioxidant function, regulation of gene expression, and maintenance of normal cell turnover.[11] Although there is still much to learn, scientists now believe that the carotenoids are especially important phytochemicals in the diet. The relationship between carotenoid consumption and health is described in more detail in the Focus on Food feature.

**CONNECTIONS** Remember from Chapter 10 that free radicals are reactive molecules with 1 or more unpaired electrons (Chapter 10, page 409).

**macular degeneration** (MA – cu – lar) A chronic disease that results from deterioration of the retina.

**macula** A portion of the retina important for sight.

## *Focus On* FOOD
### Fruits, Vegetables, Carotenoids, and Health

Recommending abundant amounts of fruits and vegetables in the diet has long been part of standard nutritional advice. But what about these foods makes them so good for our health? Most fruits and vegetables are excellent sources of many essential nutrients. They also contain other compounds (such as phytochemicals and fiber) that are not essential nutrients but important nonetheless. For example, you have already learned that dietary fiber is vitally important for many aspects of health. Similarly, many scientists believe phytochemicals such as the carotenoids in fruits and vegetables are important in achieving optimal health. Some carotenoids, such as β-carotene, can be converted to vitamin A and therefore provide this essential nutrient. However, other carotenoids cannot. Nonetheless, these nonprovitamin A carotenoids, which include lycopene, astaxanthin, zeaxanthin, and lutein, have potent antioxidant activities. Some provitamin A carotenoids, such

as β-carotene, have antioxidant functions as well.

The antioxidant properties of the carotenoids appear especially important in promoting a healthy heart and immune system and may help prevent some forms of cancer and eye disease. For example, animal studies show that lutein consumption decreases growth of mammary (breast) tumors,[12] and human epidemiologic studies suggest that increased lycopene consumption is related to lower risk of prostate cancer.[13] Higher circulating levels of lutein and zeaxanthin have also been related to decreased risk of age-related deterioration of the retina—a disease called **macular degeneration.**[14] Macular degeneration affects the **macula,** the highly sensitive portion of the retina needed for central vision.

Many studies show that consuming high amounts of carotenoid-containing fruits and vegetables is associated with

a reduction in many chronic diseases. However, dietary intervention trials have not shown a protective effect and, in fact, often suggest a negative effect of carotenoid supplements on health.[15] Thus scientists believe that the health-promoting effects of fruits and vegetables may be due to carotenoids working in synergy with each other and with other nutrients found in foods.[16]

In summary, growing evidence shows that consuming carotenoids can provide many health benefits. Although researchers are still studying the mechanisms, nutritionists recommend that we choose a variety of carotenoid-rich fruits and vegetables every day. These foods include carrots and sweet potatoes, which are high in β-carotene; tomatoes, pink grapefruit, apricots, and watermelon, which are excellent sources of lycopene; and dark green, leafy vegetables (such as spinach), which are good sources of lutein.

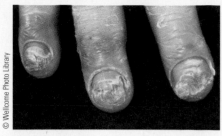

Vitamin A deficiency can cause hyperkeratosis, a condition in which skin becomes rough and scaly.

**vitamin A deficiency disorder (VADD)** A multifaceted disease resulting from vitamin A deficiency.

**xerophthalmia** (xer – o – PHTHAL – mi – a) A condition characterized by serious damage to the cornea due to vitamin A deficiency that can lead to blindness.

**Bitot's spots** A sign of vitamin A deficiency characterized by white spots on the eye; caused by buildup of dead cells and secretions.

**hyperkeratosis** (hy – per – ker – a – TO – sis) A symptom of vitamin A deficiency in which immature skin cells overproduce the protein keratin, causing rough and scaly skin.

## Vitamin A Deficiency

Primary vitamin A deficiency is uncommon in industrialized countries such as the United States. Nonetheless, secondary vitamin A deficiency can occur in people with diseases affecting the pancreas, liver, or gastrointestinal tract. An example is cystic fibrosis, which causes inadequate fat digestion and absorption. Vitamin A deficiency is also prevalent in alcoholics who have poor diets and suffer from liver damage. Excessive alcohol consumption also depletes the body's stores of vitamin A by mechanisms not well understood.

Primary vitamin A deficiency tends to be more pervasive in nonindustrialized societies, and it is often called **vitamin A deficiency disorder (VADD)**. VADD has important implications for worldwide health, especially among children. In its milder form, VADD causes night blindness, as described previously. More severe VADD damages the cornea and other portions of the eye, leading to dry eyes and scarring. This complex disease, called **xerophthalmia,** is often accompanied by the presence of **Bitot's spots,** which are accumulations of dead cells and secretions on the surface of the eye (Figure 11.5). Severe xerophthalmia can lead to blindness. VADD can also result in a condition called **hyperkeratosis,** in which skin cells produce too much of the protein keratin. This is because young, immature skin cells produce more keratin than do differentiated mature cells. An excess of keratin causes the skin to be bumpy, rough, and irritated.

Because of the importance of vitamin A in supporting a healthy immune system, people (especially children) with VADD have an increased risk of infection. Furthermore, protein energy malnutrition (PEM), which is also endemic to these parts of the world, can make vitamin A deficiency even worse. This is partly because the body cannot produce the proteins required to transport vitamin A in the circulation (such as retinol-binding protein). International programs aimed at vitamin A supplementation and food forti-

## FIGURE 11.5 Vitamin A Deficiency

Vitamin A deficiency is a major cause of blindness and xerophthalmia in many regions of the world.

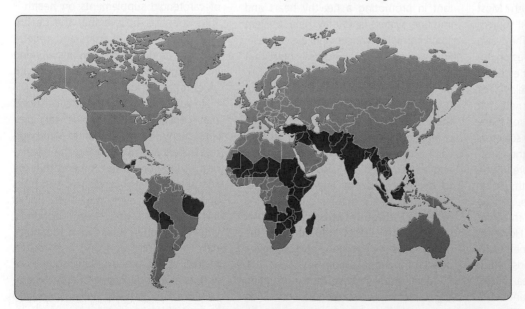

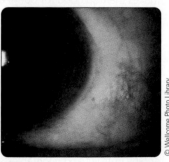

**Xerophthalmia**

**Areas of the world where blindness due to vitamin A deficiency is most common**

# *Focus On* DIET AND HEALTH
## Vitamin A and International Child Health

International efforts to decrease vitamin A deficiency have improved the health of children worldwide.

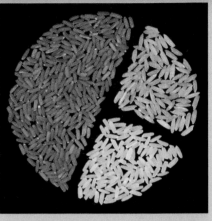

Golden Rice (upper right) and Golden Rice 2 (left) contain β-carotene, whereas plain milled rice (lower right) does not. Golden Rice and Golden Rice 2 are examples of genetically modified organisms (GMOs).

The World Health Organization estimates that between 100 and 400 million children worldwide have vitamin A deficiency disorder (VADD)—the majority being in nonindustrialized countries.[17] Health problems related to this nutrient deficiency are immensely important to overall health worldwide. Of particular concern is the effect of VADD on morbidity and mortality rates from severe illnesses, such as malaria. VADD is common in regions of the world where malaria is endemic, and vitamin A supplementation decreases both morbidity and mortality from this disease.[18] Researchers estimate that about 20% of malaria cases in children could be prevented by enhanced vitamin A status, and that doing so would prevent tens of thousands of deaths each year. Although researchers do not know precisely how vitamin A affects risk and severity of malaria, its effect on the immune system likely plays a key role. Vitamin A is important for many aspects of immunity, including the maintenance of epithelial cells such as those lining the lungs and gastrointestinal tract as well as the production of antibodies.

Current strategies to combat VADD worldwide include supplementation, food fortification, and education regarding the importance to health of vitamin A-rich foods. In addition, the World Health Organization recommends vitamin A supplementation at the earliest possible opportunity after 6 months of age in many regions of the world where VADD is common.

Another strategy being explored to reduce the incidence of VADD is the use of biotechnology or bioengineering to produce crops (genetically modified organisms, or GMOs) with higher levels of vitamin A. For example, vitamin A deficiency is common in areas of the world where people rely on rice for most of their calories. Rice is a poor source of vitamin A. In the late 1990s, however, German and Swiss researchers produced a genetically modified rice plant that could synthesize β-carotene.[19] This rice, later named Golden Rice, can provide β-carotene to the diet, whereas ordinary processed rice cannot. More recently, a strain of rice called Golden Rice 2 has been developed, and contains 20 times more β-carotene than the original Golden Rice.[20] Some people have ethical concerns about the use of GMOs such as Golden Rice. However, many scientists and public health experts hope solutions can be found to these issues and concerns so that these nutrient-dense foods can be used to prevent and treat nutritional deficiencies worldwide.[21]

fication have helped decrease the prevalence of VADD in many areas of the world, and this is described more in the Focus on Diet and Health feature.

## Vitamin A and Carotenoid Toxicity

Ingesting large doses of preformed vitamin A can lead to vitamin A toxicity, called **hypervitaminosis A.** Hypervitaminosis A can cause serious complications including blurred vision, liver abnormalities, and reduced bone strength. In addition, very high doses of either naturally occurring or synthetic vitamin A can lead to birth defects such as neurological damage and physical deformities, and high-dose carotenoid supplementation can increase risk of lung cancer in some people.

CONNECTIONS ▶ Recall that protein energy malnutrition (PEM) typically occurs in toddlers and young children and can be in the form of kwashiorkor or marasmus (Chapter 6, page 200).

**hypervitaminosis A** (hy – per – vit – a – min – O – sis) A condition in which elevated circulating vitamin A levels result in signs and symptoms of toxicity.

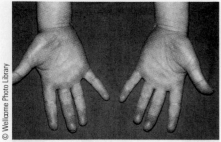

Consumption of large amounts of carotenoids can cause the skin to become orange—a condition called hypercarotenemia.

**CONNECTIONS** Dietary Reference Intakes (DRIs) are reference values for nutrient intake. When adequate information is available, Recommended Dietary Allowances (RDAs) are established, but when less information is available, Adequate Intake (AI) values are provided. Tolerable Upper Intake Levels (ULs) indicate intake levels that should not be exceeded (Chapter 2, page 36).

Consuming high amounts of carotenoids from foods and supplements can cause the skin to become yellow-orange—a condition called **hypercarotenemia.** This is because these brightly colored compounds are deposited in the skin and subcutaneous fat. Hypercarotenemia is not considered dangerous.

## Recommended Intakes for Vitamin A and the Carotenoids

The Recommended Dietary Allowances (RDAs) for vitamin A intake in men and women are 900 and 700 RAE/day, respectively. Because preformed vitamin A is found mostly in animal products, vitamin A supplementation may be necessary for vegans not consuming sufficient amounts of provitamin A carotenoids or vitamin A-fortified foods. To prevent the known toxic effects of high consumption of preformed vitamin A, a Tolerable Upper Intake level (UL) has been set at 3,000 RAE/day for adults. A complete list of the Dietary Reference Intake (DRI) values for all vitamins is provided inside the front cover of this book.

No DRIs are set for the carotenoids, because there is insufficient evidence to support their essentiality beyond their role as vitamin A precursors. Because some studies have shown that carotenoid supplements increase the risk of lung cancer in some individuals, the Institute of Medicine advises *against* supplementation for most people. However, experts recommend consuming a variety of carotenoid-containing fruits and vegetables daily.

## ESSENTIAL *Concepts*

The term *vitamin A* describes a group of compounds called retinoids. Retinoids found naturally in foods are called preformed vitamin A and include retinol, retinoic acid, and retinal. Other vitamin A–like substances found in plants and some animals are called carotenoids. Carotenoids that can be converted to vitamin A in the body are called provitamin A carotenoids. Those that cannot are called nonprovitamin A carotenoids. Vitamin A and the carotenoids have many functions, including regulation of growth, reproduction, vision, immune function, gene expression, and bone formation. The carotenoids are also potent antioxidants and protect proteins, DNA, and cell membranes from free radical damage. Good sources of preformed vitamin A are liver, fish, whole milk, and fortified foods. The carotenoids tend to be found in most brightly colored fruits and vegetables, such as cantaloupe, pumpkins, and carrots. Vitamin A deficiency can result in blindness, increased risk of infection, and even death. High intake of vitamin A can be toxic and may result in fetal malformations. High intake of the carotenoids from natural foods is not dangerous but can cause the skin to become orange. Because some studies suggest they may be harmful, consuming carotenoid supplements is not generally advised.

**CONNECTIONS** A conditionally essential nutrient is one that is nonessential in most situations but becomes essential in others (Chapter 1, page 3).

**hypercarotenemia** A condition in which carotenoids accumulate in the skin, causing it to become yellow-orange.

# Vitamin D

Vitamin D has an interesting and unique place among the nutrients. Although this vitamin is found in food, significant amounts of vitamin D can also be produced by the body. Thus many nutritional scientists consider it a conditionally essential nutrient. The 1918 discovery by Sir Edward Mellanby that a fat-soluble component of cod liver oil could strengthen bones provided the first evidence that vitamin D was an essential nutrient in some situations. Elmer V. McCullum, who discovered vitamin A, is also credited with the discovery of vitamin D.

## FIGURE 11.6    Vitamin D

Vitamin D is found in foods in two forms: ergocalciferol (vitamin $D_2$) and cholecalciferol (vitamin $D_3$).

**Vitamin D₂ (ergocalciferol)** — Form found in plant foods

**Vitamin D₃ (cholecalciferol)** — Form found in animal foods and made by the body

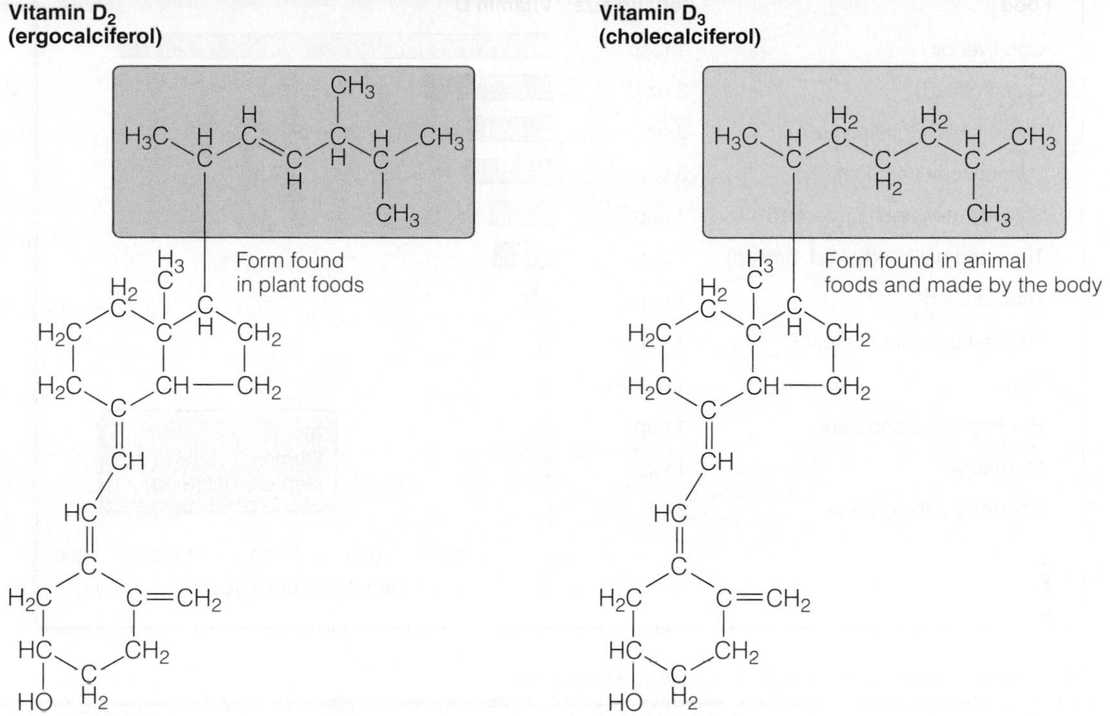

There are two forms of vitamin D in foods, and their structures are shown in Figure 11.6. **Ergocalciferol (vitamin D₂)** is found in plant sources and supplements. **Cholecalciferol (vitamin D₃)** is found in animal foods and is the form of vitamin D made by the body.

## Dietary Sources of Vitamin D

Egg yolks, butter, whole milk, fatty fish, fish oils, and mushrooms are some of the few foods that naturally contain vitamin D. However, most liquid and dried milk products as well as breakfast cereals are fortified with vitamin D, and most dietary vitamin D comes from these foods. A list of common food sources of vitamin D is provided in Figure 11.7. You can check to see whether a milk product is fortified by looking at its Nutrition Facts panel on the label. Vitamin D is quite stable and is not destroyed during food preparation, processing, and storage.

## Regulation, Metabolism, and Synthesis of Vitamin D

In the small intestine, vitamin D is packaged into micelles with the help of bile. It is then absorbed into the enterocyte, where it is incorporated into chylomicrons and circulated away from the intestine—first in the lymph and then the blood. Vitamin D not taken up by cells is delivered to the liver in chylomicron remnants. In addition to vitamin D derived from food, vitamin D can also be synthesized by the body. More specifically, vitamin D in the form of cholecalciferol (vitamin D₃) can be produced from a metabolite of cholesterol in the skin when it is exposed to ultraviolet light. For this reason, vitamin D is sometimes called "the sunshine vitamin."

**ergocalciferol** (vitamin D₂) (er – go – cal – Cl – fer – ol)  The form of vitamin D found in plant foods and vitamin-D fortified foods.

**cholecalciferol** (vitamin D₃) (cho – le – cal – Cl – fer – ol)  The form of vitamin D in animal foods and made by the human body.

## FIGURE 11.7 Good Sources of Vitamin D

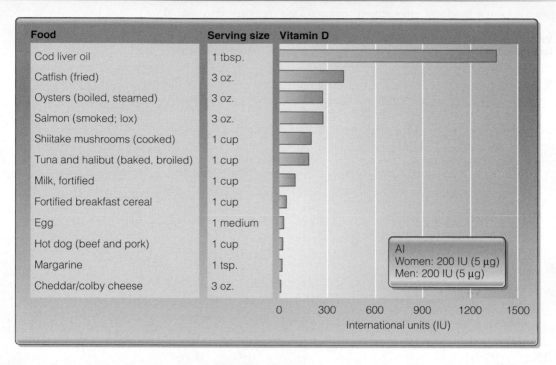

| Food | Serving size | Vitamin D |
| --- | --- | --- |
| Cod liver oil | 1 tbsp. | |
| Catfish (fried) | 3 oz. | |
| Oysters (boiled, steamed) | 3 oz. | |
| Salmon (smoked; lox) | 3 oz. | |
| Shiitake mushrooms (cooked) | 1 cup | |
| Tuna and halibut (baked, broiled) | 1 cup | |
| Milk, fortified | 1 cup | |
| Fortified breakfast cereal | 1 cup | |
| Egg | 1 medium | |
| Hot dog (beef and pork) | 1 cup | |
| Margarine | 1 tsp. | |
| Cheddar/colby cheese | 3 oz. | |

AI
Women: 200 IU (5 µg)
Men: 200 IU (5 µg)

0   300   600   900   1200   1500
International units (IU)

SOURCE: Data from Hands E. *Food Finder Vitamin and Mineral Source,* 3rd ed., Salem, OR: ESHA Research; 1995.

## Synthesis of Vitamin D

Vitamin D synthesis in the skin involves two steps, as shown in Figure 11.8. First a cholesterol metabolite called **7-dehydrocholesterol** is converted by ultraviolet light to **previtamin D₃** (also called **precalciferol**). Next, previtamin D₃ is converted to vitamin D₃ (cholecalciferol), which then diffuses into the blood and circulates to the liver. Typically, 10 to 15 minutes of sunlight

## FIGURE 11.8 Synthesis of Vitamin D in the Skin

7-Dehydrocholesterol is converted to cholecalciferol in the skin via exposure to ultraviolet (UV) rays.

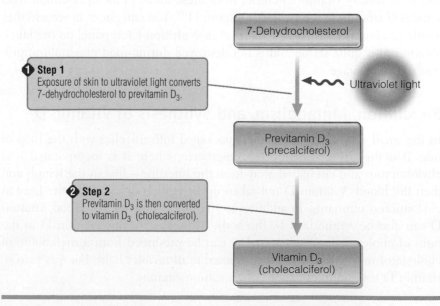

**1 Step 1**
Exposure of skin to ultraviolet light converts 7-dehydrocholesterol to previtamin D₃.

Ultraviolet light

7-Dehydrocholesterol

Previtamin D₃ (precalciferol)

**2 Step 2**
Previtamin D₃ is then converted to vitamin D₃ (cholecalciferol).

Vitamin D₃ (cholecalciferol)

**7-dehydrocholesterol** A metabolite of cholesterol that is converted to cholecalciferol (vitamin D₃) in the skin.

**previtamin D₃ (precalciferol)** An intermediate product made during the conversion of 7-dehydrocholesterol to cholecalciferol (vitamin D₃) in the skin.

three times each week is enough for the body to produce adequate amounts of cholecalciferol. Note that cholecalciferol produced in the skin is chemically the same as cholecalciferol found in animal foods.

Many environmental, genetic, and lifestyle factors can influence vitamin D synthesis. People living in areas with smog, overcast weather, or very short days may require more sun exposure to synthesize adequate vitamin D. For example, people living in Scandinavian countries (such as Sweden) experience very short periods of low-intensity sunlight during the winter and therefore produce less vitamin D during this season. Interestingly, Scandinavians are known for their high consumption of fatty fish and fish oils—both excellent natural sources of vitamin D. People living in the northern United States and Canada are also exposed to less vitamin D–producing sunlight than are those living in the south (Figure 11.9). This is especially true in the winter, when days are short and nights are long. In addition to sunlight availability, people with darker skin may need up to three times more sun exposure to produce enough vitamin D. This is because dark skin contains more melanin, a pigment that blocks synthesis of precalciferol. Sunscreen can also block the ultraviolet rays needed for vitamin D formation, and age-related changes decrease vitamin D production in the elderly.

Some, but not all, tanning machines emit the right kind of light to cause vitamin D production in the skin. However, because of the damaging effects on the skin and the increased risk for skin cancer, relying on tanning sessions to obtain enough vitamin D is not recommended. The American Academy of Dermatology and the World Health Organization both advise people to *not* use tanning beds.[22]

## Activation of Vitamin D

Regardless of whether consumed in the diet or produced in the skin, cholecalciferol must be metabolized further before the body can use it. This two-step

## FIGURE 11.9 Effect of Latitude on Sunlight Exposure

Limited exposure to sunlight results in very little vitamin D synthesis in the body.

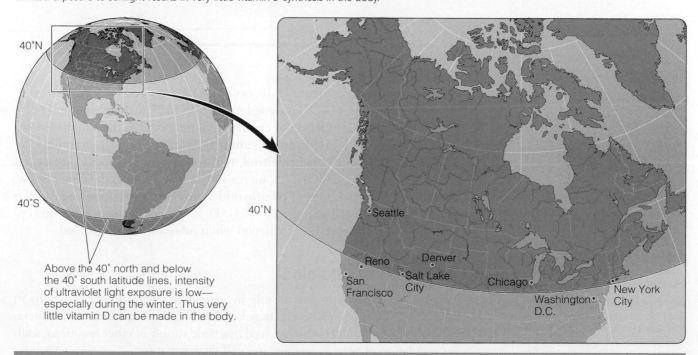

Above the 40° north and below the 40° south latitude lines, intensity of ultraviolet light exposure is low—especially during the winter. Thus very little vitamin D can be made in the body.

## FIGURE 11.10   Activation of Vitamin D

The conversion of cholecalciferol to calcitriol occurs in the liver and kidneys, and during periods of low blood calcium is stimulated by parathyroid hormone (PTH).

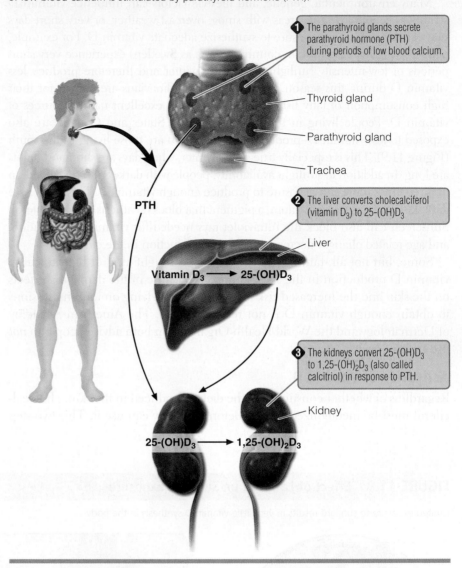

**1** The parathyroid glands secrete parathyroid hormone (PTH) during periods of low blood calcium.

Thyroid gland

Parathyroid gland

Trachea

**2** The liver converts cholecalciferol (vitamin $D_3$) to 25-(OH)$D_3$

Liver

Vitamin $D_3$ ⟶ 25-(OH)$D_3$

PTH

**3** The kidneys convert 25-(OH)$D_3$ to 1,25-(OH)$_2D_3$ (also called calcitriol) in response to PTH.

Kidney

25-(OH)$D_3$ ⟶ 1,25-(OH)$_2D_3$

process, illustrated in Figure 11.10, occurs in the liver and kidneys. First, cholecalciferol (vitamin $D_3$) is converted in the liver to **25-hydroxyvitamin D [25-(OH) $D_3$]**. Then 25-(OH) $D_3$ is circulated in the blood to the kidneys, where it is converted to **1,25-dihydroxyvitamin D [1,25-(OH)$_2$ $D_3$]**. It is the 1,25-(OH)$_2$ $D_3$ form, also called **calcitriol**, that is active in the body. Conversion of 25-(OH) $D_3$ to 1,25-(OH)$_2$ $D_3$ increases when calcium in the blood is low, stimulated by the actions of **parathyroid hormone (PTH)** produced in the parathyroid glands. Because 1,25-(OH)$_2$ $D_3$ is important for calcium absorption, the body increases its production when more calcium is needed.

## Functions of Vitamin D

Vitamin D plays an important role in regulating calcium concentrations in the blood. This requires several organs, including the small intestine, kidneys, and bone. Vitamin D is also involved in a wide variety of other functions, such as regulation of gene expression and cell differentiation.

**25-hydroxyvitamin D [25-(OH) $D_3$]** An inactive form of vitamin D that is made from cholecalciferol in the liver.

**1,25-dihydroxyvitamin D [1,25-(OH)$_2$ $D_3$], (calcitriol)** The active form of vitamin D in the body produced in the kidneys from 25-(OH) $D_3$.

**parathyroid hormone (PTH)** A hormone produced in the parathyroid glands that stimulates the conversion of 25-(OH) $D_3$ to 1,25-(OH)$_2$ $D_3$ in the kidneys.

## Calcium Homeostasis

Vitamin D [as 1,25-(OH)$_2$ D$_3$ or calcitriol] helps maintain healthy levels of calcium in the blood, assuring that calcium is always available to the body's tissues. Vitamin D increases calcium absorption in the small intestine, decreases calcium excretion in the urine, and facilitates the release of calcium from bones. This coordinated response is illustrated in Figure 11.11 and described next.

- *Small intestine*: Calcitriol is required for calcium absorption in the small intestine. More specifically, it up-regulates several genes that code for proteins required for the transport of dietary calcium into the enterocytes. In other words, vitamin D is involved in cell signaling. Without vitamin D, these proteins are not made and calcium absorption is severely limited.
- *Kidney*: Calcitriol and PTH cause the kidneys to reduce their excretion of calcium into the urine. As a result, more calcium remains in the blood.
- *Bone*: Calcitriol and PTH together stimulate bone breakdown by osteoclasts and the release of calcium into the blood.

Whereas calcium in bones is important for their structure, calcium in the blood has additional physiological functions. For example, it is needed for muscle contractions, blood pressure regulation, and the conduction of neural impulses. Without vitamin D to help maintain adequate levels of calcium in the blood, these vital functions would be impaired. Because of this close relationship between vitamin D and calcium in the body, the U.S. Food and Drug Administration (FDA) encourages vitamin D fortification of milk.

## FIGURE 11.11   Vitamin D and Calcium Homeostasis

Calcitriol helps maintain adequate calcium availability by increasing calcium absorption in the small intestine, decreasing excretion by the kidneys, and increasing resorption in bones.

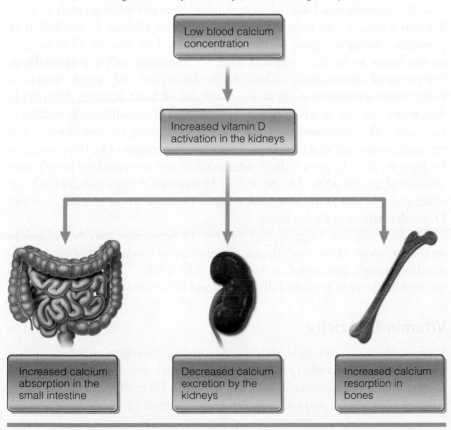

Low blood calcium concentration

Increased vitamin D activation in the kidneys

Increased calcium absorption in the small intestine

Decreased calcium excretion by the kidneys

Increased calcium resorption in bones

### Cell Differentiation and Cancer

Vitamin D also plays a critical role in stimulating immature cells to become mature, functioning cells—in other words, cell differentiation. As with vitamin A, this involves the movement of vitamin D into the nucleus of the cell and the subsequent stimulation of the genes coding for specific proteins. For example, vitamin D causes immature bone cells to become mature bone marrow cells and causes certain intestinal epithelial cells to differentiate into mature enterocytes. In this way, vitamin D plays a role in bone health and gastrointestinal function. Some evidence also suggests that vitamin D may help prevent certain types of cancers such as those of the colon, breast, skin, and prostate.[23] Although this connection warrants further investigation, vitamin D's role in cell differentiation may be involved in this relationship.

## Vitamin D Deficiency—Rickets and Osteomalacia

Vitamin D deficiency can cause inadequate mineralization and increased demineralization of bone. Vitamin D deficiency in infants and children who are in active stages of growth can result in inadequate bone mineralization—a disease called **rickets.** Although fortifying food with vitamin D has essentially eliminated rickets in the United States, an alarming number of cases have been reported during the last decade. This is described in more detail in the Focus on Life Cycle Nutrition feature. Rickets is also a significant public health concern in other parts of the world.[24] Children with rickets have slow growth and characteristically bowed legs or knocked knees caused by the bending of weak long bones that cannot support the stress of weight-bearing activities, such as walking.

In adults, vitamin D deficiency results in both poor bone mineralization and increased bone demineralization. The risk of vitamin D deficiency in adults is highest in people with dark skin, those who live in areas with little sunlight, cultures that habitually cover themselves in clothing (such as some Muslim women), and those who do not consume vitamin D–fortified milk products. Inadequate bone mineralization caused by vitamin D deficiency causes bones to become soft and weak—a condition called **osteomalacia.** Symptoms of osteomalacia include diffuse bone pain and muscle weakness. People with osteomalacia are at increased risk of bone fracture. Vitamin D deficiency can also result in demineralization of bone, ultimately leading to a disease called **osteoporosis.** Osteoporosis is a serious chronic disease, and researchers estimate that more than 28 million Americans (1 in 10 people) suffer from it. To help prevent both osteomalacia and osteoporosis, people over 50 years of age are advised to get at least 15 minutes of sun exposure each day when possible and to increase their vitamin D intake. In some cases, vitamin D supplements may be necessary.

Limited evidence suggests that vitamin D deficiency may be related to increased risk of Alzheimer's disease, a debilitating condition that causes personality changes and cognitive decline in older adults.[25] However, additional research is required to more fully understand this relationship.

## Vitamin D Toxicity

Vitamin D toxicity from food sources is uncommon. However, supplementation with high doses of vitamin D can cause calcium levels in the blood to rise, a condition called hypercalcemia. This is due to vitamin D's multiple roles in calcium absorption, excretion, and mobilization. Hypercalcemia can result in the deposit

Vitamin D deficiency causes rickets in infants and children.

© Wellcome Photo Library

**rickets** A symptom of vitamin D deficiency in young children characterized by deformed bones, especially in the legs.

**osteomalacia** (os – te – o – ma – LA – ci – a) Softening of the bones in adults that can be due to vitamin D deficiency.

**osteoporosis** (os – te – o – por – O – sis) A serious bone disease resulting in weak, porous bones.

## Focus On LIFE CYCLE NUTRITION
### Vitamin D Supplementation in Infancy

Rickets caused by vitamin D deficiency was commonly seen in the United States during the early 1900s.[26] However, the use of vitamin D–fortified milk and infant formulas nearly eradicated rickets in this country. That is, until recently.[27] In 2000 a research group at the Bowman Gray School of Medicine in North Carolina published a disturbing report documenting that the incidence of rickets was on the rise, especially in dark-skinned, breastfed babies. Not only did these babies have the characteristic bone deformities of rickets, but nearly a third of them exhibited poor growth. Public health officials were quick to investigate, to deter-mine what could be done to prevent additional cases of rickets from developing.

Evidence suggests that some infants nourished solely on their mothers' milk are at increased risk of inadequate vitamin D intake. This may be due to a lack of sun exposure in the mothers, resulting in low amounts of vitamin D in their milk. In addition, insufficient sun exposure of infants may contribute to the rise in rickets, especially those with dark skin. However, because of concerns that increased sunlight exposure may lead to skin cancer, clinicians do not recommend that parents expose their infants to additional sunlight to prevent rickets.

Instead, the American Academy of Pediatrics recommends that all breastfed infants and nonbreastfed infants not consuming at least half a liter (about 2 cups) of vitamin D–fortified infant formula daily be given vitamin D supplements (5 μg/day).[28] Supplementation should begin within the first two months of life and continue until the child is consuming sufficient vitamin D from the diet. This supplementation is especially recommended for populations at increased risk for developing rickets, such as those with increased skin pigmentation and/or little exposure to sunlight.

of calcium in soft tissues such as the heart and lungs, and can affect the function of the central nervous system. Vitamin D toxicity also promotes bone loss. Consuming very high levels of supplementary vitamin D can result in death.

## Recommended Intakes for Vitamin D

Because there are insufficient data to establish RDAs for vitamin D, AIs have been set. For some people, vitamin D is considered a conditionally essential nutrient. Thus the AI values may not apply to segments of the population with adequate sunlight exposure. Nonetheless, it is recommended that adults consume 5 μg of vitamin D each day. This is the amount in about 1 liter (4 cups) of vitamin D-fortified milk. To help decrease bone loss in the elderly, the AI increases to 10 and then 15 μg per day at 51 and 71 years of age, respectively. The Institute of Medicine has established a UL of 50 μg for vitamin D in supplemental form.

## ESSENTIAL Concepts

Vitamin D is found in fatty fish, and to a lesser extent in some plant foods. In the presence of sunlight, the body can also produce vitamin D from a cholesterol metabolite. Because of this, vitamin D is often considered a conditionally essential nutrient. The active form of vitamin D is 1,25-dihydroxyvitamin D [1,25-$(OH)_2$ $D_3$], also called calcitriol. Calcitriol is involved in maintaining calcium homeostasis by stimulating calcium absorption in the intestine, decreasing calcium loss in the urine, and increasing breakdown of bone cells for release of calcium into the blood. Vitamin D also influences the regulation of genes that are important for cell differentiation and maturation. Vitamin D deficiency causes rickets in children and osteomalacia and osteoporosis in adults, all of which result in weakened bones. Vitamin D toxicity from foods is rare, but high intake from supplements can be dangerous.

Nuts and seeds are good sources of vitamin E.

# Vitamin E

The term *vitamin E* refers to eight different compounds that all have somewhat similar chemical structures. Of these, **α-tocopherol** is the most biologically active. The essentiality of vitamin E was first recognized in 1922 when it was discovered that vitamin E–deficient animals could not reproduce. In fact, the name *tocopherol* was derived from the Greek *tokos* (childbirth) and *phero* (to bear). However, vitamin E is now known to have many other functions. For example, its potential for decreasing risk of chronic diseases such as heart disease has attracted much interest.

## Dietary Sources of Vitamin E

Although vitamin E is found in both plant and animal foods, it is especially abundant in vegetable oils, nuts, and seeds. Some dark green vegetables such as broccoli and spinach contain vitamin E as well. Figure 11.12 lists food sources of vitamin E. Vitamin E can be easily destroyed during food preparation, processing, and storage. Thus, exposing vitamin E-containing foods to as little heat as is possible during cooking can help protect this vitamin. Proper storage of foods in air-tight containers is also important.

## Regulation of Vitamin E in the Body

Absorption of vitamin E occurs in the small intestine and requires the presence of bile and the synthesis of micelles. Vitamin E is circulated in chylomicrons via the lymph and, in the blood, eventually reaches the liver. In the liver,

**α-tocopherol** (to – CO – pher – ol) The most active form of vitamin E.

## FIGURE 11.12   Good Sources of Vitamin E

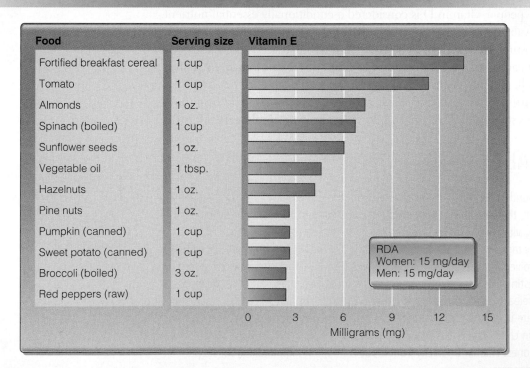

| Food | Serving size | Vitamin E |
|------|-------------|-----------|
| Fortified breakfast cereal | 1 cup | |
| Tomato | 1 cup | |
| Almonds | 1 oz. | |
| Spinach (boiled) | 1 cup | |
| Sunflower seeds | 1 oz. | |
| Vegetable oil | 1 tbsp. | |
| Hazelnuts | 1 oz. | |
| Pine nuts | 1 oz. | |
| Pumpkin (canned) | 1 cup | |
| Sweet potato (canned) | 1 cup | |
| Broccoli (boiled) | 3 oz. | |
| Red peppers (raw) | 1 cup | |

RDA
Women: 15 mg/day
Men: 15 mg/day

Milligrams (mg)

SOURCE: Data from USDA Nutrient Database, Release 16-1.

vitamin E is repackaged into VLDLs for further delivery in the body. Excess vitamin E is stored mainly in adipose tissue.

## Functions of Vitamin E

Like the carotenoids, vitamin E acts as an antioxidant preventing oxidation and free radical damage. Specifically, vitamin E typically carries out its function within membranes. Vitamin E may also protect the cell's genetic material (DNA). These functions are described in more detail next.

### Protecting Cell Membranes

As illustrated in Figure 11.13, much of the body's vitamin E is associated with various membranes. Recall that cell membranes consist of a phospholipid bilayer. In addition, many cell organelles, such as mitochondria and endoplasmic reticula, are encased in phospholipid bilayer membranes. Maintaining these membranes is vital to the stability and function of cells and their organelles, and vitamin E plays a major role. Specifically, it protects the fatty acids in the membrane's phospholipid bilayers from free radical–induced, oxidative damage. This occurs because vitamin E can donate electrons to free radicals, making them more stable. This protection is especially important in cells that are exposed to oxygen, such as those in the lungs and red blood cells. The ability of vitamin E to act as an antioxidant is enhanced in the presence of other antioxidant micronutrients, such as vitamin C and selenium.

**CONNECTIONS** Remember that phospholipids are amphipathic lipids containing both polar (phosphate group) and nonpolar (fatty acids) components (Chapter 7, page 234).

### Relationship with Cancer

Because antioxidant nutrients protect DNA from cancer-causing free radical damage, people are very interested in the possibility that vitamin E might prevent or cure cancer. However, although diets high in vitamin E are associated with decreased cancer risk, there is little experimental evidence that vitamin

## FIGURE 11.13   Vitamin E Protects Cell Membranes

Vitamin E protects cell membranes by donating electrons, thereby reducing free radicals that damage the fatty acids in phospholipid bilayers.

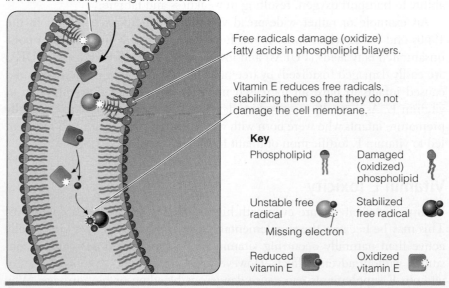

Free radicals have unpaired electrons in their outer shells, making them unstable.

Free radicals damage (oxidize) fatty acids in phospholipid bilayers.

Vitamin E reduces free radicals, stabilizing them so that they do not damage the cell membrane.

**Key**

| Phospholipid | Damaged (oxidized) phospholipid |
| Unstable free radical | Stabilized free radical |
| Missing electron | |
| Reduced vitamin E | Oxidized vitamin E |

E by itself decreases the risk of this disease.[29] Vitamin E likely interacts with other dietary components to protect the body from cancer.

### Eye Health

**Cataracts** are cloudy growths that develop on the lens of the eye, causing vision to become hazy or cloudy. The development of cataracts increases as people age, leading to impaired vision. If left untreated, cataracts can lead to blindness. Because of increased free radical damage, people who smoke and those who are exposed to excessive amounts of sunlight are at increased risk for developing cataracts. Dietary antioxidants may prevent or delay the growth of cataracts, and epidemiologic studies have shown that older people who take vitamin E supplements are at lower risk for developing them.[30] However, researchers continue to study the relationships among vitamin E, other antioxidants, and this chronic disease.

### Heart Health

Atherosclerosis, sometimes called "hardening of the arteries," can lead to heart disease or stroke from the accumulation of fatty material called plaque in the arteries. Many animal studies have shown that vitamin E slows the rate of plaque formation, and increased dietary vitamin E intake appears to be heart healthy.[31] However, recent studies suggest that vitamin E supplementation (400 mg/day) can cause an overall increased risk of mortality in people with chronic disease.[32] For this reason, although dietary vitamin E may be important in maintaining a healthy heart, taking vitamin E supplements to decrease risk of cardiovascular disease is discouraged.

## Vitamin E Deficiency

Vitamin E deficiency is uncommon, and cases have only been reported in infants fed formulas with inadequate vitamin E, people with genetic abnormalities, and in diseases causing fat malabsorption. Vitamin E deficiency is characterized by a variety of symptoms, including neuromuscular problems, loss of coordination, and muscular pain. Vitamin E deficiency also causes red blood cell membranes to weaken and rupture, a condition called **hemolytic anemia.** This is because vitamin E is especially important in protecting red blood cells from oxidative damage. Hemolytic anemia reduces the blood's ability to transport oxygen, resulting in weakness and fatigue.

An example of rather widespread vitamin E deficiency occurred in the 1960s and 1970s when some infant formulas contained high levels of polyunsaturated fatty acids (PUFAs) and low levels of vitamin E. Because PUFAs are easily damaged (oxidized) by free radicals, consumption of these formulas caused infants to have an increased need for antioxidant nutrients such as vitamin E. As a result, some babies developed hemolytic anemia, especially premature infants who were born with very low stores of vitamin E. This event led to vitamin E fortification of infant formulas.

## Vitamin E Toxicity

Vitamin E toxicity is rare even with high intakes of vitamin E supplements. This may be because the supplemental form of vitamin E is less biologically active than naturally occurring vitamin E, so that large doses can be consumed without adverse effects. However, in some people, very high doses of vitamin E supplements can cause dangerous bleeding or hemorrhage. Why

**CONNECTIONS** Recall that polyunsaturated fatty acids (PUFAs) are those that have more than one double bond between the carbons making up the fatty acid backbone (Chapter 7, page 223).

**cataract** (CAT – a – ract) A cloudy growth that develops on the lens of the eye, causing impaired vision.

**hemolytic anemia** Decreased ability of the blood to carry oxygen and carbon dioxide due to rupturing of red blood cells.

some people respond this way and others do not is not known but probably involves genetic differences. Nonetheless, caution is advised when taking vitamin E supplements.

## Recommended Intakes for Vitamin E

The RDA for vitamin E is 15 mg/day. ULs have also been established and apply to any form of supplemental vitamin E, fortified food, or a combination of both. It is recommended that total consumption of vitamin E not exceed 1,000 mg/day.

## ESSENTIAL *Concepts*

Vitamin E includes eight different compounds, although α-tocopherol is the most biologically active form. Vitamin E is found in oils, nuts, seeds, and some fruits and vegetables. Vitamin E functions mainly as an antioxidant, protecting cell membranes from free radical damage. Vitamin E may also protect the eyes from cataract formation and influence cancer risk by decreasing DNA damage. Vitamin E deficiency is rare but results in neuromuscular problems and hemolytic anemia. Vitamin E toxicity is also rare. When it does occur, it results in dangerous symptoms such as bleeding.

# Vitamin K

Vitamin K was discovered and named for its role in *koagulation* ("coagulation" in Danish) by Henrik Dam, a Danish physiologist who found that vitamin K deficiency in chickens caused excessive bleeding. Dam received a Nobel Prize in physiology or medicine in 1943 for this discovery.

The term *vitamin K* refers to three compounds that have similar structures and functions. Vitamin K found naturally in plant foods is called **phylloquinone** (vitamin $K_1$). Phylloquinone provides most of the vitamin K that we consume from foods; it is also found in some vitamin K supplements. **Menaquinone** (vitamin $K_2$) is produced by bacteria present in the large intestine. Because we cannot get enough vitamin K from this bacterial production, vitamin K is an essential nutrient. A third form of vitamin K called **menadione** (vitamin $K_3$) is not found naturally in food or made by bacteria but is produced commercially.

Many dark green vegetables, such as asparagus and broccoli, provide vitamin K to the diet.

## Dietary Sources of Vitamin K

In general, dark green vegetables such as kale, spinach, broccoli, and brussels sprouts are good sources of dietary vitamin K. Fish and legumes also provide vitamin K, and good food sources of this vitamin are listed in Figure 11.14. Excessive exposure of foods to light and heat can destroy vitamin K.

## Regulation of Vitamin K in the Body

Dietary vitamin K is absorbed, along with other fat-soluble vitamins in the small intestine, via micelles. Vitamin K is then incorporated into chylomicrons and put into lymph, eventually entering the blood. Vitamin K produced by bacteria in the large intestine is transported into epithelial cells by simple diffusion and then circulated to the liver in the blood. The liver packages

**phylloquinone** (phyll – O – quin – one) A form of vitamin K found in foods.

**menaquinone** (men – A – quin – one) A form of vitamin K produced by bacteria.

**menadione** (men – A – di – one) A form of vitamin K produced commercially.

**FIGURE 11.14   Good Sources of Vitamin K**

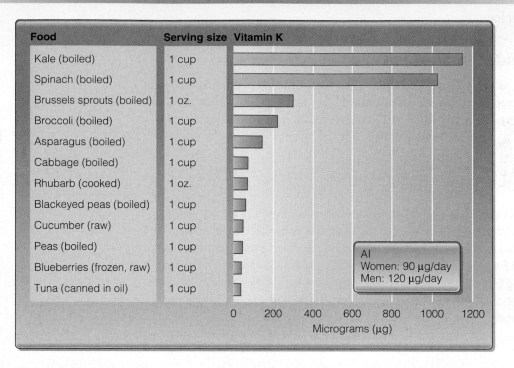

SOURCE: Data from USDA Nutrient Database, Release 16-1.

**CONNECTIONS** Recall that carboxylation reactions add a bicarbonate subunit (HCO₃) to a molecule (Chapter 10, page 399).

**coagulation** (co –ag – u – LA – tion) The process by which blood clots are formed.

**prothrombin** A clotting factor (protein) that is converted to the enzyme thrombin.

**thrombin** The enzyme that catalyzes the conversion of fibrinogen to fibrin.

**fibrinogen** (fi – BRIN – o – gen) A water-soluble protein that is converted to the water-insoluble protein fibrin.

**fibrin** An insoluble protein that forms blood clots.

both dietary and bacterially produced forms of vitamin K into lipoproteins for delivery to the rest of the body.

## Functions of Vitamin K

Vitamin K functions as a coenzyme in a variety of carboxylation reactions that add calcium to molecules. These vitamin K-dependent reactions are needed for the life-and-death process of blood clotting, which is illustrated in Figure 11.15. When a blood vessel is injured, a blood clot forms to stop the bleeding. Without this process, called **coagulation,** we might bleed to death even after a minor scrape.

For blood to coagulate and form a clot, many chemical reactions must take place. In each set of reactions, vitamin K carboxylates inactive clotting factors, which in turn bind calcium. Binding to calcium activates the clotting factors and allows the next reaction in the cascade to take place. These reactions ultimately convert the protein **prothrombin** to **thrombin,** another calcium-containing protein. Thrombin, in turn, catalyzes the conversion of **fibrinogen** to **fibrin.** Fibrin is a water-insoluble protein that forms a weblike clot to stop the bleeding. Without adequate vitamin K, this cascade of events shuts down, and clots cannot form. A drug called Coumadin® delays blood clot formation by decreasing the activity of vitamin K. This is described in more detail in the Focus on Clinical Applications feature.

Vitamin K also catalyzes the carboxylation of other proteins needed for bone and tooth formation. Only after they have been carboxylated can these proteins bind calcium. Some studies show that consuming foods high in vitamin K is associated with decreased risk for hip fracture.[33] However, further studies are needed to determine whether increased vitamin K intake results in increased bone strength.

## FIGURE 11.15    Vitamin K and Blood Clotting

Vitamin K acts as a coenzyme in carboxylation reactions involved in the formation of blood clots.

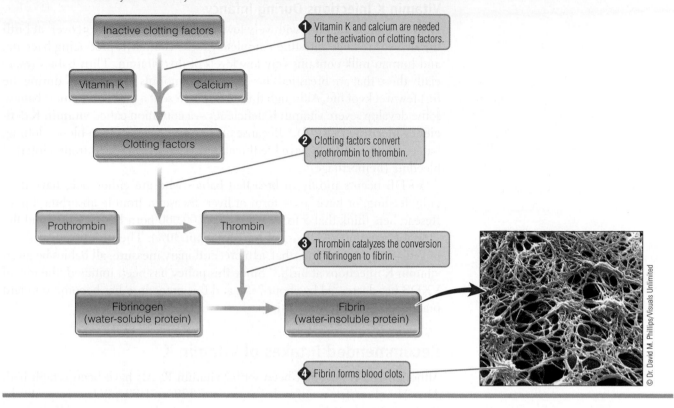

1. Vitamin K and calcium are needed for the activation of clotting factors.

2. Clotting factors convert prothrombin to thrombin.

3. Thrombin catalyzes the conversion of fibrinogen to fibrin.

4. Fibrin forms blood clots.

Inactive clotting factors → Vitamin K / Calcium → Clotting factors → Prothrombin → Thrombin → Fibrinogen (water-soluble protein) → Fibrin (water-insoluble protein)

© Dr. David M. Phillips/Visuals Unlimited

## Vitamin K Deficiency and Toxicity

Although rare in healthy adults, vitamin K deficiency appears in some infants and people with diseases that cause lipid malabsorption. In addition, prolonged use of antibiotics can kill the bacteria that normally live in the large intestine, resulting in vitamin K deficiency. The main sign of vitamin

## *Focus On* CLINICAL APPLICATIONS
### Nutrient–Drug Interactions and Vitamin K

In many situations the use of a medication can influence the metabolism or function of a nutrient. This is especially the case for vitamin K. For example, heavy long-term use of antibiotics can kill bacteria in the large intestine that normally produce vitamin K, increasing risk for vitamin K deficiency. Another example of a nutrient–drug interaction occurs with Coumadin (the generic name is warfarin), a drug often prescribed to prevent blood clots in people with certain conditions such as

cardiovascular disease. Coumadin works by decreasing the activity of vitamin K, especially in its role in blood clot formation. This is why Coumadin is often said to "thin the blood." Because blood clots can block the flow of blood to the heart or brain, the use of Coumadin can be life-saving. However, people using this drug must take special care to prevent injury, because taking Coumadin can cause serious hemorrhaging and blood loss even with a seemingly minor cut or bruise.

Because a sudden increase or decrease in vitamin K intake can interfere with the effectiveness of Coumadin, it is important to keep vitamin K intake constant. In general, clinicians advise people taking Coumadin to eat a balanced diet and to limit, to one serving each day, consumption of foods that contain very high amounts of vitamin K. These include many dark green vegetables such as kale, spinach, and broccoli.

K deficiency is excessive bleeding. Very high vitamin K intake has no known toxic effects.

### Vitamin K Injections During Infancy

Healthy infants are born with very low stores of vitamin K. Moreover, at birth the newborn's large intestine completely lacks vitamin K-producing bacteria, and human milk contains very low levels of this vitamin. Thus babies (especially those that are breastfed) have minimal amounts of vitamin K during the first few weeks of life. Although this does not present a problem to most babies, some develop severe vitamin K deficiency—a condition called **vitamin K deficiency bleeding (VKDB).*** Because of vitamin K's vital role in blood clotting, babies with VKDB can have life-threatening episodes of uncontrolled internal bleeding (hemorrhage).

VKDB occurs mostly in breastfed babies who are either sick, have difficulty feeding, or have some form of liver disease or trouble absorbing lipids. Researchers think that 5 to 10 babies per 100,000 born have VKDB, and the mortality rate for VKDB is quite high (about 30%). The American Academy of Pediatrics recommends that, as a precautionary measure, all babies be given vitamin K injections at birth.[34] Since this policy has been initiated, the rate of VKDB has decreased by about 75%, and this procedure has become standard practice in most clinical settings.

## Recommended Intakes of Vitamin K

Although RDAs have not been set for vitamin K, AIs have been established. The AI values for men and women are 120 and 90 µg/day, respectively. Because vitamin K is rarely toxic even in very high amounts, no UL is established for this vitamin.

## ESSENTIAL *Concepts*

The term *vitamin K* refers to three related compounds: phylloquinone, menaquinone, and menadione. Dark green vegetables are often good sources of vitamin K, and light and heat can destroy this vitamin in foods. In addition, intestinal bacteria produce some vitamin K, although not in quantities that meet our needs. Vitamin K is involved in calcium-requiring reactions that are needed for blood clotting and proper bone mineralization. Vitamin K deficiency causes bleeding and, perhaps, weak bones. In infants, severe vitamin K deficiency can cause a fatal condition called vitamin K deficiency bleeding, which is largely prevented by vitamin K injections.

## Summary and Overall Recommendations

The fat-soluble vitamins are involved in a wide variety of processes encompassing all the physiological systems in the body. In addition, the carotenoids may impart additional health benefits. In this chapter, you have learned about the many functions of each of these compounds. However, it is important to

**vitamin K deficiency bleeding (VKDB)** A disease that occurs in newborn infants; characterized by uncontrollable internal bleeding (hemorrhage) from inadequate vitamin K.

---

* Vitamin K deficiency bleeding (VKDB) was previously called "hemorrhagic disease of the newborn."

**TABLE 11.3    General Functions of the Fat-Soluble Vitamins and Carotenoids**

| | Vitamin A | Carotenoids[a] | Vitamin D | Vitamin E | Vitamin K |
|---|---|---|---|---|---|
| **Functions** | | | | | |
| Regulation of gene transcription | X | X | X | | |
| Immune function | X | X | | | |
| Antioxidant function | | X | | X | |
| Bone/tooth health | X | | X | | X |
| Blood health | X | | X | X | X |
| Eye health | X | X | | X | |
| Heart health | | X | | X | |

[a]Note that these are proposed functions of the carotenoids.

remember that they do not work independently of each other. This is similar to many of the B vitamins, which coordinately regulate energy metabolism in the body. For example, vitamins A, D, and E, as well as the carotenoids are involved in regulating gene transcription. Vitamin E and the carotenoids act as antioxidants, whereas vitamins A, D, and K contribute to bone health. The general overlapping functions of the fat-soluble vitamins and carotenoids are summarized in Table 11.3.

Like the water-soluble vitamins, the fat-soluble ones are found in a wide variety of foods—including those from both plant and animal sources—as shown in Table 11.4. Because of the overlapping distribution of these nutrients, it is important to consume a balance of food from all the food groups and a variety of foods from within each group. You can also help assure adequate intakes of these vitamins by following the USDA Dietary Guidelines and MyPyramid food guidance system. Tips on how you can incorporate more fruits and vegetables into your diet have been provided in the Food Matters feature. As with all the essential nutrients, choosing a diet high in foods that are nutrient dense and emphasizing moderation, variety, and balance can go a long way to help maintain optimal health in terms of the fat-soluble vitamins.

**TABLE 11.4    Major Food Sources of the Fat-Soluble Vitamins and Carotenoids**

| | Vegetables | Fruits | Milk | Organ Meats | Fish, Eggs | Nuts, Seeds |
|---|---|---|---|---|---|---|
| **Compounds** | | | | | | |
| Vitamin A (preformed) | | | X[a] | X | X | |
| Carotenoids | X | X | | | X | |
| Vitamin D | | | X[a] | | X | |
| Vitamin E | X | | | | | X |
| Vitamin K | X | | | | X | |

[a]Whole-milk and fortified milk are better sources than their reduced-fat/unfortified counterparts.

# *Food Matters*

*Working Toward the Goal*

The USDA Dietary Guidelines recommend that we choose a variety of fruits and vegetables each day, selecting from all five vegetable subgroups (dark green, orange, legumes, starchy vegetables, and other vegetables). The following selection and preparation tips will help you meet your goal.

- When grocery shopping, buy one fruit and one vegetable that you do not normally eat. This way, you will be more likely to get all the important vitamins, minerals, and phytochemicals needed by your body.
- Instead of high-fat snacks, choose dried fruits, banana chips, or vegetable chips. These snacks are nutrient dense, providing many vitamins and phytochemicals.
- Try to eat at least one salad each day, and vary what you put into it. For example, alternate between lettuce salads and spinach salads. And try not to use the same salad toppings each time. Add mushrooms and cheese for vitamin D and a spoonful of sunflower seeds for a vitamin E boost.
- Think about color when you plan a meal. Because many of the fat-soluble vitamins as well as the carotenoids tend to be found in brightly colored fruits and vegetables, extending the color palette on your plate will help you get a variety of nutrients and phytochemicals.

## Review Questions  Answers are found in Appendix G.

1. Which form of vitamin A is the most potent?
   a. Retinol
   b. Retinoic acid
   c. Retinal
   d. Lycopene
2. Preformed vitamin A is typically found in what kinds of foods?
   a. Carrots
   b. Green leafy vegetables
   c. Animal products
   d. Dried beans (legumes)
3. In general, fat-soluble vitamins are absorbed in the _____ and circulated away from the gastrointestinal tract in the _____.
   a. large intestine    blood
   b. large intestine    lymph
   c. small intestine    blood
   d. small intestine    lymph
4. Which of these vitamins is important for bone health?
   a. Vitamin A
   b. Vitamin D
   c. Vitamin K
   d. All of the above
5. Parathyroid hormone stimulates the conversion of 25-(OH) $D_3$ to 1,25-(OH)$_2$ $D_3$ when
   a. calcium levels in the blood are low.
   b. calcium levels in the blood are high.
   c. a person is consuming excessive amounts of calcium.
   d. a person is injured and bleeding.
6. Which of the following describes a major function of vitamin E?
   a. Regulation of energy metabolism during periods of fasting
   b. Carboxylation reactions
   c. Blood clotting
   d. Protection of cell and organelle membranes from free radical damage

7. Rickets is caused by a deficiency of which of the following?
   a. Vitamin A
   b. Vitamin D
   c. Vitamin E
   d. Vitamin K

8. The functions of vitamin K depend largely on which of the following minerals?
   a. Selenium
   b. Iron
   c. Calcium
   d. Magnesium

9. For which fat-soluble vitamin has the Institute of Medicine *not* established a UL?
   a. Vitamin A
   b. Vitamin D
   c. Vitamin E
   d. Vitamin K

10. Vitamin E deficiency can cause which of the following?
    a. Neuromuscular problems
    b. Loss of coordination
    c. Hemolytic anemia
    d. All of the above

11. Name two good food sources of each fat-soluble vitamin and whether food processing (such as heating and exposure to air) can influence the fat-soluble vitamin content of the food.

12. Several vitamins—both water-soluble and fat-soluble—participate in antioxidant functions in the body. Which vitamins are these, and how do they act as antioxidants? Describe, in general, how these compounds are thought to be important for cancer prevention.

13. Why is vitamin D often called the "sunshine vitamin?" Outline how vitamin D is synthesized in the skin and how it is converted to calcitriol in the body.

## Practice Calculations    Answers are found in Appendix G.

1. Using the food composition table that accompanies this text or other sources, calculate the vitamin A content of a lunch containing a hamburger on a white bun, a small order of french fries, and a large cola soft drink. What percentage of your Recommended Dietary Allowance of vitamin A does this represent? What foods might be added to or substituted for the food in this lunch to increase the vitamin A content?

2. Using the MyPyramid food guidance system (http://www.mypyramid.gov), determine the amount of foods from the meats and beans and dairy groups that are recommended for you. Then make up a sample menu that would satisfy these recommendations, and calculate the amount of vitamin D you would get if you consumed these amounts of foods.

## Media Links    A variety of study tools for this chapter are available at our website www.prenhall.pearsonedu.com/nutrition/mcguire.

Prepare for tests and deepen your understanding of chapter concepts with these online materials:

- Practice tests
- Flashcards
- Glossary
- Student lecture notebook
- Web links
- Animations
- Chapter summaries, learning objectives, and crossword puzzles

## Notes

1. Futoryan T, Gilchrest BE. Retinoids and the skin. Nutrition Reviews. 1994;52:299–310. Gerster H. Vitamin A—functions, dietary requirements and safety in humans. International Journal of Vitamins and Nutrition Research. 1997;67:71–90. Hinds TS, West WL, Knight EM. Carotenoids and retinoids: a review of research, clinical, and public health applications. Journal of Clinical Pharmacology. 1997;37:551–8. Ross AC, Gardner EM. The function of vitamin A in cellular growth and differentiation, and its roles during pregnancy and lactation. Advances in Experimental Medicine and Biology. 1994;352:187–200.

2. Jason J, Archibald LK, Nwanyanwu OC, Sowell AL, Buchanan I, Larned J, Bell M, Kazembe PN, Dobbie H, Jarvis WR. Vitamin A levels and immunity in humans. Clinical and Diagnostic Laboratory Immunology. 2002;9:616–21. Villamor E, Fawzi WW. Effects of vitamin A supplementation on immune responses and correlation with clinical outcomes. Clinical Microbiology Reviews. 2005;18:446–64.

3. Underwood B. Vitamin A deficiency disorders: international efforts to control a preventable "pox." Journal of Nutrition. 2004;134:231S–6S.

4. Gey KF, Moser UK, Jordan P, Stahelin HB, Eichholzer M, Ludin E. Increased risk of cardiovascular disease at suboptimal plasma concentrations of essential antioxidants: an epidemiological update with special attention to carotene and vitamin C. American Journal of Clinical Nutrition. 1993;57(5 Suppl):787S–97S. Mayne ST. Beta-carotene, carotenoids, and disease prevention in humans. FASEB Journal. 1996;10:690–701.

5. The Alpha-Tocopherol, Beta Carotene Cancer Prevention Study Group. The effect of vitamin E and beta carotene on the incidence of lung cancer and other cancers in male smokers. New England Journal of Medicine. 1994;330:1029–35. Baron JA, Cole BF, Mott L, Haile R, Grau M, Church TR, Beck GJ, Greenberg ER. Neoplastic and antineoplastic effects of β-carotene on colorectal adenoma recurrence: results of a randomized trial. Journal of the National Cancer Institute. 2003;95:717–22. Omenn GS, Goodman GE, Thornquist MD, Balmes J, Cullen MR, Glass A, Keogh JP, Meyskens FL, Valanis B, Williams JH, Barnhart S, Hammar S. Effects of a combination of beta carotene and vitamin A on lung cancer and cardiovascular disease. New England Journal of Medicine. 1996;334:1150–55.

6. Blinkey N, Krueger D. Hypervitaminosis A and bone. Nutrition Reviews. 2000;58:138–44. Forsyth KS, Watson RR, Gensler HL. Osteotoxicity after chronic dietary administration of 13-cis-retinoic acid, retinyl palmitate, or selenium in mice exposed to tumor initiation and promotion. Life Sciences. 1989;45:2149–56. Whiting SJ, Lemke B. Excess retinal intake may explain the high incidence of osteoporosis in northern Europe. Nutrition Reviews. 1999;57:249–50.

7. Johansson S, Melhus H. Vitamin A antagonizes calcium response to vitamin D in man. Journal of Bone Mineral Research. 2001;16:1899–905.

8. Campbell JK, Canene-Adams K, Lindshield BL, Boileau TWM, Clinton SK, Erdman JW. Tomato phytochemicals and prostate cancer risk. Journal of Nutrition. 2004;134:3486S–92S. Wertz K, Siler U, Goralczyk R. Lycopene: modes of action to promote prostate health. Archives of Biochemistry and Biophysics. 2004;430:127–34.

9. Beatty S, Nolan J, Kavanagh H, O'Donovan O. Macular pigment optical density and its relationship with serum and dietary levels of lutein and zeaxanthin. Archives of Biochemistry and Biophysics. 2004;430:70–6. Stringham JM, Hammond BR. Dietary lutein and zeaxanthin: possible effects on visual function. Nutrition Reviews. 2005;63:59–64.

10. El-Agamey A, Lowe GM, McGarvey DJ, Mortensen A, Phillip DM, Truscott G, Young AJ. Carotenoid radical chemistry and antioxidant/pro-oxidant properties. Archives of Biochemistry and Biophysics. 2004;430:37–48.

11. Chew BP, Park JS. Carotenoid action on the immune response. Journal of Nutrition. 2004;134:257S-61S. Johnson EJ. The role of carotenoids in human health. Nutrition in Clinical Care. 2002;5:56–65. Palozza P, Serini S, Nicuolo FD, Calviello G. Modulation of apoptotic signaling by carotenoids in cancer cells. Archives of Biochemistry and Biophysics. 2004;430:104–9. Sharoni Y, Danilenko M, Dubi N, Ben-Dor A, Levy J. Carotenoids and transcription. Archives of Biochemistry and Biophysics. 2004;430:89–96.

12. Chew BP, Brown CM, Park JS, Mixter PF. Dietary lutein inhibits mouse mammary tumor growth by regulating angiogenesis and apoptosis. Anticancer Research. 2003;23:3333–9. Terry P, Jain M, Miller A, Howe GR, Rohan TE. Dietary carotenoids and risk of breast cancer. American Journal of Clinical Nutrition. 2002;76:883–8.

13. Campbell JK, Canene-Adams K, Lindshield BL, Boileau TWM, Clinton SK, Erdman JW. Tomato phytochemicals and prostate cancer risk. Journal of Nutrition. 2004;134:3486S–92S. Wertz K, Siler U, Goralczyk R. Lycopene: modes of action to promote prostate health. Archives of Biochemistry and Biophysics. 2004;430:127–34.

14. Beatty S, Nolan J, Kavanagh H, O'Donovan O. Macular pigment optical density and its relationship with serum and dietary levels of lutein and zeaxanthin. Archives of Biochemistry and Biophysics. 2004;430:70–6. Stringham JM, Hammond BR. Dietary lutein and zeaxanthin: possible effects on visual function. Nutrition Reviews. 2005;63:59–64.

15. The Alpha-Tocopherol, Beta Carotene Cancer Prevention Study Group. The effect of vitamin E and beta carotene on the incidence of lung cancer and other cancers in male smokers. New England Journal of Medicine. 1994;330:1029–35. Baron JA, Cole BF, Mott L, Haile R, Grau M, Church TR, Beck GJ, Greenberg ER. Neoplastic and antineoplastic effects of β-carotene on colorectal adenoma recurrence: results of a randomized trial. Journal of the National Cancer Institute. 2003;95:717–22. Omenn GS, Goodman GE, Thornquist MD, Balmes J, Cullen MR, Glass A, Keogh JP, Meyskens FL, Valanis B, Williams JH, Barnhart S, Hammar S. Effects of a combination of beta carotene and vitamin A on lung cancer and cardiovascular disease. New England Journal of Medicine. 1996;334:1150–55.

16. Liu RH. Potential synergy of phytochemicals in cancer prevention: mechanism of action. Journal of Nutrition. 2004;134:3479S–85S.

17. World Health Organization. Micronutrient deficiencies. Combating vitamin A deficiency. Available from: http://www.who.int/nut/vad.htm.

18. Caulfield LE, Richard SA, Black RE. Undernutrition as an underlying cause of malaria morbidity and mortality in children less than five years old. American Journal of Tropical Medicine and Hygiene. 2004;71:55–63. Underwood B. Vitamin A deficiency disorders: international efforts to control a preventable "pox." Journal of Nutrition. 2004;134:231S–6S.

19. Beyer P, Al-Babili S, Ye X, Lucca P, Schaub P, Welsch R, Potrykus I. Golden rice: introducing the β-carotene biosynthesis pathway into rice endosperm by genetic engineering to defeat vitamin A deficiency. Journal of Nutrition. 2002;132:506S–10S. Hoa TTC, Al-Babili A, Schaub P, Potrykus I, Beyer P. Golden Indica and Japonica rice lines amendable to deregulation. Plant Physiology. 2003;133:161–9. Potrykus I. Golden rice and beyond. Plant Physiology. 2001;125:1157–61. Ye X, Al-Babili A, Kloti A, Zhang J, Lucca P, Beyer P, Potrykus I. Engineering the provitamin A

(β-carotene) biosynthetic pathway into (carotenoid-free) rice endosperm. Science. 2000;287:303–5.

20. Paine JA, Shipton CA, Chaggar S, Howells RM, Kennedy MJ, Vernon G, Wright SY, Hinchliffe E, Adams JL, Silverstone AL, Drake R. Improving the nutritional value of Golden Rice through increased pro-vitamin A content. Nature Biotechnology. 2005;23:482–7.

21. Thomson JA. Research needs to improve agricultural productivity and food quality, with emphasis on biotechnology. Journal of Nutrition. 2002;132:3441S–2S. Welch RM, Graham RD. Breeding for micronutrients in staple food crops from a human nutrition perspective. Journal of Experimental Botany. 2004;55:353–64.

22. American Academy of Dermatology. American Academy of Dermatology issues statement endorsing the World Health Organization's recommendation to restrict tanning bed use. Available from: http://www.aad.org/aad/Newsroom/who_endorsement.htm. American Academy of Dermatology. American Academy of Dermatology Association reconfirms need to boost vitamin D intake through diet and nutritional supplements rather than ultraviolet radiation. Available from: http://www.aad.org/aad/Newsroom/Vitamin+D+Consensus+Conf.htm. World Health Organization. Sunbeds, tanning and UV exposure. Available from: http://www.who.int/mediacentre/factsheets/fs287/en/print.html.

23. Bikle DD. Vitamin D and skin cancer. Journal of Nutrition. 2004;134:3472S–8S. Gross MD. Vitamin D and calcium in the prevention of prostate and colon cancer: new approaches for the identification of needs. Journal of Nutrition. 2005;135:326–31. Harris DM, Go VLW. Vitamin D and colon carcinogenesis. Journal of Nutrition. 2004;134: 3463S–71S. Holick MF. Sunlight and vitamin D for bone health and prevention of autoimmune diseases, cancers, and cardiovascular disease. American Journal of Clinical Nutrition. 2004;6 Suppl:1678S–88S. Welsh J. Vitamin D and breast cancer: insights from animal models. American Journal of Clinical Nutrition. 2004;80:1721S–4S.

24. Calvo MS, Whiting SJ, Barton CN. Vitamin D intake: a global perspective of current status. Journal of Nutrition. 2005;135:310–6.

25. Buchner DM, Larson EB. Falls and fractures in patients with Alzheimer-type dementia. JAMA (Journal of the American Medical Association). 1987;20:1492–5. Landfield PW, Cadwallader-Neal L. Long-term treatment with calcitriol $(1,25(OH)_2$ vit $D_3)$ retards a biomarker of hippocampal aging in rats. Neurobiology of Aging. 1998;19:469–77. Sato Y, Kanoko T, Satoh K, Iwamoto J. The prevention of hip fracture with risedronate and ergocalciferol plus calcium supplementation in elderly women with Alzheimer disease: a randomized controlled trial. Archives of Internal Medicine. 2005;165:1737–42. Sato Y, Iwamoto J, Kanoko T, Satoh K. Amelioration of osteoporosis and hypovitaminosis D by sunlight exposure in hospitalized, elderly women with Alzheimer's disease: a randomized controlled trial. Journal of Bone Mineral Research. 2005;20:1327–33.

26. Rajakumar K. Vitamin D, cod-liver oil, sunlight, and rickets: a historical perspective. Pediatrics. 2003;112:132–5.

27. Centers for Disease Control and Prevention. Severe malnutrition among young children—Georgia, January 1997–June 1999. Morbidity and Mortality Weekly Report. 2001;50:224–7. Kreiter SR, Schwartz RP, Kirkman HN, Charlton PA, Calikoglu AS, Davenport ML. Nutrition rickets in African American breast-fed infants. Journal of Pediatrics. 2000;137:153–7. Pugliese MF, Blumberg DL, Hludzinski J, Kay S. Nutritional rickets in suburbia. Journal of the American College of Nutrition. 1998;17:637–41. Sills IN, Skuza KA, Horlick MN, Schwartz MS, Rapaport R. Vitamin D deficiency rickets. Reports of its demise are exaggerated. Clinical Pediatrics (Philadelphia). 1994;33:491–3.

28. Gartner LM, Greer FR, and the Section on Breastfeeding and Committee on Nutrition. Prevention of rickets and vitamin D deficiency: new guidelines for vitamin D intake. Pediatrics. 2003;111:908–10.

29. Bostick RM, Potter JD, McKenzie DR, Sellers TA, Kushi LH, Steinmetz KA, Folsom AR. Reduced risk of colon cancer with high intakes of vitamin E: the Iowa Women's Health Study. Cancer Research. 1992;15:4230–7. Graham S, Sielezny M, Marshall J, Priore R, Freudenheim J, Brasure J, Haughey B, Nasca P, Zdeb M. Diet in the epidemiology of postmenopausal breast cancer in the New York State cohort. American Journal of Epidemiology. 1992;136:3127–37. Kline K, Yu W, Sanders BG. Vitamin E and breast cancer. Journal of Nutrition. 2004;134:3458S–62S.

30. Jacques PF, Taylor A, Moeller S, Hankinson SE, Rogers G, Tung W, Ludovico J, Willett WC, Chylack LT. Long-term nutrient intake and 5-year change in nuclear lens opacities. Archives of Ophthamology. 2005;123:517–26. Leske MC, Chylack LT, He Q, Wu SY, Schoenfeld E, Friend J, Wolfe J. Antioxidant vitamins and nuclear opacities: the longitudinal study of cataract. Ophthalmology. 1998;105:831–6. Teikari JM, Virtamo J, Rautalahti M, Palmgren J, Liesto K, Heinonen OP. Long-term supplementation with alpha-tocopherol and beta-carotene and age-related cataract. Acta Ophthalmologica Scandinavica. 1997;75:634–40.

31. Jialal I, Devaraj S. Scientific evidence to support a vitamin E and heart disease health claim: research needs. Journal of Nutrition. 2005;135:348–53. Jialal I, Fuller CJ. Effect of vitamin E, vitamin C and beta-carotene on LDL oxidation and atherosclerosis. Canadian Journal of Cardiology. 1995;11:97G–103G. Knekt P, Reunanen A, Jarvinen R, Seppanen R, Heliovaara M, Aromaa A. Antioxidant vitamin intake and coronary mortality in a longitudinal population study. American Journal of Epidemiology. 1994;139:1180–9. Lonn EM, Yusuf S. Is there a role for antioxidant vitamins in the prevention of cardiovascular diseases? An update on epidemiological and clinical trials data. Canadian Journal of Cardiology. 1997;13:957–65. Stampfer MJ, Hennekens CH, Manson JE, Colditz GA, Rosner B, Willett WC. Vitamin E consumption and the risk of coronary disease in women. New England Journal of Medicine. 1993;328:1444–9.

32. Miller ER, Pastor-Barriuso R, Dalal D, Riemersma RA, Appel LJ, Guallar E. Meta-analysis: high-dosage vitamin E supplementation may increase all-cause mortality. Annals of Internal Medicine. 2005;142:37–46.

33. Booth SL, Brow KE, Peterson JW, Cheng DM, Dawson-Hughes B, Gundberg CM, Cupples LA, Wilson PW, Kiel DP. Associations between vitamin K biochemical measures and bone mineral density in men and women. Journal of Clinical Endocrinology and Metabolism. 2004;89:4904–9. Radecki TE. Calcium and vitamin D in preventing fractures: vitamin K supplementation has powerful effect. British Medical Journal. 2005;331:108. Sasaki N, Kusano E,

Takahashi H, Ando Y, Yano K, Tsuda E, Asano Y. Vitamin K$_2$ inhibits glucocorticoid-induced bone loss partly by preventing the reduction of osteoprotegerin (OPG). Journal of Bone and Mineral Metabolism. 2005;23:41–47.

34. Committee on Fetus and Newborn (American Academy of Pediatrics). Controversies concerning vitamin K and the newborn. Pediatrics. 2003;112:191–2.

**Check out the following sources for additional information.**

35. Blount JD. Carotenoids and life-history evolution in animals. Archives of Biochemistry and Biophysics. 2004;430:10–5.

36. DeLuca HF. Overview of general physiologic features and functions of vitamin D. American Journal of Clinical Nutrition. 2004;80:S1689–96.

37. Duvall WL. Endothelial dysfunction and antioxidants. Mount Sinai Journal of Medicine. 2005;72:71–80.

38. Fairfield KM, Fletcher RH. Vitamins for chronic disease prevention in adults: scientific review. JAMA (Journal of the American Medical Association). 2002;288:3116–26.

39. Ferland G. Vitamin K. In: Present knowledge in nutrition, 8th ed. Bowman BA, Russell RM, editors. Washington, DC: ILSI Press; 2001.

40. Gibson RS. Principles of nutritional assessment. New York: Oxford University Press; 2005.

41. Gropper SS, Smith JL, Groff JL. Advanced nutrition and human metabolism, 4th ed. Belmont, CA: Thomson/Wadsworth; 2005.

42. Institute of Medicine. Dietary Reference Intakes for vitamin A, vitamin K, arsenic, boron, chromium, copper, iodine, iron, manganese, molybdenum, nickel, silicon, vanadium, and zinc. Washington, DC: National Academy Press; 2001.

43. Institute of Medicine. Dietary Reference Intakes for vitamin C, vitamin E, selenium, and the carotenoids. Washington, DC: National Academy Press; 2000.

44. Institute of Medicine. Dietary Reference Intakes: guiding principles for nutrition labeling and fortification. Washington, DC: National Academy Press, 1998.

45. Johnson IT. Micronutrients and cancer. Proceedings of the Nutrition Society. 2004;63:87–595.

46. McLaren DS, Frigg M. Sight and life manual on Vitamin A deficiency disorders. Basel, Switzerland: Sight and Life Task Force; 2001.

47. The Merck Manual of Diagnosis and Therapy, 17th ed. Beers MH, Berkow R, editors. Merck and Co. Inc.; 1999–2005. Available from: http://www.merck.com/mrkshared/mmanual/home.jsp.

48. National Institutes of Health Drug-Nutrient Interaction Task Force. Important information to know when you are taking: Coumadin® and vitamin K. Available from: http://ods.od.nih.gov/factsheets/cc/coumadin1.pdf.

49. Norman AW. Vitamin D. In: Present knowledge in nutrition, 8th ed. Bowman BA, Russell RM, editors. Washington, DC: ILSI Press; 2001.

50. Pryor WA. Vitamin E. In: Present knowledge in nutrition, 8th ed. Bowman BA, Russell RM, editors. Washington, DC: ILSI Press; 2001.

51. Raiten DJ, Picciano MF. Vitamin D and health in the 21st century: bone and beyond. Executive summary. American Journal of Clinical Nutrition. 2004;80:S1673–7.

52. Rodriguez-Amaya DB. Food carotenoids: analysis, composition and alterations during storage and processing of foods. Forum in Nutrition. 2003;56:35–7.

53. Solomons N. Vitamin A and carotenoids. In: Present knowledge in nutrition, 8th ed. Bowman BA, Russell RM, editors. Washington, DC: ILSI Press; 2001.

54. Traber MG. The ABCs of vitamin E and beta-carotene absorption. American Journal of Clinical Nutrition. 2004;80:3–4.

55. U.S. Department of Health and Human Services, and U.S. Department of Agriculture. Dietary Guidelines for Americans 2005. Washington, DC: U.S. Government Printing Office; 2005. Available from http:www.healthierus.gov/dietaryguidelines.

56. Weisberg P, Scanlon KS, Li R, Cogswell ME. Nutritional rickets among children in the United States: review of cases reported between 1986–2003. American Journal of Clinical Nutrition. 2004;80:S1697–705.

# Nutrition and Cancer

© Photodisc/Getty Images

For most people, the thought of cancer is frightening. This is not surprising, as cancer is the second leading cause of death in the United States, exceeded only by heart disease. Almost 10 million Americans have either had or currently have cancer, and cancer causes one in every four deaths.[1] However, our understanding of how cancer can be prevented and cured is greater now than ever before. For example, scientists estimate that about one third of all cancer deaths are related to nutrition, physical inactivity, and obesity.[2] In other words, a healthy diet and regular exercise might prevent almost 200,000 cancer deaths annually in the United States. In this Nutrition Matters, you will learn basic facts about cancer and the importance of nutrition and other lifestyle choices in its prevention.

## Understanding How Cancer Develops

Life is created from the joining of an egg to a sperm. This single new cell contains the basic genetic material (deoxyribonucleic acid, or DNA) that will be contained in the millions of cells that ultimately make up the human body. For this to occur, the single cell must replicate its DNA and divide into two cells. In turn, those two cells replicate their DNA and divide into four cells. This cycle of DNA replication and cell division, called the **cell cycle,** occurs continually throughout a person's lifetime as new cells are needed for growth, development, and repair of the body's tissues. In addition, old, damaged, or poorly functioning cells can undergo **apoptosis,** which is a healthy process of programmed cell death. The cell cycle is illustrated in Figure 1. The balance of cell replication via the cell cycle and cell death via apoptosis ensures that cells are healthy and functional.

> **CONNECTIONS** Recall that growth due to the division and multiplication of cells is called hyperplasia (Chapter 9, page 320).

**Cancer** occurs when the balance between cell growth and cell death is disrupted. In other words, cancer results from unregulated cell growth and division and/or an inability to initiate apoptosis. Cancer is caused by an alteration, or mutation, in specific regions of the DNA that code for the many proteins that regulate the cell cycle. Although some mutations can be inherited from our parents, most are thought to be caused by something in our lifestyle or environment.

Cancer can occur in almost any tissue type. However, it is most common in the prostate gland (prostate cancer; males only), the colon or rectum (colorectal cancer), mammary tissue (breast cancer), and the lungs (lung cancer). Of these types of cancers, the mortality rate from lung cancer is highest.[3]

## Carcinogens: The Causes of Cancer

Some cancers form spontaneously from mutations that occur during the normal process of DNA replication. However, most cancers are triggered by environmental conditions or substances. Together, these are called **carcinogens.** There are many known carcinogens, such

**cell cycle** The process by which cells grow, mature, replicate their DNA, and divide.

**apoptosis** (a – po – TO – sis) The normal process by which a cell leaves the cell cycle and dies; programmed cell death.

**cancer** A condition characterized by unregulated cell division.

**carcinogen** (car – CIN – o – gen) Compound or condition that causes cancer.

## FIGURE 1   The Cell Cycle

Healthy cells undergo a cycle of DNA replication, followed by cell division, and eventually cell death (apoptosis). Cancer can result from unregulated cell growth or insufficient apoptosis.

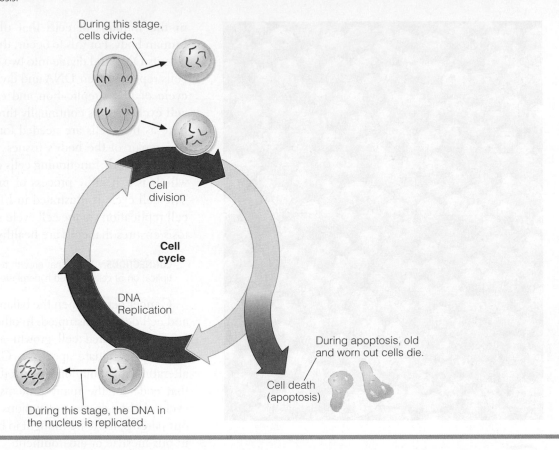

During this stage, cells divide.

Cell division

**Cell cycle**

DNA Replication

During apoptosis, old and worn out cells die.

Cell death (apoptosis)

During this stage, the DNA in the nucleus is replicated.

---

as x-rays, pollutants, and chemicals in tobacco. Exposure to chemical carcinogens tends to be highly regulated by public health agencies. However, exposure to other types of carcinogens, such as radiation and sun exposure, bacteria, and viruses, is more difficult to control. Many nutrients protect the body from the effects of these compounds.

## The Multistage Nature of Cancer

Like type 2 diabetes and cardiovascular disease, cancer is a chronic degenerative disease that can take years to develop. Many scientists believe that some cancers develop early in life but may not become evident until much later. The process by which normal, healthy cells develop into a cancerous growth occurs in stages, as illustrated in Figure 2. In the first stage, called **initiation,** normal cells become initiated cells via a mutation in the DNA. A gene that is capable of transforming normal cells into cancer cells is called an **oncogene.** Although oncogenes can be genetically inherited from a parent,

their formation is thought to be primarily caused by carcinogens. The body has many defense mechanisms to make sure that mutated cells do not continue to divide and grow. For example, antioxidant vitamins and minerals can prevent and repair some forms of DNA damage. The immune system is also involved in ridding the body of defective cells. In this way, antioxidants and the immune system can halt cancer growth at the initiation stage, preventing mutated cells from passing on their defective genes.

When damaged DNA is not repaired or the cell is not destroyed, it can enter the second stage of cancer, called **promotion.** During promotion, the cell begins to repeatedly replicate its mutated DNA and divide, ultimately

**initiation** The first stage of cancer, in which a normal gene is transformed into a cancer-forming gene (oncogene).

**oncogene** (ON – co – gene) An abnormal gene that transforms normal cells to cancer cells.

**promotion** The second stage of cancer in which the initiated cell begins to replicate itself, forming a tumor.

## FIGURE 2   The Stages of Cancer

The three stages of cancer are initiation, promotion, and progression.

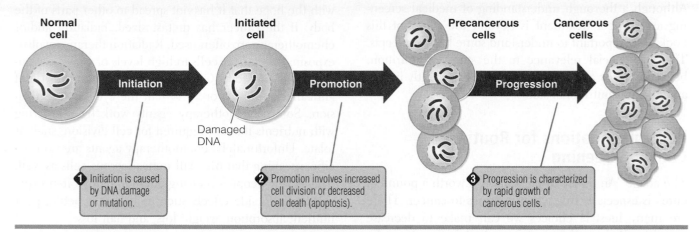

**Normal cell** → **Initiation** → **Initiated cell** (Damaged DNA) → **Promotion** → **Precancerous cells** → **Progression** → **Cancerous cells**

1. Initiation is caused by DNA damage or mutation.
2. Promotion involves increased cell division or decreased cell death (apoptosis).
3. Progression is characterized by rapid growth of cancerous cells.

forming a growth or **tumor.** Not all tumors are dangerous, and those that are not life threatening are said to be **benign.** Cancerous growths are said to be **malignant.** For example, breast tissue can develop benign noncancerous growths (consisting of fibrous tissue) or malignant cancerous growths. It is important to seek professional advice to determine if a tumor is benign or malignant. Cancerous tumors can enter a third stage of cancer called **progression.** During progression, the tumor continues to grow rapidly and invades the surrounding tissue, greatly disrupting its function. At this stage, tumors also produce substances that encourage the growth of blood vessels, ensuring a constant supply of energy, nutrients, and oxygen to the rapidly growing, cancerous cells. In addition, cancerous cells can break away and travel in the blood or lymph to other tissues, forming new tumors. The movement of cancerous cells from their original site to other locations is called **metastasis.** For example, prostate cancers often break away, or metastasize, from the prostate gland.

The process of cancer formation can be compared to what might happen if a mistake occurred in manufacturing a factory-produced product. Consider the many parts needed to build an automobile. A mistake in an inconsequential part (such as stitches in the seat coverings) would not influence the functioning of the vehicle—just as some mutations result in benign tumors. However, if the mistake affects something important like the brakes, this could be serious—similar to the initiation phase of cancer. If this mistake is not corrected, and the part is mass-produced, the vehicle would malfunction—similar to the promotion stage of cancer. Even worse, shipping these faulty parts to other automobile-producing factories would result in hundreds of defective automobiles—similar to cancer progression and metastasis.

The progression of a healthy cell to a cancerous one, and eventually metastasis, involves many steps. Thus anything that interrupts any step can halt the growth of cancer. Millions of dollars are spent annually to discover new ways to prevent and treat cancer. Scientists now know that lifestyle alterations (such as good nutrition and not smoking ), surgery, radiation, and chemotherapy, all can influence the various stages of cancer growth. These preventive lifestyle choices and cancer treatments are described in the following sections.

## ESSENTIAL *Concepts*

Cancer results from the uncontrolled growth of cells caused by an alteration or mutation in the DNA, and its formation involves several phases. During the initiation stage, the DNA is altered. This can be due to a carcinogen or simply to a mistake in DNA replication during the normal cell cycle. If the mutation in the DNA is not repaired or if the cell does not die via apoptosis, it can enter the second stage of cancer, called promotion. During this stage, the cell rapidly divides, forming a tumor. In the third stage, called progression, cancerous cells break away and lodge elsewhere in the body. This is called metastasis. Cancer prevention and treatment strategies disrupt one or more of these phases of cancer growth.

**tumor** A growth sometimes caused by cancer.

**benign tumor** A growth that has minimal physiological consequences and does not metastasize.

**malignant tumor** A cancerous growth.

**progression** The third stage of cancer in which tumor cells rapidly divide and invade surrounding tissues.

**metastasis** The spreading of cancer cells to other parts of the body via the blood or lymph.

# Cancer Screening and Treatments

Although a thorough understanding of medical screening and cancer treatment is beyond the scope of this book, it is important to understand some basic concepts. This has special relevance to the study of nutrition, because cancer and cancer treatment can greatly impair appetite and nutritional status.

## Recommendations for Routine Cancer Screening

The adage "An ounce of prevention is worth a pound of cure" is especially true when it comes to cancer. There are many lifestyle choices we can make to decrease our risks. Perhaps the most important is to get regular medical checkups to screen for cancer. It is important to detect cancer early when it is most treatable. This is especially important for people with increased risk factors for the disease. Guidelines concerning regular cancer screening are outlined in Table 1. Depending on the cancer, clinicians recommend that baseline screening begin as early as 20 years of age.

## Treatment Options

The most common methods of treating cancer are surgery, radiation therapy, and **chemotherapy.** However, which treatment or combination of treatments is used depends on many factors, such as form of cancer, stage and location of cancer, and health and age of the patient. Regardless of the treatment used, the goal is always the same—to interrupt the uncontrolled growth of the cancer cells.

In surgical treatment, the cancerous growth is removed, with the hope that it has not spread to other parts of the body. If the cancer has metastasized, radiation and/or chemotherapy are often used. Radiation therapy involves exposing the cancer cells to high levels of x-rays to damage or kill the cells. Chemotherapy involves the use of anticancer drugs in order to halt the cycle of cell division. Some chemotherapy agents work by interfering with nutrients that are required for cell division, such as folate. Unfortunately, chemotherapy agents are nonspecific, meaning that they kill noncancerous cells as well. This is why people receiving chemotherapy often experience many side effects such as loss of appetite, poor nutrient absorption, weight loss, and hair loss.

## Nutritional Implications of Cancer and Cancer Treatments

Many nutritional complications are associated with cancer and its treatment. The severity of these complications depends on the type of cancer and type of treatment. For example, cancer often results in loss of appetite and aversion to selected foods, otherwise known as anorexia. In addition, cancer can cause severe wasting of lean (muscle) tissue, a condition called **cachexia.** The causes

**chemotherapy** The use of drugs to stop the growth of cancer.

**cachexia** (ca – CHEX – i – a) A condition in which a person loses lean body mass (muscle tissue).

**TABLE 1    The American Cancer Society's Guidelines for Regular Cancer Screening**

| Type of Cancer | Sex | Age | Type of Screening Recommended |
|---|---|---|---|
| Breast | Females | 20 years | Clinical breast exams should be part of a periodic health checkups. |
| | | 40 years | Yearly mammograms should continue for as long as the woman is in good health. |
| Colon and rectum | Males and females | 50 years | Yearly exams should be done to test for blood in the feces. Colonoscopy should be part of a health exam every 10 years. |
| Cervix | Females | 21 years | Yearly Pap test after 21 years. |
| | | 30 years | Women who have had 3 normal Pap tests in a row should get screened every 2 to 3 years. |
| | | | Women at higher risk of cervical cancer should continue to get Pap tests annually. |
| | | 70 years | Women who have had 3 or more normal Pap tests in a row in the last 10 years may stop having cervical cancer screening. |
| Uterus | Females | 35 years | Women with high risk for this cancer should begin annual screening. |
| Prostate | Males | 40 years | Men at high risk should begin annual screening with blood tests and physical examinations. |
| | | 50 years | Blood tests and physical examinations should be done annually. |

SOURCE: American Cancer Society. American Cancer Society Detection Guidelines. Available from: http://www.cancer.org/docroot/PED/content/PED_2_3X_ACS_Cancer_Detection_Guidelines_36.asp.

of cancer-associated anorexia and cachexia are unclear. However, there is growing evidence that a variety of substances produced by the tumors themselves may be responsible.[4] Together, anorexia and cachexia can result in severe malnutrition. Cancer treatments can also cause nausea, diarrhea, and vomiting—which can all negatively affect nutritional status.

Although often difficult, it is important for people with cancer to maintain adequate nutrient and fluid intakes by consuming small meals and sufficient fluids throughout the day. Each person needs to develop his or her own strategies for assuring adequate nutrition, but several guidelines are recommended, including eating dry toast or crackers before getting out of bed in the morning, avoiding food prior to a treatment, sipping liquids throughout the day using a straw, and using plastic rather than metal utensils.[5]

In addition, because it is common for cancer patients to have difficulty digesting the milk sugar lactose, it may be necessary for them to avoid dairy products. Further, because chemotherapy can cause diarrhea, there is potential loss of electrolytes. Therefore, it is important to consume foods and liquids containing sodium and potassium. These include potatoes, bananas, apricot nectar, and bouillon or broth.

**CONNECTIONS** Recall that people with lactose intolerance do not produce adequate amounts of the enzyme lactase. (Chapter 5, page 140).

## ESSENTIAL *Concepts*

It is not always possible to prevent cancer, but routine screening helps detect it earlier, when it is most treatable. Therefore the American Cancer Society recommends medical screening for cancer beginning at age 20 years. When cancer is detected, there are many options for treatment, such as surgery, radiation, and chemotherapy. Radiation and chemotherapy both kill the rapidly growing cancer cells, whereas surgery is used to physically remove them. Both cancer and cancer treatment have many implications for nutritional status. For example, anorexia, cachexia, and cancer treatment often cause nausea, loss of appetite, and loss of lean body mass. Assuring adequate nutrient and fluid intake can be difficult during these times.

# Nondietary Factors Related to Risk of Cancer

Many factors influence a person's risk for developing cancer, as outlined in Table 2.[6] Some of these risk factors for cancer cannot be changed, such as genetic predisposition, sex (gender), and age. However, lifestyle risk factors such as nutrition, tobacco use, and physical activity are well within our control. Environmental factors such as sun exposure and pollutants can also affect a person's risk for cancer. Remember that having a risk factor for a disease does not mean the disease will develop. In fact, some people may have several risk factors for cancer, yet never develop it. However, being aware of risk factors and making healthy lifestyle choices can help lower the risk for developing cancer.

## Biological Risk Factors Related to Cancer

Biological (nonmodifiable) risk factors related to cancer include sex, ethnicity, and a variety of genetic factors. Although childhood cancers can occur, cancer is typically a disease diagnosed later in life—making age an important biological risk factor as well. Researchers estimate that 25% of all cancer cases are caused by biological factors.

### Sex (Gender)

Whether you are male or female can be a risk factor for some types of cancer. For example, prostate cancer is found only in men, whereas breast cancer is far more common in women. This sex difference has to do with basic anatomical and physiological differences between males and females. Other cancers, such as lung cancer, are more common in males than females. In this case, the difference is more likely due to differences in lifestyle choices, such as tobacco use, and less to biological (genetic) differences. Risk of getting some cancers, such as colorectal cancer, is similar for both sexes.

### Ethnicity (Race)

Although scientists do not know why, ethnicity or race can also influence risk for some cancers. For example, black men of African descent are at the greatest risk of developing prostate cancer. Conversely, this form of cancer is rare among Asian men. Whether this is due to genetic, lifestyle, or environmental differences is difficult to determine, although it is most likely an interaction among these factors.

### Other Genetic Factors

In addition to age, sex, and ethnicity, genetic factors can also predispose a person to various cancers. The presence of certain genes inherited from our parents can increase or decrease overall risk of developing cancer. For example, inheriting the oncogenes called *BRCA1* and *BRCA2* can increase the risk of breast cancer.[7] Conversely, we can inherit genes that *protect* against cancer.[8] The advent of genetic testing has made it possible to

## TABLE 2     Major Risk Factors for Cancer

| Risk Factors | Comments |
| --- | --- |
| **Biological (nonmodifiable) Risk Factors** | |
| Age | Most cancers are detected in people age 50 years and older. |
| Sex | Some cancers (e.g., prostate) only occur in men, whereas others (e.g., cervical) only occur in women; some cancers occur in both sexes but are more prevalent in one. |
| Ethnicity (race) | Some cancers are more common in people of a certain race. |
| Genetic predisposition | Many genetic factors can put a person at greater risk for various types of cancers. |
| Early age at first menstruation | Breast cancer is more common in women who began menstruating relatively early; note that diet may play a role in this relationship. |
| **Lifestyle and Environmental (modifiable) Risk Factors** | |
| Body weight | Being overweight or obese increases risk for many cancers (e.g., breast, colorectal, and ovarian). |
| Physical activity | Being physically inactive increases risk for many cancers (e.g., breast and colorectal). |
| Tobacco use | Cancers of the lung, colon and rectum, mouth, larynx, bladder, kidney, cervix, esophagus, and pancreas are related to tobacco use, including cigarettes, cigars, chewing tobacco and snuff; secondhand smoke can also increase risk for cancer. |
| Lactation | Having never breastfed a baby or having only breastfed for a short period of time increases risk for breast cancer. |
| Reproductive history | Breast and uterine cancers are more common in women who never have children or begin having children after the age of 30 years. |
| Alcohol consumption | Alcohol consumption over 1 to 2 drinks per day increases risk of some cancers (e.g., breast, colorectal, and mouth). |
| Dietary fat intake | Diets high in fat are related to increased risk for several cancers (e.g., breast, colorectal, and pancreatic). |
| Fruit and vegetable intake | Diets high in fruits and vegetables appear to be protective against some cancers (e.g., colorectal and bladder). |
| Red meat or processed meat | Very high intakes of red or processed meat are related to increased risk for colorectal cancer. |
| Exposure to strong sunlight or tanning beds | Skin cancer is related to unprotected exposure to strong sunlight or tanning beds. |
| Infection with certain bacteria or viruses | Some bacteria and viruses (e.g., hepatitis B and C) can cause cancer. |
| Radiation exposure | Radiation from occupational, medical, and environmental sources increases the risk for lung cancer. |
| Exposure to environmental carcinogens | Some compounds such as benzene, radon, asbestos, and lead can increase the risk for cancer. |

SOURCE: Adapted from American Cancer Society. *Cancer Facts and Figures 2006*. Atlanta, GA: American Cancer Society; 2006.

screen for these gene mutations, although this is still not widely done. As scientists learn more about the human genome, surely more cancer-related genes will be identified. Understanding the interactions among genetics, lifestyle, and environment will undoubtedly lead to better strategies for prevention and treatment.

## Nondietary Lifestyle and Environmental Factors Related to Cancer Risk

Although we have no control over the biological factors that predispose or protect us from cancer, the good news is that many cancers can be prevented by leading a healthy lifestyle and avoiding environmental carcinogens. The major lifestyle and environmental risk factors for cancer, listed in Table 2, include tobacco use (especially smoking), chemical carcinogens, radiation, infec-

tious disease, reproductive patterns, and physical activity. Many dietary variables are also related to risk of cancer.

### Smoking and Tobacco Use

Researchers estimate that 80% of all lung cancer cases are caused by tobacco products.[9] In addition, nonsmoking spouses of smokers have a 30% greater risk of developing lung cancer than do spouses of nonsmokers.[10] Of course, you do not need to be married to a smoker to be harmed by secondhand smoke. The best way to prevent lung cancer is to not smoke and to avoid exposure to secondhand smoke. Using smokeless tobacco (such as chewing tobacco) also increases the risk of cancer, especially that of the mouth. The risk for cancer of the cheek and gums increases nearly 50-fold among long-term tobacco chewers compared to nonusers.[11] The most powerful way to avoid cancer is to abstain from all types of tobacco use.

## Other Environmental and Lifestyle Factors

In addition to tobacco use, there are many other environmental carcinogens, such as high exposure to x-rays, strong ultraviolet light (from sunlight and tanning beds), and chemical carcinogens. Also, many cancers are caused by viral or bacterial infection. For example, the hepatitis B virus and the hepatitis C virus are the major causes of liver cancer.[12] Similarly, cervical cancer is often caused by the human papilloma virus,[13] and stomach cancer can be due to *Helicobacter pylori*, a bacterium that, you may recall, also causes peptic ulcers.[14] As researchers learn more about these infectious agents, cancer prevention and treatment via vaccinations (for viruses) and antibiotics (for bacteria) will likely become more available.

Many lifestyle choices can also influence cancer risk. This is especially true for breast cancer, which is related to reproductive patterns such as using oral contraceptives, never having had children, having had one's first child after the age of 30 years, and never having breastfed a baby.[15] Researchers think these factors increase risk for breast cancer in part by increasing lifetime exposure to the cancer-promoting hormone estrogen.

## ESSENTIAL *Concepts*

Cancer can be caused by biological, lifestyle, and environmental variables. Biological factors include genetic predisposition, age, ethnicity, and sex. In addition, environmental carcinogens such as cigarette smoke, radiation, and pathogens (bacteria and viruses) can also cause cancer. Nondietary lifestyle factors that influence cancer risk include a variety of reproductive patterns in women that increase the body's exposure to the cancer-promoting hormone estrogen.

# Dietary Factors Related to Cancer Risk

The National Cancer Institute and the American Cancer Society estimate that 30 to 35% of all cancers are diet related. This estimate is both staggering and hopeful, because it means that many cancers can be prevented by dietary change. Specifically, eating a diet that promotes a healthy body weight throughout life is critical to preventing cancer initiation, promotion, and progression. Consuming a diet that provides adequate amounts of essential nutrients and phytochemicals is also vital.

Scientists are still unraveling the myriad complementary and overlapping mechanisms by which nutrition influences cancer. Nonetheless, the American Cancer Society and the American Institute for Cancer Research have developed dietary guidelines that reflect our current knowledge. These are outlined in Table 3 and summarized here. Once again, the multifaceted and complex relationship between diet and risk of cancer demonstrates the importance of dietary variety and moderation in relation to health and well-being.

## Energy Balance: Energy Intake, Obesity, and Physical Activity

Obesity is a risk factor for several types of cancer, including breast cancer and colorectal cancer. Thus, it is recommended that we maintain a healthy body weight by balancing caloric intake with physical activity. Specifically, the American Cancer Society recommends that we maintain a body mass index (BMI) of less than 25 kg/m² and engage in at least 30 minutes or more of moderate physical activity, five or more days of the week. People with increased risk for breast or colon cancer are encouraged to get at least 45 minutes or more of moderate to vigorous activity, five or more days per week. Maintaining a healthy body weight in childhood and adolescence is especially important, because the increasing prevalence of obesity in children may increase the incidence of cancer in the years to come. The American Cancer Society recommends that children participate in at least 60 minutes of moderate to vigorous activity each day.

## The Macronutrients: Lipids, Protein, and Fiber

Not only is maintaining a healthy body weight important for decreasing risk for cancer, but so is the kind of foods we eat. Research suggests that a relatively low-fat diet emphasizing healthy fatty acids and sufficient dietary fiber is best when it comes to cancer prevention. In addi-

Exercising is a powerful way to help prevent cancer.

**TABLE 3    The American Cancer Society's Guidelines on Nutrition and Physical Activity for Cancer Prevention**

| Recommendation | Comments |
|---|---|
| Eat 5 or more servings of a variety of vegetables and fruits each day. | This includes fresh, cooked, frozen, and canned products. 100% fruit and vegetable juices are also excellent sources. Fruits and vegetables provide many vitamins and minerals as well as dietary fiber and a multitude of phytochemicals thought to be beneficial in cancer prevention. |
| Choose whole grains and legumes in preference to processed (refined) grains, sugars, and starches. | Legumes and whole-grain foods are nutrient dense, providing many vitamins, minerals, fiber, and phytochemicals to the diet. Legumes are also an excellent source of protein. |
| Limit consumption of red and processed meat, especially those high in fat. | Because very high amounts of red meat and processed meat are associated with increased risk of colorectal cancer, nutritionists recommend moderate consumption of these foods. Experts recommend selecting lean cuts of meat, and preparing meat by baking, broiling, or poaching rather than by frying or charbroiling. |
| Choose foods that help maintain a healthful weight. | Choose foods low in fat, calories, and sugar, and avoid large portions. Substitute vegetables, fruits, and other low-calorie foods for calorie-dense foods. |
| Adopt a physically active lifestyle. | *Adults:* engage in moderate activity for at least 30 minutes on 5 or more days of the week. *Children and adolescents:* engage in at least 60 minutes per day of moderate to vigorous physical activity. |
| If you drink alcoholic beverages, limit consumption. | Alcohol consumption increases risk for several types of cancer, including breast cancer. Women are advised to limit alcohol intake to no more than 1 drink per day, and men should limit theirs to no more than 2 drinks per day. |

SOURCE: Adapted from American Cancer Society. *Cancer Facts and Figures 2006.* Atlanta, GA: American Cancer Society; 2006.

tion, there may be cancer-causing compounds in some high-protein foods. The relationship between macronutrients and cancer is described next.

## Lipids and Fatty Acids

Epidemiologic studies show that high intake of dietary lipids is associated with increased risk for certain cancers, such as those of the breast, colon, and prostate. However, these findings are often inconsistent—especially when factors such as body weight and other dietary factors are taken into account. This is in part because people who eat high-fat diets are often overweight. Furthermore, diets high in fat are often low in essential vitamins, minerals, and dietary fiber. Therefore, it is difficult to determine which of these dietary components common in high-fat diets is actually influencing cancer risk. Nonetheless, experts recommend that adults limit their consumption of high-fat foods so that dietary lipids contribute 20 to 35% of energy.[16]

In addition to total lipids in the diet, consuming individual fatty acids may also influence risk for cancer.[5] For example, a fatty acid called conjugated linoleic acid (CLA) found in milk and beef fat is a potent anticarcinogen in laboratory animals, meaning that CLA

may inhibit cancer.[17] Omega-3 fatty acids also appear to decrease cancer risk—at least in laboratory animals.[18] However, more research is needed to determine whether these and other fatty acids can actually influence cancer risk in humans.

## High-Protein Foods

Although cell culture and animal studies provide some evidence that high protein intake may increase cancer risk, the vast majority of human studies are inconclusive. However, limited evidence suggests that very high, chronic intake of red meat (beef, lamb, and pork) or processed meats (bacon, sausage, hot dogs, ham, and cold cuts) is associated with increased risk for colorectal cancer.[19] Only people who consume 2 to 3 ounces each day are at increased risk.

Researchers do not know what component of red and processed meats influences cancer risk, but they believe iron, fat, salt, smoke residue, and/or preservatives may play a role. For example, **nitrites** sometimes added to

**nitrites** Nitrogen-containing compounds that are often added to processed meats to enhance color and flavor.

processed meats to preserve color and enhance flavor may play a role. Nitrites combine with other nitrogen-containing substances in the stomach to form **nitrosamines**—potent compounds associated with some forms of cancer.[20] Charring meat at high temperatures may also lead to the formation of **heterocyclic amines,** which have been shown to cause cancer in laboratory animals.[21] Small amounts of other carcinogens called **polycyclic aromatic hydrocarbons** are produced when meat is grilled.[22] There is also evidence that high intake of iron from red meat may be the factor related to increased cancer risk.[23] However, further research is needed to understand the physiological relationship between red meat intake and cancer risk. In the meantime, experts recommend that we choose reasonable portions and lean cuts of red or processed meats.

## Dietary Fiber

Consuming high-fiber foods, such as fruits and vegetables, has consistently been associated with decreased risk of cancer. However, as with lipids and red meat, it is often difficult to determine what component—or combination thereof—is involved in inhibiting cancer. Remember that high-fiber foods such as whole grains, fruits, and vegetables contain many other compounds known to influence health, including vitamins, minerals, and phytochemicals.

There are many hypotheses as to why high-fiber foods may decrease the risk of cancer, especially in the colon and rectum. These include dilution of potential carcinogens in the stool, decrease in the transit time of food in the gastrointestinal tract, binding of potential carcinogens in the intestine, lowering the pH of the stools, alteration of colonic microflora, and modulation of glucose regulation. Most likely dietary fiber has multiple actions, which together with other nutrients in the diet influence the cell cycle of cells lining the gastrointestinal tract and elsewhere. To ensure intake of high-fiber foods, the American Cancer Society recommends that we eat five or more servings of a variety of fruits and vegetables daily, choose whole-grain products frequently, and limit consumption of refined carbohydrates when possible.

## Micronutrients and Phytochemicals

Although the associations among energy consumption, macronutrient intake, and cancer risk are well known, many micronutrients may also influence cancer risk. These include several water-soluble vitamins, fat-soluble vitamins, and minerals. In addition, much current research is being conducted to assess the relationships between phytochemicals and cancer.

## Vitamins

Several vitamins are known to be involved in the complex processes ultimately leading to cancer. These include the antioxidants vitamins (vitamins C and E), folate, vitamin A, and vitamin D. There are many mechanisms by which these vitamins might inhibit cancer. For example, the antioxidant function of vitamin C and vitamin E may inhibit or repair free radical damage of DNA, thus inhibiting the initiation phase of cancer.[24] Similarly, vitamin A and folate are both required for cell differentiation and growth, and many vitamins are needed for maintaining a healthy immune system. Because the immune system helps rid the body of infectious organisms (such as bacteria and viruses) as well as defective cells, some vitamins may inhibit cancer indirectly by promoting a well-functioning immune response. Emerging evidence also suggests that vitamin D may help prevent colorectal cancer by influencing calcium metabolism.[25] Interestingly, sunlight exposure has been found to be related to decreased risk for some cancers, suggesting that vitamin D production in the body may be involved in preventing this disease.[26] However, results from the recently published Women's Health Initiative do not support a protective effect of vitamin D and calcium on colorectal cancer.[27] In this randomized, double-blind, placebo-controlled study, 36,282 healthy postmenopausal women were assigned to take either a combination of calcium (1,000 mg/day) and vitamin D supplements (400 IU/day) or a placebo daily. After 7 years, there was no difference in colorectal cancer risk between the study groups. However, many scientists question whether the dosage amounts were sufficient and duration of the study long enough for these conclusions to be valid.[28] Clearly, there is still much to learn about the relationships among calcium, vitamin D, and colorectal cancer risk.

> **CONNECTIONS** ▶ Free radicals are atoms or molecules with unpaired electrons, making them reactive and harmful to DNA, lipids, and proteins (Chapter 10, page 409).

Although all of these essential vitamins are collectively required for the cell cycle to function properly, it is difficult to study the role that each of these compounds plays in determining who gets cancer and who does not. What is certain is that consuming a diet high in a variety of vitamins helps protect against cancer. However, because

**nitrosamines** Nitrogen-containing chemical carcinogens that are produced from nitrites.

**heterocyclic amines** Cancer-causing compounds that can be formed when meat is cooked at high temperatures.

**polycyclic aromatic hydrocarbons** Cancer-causing compounds that can be formed when meat is grilled.

Fruits and vegetables contain many nutrients and phytochemicals that are likely important in preventing cancer.

controlled human studies using vitamin A and/or E supplements have shown increased risk for some cancers and heart disease,[29] consuming large doses of supplemental vitamins is not recommended. Instead, experts recommend that we consume at least five servings of fruits and vegetables each day and choose high-fiber foods when possible. Once again, variety is important, because no single food contains all the essential vitamins.

## Phytochemicals—Potential Cancer Fighters

In addition to vitamins, many phytochemicals may protect against cancer. Examples of potentially anticarcinogenic phytochemicals include β-carotene, lycopene, and isoflavones. Some of these are listed in Table 4. Many foods are known to contain phytochemicals and,

therefore, may help reduce cancer risk. These functional foods include tomatoes, green tea, garlic, soy products, certain spices, and cruciferous vegetables. For example, the bright red compound lycopene found in tomatoes is a potent antioxidant, and some studies show that increased consumption of tomato products decreases risk for prostate cancer.[30] Other compounds, called isothionates (found in cruciferous vegetables such as broccoli, cauliflower, cabbage, and brussels sprouts), is also a potent antioxidant.[31] Thus, consumption of cruciferous vegetables as well as the hundreds of brightly colored phytochemicals found in fruits and vegetables may also be protective. Researchers continue to actively study the roles of phytochemicals and functional foods in preventing and treating cancer.

> **CONNECTIONS** Remember that lycopene is a nonprovitamin A carotenoid with potent antioxidant properties (Chapter 11, page 437).

## Minerals

Minerals that are involved in preventing and repairing DNA damage by free radicals may also play a role in cancer prevention. These include selenium, zinc, copper, iron, and manganese. In addition, some studies suggest that getting enough calcium either from low-fat dairy products or calcium supplements may decrease the risk of colorectal cancer.[32] Researchers think calcium binds potential carcinogens in the intestinal tract, increasing their excretion in the feces. However, as previously discussed, data from the Women's Health Initiative study do not support a protective effect of calcium supplementation on cancer risk.[33] Although the impact of these min-

## TABLE 4    Potentially Anticarcinogenic Phytochemicals

| Phytochemical | Some Food Sources | Some Proposed Actions[a] |
|---|---|---|
| β-carotene | Carrots, pumpkin, sweet potatoes | Antioxidant; cell cycle; antibacterial |
| Caffeic acid phenethyl ester | Honey | Immune modulation; antioxidant |
| Capsaicin | Chili peppers | Immune modulation; apoptosis |
| Curcumin | Turmeric | Immune modulation |
| Diallyl sulfide | Garlic, onions | Modification of enzymes; antibacterial |
| Gingerol | Ginger | Antioxidant; immune modulation |
| Isoflavones | Soybeans | Estrogen antagonist |
| Isothiocyanates | Broccoli, brussels sprouts, cauliflower, cabbage | Modification of enzymes; antioxidant |
| Limonoids | Oranges, lemons, limes, grapefruit | Apoptosis |
| Lignins | Flaxseed | Estrogen antagonist |
| Lutein | Tomatoes, spinach | Antioxidant |
| Lycopene | Tomatoes, pink grapefruit | Antioxidant |
| Phenylpropanoids | Cinnamon, cloves, vanilla | Antioxidant; antibacterial |
| Polyphenols | Green tea | Immune modulation; apoptosis |
| Resveratrol | Grapes | Antioxidant; immune modulation |

[a]Note that these should be viewed as simply *proposed* actions of these compounds; in some cases, data are inconsistent and/or weak.

SOURCES: Lampe JW. Spicing up a vegetarian diet: chemopreventive effects of phytochemicals. *American Journal of Nutrition*. 2003;78:579S–83S. Nishino H, Murakoshi M, Mou XY, Wada S, Masuda M, Ohsaka Y, Satomi Y, Jinno K. Cancer prevention by phytochemicals. *Oncology*. 2005;69(Suppl 1):38–40. Rafter JJ. Scientific basis of biomarkers and benefits of functional foods for reduction of disease risk: cancer. *British Journal of Nutrition*. 2002;88:S219–24. Surh Y-J. Cancer chemoprevention with dietary phytochemicals. *Nature Reviews*. 2003;3:768–80. Talalay P, Fahey JW. Phytochemicals from cruciferous plants protect against cancer by modulating carcinogen metabolism. *Journal of Nutrition*. 2001;131:3027S–33S.

Studies show that increased calcium intake is associated with decreased risk for colorectal cancer.

erals on cancer risk is not fully understood, most likely they modulate the various stages of cancer. Once again, consuming adequate amounts of nutrient-dense foods such as fruits, vegetables, whole-grain products, and low-fat dairy foods is a prudent way to assure adequate intakes of all these minerals.

## Alcohol

Excessive alcohol consumption can contribute to an increased risk of several types of cancer, especially of the mouth, liver, and breast.[34] Doctors recommend that alcohol consumption be limited to no more than 2 drinks per day for men and 1 drink per day for women. A drink is defined as 12 ounces of beer, 5 ounces of wine, or 1.5 ounces of distilled spirits such as gin or vodka.

## Diet and Cancer: Future Directions

There is still much to learn about how dietary choices influence risk of cancer. This is partially due to the inherent difficulties in studying the relationship between nutrition and the development of this disease. In addition, many genetic factors can modulate the relationship between dietary intake and cancer risk. In other words, genetic factors interact with diet to influence risk of cancer. The medical community hopes that as we learn more about genes, lifestyle, and other factors that influence cancer risk, it will be possible to offer specialized dietary advice to people who are at greatest risk for developing certain types of cancers. Perhaps more than any other chronic disease, cancer holds the greatest hope in the area of nutritional prevention and treatment.

## ESSENTIAL *Concepts*

Obesity and physical inactivity are major risk factors for several types of cancer. In addition, many macronutrients and micronutrients are likely important in preventing cancer. High consumption of dietary fiber is related to decreased risk of cancer, whereas high consumption of alcohol, red meat, or processed meat is related to increased risk of cancer. To help prevent cancer, consume adequate amounts of whole-grain and high-fiber foods. If you drink, do so in moderation. It is also recommended that we choose reasonable portions of red or processed meats. Several vitamins, minerals, and phytochemicals are important for promoting a healthy cell cycle and immune system. These include the antioxidant vitamins and minerals, folate, vitamin D, and a growing list of phytochemicals. Consuming a variety of fruits and vegetables is highly recommended to obtain adequate amounts of these compounds.

## Notes

1. American Cancer Society. Cancer facts and figures 2006. Atlanta, GA: American Cancer Society; 2006.

2. World Cancer Research Fund/American Institute for Cancer Research. Food, nutrition and the prevention of cancer: a global perspective. Washington, DC: World Cancer Research Fund/American Institute for Cancer Research; 1997.

3. American Cancer Society. Cancer facts and figures 2006. Atlanta, GA: American Cancer Society; 2006.

4. Tisdale MJ. Tumor-host interactions. Journal of Cellular Biochemistry. 2004;93:871–7. Tisdale MJ. Cancer cachexia. Langenbecks Archives of Surgery. 2004;389: 299–305.

5. American Institute for Cancer Research. CancerResource. Washington, DC: American Institute for Cancer Research; 2000. Available from: http://www.aicr.org.

6. American Cancer Society. Cancer facts and figures 2006. Atlanta, GA: American Cancer Society; 2006.

7. Nelson HD, Huffman LH, Fu R, Harris EL. Genetic risk assessment and BRCA mutation testing for breast and ovarian cancer susceptibility: systematic evidence review for the U.S. preventive services task force. Annals of Internal Medicine. 2005;143:362–79. U.S. Preventive Services Task Force. Genetic risk assessment and BRCA mutation testing for breast and ovarian cancer susceptibility: recommendation statement. Annals of Internal Medicine. 2005;143:355–61.

8. Kim W-J, Quan C. Genetic and epigenetic aspects of bladder cancer. Journal of Cellular Biochemistry. 2005;95:24–33.

9. Doll R, Peto R. The causes of cancer. New York: Oxford University Press; 1981; U.S. Department of Health and Human Services. Reducing the health consequences of

smoking: 25 years of progress. A report of the surgeon general. Rockville, MD: U.S. Department of Health and Human Services, Public Health Service, Centers for Disease Control, Center for Chronic Disease Prevention and Health Promotion, Office of Smoking and Health; 1989.

10. U.S. Environmental Protection Agency. Respiratory health effects of passive smoking: lung cancer and other disorders. Washington, DC: U.S. Environmental Protection Agency (EPA/600/6-90/006F); 1992.

11. U.S. Department of Health and Human Services. The health consequences of using smokeless tobacco: a report of the advisory committee to the surgeon general. Atlanta, GA: U.S. Department of Health and Human Services, National Institutes of Health, National Cancer Institute; 1986.

12. Pisani P, Parkin DM, Munoz N, Ferlay J. Cancer and infection: estimates of the attributable fraction in 1990. Cancer Epidemiology, Biomarkers and Prevention. 1997;6:387–400.

13. International Agency for Research on Cancer. Human papillomaviruses. IARC monographs on the evaluation of carcinogenic risks to humans. Vol. 64. Lyon, France: IARC Press; 1995. Walboomers JM, Jacobs MV, Manos MM, Bosch FX, Kummer JA, Shah KV, Snijders PJ, Peto J, Meijer CJ, Munoz N. Human papillomavirus is a necessary cause of invasive cervical cancer worldwide. Journal of Pathology. 1999;189:12–19.

14. Plummer M, Franceschi S, Munoz N. Epidemiology of gastric cancer. International Agency for Research on Cancer Scientific Publications. 2004;157:311–26.

15. American Academy of Pediatrics. Breastfeeding and the use of human milk. Pediatrics. 2005;115:496–506. Jernstrom H, Lubinski J, Lynch HT, Ghadirian P, Neuhausen S, Isaacs C, Weber BL, Horsman D, Rosen B, Foulkes WD, Friedman E, Gershon-Baruch R, Ainsworth P, Daly M, Garber J, Olsson H, Sun P, Narod SA. Breast-feeding and the risk of breast cancer in BRCA1 and BRCA2 mutation carriers. Journal of the National Cancer Institute. 2004;96:1094–8.

16. Institute of Medicine. Dietary Reference Intakes for energy, carbohydrate, fiber, fat, fatty acids, cholesterol, protein, and amino acids (macronutrients). Washington, DC: National Academies Press; 2005.

17. Field CJ, Schley PD. Evidence for potential mechanisms for the effect of conjugated linoleic acid on tumor metabolism and immune function: lessons from n-3 fatty acids. American Journal of Clinical Nutrition. 2004;79:1190S–8S. Lee KW, Lee HJ. Role of the conjugated linoleic acid in the prevention of cancer. Critical Reviews in Food Science and Nutrition. 2005;45:135–44.

18. Hardman WE. (n-3) fatty acids in cancer therapy. Journal of Nutrition. 2004;134:3427S-30S. Larsson SC, Kumlin M, Ingelman-Sundberg M, Wolk A. Dietary long-chain n-3 fatty acids for the prevention of cancer: a review of potential mechanisms. American Journal of Clinical Nutrition. 2004;79:935–45.

19. Chao A, Thun MJ, Connell CJ, McCullough ML, Jacobs EJ, Flanders D, Rodriguez C, Sinha R and Calle EE. Meat consumption and risk of colorectal cancer. JAMA (Journal of the American Medical Association). 2005;293:172–82. Cross AJ, Sinha R. Meat-related mutagens/carcinogens in the etiology of colorectal cancer. Environmental and Molecular Mutagenesis. 2004;44:44–55. Paik DC, Saborio

DV, Oropeza R, Freeman HP. The epidemiologic enigma of gastric cancer rates in the U.S.: was grandmother's sausage the cause? International Journal of Epidemiology. 2001;30:181–2. Palli D, Saieva C, Coppi C, del Guidice G, Magagnotti C, Nesi G, Orsi F, Airoldi L. O6-alkylguanines, dietary N-nitroso compounds, and their precursors in gastric cancer. Nutrition and Cancer. 2001;39:42–49.

20. Knekt P, Jarvinene R, Dich J, Hakulinen T. Risk of colorectal and other gastro-intestinal cancers after exposure to nitrate, nitrite, and N-nitroso compounds: a follow-up study. International Journal of Cancer. 1999;80:852–6. Silvester KR, Bingham SA, Pollock JR, Cummings JH, O'Neill IK. Effect of meat and resistant starch on fecal excretion of apparent N-nitroso compounds and ammonia from the human large bowel. Nutrition and Cancer. 1997;29:13–23.

21. Shut HA, Snyderwise EG. DNA adducts of heterocyclic amine food mutagens: implications for mutagenesis and carcinogenesis. Carcinogenesis. 1999;20:353–68.

22. Van Maanen JM, Moonen EJ, Maas LM, Kleinjans JC, van Schooten FJ. Formation of aromatic DNA adducts in white blood cells in relation to urinary excretion of 1-hydroxypyrene during consumption of grilled meat. Carcinogenesis. 1994;15:2263–8.

23. Bingham SA, Hughes R, Cross AJ. Effect of white versus red meat on endogenous N-nitrosation in the human colon and further evidence of a dose response. Journal of Nutrition. 2002;132:3522–5.

24. Bowen DJ, Beresford SAA. Dietary interventions to prevent disease. Annual Review of Public Health. 2002;23:255–86. Kris-Etherton PM, Hecker KD, Bonanome A, Coval SM, Binkowski AM, Hilpert KF, Griel AE, Etherton TD. Bioactive compounds in foods: their role in the prevention of cardiovascular disease and cancer. American Journal of Medicine. 2002;113:71S–88S. Fairfield KM, Fletcher RH. Vitamins for chronic disease prevention in adults. JAMA (Journal of the American Medical Association). 2002;287:3116–26. Jacob RA, Aiello GM, Stephensen CB, Blumberg JB, Milbury PE, Wallock LM, Ames BN. Moderate antioxidant supplementation has no effect on biomarkers of oxidant damage in healthy men with low fruit and vegetable intakes. Journal of Nutrition. 2003;133: 740–3. Padayatty SJ, Katz A, Wang Y, Eck P, Kwon O, Lee J-H, Chen S, Corpe C, Dutta A, Dutta SK, Levine M. Vitamin C as an antioxidant: evaluation of its role in disease prevention. Journal of the American College of Nutrition. 2003;22:18–35.

25. Gross MD. Vitamin D and calcium in the prevention of prostate and colon cancer: new approaches for the identification of needs. Journal of Nutrition. 2005;135:326–31. Harris DM, Go VLW. Vitamin D and colon carcinogenesis. Journal of Nutrition. 2004;134:3463S–71S. Levine AJ, Harper JM, Ervin CM, Chen Y-H, Harmon E, Xue S, Lee ER, Frankel HD, Haile RW. Serum 25-hydroxyvitamin D, dietary calcium intake, and distal colorectal adenoma risk. Nutrition and Cancer. 2001;39:35–41.

26. Giovannucci E. The epidemiology of vitamin D and cancer incidence and mortality: a review (United States). Cancer Causes and Control. 2005;16:83–95.

27. Wactawski-Wende J, Kotchen JM, Anderson GL, et al. Calcium plus vitamin D supplementation and the risk of colorectal cancer. New England Journal of Medicine. 2006;354:684–96.

28. Forman M, Levin B. Calcium plus vitamin D$_3$ supplementation and colorectal cancer in women. New England Journal of Medicine. 2006;354:752–54.

29. Gey KF, Moser UK, Jordan P, Stahelin HB, Eichholzer M, Ludin E. Increased risk of cardiovascular disease at suboptimal plasma concentrations of essential antioxidants: an epidemiological update with special attention to carotene and vitamin C. American Journal of Clinical Nutrition. 1993;57(5 Suppl):787S–97S. Mayne ST. Beta-carotene, carotenoids, and disease prevention in humans. FASEB Journal: official publication of the Federation of American Societies for Experimental Biology. 1996;10:690–701. Miller ER, Pastor-Barriuso R, Dalal D, Riemersma RA, Appel LJ, Guallar E. Meta-analysis: high-dosage vitamin E supplementation may increase all-cause mortality. Annals of Internal Medicine. 2005;142:37–46.

30. Campbell JK, Canene-Adams K, Lindshield BL, Boileau TWM, Clinton SK, Erdman JW. Tomato phytochemicals and prostate cancer risk. Journal of Nutrition. 2004;134:3486S–92S. Heber D, Lu Q-Y, Go VLW. Role of tomatoes, tomato products and lycopene in cancer prevention. Advances in Experimental Medicine and Biology. 2001;492:29–37.

31. Milner JA. Mechanisms by which garlic and allyl sulfur compounds suppress carcinogen bioactivation. Advances in Experimental Medicine and Biology. 2001;492:69–81.

32. Grau MV, Baron JA, Sandler RS, Haile RW, Beach ML, Church TR, Heber D. Vitamin D, calcium supplementation, and colorectal adenomas: results of a randomized trial. Journal of the National Cancer Institute. 2003;95:1765–71. Gross MD. Vitamin D and calcium in the prevention of prostate and colon cancer: new approaches for the identification of needs. Journal of Nutrition. 2005;135:326–31. Moorman PG, Terry PD. Consumption of dairy products and the risk of breast cancer: a review of the literature. American Journal of Clinical Nutrition. 2004;80:5–14.

33. Wactawski-Wende J, Kotchen JM, Anderson GL, et al. Calcium plus vitamin D supplementation and the risk of colorectal cancer. New England Journal of Medicine. 2006;354:684–96.

34. Blot WJ. Alcohol and cancer. Cancer Research. 1992;52:2119S–23S. Mann RE, Smart RG, Govoni R. The epidemiology of alcoholic liver disease. Alcohol Research and Health. 2003;27:209–19.

**Check out the following sources for additional information.**

35. Bohnsack BL, Hirschi KK. Nutrient regulation of cell cycle progression. Annual Review of Nutrition. 2004;24:433–53.

36. Chao A, Thun MJ, Connell CJ, McCullough ML, Jacobs EJ, Flanders D, Rodriguez C, Sinha R, Calle EE. Meat consumption and risk of colorectal cancer. JAMA (Journal of the American Medical Association). 2005;293:172–82.

37. Donaldson MS. Nutrition and cancer: a review of the evidence for an anti-cancer diet. Nutrition Journal. 2004; 3:19–21.

38. Forman MR, Hurstin SD, Umar A, Barrett JC. Nutrition and cancer prevention: a multidisciplinary perspective on human trials. Annual Review of Nutrition. 2004;24:223–54.

39. Friso S, Choi S-W. Gene-nutrient interactions and DNA methylation. Journal of Nutrition. 2002;132:2382S–7S.

40. Go VLW, Song DA, Butrum R. Diet, nutrition and cancer prevention: where are we going from here? Journal of Nutrition. 2001;13:3121S–6S.

41. Kim Y-I. Nutrition and cancer. In: Present knowledge in nutrition, 8th ed. Bowman BA, Russell RM, editors. Washington, DC: ILSI Press; 2001.

42. Kris-Etherton PM, Lefevre M, Beecher GR, Gross MD, Keen CL, Etherton TD. Bioactive compounds in nutrition and health-research methodologies for establishing biological function: the antioxidant and anti-inflammatory effects of flavonoids on atherosclerosis. Annual Review of Nutrition. 2004;24:511–38.

43. Kritchevsky D. Diet and cancer: what's next? Journal of Nutrition. 2003;133:3827S–9S.

44. Milner JA. Incorporating basic nutrition science into health interventions for cancer prevention. Journal of Nutrition. 2003;133:3820S–6S.

45. Institute of Medicine. Dietary Reference Intakes for vitamin C, vitamin E, selenium and the carotenoids. Washington, DC: National Academy Press; 2000.

46. Stover PJ. Nutritional genomics. Physiological Genomics. 2004;16:161–5.

47. Temple NJ, Balay-Karperien AL. Nutrition in cancer prevention: an integrated approach. Journal of the American College of Nutrition. 2002;21:79–83.

48. van Gils CH, Peeters PHM, Bueno-de-Mesquite HB, Boshuizen HC, Lahmann PH, Clavel-Chapelon F, Thiebaut A, Kesse E, Sieri S, Palli D, Tumino R, Panico S, Vineis P, Gonzalez CA, Ardanaz E, Sanchez M-J, et al. Consumption of vegetables and fruits and risk of breast cancer. JAMA (Journal of the American Medical Association). 2005;293:183–93.

49. Wargovich MJ, Cunningham JE. Diet, individual responsiveness and cancer prevention. Journal of Nutrition. 2003;133:2400S–3S.

50. Willett WC. Diet and cancer. JAMA (Journal of the American Medical Association). 2005;293:233–4.

51. Wu AH. Soy and risk of hormone-related and other cancers. In: Nutrition and cancer prevention: new insights into the role of phytochemicals. New York: Kluwer Academic/Plenum Publishers; 2001.

## Chapter 12

# The Trace Minerals

*I*n addition to organic molecules such as protein, carbohydrates, lipids, and vitamins, our bodies are also made of inorganic matter. These inorganic substances include minerals and water, which together constitute over 60% of the body's weight. Minerals make up the structure of our bones and teeth, and participate in hundreds of chemical reactions. Water provides the ideal environment for many chemical processes and is the main constituent of our blood, serving as a transportation superhighway. In this chapter you will learn about some of the minerals that are essential in very small quantities for life. These minerals, such as selenium, zinc, iron, and iodine, are called the trace minerals. Although we need only minute amounts of them, they play vital roles in health. Other essential minerals that are needed in larger quantities—the major minerals—include calcium and sodium. The major minerals and water are presented in Chapter 13.

Photos © Matthew Farruggio

# The Trace Minerals: An Overview

In nutrition, the term **mineral** is used to describe inorganic atoms or molecules, other than water. Because they are required by the body in very small amounts, these minerals are micronutrients. All minerals are essential nutrients in that we must get them from the diet. This is because the body cannot make minerals from other compounds, as is the case for many macronutrients. Minerals can be neither created nor destroyed; even if you completely combust (burn) a food, the minerals will remain as ash.

The periodic table depicts all the atoms that make up our world. Although many atoms are listed in the periodic table, only some of them are minerals, and only a few of these are essential to life (Figure 12.1). Essential minerals are classified as major minerals or **trace minerals,** depending on how much we need. Major minerals are required in amounts greater than 100 mg/day, whereas less than 100 mg of each trace mineral is required daily. Sometimes trace minerals are called "microminerals" or "trace elements." The body needs at least eight trace minerals. Other trace minerals, such as fluoride, are not technically essential nutrients but may influence health in general. Some trace minerals are very difficult to measure, and scientists are still learning how the body uses them. An overview of general characteristics of the trace minerals is presented in Table 12.1.

**mineral** Inorganic substance, other than water, that is required by the body in small amounts.

**trace mineral** An essential mineral that is required in amounts less than 100 mg daily.

## FIGURE 12.1  The Periodic Table

Six major minerals and at least eight trace minerals are vital for health.

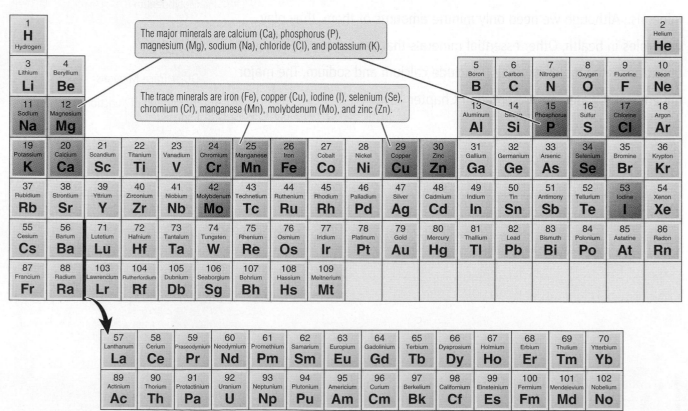

The major minerals are calcium (Ca), phosphorus (P), magnesium (Mg), sodium (Na), chloride (Cl), and potassium (K).

The trace minerals are iron (Fe), copper (Cu), iodine (I), selenium (Se), chromium (Cr), manganese (Mn), molybdenum (Mo), and zinc (Zn).

**TABLE 12.1   General Characteristics of the Trace Minerals**

| | General Characteristics |
|---|---|
| **Food Sources** | • Amount often dependent on mineral content of soil<br>• Found in all food groups<br>• Whole-grain products tend to contain more than milled cereal products<br>• Amount not influenced by cooking |
| **Digestion** | • Very little needed |
| **Absorption** | • Occurs mostly in small intestine, but also in stomach<br>• Bioavailability often influenced by form, nutritional status, and interactions with other dietary components |
| **Circulation** | • Via blood |
| **Functions** | • Most are cofactors for enzymes, some of which are involved in redox reactions<br>• Some are components of nonenzymatic proteins<br>• Some have structural roles (such as mineralization) |
| **Toxicity** | • Rare; associated with excess supplement intake or environmental exposure |

## Bioavailability and Regulation of Trace Minerals

Trace minerals are found in a wide variety of plant- and animal-based foods, and their bioavailability can be influenced by many factors, such as genetics, nutritional status, and interactions with other components of foods. Normal aging processes can also affect mineral bioavailability. Trace minerals are absorbed primarily in the small intestine and circulated to the liver via the blood. The amount of most trace minerals in the body is regulated, although the site of regulation varies. For example, iron absorption in the small intestine is adjusted to maintain healthy iron status. For other minerals, such as iodine and selenium, the site of regulation is the kidneys, and excess amounts are excreted in the urine. The regulation site of others, such as copper and manganese, is the liver, which incorporates excess minerals into the bile and subsequently eliminates them in the feces. Although there are clear exceptions (such as iron and iodine), both deficiencies and toxicities of the trace minerals are rare except in genetic disorders and environmental exposure.

## Functions of Trace Minerals in the Body

Trace minerals participate in many chemical reactions by serving as cofactors. For example, selenium is part of the quaternary structure of an enzyme that protects the body from free radical damage. When a mineral is part of an enzyme in this way it is called a cofactor, and the enzyme is called a **metalloenzyme.** The function of a cofactor is somewhat similar to that of a coenzyme, a role common to many vitamins. Binding of the mineral cofactor to the enzyme activates it, allowing it to bind to its substrate and carry out its function, as shown in Figure 12.2. Other times, trace minerals are components of larger nonenzymatic molecules, such as hemoglobin. Some trace minerals also provide structure to mineralized tissues. An example is fluoride, which provides strength to bones and teeth.

**CONNECTIONS** Recall that a free radical is a reactive molecule with 1 or more unpaired electrons. Free radicals are destructive to cell membranes, DNA, and proteins (Chapter 10, page 409).

**metalloenzyme** An enzyme that contains a mineral cofactor.

## FIGURE 12.2 Cofactors Assist Enzymes

Many minerals act as cofactors, helping enzymes to function. Enzymes with mineral cofactors are called metalloenzymes.

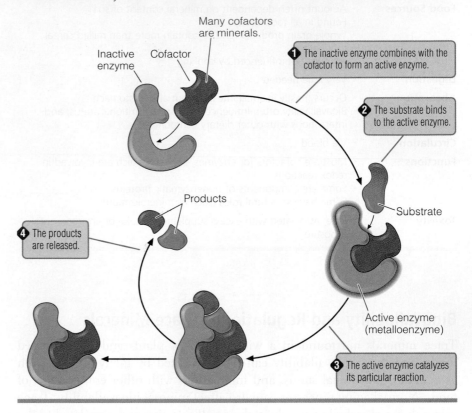

Many cofactors are minerals.

Inactive enzyme

Cofactor

**1** The inactive enzyme combines with the cofactor to form an active enzyme.

**2** The substrate binds to the active enzyme.

Substrate

Products

**4** The products are released.

Active enzyme (metalloenzyme)

**3** The active enzyme catalyzes its particular reaction.

## ESSENTIAL *Concepts*

Trace minerals are inorganic atoms or compounds the body needs in very small quantities. The bioavailability of trace minerals is influenced by many factors, such as nutritional status and other components of foods. In general, trace minerals serve functional roles as cofactors of enzymes or components of nonenzyme molecules such as hemoglobin. Some trace minerals also provide structure to the body's mineralized tissues.

# Iron

Of all the trace minerals, **iron (Fe)** is likely the most studied. Nonetheless, iron deficiency remains the most common nutrient deficiency worldwide. Once iron was thought to only affect the blood's ability to transport oxygen, but scientists now know that iron has many functions. For example, it affects cognitive development in infants and children, intellectual performance, and pregnancy outcome. Iron is also important for optimal immune function and body temperature regulation.

## Dietary Sources of Iron

Iron is in a wide variety of foods, and some examples of iron-rich foods are listed in Figure 12.3. Iron (Fe) in the diet exists in two forms—**heme iron** and **nonheme iron.** The term *heme iron* refers to iron that is bound to a heme

**iron (Fe)** A trace mineral needed for oxygen and carbon dioxide transport, energy metabolism, removal of free radicals, and synthesis of DNA.

**heme iron** Iron that is a component of a heme group; heme iron includes hemoglobin in blood, myoglobin in muscles, and cytochromes in mitochondria.

**nonheme iron** Iron that is not attached to a heme group.

**FIGURE 12.3** Good Sources of Iron

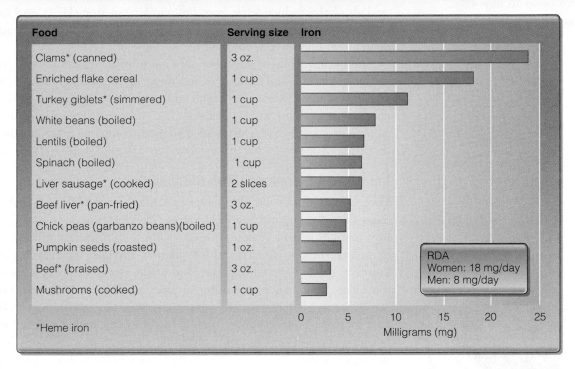

| Food | Serving size | Iron |
|------|-------------|------|
| Clams* (canned) | 3 oz. | |
| Enriched flake cereal | 1 cup | |
| Turkey giblets* (simmered) | 1 cup | |
| White beans (boiled) | 1 cup | |
| Lentils (boiled) | 1 cup | |
| Spinach (boiled) | 1 cup | |
| Liver sausage* (cooked) | 2 slices | |
| Beef liver* (pan-fried) | 3 oz. | |
| Chick peas (garbanzo beans)(boiled) | 1 cup | |
| Pumpkin seeds (roasted) | 1 oz. | |
| Beef* (braised) | 3 oz. | |
| Mushrooms (cooked) | 1 cup | |

RDA
Women: 18 mg/day
Men: 8 mg/day

0    5    10    15    20    25
Milligrams (mg)

*Heme iron

SOURCE: Data from USDA Nutrient Database, Release 17.

group. Recall from Chapter 6 that heme is an iron-containing prosthetic group that is a component of several complex proteins such as hemoglobin. Heme is also a component of myoglobin and cytochromes, found in muscles and mitochondria, respectively. Excellent sources of heme iron include shellfish, beef, poultry, and organ meats such as liver. Nonheme iron is found in green leafy vegetables, mushrooms, and legumes, and typically accounts for more than 85% of dietary iron. Recall that certain foods (such as rice, corn meal, and breads) fortified with iron can be labeled as being "enriched" cereal products. These products as well as other iron-fortified foods are also good sources of this mineral. Whether iron is in its heme or nonheme form greatly affects its bioavailability.

## Bioavailability of Iron

The bioavailability of iron is complex and influenced by many factors, including its form, a person's iron status, and the presence or absence of other dietary components. Perhaps the major factor is its form. For example, the bioavailability of heme iron is two to three times greater than that of nonheme iron. Absorption of heme iron is high, being most affected by iron status, whereas many factors can influence absorption of nonheme iron. One of the most important factors affecting nonheme iron absorption is its ionic state. Nonheme iron is found in two ionic forms in foods: the more oxidized **ferric iron ($Fe^{3+}$)** and the more reduced **ferrous iron ($Fe^{2+}$)**. Of these, the ferrous form is more readily absorbed.

### Enhancers of Nonheme Iron Bioavailability

One of the best-known enhancers of iron absorption is vitamin C (ascorbic acid), which converts ferric iron ($Fe^{3+}$) to ferrous iron ($Fe^{2+}$) in the intestinal lumen. For this to occur, vitamin C donates an electron to ferric iron ($Fe^{3+}$),

**CONNECTIONS** Cytochromes are involved in the electron transport chain and therefore in ATP production (Chapter 8, page 278).

Remember that the nutrients that must be added to "enriched foods" are iron, thiamin (vitamin $B_1$), riboflavin (vitamin $B_2$), niacin (vitamin $B_3$), and folate (Chapter 10, page 382).

Shellfish and beef are excellent sources of iron.

**ferric iron ($Fe^{3+}$)** The oxidized form of iron.

**ferrous iron ($Fe^{2+}$)** The reduced form of iron.

**CONNECTIONS** ▶ Recall that vitamin C is an antioxidant, because it donates electrons to free radicals (Chapter 10, page 409).

reducing it to its ferrous ($Fe^{2+}$) form. Thus consuming vitamin C in a meal that contains nonheme iron enhances the bioavailability of the iron. Stomach acid also helps reduce ferric iron ($Fe^{3+}$) to ferrous iron ($Fe^{2+}$), and some studies suggest that the chronic use of antacids to neutralize stomach acidity can decrease nonheme iron absorption.[1]

Another factor that enhances nonheme iron bioavailability is a compound in meat, poultry, and seafood called **meat factor.** Consuming even a small amount of meat factor along with nonheme iron-containing grains or vegetables can increase iron's bioavailability. Although the exact nature of meat factor is not known, recent research suggests it may be a carbohydrate found in muscle tissue.[2]

### Inhibitors of Nonheme Iron Bioavailability

Whereas vitamin C and meat factor enhance bioavailability of nonheme iron, other dietary components are inhibitory. These compounds, called **chelators,** bind nonheme iron, making it unavailable for absorption. Chelators of nonheme iron include phytates and polyphenols. **Phytates** are complex substances found in many vegetables, grains, and seeds. Because phytates are often lost during preparation and processing, unprocessed bran, oats, and fiber-rich foods are especially phytate rich, whereas milled cereal products are not. **Polyphenols** are found in some vegetables (such as spinach), tea, coffee, and red wine. The negative effect of phytates and polyphenols on iron bioavailability is relatively strong. For example, consuming 1 cup of coffee or tea with a meal can decrease nonheme iron absorption by 40 to 70%.[3] Moreover, although some foods (such as spinach) contain relatively high amounts of iron, they really are not especially good sources of iron, because of the phytates and/or polyphenols they contain. Note that simultaneously consuming vitamin C or meat factor can partly counteract the inhibitory effects of phytates and polyphenols.

Consuming vitamin C along with nonheme iron enhances its absorption.

## Iron Absorption, Circulation, and Storage

Although iron deficiency causes many problems, iron toxicity is perhaps more dangerous. Thus humans have complex homeostatic mechanisms to regulate how much iron is stored in the body, making sure there is neither too little nor too much. Except for bleeding associated with menstruation, injury, or childbirth, the body loses very little iron once it is absorbed. The major mechanism by which the amount of iron is regulated in the body is the alteration of its absorption. How the body regulates iron absorption and how it circulates and stores iron are described next.[4]

### Iron Absorption

As with other nutrients, absorption of iron involves transport across two membranes in the intestinal enterocyte: the brush border membrane that faces the intestinal lumen and the basolateral membrane that faces the submucosa. The transport of heme and nonheme iron across these membranes involves different mechanisms. Whereas heme iron can be transported across both membranes without chemical modification, nonheme iron must be reduced to its ferrous form ($Fe^{2+}$) prior to transport. This is why reducing substances such as vitamin C increase bioavailability of nonheme iron. Several membrane-bound iron transport proteins move iron from the intestinal lumen into the enterocyte.

Once inside the enterocyte, ferrous iron ($Fe^{2+}$) is bound to the protein **ferritin.**[5] The binding of iron to ferritin in the enterocyte is often temporary, and

**meat factor** An unidentified compound, found in meat, that increases the absorption of nonheme iron.

**chelator** (CHE – la – tor) A substance that binds compounds in the gastrointestinal tract, making them unavailable for absorption.

**phytates** (PHY – tates) Phosphorus-containing compounds often found in the outer coating of a kernel of grain, vegetables, and legumes.

**polyphenols** (pol – y – PHEN – ols) Organic compounds found in some foods.

**ferritin** (FER – ri – tin) A protein important for iron absorption and storage in the body.

the role of the enterocyte is somewhat like a gate in an airport. The passengers (like iron) must wait in the gate (like the enterocyte) until the airplane (like the body) is ready to board. However, if the airplane is overbooked, not all the passengers will be allowed to board. Similarly, ferritin temporarily holds or stores iron in the enterocyte until it receives a signal from the body that the iron should be released for further transport across the basolateral membrane. When the body needs more iron, both ferrous iron ($Fe^{2+}$) and heme iron are transported across the basolateral membrane into the submucosa for uptake into the blood.

CONNECTIONS Remember that an entero-cyte is the absorptive cell lining the intestinal mucosa (Chapter 4, page 104).

***Effect of Iron Status on Iron Absorption*** Iron status can greatly influence how much iron is absorbed, such that it increases during iron deficiency and decreases during periods of iron excess. These mechanisms involve altering the production of the iron transport proteins in the brush border cell membrane as well as of the ferritin in the enterocyte. This is illustrated in Figure 12.4.

During iron deficiency, the body increases production of the iron transport proteins while decreasing production of ferritin. In this way, iron transport into the enterocyte is increased, and there is less ferritin to bind the iron within the enterocyte. Together, these actions increase movement of iron from the intestinal lumen through the enterocyte, and ultimately into the circulation. Conversely, if the body has adequate or excess iron, production of iron transport proteins decreases and production of ferritin increases. This causes less iron to be transported into the enterocyte, and that which is transported is more likely to be retained in the cell. These iron-rich enterocytes are eventually sloughed off into the intestinal lumen and eliminated in the feces. The presence of iron in the feces can cause stools to be black in color. Enterocytes lining the small intestine thus serve as important regulators of iron status, helping prevent both iron deficiency and toxicity.

## FIGURE 12.4 Effect of Iron Status on Iron Absorption

The body increases iron absorption when it needs iron and decreases iron absorption when it has excess amounts.

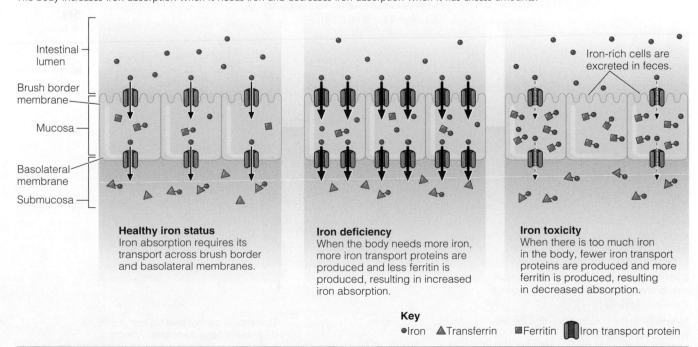

**Healthy iron status**
Iron absorption requires its transport across brush border and basolateral membranes.

**Iron deficiency**
When the body needs more iron, more iron transport proteins are produced and less ferritin is produced, resulting in increased iron absorption.

**Iron toxicity**
When there is too much iron in the body, fewer iron transport proteins are produced and more ferritin is produced, resulting in decreased absorption.

Iron-rich cells are excreted in feces.

**Key**
●Iron   ▲Transferrin   ■Ferritin   ▊Iron transport protein

### Iron Circulation, Uptake into Cells, and Storage

As shown in Figure 12.4, once iron enters the blood it binds to **transferrin,** a protein produced in the liver. Transferrin delivers iron to all body cells to proteins on cell membranes called **transferrin receptors.** As shown in Figure 12.5, the number of transferrin receptors on a cell membrane is regulated by the cell's needs. For example, if enough iron is in the cell, the number of transferrin receptors decreases, inhibiting uptake of additional iron. Cells needing iron produce more receptors so they can bind or "capture" more iron for uptake.

Excess iron is stored in the liver, bone marrow, and spleen. There are two iron storage compounds in these tissues—ferritin and **hemosiderin.** These large proteins contain about a third of a healthy person's total iron. As you have learned, ferritin is also produced in enterocytes where it helps regulate iron absorption. Note that small amounts of the storage form of ferritin present in liver, bone, and spleen are continually released into the blood, where its concentration is proportional to the amount of iron stored in the body. Because of this, blood ferritin is often measured to assess iron status. Whereas ferritin is considered the main storage form of iron, hemosiderin is needed for more long-term storage. Hemosiderin protects the body from iron toxicity when intake is chronically high.

## Functions of Iron

The body's iron-containing compounds are classified as being either for storage or functional use. Functional iron is contained in a variety of heme proteins and nonheme proteins. Heme proteins include hemoglobin, myoglobin, and the cytochromes. Nonheme proteins include enzymes for which iron is a cofactor.

### FIGURE 12.5 Regulation of Iron Uptake into Cells

Iron uptake into cells is regulated by the abundance of transferrin receptors on their cell membranes. Tissue A would be able to take up iron, whereas Tissue B would not.

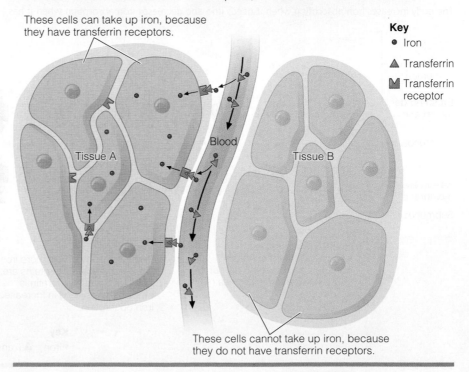

These cells can take up iron, because they have transferrin receptors.

**Key**
- ● Iron
- ▲ Transferrin
- Ϻ Transferrin receptor

Blood

Tissue A

Tissue B

These cells cannot take up iron, because they do not have transferrin receptors.

---

**transferrin** (trans – FER – rin) A protein important for iron transport in the blood.

**transferrin receptor** Protein found on cell membranes that binds transferrin, allowing the cell to take up iron.

**hemosiderin** (he – mo – SID – er – in) An iron-storing protein.

## Oxygen and Carbon Dioxide Transport: Hemoglobin

**Hemoglobin** is the most abundant protein in red blood cells, containing four protein subunits and four iron-containing heme groups (Figure 12.6). As hemoglobin travels through the blood vessels of the lungs, it comes in contact with oxygen, which then attaches to iron atoms in hemoglobin. As the oxygenated blood then circulates from the lungs to other tissues in the body, hemoglobin delivers oxygen to cells that need it for aerobic energy metabolism. Once oxygen is released, hemoglobin picks up carbon dioxide, a waste product of energy metabolism. Deoxygenated, carbon dioxide-rich blood is circulated back to the lungs, where the carbon dioxide is eliminated when the person exhales. Without sufficient hemoglobin, oxygen availability to tissues decreases, which affects energy metabolism. This can result in lack of energy and fatigue. Red blood cells contain about two thirds of the total body iron and have a normal life span of about 120 days. However, when red blood cells are degraded, most of the iron is recycled or stored for later use.

## Oxygen Reservoir: Myoglobin

**Myoglobin** is another iron-containing protein that contains heme. However, unlike hemoglobin, which is found in blood, myoglobin is found in muscle cells. It consists of a single heme group and a single protein subunit. Myoglobin acts as a reservoir of oxygen, releasing oxygen to muscle cells when needed for ATP production and, ultimately, for muscle contraction.

## Cellular Energy Metabolism

In addition to delivering oxygen to cells, iron is also needed for other aspects of energy metabolism (that is, ATP production). For example, it is a basic component of the **cytochromes,** which are heme-containing protein complexes that function in the electron transport chain. Cytochromes serve as electron carriers, ultimately allowing the conversion of adenosine diphosphate (ADP)

**CONNECTIONS** Recall that a prosthetic group such as heme is a nonprotein component of a protein and is important for the quaternary structure of a protein (Chapter 6, page 183).

Recall that oxygen is the final electron acceptor in the electron transport chain, and is thus essential for ATP production (Chapter 8, page 278).

---

**FIGURE 12.6** **Hemoglobin Contains Iron**

Hemoglobin is a complex protein containing 4 atoms of iron in its quaternary structure.

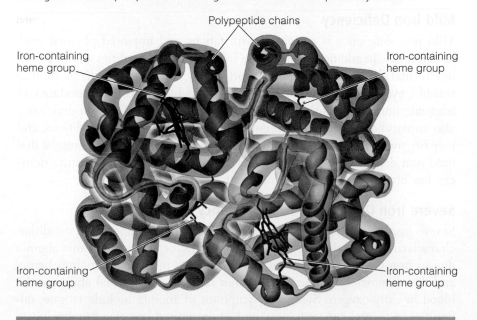

Polypeptide chains

Iron-containing heme group

Iron-containing heme group

Iron-containing heme group

Iron-containing heme group

---

**hemoglobin** (HE – mo – glob – in) A complex protein composed of four iron-containing heme groups and four protein subunits needed for oxygen and carbon dioxide transport in the body.

**myoglobin** (MY – o – glob – in) A heme protein found in muscle.

**cytochrome** A heme protein complex that is part of the electron transport chain in mitochondria.

to adenosine triphosphate (ATP). Iron also serves as a cofactor for a variety of nonheme-containing enzymes involved in the electron transport chain, citric acid cycle, and gluconeogenesis.

## Other Roles of Iron

Iron is also a cofactor of many other important enzymes that help metabolize drugs and remove toxins from the body. Activity of these enzymes, collectively called the **cytochrome P450** enzymes, is thought to influence risk for many chronic diseases such as cancer and cardiovascular disease.[6] Iron is also a cofactor for antioxidant enzymes that stabilize free radicals. In this way, iron is important for protecting DNA, cell membranes, and proteins from oxidative damage that can be harmful to cells. Iron is also a cofactor for an enzyme needed for DNA synthesis. This is one reason why iron is so important for optimal growth and development.

## Iron Deficiency

Iron deficiency is the most common nutritional deficiency in the United States and the world.[7] Because iron requirements increase during growth and development, iron deficiency is typically seen in infants, growing children, and pregnant women. Iron is lost in the blood each month during the menstrual cycle, and women of child-bearing age are also at increased risk for iron deficiency. Interestingly, the eating of nonnutritive substances such as dirt and clay, often referred to as **pica** or geophagia, is also thought to be associated with iron deficiency.[8] Pica is most prevalent in pregnant women. Whereas some researchers believe pica *results* in iron deficiency, others believe iron deficiency may somehow *cause* pica. A better understanding of this relationship requires further study.

Although iron deficiency was once thought to cause only anemia, scientists now know that it can influence many aspects of health. Thus experts are interested in understanding the entire spectrum of iron deficiency—from depletion of iron stores and mild iron deficiency to severe iron deficiency. Several methods are used to assess iron status, and you can read more about these in the Focus on Clinical Applications feature.

## Mild Iron Deficiency

Mild iron deficiency is associated with fatigue and impaired physical work performance. In addition, it can cause behavioral abnormalities and impaired intellectual abilities in children.[9] Unfortunately, some of these effects are irreversible, even after iron supplementation. This highlights the importance of adequate iron intake during the very early years of life. Mild iron deficiency also impairs body temperature regulation,[10] especially in cold conditions, and may negatively influence the immune system.[11] Some studies also suggest that mild iron deficiency during pregnancy increases the risk of premature delivery, low birth weight, and maternal mortality.[12]

## Severe Iron Deficiency: Iron Deficiency Anemia

Severe iron deficiency causes microcytic hypochromic anemia, a condition characterized by small, pale red blood cells. Microcytic hypochromic anemia due to iron deficiency is caused by the inability to produce enough heme, and thus hemoglobin. The term *anemia* refers to a decreased ability of the blood to carry oxygen. Signs and symptoms of anemia include fatigue, difficulties in mental concentration, and compromised immune function.

CONNECTIONS Recall that microcytic hypochromic anemia can also be caused by vitamin $B_6$ deficiency (Chapter 10, page 397).

**cytochrome P450** An iron-containing enzyme that helps stabilize free radicals.

**pica** (PI – ca) An abnormal eating behavior that involves consuming nonfood substances such as dirt or clay.

# *Focus On* CLINICAL APPLICATIONS

## Assessing Iron Status

There are several ways to assess a person's iron status. In general, these tests fall into three categories depending on whether they assess iron stores or are meant to diagnose mild iron deficiency or severe iron deficiency. When iron intake is inadequate, the body begins to use stored iron, causing iron stores to eventually become depleted. The depletion of iron stores is called mild iron deficiency or impaired iron status, and can eventually progress into more severe iron-deficiency anemia. Clinicians use a variety of biochemical measurements to assess impaired iron status and more severe iron deficiency.

- *Serum ferritin concentration.* Much of the iron in the body is stored as ferritin, and serum ferritin levels in the blood are correlated with the amount of stored iron. Because serum ferritin levels drop long before anemia develops, measuring ferritin concentrations allows detection of iron deficiency in its early stages. Serum ferritin levels lower than 12 µg/L indicate depleted iron stores.

- *Total iron-binding capacity (TIBC).* Iron is transported in the blood attached to the protein transferrin, and each transferrin molecule can hold 2 atoms of iron. Total iron-binding capacity (TIBC) is a laboratory measurement related to the total number of *free* iron-binding sites on transferrin. When iron stores are being depleted, TIBC is increased, meaning there is insufficient iron available to saturate the transferrin molecule. A TIBC greater than 400 µg/dL is indicative of iron depletion.

- *Serum transferrin saturation.* In well-nourished people, most of the transferrin in the blood contains (or is "saturated" with) iron. However, during early stages of iron deficiency some transferrin does not contain iron. In this way, serum transferrin saturation reflects the *percentage* of transferrin that contains iron, and low levels of serum transferrin saturation reflect mild iron deficiency. Serum transferrin saturation is calculated as serum iron ÷ TIBC × 100. As with TIBC, low levels of transferrin saturation can be detected before iron deficiency anemia develops, and a value of less than 16% indicates mild iron deficiency.

- *Hemoglobin concentration.* As iron deficiency progresses, adequate amounts of hemoglobin can no longer be made, and its concentration in the blood decreases. However, because other nutrient deficiencies (such as protein or vitamin $B_6$) can also inhibit hemoglobin synthesis, low hemoglobin concentrations cannot by themselves be used to diagnose iron deficiency. Instead, they are usually used as an inexpensive and easy way to test for any type of anemia. Because hemoglobin concentration remains normal during mild iron deficiency, this measure can only be used to test for more severe forms of iron deficiency. A value of less than 13 g/dL or 12 g/dL indicates iron deficiency for men and women, respectively.

- *Hematocrit.* Hematocrit is the percentage of total blood that is made of red blood cells (Figure 12.7). Hematocrit values decrease during iron deficiency, because microcytic anemia causes red blood cells to be small. This measure is an easy and inexpensive way to test for iron deficiency anemia. However, as with hemoglobin concentration, a low hematocrit can be caused by many other nutritional deficiencies. Nonetheless, it is often used as an initial screen of iron status. A hematocrit of less than 39% or 36% for men and women, respectively, indicates anemia.

## FIGURE 12.7 Hematocrit

Hematocrit is a measure of the portion of blood consisting of red blood cells. Iron deficiency anemia lowers hematocrit.

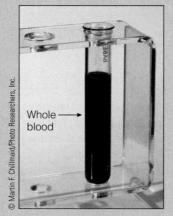
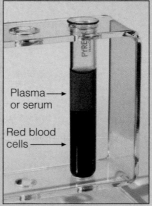

© Martin F. Chillmaid/Photo Researchers, Inc.

Whole blood

Plasma or serum

Red blood cells

100%

0%

The percentage of whole blood that is composed of red blood cells is called the "hematocrit."

## Basics of Iron Supplementation

Iron supplementation is sometimes needed when diet alone cannot maintain iron status or during specific periods of the life cycle (such as pregnancy) when iron requirements are especially high. Supplemental iron is available in two general forms: ferrous iron ($Fe^{2+}$) and ferric iron ($Fe^{3+}$). Of these, the ferrous forms are best absorbed.[13] You can find these types of supplements by looking for the terms "ferrous fumarate," "ferrous sulfate," and "ferrous gluconate" on product labels.

## Iron Toxicity

Although iron toxicity (also called iron overload) from foods is rare, it is sometimes seen with overdoses of medicinal or supplemental iron. In fact, accidental consumption of toxic levels of iron is one of the most common causes of childhood poisoning in the United States. Symptoms of iron toxicity include vomiting, diarrhea, constipation, and black stools. In severe cases, it can cause death. This is because excess iron is deposited in soft tissues such as the liver, heart, and muscles, impairing their function. Some researchers have also suggested that excessive iron intake from supplements can lead to cardiovascular disease and some forms of cancer.[14] However, results of studies are inconsistent, and more research is needed before conclusions can be made concerning these hypotheses. Iron toxicity can also be caused by genetic abnormalities. One example, called **hereditary hemochromatosis,** is described in the Focus on Clinical Applications feature.

## Recommended Intakes for Iron

**hereditary hemochromatosis** (he – ma – to – chrom – a – TO – sis) A genetic abnormality resulting in increased absorption of iron in the intestine.

Iron requirements are highest during periods of rapid growth and development and are influenced by the amount of lean body mass (muscle) a person has. Because women of child-bearing age lose iron each month during

# Focus On CLINICAL APPLICATIONS
## Hereditary Hemochromatosis

Although iron toxicity can be caused by consuming large amounts of iron from medicine or supplements, it can also be caused by a genetic abnormality called hereditary hemochromatosis. Although this condition has long been recognized in the medical community, its true cause was only discovered in 1996.[15] Scientists now know that people with this disease have a defect in the gene (DNA) that codes for one of the transport proteins required for regulating intestinal iron absorption. As a result, greater than normal amounts of iron are absorbed into the blood. The gene for hereditary hemochromatosis is commonly found in people of northern European ancestry, and its symptoms include arthritis, chronic fatigue, lack of normal menstruation in women, and abnormal liver function. Researchers think that 4 of every 1,000 Caucasians has hereditary hemochromatosis, although not all people with the defective gene develop symptoms. In severe cases, hereditary hemochromatosis can lead to liver failure, liver cancer, and increased risk for diabetes and heart disease.

Until recently, the only way to test for hereditary hemochromatosis was by measuring levels of iron in the liver. This required a small sample of liver tissue to be taken. However, this disease can now be diagnosed by genetic testing of the DNA in blood.

Although hereditary hemochromatosis cannot be prevented or cured, it can be treated by frequent removal of blood. It is also highly recommended that people with this condition avoid iron-containing supplements and vitamin C supplements (especially with food). In addition, they are advised to consume only small portions of iron-rich foods. Perhaps genetic engineering will someday allow the defective genes causing hereditary hemochromatosis to be replaced by normal ones.

menstruation, requirements for females in this age group are relatively high. Recommended Dietary Allowances (RDAs) for men and women are 8 and 18 mg/day, respectively. During pregnancy the recommended intake increases to 27 mg/day. A serving (3 oz.) of beef contains only about 3 mg of iron, so it is difficult for some women to meet the RDA, especially during pregnancy. For this reason, iron supplements are often recommended. To prevent gastrointestinal distress from iron excess, a Tolerable Upper Intake Level (UL) has been set at 45 mg/day. A complete list of the Dietary Reference Intake (DRI) values for all trace minerals is provided inside the front cover of this book.

### Special Recommendations for Vegetarians and Endurance Athletes

Vegetarians may have difficulty consuming the needed amount of iron, because meat provides substantial amounts of highly bioavailable iron, and nonheme iron found in plant foods has low bioavailability. In fact, the Institute of Medicine estimates that dietary iron requirements for vegans consuming no animal-based foods are 80% higher than requirements for nonvegetarians. To take advantage of the positive effect of meat factor on iron absorption, vegetarians who eat fish should try to consume it together with nonheme iron sources in the same meal. For example, consuming even small amounts of fish along with iron-fortified pasta can increase the amount of iron absorbed. Both vegetarians and vegans may require iron supplements.

Athletes engaged in endurance sports such as long-distance running may also have increased iron requirements. There are many possible reasons for this, including increased blood loss in feces and urine, and chronic rupture of red blood cells within the feet. The Institute of Medicine suggests that iron requirements for endurance athletes may be increased by as much as 70%.

## ESSENTIAL *Concepts*

Two forms of iron are found in foods: heme iron and nonheme iron. Heme iron is more readily absorbed in the intestine, whereas nonheme iron bioavailability is influenced by many factors. Good sources of heme iron include meat and seafood. Nonheme iron is found in some vegetables, enriched cereal products, and fortified foods. Regulation of iron homeostasis occurs primarily in the small intestine, such that poor iron status increases iron absorption. Iron is a component of hemoglobin and myoglobin and is needed for oxygen transport, and thus aerobic metabolism. Iron is also associated with the electron transport chain and is involved in protecting the body from toxins and free radicals. Iron deficiency can cause fatigue, decreased work performance, and impaired intellectual abilities. Severe iron deficiency causes microcytic hypochromic anemia. Iron toxicity resulting from genetic disorders and overconsumption of iron supplements or medicine can be fatal.

## Copper

**Copper (Cu)** is present in two forms: its oxidized **cupric** form ($Cu^{2+}$) and its reduced **cuprous** form ($Cu^+$). Note that, as with iron, the ending "-ous" represents the more reduced form, whereas "-ic" represents the more oxidized form. Copper is a cofactor for several enzymes involved in a wide variety of processes, such as ATP production and protection from free radicals. As you will see, copper and iron share many similarities in terms of food sources, absorption, and functions in the body.[16]

**CONNECTIONS** Recall that Dietary Reference Intakes (DRIs) are reference values for nutrient intakes. Recommended Dietary Allowances (RDAs) represent values sufficient to meet the needs of about 97% of the population. When less information is available, Adequate Intake (AI) values are provided. Tolerable Upper Intake Levels (ULs) indicate intake levels that should not be exceeded (Chapter 2, page 36).

**copper (Cu)** An essential trace mineral that acts as a cofactor for nine enzymes involved in redox reactions.

**cupric ion ($Cu^{2+}$)** The more oxidized form of copper.

**cuprous ion ($Cu^+$)** The more reduced form of copper.

## Dietary Sources and Bioavailability of Copper

Although organ meats such as liver are likely the best sources of copper, this mineral is also found in shellfish, whole-grain products, mushrooms, nuts, and legumes (Figure 12.8). Copper bioavailability can decrease in response to heavy use of antacids, which cause copper to form insoluble complexes that cannot be absorbed. High amounts of iron in the diet can also decrease copper absorption. This is because iron and copper compete for the same transport proteins in enterocyte cell membranes.

## Absorption, Metabolism, and Regulation of Copper

Copper is absorbed in the small intestine, and to a lesser extent the stomach, and absorption is influenced by copper status. As with iron, copper absorption increases during periods of copper deficiency and decreases when copper status is adequate. However, the mechanisms by which this occurs are poorly understood. On absorption, copper circulates in the blood to the liver, where it is bound to its transport protein **ceruloplasmin.** Excess copper is not stored to any great extent. Instead, the liver incorporates it into bile and eliminates it in the feces.

**ceruloplasmin** (ce – RU – lo – plas – min) The protein that transports copper in the blood.

**cytochrome c oxidase** A copper-containing enzyme needed in the electron transport chain.

**CONNECTIONS** Oxidation is the loss of electrons, whereas reduction is the gain of electrons (Chapter 3, page 67).

## Functions of Copper

Copper is a cofactor for at least nine metalloenzymes involved in reduction-oxidation (redox) reactions important in ATP production, iron metabolism, neural function, antioxidant function, and connective tissue synthesis. For example, copper serves as a cofactor for the enzyme **cytochrome c oxidase,**

## FIGURE 12.8   Good Sources of Copper

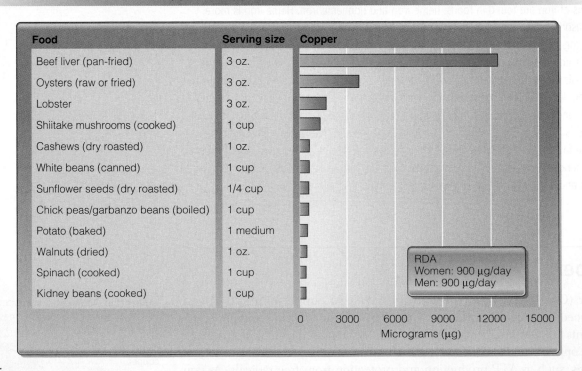

| Food | Serving size | Copper |
|---|---|---|
| Beef liver (pan-fried) | 3 oz. | |
| Oysters (raw or fried) | 3 oz. | |
| Lobster | 3 oz. | |
| Shiitake mushrooms (cooked) | 1 cup | |
| Cashews (dry roasted) | 1 oz. | |
| White beans (canned) | 1 cup | |
| Sunflower seeds (dry roasted) | 1/4 cup | |
| Chick peas/garbanzo beans (boiled) | 1 cup | |
| Potato (baked) | 1 medium | |
| Walnuts (dried) | 1 oz. | |
| Spinach (cooked) | 1 cup | |
| Kidney beans (cooked) | 1 cup | |

RDA
Women: 900 µg/day
Men: 900 µg/day

Micrograms (µg)

SOURCE: Data from USDA Nutrient Database, Release 17.

which combines electrons, hydrogen ions, and oxygen to form water in the electron transport chain. Thus copper is vital for ATP production, and copper levels tend to be highest in tissues with high metabolic activity. Copper is also a cofactor for the enzyme **superoxide dismutase,** which converts the superoxide free radical ($O_2^-$) to the less harmful hydrogen peroxide molecule ($H_2O_2$). The synthesis of norepinephrine (a neurotransmitter) and collagen (a protein needed for connective tissue) also requires copper.

## Copper Deficiency and Toxicity

Copper deficiency is rare but occurs occasionally in hospitalized patients and preterm infants receiving improper nutrition support. People consuming large amounts of antacids can also develop copper deficiency because of decreased copper absorption. Signs and symptoms of deficiency include defective connective tissue, anemia, and neural problems. In infants and young children, copper deficiency causes bones to weaken. Mild copper deficiency has also been associated with poor blood glucose regulation[17] and depressed immune function.[18] However, further studies are needed to better understand these relationships.

Because copper toxicity is rare in humans, very little is known about it. However, there have been situations in which people have consumed large doses of copper from drinking water or contaminated soft drinks.[19] In these cases, copper toxicity caused cramping, nausea, and diarrhea. In addition, copper toxicity may cause liver damage.

## Recommended Intakes for Copper

An RDA for copper has been established at 900 μg/day. To prevent possible liver damage due to copper toxicity, a UL of 10 mg (10,000 μg)/day has been established.

## ESSENTIAL *Concepts*

Copper is similar to iron in its regulation and use in the body. Homeostasis of this mineral occurs via alterations in absorption, such that it increases during copper deficiency. Excess copper is excreted in the bile. Copper is a cofactor for enzymes involved in redox reactions and is important for energy metabolism, neural function, and antioxidant reactions. Good sources of copper include liver, shellfish, mushrooms, and nuts. Copper deficiency, which is rare, causes connective tissue and bones to weaken, anemia, and neural problems. Copper toxicity results in gastrointestinal distress and perhaps liver damage.

# Iodine (Iodide)

**Iodine (I)** is needed for only one function in the body. That is, it is an essential component of the hormones produced by the thyroid gland. Thyroid hormones regulate growth, reproduction, and energy metabolism. In addition, they influence the immune system and neural development. Technically, most iodine in the body is in the anion form of **iodide (I⁻)**. However, to be consistent with much of the nutrition literature we refer to this mineral as "iodine."

**superoxide dismutase** A copper-containing enzyme that reduces the superoxide free radical to form hydrogen peroxide.

**iodine (I)** An essential trace mineral that is a component of the thyroid hormones.

**iodide (I⁻)** The most abundant form of iodine in the body.

© Food Collection/Stockfood America

Because of frequent use of seafood and sea-weed (nori), Asian cuisines are often high in iodine.

**thyroid-stimulating hormone (TSH)** A hormone produced in the pituitary gland that stimulates uptake of iodine by the thyroid gland.

**goitrogens** (GOIT – ro – gens) Compounds found in some vegetables that decrease iodine utilization by the thyroid gland.

## Dietary Sources of Iodine

The iodine content of foods frequently depends on the iodine content of the soil and water used to grow them (Figure 12.9). Ocean fish and mollusks tend to contain high amounts of iodine, because they concentrate iodine found in seawater into their tissues. Seaweed (such as nori), which is used in many Asian cuisines, also contains iodine, and dairy products are excellent sources, because iodine is used in milk processing. However, most of the iodine we consume comes from iodized salt. There are two kinds of table salt available—noniodized and iodized. Iodized salt has been fortified with iodine. In fact, about half of the salt that Americans put on their food is iodized. You can read more about the iodine fortification of salt in the Focus on Food feature on page 498.

## Absorption, Metabolism, and Regulation of Iodine

Iodine is highly bioavailable, being almost completely absorbed in the small intestine and, to a lesser extent, the stomach. Once in the blood, iodine is rapidly taken up by the thyroid gland and incorporated into the thyroid hormones. **Thyroid-stimulating hormone (TSH),** which is produced by the pituitary gland located at the base of the brain, regulates iodine uptake by the thyroid gland (Figure 12.10). During periods of iodine deficiency, TSH release increases, in turn increasing uptake of iodine by the thyroid gland. The opposite happens when iodine status is satisfactory. Excess iodine is excreted in the urine.

Compounds called **goitrogens** can inhibit iodine uptake by the thyroid gland. Goitrogens are found in cassava (a root eaten worldwide) and in cru-

## FIGURE 12.9 Good Sources of Iodine

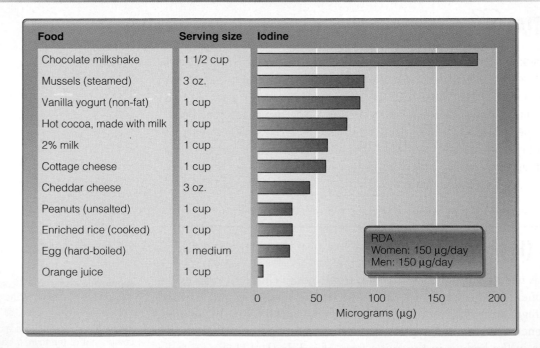

| Food | Serving size | Iodine |
|------|-------------|--------|
| Chocolate milkshake | 1 1/2 cup | |
| Mussels (steamed) | 3 oz. | |
| Vanilla yogurt (non-fat) | 1 cup | |
| Hot cocoa, made with milk | 1 cup | |
| 2% milk | 1 cup | |
| Cottage cheese | 1 cup | |
| Cheddar cheese | 3 oz. | |
| Peanuts (unsalted) | 1 cup | |
| Enriched rice (cooked) | 1 cup | |
| Egg (hard-boiled) | 1 medium | |
| Orange juice | 1 cup | |

RDA
Women: 150 µg/day
Men: 150 µg/day

Micrograms (µg)

SOURCE: Data from Hands E. *Food Finder—Vitamin and Mineral Source Guide,* 3rd ed. Salem, OR: ESHA Research, 1995.

## FIGURE 12.10 Regulation of Iodine Uptake by the Thyroid Gland

Thyroid-stimulating hormone (TSH), made in the pituitary gland, stimulates iodine uptake by the thyroid gland and production of thyroid hormones ($T_3$ and $T_4$). TSH release increases during iodine deficiency.

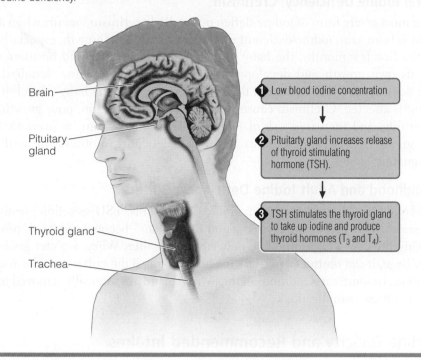

Brain

Pituitary gland

Thyroid gland

Trachea

**1** Low blood iodine concentration

**2** Pituitarty gland increases release of thyroid stimulating hormone (TSH).

**3** TSH stimulates the thyroid gland to take up iodine and produce thyroid hormones ($T_3$ and $T_4$).

ciferous vegetables such as cabbage, cauliflower, and brussels sprouts. The term *goitrogen* refers to the fact that they can potentially cause a disease called *goiter*, which is described in detail below. Consuming goitrogens typically poses no problem except in situations of very low iodine intake or in people who have thyroid dysfunction.

## Functions of Iodine

Iodine is an essential component of the thyroid hormones **thyroxine** ($T_4$) and **triiodothyronine** ($T_3$). Thyroxine, which contains 4 iodine atoms, is produced in the thyroid gland and is converted to the more active form triiodothyronine, which contains 3 iodine atoms. $T_3$ and $T_4$ help regulate energy metabolism, growth, and development. They are especially critical for proper development of the brain, spinal cord, and skeleton during fetal growth. Because thyroid hormones are involved in regulating energy metabolism, a deficiency causes severe fatigue and lethargy. However, the mechanisms by which $T_3$ and $T_4$ influence such a vast number of physiological systems are not completely understood.

## Iodine Deficiency

Iodine deficiency is a significant public health problem worldwide. Countries where it is most prevalent are those without iodized salt and those that are not bordered by an ocean or sea.[20] Manifestations of iodine deficiency take on many forms, which are collectively called **iodine deficiency disorders**

Cruciferous vegetables contain compounds called goitrogens that can decrease iodine absorption.

© Aurora Photos/StockFood Munich/StockFood America

**thyroxine ($T_4$)** The less active form of thyroid hormone; contains 4 atoms of iodine.

**triiodothyronine ($T_3$)** The more active form of thyroid hormone; contains 3 atoms of iodine.

**iodine deficiency disorders (IDDs)** A broad spectrum of conditions caused by inadequate iodine.

Cretinism occurs when an iodine deficient mother becomes pregnant, causing her child to experience iodine deficiency *in utero.*

© Wellcome Photo Library

**cretinism** A form of IDD that affects babies born to iodine-deficient mothers.

**goiter** (GOIT – er) A form of IDD that affects children and adults; characterized by an enlarged thyroid gland.

(IDDs). Which type of IDD a person has depends on various factors, including severity of the deficiency and its timing in the life cycle. The two most well-studied forms of IDD are cretinism and goiter.

### Fetal Iodine Deficiency: Cretinism

The most severe form of iodine deficiency, called **cretinism,** occurs when a baby is born to an iodine-deficient mother.[21] During fetal growth, especially in the first few months, the baby relies on the mother's thyroid hormones for its own growth and development. If the mother is iodine deficient, she does not produce sufficient thyroid hormones, and the baby's development is affected. Cretinism causes severe mental retardation, poor growth, infertility, and increased risk of mortality. This disease can be prevented by giving iodine supplements to iodine-deficient mothers in early pregnancy.

### Childhood and Adult Iodine Deficiency: Goiter

When iodine deficiency occurs in children and adults, TSH secretion greatly increases. This causes the thyroid gland to grow and become enlarged, producing the classic sign of iodine deficiency—a **goiter.** When a goiter grows very large, it can obstruct a person's trachea, making it difficult to breathe and swallow. In some cases, portions of the thyroid gland are surgically removed to prevent these complications.

## Iodine Toxicity and Recommended Intakes

Although most people tolerate excess iodine intake, iodine toxicity has been documented from high intakes from food, water, and supplements. Iodine toxicity can take several forms, including both underactive thyroid activity (hypothyroidism) and overactive thyroid activity (hyperthyroidism). In fact, it can actually cause goiters to form in some people. The RDA for iodine intake is 150 μg/day. To prevent iodine toxicity, a UL of 1,100 μg/day has been established.

## *Focus On* FOOD

### Iodine Deficiency and Iodine Fortification of Salt

Although iodine deficiency is now quite rare in the United States, this was not always the case. Prior to the fortification of salt with iodine in the 1920s, goiter was a major public health problem in some regions of the country. Specifically, it was common in regions that spanned the Great Lakes and Rocky Mountains, once referred to as the "Goiter Belt." A survey done in Michigan in the early 1920s (prior to iodination of salt) found that about 39% of schoolchildren had visible goiters.[22] When it was discovered that iodine deficiency was the cause of goiter, statewide campaigns for goiter prevention were launched promoting iodized salt as the hopeful cure. Within four years, there was a 75% reduction in childhood goiter,[23] and by the 1950s, fewer than 1% of children were found to have goiter.[24] Clearly, consuming iodized salt has drastically improved the health of many people. Today, U.S. consumers have the choice to purchase noniodized or iodized salt—the difference in cost typically being less than 10¢.

This success story is a prime example of how scientists can team up with public health officials and food manufacturers to improve the well-being of millions of people worldwide via better nutrition. Currently, some charitable organizations, such as Kiwanis International, are working with international health organizations to eradicate all iodine deficiency disorders (IDDs) around the world.

# ESSENTIAL *Concepts*

Iodine is found in fortified salt (iodized salt), shellfish, and dairy products. Its bioavailability is typically high, and excess iodine is excreted in the urine. Iodine is a component of the thyroid hormones thyroxine ($T_4$) and triiodothyronine ($T_3$),which are important for growth, development, brain function, and energy metabolism. However, the presence of goitrogens in some vegetables can decrease incorporation of iodine into $T_3$ and $T_4$. The many forms of iodine deficiency are collectively called iodine deficiency disorders (IDDs). Cretinism is the form of IDD that affects infants born to iodine-deficient mothers and causes mental retardation, poor growth, and delayed development. IDD in children and adults can cause goiters, a condition characterized by an enlarged thyroid gland. Goiters can be serious if they obstruct a person's trachea. Iodine toxicity causes both hypo- and hyperthyroidism.

Goiter is a result of compensatory growth of the thyroid gland in response to iodine deficiency.

# Selenium

The essentiality of **selenium (Se)** was only discovered in the 1950s. Since that time much has been learned about how the body uses this trace mineral. Selenium is thought to be involved in decreasing risk for cancer, protecting the body from toxins and free radicals, slowing the aging process, and enhancing immunity. Because of this dynamic group of functions, many researchers continue to actively study this important nutrient.

## Dietary Sources of Selenium

Foods contain several forms of selenium, but typically it is associated with the amino acid methionine. Usually methionine contains sulfur. However, when it contains selenium it is called **selenomethionine.** The best sources of selenium are nuts, seafood, and meats (Figure 12.11). In addition, although it does not provide a large amount of selenium to the diet, garlic is also a good source. Grains can contain selenium, but their content is highly dependent on that of the soil in which they grow. For example, in some areas of China where the soil selenium content is very low, grains contain negligible amounts. Conversely, in some portions of the western United States where the soil selenium content is very high, grains can contain toxic levels. Grazing animals have been known to develop selenium toxicity from eating grass grown in soils that contain high levels of selenium. Water can also be a source of this mineral.

## Absorption, Metabolism, and Regulation of Selenium

The bioavailability of selenium in foods is high, and absorption of this mineral in the intestine is not regulated. Therefore, almost all selenium that is consumed enters the blood. In the body, selenium is taken up by cells, incorporated into selenomethionine, and used to make proteins. Proteins that contain selenomethionine are called **selenoproteins.** Metabolism of selenoproteins releases selenium that can then be inserted into other proteins. In this way, the body stores a relatively large amount of selenium in muscles as selenomethionine and selenoproteins. Blood concentrations of selenium are maintained by increasing or decreasing the amount excreted in the urine.

**selenium (Se)** An essential trace mineral that is important for redox reactions, thyroid function, and activation of vitamin C.

**selenomethionine** The amino acid methionine that has been altered to contain selenium instead of sulfur.

**selenoprotein** A protein that contains selenomethionine instead of sulfur-containing methionine.

## FIGURE 12.11 Good Sources of Selenium

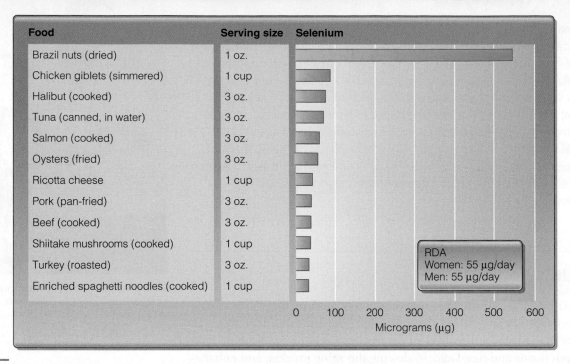

| Food | Serving size | Selenium |
|------|-------------|----------|
| Brazil nuts (dried) | 1 oz. | |
| Chicken giblets (simmered) | 1 cup | |
| Halibut (cooked) | 3 oz. | |
| Tuna (canned, in water) | 3 oz. | |
| Salmon (cooked) | 3 oz. | |
| Oysters (fried) | 3 oz. | |
| Ricotta cheese | 1 cup | |
| Pork (pan-fried) | 3 oz. | |
| Beef (cooked) | 3 oz. | |
| Shiitake mushrooms (cooked) | 1 cup | |
| Turkey (roasted) | 3 oz. | |
| Enriched spaghetti noodles (cooked) | 1 cup | |

RDA
Women: 55 μg/day
Men: 55 μg/day

Micrograms (μg)

SOURCE: Data from USDA Nutrient Database, Release 17.

## Functions of Selenium

There are at least 14 selenoproteins in the body. One group of enzymatic selenoproteins called **glutathione peroxidases** has redox functions, protecting against oxidative damage.[25] Other selenoproteins are needed for thyroid function and for vitamin C metabolism. Because selenium has many protective effects, scientists are interested in its potential importance to the immune system,[26] as well as its role in cancer prevention.[27] However, more research is required to understand the role of selenium in cancer and disease prevention.

## Selenium Deficiency, Toxicity, and Recommended Intakes

Severe selenium deficiency causes **Keshan disease,** which was first documented in the Keshan region of China. This part of Asia has very low soil selenium levels. Keshan disease mostly affects children and can be fatal, because it causes serious heart problems.[28] Consuming high amounts of selenium can cause a garlic-like odor of the breath, nausea, vomiting, diarrhea, and brittleness of the teeth and fingernails. The RDA for selenium is 55 μg/day. A UL for selenium has been established at 400 μg/day.

## ESSENTIAL *Concepts*

Food sources of selenium include nuts, seafood, and meats. Cereal products can also provide selenium, although the amount is dependent on soil selenium levels. Selenium absorption is efficient, allowing most dietary selenium to enter the blood. Excess selenium is excreted in the urine. Selenium is a component of at least 14 proteins, called selenoproteins, some of which

**glutathione peroxidases** A group of selenoprotein enzymes that have redox functions in the body.

**Keshan disease** A disease resulting from selenium deficiency.

are involved in redox reactions, protecting the body from oxidative damage. Others regulate thyroid metabolism and activation of vitamin C. Selenium deficiency can cause Keshan disease, especially in children, resulting in sometimes fatal heart problems. Selenium toxicity causes a garlic-like odor of the breath, gastrointestinal upset, and brittle teeth and fingernails.

# Chromium

**Chromium (Cr)** was designated an essential nutrient because its deficiency in animals caused a diabetic-like state. Later, it was discovered that chromium is needed for glucose homeostasis and insulin function in humans as well.

## Dietary Sources, Bioavailability, and Regulation of Chromium

Chromium is found in a variety of foods. However, it is difficult to list those that are especially good sources, partly because the chromium content of a food can vary greatly depending on its content in soil. In addition, chromium is found in such small quantities that it can be difficult to accurately measure. Nonetheless, whole-grain products, fruits, and vegetables tend to be sources of chromium, whereas refined cereals, meat products, and dairy foods contain very little chromium. However, some processed meats have relatively high amounts of chromium, as do some beers and wines.

Whole grain foods can provide chromium to the diet.

Chromium bioavailability is increased by vitamin C and acidic medications such as aspirin. Conversely, antacids can decrease chromium bioavailability. Regardless of chromium intake, very little (< 2%) is absorbed—most is excreted in the feces. Chromium that is absorbed is transported in the blood to the liver, and excess chromium is excreted in the urine. Although scientists do not know why, consuming large amounts of simple sugars can increase urinary chromium excretion.[29] Excess chromium is stored mainly in the liver.

## Functions of Chromium

Chromium is needed for the hormone insulin to function properly in the body and appears to be especially important in regulating its function in people with type 2 diabetes.[30] Chromium is also required for normal growth and development in children. In addition, it increases lean mass and decreases fat mass—at least in laboratory animals.[31] Because of this, chromium in a form called **chromium picolinate** has been widely marketed as an ergogenic aid for athletes.[32] Other types of chromium supplements are promoted as products that help regulate blood glucose. In fact, estimated sales of chromium-containing supplements in 2002 were $85 million, representing 5.6% of the total mineral-supplement market.[33] However, most controlled studies investigating the effect of these supplements on athletic performance or blood glucose regulation have shown no beneficial outcomes.[34]

**CONNECTIONS** Remember that insulin is released from the pancreas in response to elevated blood glucose concentrations. Insulin is needed for some cells to take up and use glucose (Chapter 5, page 143).

## Chromium Deficiency and Toxicity

Chromium deficiency is only seen in hospitalized patients receiving inadequate nutrition support, resulting in elevated blood glucose levels, decreased sensitivity to insulin, and weight loss. Chromium toxicity is also rare, even

**chromium (Cr)** An essential trace mineral needed for proper insulin function.

**chromium picolinate** A form of chromium taken as an ergogenic aid by some athletes.

when supplemental chromium is consumed. This is probably due to low chromium absorption. However, toxic levels are sometimes seen when people are exposed to high levels of industrially released chromium. For example, when stainless steel is heated to a very high temperature during welding, chromium can be released into the air. Environmental exposure of this kind causes skin irritations and increased risk for lung cancer. These effects are never seen when chromium is obtained from the diet.

## Recommended Intakes for Chromium

There is insufficient information for establishing RDAs for chromium. However, AIs have been set. These recommendations are quite variable (20 to 35 μg/day), being higher for men than women. Because there are no known adverse effects of high intake of chromium from diet or supplements, the Institute of Medicine has not established ULs for this mineral.

## ESSENTIAL *Concepts*

Chromium is found in whole-grain products, fruits, and vegetables, and its bioavailability is increased by vitamin C. Excess chromium is excreted in the urine. Chromium is needed for the function of the hormone insulin and for optimal growth and development of children. Although chromium supplementation can increase growth of muscles and decrease adipose tissue in animals, these effects have not been seen in humans. Nonetheless, chromium supplementation is common in athletes. Chromium deficiency is rare but causes high blood glucose concentrations, decreased insulin sensitivity, and weight loss. Chromium toxicity from foods is rare, although it is sometimes caused by environmental exposure, resulting in severe skin irritations and increased risk for lung cancer.

# Manganese

**Manganese (Mn)** is a necessary cofactor of many enzymes in the body. As with chromium, manganese deficiency is rare and toxicity is typically found only from environmental exposure.

## Dietary Sources, Regulation, and Functions of Manganese

The best sources of manganese include whole-grain products, pineapples, nuts, and legumes. Dark green, leafy vegetables such as spinach are also high in manganese, and water can contain significant amounts. You may have noticed that nuts and seeds are good sources of several of the trace minerals. For this and other reasons, the Dietary Guidelines for Americans recommends that nuts be consumed as part of a healthy diet. You can read more about reaching this goal by reading the Food Matters feature. Regardless of manganese intake, very little (< 10%) is absorbed. Excess manganese is typically not excreted in the urine but is delivered to the liver, where it is incorporated into bile and excreted in the feces.

Manganese is a cofactor for numerous metalloenzymes. For example, it is needed for enzymes involved in glucose production (gluconeogenesis) and

© Photodisc/Getty Images

Pineapples are excellent sources of the trace mineral manganese.

**manganese (Mn)** An essential trace mineral that is a cofactor for enzymes needed for bone formation and the metabolism of carbohydrates, proteins, and cholesterol.

bone formation. Manganese also binds ADP and ATP in a variety of reactions, being important for some aspects of energy metabolism. Like copper, manganese is a cofactor for superoxide dismutase, an enzyme that protects cells from free radicals.

## Manganese Deficiency and Toxicity

Because manganese intake is almost always sufficient, manganese deficiency is rare. Severe manganese deficiency results in scaly skin, poor bone formation, and growth faltering. Manganese toxicity from foods and supplements is uncommon, but exposure to high levels of manganese in the air as a result of mining can cause serious neural problems. Manganese toxicity can also occur in people with liver disease and in those consuming water with high manganese levels.

## Recommended Intakes for Manganese

AIs for manganese have been set at 2.3 and 1.8 mg/day for men and women, respectively. To prevent nerve damage caused by manganese toxicity, a UL has been established at 11 mg/day. This UL includes manganese intake from food, water, and supplements.

*Working Toward the Goal*

The USDA Dietary Guidelines for Americans and the MyPyramid food guidance system recommend that we consume a variety of nuts and seeds in moderation as part of a nutritious diet. This is partly because they tend to be excellent sources of trace minerals. The following selection and preparation tips can help you meet your goal.

- Include nuts and seeds in main dishes. However, because nuts and seeds tend to be high in fat, they should be used to *replace* meat or poultry, not used *in addition* to them.
- Use basil and pine nut pesto for pasta or as a nutritious and tasty spread on bread and crackers.
- Add slivered almonds to steamed vegetables.
- Add toasted peanuts or cashews to a vegetable stir-fry.
- Sprinkle a few nuts on top of low-fat ice cream or frozen yogurt.
- Add toasted walnuts or pecans to a green salad instead of cheese or meat.
- Instead of snacking on chips, try sunflower or pumpkin seeds.
- Try toasted hazel nuts on your oatmeal for a crunchy and delicious breakfast.
- Choose a slice of pecan pie instead of a less nutrient-dense variety (such as chocolate cream).

Manganese is found in whole-grain products, nuts, legumes, and some fruits and vegetables. Water can also be a good source. Very little manganese is absorbed, and excess is excreted in the bile. Manganese works with several enzymes in the body involved in energy metabolism, bone formation, gluconeogenesis, and antioxidant function. Manganese deficiency causes dry skin, weak bones, and poor growth. Toxicity from foods is rare, but environmental exposure can cause serious problems.

# Molybdenum

**Molybdenum (Mo)** is an essential nutrient for humans because it is a cofactor for several important metalloenzymes. Because the body requires very little molybdenum and it is present in a variety of foods, molybdenum deficiency is almost unheard of. Thus there is little concern among health professionals about molybdenum as a dietary component.

## Dietary Sources and Functions of Molybdenum

The molybdenum content of foods varies greatly depending on its content in soils. In general, legumes (peas and lentils) as well as grains and nuts are good sources of this mineral. Molybdenum appears to be almost completely absorbed in the intestine, where it is circulated to the liver in the blood. Molybdenum functions in redox reactions and is a cofactor for several enzymes in the body. These enzymes are involved in the metabolism of sulfur-containing amino acids (such as methionine and cysteine) as well as purines that make up DNA and RNA. Molybdenum-requiring enzymes are also involved in detoxifying drugs in the liver.

## Molybdenum Deficiency, Toxicity, and Recommended Intake

Only one case of molybdenum deficiency has been documented. In this case, molybdenum deficiency was accompanied by abnormal heart rhythms, headache, and visual problems.[35] There are no known adverse effects of high intakes of molybdenum in humans. However, in experimental animals toxicity can seriously harm reproduction. The RDA for molybdenum is 45 $\mu$g/day. To prevent potential reproduction problems, a UL of 2,000 $\mu$g/day has been established.

In general, legumes, grains, and nuts are good sources of molybdenum. Molybdenum is a cofactor for enzymes involved in the metabolism of sulfur-containing amino acids and the purines important for protein, DNA, and RNA structures. Thus this mineral is needed for protein synthesis and cellular growth. Deficiency is extremely rare but causes headaches, abnormal cardiac function, and visual difficulties. Toxicity has never been documented in humans but causes reproductive failure in laboratory animals.

CONNECTIONS ▶ Purines, together with pyrimidines, make up DNA and RNA macromolecules (Chapter 10, page 402).

**molybdenum (Mo)** (mo – LYB – de – num) An essential trace mineral that is a cofactor for enzymes needed for amino acid and purine metabolism.

# Zinc

Although **zinc (Zn)** is a cofactor for over 300 enzymes, its essentiality was only established in the 1960s. The mechanisms by which zinc influences human health are still being studied. However, researchers know that it is needed for growth, reproduction, immunity, and protein synthesis.

## Dietary Sources and Bioavailability of Zinc

Zinc is found in high concentrations in shellfish, meat, organ meats, dairy products, legumes, and chocolate (Figure 12.12). It is also often added to fortified cereals and grain products. Zinc bioavailability is influenced by a variety of dietary factors, many of which are similar to those that influence iron absorption.[36] For example, phytates bind zinc and form insoluble complexes that are difficult to absorb. Thus, phytate consumption can decrease zinc bioavailability. In addition, bioavailability of zinc from animal sources is greater than from plant sources, and acidic substances increase its absorption. Because zinc, iron, and calcium share transport proteins in the intestinal cell, high intakes of iron and calcium can decrease zinc absorption.

## Absorption, Metabolism, and Regulation of Zinc

As with iron, the zinc content of blood is highly regulated at the small intestine. Zinc absorption requires at least two proteins. The first protein transports zinc into the enterocyte, and the second binds the mineral within the enterocyte. Recall that ferritin in the enterocyte binds excess iron, blocking

Chocolate is a good source of zinc. Remember that all foods can be part of a healthy diet.

**zinc (Zn)** An essential trace mineral involved in gene expression, immune function, and cell growth.

**FIGURE 12.12** Good Sources of Zinc

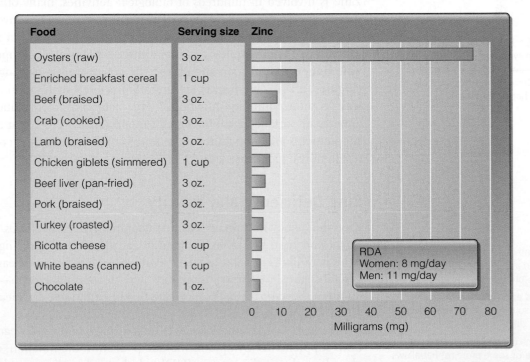

| Food | Serving size | Zinc |
|---|---|---|
| Oysters (raw) | 3 oz. | |
| Enriched breakfast cereal | 1 cup | |
| Beef (braised) | 3 oz. | |
| Crab (cooked) | 3 oz. | |
| Lamb (braised) | 3 oz. | |
| Chicken giblets (simmered) | 1 cup | |
| Beef liver (pan-fried) | 3 oz. | |
| Pork (braised) | 3 oz. | |
| Turkey (roasted) | 3 oz. | |
| Ricotta cheese | 1 cup | |
| White beans (canned) | 1 cup | |
| Chocolate | 1 oz. | |

RDA
Women: 8 mg/day
Men: 11 mg/day

Milligrams (mg)

SOURCE: Data from USDA Nutrient Database, Release 17.

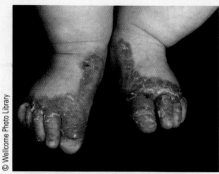

**Acrodermatitis enteropathica is a genetic disorder that causes decreased zinc absorption.**

**CONNECTIONS** The Human Genome Project was a study undertaken to understand the genes making up the human genome (Chapter 6, page 185).

**CONNECTIONS** Recall that in the second step of protein synthesis, called gene transcription, the information contained in genes (DNA) is transferred to mRNA molecules in the nucleus (Chapter 6, page 179).

**metallothionine** (me – tall – o – THI – o – nine) A protein in the enterocyte that regulates zinc absorption and elimination.

**acrodermatitis enteropathica** A genetic abnormality resulting in decreased absorption of dietary zinc.

**zinc finger** Zinc containing three-dimensional structure of some proteins that allows them to regulate gene expression.

its absorption during periods of high iron status. Similarly, a protein called **metallothionine** binds zinc in the enterocyte, making it unavailable for transport across the basolateral membrane. To facilitate zinc absorption, synthesis of metallothionine decreases during periods of zinc deficiency. Conversely, synthesis of metallothionine increases during periods of zinc excess. Because intestinal cells are continually sloughed off, excess zinc bound to metallothionine is excreted in the feces. Zinc absorption is also influenced by genetic factors, and is discussed next.

## Acrodermatitis Enteropathica

One genetic condition that decreases zinc absorption is **acrodermatitis enteropathica.** This disease is characterized by severe zinc deficiency even when zinc is adequate in the diet, and is usually seen in the first few months of life. Babies with acrodermatitis enteropathica fail to grow properly and have severely red and scaly skin especially around the scalp, eyes, and feet.[37] They also suffer from diarrhea. Untreated acrodermatitis enteropathica is fatal. Researchers estimate that 1 in 500,000 babies are born with this disease.

Although researchers have long known that acrodermatitis enteropathica is due to an inability to absorb zinc, more specific clues to its cause were not available until the Human Genome Project was completed. It is now known that people with this disease have mutations in a gene that codes for a protein needed to transport zinc into the enterocyte. The defective protein is less able to transport zinc, causing it to be lost in the feces. Treatment of acrodermatitis enteropathica requires life-long zinc supplementation. When high intakes of zinc are consumed, adequate amounts of zinc can be absorbed, even in the presence of the faulty transport protein.

## Functions of Zinc

Zinc is involved in hundreds of biological activities, many of which require zinc as a cofactor. For example, it is needed for enzymes involved in RNA synthesis. Zinc also stabilizes portions of many proteins that regulate gene expression. These proteins contain three-dimensional components called **zinc fingers** that turn on and off specific genes, thus regulating transcription (Figure 12.13). Zinc fingers are important in cell maturation and growth as well as proper immune function.[38] Zinc acts as a potent antioxidant,[39] and appears important in stabilizing cell membranes.[40] Note that although zinc supplements are often touted as helping "cure" the common cold, most studies do not support this claim.[41]

## Zinc Deficiency and Toxicity

Zinc deficiency in humans was first documented in the 1960s in portions of Egypt and Iran, where people tend to eat plant-based diets high in phytates (which inhibit zinc absorption).[42] Mild zinc deficiency appears to decrease appetite, thus increasing morbidity and decreasing growth—especially in children. Severe zinc deficiency causes skin irritations, diarrhea, and delayed sexual maturation.

Although dietary zinc toxicity is uncommon, high levels can be obtained from supplements. Symptoms include poor immune function, depressed levels of high-density lipoprotein (HDL) cholesterol, and impaired copper status. Nausea, vomiting, and loss of appetite have also been reported.

## Recommended Intake for Zinc

RDAs for zinc have been set at 11 and 8 mg/day for men and women, respectively. The Institute of Medicine suggests that vegetarians, particularly vegans, should consume up to 50% more zinc than nonvegetarians. A UL for zinc has been set at 40 mg/day.

## ESSENTIAL *Concepts*

Zinc is found in shellfish, fortified cereals, meat, legumes, and chocolate, and its absorption is influenced by many factors. For example, phytates decrease absorption, and animal sources are more bioavailable than plant sources. Zinc absorption is highly regulated, increasing during zinc deficiency and decreasing during toxicity. Acrodermatitis enteropathica is a genetic defect resulting in decreased zinc absorption. This disease is caused by the production of a faulty protein needed for zinc transport into the enterocyte. Zinc is a cofactor for enzymes needed for RNA synthesis and is a vital component of proteins (zinc fingers) that modulate gene expression. It is also important for cell maturation and immune function. Zinc deficiency causes poor appetite, growth failure, skin irritations, diarrhea, delayed physical maturation, and infertility. Zinc toxicity is rare but can influence immune function, lipid metabolism, and appetite.

## Fluoride

**Fluoride (F⁻)** is not an essential nutrient. That is, it is not needed for growth, reproduction, or maintenance of basic body functions. However, it has long been recognized that dietary fluoride strengthens bones and teeth. Therefore, knowing about sources of fluoride and how it works in the body is important. Note that the nonionic form of fluoride is fluor*ine*, which is a poisonous gas.

### Dietary Sources, Bioavailability, and Regulation of Fluoride

Very few foods are good sources of fluoride, although potatoes, tea, and legumes contain some fluoride, as does fish with bones intact. In addition, swallowing fluoridated toothpaste can contribute fluoride to the body. Because foods provide little fluoride, many communities add it to their drinking water. The American Dental Association recommends that water be fluoridated to contain 1 to 2 parts fluoride per million parts water (1 to 2 ppm).[43] You can find out whether your town has fluoridated water by contacting a local dentist or the city water department. Bottled water typically contains little fluoride and is therefore a poor source of this mineral.

The gastrointestinal tract provides very little fluoride regulation. In fact, almost all fluoride consumed is absorbed in the small intestine and then circulates in the blood to the liver and then bones and teeth, where it is actively taken up. Fluoride remaining in the blood is excreted in the urine.

### Functions of Fluoride

Fluoride affects the health of bones and teeth in several ways. First, it strengthens them by being incorporated into their basic mineral matrix. Bones and teeth that contain fluoride are stronger than those without it, being especially

**FIGURE 12.13   Zinc Fingers**

Special zinc-containing portions of proteins, called zinc fingers, modulate gene expression and, ultimately, protein synthesis.

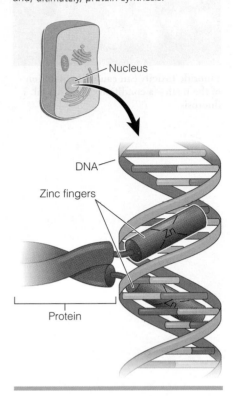

Nucleus

DNA

Zinc fingers

Protein

**fluoride (F⁻)** A nonessential trace mineral that strengthens bones and teeth.

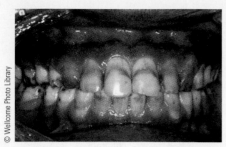

Fluoride toxicity can cause discoloration of the teeth—a condition called dental fluorosis.

resistant to bacterial breakdown and cavities (also called caries). Consuming fluoride also appears to stimulate maturation of osteoblasts, the cells that build new bone. Second, *topical* (not dietary) application of fluoride-containing toothpaste and fluoride treatments works directly on the cavity-causing bacteria in the mouth to decrease acid production. The damaging effects of acids produced by these bacteria are largely responsible for tooth decay. Thus lower acid production means fewer cavities.

## Fluoride Deficiency, Toxicity, and Recommended Intake

Aside from the beneficial effects of fluoride on tooth and bone strength, there are no known biologically important effects of this mineral. Thus no deficiency symptoms are known. However, fluoride toxicity is well documented. Signs and symptoms include gastrointestinal upset, excessive production of saliva, watering eyes, heart problems, and in severe cases coma. In addition, very high fluoride intake causes pitting and mottling (discoloration) of teeth, called **dental fluorosis,** and a weakening of the skeleton called **skeletal fluorosis.** Fluoride toxicity is a special concern in small children, who sometimes swallow large amounts of toothpaste on a daily basis. Thus parents should carefully monitor tooth-brushing routines. Because fluoride is not technically considered an essential nutrient, no RDAs are established. However, AIs have been set at 4 and 3 mg/day for men and women, respectively. The UL is 10 mg/day.

## ESSENTIAL *Concepts*

Fluoride is not an essential nutrient. However, increased fluoride intake decreases tooth decay and strengthen bones. Fluoride is not found naturally in many foods. Instead, we tend to consume it from the water we drink. Fluoride increases the strength of both teeth and bones by becoming incorporated in the mineral matrix of these tissues. In addition, fluoride stimulates osteoblasts, which build bone. Topical application of fluoride (such as in toothpaste) decreases bacterial production of acid in the mouth. Because these acids cause cavities, the use of fluoridated toothpaste enhances dental health. Fluoride toxicity can discolor teeth (dental fluorosis) and weaken bones (skeletal fluorosis).

## Other Trace Minerals

In addition to the essential trace minerals already covered in this chapter, others may influence human health. These include nickel, aluminum, silicon, vanadium, arsenic, and boron. The essentiality of these minerals has been neither confirmed nor denied, and requires further study. In fact, most of what is known about them is based on animal studies, not human studies. Therefore, we will simply mention each of these minerals and comment on possible biological functions as well as food sources. Although there are no recommended intakes for these minerals, some of them have Tolerable Upper Intake Levels (ULs), which are listed inside the front cover of this book.

- *Nickel (Ni)* may be involved in protein and lipid metabolism, redox reactions, and gene regulation. It may also influence calcium metabolism, especially in terms of bone formation. Nickel is found in legumes, nuts, and chocolate.
- *Aluminum (Al)* may be important for reproduction, bone formation, DNA synthesis, and behavior. It is found in many baked goods, grains, some vegetables, tea, and many antacids.

**dental fluorosis** Discoloration and pitting of teeth caused by excessive fluoride intake.

**skeletal fluorosis** Weakening of the bones caused by excessive fluoride intake.

- *Silicon (Si)* is important for bone growth in laboratory animals. Dietary sources include whole-grain products. In addition, silicon is sometimes used as a food additive.
- *Vanadium (V)* acts much like insulin in some laboratory animals and cell culture systems. It also influences cell growth and differentiation and may affect thyroid hormone metabolism. Sources include mushrooms, shellfish, and black pepper.
- *Arsenic (As)* may be a cofactor for a variety of enzymes and may be important for DNA synthesis. Seafood and whole grains are good sources of arsenic.
- *Boron (B)* is especially important for reproduction in laboratory animals, and deficiency causes severe fetal malformations. Sources of boron include fruits, leafy vegetables, legumes, and nuts.

## ESSENTIAL *Concepts*

In addition to the trace minerals known to be essential for life, many others may influence health. These include nickel, aluminum, silicon, vanadium, arsenic, and boron. Although scientists do not know whether these minerals are important for human health, there is evidence that they influence the health of other animals. Further research is needed to understand whether these minerals are vital for humans or whether they, like fluoride, are involved in nonessential but important functions.

# Integration of Functions and Food Sources

The functions of the trace minerals are diverse. Aside from participating as cofactors for hundreds of metalloenzymes, trace minerals also regulate gene expression and energy metabolism, protect against free radicals via redox systems, and are important for blood health, immune function, and bone health (Table 12.2). Thus it is important that we consume sufficient—albeit small—amounts of each of these essential minerals on a regular basis. As with the vitamins, no single food group can supply all the trace minerals. Therefore, consumption of a balanced diet featuring a wide variety of foods is required to assure adequate intake of each mineral. This is shown in Table 12.3.

**TABLE 12.2    General Functions of the Essential Trace Minerals**

|  | Minerals | | | | | | | |
|---|---|---|---|---|---|---|---|---|
|  | Iron | Copper | Iodine | Selenium | Chromium | Manganese | Molybdenum | Zinc |
| **Functions** | | | | | | | | |
| Cofactor (metalloenzyme) | X | X |  | X |  | X | X | X |
| Regulation of gene transcription |  |  |  |  |  |  |  | X |
| Immune function |  |  |  | X |  |  |  | X |
| Antioxidant (redox) function | X | X |  | X |  | X | X | X |
| Bone/tooth health |  |  |  |  |  | X |  |  |
| Blood health | X |  |  |  |  |  |  |  |
| Energy metabolism[a] | X | X | X | X | X | X |  |  |

[a]Note that these minerals are not themselves energy-yielding nutrients but are involved in energy metabolism via other mechanisms.

## TABLE 12.3  Major Food Sources of the Essential Trace Minerals

| | Types of Foods | | | | | | | | | |
|---|---|---|---|---|---|---|---|---|---|---|
| | Grains^a | Vegetables | Fruits | Dairy | Meat | Organ Meats | Seafood | Nuts, Seeds | Legumes | Mushrooms |
| **Trace Minerals** | | | | | | | | | | |
| Iron | X^b | X | | | X | X | X | X | X | X |
| Copper | X | | | | | X | X | X | X | X |
| Iodine | | | | X | | | X^c | | | |
| Selenium | X^c | | | | X | X | X | X | | |
| Chromium | X^c | X | X | | | | | | | |
| Manganese | X | X | X | | | | | X | X | |
| Molybdenum | X^c | | | | | | | X | X | |
| Zinc | X | | | X | X | X | X | | X | |

^a Whole-grain products tend to contain more trace minerals than milled products.
^b Especially enriched cereal products.
^c Concentration highly dependent on mineral content of soil or water.

## Review Questions  Answers are found in Appendix G.

1. Which of the following would be expected in iron deficiency?
   a. Increased iron transport protein production and decreased ferritin production
   b. Decreased iron transport protein production and decreased ferritin production
   c. Elevated hematocrit
   d. Increased transferrin saturation

2. Vitamin C increases the bioavailability of nonheme iron by
   a. binding to ferritin in the intestinal cell.
   b. increasing the activity of the digestive enzymes in the small intestine.
   c. reducing ferric iron to ferrous iron.
   d. helping iron bind to chelators in the intestinal tract.

3. Ceruloplasmin transports _____ in the blood.
   a. cobalt
   b. copper
   c. selenium
   d. molybdenum

4. Thyroid-stimulating hormone (TSH) causes
   a. the uptake of iodine by the thyroid gland.
   b. the excretion of iodine in the urine.
   c. increased absorption of iodine in the small intestine.
   d. the release of iodine from the thyroid gland into the blood.

5. Glutathione peroxidases require which of the following trace minerals?
   a. Manganese
   b. Iron
   c. Fluoride
   d. Selenium

6. Fluoride is *not* considered an essential nutrient, because
   a. the body does not use it for any processes.
   b. it is not important for ATP production.
   c. it is not required for growth, reproduction, or basic body functions.
   d. we do not consume any fluoride from food or water.

7. The hormone _____ requires chromium for optimal function.
   a. gastrin
   b. insulin
   c. thyroid hormone
   d. glucagon

8. Regulation of the amount of manganese in the body is via its loss in the _____.
   a. urine
   b. blood
   c. sweat
   d. feces

9. Which of the following best describes acrodermatitis enteropathica?
   a. A primary zinc deficiency caused by inadequate zinc in the diet
   b. A genetic disease that inhibits zinc excretion
   c. A genetic disease that results in zinc toxicity
   d. A genetic disease that causes inadequate zinc absorption

10. The trace minerals are found mainly in which of the following food groups?
   a. Grains and cereals
   b. Meat
   c. Nuts and seeds
   d. All of the above

11. The mineral content of foods can vary greatly. For example, wheat produced in one region of the United States can have very high levels of selenium, whereas that raised elsewhere can have very low amounts. Why is this?

**12.** Are there any trace minerals that a vegetarian may have trouble consuming enough of? If so, what foods might she or he need to consume more of?

**13.** What distinguishes a person with cretinism from a person with goiter? Why, do you think, are infants not born with goiter?

## Practice Calculations   Answers are found in Appendix G.

**1.** Using the food composition table that accompanies this book, calculate the chromium content of a lunch containing a hamburger on a nonenriched white bun, a small order of french fries, and a large cola soft drink. What percentage of your Recommended Dietary Allowance of iron does this represent? What foods might be added to or substituted for the food in this lunch to increase the iron content?

**2.** Using the MyPyramid food guidance system (http://www.mypyramid.gov), determine the amount of foods from the meats, beans, and dairy groups that are recommended for you. Then make up a sample menu that would satisfy these recommendations, and calculate the amount of magnesium you would get if you consumed these amounts of foods.

## Media Links   A variety of study tools for this chapter are available at our website, www.thomsonedu.com/nutrition/mcguire.

Prepare for tests and deepen your understanding of chapter concepts with these online materials:

- Practice tests
- Flashcards
- Glossary
- Student lecture notebook
- Web links
- Animations
- Chapter summaries, learning objectives, and crossword puzzles

## Notes

**1.** Pruchnicki MC, Coyle JD, Hoshaw-Woodard S, Bay WH. Effect of phosphate binders on supplemental iron absorption in healthy subjects. Journal of Clinical Pharmacology. 2002;42:1171–6.

**2.** Chul Huh E, Hotchkiss A, Broulette J, Glahn RP. Carbohydrate fractions from cooked fish promote iron uptake by Caco-2 cells. Journal of Nutrition. 2004;134:1681–9.

**3.** Lopez MAA , Martos FC. Iron availability: an updated review. International Journal of Food Sciences and Nutrition. 2004;55:597–606. Zijp IM, Korver O, Tijburg LBM. Effect of tea and other dietary factors on iron absorption. Critical Reviews in Food Science and Nutrition. 2000;40:371–98.

**4.** Miret S, Simpson RJ, McKie AT. Physiology and molecular biology of dietary iron absorption. Annual Review of Nutrition. 2003;23:283–301.

**5.** Theil EC. Iron, ferritin, and nutrition. Annual Review of Nutrition. 2004;24:327–43.

**6.** Coon MJ. Cytochrome P450: nature's most versatile biological catalyst. Annual Review of Pharmacology and Toxicology. 2005;45:1–25. Masson LF, Sharp L, Cotton SC, Little J. Cyrochrome P-450 1A1 gene polymorphisms and risk of breast cancer: a HuGE review. American Journal of Epidemiology. 2005;161:901–15.

**7.** Centers for Disease Control and Prevention. Recommendations to prevent and control iron deficiency in the United States. Mortality and Morbidity Weekly Report. 1998;47:1–29. Looker AC, Dallman PR, Carroll MD, Gunter EW, Johnson CL. Prevalence of iron deficiency in the United States. JAMA (Journal of the American Medical Association). 1997;277:973–6. Stoltzfus RJ. Defining iron-deficiency anemia in public health terms: reexamining the nature and magnitude of the public health problem. Journal of Nutrition. 2001;131:565S–7S.

**8.** Rose EA, Porcerelli JH, Neale AV. Pica: common but commonly missed. Journal of the American Board of Family Practice. 2000;13:353–8. Singhi S, Ravishanker R, Singhi P, Nath R. Low plasma zinc and iron in pica. Indian Journal of Pediatrics. 2003;70:139–43.

**9.** Black MM. Micronutrient deficiencies and cognitive functioning. Journal of Nutrition. 2003;133:3927S–31S. Bryan J, Osendarp S, Hughes D, Calvaresi E, Baghurst K, van Klinken J-W. Nutrients for cognitive development in school-aged children. Nutrition Reviews. 2004;62:295–306.

**10.** Rosenzweig PH, Volpe SL. Iron, thermoregulation, and metabolic rate. Critical Reviews in Food Science and Nutrition. 1999;39:131–48.

11. Cunningham-Rundles S, McNeeley DF, Moon A. Mechanisms of nutrient modulation of the immune response. Journal of Allergy and Clinical Immunology. 2005;115:1119–28. Failla ML. Trace elements and host defense: recent advances and continuing challenges. Journal of Nutrition. 2003;133:1443S–7S.

12. Gambling L, Danzeisen R, Fosset C, Andersen HS, Dunford S, Srai SKS, McArdle HJ. Iron and copper interactions in development and the effect on pregnancy outcome. Journal of Nutrition. 2003;133:1554S–6S.

13. Hoffman R, Benz E, Shattil S, Furie B, Cohen H, Silberstein L, McGlave P. Hematology: Basic principles and practice, 3rd ed. New York: Churchill Livingstone, Harcourt Brace; 2000.

14. Nelson RL, Davis FG, Persky V, Becker E. Risk of neoplastic and other diseases among people with heterozygosity for hereditary hemochromatosis. Cancer. 1995;76:875–79.

15. Franchini M, Veneri D. Hereditary hemochromatosis. Hematology. 2005;10:145–9. Robson KJH, Merryweather-Clarke AT, Cadet E, Viprakasit V, Zaahl MG, Pointon JJ, Weatherall DJ, Rochette J. Recent advances in understanding haemochromatosis: a transition state. Journal of Medical Genetics. 2004;41:721–30.

16. Puig S, Thiele D. Molecular mechanisms of copper uptake and distribution. Current Opinion in Chemical Biology. 2002;6:171–80.

17. Cooper GJS, Chan Y-K, Dissanayake AM, Leahy FE, Koegh GF, Frampton CM, Bamble GD, Brunton DH, Baker JR, Poppitt SD. Demonstration of a hyperglycemia-driven pathogenic abnormality of copper homeostasis in diabetes and its reversibility by selective chelation. Diabetes. 2005;54:1468–76.

18. Failla ML. Trace elements and host defense: recent advances and continuing challenges. Journal of Nutrition. 2003;133:1443S–7S. Turnlund JR, Jacob RA, Keen CL, Strain JJ, Kelley DS, Domek JM, Keyes WR, Ensunsa JL, Lykkesfeldt J, Coulter J. Long-term high copper intake: effects on indexes of copper status, antioxidant status, and immune function in young men. American Journal of Clinical Nutrition. 2004;79:1037–44.

19. Chuttani H, Gupta P, Gulati S, Gupta D. Acute copper sulfate poisoning. American Journal of Medicine. 1965;39:849–54. Bremner I. Manifestations of copper excess. American Journal of Clinical Nutrition. 1998;67:1069S–73S.

20. Zimmermann MB. Assessing iodine status and monitoring progress of iodized salt programs. Journal of Nutrition. 2004;134:1673–7.

21. Cao X-Y, Jiang X-M, Dou Z-H, Rakeman MA, Zhang M-L, O'Donnell K, Ma T, Amette K, DeLong N, DeLong GR. Timing of vulnerability of the brain to iodine deficiency in endemic cretinism. New England Journal of Medicine. 1994;331:1739–44. Miles M. Goitre, cretinism and iodine in South Asia: historical perspectives on a continuing scourge. Medical History. 1998;42:46–67. Pharoah PO, Buttfield IH, Hetzel BS. Neurological damage to the fetus resulting from severe iodine deficiency during pregnancy. Lancet. 1971;1:308–10.

22. Olin RM. Iodine deficiency and prevalence of simple goiter in Michigan. JAMA (Journal of the American Medical Association). 1924;82:1328–32.

23. Kimball OP. The efficiency and safety of the prevention of goiter. JAMA (Journal of the American Medical Association). 1928;91:454–60. Kimball OP. Prevention of goiter in Michigan and Ohio. JAMA (Journal of the American Medical Association). 1937;108:860–4.

24. Altland JK, Brush BE. Goiter prevention in Michigan, results of thirty years' voluntary use of iodized salt. Journal of the Michigan Medical Society. 1952;51:985–9. Brush BE, Altland JK. Goiter prevention with iodized salt: results of a thirty-year study. Journal of Clinical Endocrinology and Metabolism. 1952;12:1380–8.

25. Valko M, Izakovic M, Mazur M, Rhodes CJ, Telser J. Role of oxygen radicals in DNA damage and cancer incidence. Molecular and Cellular Biochemistry. 2004;266:37–56.

26. Baum MK, Shor-Posner G, Lai S, Zhang G, Lai H, Fletcher MA, Sauberlich H, Page JB. High risk of HIV-related mortality is associated with selenium deficiency. Journal of Acquired Immune Deficiency Syndromes. 1997;15:370–1. Look MP, Rockstroh JK, Rao GS, Kreuzer KA, Spengler U, Sauerbruch T. Serum selenium versus lymphocyte subsets and markers of disease progression and inflammatory response in human immunodeficiency virus-1 infection. Biological Trace Element Research. 1997;56:31–41. Romero-Alvira D, Roche E. The keys of oxidative stress in acquired immune deficiency. Medical Hypotheses. 1998;51:169–73.

27. Combs GF, Clark LC, Turnbull BW. An analysis of cancer prevention by selenium. Biofactors. 2001;14:153–9. El-Bayoumy K, Richie JP, Boyiri T, Komninou D, Prokopczyk B, Trushin N, Kleinman W, Cox J, Pittman B, Colosimo S. Influence of selenium-enriched yeast supplementation on biomarkers of oxidative damage and hormone status in healthy adult males: a clinical pilot study. Cancer Epidemiology, Biomarkers and Prevention. 2002;11:1459–65. Hartwig A, Blessing H, Schwerdtle T, Walter I. Modulation of DNA repair processes by arsenic and selenium compounds. Toxicology. 2003;193:161–9. Johnson IT. Micronutrients and cancer. Proceedings of the Nutrition Society. 2004;63:587–95. Shen C-L, Song W, Pence BC. Interactions of selenium compounds with other antioxidants in DNA damage and apoptosis in human normal keratinocytes. Cancer Epidemiology, Biomarkers and Prevention. 2001;10:385–90.

28. Thomson CD. Assessment of requirements for selenium and adequacy of selenium status: a review. European Journal of Clinical Nutrition. 2004;58:391–402.

29. Kozlovsky AS, Moser PB, Reiser S, Anderson RA. Effects of diets high in simple sugars on urinary chromium losses. Metabolism. 1986;35:515–8.

30. Cefalu WT, Hu FB. Role of chromium in human health and in diabetes. Diabetes Care. 2004;27:2741–51. Hopkins Jr. LL, Ransome-Kuti O, Majaj AS. Improvement of impaired carbohydrate metabolism by chromium (III) in malnourished infants. American Journal of Clinical Nutrition. 1968;21:203–11. Mertz W. Interaction of chromium with insulin: a progress report. Nutrition Reviews. 1998;56:174–7. Porte Jr. D, Sherwin RS, Baron A, editors. Ellenberg and Rifkin's Diabetes mellitus, 6th ed. New York: McGraw-Hill; 2003.

31. McNamara JP, Valdez F. Adipose tissue metabolism and production responses to calcium proprionate and chromium proprionate. Journal of Dairy Science. 2005;88:2498–507. Page TG, Southern LL, Ward TL, Thompson DLJ. Effect of chromium picolinate on growth and serum and carcass

traits of growing-finishing pigs. Journal of Animal Science. 1993;71:656–62.

32. Lukaski HC. Chromium as a supplement. Annual Review of Nutrition. 1999;19:279–302.

33. Nutrition Business Journal. NJ's Supplement Business Report 2003. San Diego, CA: Penton Media Inc.; 2003.

34. Pittler MH, Stevinson C, Ernst E. Chromium picolinate for reducing body weight: meta-analysis of randomized trials. International Journal of Obesity and Related Metabolic Disorders. 2003;27:522–9. Vincent JB. The potential value and toxicity of chromium picolinate as a nutritional supplement, weight loss agent and muscle development agent. Sports Medicine. 2003;33:213–30.

35. Abumrad NN, Schneider AJ, Steel D, Rogers LS. Amino acid intolerance during total parenteral nutrition reversed by molybdate therapy. American Journal of Clinical Nutrition. 1981;34:2551–9.

36. Ford D. Intestinal and placental zinc transport pathways. Proceedings of the Nutrition Society. 2004;63:21–9. Lonnerdal B. Dietary factors influencing zinc absorption. Journal of Nutrition. 2000;130:1378S–83S.

37. Kury S, Dreno B, Bezieau S, Giraudet S, Kharfi M, Kamoun R, Moisan JP. Identification of *SLC39A4*, a gene involved in acrodermatitis enteropathica. Nature Genetics. 2002;31: 239–40. Wang K, Zhou B, Kuo YM, Zemansky J, Gitschier J. A novel member of a zinc transporter family is defective in acrodermatitis enteropathica. American Journal of Human Genetics. 2002;71:66–73.

38. MacDonald RS. The role of zinc in growth and cell proliferation. Journal of Nutrition. 2000;130:1500S–8S.

39. Powell SR. The antioxidant properties of zinc. Journal of Nutrition. 2000;130:1447S–54S.

40. O'Dell BL. Role of zinc in plasma membrane function. Journal of Nutrition. 2000;130:1432S–6S.

41. Jackson JL, Lesho E, Peterson C. Zinc and the common cold: a meta-analysis revisited. Journal of Nutrition. 2000;130:1512S–5S.

42. Hambidge M. Human zinc deficiency. Journal of Nutrition. 2000;130:1344S–9S.

43. American Dental Association. American Dental Association supports fluoridation. Available from: http://www.ada.org/ prof/resources/positions/statements/fluoride3.asp. American Dental Association. Statement on the effectiveness of community water fluoridation. Available from: http://www.ada .org/prof/resources/positions/statements/ fluoride_community_effective.asp.

**Check out the following sources for additional information.**

44. Borek C. Antioxidant health effects of aged garlic extract. Journal of Nutrition. 2001;131:1010S–15S.

45. Davidsson L. Approaches to improved iron bioavailability from complementary foods. Journal of Nutrition. 2003;133:1560S–2S.

46. Dibley MJ. Zinc. In: Present knowledge in nutrition, 8th ed. Bowman BA, Russell RM, editors. Washington, DC: ILSI Press; 2001.

47. Failla ML, Johnson MA, Prohaska JR. Copper. In: Present knowledge in nutrition, 8th ed. Bowman BA, Russell RM, editors. Washington, DC: ILSI Press; 2001.

48. Feldman EB. The scientific evidence for a beneficial health relationship between walnuts and coronary heart disease. Journal of Nutrition. 2002;132:1062S–1S.

49. Finley JW. Does environmental exposure to manganese pose a health risk to healthy adults? Nutrition Reviews. 2004;62:148–53.

50. Gibson RS. Principles of nutritional assessment, 2nd ed. New York: Oxford University Press; 2005.

51. Hurrell RF. Influence of vegetable protein sources on trace element and mineral bioavailability. Journal of Nutrition. 2003;133:2973S–7S.

52. Institute of Medicine. Dietary Reference Intakes for calcium, phosphorus, magnesium, vitamin D, and fluoride. Washington, DC: National Academy Press; 1997.

53. Institute of Medicine. Dietary Reference Intakes for vitamin C, vitamin E, selenium, and carotenoids. Washington, DC: National Academy Press; 2000.

54. Institute of Medicine. Dietary Reference Intakes for vitamin A, vitamin K, arsenic, boron, chromium, copper, iodine, iron, manganese, molybdenum, nickel, silicon, vanadium, and zinc. Washington, DC: National Academy Press; 2001.

55. Kader AA, Perkins-Veazie P, Lester GE. Nutritional quality of fruits, nuts, and vegetables and their importance in human health. Available from: www.ba.ars.usda.gov/hb66/ 025nutrition.pdf.

56. Kris-Etherton PM, Zhao G, Binkoski AE, Coval SM, Etherton TD. The effects of nuts on coronary heart disease risk. Nutrition Reviews. 2001;59:103–11.

57. Kris-Etherton PM, Hecker KD, Bonanome A, Coval SM, Binkoski AE, Hilpert KF, Griel AE, Etherton TD. Bioactive compounds in foods: their role in the prevention of cardiovascular disease and cancer. American Journal of Medicine. 2002;113:71S–88S.

58. Nielsen FH. Boron, manganese, molybdenum, and other trace elements. In: Present knowledge in nutrition, 8th ed. Bowman BA, Russell RM, editors. Washington, DC: ILSI Press; 2001.

59. Sharp P. The molecular basis of copper and iron interactions. Proceedings of the Nutrition Society. 2004;63:563–9.

60. Shay NF, Mangian HF. Neurobiology of zinc-influenced eating behavior. Journal of Nutrition. 2000;130(55 Suppl):1493S–9S.

61. Simopoulos AP. The Mediterranean diets: what is so special about the diet of Greece? The scientific evidence. Journal of Nutrition. 2001;131:3065S–73S.

62. St.-Onge M-P. Dietary fats, teas, dairy, and nuts: potential functional foods for weight control? American Journal of Clinical Nutrition. 2005;81:7–15.

63. Stanbury JB, Dunn JT. Iodine and the iodine deficiency disorders. In: Present knowledge in nutrition, 8th ed. Bowman BA, Russell RM, editors. Washington, DC: ILSI Press; 2001.

64. Stoecker BJ. Chromium. In: Present knowledge in nutrition, 8th ed. Bowman BA, Russell RM, editors. Washington, DC: ILSI Press; 2001.

65. Sunde RA. Selenium. In: Present knowledge in nutrition, 8th ed. Bowman BA, Russell RM, editors. Washington, DC: ILSI Press; 2001.

66. Yip R. Iron. In: Present knowledge in nutrition, 8th ed. Bowman BA, Russell RM, editors. Washington, DC: ILSI Press; 2001.

Chapter 13

# The Major Minerals and Water

ike the trace minerals, the major minerals play both structural and functional roles in the body. They act as cofactors, regulate gene expression, are involved in the formation of bones and teeth, and influence every physiological system in the body. In addition, several of the major minerals are involved in water balance. Indeed, it is impossible to provide a simple summary of how each mineral is necessary for good health, because each one has such diverse roles. In this chapter, we discuss the fundamentals related to each major mineral in terms of dietary sources, regulation, function, deficiency, toxicity, and requirements. In addition, we discuss the importance and regulation of water in the body.

Photos © Matthew Farruggio

# The Major Minerals: an Overview

Major minerals (sometimes called macrominerals) are those required in amounts greater than 100 mg/day. The body needs six major minerals, which include calcium, phosphorus, magnesium, sodium, chloride, and potassium. An overview of the food sources, regulation, functions, and toxicities of the major minerals is provided in Table 13.1.

## Sources, Bioavailability, and Regulation of Major Minerals

Major minerals are found mostly in seafood, meat, and dairy products, although some vegetables, fruits, and legumes can be good sources as well. Whereas absorption of some of the major minerals is not regulated, the bioavailability of others can be influenced by genetics, nutritional status, and interactions with other compounds in foods. In addition, the absorption of some major minerals decreases as we age. The primary site of major mineral absorption is the small intestine, although some minerals are absorbed in the large intestine. All major minerals circulate directly from the gastrointestinal tract to the liver via the blood. The amount of each major mineral in the body is regulated, either by the small intestine or the kidneys. For example, circulating concentrations of calcium are regulated by the small intestine so that calcium absorption increases when calcium status is inadequate. For other minerals, such as sodium, circulating levels are regulated by the kidneys, so that excess amounts are excreted in the urine. For most major minerals, toxicities are rare except in genetic disorders and after consumption of large amounts in supplemental or medicinal forms.

## General Functions of Major Minerals

Major minerals have many structural and functional roles in the body. For example, calcium and phosphorus are vitally important for the structure of bones and teeth, whereas others participate in chemical reactions by serving as cofactors. Major minerals are also required for energy metabolism, nerve

### TABLE 13.1   General Characteristics of the Major Minerals

| | |
|---|---|
| **Food Sources** | • Seafood, meat, and dairy products tend to be best sources<br>• Vegetables and legumes are sometimes good sources<br>• Whole-grain products are better sources than milled products |
| **Digestion** | • Very little needed |
| **Absorption** | • Occurs mostly in small intestine but sometimes also in large intestine<br>• Bioavailability sometimes influenced by nutritional status and interactions with other dietary components |
| **Circulation** | • Via blood |
| **Functions** | • Many are cofactors for enzymes, some of which are involved in energy metabolism<br>• Some have major structural roles such as maintaining bone and tooth health<br>• Involved in nerve and muscle function<br>• Electrolytes involved in fluid balance |
| **Toxicity** | • Rare and usually associated with excess supplemental or medicinal intake |

function, and muscle contraction. In addition, the electrolytes (sodium, chloride, and potassium) are essential for maintaining a healthy balance of fluids in the body's various compartments. As with the trace minerals, it is important to remember that not all the functions of each major mineral are known yet.

## ESSENTIAL *Concepts*

Major minerals are essential nutrients that are important for maintaining the structure of our skeletal system and play many functional roles in the body. For example, some act as cofactors for enzymes and are involved in nerve and muscle function, as well as energy metabolism. In addition, electrolytes help maintain fluid balance in the body. The body maintains its availability of the major minerals via alterations in absorption by the intestine and excretion by the kidneys. Animal products provide a good source of the major minerals, although some are also found in certain fruits, vegetables, and legumes.

# Calcium

**Calcium (Ca)** is the most abundant mineral in the body, making up about 1 kg (2.2 pounds) of a person's weight. Although over 99% of the body's calcium is in the skeleton, calcium is also present in the blood and other tissues, where it serves functions vital to life.

## Dietary Sources and Bioavailability

Calcium is found in both plant and animal foods, although the best sources are dairy products (Figure 13.1). Other good sources of calcium include dark green leafy vegetables such as collards and spinach, salmon and sardines (with bones), and some legumes. Calcium-fortified foods such as breakfast cereals, orange juice, and soy products can also provide considerable amounts of calcium. It is relatively easy to assess the calcium content of most foods, because the Nutrition Facts panels on food labels must contain this information, and foods fortified with calcium are often labeled as such.

Many forms of calcium supplements are available, and in general all are relatively well utilized by the body.[1] At one time, there was concern regarding some "natural" forms of calcium such as oyster shell and bonemeal, because they contained high levels of lead. However, changes in manufacturing processes now ensure that these products are safe to consume.[2] Calcium supplements are best absorbed when taken with a meal and in doses of no more than 500 mg at a time.

Many factors can influence the bioavailability of calcium. For example, its absorption can be hindered by the presence of oxalates and phytates, which bind calcium in the intestine. Oxalates are found in cocoa, tea, and many vegetables, including snap beans, brussels sprouts, collards, and spinach. Phytates are commonly found in whole grains, raw beans, and nuts. Although some foods such as spinach contain high amounts of calcium, much of it cannot be absorbed because of the oxalates and/or phytates. In fact, a person would have to consume over eight servings (8 cups) of spinach to absorb the same amount of calcium available in one serving (1 cup) of milk. Nondietary factors such as age can also affect calcium bioavailability. Calcium absorption tends

**CONNECTIONS** Recall that the term *fortified* refers to any food to which a nutrient or combination of nutrients has been added (Chapter 10, page 382).

**CONNECTIONS** Recall that *bioavailability* refers to the extent to which a nutrient is absorbed in the gastrointestinal tract (Chapter 4, page 109).

**calcium (Ca)** A major mineral found in the skeleton and needed for blood clotting, muscle and nerve function, and energy metabolism.

## FIGURE 13.1 Good Sources of Calcium

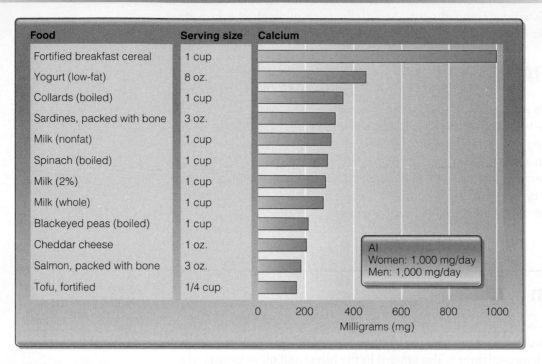

| Food | Serving size | Calcium |
|---|---|---|
| Fortified breakfast cereal | 1 cup | |
| Yogurt (low-fat) | 8 oz. | |
| Collards (boiled) | 1 cup | |
| Sardines, packed with bone | 3 oz. | |
| Milk (nonfat) | 1 cup | |
| Spinach (boiled) | 1 cup | |
| Milk (2%) | 1 cup | |
| Milk (whole) | 1 cup | |
| Blackeyed peas (boiled) | 1 cup | |
| Cheddar cheese | 1 oz. | |
| Salmon, packed with bone | 3 oz. | |
| Tofu, fortified | 1/4 cup | |

AI
Women: 1,000 mg/day
Men: 1,000 mg/day

Milligrams (mg)

SOURCE: Data from USDA Nutrient Database, Release 17.

You would have to eat about 8 servings of spinach to obtain the amount of calcium available from a single serving of milk.

**calcitonin** (cal – ci – TO – nin) A hormone produced in the thyroid gland in response to high blood calcium levels.

**calbindin** A transport protein made in enterocytes that assists in calcium absorption; synthesis is stimulated by vitamin D.

to decrease around the age of 50 years, making it difficult for older people to satisfy their need for calcium.[3]

## Regulation of Calcium Homeostasis

The amount of calcium in the blood is tightly regulated, and the homeostatic mechanisms regulating blood calcium are complex.[4] Calcium homeostasis involves a well-orchestrated system that includes three hormones: calcitriol (vitamin D), parathyroid hormone (PTH), and **calcitonin.** These hormones work together to maintain blood calcium levels within a healthy range. This relationship is illustrated in Figure 13.2 and described next.

### Responses to Low Blood Calcium

When blood calcium levels are low, the parathyroid gland releases a hormone called parathyroid hormone (PTH). PTH then circulates to the kidneys, where it stimulates the conversion of the inactive form of vitamin D to its active form, calcitriol [25 (OH) $D_3 \rightarrow$ 1,25 (OH)$_2$ $D_3$]. Recall from Chapter 11 that calcium absorption, in part, requires the assistance of several vitamin D–dependent calcium transport proteins, one of which is called **calbindin,** that carry calcium into the enterocyte. Calbindin-mediated calcium absorption is an energy-requiring active transport system.

Both PTH and vitamin D act on the kidneys to decrease calcium loss in the urine. Recall that this process is called resorption (or reabsorption). Whereas *absorption* refers to the movement of substances from the gastrointestinal tract into the blood, *resorption* refers to the movement of substances from other tissues (such as kidneys and bone) into the blood. Thus anything that increases

## FIGURE 13.2   Regulation of Blood Calcium

Blood calcium concentration is regulated by three hormones: parathyroid hormone (PTH), calcitriol (vitamin D), and calcitonin.

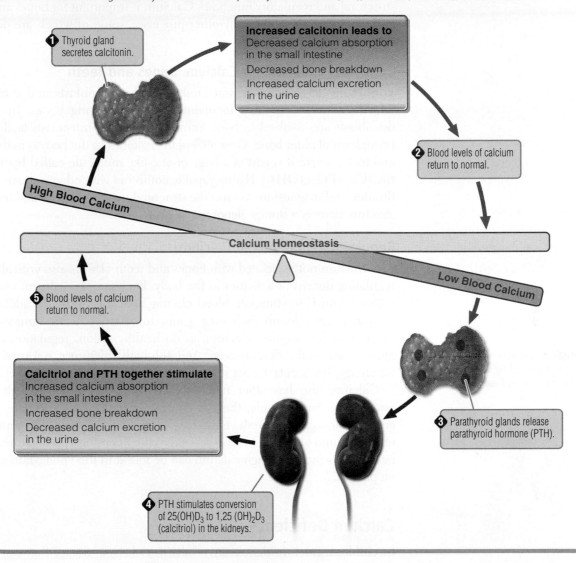

**1** Thyroid gland secretes calcitonin.

**Increased calcitonin leads to**
Decreased calcium absorption in the small intestine
Decreased bone breakdown
Increased calcium excretion in the urine

**2** Blood levels of calcium return to normal.

High Blood Calcium

**Calcium Homeostasis**

Low Blood Calcium

**5** Blood levels of calcium return to normal.

**Calcitriol and PTH together stimulate**
Increased calcium absorption in the small intestine
Increased bone breakdown
Decreased calcium excretion in the urine

**3** Parathyroid glands release parathyroid hormone (PTH).

**4** PTH stimulates conversion of 25(OH)D$_3$ to 1,25 (OH)$_2$D$_3$ (calcitriol) in the kidneys.

absorption or resorption of a nutrient, such as calcium, raises blood levels of the nutrient. In addition, PTH, often aided by vitamin D, also stimulates the breakdown of bone and resorption of its calcium into the blood. Collectively, increased calcium absorption at the small intestine and calcium resorption at the kidneys and bone increase blood calcium concentration.

### Responses to High Blood Calcium

When blood calcium is high, the parathyroid gland produces less PTH, ultimately decreasing the activation of vitamin D to calcitriol in the kidneys. Less activation of vitamin D means reduced calcium absorption in the small intestine. In addition, the thyroid gland responds to elevated blood calcium levels by releasing a hormone called calcitonin. Calcitonin decreases calcium resorption from bone and calcium absorption in the small intestine while increasing calcium loss in the urine. Together, these processes lower blood calcium back to normal.

**CONNECTIONS** In Chapter 11 you learned that vitamin D is both a nutrient and a prohormone, being converted to calcitriol (Chapter 11, page 448).

## Functions of Calcium

Calcium is the most abundant mineral in the body, participating in a variety of structural and regulatory functions. Calcium is important for bones and teeth, and is also required for many other processes, some of which are described here.

### Structural Functions of Calcium: Bones and Teeth

Bone is made of two different kinds of cells: **osteoblasts** and **osteoclasts.** Together, these cells help to maintain healthy, strong bones. In general, osteoblasts are involved in bone formation, whereas osteoclasts facilitate the breakdown of older bone. Over 99% of the calcium in the body is in the bones and teeth, where it is part of a large crystal-like molecule called **hydroxyapatite $[Ca_{10}(PO_4)_6(OH)_2]$.** Hydroxyapatite combines with other minerals such as fluoride and magnesium to form the structural matrix of bones and teeth, and also functions as a storage depot for calcium.

### Regulatory Functions of Calcium

The calcium not associated with bones and teeth (1%) is also critical for life, regulating dozens of activities in the body. For example, calcium cooperates with vitamin K to stimulate blood clotting after an injury.[5] In addition, the body requires calcium each time a muscle contracts.[6] The transmission of nerve impulses requires calcium, as do healthy vision, regulation of blood glucose, and cell differentiation.[7] And although calcium is not used directly for energy, it is a cofactor for several enzymes needed for energy metabolism.

Calcium also has other functions that scientists are just beginning to understand. For example, there is growing evidence that adequate calcium consumption can help reduce the risk for cardiovascular disease, some forms of cancer, and obesity. This is described in the Focus on Food feature. Tips to help you consume enough calcium can be found in the Food Matters feature on page 522.

## Calcium Deficiency

In children, calcium deficiency results in rickets, a disease that can also be caused by vitamin D deficiency (see Chapter 11). Children with rickets have poor bone mineralization and characteristically "bowed" bones, especially in the legs. In adults, calcium deficiency can cause **osteopenia,** the moderate loss of bone mass. In older adults, calcium deficiency can cause resorption of calcium out of bones, resulting in a condition called osteoporosis. Normal and osteoporotic bone is shown in Figure 13.3. Having osteoporosis increases the risk of bone fractures, which can be serious in older people. The National Osteoporosis Foundation estimates that osteoporosis is a major health threat for 44 million Americans, or 55% of people 50 years of age and older.[8] Although adequate calcium intake is needed to maintain healthy bones, results from the Women's Health Initiative suggest only small improvements in bone health when calcium and vitamin D supplements are consumed.[9] More detail on the relationship between diet and osteoporosis is presented in the Nutrition Matters after this chapter.

Calcium deficiency also affects other tissues. Because of calcium's role in muscle contraction and nerve function, low blood calcium can cause muscle pain, muscle spasms, and tingling of the hands and feet. More serious calcium deficiency causes muscles to tighten and become unable to relax, a condition called **tetany.** However, tetany is typically not caused by dietary

**CONNECTIONS** Remember that cell differentiation is the process by which immature cells become mature, functioning cells (Chapter 11, page 440).

**osteoblast** A bone cell that promotes bone formation.

**osteoclast** A bone cell that promotes bone breakdown.

**hydroxyapatite $[Ca_{10}(PO_4)_6(OH)_2]$** The mineral matrix of bones and teeth.

**osteopenia** (os – te – o – PE – ni – a) A disease characterized by moderate bone demineralization.

**tetany** A condition in which muscles tighten and are unable to relax.

## *Focus On* FOOD
### Milk Consumption and Chronic Disease

The relationship between dairy products and good health has been part of standard nutritional advice for decades. Recent research has highlighted how true this is, as some studies show that consuming high amounts of low-fat dairy products is associated with decreased risk for many chronic degenerative diseases, including osteoporosis, heart disease, cancer, and obesity.

How dairy foods decrease the risk for these diseases is not completely understood, although calcium appears to be a key player. The protective effect of calcium consumption on bone disease is well accepted by medical experts and is reviewed in the Nutrition Matters following this chapter. However, the relationships between calcium and other chronic diseases have only recently been discovered. For example, increased calcium consumption from foods and supplements can lower blood pressure in some people.[10] There are also proteins in dairy products that can lower blood pressure in some people.[11] This is important, because high blood pressure is a major risk factor for cardiovascular disease. Because high blood pressure is estimated to cause approximately 25,000 deaths per year in the United States, the potential benefit of dairy consumption is significant.

In addition, increased consumption of dairy products may decrease risk for breast, prostate, and colon cancer.[12] One study suggested that women who drink at least three glasses of milk each day have half the incidence of breast cancer than women who do not drink milk.[13] Similarly, consuming even one additional serving of low-fat dairy foods is associated with a 40 to 50% reduction of colon cancer risk.[14] Although it is not known which factors in milk provide these health benefits, calcium, vitamin D, and conjugated linoleic acid (CLA—a fatty acid) are likely candidates.[15]

Considerable research also shows that consuming low-fat dairy products can help prevent obesity, especially in children, and can help overweight adults lose weight. Michael Zemel at the University of Tennessee has led the way with this research. For example, his research has shown that obese people consuming three to four daily servings of milk, yogurt, or cheese can lose over 50% more body fat than those who consume one or fewer servings of dairy products.[16] The effect of dairy foods on weight maintenance is probably linked to the high calcium content of these foods, although other components may also be important. Not all studies have supported the beneficial effect of dairy consumption on weight loss, and research continues in this area.[17] Nonetheless, by most accounts milk is good for you!

calcium insufficiency alone, but by a combination of factors such as disease and medication.

## Calcium Toxicity

Consuming too much calcium can cause calcium to be deposited in soft tissues such as muscle and kidney. High calcium intakes are also associated with impaired kidney function and can interfere with the bioavailability of other

### FIGURE 13.3    Normal and Osteoporotic Bone

Calcium deficiency in adults can cause bones to weaken and ultimately become osteoporotic.

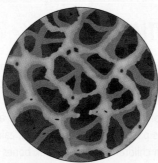

Normal bone                    Osteoporotic bone

*Food Matters*

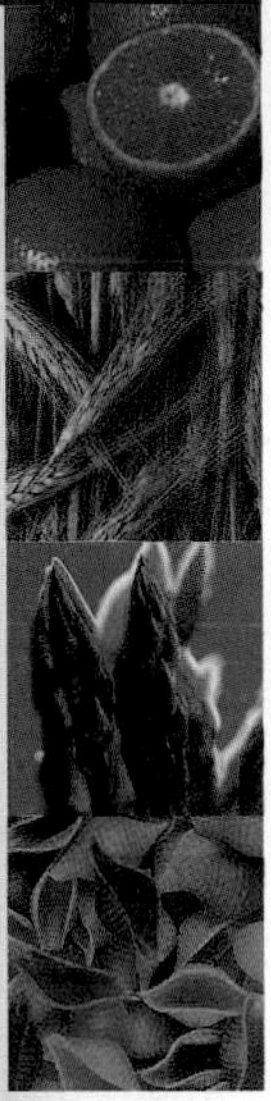

*Working Toward the Goal*

The USDA Dietary Guidelines for Americans recommend that we consume 3 cups per day of fat-free or low-fat milk or equivalent milk products. The following selection and preparation tips will help you meet your goal.

- When grocery shopping, visit the cheese section and buy a variety that you do not usually purchase.
- Buy single-serving containers of plain or flavored milk. They make quick and appealing snacks, and are more nutrient dense than soft drinks.
- Use fat-free or low-fat milk instead of water when making instant oatmeal or other hot cereals.
- Rather than having a candy bar or bag of chips, snack on fat-free or low-fat yogurt or cheese.
- Make a dip for fruits or vegetables from yogurt or low-fat cream cheese.
- Experiment with making calcium-rich smoothies by blending various fruits with yogurt or reduced-fat ice cream.
- Top cut-up fruit with flavored yogurt for a quick dessert.
- If you drink cappuccinos or lattes, ask that they be made with fat-free milk.

nutrients, such as iron and zinc.[18] These effects of high calcium intake can occur from foods, but they are more commonly associated with consumption of high doses of calcium supplements or certain diseases.

## Recommended Intakes for Calcium

There are insufficient data to generate Recommended Dietary Allowances (RDAs) for calcium. Therefore, an Adequate Intake (AI) level has been set at 1,000 mg/day for adults. Because calcium absorption tends to decrease with age, the AI increases to 1,200 mg/day at 51 years of age. To decrease the risk of calcium toxicity, a Tolerable Upper Intake Level (UL) has been set at 2,500 mg/day. A complete list of the Dietary Reference Intake (DRI) values for all major minerals is provided inside the front cover of this book.

**CONNECTIONS** The Dietary Reference Intakes (DRIs) are reference values for nutrient intake. When adequate information is available, Recommended Dietary Allowances (RDAs) are established to reflect intakes that meet the needs of about 97% of the population. When less information is available, Adequate Intake (AI) values are provided. Tolerable Upper Intake Levels (ULs) indicate intakes that should not be exceeded (Chapter 2, page 36).

## ESSENTIAL *Concepts*

Calcium plays major structural and functional roles in the body. As a component of hydroxyapatite, calcium is important for the structure of bones and teeth. Calcium is also needed for blood clotting, nerve and muscle function, and metabolism. Blood calcium concentrations are highly regulated, mostly at the levels of the small intestine and bone. This involves parathyroid hormone (PTH), calcitriol (vitamin D), and calcitonin. Calcium deficiency in children can cause weak bones and can result in rickets. In adults, calcium deficiency can cause osteopenia and osteoporosis. Very high calcium intake can cause the mineralization of soft tissues such as muscle and kidney. Although both plant and animal foods provide calcium, dairy products provide most of this mineral in the diet.

# Phosphorus

Because **phosphorus (P)** is a component of the phospholipid bilayers that make up cell membranes (animals) and cell walls (plants), it is found in practically all foods. Like calcium, phosphorus has both structural and functional roles in the body. Phosphorus is also critical for energy metabolism.

## Dietary Sources and Bioavailability

Perhaps more than any other nutrient, phosphorus is naturally abundant in food, especially those rich in protein. Good sources include dairy products, meat (including poultry), seafood, nuts, and seeds. Bioavailability of phosphorus is high in most foods, although phosphorus in seeds and grains is more difficult to absorb than that in animal foods. This is because seeds and grains contain phytates, which decrease phosphorus absorption.

In addition to the phosphorus found naturally in foods, dietary phosphorus can also be obtained from processed foods and some soft drinks. Phosphorus is added to foods to enhance moisture retention, smoothness, and taste. A typical cola drink (12 ounces) contains about 50 mg of phosphorus. You can find out if a soft drink contains added phosphorus by looking for the ingredient "phosphoric acid" on the label. Reliance on soft drinks to supply dietary phosphorus is not recommended, because they typically do not contain any other essential nutrients. In other words, they have low nutrient densities.

> **CONNECTIONS** Recall that the term *nutrient density* refers to the relative ratio of nutrients to energy in a food (Chapter 2, page 44).

## Metabolism and Regulation of Phosphorus in the Body

Phosphorus is readily absorbed in the small intestine by both vitamin D–dependent active transport and by simple diffusion.[19] Like calcium, the regulation of blood phosphorus concentrations is controlled mainly by calcitriol (vitamin D), PTH, and calcitonin. When blood phosphorus levels are low, calcitriol together with PTH increase phosphorus absorption in the small intestine and phosphorus resorption from bones. Together, these actions return blood phosphorus to normal.[20] During periods of high blood phosphorus, calcitonin stimulates the bone-building activity of osteoblasts, which take up phosphorus from the blood—lowering blood phosphorus to normal levels.

> **CONNECTIONS** Active transport is an energy (ATP)-requiring process. Simple diffusion does not require energy and transports substances from a region of higher concentration to that of a lower concentration (Chapter 3, pages 79–80).

> **phosphorus (P)** A major mineral needed for cell membranes, bone and tooth structure, DNA, RNA, ATP, lipid transport, and a variety of reactions in the body that require phosphorylation.

## Functions of Phosphorus

Recall that cell membranes are made from *phospho*lipids, which consist of phosphorus-containing polar head groups. Therefore, a primary role of phosphorus in the body is its function as a component of cell membranes. Phospholipids also surround lipoproteins and are therefore important for the transport of lipids in the body. Phosphorus is a component of our genetic material (DNA) and adenosine triphosphate (ATP) and is also needed for protein synthesis and energy metabolism. In addition, phosphorus-containing

> **CONNECTIONS** Phospholipids are made of a glycerol backbone bonded to two fatty acids and a polar head group via a phosphate linkage (Chapter 7, page 234).
>
> Remember that lipoproteins (such as low-density lipoproteins or LDLs) transport lipids in the blood (Chapter 7, page 245).

compounds help maintain blood pH (acid–base balance) by acting as buffers that accept and donate hydrogen ions ($H^+$).

Phosphorus is also involved in hundreds of metabolic reactions in the body. In these reactions, phosphate groups are transferred from one molecule to another, producing "phosphorylated" molecules. In fact, some molecules remain inactive until they are phosphorylated. For example, the enzyme needed to break down glycogen into its glucose subunits must be phosphorylated before it can work.

Phosphorus, along with calcium, is also required to form hydroxyapatite needed for bones and teeth. Hydroxyapatite crystals contain a ratio of calcium to phosphorus of 2:1. In other words, there are 2 atoms of calcium to each atom of phosphorus. In this way, phosphorus serves an important structural function in the body.

## Phosphorus Deficiency, Toxicity, and Recommended Intakes

Phosphorus deficiency results in loss of appetite, anemia, muscle weakness, poor bone development and, in extreme cases, death. Because phosphorus is in so many foods and is a common food additive, phosphorus deficiency is rare. Phosphorus toxicity, however, is more common, resulting in mineralization of soft tissues, especially the kidneys. The RDA for phosphorus in adults is 700 mg/day. Because of the potentially toxic effects of high doses of phosphorus, a UL of 4,000 mg/day has been established.

## ESSENTIAL *Concepts*

Phosphorus is a component of phospholipids, which make up cell membranes and lipoproteins. Phosphorus is also found in bone, DNA, RNA, and ATP and can activate or inactivate a variety of enzymes. Thus phosphorus has both structural and functional roles in the body. Regulation of blood levels of phosphorus occurs via the actions of parathyroid hormone (PTH) and vitamin D (calcitriol) at the small intestine and bone. Phosphorus deficiency is rare, but can be fatal. Toxic intakes of phosphorus can damage the kidneys. Good sources of this mineral include dairy products, meat, seafood, nuts, and seeds.

# Magnesium

**Magnesium (Mg)** is important for many physiological processes, including energy metabolism and enzyme function. More recently, magnesium has gained significant attention because of its possible associations with several chronic diseases, such as heart disease and type 2 diabetes.

## Dietary Sources and Bioavailability

Good sources of magnesium include green leafy vegetables, seafood, legumes, nuts, dairy products, chocolate, and unprocessed (brown) rice. In general, foods made with whole-grain products contain more magnesium than do refined-grain products. Bioavailability of magnesium from the diet is variable, and it is likely that several dietary factors influence its absorption. For exam-

**magnesium (Mg)** A major mineral needed for stabilizing enzymes and ATP and as a cofactor for many enzymes.

ple, some studies suggest that diets high in calcium or phosphorus decrease the bioavailability of magnesium.[21] However, the factors affecting magnesium absorption are not well understood.

## Metabolism and Regulation of Magnesium in the Body

Blood levels of magnesium are regulated both by the intestine and the kidneys. Like most minerals, magnesium absorption in the small intestine and resorption by the kidneys increase when blood levels are low. The opposite occurs when circulating magnesium levels are high. The kidneys appear to be the main regulators of magnesium levels in the body, although the mechanisms involved in this remain unclear.

## Functions of Magnesium

The majority of magnesium in the body is associated with bone where it provides structure. Because magnesium is typically found in the body as its positively charged, cation form—$Mg^{2+}$—another role of $Mg^{2+}$ in the body is to stabilize enzymes and to neutralize negatively charged ions (that is, anions). Among its substrates are ATP and ADP, which are typically found in the body as the negatively charged $ATP^{4-}$ and $ADP^{3-}$. Because magnesium helps stabilize these high-energy compounds, it is vital for energy metabolism. Magnesium participates in over 300 reactions—most notably those involved in DNA and RNA metabolism—and also influences nerve and muscle function, especially in heart tissue.[22]

## Magnesium Deficiency and Toxicity

Magnesium deficiency is rare but sometimes seen in alcoholics because of dietary inadequacy (primary malnutrition) and poor overall nutrient absorption (secondary malnutrition). Severe magnesium deficiency causes abnormal nerve and muscle function, especially in cardiac tissue. Because of this, there is much interest in the possibility that mild magnesium deficiency may increase risk for cardiovascular disease.[23] Some research also suggests that magnesium deficiency may predispose people to type 2 diabetes and even migraine headaches.[24] However, controlled clinical studies have not provided convincing evidence that higher intakes of magnesium decrease the incidence of these conditions.

Magnesium toxicity only appears in people taking large doses of supplemental magnesium or medications containing magnesium, such as "milk of magnesia" used to treat heartburn, indigestion, and constipation. Toxicity results in intestinal distress (diarrhea, cramping, nausea) and, when severe, can cause the heart to stop beating. Thus consumption of magnesium supplements should be carefully monitored.

## Recommended Intakes for Magnesium

The RDAs for magnesium for men and women are 420 and 320 mg/day, respectively. The UL for magnesium is 350 mg/day. This UL is unusual, because it is actually similar to or less than the RDA values. This is because the UL does not apply to intake of magnesium from foods—only from supplemental and medicinal forms.

## ESSENTIAL *Concepts*

Magnesium plays many regulatory roles in the body. It is required for energy metabolism, is a cofactor for hundreds of enzymes, and is needed for nerve and muscle function. Magnesium deficiency is rare but can result in poor nerve and muscle action, especially in heart tissue. Magnesium toxicity from supplements or some medications can cause intestinal upset and, when severe, cardiac arrest. Magnesium is found in seafood, legumes, nuts, chocolate, and unprocessed grains.

# Sodium and Chloride

**CONNECTIONS** Recall that ionic bonds are chemical bonds between positively and negatively charged atoms (ions) (Chapter 3, page 67).

Sodium (Na) and chloride (Cl) are almost always found together in foods, and in many ways have similar functions in the body. This is because they join together via ionic bonds to form a salt—sodium chloride (NaCl), better known as "table salt." As such, these minerals are added freely to foods during processing, cooking, and during a meal. Because sodium and chloride are found together in foods and both play major roles in fluid balance, they are discussed together here.

## Dietary Sources and Bioavailability

Foods that contribute sodium to the diet are also good sources of chloride. The sodium content of foods is easily determined by using Nutrition Facts panels, and foods with low sodium content are often labeled as such. Because increased sodium consumption can lead to hypertension in some people, the following unqualified health claim is sometimes found on low-sodium foods: "Diets low in sodium may reduce the risk of high blood pressure, a disease associated with many factors." The following terms are also used on food packaging to describe their salt content.

- Salt free—less than 5 mg sodium per serving
- Very low salt—less than 35 mg sodium per serving
- Low salt—less than 140 mg sodium per serving

Indeed, table salt is added to many foods and provides the majority of both sodium and chloride to the American diet. One teaspoon of salt contains more than 2,000 mg of each of these minerals. Other food additives, such as monosodium glutamate, which is commonly used in Asian cuisine, also contain sodium. In general, unprocessed foods such as fresh fruits and vegetables contain small amounts of sodium and chloride, whereas manufactured and highly processed foods such as fast foods, frozen entrees, and savory snacks contain larger amounts. Some meats, dairy products, poultry, and seafood naturally contain moderate amounts of both sodium and chloride. In addition, we tend to add salt-containing condiments, such as soy sauce and ketchup, to our foods. Here is the sodium content of some commonly eaten foods (keep in mind that the AI for sodium is 1500 mg per day).

**sodium (Na)** A major mineral important for regulating fluid balance, nerve function, and muscle contraction.

**chloride (Cl)** A major mineral important for regulating fluid balance, protein digestion in the stomach (via HCl), and carbon dioxide removal by the lungs.

- ½ cup of fresh or frozen vegetables cooked without salt—1 to 70 mg
- 1 cup of milk—120 mg
- 1 fast food taco—802 mg
- 1 tablespoon soy sauce—902 mg
- 3 ounces of ham—1,128 mg

The bioavailability of both sodium and chloride is very high, except in conditions of intestinal malabsorption, which can occur with gastrointestinal infections.

## Regulation of Sodium and Chloride in the Body

Sodium and chloride are efficiently absorbed, mostly in the small intestine. There are several transport mechanisms, but in general sodium is absorbed first, with chloride following close behind.[25] One mechanism by which sodium is transported into the enterocyte requires that glucose be cotransported with it. Without glucose sodium cannot be absorbed. This is why sports drinks designed to replace salt lost during athletic events were originally formulated to contain both glucose and salt—without the glucose, the salt would not be absorbed. However, many "sports drinks" currently on the market are made with high-fructose corn syrup, not glucose. Sodium is also actively absorbed in the colon prior to elimination in the feces, resulting in water absorption as well. Without this, large amounts of water would be lost in the feces, causing potentially dangerous diarrhea and dehydration.

**Many condiments contain sodium.**

### Regulation of Sodium in the Blood

The sodium and chloride concentrations in the blood are carefully regulated, as shown in Figure 13.4.[26] When the concentration of blood sodium decreases, a condition called **hyponatremia** develops. Hyponatremia can be caused by low intake of salt or when sodium chloride is lost from the body via excessive sweating. It can also be caused by excessive water consumption not accompanied by increased sodium intake. Hyponatremia stimulates the adrenal glands to secrete the hormone **aldosterone,** which in turn causes the kidneys to retain (resorb) sodium. As a result, blood sodium levels are restored to normal. Low blood pressure also increases aldosterone release from the adrenal glands by stimulating the kidneys to produce an enzyme called **renin.** In the blood, renin converts the liver-derived protein **angiotensinogen** into **angiotensin I.** Angiotensin I is then converted to **angiotensin II** in the lungs. Angiotensin II, in turn, stimulates the adrenal glands to release additional aldosterone.

When the concentration of blood sodium increases (**hypernatremia**), aldosterone and renin secretion decrease, in turn decreasing salt resorption by the kidneys. In this way, renin, angiotensin II, and aldosterone work together to maintain healthy levels of blood sodium. This is often referred to as the renin-angiotensin-aldosterone system.

## Functions of Sodium and Chloride

Sodium and chloride, the most abundant ions in the blood, are the body's principal electrolytes. When the ionic bond of a NaCl molecule dissociates in water, sodium is released as a cation ($Na^+$), whereas chloride is released as an anion ($Cl^-$). Both ions play major roles in fluid balance. Because water naturally moves to areas that have high $Na^+$ and/or $Cl^-$ concentrations, the body can maintain fluid balance by selectively moving these electrolytes where more water is needed. This process is described in more detail later in this chapter. Sodium is also important for nerve function and muscle contraction, both of which also involve potassium ($K^+$). In addition, chloride is needed for production of hydrochloric acid (HCl) in the stomach, for removal of carbon dioxide ($CO_2$) by the lungs, and for optimal immune function.

**CONNECTIONS** Remember that electrolytes are substances that produce charged particles, or ions, when dissolved in fluids. The terms *electrolyte* and *ion* are often used interchangeably (Chapter 3, page 67).

**hyponatremia** Low blood sodium concentration.

**aldosterone** (al – DO – ster – one) A hormone produced by the adrenal glands in response to low blood sodium concentration (hyponatremia) and angiotensin II.

**renin** (RE – nin) An enzyme produced in the kidneys in response to low blood pressure; converts angiotensinogen to angiotensin I.

**angiotensinogen** (an – gi – o – ten – SIN – o – gen) An inactive protein made by the liver that is converted by renin into angiotensin I.

**angiotensin I** The precursor of angiotensin II.

**angiotensin II** A protein derived from angiotensin I in the lungs; stimulates aldosterone release.

**hypernatremia** High blood sodium concentration.

## FIGURE 13.4  Regulation of Blood Sodium

Blood sodium concentration is regulated by the kidneys via aldosterone, angiotensin II, and renin.

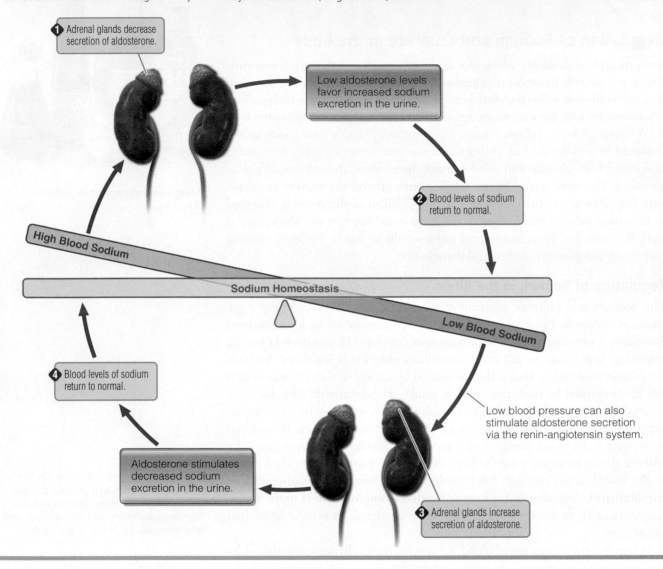

## Sodium and Chloride Deficiencies

Sodium and chloride deficiencies are rare. However, they can occur in infants and small children with diarrhea and/or vomiting. These conditions result in loss of sodium and chloride through the gastrointestinal tract (diarrhea) or loss of nutrients before they even enter the intestine (vomiting). Diarrhea and vomiting can be life threatening because of the rapid loss of both electrolytes and the accompanying water. Less severe sodium and chloride deficiencies can occur in athletes, especially those involved in endurance sports such as marathon running. Symptoms of electrolyte deficiency include nausea, dizziness, muscle cramps, and, in severe cases, coma.

## Overconsumption of Sodium Chloride (Salt)

There are no "toxic" effects of high salt intake, although high intakes of sodium chloride are associated with increased blood pressure in some people, a major risk factor for heart disease and stroke.[27] This is because the sodium concentration of the blood is one of the major regulators of blood volume, with higher sodium causing greater blood volume—and thus increased blood pressure. The relationships among sodium, blood volume, and blood pressure are described in more detail later in this chapter. In reality, only a small percentage of people are sensitive to the blood pressure–raising effects of salt. The most susceptible are older people, African Americans, and individuals with hypertension, diabetes, or chronic kidney disease. The relationship between sodium and hypertension is described in more detail in the Focus on Food feature.

## Recommended Intakes for Sodium and Chloride

Despite the long-standing interest in the health effects of salt, data are insufficient to establish RDAs for sodium and chloride. However, AIs have been set at 1,500 and 2,300 mg/day for sodium and chloride, respectively. Recognizing that typical salt intake is significantly higher than this, the Dietary Guidelines for Americans recommend that we "choose and prepare foods with less salt." Because high intakes of salt are associated with increased risk for high blood pressure in some people, ULs of 2,300 and 3,600 mg/day for sodium and chloride, respectively, have been established. No ULs are set for infants, but it is recommended that salt not be added to foods during the first year of life.

## *Focus On* FOOD
### Salt: Is it Really So Bad?

Sodium and chloride are essential nutrients required for the basic functions of life. Therefore, recommendations to restrict salt intake may seem confusing. Even the U.S. Department of Agriculture's (USDA) Dietary Guidelines for Americans recommend that we consume less salt. How can something so vital for health be so bad?

Sodium helps maintain blood volume, which is directly related to blood pressure—when blood volume increases, so does blood pressure. High blood pressure is also called hypertension. It has long been known from epidemiologic studies that people who consume large amounts of salt tend to have higher blood pressure than those who consume less salt. People with hypertension also tend to be overweight and smoke. Approximately 50 million Americans (25% of the U.S. adult population) have high blood pressure.[28] Understanding what causes hypertension is important,

because of its associated risk for cardiovascular disease. Thus, public health organizations have long recommended that we strive to decrease blood pressure by weight loss, by not smoking, and by salt restriction.

However, it is clear that not all people with hypertension benefit from a low-salt diet. In fact, whereas salt restriction decreases blood pressure in some people (referred to as "salt sensitive"), in a small portion of the population a low-salt diet actually increases blood pressure. For others, salt intake does not appear to influence blood pressure one way or another; these people are "salt insensitive." This wide variation in how people respond to salt intake may be due to factors such as genetics, exercise, and the responsiveness of the renin-angiotensin-aldosterone system. Although there is no easy way to determine if a person is salt sensitive, certain subgroups of the population have high prevalences of

salt-sensitive people. These include the elderly, African-Americans, and people with hypertension, diabetes, or chronic kidney disease.

Even a small drop in blood pressure can help reduce a person's risk for cardiovascular disease. Therefore, although reducing salt intake may not influence blood pressure in most individuals, public health agencies continue to recommend that people limit their salt intake. Perhaps medical professionals will someday be able to assess a person's sensitivity to salt and make individualized dietary recommendations based on their genetic makeup. Until that time, it is recommended that sodium intake not exceed 2,300 mg per day for the general population, and the Institute of Medicine recommends that individuals with increased likelihood of being salt sensitive consume considerably less than this amount.[29]

### Special Recommendations for Athletes and High Temperatures

Because salt is lost in perspiration, the AIs are not meant to apply to highly active individuals, such as endurance athletes. In addition, people living in very warm climates may require additional salt in their diets. However, the Institute of Medicine has not yet established AI recommendations for these groups.

## ESSENTIAL *Concepts*

Sodium and chloride are found together in foods and often work together in the body. Both minerals, when dissolved in water, become ions important in maintaining fluid balance. Sodium is needed for nerve and muscle function, whereas chloride (as HCl) is required for digestion in the stomach, carbon dioxide removal, and immune function. Deficiencies of sodium and chloride are uncommon, except in sick infants and children, and endurance athletes. Symptoms include nausea, dizziness, and muscle cramps. Although high intakes of salt have long been associated with increased risk for heart disease, we now know this is true only in a small segment of the population. These salt-sensitive individuals should be careful to limit their salt intakes. We consume most of the sodium and chloride in our diets in the form of table salt (NaCl).

# Potassium

**CONNECTIONS** Extracellular fluids are those outside of cells, whereas intracellular fluids are those within cells (Chapter 3, page 77).

Whereas sodium is the most abundant cation in the extracellular fluids, **potassium (K)** is the most abundant cation in intracellular fluids. And unlike sodium, which can increase blood pressure in some people, potassium lowers blood pressure.

## Dietary Sources and Bioavailability

Dietary sources of potassium include legumes, potatoes, seafood, dairy products, meat, and a variety of fruits and vegetables such as sweet potatoes and bananas. The bioavailability of potassium from these foods is high and is not influenced by other factors.

Potatoes are an excellent source of potassium.

## Regulation and Functions of Potassium in the Body

As with sodium and chloride, potassium absorption occurs with great efficiency in the small intestine, and to a lesser extent in the colon. And as with sodium and chloride, regulation of blood potassium levels is achieved mostly by the kidneys. In other words, when blood potassium is elevated, the kidneys excrete more potassium. The opposite is true when blood potassium is low. However, in contrast to what occurs with sodium and chloride, aldosterone causes the kidneys to *increase* potassium excretion in the urine. Thus aldosterone causes blood levels of sodium and chloride to increase while simultaneously causing blood levels of potassium to decrease.

The potassium cation ($K^+$) is an important electrolyte, working with sodium and chloride to maintain proper fluid balance in the body. In addition, potassium is critical for muscle function (especially in heart tissue), nerve function, and energy metabolism. Whereas increased sodium intake

**potassium (K)** A major mineral important in fluid balance, muscle and nerve function, and energy metabolism.

causes a rise in blood pressure in salt-sensitive individuals, consuming high amounts of potassium can decrease blood pressure in some people.

## Potassium Deficiency, Toxicity, and Recommended Intake

Potassium deficiency is rarely seen because of the abundance of the mineral in the diet, although it can result from diarrhea and vomiting. Heavy use of certain diuretics can also result in excessive potassium loss in the urine. **Diuretics** are drugs used to lower blood pressure by helping the body eliminate water. This reduces blood volume and helps decrease blood pressure. However, when the body excretes excessive amounts of water, it also loses electrolytes. This can lead to low blood potassium, a condition called **hypokalemia.** People with eating disorders involving vomiting, such as bulimia nervosa, are at increased risk for hypokalemia. Potassium deficiency causes muscle weakness, constipation, irritability, and confusion. Recent studies also suggest that it may cause insulin resistance.[30] In severe cases, potassium deficiency can cause irregular heart function, muscular paralysis, decreased blood pressure, and difficulty breathing.

Potassium toxicity is rare from foods but can occur with potassium supplementation. Because high levels of potassium can cause cardiac arrest, potassium toxicity can be life threatening.[31] Less severe symptoms include weakness, tingling in the feet and hands, and muscular paralysis.

As with the other electrolytes, no RDAs have been established for potassium, although an AI has been set at 4,700 mg/day for adults. No ULs have been set for this mineral.

> **CONNECTIONS** Bulimia nervosa is an eating disorder characterized by bingeing and purging—often vomiting. (Chapter 9 Nutrition Matters, page 364).

## ESSENTIAL *Concepts*

Potassium, an important cation in the body, cooperates with sodium and chloride to maintain fluid balance. Potassium is also needed for muscle function (especially in the heart), nerve impulses, and energy metabolism. As with the other electrolytes, potassium deficiency is rare except during illness and in some eating disorders, and can cause cardiac disturbances, weakness, and confusion. Excessive potassium intake from supplements can cause cardiac arrest. Potassium is found in legumes, potatoes, seafoods, a variety of fruits and vegetables, dairy products, and meat.

# Water: The Essence of Life

Water ($H_2O$), the most abundant molecule in the human body, is truly the essence of life. Water is found both inside and outside of cells, and although some water is made in the body during metabolism, it is an essential nutrient. Classified as a macronutrient, water acts as a biological solvent, serves as a chemical reactant in biochemical reactions, and helps regulate body temperature. Water balance within different compartments of the body is vital for health and is regulated primarily by the movement of electrolytes.

## Distribution of Water in the Body

Water accounts for 50 to 65% of total adult body weight. Because muscle has more water associated with it than does fat, the total volume of body water is

**diuretic** (di – ur – E – tic) A substance or drug that causes water loss from the body.

**hypokalemia** Low blood potassium concentration.

larger in people with greater lean tissue compared to people with more body fat. For this reason, males generally have more water than females.

Water is a principal component of biological fluids in the body; its distribution is shown in Figure 13.5. Recall that fluids can be categorized as intracellular fluid (inside cells) or extracellular fluid (outside cells). Extracellular fluid includes interstitial fluid (between cells) and intravascular fluid (blood and lymph). Plasma, which is the fluid component of blood, is an example of an intravascular fluid. For example, a typical 70-kg man has about 42 liters of water in his body—28 liters inside of cells (intracellular fluid) and 14 liters outside of cells (extracellular fluid). Most of the extracellular fluid is in interstitial space (11 L), with the remainder as intravascular fluid (3 L).

## Fluid Balance and the Electrolytes

The amount of water in the various spaces is highly regulated and, as previously mentioned, is dependent on several essential minerals—namely sodium, chloride, and potassium. Plasma proteins such as albumin also help maintain a healthy fluid balance between intravascular and interstitial compartments.

Intracellular and extracellular fluids each contain specific concentrations of various electrolytes. For example, sodium ($Na^+$) and chloride ($Cl^-$) ions are predominant in extracellular fluid, whereas potassium ($K^+$) and phosphate ($PO_4^-$) ions are predominant in intracellular fluid. The concentrations of these electrolytes must be maintained within certain ranges for cells to function properly. Recall that cell membranes control the movement of most substances into and out of cells. For example, $Na^+$ and $Cl^-$ cannot cross cell membranes

CONNECTIONS Recall that proteins, such as albumin, in the blood help move water from the interstitial space back into the intravascular space in capillary beds (Chapter 6, page 190).

## FIGURE 13.5 Fluid Compartments

Water in the body can be characterized as being intracellular, extracellular, interstitial, and intravascular.

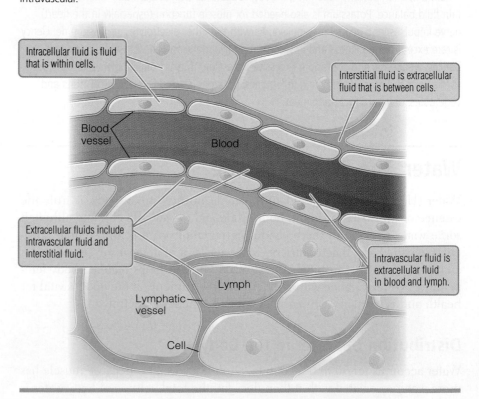

Intracellular fluid is fluid that is within cells.

Interstitial fluid is extracellular fluid that is between cells.

Blood vessel

Blood

Extracellular fluids include intravascular fluid and interstitial fluid.

Intravascular fluid is extracellular fluid in blood and lymph.

Lymph

Lymphatic vessel

Cell

passively but instead need help from membrane-bound transport proteins, or "pumps," and the input of energy (ATP). Thus movement of electrolytes into and out of cells is an active transport process. However, water is unique in that it passes freely across cell membranes, making this a passive transport process. The body can couple the active pumping of ions across cell membranes with the passive movement of water. In this way, fluid movement and balance are maintained in the various compartments at appropriate levels.

## Movement of Water Across Membranes Occurs via Osmosis

Just as the body maintains the concentrations of other substances (such as blood glucose) within specific ranges, it also tightly regulates the amounts of water within the various fluid compartments. This is done by a process called osmosis, which is a special form of simple diffusion. Osmosis occurs when a cell membrane is selective in terms of the types of molecules that are allowed across it. This sort of selective membrane is called a **semipermeable membrane.** Cell membranes are semipermeable, allowing the passive movement of some substances (such as water) but not others (such as ions).

Osmosis relies on a basic rule of nature that water flows from a region of low solute concentration to a region of high solute concentration. A solution with a low solute concentration is said to have **low osmotic pressure,** whereas a solution with a high solute concentration is said to have **high osmotic pressure.** For example, watch what happens when cucumber slices are salted and left to sit. Eventually, the salt "draws" water out of the cucumber slices via osmosis. In general, water continues to move from one region to another until the concentration of solutes is the same in both regions. Another example of osmosis is illustrated in Figure 13.6.

How are ions involved in the process of osmosis and fluid balance in the body? The active transport of ions across a cell membrane (let's say from region A to region B) causes the concentration of ions (that is, solutes) to build up on one side of the membrane (region B). When this occurs, water moves

**CONNECTIONS** Simple diffusion is a form of passive transport whereby substances cross cell membranes without the assistance of a transport protein or energy (Chapter 3, page 79).

**CONNECTIONS** A solute is a substance dissolved in a liquid (Chapter 3, page 69).

**semipermeable membrane** A barrier that allows passage of some, but not all, molecules across it.

**low osmotic pressure** When a solution has a small amount of solutes dissolved in it.

**high osmotic pressure** When a solution has a large amount of solutes dissolved in it.

## FIGURE 13.6  Osmosis

Osmosis occurs when water moves from a region of low osmotic pressure (low solute concentration) to a region of high osmotic pressure (high solute concentration).

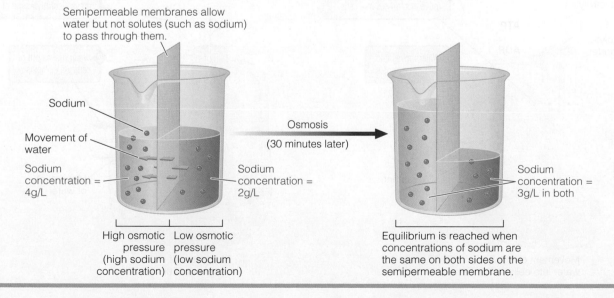

Semipermeable membranes allow water but not solutes (such as sodium) to pass through them.

Sodium

Movement of water

Sodium concentration = 4g/L

Osmosis
(30 minutes later)

Sodium concentration = 2g/L

Sodium concentration = 3g/L in both

High osmotic pressure (high sodium concentration)  Low osmotic pressure (low sodium concentration)

Equilibrium is reached when concentrations of sodium are the same on both sides of the semipermeable membrane.

passively via osmosis from the region of low osmotic pressure (region A) to the region of high osmotic pressure (region B). In other words, fluid balance shifts toward region B. The active movement of ions (such as sodium, chloride, and potassium) across cell membranes maintains a healthy fluid balance among fluid compartments. This is shown in Figure 13.7.

You can compare the process of fluid balance regulation via osmosis to what might happen on a class trip at a local elementary school. Imagine that 50 children and 50 mothers all need to fit evenly into two school buses. Let's say that at first all the children and mothers load into bus 1. However, if 25 of the mothers "actively" get off bus 1 and move into bus 2, their children will "passively" follow. This is somewhat similar to how the body passively moves water (like the children) via the directed, active movement of electrolytes (like the mothers). Where one goes, the other will follow.

An example of how the body uses osmosis to selectively move water is the active transport of sodium into the cells that line the colon, causing water to be absorbed as well. Without this absorption, excessive amounts of water would be lost in the feces. An example of how osmosis may have negative health consequences is the regulation of blood volume and blood pressure. Recall that high salt (sodium chloride) intake can lead to increased blood volume and blood pressure in some people. This is because salt-sensitive people are unable to excrete excess sodium, resulting in high levels of sodium in the blood. This, in turn, causes water to move into the intravascular space, increasing blood volume—and, in turn, blood pressure.

## The Functions of Water

Water plays an active role in many processes, including hundreds of chemical reactions. Water also helps keep the body at a constant temperature, even

## FIGURE 13.7 Maintaining Fluid Balance Across Membranes

Active transport of electrolytes across membranes causes the movement of water across the membranes as well; water flows from a region of low osmotic pressure to one of high osmotic pressure via the process of osmosis.

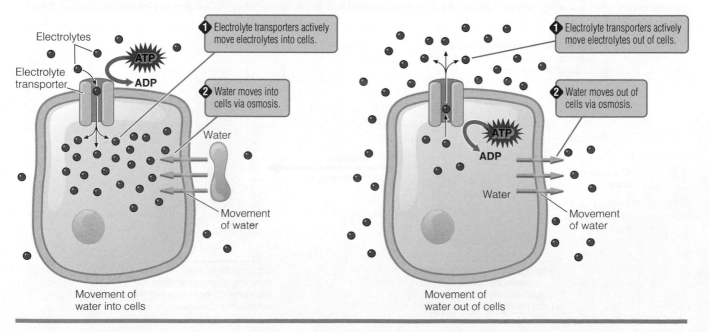

when the environment is very cold or hot. Water also provides protection and serves as an important solvent and lubricant.

## Water as a Solvent, Transport Medium, and Lubricant

Water is the primary solvent in blood, saliva, and gastrointestinal secretions. For example, blood is a solution consisting of water and a variety of dissolved solutes including nutrients and metabolic waste products, such as carbon dioxide and urea. Substances dissolved in blood can move from inside the blood vessel out into the watery environment within and around tissues and cells, delivering important nutrients and allowing for removal of waste products.

Water is also a lubricant. This is especially true in the gastrointestinal tract, respiratory tract, skin, and reproductive system, which produce important secretions such as digestive juices, mucus, sweat, and reproductive fluids, respectively. The ability of the body to incorporate water into these secretions is vital for health. For example, water is needed for producing functional mucus in the lungs. Mucus both protects the lungs from environmental toxins and pathogens and lubricates the sensitive lung tissue so that it can remain moist and supple. When insufficient water is available or the ability to regulate water balance across membranes is diminished, the ability to produce secretions such as mucus becomes limited. As a result, the secretions become ineffective or even counterproductive. An example of this occurs with the disease **cystic fibrosis,** presented in more detail in the Focus on Clinical Applications feature.

**CONNECTIONS** A solvent is a component of a solution in which a solute dissolves (Chapter 3, page 69).

**cystic fibrosis** A genetic (inherited) disease in which a defective chloride transporter results in the inability of the body to transport chloride out of cells.

# Focus On CLINICAL APPLICATIONS
## Electrolytes, Fluid Balance, and Cystic Fibrosis

Secretions are produced by a variety of tissues, including the salivary glands (saliva), gastrointestinal tract (gastric juices), and lungs (mucus). However, sometimes the body is unable to produce these secretions, and this can have serious consequences. An example is cystic fibrosis, which affects approximately 3,200 babies in the United States annually. Cystic fibrosis is most common in people of northern European descent, and is quite rare in African- and Asian-Americans. Researchers estimate that about 30,000 U.S. adults and children have cystic fibrosis.[32]

Cystic fibrosis is caused by a genetic defect that results in the production of an abnormal protein that typically transports chloride ions ($Cl^-$) across cell membranes.[33] In cells that produce secretions, such as in the lungs and

pancreas, the ability to actively pump $Cl^-$ from the intracellular space across the cell membrane into the extracellular space is crucial for incorporating water into the secretions produced there. This is because water moves from a region of low $Cl^-$ concentration (low osmotic pressure) to one of high $Cl^-$ concentration (high osmotic pressure). Thus, if a cell cannot actively pump $Cl^-$ from the intracellular compartment to the extracellular compartment, water will not be incorporated into its secretion.

In the case of cystic fibrosis, this transport system is disrupted because of the production of a faulty $Cl^-$ transport protein, resulting in thick, sticky secretions. This is especially true in the lungs and the pancreas. As a result, people with cystic fibrosis have significant—often life threatening—respiratory difficulties, and may have profound digestive

complications. Cystic fibrosis also causes the sweat to be concentrated and very salty, and is often first suspected when parents kiss their baby's skin and find that it tastes salty. Other complications of cystic fibrosis include blockage of the bowel, delayed growth, coughing, frequent respiratory infection, and liver damage.

There is currently no way to prevent or cure cystic fibrosis. However, scientists now know precisely what genes contain the mutations responsible for this disease and are working to develop ways to insert copies of the normal gene into the cells of cystic fibrosis patients. In the meantime, clinicians recommend that people with cystic fibrosis maintain a healthy diet, engage in regular exercise, consume adequate fluids, and use nutritional supplements to help ensure optimal health.

## Hydrolysis and Condensation Reactions

Water is a common reactant in many chemical reactions. More specifically, hydrolysis reactions occur when chemical bonds in a molecule are broken by the addition of a water molecule. You have learned about many kinds of hydrolysis reactions in the preceding chapters, such as those required for breaking down glycogen and triglycerides. Conversely, chemical reactions that produce water are called condensation reactions. The typical adult produces about 250 mL of water each day from condensation reactions; more physically active people produce twice this amount.

## Water Regulates Body Temperature

When energy-yielding nutrients such as amino acids, glucose, and fatty acids are metabolized, energy is released. Some of this energy becomes heat and helps maintain the internal body temperature at a comfortable 98.6°F. However, excess heat generated by metabolism must be removed from the body so the body's temperature does not rise. Hot environments can also raise the body's internal temperature.

Water plays a key role in the body's cooling system, because a given amount of water can "hold" or retain a surprising amount of heat. The term **specific heat** refers to the amount of energy it takes to increase the temperature of 1 gram of a substance 1 degree C. Water has a high specific heat, meaning that changing its temperature takes a large amount of energy. You have probably experienced this unique property of water if you have visited a swimming pool in the heat of the day or the cool of the evening. Although the air temperature fluctuates around the clock, the temperature of the water remains somewhat constant. In contrast, metals such as those found in pots and pans tend to have low specific heats—it takes very little change in the environmental temperature to change the temperature of the metal. Because water can handle so much energy without heating up, our bodies can maintain a relatively stable internal temperature even when metabolic rates are high or the environment is hot. Additional heat energy contained in our sweat is eliminated via the cooling process of evaporation when we perspire. To facilitate heat loss via perspiration, the small blood vessels beneath the skin become larger (dilate), allowing fluids to be more easily lost from the body.

# Water Insufficiency: Dehydration

The condition of water insufficiency is called **dehydration,** and can lead to serious consequences. Compared to other classes of macronutrients, the body is the least tolerant of water loss. For example, as little as a 2% loss of a person's body weight in water can lead to many complications, listed here.

- ↓ Cognitive function
- ↓ Motor control and ability to engage in exercise
- ↓ Short-term and long-term memory
- ↓ Attention span
- ↓ Ability to maintain core temperature, especially when exercising
- ↑ Urinary tract infections
- ↑ Fatigue

These complications are even more severe in children and the elderly, who have little tolerance for the negative consequences of dehydration.[34] In the elderly, signs of dehydration can be mistaken for signs of dementia. In addition, athletes who lose large amounts of water in sweat are at increased risk for

**specific heat** The energy required to raise the temperature of a substance.

**dehydration** A condition in which the body has an insufficient amount of water.

dehydration. Ensuring adequate water intake in infants, children, the elderly, and athletes is very important.

## The Body's Response to Dehydration

When water intake is inadequate to replace water loss, the body tries to retain all the fluid it has. This is illustrated in Figure 13.8. The pituitary gland at the base of the brain responds to dehydration (in other words low blood volume) by releasing **antidiuretic hormone (ADH)** (also called vasopressin). ADH circulates in the blood to the kidneys, where it stimulates water conservation by decreasing urine production and output. ADH also causes blood vessels to construct, raising blood pressure.

---

### FIGURE 13.8 Regulation of Blood Volume and Blood Pressure

Blood volume and pressure are regulated by the brain and kidneys via aldosterone (produced in the adrenal glands) and antidiuretic hormone (ADH, produced in the pituitary gland).

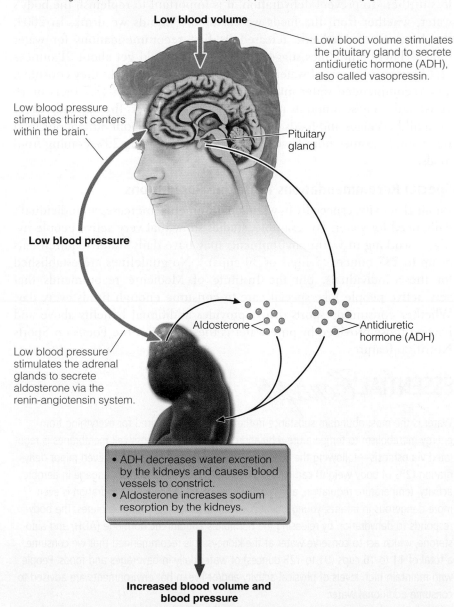

**Low blood volume**

Low blood volume stimulates the pituitary gland to secrete antidiuretic hormone (ADH), also called vasopressin.

Low blood pressure stimulates thirst centers within the brain.

Pituitary gland

**Low blood pressure**

Aldosterone

Antidiuretic hormone (ADH)

Low blood pressure stimulates the adrenal glands to secrete aldosterone via the renin-angiotensin system.

- ADH decreases water excretion by the kidneys and causes blood vessels to constrict.
- Aldosterone increases sodium resorption by the kidneys.

**Increased blood volume and blood pressure**

**antidiuretic hormone (ADH)** (also called vasopressin) A hormone produced in the pituitary gland and released during periods of low blood volume; stimulates the kidneys to decrease urine production, thus conserving water.

Our requirement for water can be met by both the foods we eat and the beverages we drink.

At the same time, low blood pressure caused by dehydration stimulates the kidneys to produce the enzyme renin, as previously described. Renin ultimately causes the adrenal glands to release aldosterone, which stimulates the kidneys to retain sodium and chloride. Increased sodium and chloride resorption by the kidneys further conserves water via osmosis. Angiotensin II (produced via renin) works at the level of the brain to stimulate additional ADH release as well as the sensation of thirst—increasing blood volume even more. Recall that the initial stimulus in this scenario was a decrease in blood volume. Thus the combined actions of ADH and the renin-angiotensin-aldosterone system represent an integrated physiological system with a negative feedback loop—the end result (increased blood volume) being the opposite of the initial stimulus (decreased blood volume).

## Recommendations for Water Intake

Every day we lose water in our perspiration (500 mL), breath (300 mL), urine (1,500 mL), and feces (150 mL). Heavy exercise can increase water loss further. To prevent dehydration, it is important to replenish the body's water, whether from the foods we eat or the liquids we drink. In 2004, the Institute of Medicine released its first recommendations for water intake. These guidelines suggest that women should get about 91 ounces (2.7 liters or 11 cups) of water daily in the foods and drinks they consume. The recommended water intake for males is 125 ounces (3.7 liters or 16 cups) daily. These amounts may sound high. However, they include water from all beverages and foods consumed. In general, about 80% of our total water intake comes from beverages, with the remaining 20% coming from foods.

### Special Recommendations for Active Populations

Physical activity, especially in warm environments, increases an individual's daily need for water. For example, studies show that very active people living or working in warm environments may have daily water requirements of up to 237 ounces (7 liters or 30 cups).[35] No guidelines are established for these individuals, but the Institute of Medicine recommends that very active people take special care to consume enough fluids every day. Whether consuming sports drinks provides additional benefits above and beyond those provided by pure water is the subject of the Focus on Sports Nutrition feature.

## ESSENTIAL Concepts

Water is the most abundant substance in the body and is required for everything from energy metabolism to temperature regulation. Its movement across cell membranes is regulated via osmosis—following the active transport of various electrolytes. Even minor dehydration (2% of body weight) can influence cognitive function, ability to engage in aerobic activity, temperature regulation, and risk of urinary tract infections. Dehydration is even more dangerous in infants, young children, the elderly, and endurance athletes. The body responds to dehydration by releasing the hormones antidiuretic hormone (ADH) and aldosterone, which act to conserve water at the kidneys. It is recommended that we consume a total of 11 to 16 cups (91 to 125 ounces) of water daily in beverages and foods. People who maintain high levels of physical activity and/or live in hot environments are advised to consume additional water.

# *Focus On* SPORTS NUTRITION
## Sports Drinks: Are They Better Than Water?

Although once found only in athletic shops, sports drinks are now sold in grocery stores, gas stations, and convenience markets. Consequently, sports drinks developed to help athletes rehydrate are now being consumed by children, teens, and adults—whether or not they are engaged in athletic pursuit. But are there any benefits to drinking sports drinks over simple, pure water?

To answer this question, first consider the ingredients and purpose of sports drinks. Basically, they are made from water, sugar, salt (NaCl), and flavoring (see the table for a simple recipe for a homemade sports drink). The original purpose of these drinks was to rehydrate athletes and provide important electrolytes ($Na^+$ and $Cl^-$) lost via perspiration, especially in hot environments. For example, a 150-pound runner can lose ½ pound (8 ounces, or 1 cup) of sweat per mile. Thus, over the course of a marathon (26.2 miles) a runner might lose 8 to 9 pounds of sweat containing a significant amount of water and electrolytes. Consuming sports drinks dur-

| Recipe for a Sports Drink (makes 20 ounces) | |
| --- | --- |
| Ingredient | Amount |
| Sugar | 3 tablespoons |
| Salt | ⅛ teaspoon |
| Unsweetened drink mix (such as Kool-Aid®) | ⅓ packet or ⅓ teaspoon, any flavor |
| Water | 20 ounces |

ing such events provides these needed nutrients to high-intensity athletes. The glucose in sports drinks both facilitates electrolyte absorption and provides a small amount of energy. Using sports drinks instead of water during endurance training and competition can help an athlete prevent the serious complications of dehydration as well as provide some needed energy.

The question remains as to whether sports drinks are, in general, healthy alternatives to water for most people. Clearly, for sedentary people water is the drink of choice for simple hydration. Similarly, there is no evidence that

consuming sports drinks provides a competitive edge to most athletes in training or competition—unless the activity is of high intensity or in a hot environment.[36] One special case is that of the exercising child. Compared to adults, children are at increased risk for dehydration during exercise. Because they often do not automatically get thirsty during prolonged exercise, they must be encouraged to take a water break. Children are also more sensitive to hot environments than are adults.

To prevent dehydration in children, the American Academy of Pediatrics recommends that special care be taken to restrict the amount of time young athletes train and compete, especially in warm environments.[37] The American Academy of Pediatrics also recommends that children be well hydrated prior to physical activity and that they be encouraged to consume 5 to 9 ounces (0.6 to 1.1 cups) of either cold tap water or a sports drink every 20 minutes, even if they are not thirsty. Because children tend to prefer sports drinks over water, providing them with this choice helps ensure healthy hydration.

# Integration of Functions and Food Sources

As you have seen, the body uses the major minerals for many purposes, including both functional and structural roles. Like the trace minerals, the major minerals participate as cofactors for enzymes, regulate energy metabolism, are important for a healthy immune system, and are involved in glucose homeostasis. In addition, several of the major minerals play important parts in regulating fluid balance in the body as well as in nerve and muscle function. Several major minerals also make up most of our bones and teeth and therefore are important for their strength and functions. The overlapping roles of the major minerals in these activities is summarized in Table 13.2.

It is important that we consume sufficient amounts of each of these essential minerals from the diet, and as with the other micronutrients, no single food group can supply all of them. Consuming a balanced diet featuring a wide variety of foods is required, as shown in Table 13.3.

**TABLE 13.2    General Functions of the Essential Major Minerals**

| | Minerals | | | | |
|---|---|---|---|---|---|
| | Calcium | Phosphorus | Magnesium | Sodium/Chloride | Potassium |
| **Functions** | | | | | |
| Cofactor | X | | | | |
| Immune function | | | | X | |
| Bone and tooth health | X | X | X | | |
| Blood health | X | X | | | |
| Energy metabolism[a] | X | X | X | | X |
| Fluid balance | | | | X | X |
| Nerve and muscle function | X | X | X | X | X |

[a]Note that these minerals are not, themselves, energy-yielding nutrients but are involved in energy metabolism via other roles, such as activation of enzymes or as a cofactor.

**TABLE 13.3    Major Food Sources of the Essential Major Minerals**

| | Food Groups | | | | | | | |
|---|---|---|---|---|---|---|---|---|
| | Grains[a] | Vegetables | Fruits | Dairy | Meat | Seafood | Nuts, Seeds | Legumes |
| **Major Minerals** | | | | | | | | |
| Calcium | | X | | X | | X[b] | | X |
| Phosphorus | | | | X | X | X | X | |
| Magnesium | X | X | | X | | X | X | X |
| Sodium and Chloride | | | | X | X | X | | |
| Potassium | | X | X | X | X | X | | X |

[a]Whole-grain products tend to contain more major minerals than do milled products, unless they are fortified.
[b]Packed with bones.

# Review Questions   Answers are found in Appendix G.

1. Calcitonin is released by the _____ during periods of _____ blood calcium concentration.
   a. parathyroid glands      high
   b. thyroid gland           high
   c. parathyroid glands      low
   d. thyroid gland           low

2. Which of the following can influence calcium bioavailability?
   a. Oxalates
   b. Phytates
   c. Aging
   d. All of the above

3. Which of the major minerals is a component of cell membranes?
   a. Potassium
   b. Sodium
   c. Calcium
   d. Phosphorus

4. Hyponatremia stimulates the release of _____.
   a. parathyroid hormone
   b. aldosterone
   c. growth hormone
   d. insulin

5. Positively charged ions, such as $Na^+$, are called _____ .
   a. electrolytes
   b. cations
   c. anions
   d. both a and c

6. Increased circulating levels of _____ are associated with decreased blood pressure in some people.
   a. sodium
   b. chloride
   c. manganese
   d. potassium

7. Which of the following contains most of the fluid within the body?
   a. Interstitial space
   b. Intravascular space
   c. Circulatory space
   d. Intracellular space

8. During osmosis, active transport of electrolytes causes the passive transport of _____.
   a. water
   b. sodium and potassium
   c. plasma proteins
   d. urea

9. Cystic fibrosis is a genetic condition caused by the inability of cells to move _____ across cell membranes.
   a. sodium
   b. potassium
   c. chloride
   d. sulfate

10. Dehydration causes the release of _____ from the _____.
    a. aldosterone                    brain
    b. angiotensinogen                kidneys
    c. antidiuretic hormone           pituitary gland
    d. antidiuretic hormone           liver

11. The USDA Dietary Guidelines for Americans recommend that we "choose and prepare foods with less salt." List five suggestions for daily dietary changes that can help accomplish this goal.

12. Sodium and chloride are important electrolytes in maintaining fluid balance. Describe how they function in the process of osmosis.

13. Characterize the mechanisms by which phosphorus deficiency may cause weakness and fatigue.

## Practice Calculations   Answers are found in Appendix G.

1. Using the food composition table that accompanies this text or other sources, calculate the sodium content of a lunch containing a hamburger on a white bun, a small order of french fries, and a large cola soft drink. What percentage of your AI of sodium does this represent? How does this value compare to what is recommended in the USDA Dietary Guidelines for Americans? What foods could be substituted for the food in this lunch to decrease the sodium content?

2. Using the MyPyramid food guidance system (http://www.mypyramid.gov), determine the amount of foods from the fruits and vegetables groups that are recommended for you. Then make up a sample menu that would satisfy these recommendations, and calculate the amount of potassium you would get if you consumed these amounts of foods.

## Media Links   A variety of study tools for this chapter are available at our website, www.thomsonedu.com/nutrition/mcguire.

Prepare for tests and deepen your understanding of chapter concepts with these online materials:

- Practice tests
- Flashcards
- Glossary
- Student lecture notebook
- Web links
- Animations
- Chapter summaries, learning objectives, and crossword puzzles

## Notes

1. Anderson JJB. Nutritional biochemistry of calcium and phosphorus. Journal of Nutritional Biochemistry. 1991;2:300–7. Sheikh MS, Santa Ana CA, Nicar MJ, Schiller LR, Fordtran JS. Gastrointestinal absorption of calcium from milk and calcium salts. New England Journal of Medicine. 1987;317:532–6. Whiting S. Safety of some calcium supplements questioned. Nutrition Reviews. 1994;52:95–7.

2. Scelfo GM, Flegal R. Lead in calcium supplements. Environmental Health Perspectives. 2004;108:309–13.

3. Heaney RP, Recker RR, Stegman MR, Moy AJ. Calcium absorption in women: Relationships to calcium intake, estrogen status, and age. Journal of Bone Mineral Research. 1989;469–475.

4. Hoenderop JG, Nilius B, Bindels RJ. Calcium absorption across epithelia. Physiological Reviews. 2005;85:373–422.

5. Mihalyi E. Review of some unusual effects of calcium binding to fibrinogen. Biophysical Chemistry. 2004;112:31–40.

6. Chin ER. Role of $Ca^{2+}$/calmodulin-dependent kinases in skeletal muscle plasticity. Journal of Applied Physiology. 2005;99:414–23. Thorneloe KS, Nelson MT. Ion channels in smooth muscle: regulators of intracellular calcium and contractility. Canadian Journal of Physiology and Pharmacology. 2005;83:215–42.

7. Barnes S, Kelly ME. Calcium channels at the photoreceptor synapse. Advances in Experimental Medicine and Biology. 2004;514:465–76. Borge PD, Moibi J, Greene SR, Trucco M, Young RA, Gao Z, Wolf BA. Insulin receptor signaling and sarco/endoplasmic reticulum calcium ATPase in β-cells. Diabetes. 2002;51:S427–33. French RJ, Zamponi GW. Voltage-gated sodium and calcium channels in nerve, muscle, and heart. IEEE Transactions in Nanobioscience. 2005;4:58–69. Moreau M, Leclerc C. The choice between epidermal and neural fate: a matter of calcium. International Journal of Developmental Biology. 2004;48:75–84. Senin II, Koch KW, Akhtar M, Philippov, PP. Ca²⁺-dependent control of rhodopsin phosphorylation: recoverin and rhodopsin kinase. Advances in Experimental Medicine and Biology. 2002;514:69–99.

8. National Osteoporosis Foundation. Fast facts. Available from: http://www.nof.org/osteoporosis/diseasefacts.htm.

9. Jackson RD, LaCroix AZ, et al. Calcium plus vitamin D supplementation and the risk of fractures. New England Journal of Medicine. 2006;354:669–83.

10. Allender PS, Cutler JA, Follman D, Cappuccio FP, Pryer J, Elliott P. Dietary calcium and blood pressure. Annals of Internal Medicine, 1996;124:825–31. Bucher HC, Cook RJ, Guyatt GH, Lang JD, Cook DJ, Hatala R, Hunt DL. Effects of dietary calcium supplementation on blood pressure. JAMA (Journal of the American Medical Association). 1996;275:1016–22.

11. Elwood PC, Pickering JE, Hughes J, Fehili AM, Ness AR. Milk drinking, ischaemic heart disease and ischaemic stroke II. Evidence from cohort studies. European Journal of Clinical Nutrition. 2004;58:718–24. FitzGerald RJ, Murray BA, Walsh DJ. Hypotensive peptides from milk proteins. Journal of Nutrition. 2004;134:980S–8S. Hoolihan L. Beyond calcium. Nutrition Today. 2004;39:69–77.

12. Chan JM, Giovannucci EL. Dairy products, calcium, and vitamin D and risk of prostate cancer. Epidemiologic Reviews. 2001;23:87–92. Cho E, Smith-Warner SA, Spiegelman D, Beeson WL, van den Brandt PA, Colditz GA, Folsom AR, Fraser GE, Freudenheim JL, Giovannucci E, Goldbohm RA, Graham S, Miller AB, Pietinen P, Potter JD, Rohan TE, Terry P, Toniolo P, Virtanen MJ, Willett WC, Wolk A, Wu K, Yaun SS, Zeleniuch-Jacquotte A, Hunter DJ. Dairy foods, calcium, and colorectal cancer: a pooled analysis of 10 cohort studies. Journal of the National Cancer Institute. 2004;96:1015–22. Hjartaker A, Laake P, Lund E. Childhood and adult milk consumption and risk of premenopausal breast cancer in a cohort of 48,844 women. International Journal of Cancer. 2001;93:888–93. Holt PR, Atillasoy EO, Gilman J, Guss J, Moss SF, Newmark H, Fan K, Yang K, Lipkin M. Modulation of abnormal colonic epithelial cell proliferation and differentiation by low-fat dairy foods: a randomized controlled trial. JAMA (Journal of the American Medical Association). 1998;12:1074–9. Holt PR, Wolper C, Moss SF, Yang K, Lipkin M. Comparison of calcium supplementation or low-fat dairy foods on epithelial cell proliferation and differentiation. Nutrition and Cancer. 2001;41:150–5.

13. Knekt P, Jarvinen R, Seppanen R, Pukkala E, Aromaa A. Intake of dairy products and the risk of breast cancer. British Journal of Cancer. 1998;73:687–91.

14. Wu K, Willett WC, Fuchs CS, Colditz GA, Giovannucci EL. Calcium intake and risk of colon cancer in women and men. Journal of the National Cancer Institute. 2002;94:437–46.

15. Hoolihan L. Beyond calcium. Nutrition Today. 2004;39:69–77. Parodi PW. Cows' milk fat components as potential anticarcinogenic agents. Journal of Nutrition. 1997;27:1055–60. Parodi PW. A role for milk proteins in cancer prevention. Australian Journal of Dairy Technology. 1998;53:37–47.

16. Zemel MB. Calcium and dairy modulation of obesity risk. Obesity Research. 2005;13:192–3. Zemel MB. Regulation of adiposity and obesity risk by dietary calcium: mechanisms and implications. Journal of the American College of Nutrition. 2002;21:146S–151S.

17. Barr SI. Increased dairy product or calcium intake: is body weight or composition affected in humans? Journal of Nutrition. 2003;133:245S–8S. Gunther CW, Legowski PA, Lyle RM, McCabe GP, Eagan MS, Peacock M, Teegarden D. Dairy products do not lead to alterations in body weight or fat mass in young women in a 1-y intervention. American Journal of Clinical Nutrition. 2005;81:751–6. Parikh SJ, Yanovski JA. Calcium intake and adiposity. American Journal of Clinical Nutrition. 2003;77:281–7.

18. Hallberg L, Rossander-Hulten L, Brune M, Gleerup A. Calcium and iron absorption: mechanism of action and nutritional importance. European Journal of Clinical Nutrition. 1992;46:317–27. Institute of Medicine. Dietary Reference Intakes for calcium, phosphorus, magnesium, vitamin D, fluoride. Washington, DC: National Academy Press; 1997.

19. Tenenhouse HS. Regulation of phosphorus homeostasis by the type IIA Na/phosphate cotransporter. Annual Review of Nutrition. 2005;25:197–214.

20. Takeda E, Taketani Y, Sawada N, Sato T, Yamamoto H. The regulation and function of phosphate in the human body. Biofactors. 2004;21:345–55.

21. Greger JL, Smith SA, Snedeker SM. Effect of dietary calcium and phosphorus levels on the utilization of calcium, phosphorus, magnesium, manganese, and selenium by adult males. Nutrition Research. 1981;1:315–25.

22. Gums JG. Magnesium in cardiovascular and other disorders. American Journal of Health-System Pharmacy. 2004;61:1569–76. Heaney RP. Phosphorus nutrition and the treatment of osteoporosis. Mayo Clinic Proceedings. 2004;79:91–7. Nieves JW. Osteoporosis: the role of micronutrients. American Journal of Clinical Nutrition. 2005;81:1232S–9S. Sarko J. Bone and mineral metabolism. Emergency Medicine Clinics of North America. 2005;23:703–21.

23. Alghamdi AA, Al-Radi OO, Latter DA. Intravenous magnesium for prevention of atrial fibrillation after coronary artery bypass surgery: a systematic review and meta-analysis. Journal of Cardiac Surgery. 2005;20:293–9. Bobkowski W, Nowak A, Durlach J. The importance of magnesium status in the pathophysiology of mitral valve prolapse. Magnesium Research. 2005;18:35–52. Weglicki W, Quamme G, Tucker K, Haigney M, Resnick L. Potassium, magnesium, and electrolyte imbalance and complications in disease management. Clinical and Experimental Hypertension. 2005;27:95–112.

24. Bianchi A, Salomone S, Caraci J, Pizza V, Bernardini R, D'Amato CC. Role of magnesium, coenzyme Q10, riboflavin, and vitamin B₁₂ in migraine prophylaxis. Vitamins and Hormones. 2004;69:297–312. Durlach J, Pages N, Bac P, Bara M, Guiet-Bara A. Headache due to photosensitive magnesium depletion. Magnesium Research. 2005;18:109–22. Guerrero-Romero F, Rodriguez-Moran M. Complementary

therapies for diabetes: the case for chromium, magnesium, and antioxidants. Archives of Medical Research. 2005;36:250–7. Schulze MB, Hu FB. Primary prevention of diabetes: what can be done and how much can be prevented? Annual Review of Public Health. 2005;26:445–67.

25. Gamba G. Molecular physiology and pathophysiology of electroneutral cation-chloride cotransporters. Physiological Reviews. 2005;85:423–93.

26. von Bohlen, Halbach O. The renin-angiotensin system in the mammalian central nervous system. Current Protein and Peptide Science. 2005;6:355–71.

27. Conlin PR. Interactions of high salt intake and the response of the cardiovascular system to aldosterone. Cardiology in Review. 2005;13:118–24. Lawlor DA, Smith GD. Early life determinants of adult blood pressure. Current Opinion on Nephrology and Hypertension. 2005;14:259–64. Lang F, Capasso G, Schwab M, Waldegger S. Renal tubular transport and the genetic basis of hypertensive disease. Clinical and Experimental Nephrology. 2005;9:91–9. Meneton P, Jeunemaitre X, de Wardener HE, MacGregor GA. Links between dietary salt intake, renal salt handling, blood pressure, and cardiovascular diseases. Physiological Reviews. 2005;85:679–715.

28. U.S. Department of Health and Human Services, Centers for Disease Control and Prevention, National Center for Health Statistics. Health, United States, 2004. Available from: http://www.cdc.gov/nchs/data/hus/hus04acc.pdf.

29. Institute of Medicine. Dietary Reference Intakes for water, potassium, sodium, chloride, and sulfate. Washington, DC: National Academies Press; 2005.

30. McCarty MF. Acid-base balance may influence risk for insulin resistance syndrome by modulating cortisol output. Medical Hypotheses. 2005;64:380–4. Stumvoll M, Goldstein BJ, van Haeften TW. Type 2 diabetes: principles of pathogenesis and therapy. Lancet. 2005;365:1333–46.

31. Kallen RJ, Rieger CHL, Cohen HS, Suter MA, Ong RT. Near-fatal hyperkalemia due to ingestion of salt substitute by an infant. JAMA (Journal of the American Medical Association). 1976;235:2125–26.

32. Goss CH, Rosenfeld M. Update on cystic fibrosis epidemiology. Current Opinion in Pulmonary Medicine. 2004;10: 510–4. Cystic Fibrosis Foundation. About cystic fibrosis. 2005. Available from: http://www.cff.org/about_cf/what_is_cf/.

33. Griesenbach U, Geddes DM, Alton EW. Advances in cystic fibrosis gene therapy. Current Opinion in Pulmonary Medicine. 2004;10:542–6.

34. Sentongo TA. The use of oral rehydration solutions in children and adults. Current Gastroenterology Reports. 2004;6:307–13.

35. Ruby BC, Shriver TC, Zderic TW, Sharkey BJ, Burks C, Tysk S. Total energy expenditure during arduous wildfire suppression. Medicine and Science in Sports and Exercise. 2002;34:1048–54.

36. Jeukendrup AE, Jentjens RL, Moseley L. Nutritional considerations in triathlon. Sports Medicine. 2005;35:163–81. von Duvillard SP, Braun WA, Markofski M, Beneke R, Leithauser R. Fluids and hydration in prolonged endurance performance. Nutrition. 2004;20:651–6.

37. American Academy of Pediatrics, Committee on Sports Medicine and Fitness. Climatic heat stress and the exercising child and adolescent. Pediatrics. 2000;106:158–9.

**Check out the following sources for additional information.**

38. Anderson JJB, Sell ML, Gerner SC, Calvo MS. Phosphorus. In: Present knowledge in nutrition, 8th ed. Bowman BA, Russell RM, editors. Washington, DC: ILSI Press; 2001.

39. Fleet JC, Cashman KD. Magnesium. In: Present knowledge in nutrition, 8th ed. Bowman BA, Russell RM, editors. Washington, DC: ILSI Press; 2001.

40. Gibson RS. Principles of nutritional assessment. New York: Oxford University Press; 2005.

41. Gropper SS, Smith JL, Groff JL. Advanced nutrition and human metabolism, 4th ed. Belmont, CA: Wadsworth; 2005.

42. Institute of Medicine. Dietary Reference Intakes for calcium, phosphorus, magnesium, vitamin D, and fluoride. Washington, DC: National Academy Press, 1997.

44. Institute of Medicine. Dietary Reference Intakes; guiding principles for nutrition labeling and fortification. Washington, DC: National Academy Press; 1998.

45. Preuss HG. Sodium, chloride and potassium. In: Present knowledge in nutrition, 8th ed. Bowman BA, Russell RM, editors. Washington, DC: ILSI Press; 2001.

46. U.S. Department of Health and Human Services and U.S. Department of Agriculture. Dietary Guidelines for Americans 2005. Washington, DC: U.S. Government Printing Office; 2005. Available from: http:www.healthierus.gov/dietaryguidelines.

47. Weaver C. Calcium. In: Present knowledge in nutrition, 8th ed. Bowman BA, Russell RM, editors. Washington, DC: ILSI Press; 2001.

# Nutrition and Bone Health

© Yoav Levy/Phototake, Inc.

The skeletal system makes up the basic architecture of the body. Although people tend to think of the skeleton as being inactive, bone is a living tissue. Bones not only provide structure but are the site of red and white blood cell production. Bones also produce platelets, which we need for blood clotting. Bone development begins early in fetal development and continues throughout the life span. It is important to maintain healthy bones, especially today as we enjoy longer and more active lives. This is because weak bones can lead to many negative health consequences in later life. Indeed, *Healthy People 2010* has identified the prevention of bone disease, or osteoporosis, as one of its major focus areas.[1]

## Bone Structure and Growth Through the Life Cycle

Development of bones and the skeletal system begins early in life—about the fifth week of gestational development. At first, fetal bones are soft and rubbery, being made of tissue called **cartilage.** Around the seventh week of pregnancy, cartilage is slowly replaced by hard, mineralized bone. The process of converting cartilage to bone, called **ossification,** continues throughout fetal life and childhood. In fact, ossification is not complete until early adult life.

## The Basic Structure of Bones

Bones have several components. On their outer surface, bones are covered by a thin membrane called the **periosteum,** made of connective tissue and hundreds of small blood vessels and nerves. Beneath the periosteum is a layer of hard, dense bone called **cortical bone** (also called compact bone), which surrounds another layer called **trabecular bone.** Trabecular bone is not as dense as cortical bone, instead consisting of a lattice-like structure. The major function of the cortical bone is to provide strength, whereas the trabecular bone contains the **bone marrow** that produces red blood cells, white blood cells, and platelets. Different bones in the body have different amounts of cortical and trabecular bone. For example, the long bones of the forearm have less trabecular bone, compared to the vertebral column (backbone) and pel-

**cartilage** (CAR – ti – lage) The soft, nonmineralized precursor of bone.

**ossification** The process by which minerals are added to cartilage, ultimately resulting in bone formation.

**periosteum** (per – i – O – ste – um) The outer covering of bone, consisting of blood vessels, nerves, and connective tissue.

**cortical bone** (compact bone) The dense, hard layer of bone found directly beneath the periosteum.

**trabecular bone** (tra – BE – cu – lar) The inner, less dense layer of bone that contains the bone marrow.

**bone marrow** The soft, spongy, inner part of bone that makes red blood cells, white blood cells, and platelets.

vis, which are mostly cortical bone. Trabecular bone is also abundant in the ends of long bones. Positioned in the very core of each bone are large blood vessels, lymphatic vessels, and nerves that allow hormones, nutrients, and neural impulses to be transported and transmitted to and from bone tissue.

## Bones Are Made of Osteoblasts and Osteoclasts

There are two basic types of bone cells: osteoblasts and osteoclasts. Osteoblasts are bone-forming cells responsible for depositing the many minerals and proteins needed for bone maintenance, growth, and repair. Osteoclasts work in an opposite fashion, breaking down old bone to make way for new bone. Osteoblasts and osteoclasts work together to build and break down bone as needed. During periods of calcium deficiency, the body maintains healthy blood calcium levels by stimulating osteoclasts and, thus, bone breakdown. In other words, calcium resorption from bones is stimulated during these times. This helps maintain blood calcium at healthy levels. The actions of osteoblasts and osteoclasts are orchestrated by a variety of hormones and nutrients, including vitamin D, as described in Chapters 11 and 13.[2]

> **CONNECTIONS** Recall that the term *resorption* refers to the movement of substances from nongastrointestinal tissues (such as kidneys and bone) into the blood (Chapter 4, page 119).

## Bone Remodeling and Growth During the Life Cycle

The growth and maintenance of bone are dynamic processes involving both the building of new bone by osteoblasts and the breakdown of old bone by osteoclasts. This cycle of bone breakdown and rebuilding is called **bone remodeling** (or bone turnover) and can be compared to the remodeling of a house. For example, it would be impossible to remodel a house without first removing older walls, electrical wiring, insulation, and other materials. This is similar to what occurs during the life of a bone—osteoclasts break down and remove old bone, and osteoblasts rebuild new bone. In this way, osteoclasts and osteoblasts work together to remodel bones.

**FIGURE 1**    Changes in Bone Mass During the Life Cycle

Women are at greater risk of osteoporosis than men, because women are more likely to have bone masses below the threshold for osteoporosis risk in later life.

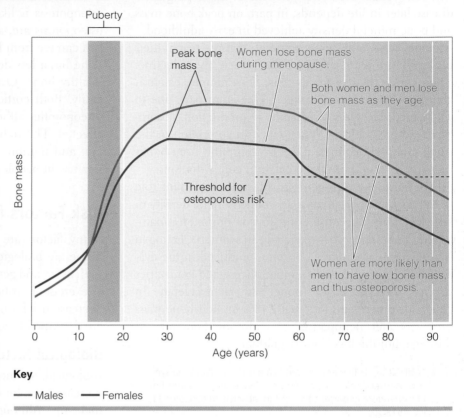

**Key**

— Males    — Females

During periods of growth, such as infancy, childhood, and puberty, the actions of the osteoblasts exceed those of the osteoclasts—lengthening and reshaping bones. In other words, bone growth is greater than bone loss. Most bone growth is complete before age 20, although a small amount (10%) of bone is laid down between 20 and 30 years of age. During this period of bone growth, the total amount of minerals contained in bones, called **bone mass,** increases. Bone mass is at its highest level at about 30 years of age (Figure 1)—this maximum amount of bone is referred to as **peak bone mass.** Peak bone mass is the greatest total amount of bone that a person will have in his or her lifetime. People with larger frames tend to have higher peak bone masses than do people with smaller frames, and women generally have somewhat lower peak bone masses than do men.

**bone remodeling** (bone turnover) The process by which older and damaged bone is removed and replaced by new bone.

**bone mass** The total amount of bone mineral in the body.

**peak bone mass** The greatest amount of bone mineral that a person has during his or her life.

Another measure of the body's bone health is **bone mineral density,** which is the amount of mineral in a given area of bone. The greater the bone mineral density, the stronger the bone. The risk for developing bone disease later in life depends, in part, on peak bone mass and bone mineral density achieved in early adulthood.

Bone mass and bone mineral density reach their peaks by 30 years of age, after which they decline. There are many reasons for this. For example, calcium absorption decreases with age.[3] This is, in part, due to age-related decreases in vitamin D production, absorption, and metabolism. When calcium absorption in the small intestine declines, calcium resorption from bones increases to maintain blood calcium concentrations.

The hormone estrogen also plays an important role in the decline of bone mass and bone mineral density in later life. This is because estrogen is important for maintaining bone strength—especially in women.[4] Estrogen is a female reproductive hormone produced in the ovaries. When a woman reaches menopause and her ovaries stop producing estrogen, bone loss can accelerate. In general, women tend to lose 1 to 3% of their bone mass each year after menopause.[5] Men also lose bone mass as they age, but the loss is more gradual.

CONNECTIONS Remember that vitamin D stimulates the synthesis of several calcium transport protein in the small intestine, and is therefore important for calcium absorption (Chapter 11, page 451).

## ESSENTIAL *Concepts*

Bones are complex, having many components. Cortical bone is hard and dense, whereas trabecular bone is less compact. The center of bones contains bone marrow, which produces blood cells and platelets. Bones also contain blood vessels and nerves. Osteoblasts are cells that build bone by depositing minerals and proteins, whereas osteoclasts break down bone. The dynamic process by which bones are built and broken down is called bone remodeling or bone turnover. Peak bone mass—total bone mass at its greatest—occurs at about 30 years of age. Bone mineral density is the amount of minerals in a given area of bone. Low peak bone mass and bone mineral density are related to increased risk for osteoporosis in later life.

## Osteoporosis and Health

When the bone breakdown activities of osteoclasts exceed the bone-building activities of osteoblasts, bone mass declines. This is a normal part of aging and does not always lead to unhealthy bones. However, if bone mass is low when this process begins, even moderate bone demineralization can lead to a condition called osteopenia. As more bone is lost, the entire matrix of the bone tissue also begins to break down. When this occurs, osteoporosis is likely to result. Bones of a person with osteoporosis are, as the name would imply, porous. As you can see from Figure 2, the matrix of an osteoporotic bone has a less dense lattice system than does that of a healthy bone. Osteoporotic bones are weak and break easily. Both cortical and trabecular bone can become osteoporotic, although trabecular bone is much more affected. The trabecular-rich bones of the pelvis, vertebra, and portions of long bones are at greatest risk for fracture in people with osteoporosis.

## Risk Factors for Osteoporosis

Many factors are associated with risk for osteoporosis. Some are biological risk factors that cannot be altered, such as sex and genetics. However, lifestyle factors—such as exercise, nutrition, and smoking—can be modified to decrease risk for bone disease. Some of these risk factors are described here.

### Biological Factors Related to Osteoporosis

Biological, or nonmodifiable, factors associated with increased risk for osteoporosis are related to our genetic make-up.[6] Although these factors cannot be changed, it is important to be aware of the biological risk factors you have for osteoporosis, some of which are listed here.

- *Sex:* Women are at greater risk for osteoporosis than are men.
- *Age:* Rate of bone loss increases after about 40 years of age.
- *Body size:* Small people tend to have lower peak bone masses and are at greater risk for bone disease. In general, people with body mass indices (BMIs) of less than 19 kg/m$^2$ are at increased risk for osteoporosis. Of course, body size is influenced by both biological and lifestyle factors (such as nutrition).
- *Ethnicity:* People of northern European or Asian descent are at greatest risk; those of African or Hispanic descent are at lowest risk.
- *Genetics:* Children of parents who have osteoporosis are at greater risk than those of parents who do not.

Although genetics does not ultimately determine who develops or does not develop osteoporosis, those with a family history of it are at especially high risk. Thus it

**bone mineral density** The amount of bone mineral per unit area.

**FIGURE 2     Normal and Osteoporotic Bone**

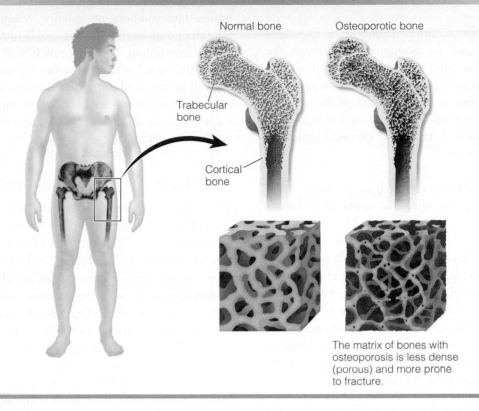

The matrix of bones with osteoporosis is less dense (porous) and more prone to fracture.

is important to do all you can to decrease risk factors related to lifestyle.

## Lifestyle Factors Related to Osteoporosis

Although we may not be able to alter our biological risk for osteoporosis, we control many other factors that influence risk for bone disease.

***Smoking***     Smoking is associated with increased risk for osteoporosis.[7] Although the reasons for this are not completely understood, smoking may weaken bones by decreasing estrogen production. However, because smoking is also related to bone loss in men, there are probably other mechanisms by which it influences bone health.

***Medications***     Chronic use of some medications (such as thyroid hormones) can damage bones and increase bone loss. Similarly, some diuretics (such as Lasix®) increase calcium loss in the urine, leading to weak bones.[8] Other medications, such as corticosteroid therapy, can also decrease bone strength and thus increase risk for osteoporosis.[9] It is important to ask your health care provider what effects, if any, your medications might have on bone health. This is especially important as we age.

***Circulating Estrogen Concentrations***     Because the hormone estrogen stimulates the bone-building actions of osteoblasts, low circulating levels of estrogen can increase the risk for osteoporosis. In women, most of their estrogen is produced by the ovaries. The surgical removal of the ovaries (called an ovariectomy or oophorectomy) before the age of 45 without **hormone replacement therapy (HRT)** increases risk for bone disease. Also, women with eating disorders often stop menstruating—this also lowers estrogen levels and increases risk for osteoporosis.[10] The use of estrogen in hormone replacement therapy after menopause helps maintain bone strength in later life.[11]

***History of Eating Disorders***     Women and men with a history of either anorexia nervosa or bulimia are at increased risk for lower bone density, especially in the lower back and hips.[12] This is partly due to the fact that eating disorders often disrupt the reproductive system, lowering estrogen levels in women. In addition, poor

**hormone replacement therapy (HRT)** Medication containing estrogen and progesterone, sometimes taken by women after having their ovaries removed or after menopause.

nutrition is common in both males and females with eating disorders and this can lead to bone disease.

***Alcohol Consumption***    Chronic alcoholism is associated with an increased risk for osteoporosis, although the mechanisms by which this occurs are not well understood.[13] Although alcohol probably affects bone remodeling directly, researchers think that much of the negative influence of alcoholism on bone health is due to chronic consumption of a poor diet (lacking especially calcium, vitamin D, and protein).

***Physical Activity***    Regular physical activity, especially weight-bearing exercise, increases bone density and decreases risk for osteoporosis.[14] This is because weight-bearing exercise increases the overall rate of bone remodeling—leading to increased bone density. Examples of bone-strengthening exercises include weight training, walking, running, dancing, and tennis. As little as one hour of these activities each week can make a big difference in bone strength. These exercises result in healthier bones in both children and the elderly, re-enforcing the idea that one is never too young nor too old to benefit from physical activity.

***Nutritional Factors***    A well-balanced diet may be one of the most important ways to prevent osteoporosis, because many nutrients are needed for healthy bones.[15] At the forefront of these nutrients is calcium. Chronically low calcium intake can decrease peak bone mass in young adulthood and increase bone loss in later life. This is because calcium is the largest component of bones, and without it they develop improperly and are weak. Phosphorus and magnesium are also important throughout the life span for building and maintaining healthy bones.

Protein and vitamins C, D, and K are also required for the bone building and remodeling processes.[16] Protein is needed not only for structural purposes within the bone matrix but also for other processes related to calcium homeostasis and bone remodeling. Vitamin D is needed for calcium absorption in the small intestine and bone cell differentiation, and vitamins C and K are involved in healthy bone-building activities. In addition, although it is not an essential nutrient, fluoride consumption can increase bone strength.[17] Thus, it is important to consume adequate amounts of all these nutrients throughout the life span.

## Type 1 and Type 2 Osteoporosis

There are two types of osteoporosis. **Type 1 osteoporosis**, also called postmenopausal osteoporosis, typically occurs in women who are between 50 and 60 years of age. This type of osteoporosis is directly linked to the negative effects of decreased estrogen concentrations that occur at menopause. Type 1 osteoporosis most dramatically affects trabecular bone. **Type 2 osteoporosis** occurs in both men and women, and tends to be diagnosed later in life (70 to 75 years of age). Type 2 osteoporosis is a result of breakdown of both cortical and trabecular bone and is due to a combination of dietary and age-related factors. People with either form of osteoporosis can lose significant height during old age and experience severe pain, especially in the vertebra. In addition, some people develop a curvature in the upper spine called **kyphosis** or dowager's hump. Kyphosis can be problematic, because

Osteoporosis can lead to kyphosis (dowager's hump).

© Martin Rotker/Phototake, Inc.

© Topham/The Image Works

**Weight-bearing exercise can strengthen bones.**

**Type 1 osteoporosis**  The form of osteoporosis that occurs in women, caused by hormone-related bone loss.

**Type 2 osteoporosis**  The form of osteoporosis that occurs in men and women caused by age-related and lifestyle factors.

**kyphosis** (dowager's hump)  A curvature of the upper spine, caused by osteoporosis.

the bending of the spine can decrease the volume of the chest cavity, resulting in difficulty breathing, abdominal pain, decreased appetite, and premature satiety.

## Impact of Osteoporosis on Geriatric Health and Well-Being

Osteoporosis is a serious chronic degenerative disease, with many personal, societal, and economic implications. Researchers estimate that more than 28 million Americans (1 in 10 people) suffer from this disease, which is associated with 1.5 million bone fractures annually.[18] Approximately one in four people who fracture a hip die within a year. Because bone fractures often result in hospitalization and loss of the ability to live independently, preventing osteoporosis means enhanced health and well-being as we age. The National Osteoporosis Foundation estimates that health care costs attributable to osteoporosis are currently about $14 million each year, and these costs are expected to reach more than $60 billion by the year 2020.

## Screening and Treatment for Osteoporosis

Although *preventing* osteoporosis is best, it is also important for people at increased risk for osteoporosis to be screened on a regular basis. This is because some medications and lifestyle changes can help slow the progression of bone loss and thus osteoporosis. Clinicians can measure both bone mineral density and mass, using a variety of instruments. One such instrument is called a dual-energy x-ray absorptiometer (DEXA), which uses very low amounts of x-rays to produce a detailed picture of the various components of the body. A printout from a DEXA scan is provided in Figure 3. DEXA testing allows the rapid, painless, and accurate measurement of bone mass and bone density and can distinguish between different areas of the body. Recent reforms in national Medicare policies now help assure that the costs associated with DEXA measurements are covered for many people. The National Osteoporosis Foundation recommends DEXA testing for the following people.[19]

- All women aged 65 years and older
- Younger postmenopausal women with one or more risk factors (other than being white, postmenopausal, and female) including estrogen deficiency, steroid therapy, or thyroid dysfunction
- Postmenopausal women who have already had a fracture

## FIGURE 3   Dual-Energy X-Ray Absorptiometry (DEXA) Scan

DEXA scans are used to monitor and assess bone health—especially as we age.

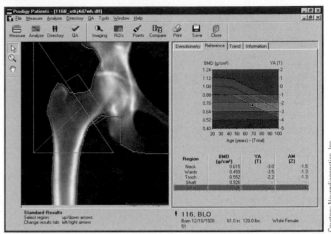

**CONNECTIONS** Recall that dual energy x-ray absorptiometry (DEXA) is also used to measure a person's body fat and lean mass (Chapter 2, page 32).

Currently, the most effective way to treat osteoporosis is the use of hormone replacement therapy (HRT), which is a combination of the hormones estrogen and progesterone.[20] However, HRT may increase the risk of other health problems, such as vaginal bleeding, increased risk for some cancers, and increased risk for cardiovascular disease.[21] Women considering this form of therapy should consult a doctor to evaluate whether HRT is right for them. Some studies suggest that consuming **phytoestrogens**, which are estrogen-like phytochemicals found primarily in soy products, may also help maintain bone health after menopause.[22]

Other medications, such as **selective estrogen receptor modulators (SERM)** and **bisphosphonates** can also help slow bone loss. SERMs mimic the beneficial effects of estrogen on bone density without some of the risks associated with HRT. Bisphosphonates work by inhibiting bone breakdown, especially in the spine and hip regions. An example of a bisphosphonate is Fosamax®, which is commonly prescribed to people with osteopenia or osteoporosis.

**phytoestrogen** An estrogen-like phytochemical.

**selective estrogen receptor modulator (SERM)** A drug that does not contain the hormone estrogen but causes estrogen-like effects in the body.

**bisphosphonates** A class of drugs sometimes taken to help reduce bone loss.

## ESSENTIAL *Concepts*

Osteoporosis is caused by severe bone demineralization (especially trabecular bone), resulting in weak and fragile bones. Both biological and lifestyle factors can influence risk of osteoporosis. Being female, aging, having a small body frame, and having one or more genetic risk factors can predispose a person to osteoporosis. Smoking, some medications, a history of eating disorders, excessive alcohol consumption, and low estrogen levels can also increase risk for this disease. Conversely, a well-balanced diet coupled with weight-bearing exercise can greatly decrease a person's risk. Type 1 osteoporosis tends to occur in women after menopause and is due to declining levels of estrogen, whereas type 2 osteoporosis affects both men and women. Kyphosis or dowager's hump is characterized by a curvature of the upper spine and can result in significant health issues such as difficulty breathing and abdominal pain. Screening for osteoporosis is important as we age, and some medications can slow bone loss during this time.

# Nutritional Recommendations Related to Bone Health

The Dietary Reference Intakes (DRIs) for many of the nutrients required for bone health have been set at levels that both promote healthy bone growth during earlier life and minimize bone loss in later life. These nutrients include calcium, phosphorus, magnesium, vitamin D, vitamin C, vitamin K, and protein. Obtaining recommended levels of these nutrients is important throughout the life span, especially in individuals with increased risk factors for osteoporosis. You can review the dietary recommendations for these nutrients by looking at the DRI tables provided inside the front cover of this book.

## Foods to Consume Only in Moderation

Some substances found in foods such as fiber, oxalates, and phytates can decrease absorption of calcium. Phytates are commonly found in whole grains and nuts, and oxalates are present in many vegetables, especially green leafy varieties such as spinach and collards. People with osteopenia or osteoporosis should not consume large amounts of these compounds on a daily basis. However, because these foods are nutrient rich it is prudent to continue to eat them in moderate amounts.

Because excessive alcohol consumption is associated with increased risk of osteoporosis, clinicians recommend that alcohol consumption be kept moderate. This is generally defined as less than one drink/day for women and less than two drinks/day for men. Studies also show a weak relationship between caffeine intake and risk for bone disease.[23] However, typical coffee consumption of 2 to 3 cups each day appears to pose no risk to bone health.[24]

There is also concern that consuming diets very high in phosphorus may have a detrimental effect on bone. This is because studies using animal models suggest that very high phosphorus intake, especially coupled with low calcium intake, can increase bone loss. However, after careful review of the literature the Institute of Medicine concluded there is little or no evidence that consuming an otherwise adequate diet that is high in phosphorus influences bone health in humans.[25]

## Foods to Encourage

Obtaining the needed bone-building nutrients requires the consumption of a variety of foods. To help reach this goal for calcium, vitamin D, and protein, the Dietary Guidelines for Americans and accompanying MyPyramid food guidance system recommend that we consume 2 to 3 cups of milk (or equivalent dairy product) each day. This recommendation is partly based on decreasing the risk for bone disease in later life. You can find out what your personal recommendation for dairy intake is by visiting the MyPyramid website (www.mypyramid.gov). If lactose intolerance is a problem, lactose-free milk or yogurt is suggested as a good alternative.

Comparing the calcium and protein contents of foods, using Nutrition Facts panels, is also a relatively easy way to choose foods that supply these nutrients. This is because the calcium and protein contents of foods are required components of food labeling. In addition, the U.S. Food and Drug Administration (FDA) has approved the following unqualified health claim that can be included on packaging when appropriate: "Regular exercise and a healthy diet with enough calcium helps teens and young women maintain good bone health and may reduce their high risk of osteoporosis later in life." Foods bearing this claim must be good sources of readily absorbed calcium. Calcium supplements are often recommended to prevent osteoporosis. However, data from the Women's Health Initiative suggest that long-term (7 year) supplementation with 1000 mg of calcium and 400 IU of vitamin D has only small effects on bone density and risk of bone fracture.[26] Hence, the benefits of calcium supplements are somewhat controversial.

Vitamin D is found in oily fish and, on exposure to sunlight, is made from cholesterol in the skin. To help prevent bone loss in older individuals, people over the age of 50 are advised to increase their vitamin D consumption and get at least 15 minutes of sun exposure each day when possible.[27] Vitamin D supplements may also be required in the elderly.

To obtain enough vitamin C, experts recommend that we consume 1 to 4 cups of vegetables and 1 to 2½ cups of fruits each day. Good sources of vitamin C are red peppers, citrus fruits, broccoli, strawberries, and brussels sprouts. Fish and dark green leafy vegetables such as kale, spinach, broccoli, and brussels sprouts provide the highest amounts of vitamin K in the diet. Legumes are also good sources of this nutrient. Foods rich in magnesium include seafood, legumes, nuts, and unprocessed (brown) rice, whereas phosphorus is found in almost all foods—especially those high in protein.

## ESSENTIAL *Concepts*

Many nutrients are important for building and maintaining healthy bones throughout life. These include calcium, magnesium, phosphorus, protein, and vitamins C, D, and K. Calcium, vitamin D, and protein are found in abundant amounts in milk and dairy product. Experts also recommend that elderly people be exposed to at least 15 minutes of sunlight daily to obtain (make) enough vitamin D. Magnesium is found in seafood, legumes, nuts, and brown rice. Phosphorus is found in almost all foods, especially good sources of protein. Broccoli and brussels sprouts are good sources of both vitamins C and K. Vitamin C is also found in citrus fruits and a variety of other fruits and vegetables, whereas legumes and fish are good sources of vitamin K.

## Notes

1. U.S. Department of Health and Human Services. Healthy People 2010: understanding and improving health, 2nd ed. Washington, DC: U.S. Government Printing Office; 2000.

2. Phan TC, Xu J, Zheng MH. Interaction between osteoblast and osteoclast: impact in bone disease. Histology and Histopathology. 2004;19:1325–44. Pogoda P, Priemel M, Rueger JM, Amling M. Bone remodeling: new aspects of a key process that controls skeletal maintenance and repair. Osteoporosis International. 2004;16(Suppl 2):S18–24. Takeda S. Central control of bone remodeling. Biochemical and Biophysical Research Communications. 2005;328: 697–9. Tanaka Y, Nakayamada S, Okada Y. Osteoblasts and osteoclasts in bone remodeling and inflammation: current drug targets. Inflammation and Allergy. 2005;4:325–8. Xing L, Boyce BF. Regulation of apoptosis in osteoclasts and osteoblastic cells. Biochemical and Biophysical Research Communications. 2005;328:709–20.

3. Bullamore JR, Wilkinson R, Gallagher JC, Nordin BE, Marshall DH. Effect of age on calcium absorption. Lancet. 1970;2:535–7.

4. Bonnelye E, Aubin JE. Estrogen receptor-related receptor alpha: a mediator of estrogen response in bone. Journal of Clinical Endocrinology and Metabolism. 2005;90:3115–21. Seeman E. Estrogen, androgen, and the pathogenesis of bone fragility in women and men. Current Osteoporosis Reports. 2004;2:90–6. Syed F, Khosla S. Mechanisms of sex steroid effects on bone. Biochemical and Biophysical Research Communications. 2005;328:688–96.

5. Institute of Medicine. Dietary Reference Intakes for calcium, phosphorus, magnesium, vitamin D, and fluoride. Washington, DC: National Academy Press; 1997.

6. Guthrie JR, Dennerstein L, Wark JD. Risk factors for osteoporosis: a review. Medscape Women's Health. 2000;5:E1.

7. Kapoor D, Jones TH. Smoking and hormones in health and endocrine disorders. European Journal of Endocrinology. 2005;152:491–9. Tanko LB, Christiansen C. An update on the antiestrogenic effect of smoking: a literature review with implications for researchers and practitioners. Menopause. 2004;11:104–9.

8. Lakotos P. Thyroid hormones: beneficial or deleterious for bone? Calcified Tissue International. 2003;73:205–9. Pack AM, Gidal B, Vazquez B. Bone disease associated with antiepileptic drugs. Cleveland Clinic Journal of Medicine. 2004;71Suppl2:S42–8. Roberts CG, Ladenson PW. Hypothyroidism. Lancet. 2004;363:793–803.

9. Cranney A, Adachi JD. Corticosteroid-induced osteoporosis: a guide to optimum management. Treatments in Endocrinology. 2002;1:271–9.

10. Abrams SA, Silber TJ, Esteban NV, Vieira NE, Stuff JE, Meyers R, Majd M, Yergey AL. Mineral balance and bone turnover in adolescents with anorexia nervosa. Journal of Pediatrics. 1993;123:326–31.

11. Stevenson JC. Hormone replacement therapy: review, update, and remaining questions after the Women's Health Initiative Study. Current Osteoporosis Reports. 2004;2:12–6.

12. Abrams SA, Silber TJ, Esteban NV, Vieira NE, Stuff JE, Meyers R, Majd M, Yergey AL. Mineral balance and bone turnover in adolescents with anorexia nervosa. Journal of Pediatrics. 1993;123:326–31.

13. Laitinen K , Valimaki M. Alcohol and bone. Calcified Tissue International. 1991;49:S70–3.

14. Chilibeck PD. Exercise and estrogen or estrogen alternatives (phytoestrogens, bisphosphonates) for preservation of bone mineral in postmenopausal women. Canadian Journal of Applied Physiology. 2004;29:59–75. Iwamoto J, Takeda T, Sato Y. Effect of treadmill exercise on bone mass in female rats. Experimental Animals. 2005;65:1–6. Jee WS, Tian XY. The benefit of combining non-mechanical agents with mechanical loading: a perspective based on the Utah Paradigm of Skeletal Physiology. Journal of Musculoskeletal and Neuronal Interactions. 2005;5:110–18.

15. Prentice A. Diet, nutrition and the prevention of osteoporosis. Public Health Nutrition. 2004;7:227–43.

16. Bugel S. Vitamin K and bone health. Proceedings of the Nutrition Society. 2003;62:839–43. Ginty F. Dietary protein and bone health. Proceedings of the Nutrition Society. 2003;62:867–76. Effects of vitamin $K_2$ on osteoporosis. Current Pharmaceutical Design. 2004;10:2557–76. Montero-

Odasso M, Duque G. Vitamin D in the aging musculoskeletal system: an authentic strength preserving hormone. Molecular Aspects of Medicine. 2005;26:203–19.

17. Duque G. Anabolic agents to treat osteoporosis in older people: is there still place for fluoride? Fluoride for treating postmenopausal osteoporosis. Journal of the American Geriatric Society. 2001;49:1387–9.

18. Riggs BL, Melton LJ III. The worldwide problem of osteoporosis: Insights afforded by epidemiology. Bone. 1995;17:505S–11S.

19. National Osteoporosis Foundation. Physician's Guide to Prevention and Treatment of Osteoporosis. Washington, DC: National Osteoporosis Foundation; 2003.

20. Stevenson JC. Hormone replacement therapy: review, update, and remaining questions after the Women's Health Initiative Study. Current Osteoporosis Reports. 2004;2:12–6.

21. Compston JE. The risks and benefits of HRT. Journal of Musculoskeletal and Neuronal Interactions. 2004;4:187–90. Nelson H. Postmenopausal osteoporosis and estrogen. American Family Physician. 2003;68:606–12.

22. Branca F. Dietary phyto-estrogens and bone health. Proceedings of the Nutrition Society. 2003;62:877–87. Harkness L. Soy and bone. Where do we stand? Orthopaedic Nursing. 2004;23:12–17. Weaver CM, Cheong JM. Soy isoflavones and bone health: the relationship is still unclear. Journal of Nutrition. 2005;135:1243–7.

23. Barger-Lux MJ, Heaney RP. Caffeine and the calcium economy revisited. Osteoporosis International. 1995;5:97–102. Barger-Lux MJ, Heaney RP, Stegman MR. Effects of moderate caffeine intake on the calcium economy of premenopausal women. American Journal of Clinical Nutrition. 1990;52:722–5.

24. Institute of Medicine. Dietary Reference Intakes for calcium, phosphorus, magnesium, vitamin D, and fluoride. Washington, DC: National Academy Press, 1997.

25. Institute of Medicine. Dietary Reference Intakes for calcium, phosphorus, magnesium, vitamin D, and fluoride. Washington, DC: National Academy Press; 1997.

26. Jackson RD, LaCroix AZ, et al. Calcium plus vitamin D supplementation and the risk of fractures. New England Journal of Medicine. 2006;354:669–83.

27. Dawson-Hughes B. Racial/ethnic considerations in making recommendations for vitamin D for adults and elderly men and women. American Journal of Clinical Nutrition. 2004;80:1763–6. Holick MF. Sunlight and vitamin D for bone health and prevention of autoimmune diseases, cancers, and cardiovascular disease. American Journal of Clinical Nutrition. 2004;80:1678S–8S. Institute of Medicine. Dietary Reference Intakes for calcium, phosphorus, magnesium, vitamin D, and fluoride. Washington, DC: National Academy Press; 1997.

**Check out the following sources for additional information.**

28. Anderson JJB, Sell ML, Gerner SC, Calvo MS. Phosphorus. In: Present knowledge in nutrition, 8th ed. Bowman BA, Russell RM, editors. Washington, DC: ILSI Press; 2001.

29. Fleet JC, Cashman KD. Magnesium. In: Present knowledge in nutrition, 8th ed. Bowman BA, Russell RM, editors. Washington, DC: ILSI Press; 2001.

30. Goulding A, Rockell JE, Black RE, Grant AM, Jones IE, Williams SM. Children who avoid drinking cow's milk are at increased risk for prepubertal bone fractures. Journal of the American Dietetic Association. 2004;104:250–3.

31. Greenblatt D. Treatment of postmenopausal osteoporosis. Pharmacotherapy. 2005;25:574–84.

32. Gropper SS, Smith JL, Groff JL. Advanced nutrition and human metabolism, 4th ed. Belmont, CA: Wadsworth; 2005.

33. Institute of Medicine. Dietary Reference Intakes for water, potassium, sodium, chloride, and sulfate. Washington, DC: National Academies Press; 2005.

34. Nieves JW. Osteoporosis: the role of micronutrients. American Journal of Clinical Nutrition. 2005;81:1232S–9S.

35. Preuss HG. Sodium, chloride and potassium. In: Present knowledge in nutrition, 8th ed. Bowman BA, Russell RM, editors. Washington, DC: ILSI Press; 2001.

36. Raisz LG. Clinical practice. Screening for osteoporosis. New England Journal of Medicine. 2005;353:164–71.

37. Weaver C. Calcium. In: Present knowledge in nutrition, 8th ed. Bowman BA, Russell RM, editors. Washington, DC: ILSI Press; 2001.

38. Davis SR, Dinatale I, Rivera-Woll L, Davison S. Postmenopausal hormone therapy: from monkey glands to transdermal patches. Journal of Endocrinology. 2005;185:207–22.

## Chapter 14

# Life Cycle Nutrition

From start to finish, the human life cycle is a process of continuous change. Birth, growth, maturation, aging, and death are all part of this natural progression of life. With each stage, the body changes in terms of size, proportion, and composition. Because of this nutritional requirements vary enormously. For example, nutritional needs during periods of growth differ vastly from those associated with the later stages of life, when the body is often in a state of physical decline.

A healthy diet is essential during all stages of life. In fact, poor nutritional status early in life can influence health at later stages. Thus, food choices made today may have a far greater effect on health than you might think. This chapter surveys the influence of growth, development, and aging on nutritional requirements across the human life span. The continuum of life encompasses infancy, childhood, adolescence, adulthood, and the special life stages of pregnancy and lactation.

Photos © Matthew Farruggio

# An Overview of the Human Life Span

Cells form, mature, carry out specific functions, die, and are replaced by new cells. In many ways, the life cycle of cells mimics life itself. That is, after a new human is conceived and born, the next 70 to 80 years are characterized by periods of growth and development, physical decline, and eventually death. The ability to reproduce enables life to be renewed by passing genetic material (DNA) on to the next generation.

## Growth and Development

Growth and development are important physiological events that take place throughout the life cycle. **Growth,** an increase in body size, occurs when the number and/or size of cells increase. As illustrated in Figure 14.1, an increase in number of cells is called hyperplasia, whereas an increase in cell size is called hypertrophy. The highest rates of growth occur during infancy, childhood, and adolescence. Growth rates are also high during pregnancy and lactation.

Height and weight are the most common measures used to assess growth. The U.S. National Center for Health Statistics (NCHS), which is part of the Centers for Disease Control and Prevention (CDC), has compiled reference standards into growth charts.[1] These charts indicate expected growth for well-nourished infants, children, and adolescents. Growth charts include percentile curves that represent growth patterns from birth through 20 years of age. To evaluate adequacy of growth, a measurement (such as weight) can be plotted in reference to age and sex. For example, if a child's weight falls at the 60th percentile, 40% of healthy children of similar age and sex weigh more,

**FIGURE 14.1  Types of Growth**

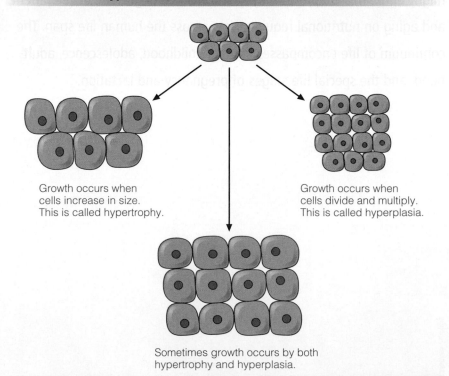

Growth occurs when cells increase in size. This is called hypertrophy.

Growth occurs when cells divide and multiply. This is called hyperplasia.

Sometimes growth occurs by both hypertrophy and hyperplasia.

**growth** An increase in size and/or number of cells.

and 60% weigh less. In this way, growth can be monitored over time and used as a general indicator of health.

**Development** is the acquisition of increased complexity of function, such as the ability to walk during late infancy. Although the rate of development varies, the pattern tends to be predictable. For example, infants usually crawl before they walk. However, some infants walk as early as 10 months, whereas others may not take their first step until several months later. As with growth, it is also important to monitor development, because a major delay could indicate a problem such as illness or poor nutrition.

Growth and development continue steadily throughout infancy and childhood, increasing markedly during adolescence. Adolescence begins when hormonal changes trigger the physical transformation of a child into an adult. Most dramatic is the maturation of reproductive organs and the ability to reproduce. Once physical maturity is reached, the rate of growth and development begins to slow. The rate of **cell turnover,** a process of cell formation and breakdown, reaches equilibrium at this time. However, with increasing age the rate at which new cells form decreases, resulting in a loss of some body tissue. As cells die, the remaining cells become less effective at carrying out their functions. Gradually, the characteristic physical changes associated with aging, called **senescence,** become apparent.

## Stages of the Life Span

Classifying the stages of life in terms of growth and development offers a meaningful way to discuss nutritional requirements (Figure 14.2). Physiological changes that accompany growth and development affect body size and com-

**development** Change in the complexity and/or attainment of a function.

**cell turnover** The cycle of cell formation and cell breakdown.

**senescence** The process of aging, during which function diminishes.

## FIGURE 14.2 Stages in the Human Life Span

Growth and development at different stages of the life cycle affect body size and composition, which in turn influence nutrient and energy requirements.

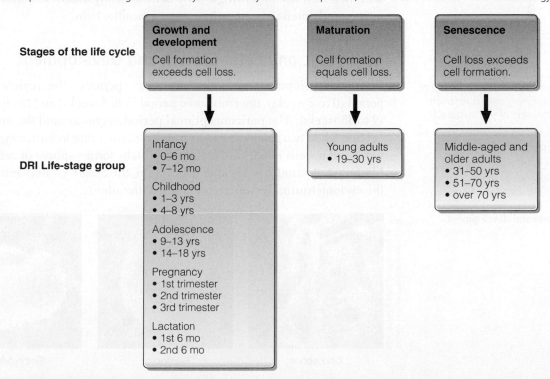

position, which in turn influence nutrient and energy requirements. For this reason, the Dietary Reference Intakes (DRIs) specify recommended nutrient and energy intakes for each life stage group, including infancy (ages 0 to 6 months and ages 7 to 12 months), early childhood (ages 1 to 3 years and ages 4 to 8 years), adolescence (ages 9 to 13 years and 14 to18 years), and young, middle, and older adults (ages 19 to 30 years, ages 31 to 50 years, ages 51 to 70 years, and over 70 years). Along with these stages, the DRI recommendations consider the special conditions of pregnancy and lactation.

## ESSENTIAL *Concepts*

Life is a continuum that begins with birth and is marked by periods of growth and development. These changes coincide with stages of the life span that include infancy, childhood, adolescence, adulthood, and the special life stages of pregnancy and lactation. Changes in body size and composition can influence nutrient requirements. Growth is an increase in body size, whereas development is the acquisition of increased complexity of function. Growth occurs when the number (hypertrophy) and/or the size (hyperplasia) of cells increase. When physical maturity is reached, the rate of cell turnover (cell formation and breakdown ) is in equilibrium. With increasing age, the rate of new cell formation slows, resulting in a decline in physiological function, called senescence.

# Prenatal Development

Although it is important for all women to have their nutritional needs met, nutrition is particularly important during pregnancy. This is because poor nutritional status before and during pregnancy can have serious and long-term effects on the unborn child. A pregnant woman with poor nutritional status is at increased risk for having a baby born too early and/or too small. For this reason, early prenatal care helps ensure a healthy baby.

## Embryonic and Fetal Growth and Development

Prenatal development is divided into three periods—the periconceptional period (0 to 2 weeks), the embryonic period (3 to 8 weeks) and the fetal period (9 to 38 weeks). The **periconceptional period** begins around the time of conception when two gametes, an ovum and a sperm, unite to form a **zygote**. After fertilization, cells rapidly divide, and eventually form a sphere of cells called a **blastocyst.** Around 2 weeks after conception, the blastocyst implants itself into the endometrium, the innermost lining of the uterus.

**periconceptional period** Prenatal stage of development beginning with conception through the formation of the blastocyst.

**zygote** (ZY – gote) An ovum that has been fertilized by a sperm.

**blastocyst** (BLAS – to – cyst) Early stage of embryonic development.

Stages of prenatal development.

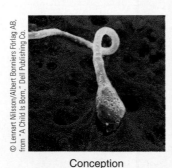
Conception

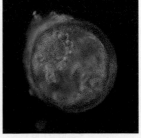
Blastocyst

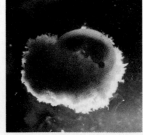
Embryo/fetus

© Lennart Nilsson/Albert Bonniers Förlag AB, from "A Child Is Born," Dell Publishing Co.

During the **embryonic period,** which extends through the eighth week of pregnancy, cells rapidly divide. These cells begin to differentiate, giving rise to tissues and organs. Each tissue follows a specific timetable in terms of development. If a critical nutrient is lacking at this time, a tissue or organ may not form properly. The term **critical period** is often used to describe the time when an organ undergoes rapid growth and development and is most vulnerable to environmental insults that could disrupt this process. Although critical periods are most likely to occur during the early stages of pregnancy, they can also occur at later stages. Substances that disrupt the normal course of cell growth and development are called **teratogens.** The harmful effects of teratogens are usually apparent at birth, although some may not be detected until much later. For example, the drug diethylstilbestrol (DES), which was prescribed to pregnant women during the 1950s and 1960s to prevent miscarriages or premature births, was at that time thought to be safe. However, 10 to 12 years later DES was linked to a rare form of vaginal cancer in daughters born to mothers who had taken DES. It soon became clear that DES was teratogenic, and by the early 1970s the drug was no longer prescribed. There are many teratogens, but perhaps the most familiar is alcohol. Women who consume alcohol while pregnant are at increased risk of having a baby born with fetal alcohol syndrome. You can read more about the effects of alcohol and other teratogens in the Focus on Clinical Applications feature on page 562.

By the end of the embryonic period, the basic structures of all major body organs are formed, and the embryo is now referred to as a fetus. However, for the fetus to survive outside the womb, much additional growth and development is needed. Fetal weight increases almost 500-fold during this period. At term, the fetus weighs approximately 7 to 8 pounds (3.2 to 3.6 kg) and is roughly 20 inches (51 cm) long. These progressive stages of prenatal development and critical periods of organ formation are illustrated in Figure 14.3.

The terms *embryonic period* and **fetal period** refer to stages of prenatal development. However, pregnancy is more commonly described in terms of trimesters. The first trimester is the time from conception to the end of week 13 and includes the entire embryonic period as well as part of the fetal period. The second trimester is from week 14 to the end of week 26, and the third trimester is from week 27 to the end of pregnancy.

## Duration of Pregnancy

A variety of terms are used to describe the length of pregnancy or what is called gestation. **Gestational age** is determined by counting the number of weeks between the first day of a woman's last normal menstrual period and birth. Because most women do not know exactly when conception takes place, calculating length of gestation using this method provides a more reliable measure of pregnancy duration.[2] Therefore, based on gestational age the average length of pregnancy is about 40 ±2 weeks.

## The Formation of the Placenta

After implantation, cells from the embryo invade the endometrial lining and join with maternal tissues to form the **placenta**. Although the placenta develops early in pregnancy, it takes several weeks before it is fully functional,

**embryonic period** The stage of development that extends through the eighth week of pregnancy.

**critical period** Period in development when cells and tissue rapidly grow and differentiate to form body structures.

**teratogen** (te – RAT – o – gen) Environmental agent that can alter normal cell growth and development, causing a birth defect.

**fetal period** Period of gestation after week 8 through birth.

**gestational age** Length of pregnancy, determined by counting the number of weeks between the first day of a woman's last normal menstrual period and birth.

**placenta** An organ made up of fetal and maternal tissue that supplies nutrients and oxygen to the fetus.

## FIGURE 14.3 Stages of Prenatal Development and Critical Periods of Organ Formation

Malformations resulting from exposure to teratogens are highly dependent on both the agent and the gestational age at which the exposure occurs.

Prenatal development is divided into 3 periods—the periconceptional period (0–2 weeks), the embryonic period (3–8 weeks), and the fetal period (9–38 weeks).

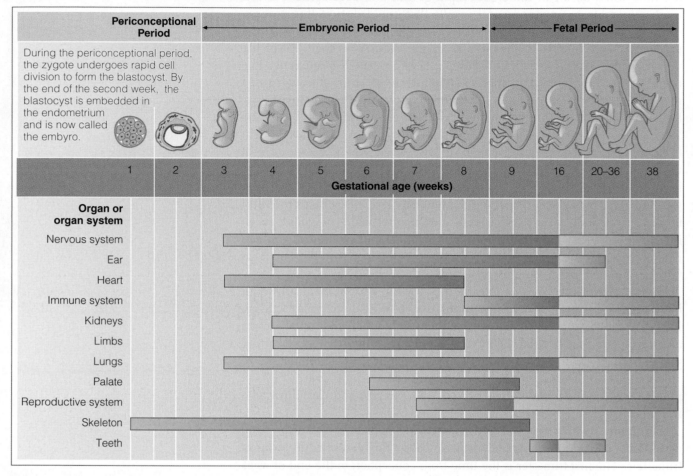

| | Periconceptional Period | Embryonic Period | Fetal Period |
|---|---|---|---|
| During the periconceptional period, the zygote undergoes rapid cell division to form the blastocyst. By the end of the second week, the blastocyst is embedded in the endometrium and is now called the embryro. | | | |

Gestational age (weeks): 1, 2, 3, 4, 5, 6, 7, 8, 9, 16, 20–36, 38

**Organ or organ system**
- Nervous system
- Ear
- Heart
- Immune system
- Kidneys
- Limbs
- Lungs
- Palate
- Reproductive system
- Skeleton
- Teeth

**Critical Periods of Development**

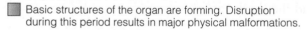

Basic structures of the organ are forming. Disruption during this period results in major physical malformations.

Basic structures of the organ are already established. Disruption during this period results in impaired growth, physiological defects, or functional deficiencies.

SOURCES: Office of Children's Health Protection (OCHP) and U.S. Environmental Protection Agency (EPA). Critical periods in development. (Paper.) Series on children's health and the environment. Washington, DC: February 2003. Dietert RR, Etzel RA, Chen D, Halonen M, Holladay SD, Jarabek AM, Landreth K, Peden DB, Pinkerton K, Smialowicz RJ, Soetis T. Workshop to identify critical windows of exposure for children's health: immune and respiratory systems work group summary. Environmental Health Perspectives. June 2000;108(Suppl 3):483–90.

weighing between 1 and 2 pounds at term. This highly vascularized structure (see Figure 14.4) has important functions.

- The placenta transfers nutrients, hormones, oxygen, and other substances from the maternal blood to the fetus.
- Metabolic waste products formed by the fetus pass through the placenta and into the mother's blood, and later are excreted by the mother's kidneys and lungs.
- The placenta is a source of several hormones that serve a variety of functions during pregnancy.

Although placental membranes prevent the fetal and maternal blood from mixing, the exchange of gases, nutrients, and waste products is quite efficient. Unfortunately, many potentially harmful substances can also cross the placenta from the mother's blood into the fetal circulation. For this reason, pregnant women must be particularly careful about consuming medications or other substances that could potentially harm the embryo/fetus.

LBW infants often require care in neonatal intensive care units. These units provide care to infants who are critically ill.

## Gestational Age and Birth Weight

Gestational age and birth weight are two important predictors of infant health. As previously mentioned, an infant's gestational age is determined by counting the number of weeks between the first day of a woman's last normal menstrual period and birth. Babies born with gestational ages between 37 and 42 weeks are considered **full-term infants**, whereas those born with gestational ages less than 37 weeks are called **preterm infants** (or premature infants). The earlier a baby is born, the greater the risk for complications that can affect survival and long-term health. This is because organs may not be fully developed and may therefore be unable to sustain life outside the womb.

Babies weighing less than 5 pounds, 8 ounces (2,500 g) at birth are called **low birth weight (LBW)** infants. Low birth weight infants are small because they are either preterm or have experienced slow growth in utero, known as **intrauterine growth retardation (IUGR)**. Babies who experience IUGR are often referred to as being **small for gestational age (SGA),** defined as having a birth weight below the 10th percentile for gestational age. Infants born with

**full-term infants** Babies born with gestational ages between 37 and 42 weeks.

**preterm** (premature) **infant** A baby born at 37 weeks or less gestation.

**low birth weight (LBW)** An infant that weighs less than 2,500 g at birth.

**intrauterine growth retardation (IUGR)** Slow or delayed growth in utero.

**small for gestational age (SGA)** An infant that is below the 10th percentile for weight for gestational age.

## FIGURE 14.4   Structure and Functions of the Placenta

The placenta forms early in pregnancy and is made up of fetal and maternal tissue.

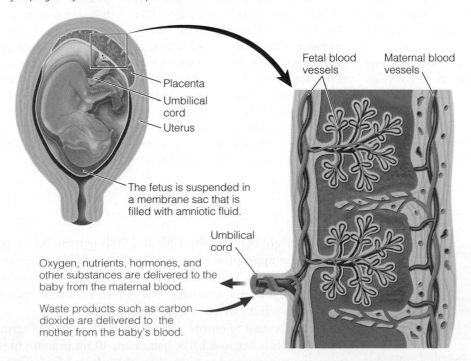

Placenta
Umbilical cord
Uterus

The fetus is suspended in a membrane sac that is filled with amniotic fluid.

Fetal blood vessels
Maternal blood vessels

Umbilical cord

Oxygen, nutrients, hormones, and other substances are delivered to the baby from the maternal blood.

Waste products such as carbon dioxide are delivered to the mother from the baby's blood.

Placenta

# *Focus On* CLINICAL APPLICATIONS

## Teratogens and Birth Defects

Although most pregnancies result in the birth of a healthy baby, approximately 3 to 4% of babies (150,000 per year) in the United States are born with a birth defect.[3] A birth defect is an abnormality related to a structure and/or function, and sometimes results in mental and/or physical impairment. Some birth defects are mild, but others can be severe and life threatening. Birth defects occur for many reasons, including genetics, environment, lifestyle, or a combination of these factors. Approximately 60% of birth defects have no known cause.

Scientists do know that some birth defects are caused by compounds called teratogens. The term *teratogen* is used to describe a broad group of environmental agents that negatively affect the development of the unborn child. Teratogens include chemicals, drugs, infections, and radiation, and are responsible for about 4 to 5% of all birth defects.[4] Although most teratogens exert damage only at certain times during the pregnancy (that is, critical periods), other teratogens such as alcohol can be damaging throughout the entire pregnancy.

During the 1950s, physicians became alarmed when thousands of European babies were born with stunted development of their arms and/or legs. It soon became apparent that a drug called thalidomide given to pregnant women to alleviate morning sickness was responsi-

ble for this devastating birth defect.[5] Yet women who took thalidomide during the later stages of pregnancy had children with normal limb development, because the critical period for forming arms and legs occurs early in pregnancy. From this tragedy, the medical community learned that some drugs have serious effects on the unborn child, and thalidomide was banned from use. Nonetheless, over 12,000 babies were born with birth defects caused by thalidomide.

Today, most drugs are tested for teratogenic effects. For example, the drug isotreinoin (Accutane™) is often prescribed for the treatment of severe acne. However, it is highly teratogenic and should not be taken by women who are pregnant or who could become pregnant.[6] Even taking small amounts of isotreinoin for a short period can put a fetus at high risk for deformities. In fact, women must agree to use contraception and to undergo regular pregnancy testing before they are prescribed this drug.

Another well-known teratogen is alcohol. About 1 to 2 babies per 1,000 live births have alcohol-related health problems, commonly referred to as fetal alcohol syndrome (FAS).[7] Characteristics associated with FAS include a small head circumference, unusual facial characteristics, and other physical deformities. In addition, many of these infants are developmentally delayed. A less severe form of FAS, called fetal alcohol effect

In the 1950s, many pregnant women who were prescribed the drug thalidomide gave birth to babies born with deformities of the arms and/or legs.

© Leonard McCombe/Time Life Pictures/Getty Images

(FAE) is associated with learning and behavior problems, which are often not apparent until later in life. Fortunately, FAS and FAE are preventable. That is, if a woman does not consume alcohol while pregnant, there is no risk of having a baby with FAS or FAE. Because no amount of alcohol is considered safe during pregnancy, the best recommendation is to totally avoid drinking alcohol during this time.

---

**appropriate for gestational age (AGA)** An infant that weighs between the 10th and the 90th percentile for weight for gestational age.

**large for gestational age (LGA)** An infant that weighs more than the 90th percentile for weight for gestational age.

CONNECTIONS ▶ Recall that infant mortality rate is the number of infant deaths during the first year of life per 1,000 live births (Chapter 1, page 20).

a birth weight between the 10th and 90th percentiles for gestational age are classified as **appropriate for gestational age (AGA),** whereas those with a birth weight above the 90th percentile are classified as **large for gestational age (LGA).** Figure 14.5 illustrates how growth charts are used to classify infants according to birth weight and gestational age.

It is important to ensure that babies are born at full term and at a healthy weight. This is because LBW babies are 40 times more likely to die before 1 year of age compared to normal-weight infants.[8] In fact, together, premature births and LBW are the leading causes of infant mortality. Moreover, poor growth during the prenatal period may have profound long-term effects. This

**FIGURE 14.5    Classification of Infants Based on Gestational Age and Birth Weight**

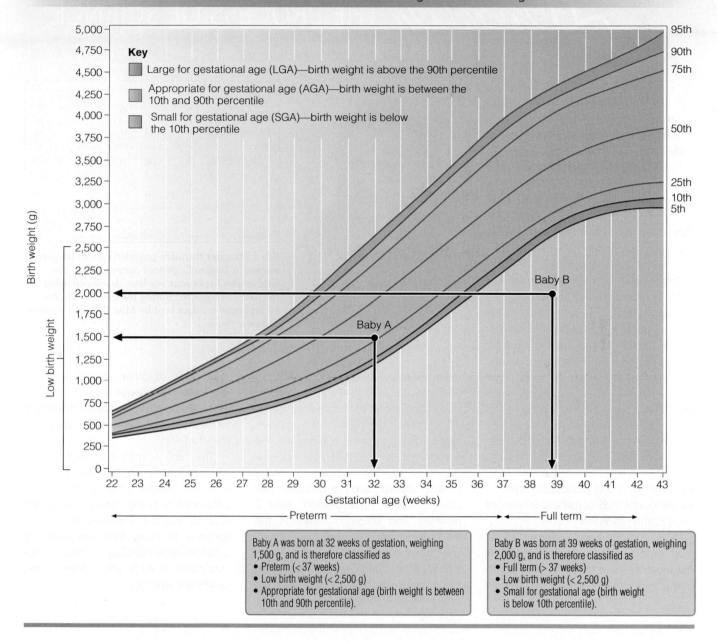

Baby A was born at 32 weeks of gestation, weighing 1,500 g, and is therefore classified as
- Preterm (< 37 weeks)
- Low birth weight (< 2,500 g)
- Appropriate for gestational age (birth weight is between 10th and 90th percentile).

Baby B was born at 39 weeks of gestation, weighing 2,000 g, and is therefore classified as
- Full term (> 37 weeks)
- Low birth weight (< 2,500 g)
- Small for gestational age (birth weight is below 10th percentile).

is known as the **fetal origins hypothesis.** Evidence suggests that nutrient shortages in the womb may cause permanent changes in the structure and/or function of organs and tissues, predisposing individuals to certain chronic diseases later in life.[9] You can read more about the fetal origins hypothesis in the Focus on the Process of Science feature.

## ESSENTIAL *Concepts*

After conception, a zygote undergoes cell division and becomes a blastocyst. Certain cells of the blastocyst form the fetus, and others join with maternal tissues to form the placenta. The placenta transfers nutrients, gases, hormones, and other products between the mother and the fetus. By the end of the embryonic period the basic

**fetal origins hypothesis** A theory suggesting that conditions during gestational development can alter risk for chronic diseases later in life.

# *Focus On* THE PROCESS OF SCIENCE
## The Fetal Origins Hypothesis

Courtesy MRC Epidemiology Resource Centre, Southampton General Hospital

In the early 1900s, lowering the infant mortality rates among England's poorest citizens was a goal of health officials. To accomplish this, a midwife named Ethel Margaret Burnside was assigned to assist pregnant women in these communities. Nurses made regular home visits during which they weighed and measured each baby at birth, and then again at 1 year of age. Throughout the years, information on thousands of infants was carefully recorded.

Years later, the data recorded by Burnside became of great interest to epidemiologists David Barker and Clive Osmond of Southampton University in England. Regions that had the highest neonatal deaths in the early 1900s were the same regions that later had the highest rates of heart disease. Curious to see if these events were related, Barker and Osmond analyzed the birth records of 13,249 men born between 1911 and 1945 and found that being born with LBW was related to increased risk for heart disease later in life. Barker and Osmond proposed that prenatal conditions had somehow influenced adult health years later.

This concept, now referred to as fetal programming, suggests that less than optimal conditions in the womb may alter fetal development, resulting in

Ethel Margaret Burnside provided care to pregnant women in England's poorest communities. The fetal origins hypothesis was later developed using the data she collected during the early 1900s. An excerpt from a ledger kept by Miss Burnside's nurses is shown above.

increased risk of certain chronic diseases later in life. Barker and Osmond's hypothesis, called the fetal origins hypothesis, or the Barker hypothesis, has now been demonstrated in many populations and animal models.[10] Considerable evidence links poor prenatal nutrient availability to adult diseases such as cardiovascular disease, stroke, hypertension, type 2 diabetes, and obesity.[11] Although the mechanisms are not well understood, scientists think that when nutrients are limited, fetal blood may be shunted to the brain to protect it, thus reducing

blood flow to other organs. Although this adaptive response may be beneficial to the fetus in the short term, it seems to have negative consequences during later stages of life.

The fetal origins hypothesis is not without controversy, because the origin of chronic disease is highly complex and influenced by many factors that are difficult to sort out. However, as scientists continue to study fetal programming, a better understanding of how various conditions in early life influence later health will emerge.

structures of all major body organs have been formed. Substances and conditions that disrupt normal growth and development are called teratogens. The fetal period follows the embryonic period. Babies with a gestational age of 37 weeks or greater are full term, whereas those with a gestational age less than 37 weeks are preterm. Most preterm infants are born with low birth weight (LBW), which can also be caused by intrauterine growth retardation (IUGR). Babies who experience IUGR are small for gestational age (SGA). Gestational age is determined by counting the number of weeks between the first day of a woman's last normal menstrual period and birth. Infants born with a birth weight between the 10th and 90th percentiles for gestational age are classified as appropriate for gestational age (AGA), whereas those with a birth weight above the 90th percentile are classified as large for gestational age (LGA).

| TABLE 14.1 | Components of Weight Gain During Pregnancy |
|---|---|
| **Component** | **Approximate Weight Gain (lb.) at 38 Weeks Gestation** |
| Fetus | 7 to 8 |
| Placenta | 1½ to 2 |
| Uterus and supporting structures | 2½ to 3 |
| Maternal adipose stores | 7 to 8 |
| Mammary tissue | 1 to 2 |
| Extracellular fluids (blood and amniotic fluid) | 6 to 7 |
| *Total weight gain* | 25 to 30 |

SOURCE: King JC. Physiology of pregnancy and nutrient metabolism. *American Journal of Clinical Nutrition*. 2000;71(5 Suppl): 1218S–25S.

# Health Recommendations for Pregnancy

Many factors can help decrease a woman's risk of having a preterm and/or LBW baby. The three most important factors largely within a woman's control are adequate weight gain, eating a healthy diet, and not smoking.

## Maternal Weight Gain

Adequate weight gain during pregnancy is an important determinant of fetal growth and development. Table 14.1 lists the components of weight gain associated with pregnancy. Current weight-gain guidelines for pregnant women, developed by the Institute of Medicine (IOM),[12] are based on maternal prepregnancy body mass index (BMI). These recommendations are summarized in Table 14.2. Women who gain weight within their BMI range are likely to deliver full-term babies with a healthy birth weight. For example, it is recommended that normal-weight women (prepregnancy BMI between 19.8 and 26.0 $kg/m^2$) gain 25 to 35 pounds, whereas overweight women (prepregnancy BMI > 26.0 to 29.0 $kg/m^2$) are encouraged to gain less—between 15 and 25 pounds. Obese women (prepregnancy BMI > 29.0 $kg/m^2$) should gain a minimum of 15 pounds, whereas underweight women (prepregnancy BMI <19.8 $kg/m^2$) are encouraged to gain between 28 and 40 pounds. Teens benefit from higher weight gains during pregnancy and should gain at the upper end of the recommended range based on their prepregnancy BMI. This is because teens are still growing, increasing the risk of having preterm and LBW infants. Regardless of prepregnancy BMI, weight loss during pregnancy is never recommended.

**CONNECTIONS** Recall that body mass index ($kg/m^2$) is a measure of adiposity (Chapter 2, page 32).

| TABLE 14.2 | Recommended Weight Gain Ranges for Pregnant Adult Women | |
|---|---|---|
| **Prepregnancy Weight-for-Height Category (BMI kg/m²)** | **Recommended Total Weight Gain During Pregnancy** | |
| | **kg** | **lb.** |
| Low (BMI <19.8) | 12.5–18 | 28–40 |
| Normal (BMI 19.8–26) | 11.5–16 | 25–35 |
| High (BMI >26–29) | 7.0–11.5 | 15–25 |
| Obese (BMI >29) | At least 6.8 kg | At least 15 lb. |

SOURCE: Institute of Medicine/National Academies of Science (NAS/IOM). *Nutrition During Pregnancy*. Washington, DC: National Academies Press; 1990.

# Maternal Nutrient and Energy Requirements

Pregnant women experience a variety of physiological changes, many of which affect nutrient requirements. These changes are described in Table 14.3. Dietary recommendations for pregnant women are based on extensive research and are intended to promote optimal health in the mother and unborn child. However, the MyPyramid food guidance system is not intended to be a tool for planning diets during pregnancy or lactation. Rather, obstetricians recommend that pregnant and lactating women work directly with a health care provider. The Dietary Guidelines for Americans 2005 provides key recommendations that specifically address issues relevant to pregnant women, and these are featured in the Food Matters.

## Recommended Energy Intake

Adequate weight gain during pregnancy requires adequate energy intake. During pregnancy, additional energy is needed to support the growth of the fetus and placenta, as well as other maternal tissues. Resting energy expenditure increases during pregnancy because of added physiological demands on the mother. For example, the heart and lungs must work harder to deliver nutrients and oxygen to the fetus.

The energy demands of pregnancy are quite high—about 60,000 kcal over the course of pregnancy. Although very little extra energy is needed during the first trimester, most women are advised to increase their energy intake above nonpregnancy Estimated Energy Requirements (EERs) by an additional 348 kcal/day and 452 kcal/day, during the second (14 to 26 weeks) and third (week 27 to the end of pregnancy) trimesters of pregnancy, respectively. However, young or underweight women may have higher energy requirements.

**CONNECTIONS** Resting energy expenditure (REE) is the energy expended for resting metabolic activity over a 24-hour period (Chapter 9, page 324).

---

### TABLE 14.3 Physiological Changes During Pregnancy

**Cardiovascular System**
- Heart enlarges slightly.
- Heart rate increases.
- Cardiac output increases.
- Blood pressure decreases during the first half of pregnancy and returns to nonpregnant values during the second half of pregnancy.
- Plasma volume increases.
- Red blood cell volume increases.
- Respiratory rate increases.
- Oxygen consumption increases.

**Gastrointestinal Tract and Food Intake**
- Appetite increases.
- Altered sense of taste and smell.
- Thirst increases.
- Gastrointestinal motility decreases.
- Efficiency in nutrient absorption increases.
- Gastroesophageal reflux becomes more common.

**Renal System**
- Filtration rate increases.
- Sodium retention increases.

**Energy Metabolism and Energy Balance**
- Basal metabolic rate (BMR) increases.
- Body temperature increases.
- Body weight increases.
- Fat mass increases.
- Lean mass increases.
- Total body water increases.

## *Food Matters*

*Working Toward the Goal*

The USDA Dietary Guidelines recommend that women pay special attention to food choices during pregnancy. The following food selection and preparation tips will help you meet this goal.

- Consume folate-rich foods such as orange juice, lentils, and dark green leafy vegetables.
- Consume additional folic acid from fortified and enriched foods such as breakfast cereals, whole-grain bread, and whole-wheat pasta.
- Eat foods that provide heme iron such as lean meat and/or consume iron-rich plant foods such as breakfast cereals, whole-grain bread, and lentils.
- To enhance iron content of foods, use cast-iron cookware.
- To enhance iron absorption, consume vitamin C-rich foods and iron-rich plant foods together.
- Eat foods that are good sources of calcium such as milk, cheese, yogurt, calcium-fortified soy products, and calcium-fortified orange juice.
- If eating deli meats and frankfurters, only eat those that have been reheated to steaming hot.
- Ensure appropriate weight gain as specified by a health care provider by eating nutrient-dense foods.
- Abstain from drinking alcohol.

## Recommended Macronutrient Intakes

To satisfy energy requirements, it is important to get enough carbohydrate, protein, and fat during pregnancy. If the pregnancy is progressing normally, carbohydrates should remain the primary energy source (45 to 65% of total calories) throughout. The RDA for carbohydrate during pregnancy is 175 g/day, which provides adequate amounts of glucose for both the mother and the fetus.

During pregnancy, additional protein is needed for forming fetal and maternal tissues. The recommendation (RDA) is for pregnant women to increase their protein intake by about 25 g per day, or approximately 70 g of total protein daily. This amount is easily obtained by eating a variety of high-quality protein sources such as meat, dairy products, and eggs. For example, 3 ounces of meat or 2 cups of yogurt provide approximately 25 g of protein. Consistent with the Acceptable Macronutrient Distribution Ranges (AMDRs), protein should continue to provide 10 to 35% of total calories.

Pregnant women who follow a vegan diet must plan their meals carefully to ensure adequate intake of essential amino acids. Plant foods that provide relatively high amounts of protein include tofu and other soy-based products and legumes such as dried beans and lentils. To ensure that all essential amino acids are consumed in adequate amounts, it is important to eat a variety of these and other foods.

Dietary fat is also an important source of energy during pregnancy, with a recommended intake of 20 to 35% of total calories. Although there are no DRI values for total fat, AIs have been established for the essential fatty acids. During pregnancy, the AIs for linoleic and linolenic fatty acids are 13 and 1.4 g per day, respectively. Linoleic acid and linolenic acid are important parent compounds used by the body to form other fatty acids. For example, linoleic acid is converted to arachidonic acid, whereas linolenic acid is converted to docosahexaenoic acid (DHA) and eicosapentaenoic acid (EPA). All these fatty acids are critical for fetal growth and development, but DHA is particularly important for brain development and formation of the retina.

To ensure adequate intake of essential fatty acids during pregnancy, women should eat fish several times per week and use omega-3-rich canola oil and flaxseed oil. Because some types of fish contain high levels of mercury, the Food and Drug Administration (FDA) and the Environmental Protection Agency (EPA) advise pregnant women to avoid eating shark, swordfish, king mackerel, and tilefish.[13] Nursing mothers and young children are also advised not to eat these fish.

## Recommended Micronutrient Intakes

Not only do energy requirements increase during pregnancy, but it is also important for women to have adequate intakes of vitamins and minerals. Vitamins and minerals are needed for the synthesis of maternal and fetal tissues and for many reactions involved in energy metabolism. With few exceptions, the requirements for most micronutrients increase during pregnancy; recommended intakes are listed inside the front cover of this book.

Although the requirements for most vitamins increase during pregnancy, some vitamins such as vitamin A increase only slightly, whereas others (vitamins D, E, and K) do not increase at all. In addition, excessively high intakes of preformed vitamin A during pregnancy have been associated with fetal malformations. In other words, large doses of vitamin A can be teratogenic. For this reason, pregnant women should not exceed the UL (3,000 μg/day) of vitamin A from either foods or supplements. Foods that contain large amounts of preformed vitamin A include beef and chicken liver. However, it is not necessary for pregnant women to limit their intake of beta-carotene, or other provitamin A carotenoids found in plant foods.

Surprisingly, the recommended intake of calcium does not increase during pregnancy. Although extra calcium is needed for the fetus to grow and develop properly, changes in maternal physiology are able to accommodate these needs without increasing dietary intake. For example, calcium absorption increases and urinary calcium loss decreases. Therefore, the AI for calcium during pregnancy (1,000 mg/day) is the same as that for nonpregnant women.

Because iron is needed for the formation of hemoglobin and for the growth and development of the fetus and the placenta, the RDA for iron increases substantially during pregnancy to 27 mg/day. Most well-planned diets provide women with approximately 15 to 18 mg of iron daily. For example, 6 ounces of lean beef provide 4 to 5 mg of heme iron, whereas 1 cup of iron-fortified breakfast cereal has between 10 and 15 mg of nonheme iron. However, the bioavailability of iron in meat is higher than plants. Many pregnant women have difficulty meeting the recommended intake for iron by diet alone. In

CONNECTIONS ▶ Recall that the body is unable to make the essential fatty acids linoleic acid and linolenic acid, and therefore these fatty acids must be supplied by the diet. Docosahexaenoic acid (DHA) is derived from linolenic acid (Chapter 7, page 227).

addition, about 12% of women enter pregnancy with impaired iron status. For these reasons, iron supplementation is often encouraged during the second and third trimesters of pregnancy, when iron requirements are the greatest.[14]

Because vegans eliminate all foods of animal origin, efforts must be made to get adequate amounts of vitamin $B_{12}$, vitamin $B_6$, iron, calcium, and zinc. Pregnant women who are vegans should include additional servings of grains, legumes, nuts, and calcium-fortified foods such as tofu and soy milk. Foods that provide vitamin $B_{12}$, such as nutritional yeast and vitamin $B_{12}$-fortified cereals, are also vital.

*Maternal Diet and Neural Tube Defects*   Not only is it important for a woman to eat a healthy diet during pregnancy, but her diet around the time of conception can affect the health of her unborn child, as well. For example, the B vitamin folate plays an important role in cell division, which is critical to the development of the nervous system. For this reason, women with poor folate status in early pregnancy are at increased risk of having a baby with a neural tube defect (NTD), a birth defect affecting the spinal cord and/or brain. The critical period for the formation of the neural tube occurs early in pregnancy—within the first 21 days. Thus the neural tube may already be formed before a woman realizes she is pregnant. Because of the importance of folate in neural tube development, an RDA of 600 μg/day has been set for pregnancy. It is also recommended that women capable of becoming pregnant consume 400 μg/day of folic acid as a supplement or in fortified foods in addition to consuming folate naturally occurring in foods.

In 1996, the FDA, recognizing the importance of folate in preventing NTDs, approved enrichment of grain products such as corn grits, pasta, and bread with folic acid. Fortification levels are approximately 0.43 mg to 1.4 mg per pound of product. In addition, food manufacturers are allowed to make claims on the label that adequate intake of folate may reduce the risk of NTDs. Since this nationwide effort, folate status in the United States has improved, and the incidence of NTDs has decreased by 19%.[15] Although folate fortification efforts have been successful at decreasing the incidence of neural tube defects, not all NTDs can be prevented by increased folate intake. Rather, some NTDs are caused by genetic mutations that alter folate metabolism and folate transfer across the placenta.

## Impact of Maternal Smoking

There is overwhelming evidence that smoking during pregnancy increases the risk of having a preterm and/or LBW baby.[16] Thus it is important that pregnant women not smoke. Furthermore, some studies show that secondhand smoke can also harm the fetus.[17]

Tobacco contains thousands of chemical compounds, many of which damage the fetus and the placenta. In addition, smoking causes maternal blood vessels to constrict, reducing blood flow from the uterus to the placenta. As a result, nutrient and oxygen availability to the fetus decreases. Smoking also increases the risk of premature detachment of the placenta, which can result in a **miscarriage,** or the death of a fetus during the first 20 weeks of a pregnancy. If a woman smokes, she is advised to quit as early in the pregnancy as possible.

CONNECTIONS  Recall that vegans do not eat animal-derived foods, including meat, dairy products, and eggs (Chapter 6, page 199).

**miscarriage** Death of a fetus during the first 20 weeks of pregnancy.

# ESSENTIAL *Concepts*

Adequate weight gain, a healthy diet, and smoking can influence birth weight and gestation length. Energy requirements increase during pregnancy, as do those for macro- and most micronutrients. Energy is needed during pregnancy to support the growth of the fetus, placenta, and maternal tissues. Women are advised to increase their energy intake above nonpregnancy Estimated Energy Requirements by an additional 348 kcal/day and 452 kcal/day during the second and third trimesters of pregnancy, respectively. Carbohydrates should remain the primary energy source throughout pregnancy. Additional protein (25 g/day) is needed for the formation of fetal and maternal tissues. Dietary fat should provide 20 to 35% of total calories. Although extra calcium is needed for the fetus to grow and develop, changes in maternal physiology can accommodate these needs without increasing dietary intake. Iron is needed for forming hemoglobin and for the growth and development of the fetus and the placenta. The RDA for iron during pregnancy increases to 27 mg/day. Folate plays an important role in cell division and is critical to the development of the nervous system. Women with poor folate status are at increased risk of having a baby born with a neural tube defect. The RDA for folate during pregnancy is 600 μg/day. Smoking during pregnancy increases the risk of having a preterm and/or LBW baby.

# Nutrition During Lactation

During pregnancy, a woman's body undergoes many physiological changes that prepare the breasts (mammary glands) for milk production. The nutrient composition of human milk is ideally suited to the growth and development of the baby. In addition, human milk provides immunologic protection against pathogenic viruses and bacteria. It is not surprising that breastfed babies tend to get sick less often than babies fed formula. Moreover, breastfeeding is beneficial to the mother as well. As with pregnancy, nutrition is important during lactation, when the woman is nourishing herself and providing milk to feed her baby.

## Anatomy and Physiology of Lactation

During pregnancy, many hormones, including estrogen and progesterone, prepare the mammary glands for milk production, a process called **lactation.** These hormones stimulate an increase in the number of milk-producing cells and an expansion of ducts that transport milk in the breast. As illustrated in Figure 14.6, milk production takes place in specialized structures of the breast called **alveoli.** Each alveolus (the singular of *alveoli*) is made up of secretory cells, which produce milk. When stimulated, the secretory cells release the milk into a network of **mammary ducts** that eventually lead to the nipple. Women are encouraged to nurse their babies soon after delivery, because suckling initiates the process of **lactogenesis,** the onset of milk production.

The hormones **prolactin** and **oxytocin** are both produced in the pituitary gland and regulate milk production and the release of milk from alveoli into the mammary ducts. When a baby suckles, nerves in the nipple are stimulated and signal the hypothalamus, which in turn signals the pituitary gland to release both of these hormones (Figure 14.7). Prolactin stimulates the secretory cells in the mammary gland to synthesize milk, whereas oxytocin

**lactation** The production and release of milk.

**alveolus** (plural, *alveoli*) A cluster of milk-producing cells that make up the mammary glands.

**mammary duct** Structure that transports milk from the secretory cells toward the nipple.

**lactogenesis** (lac – to – GEN – e – sis) The onset of milk production.

**prolactin** (pro – LAC – tin) A hormone produced in the pituitary gland that stimulates the production of milk in alveoli.

**oxytocin** (ox – y – TO – cin) A hormone produced in the pituitary gland that stimulates the movement of milk into the mammary ducts.

## FIGURE 14.6 Anatomy and Physiology of the Human Breast

The human breast, or mammary gland, is a complex organ composed of many different types of tissue that produce and secrete milk.

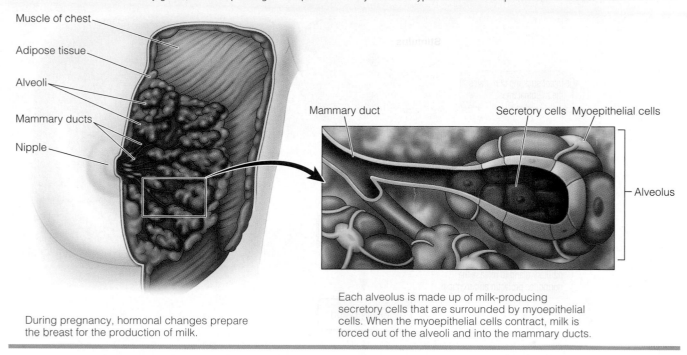

Muscle of chest
Adipose tissue
Alveoli
Mammary ducts
Nipple

Mammary duct    Secretory cells  Myoepithelial cells

Alveolus

During pregnancy, hormonal changes prepare the breast for the production of milk.

Each alveolus is made up of milk-producing secretory cells that are surrounded by myoepithelial cells. When the myoepithelial cells contract, milk is forced out of the alveoli and into the mammary ducts.

causes small muscles (called **myoepithelial cells**) that surround the alveoli to contract. This muscular contraction forces the milk out of the alveoli and into the mammary ducts. This active release of milk is called milk **let-down.** As the baby suckles, the milk moves through the mammary ducts, toward the nipple, and into the baby's mouth. Anxiety, stress, and fatigue can interfere with the milk let-down reflex, sometimes making breastfeeding challenging. For this reason, it is important for women to seek physical and emotional support during this time.

## Milk Production: A Matter of Supply and Demand

Milk production is regulated by many physiological factors, although breast size has no effect. The amount of milk produced is determined largely by how much milk the infant consumes. Women who exclusively breastfeed, meaning that human milk is the sole source of infant feeding, produce more milk than those who supplement breastfeeding with formula feeding. On average, women produce 26 ounces (3⅓ cups) of milk per day during the first 6 months postpartum, and 20 ounces (2½ cups) per day in the second 6 months. Because newborns have small stomachs and can only consume small amounts of milk at each feeding, many mothers breastfeed as frequently as every two to three hours. However, as the baby grows the amount of milk consumed increases, reducing the need to breastfeed as frequently. The American Academy of Pediatrics (AAP) recommends that women nurse at least 8 to 12 times each day and feed on demand rather than schedule feedings.

Breastfeeding requires proper positioning of the baby at the breast, the ability of the baby to latch onto the nipple, and the ability of the baby to coor-

**Most women find it convenient to breast-feed their babies. It is recommended that mothers breastfeed on demand.**

**myoepithelial cells** Muscle cells that surround alveoli and that contract, forcing milk into the mammary ducts.

**let-down** The movement of milk through the mammary ducts toward the nipple.

## FIGURE 14.7 Neural and Hormonal Regulation of Lactation

Milk production is regulated by "supply and demand."

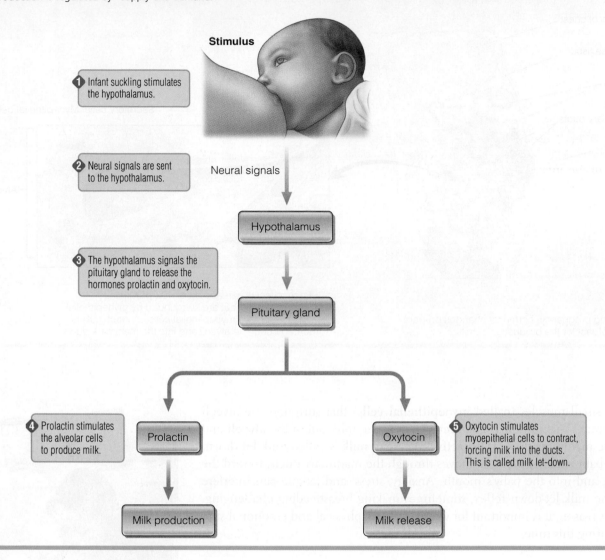

**Stimulus**

1 Infant suckling stimulates the hypothalamus.

2 Neural signals are sent to the hypothalamus.

Neural signals

Hypothalamus

3 The hypothalamus signals the pituitary gland to release the hormones prolactin and oxytocin.

Pituitary gland

4 Prolactin stimulates the alveolar cells to produce milk.

Prolactin

5 Oxytocin stimulates myoepithelial cells to contract, forcing milk into the ducts. This is called milk let-down.

Oxytocin

Milk production

Milk release

dinate sucking and swallowing. Often a mother is concerned about whether her infant is receiving enough milk. The best indicators of adequate milk production are infant weight gain and the frequency of wet diapers. Under most conditions a baby needs no additional source of nourishment other than human milk for the first six months of life. Most problems associated with breastfeeding can be resolved with the assistance of a medical professional such as a pediatrician or lactation specialist. All new mothers should seek help if they experience problems with breastfeeding.

## Human Milk Is Beneficial for Babies

Shortly after the mother gives birth, her breasts produce a special substance called **colostrum**, which nourishes the newborn and helps protect the baby from disease. This is, in part, because colostrum contains an abundance of immunoglobulins, which are proteins that fight infections. Consuming colostrum also helps eliminate **meconium** from the baby's intestinal tract. Meconium is a dark green, feces-like substance that collects in the large intestine of the fetus while it is developing in utero.

**colostrum** (co – LOS – trum) The first secretion from the breasts after birth that provides nourishment and immunological protection to newborns.

**meconium** (me – CO – ni – um) Dark green, feces-like matter that collects in the intestine of an unborn baby.

Over the next several days, the mammary glands stop producing colostrum and begin to produce mature milk. Like colostrum, mature human milk has many important immunological benefits. For example, some components help inhibit the growth of disease-causing pathogens, whereas others promote the development of the infant's immature gastrointestinal tract. This may help reduce the risk of developing food allergies later in life. Not only does breastfeeding reduce the incidence of infectious diseases, but some studies suggest that infants fed human milk are less likely to develop type 1 and type 2 diabetes, certain types of cancer, asthma, and obesity in later life.[18]

## Composition of Human Milk

The nutrient composition of human milk perfectly matches the nutritional needs of the infant. In addition to nutrients, human milk contains enzymes and other compounds that make certain nutrients easier to digest and absorb. For example, the presence of lipase in the milk helps the infant digest triglycerides. In addition, human milk contains lactoferrin, a multifunctional glycoprotein that enhances iron absorption and has antimicrobial activity that protects the infant from certain infectious diseases.

The protein content of human milk is relatively low compared to cow milk, and is considered ideal for infant growth and development. The high protein content of cow milk makes it unsuitable for infants. In addition to the right amount of protein, the protein in human milk is easily digested. In fact, before cow milk can be used to make formula, the protein content must be adjusted.

Over half the calories in human milk comes from lipids. Some lipids are synthesized by the mammary gland, and others come from the maternal diet. Although there are dozens of fatty acids in human milk, scientists are particularly interested in DHA because of its important role in brain and eye development during infancy. Some studies have shown improved cognitive performances in breastfed infants compared to formula-fed infants.[19] When DHA is added to infant formula, these differences disappear. Only in the last few years have some manufacturers started adding DHA to infant formula. Human milk also contains large amounts of cholesterol, which is an important component of cell membranes. Currently, cholesterol is not added to infant formula, but there is interest in understanding whether it should be included.

The primary carbohydrates in human milk are lactose and oligosaccharides. Not only is lactose an important source of energy, it also facilitates the absorption of other nutrients such as calcium, phosphorus, and magnesium. In addition to providing a source of energy, oligosaccharides may have functional roles such as inhibiting the growth of harmful bacteria in the infant's gastrointestinal tract.

CONNECTIONS  Recall that oligosaccharides are carbohydrates that consist of 3 to 10 monosaccharides (Chapter 5, page 130).

## Energy and Nutrient Requirements During Lactation

The amount of additional energy needed by the mother during lactation depends on whether she is exclusively breastfeeding or feeding a combination of human milk and formula. Because women tend to produce more milk during the first six months of lactation compared to the second 6 months, additional energy required for milk production is approximately 500 kcal/day and 400 kcal/day, respectively, during these periods. However, some of the energy needed for milk production during the first six months comes from the mobilization of maternal body fat. As the following formulas show, energy intake recommendations for lactation are based on nonpregnant, nonlactating

Estimated Energy Requirements (EERs), energy expenditure associated with milk production, and energy "saved" by mobilization of body fat stores.

### Estimated Energy Requirement (EER) Calculations

- 0 to 6 months of lactation = Adult EER + milk energy output − weight loss
  Adult EER + 500 kcal − 170 kcal from mobilized body fat
  Adult EER + 330 kcal/day

- 6 to 12 months of lactation = Adult EER + milk energy output
  Adult EER + 400 kcal
  Adult EER + 400 kcal/day

Thus the Institute of Medicine (IOM) recommends an additional 330 kcal/day above nonpregnant Estimated Energy Requirements (EERs) during the first six months of lactation and an additional 400 kcal/day in the second 6 months. Equations for calculating adult EERs are provided in Appendix B.

Recommendations for micronutrient intakes during lactation are similar to those during pregnancy, although some (such as vitamin A) are greater and others (such as folate) are less. These are listed inside the front cover of this book. Of special interest is the recommendation for vitamin C, which increases substantially in lactation. Because a relatively large amount of vitamin C is secreted in milk, the RDA increases from 85 mg/day during pregnancy to 120 mg/day during lactation. In addition, it is important that lactating women take in enough fluid. Although increased fluid intake does not increase milk production, a lack of fluid can decrease milk volume. The AI for total fluids for lactating women is 3.8 liters (13 cups) per day. This includes fluids in both foods and beverages. Recommended nutrient intakes for nonpregnant, pregnant, and lactating women are compared in Figure 14.8.

## Breastfeeding Is Beneficial for Mothers

Although most people are aware that breastfeeding is beneficial for infants, many people are unaware that it is also beneficial for mothers. Breastfeeding shortly after giving birth stimulates the uterus to contract, helping to minimize blood loss and shrink the uterus to its prepregnant size. Some women also find that breastfeeding helps them return to their prepregnant weight more easily.[20]

Women who breastfeed also gain long-term health benefits. For example, breastfeeding delays the return of menstrual cycles. This period of time between the birth of the baby and the first menses, called **postpartum amenorrhea,** allows iron stores to recover. Although the duration of postpartum amenorrhea varies greatly, menstruation tends to resume around 6 months postpartum in breastfeeding women, and around 6 weeks postpartum in nonbreastfeeding women. Because it is possible to ovulate without menstruating, gynecologists recommend that women use contraception until they wish to become pregnant again. In addition to prolonging postpartum amenorrhea, breastfeeding also reduces a woman's risk of developing breast cancer, ovarian cancer, and possibly osteoporosis later in life.[21]

## ESSENTIAL *Concepts*

Physiological changes during pregnancy prepare the breasts for producing milk. Milk production takes place in mammary secretory cells and is regulated in part by the hormones prolactin and oxytocin. Nursing stimulates the hypothalamus, which in turn signals release

**postpartum amenorrhea** (a – men – or – RHE – a) The period after giving birth and before the return of menses.

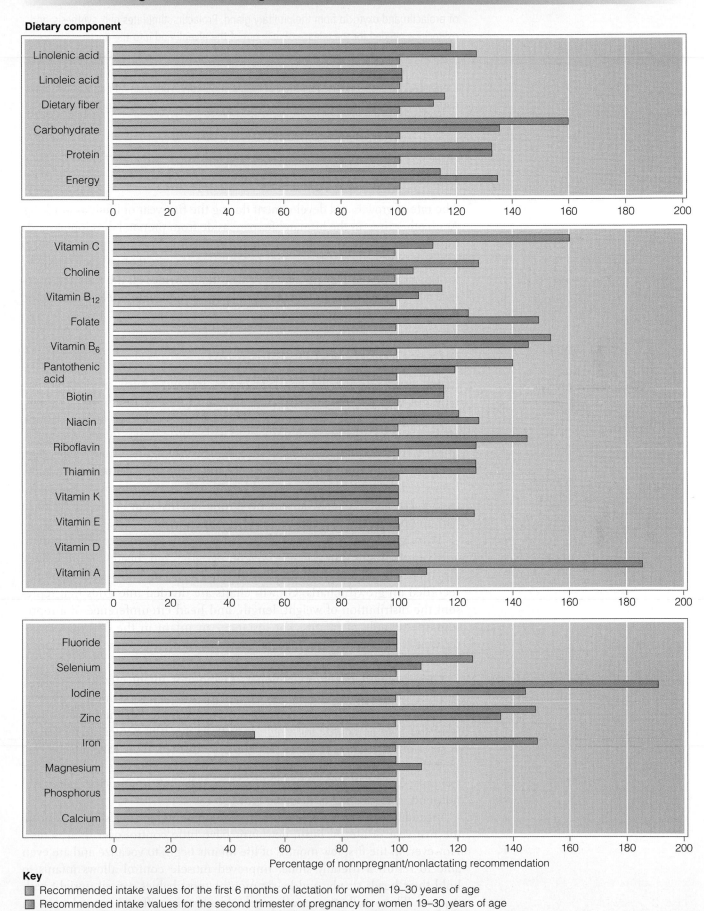

**FIGURE 14.8** Comparison of Recommended Energy and Nutrient Intakes for Nonpregnant, Pregnant, and Lactating Women

Percentage of nonpregnant/nonlactating recommendation

**Key**

Recommended intake values for the first 6 months of lactation for women 19–30 years of age
Recommended intake values for the second trimester of pregnancy for women 19–30 years of age
Recommended intake values for nonpregnant/nonlactating women 19–30 years of age

of prolactin and oxytocin from the pituitary gland. Prolactin stimulates milk synthesis, and oxytocin is needed for the release of milk out of the alveoli and into the ducts—a process called milk let-down. To maintain lactation, it is important for the mother to breastfeed her infant regularly. Like pregnancy, lactation requires additional nutrients and energy. Breastfeeding is the preferred method of nourishing infants, because human milk has many nutritional and immunological benefits. Breastfeeding also has health benefits for women.

# Nutrition During Infancy

The rate of growth and development during the first year of life is astonishing. At no other time in the human life span, aside from prenatal development, do these processes occur so rapidly. Therefore feeding becomes more challenging, and it is important for parents to be aware of signs that indicate readiness for this next phase. During infancy, a diet that provides all the essential nutrients and an environment that is safe, secure, and engaging help build a solid foundation for the next stages of life.

## Infant Growth and Development

Infant growth during the first year of life follows a fairly predictable pattern. For example, within a few days after birth, infants lose 5 to 6% of their body weight, because of a loss of fluids and some breakdown of body tissues. However, most infants regain this weight within two weeks. By 4 to 6 months, infant weight doubles, and length increases by 20 to 25%. Growth rates then decrease slightly from 6 to 12 months. By end of the first year of life, a healthy baby's weight will almost triple, and length will increase by 50%. Equations used to estimate energy requirements during the first year of life are based in part on body weight and can be found in Appendix B.

Throughout the first year of life, it is important to monitor infant growth and development. An infant's weight, length, and head circumference are routinely measured during health checkups, and this information is recorded on growth charts. Growth charts are divided into grids that represent the distribution of weight, length, and head circumference of a representative sample of infants. For instance, an infant in the 60th percentile for weight at a given age is heavier than 60% of the reference population of infants (Figure 14.9).

Growth charts enable practitioners to monitor infant growth over time. Because infants tend to follow a consistent growth pattern, a dramatic change in percentile could indicate a problem. For example, an infant who drops from the 50th percentile for weight at birth to the 10th percentile for weight at six months is growing more slowly than expected. This can indicate that something is wrong and should be evaluated. Poor nursing technique, for example, can result in an infant not receiving adequate amounts of energy and nutrients, thus slowing or delaying growth.

In addition to growth, significant developmental changes occur during the first year of life. At birth, newborns have little control over their bodies. However, in the first few months of life infants begin to vocalize and are even able to return a friendly smile. Improved muscle control allows infants to hold their heads steady, and by six months most babies can sit upright with support. These and other developmental milestones affect how and what an infant should be fed. Within this first year of life, infants progress from a diet

## FIGURE 14.9 Growth During the First Three Years of Life

Weight and length can be monitored during infancy and childhood using growth charts.

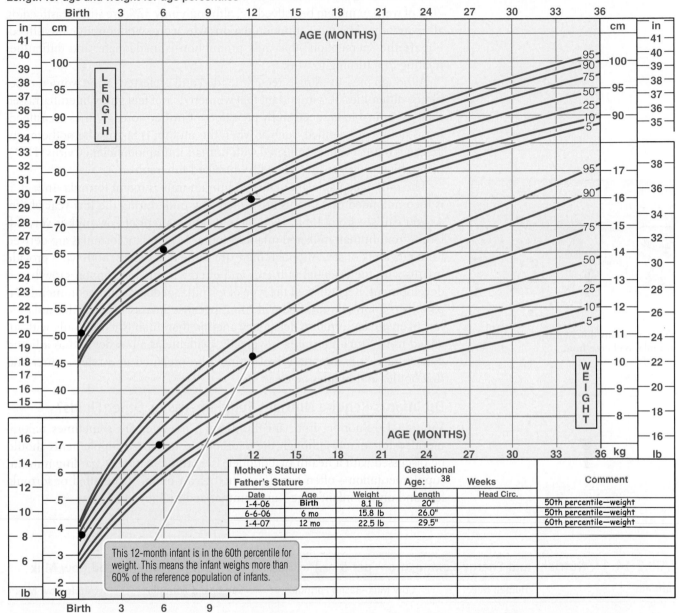

**Birth to 36 months: Girls**
**Length-for-age and Weight-for-age percentiles**

| Mother's Stature | | | Gestational Age: 38 Weeks | | Comment |
|---|---|---|---|---|---|
| Father's Stature | | | | | |
| Date | Age | Weight | Length | Head Circ. | |
| 1-4-06 | Birth | 8.1 lb | 20" | | 50th percentile—weight |
| 6-6-06 | 6 mo | 15.8 lb | 26.0" | | 50th percentile—weight |
| 1-4-07 | 12 mo | 22.5 lb | 29.5" | | 60th percentile—weight |

This 12-month infant is in the 60th percentile for weight. This means the infant weighs more than 60% of the reference population of infants.

By 4 to 6 months, weight doubles and length increases by approximately 25%.

By 12 months, weight triples and length increases by approximately 50%.

Source: Developed by the National Center for Health Services in collaboration with the National Center for Chronic Disease Prevention and Health Promotion (2000). http://www.cdc.gov/growthcharts

that consists solely of human milk and/or infant formula to being able to feed themselves a variety of foods.

## Infant Feeding Recommendations

In the late 1950s, fewer than 20% of babies were breastfed, partly because of the belief that infant formula was superior to human milk. Fortunately, this trend has now been reversed. Over the past 30 years, the number of women in the United States that breastfeed has steadily increased.[22] This is, in part, a response to compelling scientific evidence that human milk is ideally suited for optimal growth and development of infants. The AAP recommends exclusive breastfeeding for the first four to six months of life and the continuation of breastfeeding for the second six months until at least one year.[23] Currently 70% of women initiate breastfeeding, although only 32% are still breastfeeding at six months. The most common reasons why women discontinue breastfeeding are the "perception" that milk production is inadequate, and difficulties relating to returning to work.

Although breastfeeding is recommended and preferred, some women experience difficulties. It is important that women do not feel guilty or embarrassed if they are unable to successfully breastfeed. Furthermore, at times breastfeeding is not recommended, such as when the mother is taking chemotherapeutic drugs to treat cancer, infected with human immunodeficiency virus (HIV), using illicit drugs, or infected with untreated tuberculosis.

The only acceptable alternative to human milk is infant formula. In fact, it is recommended that infants not be fed cow milk during the first year of life. As you can see from Table 14.4, the nutrient content of cow milk is very different from human milk and infant formula. Most infant formulas are derived from cow milk or soy. Although there are some differences, both types provide infants with the essential nutrients and energy needed to support growth and development. Regardless of the type of formula, pediatricians recommend that parents use formula fortified with iron, because infants fed such formula are less likely to develop iron deficiency anemia than infants fed formula without added iron.[24] Infant formula is typically available as a powder and as a liquid concentrate. It is important that formula be prepared in a safe manner and that bottles and nipples be clean.

There are many different types of infant formula to choose from.

## Do Infants Require Nutrient Supplements or Other Fluids?

Vitamin D, fluoride, iron, and fluids such as water are sometimes recommended for infants although the decisions to use these supplements should be discussed with a health practitioner. Recommendations regarding nutrient supplementation—often based on whether an infant is breastfed or formula fed, and on the infant's age—are summarized in Table 14.5.

| TABLE 14.4 | Nutrient and Energy Composition (per 5 oz.) of Human Milk, Infant Formula, and Cow Milk | | | |
|---|---|---|---|---|
| Nutrient | Human Milk[a] | Cow Milk–Based Formula[b] | Soy-Based Formula[c] | Whole Cow Milk[d] |
| Energy (kcal) | 105[e] | 100 | 100 | 92 |
| Protein (g) | 2 | 2 | 3 | 5 |
| Fat (g) | 6 | 6 | 5 | 5 |
| Carbohydrate (g) | 9 | 11 | 9 | 7 |
| Cholesterol (mg) | 21 | 4 | 0 | 15 |
| Iron (g) | 0.1 | 1.8 | 1.8 | 0 |
| Calcium (mg) | 45 | 78 | 83 | 173 |
| Vitamin A (IU) | 331 | 300 | 300 | 156 |
| Vitamin D (IU) | 3 | 60 | 60 | 62 |
| Vitamin C (mg) | 6 | 9 | 12 | 0 |
| Folate (μg) | 7 | 15 | 16 | 8 |

[a]Picciano MF. Representative values for constituents of human milk. *Pediatric Clinics of North America.* 2001;48:1–3.
[b]Infant formula, Similac™, with iron, ready-to-serve. Data from USDA Standard Reference Release 17.
[c]Infant formula, Prosobee™ with iron, ready-to-serve. Data from USDA Standard Reference Release 17.
[d]Milk, whole, 3.25% milkfat. Data from USDA Standard Reference Release 17.
[e]Human milk averages 20–22 kcal/oz.

## TABLE 14.5 Recommended Nutrient Supplementation During Infancy (0–12 months)

| | Infant-Feeding Method | |
|---|---|---|
| Nutrient | Exclusive Human Milk | Iron-Fortified Infant Formula |
| Vitamin D | Vitamin D supplements (200 IU/day) recommended beginning during the first 2 months of life and continuing until the infant is consuming at least 16 oz. of infant formula per day | Not needed, because all infant formulas manufactured in the United States meet standards for vitamin D |
| Fluoride | Fluoride supplements (0.25 mg/day) starting at age 6 months if local water has a fluoride concentration less than 0.3 ppm | Not needed if formula is prepared with local water that has a fluoride concentration of at least 0.3 ppm. If local water isn't fluoridated, supplements (0.25 mg/day) recommended starting at 6 months |
| Iron | Iron supplements (1 mg/day per kilogram of body weight) for infants exclusively breastfed during the second 6 months of life | Not needed if iron-fortified formula contains at least 1 mg iron/100 kcal |
| | | Iron supplements (1 mg per kilogram of body weight) for infants not fed iron-fortified infant formula |
| Fluids | Additional fluids not needed unless infant has excessive fluid loss due to vomiting and/or diarrhea | Additional fluids not needed unless infant has excessive fluid loss due to vomiting and/or diarrhea |

*Vitamin D*    In recent years, several cases of rickets in breastfed infants have been reported nationwide, raising concerns about the adequacy of vitamin D content in human milk. Although scientists have long assumed that babies receive enough vitamin D from their diet and/or exposure to sunlight, apparently this is not always so. Factors such as skin color, cloud cover, smog, and use of sunscreen can interfere with vitamin D synthesis. For example, sunscreens with a sun protection factor (SPF) of 8 or greater block the ultraviolet (UV) rays needed for vitamin D synthesis in skin. As a result, breastfed infants who do not get adequate sunlight exposure are at increased risk for developing vitamin D deficiency or rickets. Because it is important to use sunscreen to help prevent skin cancer, the National Academy of Sciences and the AAP recommend that breastfed infants receive vitamin D supplements (200 IU/day) beginning during the first two months of life and continuing until they are consuming at least 16 ounces (500 mL) of infant formula per day.[25] Because all infant formulas manufactured in the United States meet standards for vitamin D, there is little risk of vitamin D deficiency or rickets in formula-fed infants.

*Fluoride*    Fluoride is important for forming teeth and can help prevent dental caries later in life. The National Academy of Sciences and the AAP recommend that the decision to provide fluoride supplements for breastfed infants starting at 6 months of age be made on the basis of the fluoride concentration in the water supply.[26] If the local water supply has a fluoride concentration of at least 0.3 parts per million (ppm), fluoride supplements are not necessary. Parents should check with the local water department to find out the fluoride content of the drinking water in their community. If purified bottled water is used to prepare infant formula, fluoride supplements are recommended. Recommended daily dosage of fluoride supplements is 0.25 mg for children between 6 months and 3 years of age.

*Iron*    Iron is another nutrient that is sometimes given to infants in the form of supplements. A full-term infant is born with a substantial iron reserve that helps meet his or her need for iron during the first six months of life. After this time infants receiving iron-fortified formula that contain at least 1 mg iron per 100 kcal formula are likely to maintain adequate iron intakes. Although

CONNECTIONS ▶ Rickets is a disease in children that causes bones to become soft and bend (Chapter 11, page 452; Chapter 13, page 520).

human milk contains less iron than infant formula, it is more readily absorbed. The AAP and CDC recommend iron supplementation for infants who continue to be exclusively breastfed during the second six months of life.[27] It is suggested that these infants be given 1 mg of iron per kilogram of body weight daily until complementary iron-rich foods are introduced. Iron supplements are generally recommended for preterm and/or LBW infants. For breastfed infants who were preterm or LBW, a supplement of 2 to 4 mg of iron per kilogram of body weight daily is recommended starting at one month after birth and continuing through the first year of life.

*Water*    Water requirements during the first six months of life are likely met if adequate amounts of breast milk and/or formula are provided, so infants do not need additional fluids. However, when vomiting and diarrhea cause excess fluid loss, water replacement is necessary to prevent dehydration. During these situations, it is important that parents contact their health care provider. The AAP recommends plain water rather than juice during these times.[28] Although over-the-counter fluid replacement products can be given, they are not usually needed.

## Infant Feeding Practices: Introducing Complementary Foods

The AAP recommends introducing nonmilk complementary foods when an infant is between 4 and 6 months of age.[29] Until this time younger infants are not physiologically or physically ready for foods other than human milk and/or infant formula. Furthermore, early introduction of complementary foods can increase the risk of developing food allergies.[30] Signs that an infant is ready for complementary foods include sitting up with support and good head and neck control.

Because iron status begins to decline at four to six months of life, pediatricians recommend that the first complementary foods be iron rich, such as iron-fortified cereal. Rice cereal, or other single-grain cereals, can be mixed with human milk or infant formula to give it a smooth, soft consistency. At first, some infants find it difficult to consume food from a spoon. However, in time most infants become quite skilled at it. Once spoon-feeding is well established, the consistency of the food can be made thicker and more challenging. Once an infant is accustomed to cereal, pureed vegetables, fruits, and meats can be introduced. For now, complementary foods are considered "extra," because infants still need regular feedings of human milk and/or formula.

It is important for parents to introduce new foods into the infant's diet gradually. After a new food is given, parents should wait three to four days to make sure the food is tolerated and there are no adverse reactions. Signs and symptoms associated with allergic reactions often include a rash, diarrhea, runny nose, and in more severe cases difficulty breathing. To minimize the risk of developing food allergies, foods with a high allergic potential such as cow milk, egg white, peanut butter, and wheat should be delayed until the infant is 1 year old.

Infants should continue to be fed human milk and/or iron-fortified formula throughout the first year of life. Cow milk, goat milk, and soy milk are not recommended until after a baby's first birthday, primarily because these types of milk are low in iron. Consuming large amounts of milk can displace iron-containing foods, which increases an infant's risk of developing iron deficiency anemia. In addition, many pediatricians caution parents about giving infants

© Myrleen Ferguson Cate/PhotoEdit

**Iron-fortified cereal is most often recommended as a "first food."**

## TABLE 14.6  Developmental Skills and Feeding Practices During Infancy

| Age | Developmental Skills | Appropriate Foods and Feeding Methods |
| --- | --- | --- |
| 0–1 months | Startle reflex, able to suck and swallow | • Human milk and/or iron-fortified formula |
| 2–3 months | Able to support head, interactive, reaches toward objects | • Human milk and/or iron-fortified formula |
| 4–6 months | Able to roll over to back, can sit with support, vocalizing, props self using forearms, grasps objects, able to move tongue from side to side, teething and eruption of upper and lower teeth | • Human milk and/or iron-fortified formula<br>• Slow introduction to iron-fortified infant cereal begins sometime between ages of 4 and 7 months |
| 7–9 months | Begins to sit without support, holds objects, transfers objects between hands, pulls to stand, improved mouth control and can begin to drink from a cup with assistance, can pick up small pieces of food and place them in mouth | • Human milk and/or iron-fortified formula<br>• Iron-fortified infant cereal<br>• 100% fruit juice<br>• Pureed or mashed vegetables and fruits<br>• Soft pieces/mashed food such as meat<br>• Dry cereal |
| 10–12 months | Begins to walk with assistance, crawls, refined grasp, mature chewing, improved ability to drink from a cup, develops skills needed for self-feeding | • Human milk and/or iron-fortified formula<br>• Iron-fortified cereal<br>• Soft pieces of fruits and vegetables<br>• Bread, crackers, and dry cereal<br>• Soft pieces of meat<br>• 100% fruit juice |

SOURCE: American Academy of Pediatrics. Available from: http://www.brightfutures.org/nutrition/pdf/infancy.pdf.

too much fruit juice. Fruit juice was once recommended for infants because it provides a source of vitamin C and additional water. However, current recommendations by the AAP discourage introducing fruit juice into the diet of infants before 6 months of age.[31] It is also recommended that older infants not be given more than 4 to 6 ounces of fruit juice a day. Because fruit juice tastes good, it can lead to overconsumption, which in turn can lead to excess energy intake, dental caries, diarrhea, flatulence, and abdominal cramps.

Older infants (9 to 10 months of age) are more adept at chewing, swallowing, and manipulating food in their hands, and therefore are ready to move on to the next stages of feeding. Table 14.6 provides a list of appropriate foods that can be given to infants at this age. Certain foods should be avoided during the first year of life because they pose a risk for choking and therefore are unsafe. These include such foods as popcorn, peanuts, whole grapes, pieces of hot dogs, and hard candy. In addition, the AAP recommends not feeding honey to infants and young children under 2 years of age, because it can contain spores that cause botulism. Because infants and children are small, even very low exposure to these spores can make them sick. Approximately 250 cases of infant botulism occur in the United States each year.[32] Other sources that can harbor botulism spores are soil, corn syrup, and improperly canned foods.

**CONNECTIONS** Botulism is a serious form of foodborne illness caused by consuming foods that contain the bacterium *Clostridium botulinum* (Nutrition Matters, page 208).

# ESSENTIAL *Concepts*

Infants grow and development rapidly during the first year of life. By 6 months of age, weight has doubled, and by 12 months, it has tripled. The American Academy of Pediatrics recommends exclusively breastfeeding during the first four to six months of life. Although human milk is recommended, infant formula is an alternative derived either from cow milk (cow-based formula) or soybeans (soy-based formula). In communities without fluoridated water, fluoride supplements may be necessary after 6 months of age. Iron supplements are recommended for infants who are exclusively breastfed during the second six months of life. Vitamin D drops are recommended for breastfed infants after 2 months of age. Human milk and/or infant formula should be the primary source of nutrients and energy throughout the first year of life. However, nonmilk complementary foods can begin sometime between 4 to 6 months of age, depending on an infant's readiness. The first solid food often fed to infants is iron-fortified cereal. Older infants should be given a wider variety of foods, although they should be chosen carefully to pose minimal risk for choking.

# Nutrition During Childhood

The life stage group referred to as "childhood" spans the ages of 1 through 8 years. Over this period of time, many physical, cognitive, psychological, and developmental changes take place. It is a time of growing independence as children gain the ability and confidence to function on their own. At this age, children become more opinionated, often expressing their likes and dislikes. Indeed, feeding children can now be challenging for parents. Childhood is also a time when attitudes about food are being formed, and parents play an important role in helping children develop a healthy relationship with food. How parents deal with the challenges of feeding children is very important.

## Childhood Growth and Development

Childhood is divided into stages, which include toddlerhood (ages 1 to 3 years) and early childhood (ages 4 to 8 years). These periods of the life span are characterized by a steady rate of growth, although it is considerably slower than in infancy. Eating habits change and appetite often decreases, a concern for many parents. It is important for parents to provide children with enough nutritious foods for the child to get the nutrients and energy needed for growth and development.

As in infancy, sex-specific growth charts are used to monitor weight and height throughout childhood. In addition, BMI is used to assess weight in children over 2 years of age. Children with BMIs between the 85th and 95th percentile are considered at risk for being overweight, whereas children with BMIs at or above the 95th percentile are considered overweight. At the other end of the spectrum, children at the 5th percentile or below are classified as underweight. Typically, children are considered at risk for undernutrition if they are below the 3rd percentile.

It can be difficult to determine if a child is overweight, because it is sometimes not clear if he or she is actually overweight or simply in a transient part of the growth cycle. However, excess weight for height that persists throughout childhood is of concern. According to recent estimates by the CDC, approximately 15% of children (6 to 11 years of age) and 15% of adolescents (12 to 19 years of age) are at or above the 95th percentile for BMI.[33] This is two to three times higher than 20 years ago. You can read more about the prevalence and health concerns of overweight children in the Focus on Diet and Health feature.

## Feeding Behaviors in Children

"Please, just one bite," begs an anxious parent. This all too familiar plea makes clear that feeding children is not always easy. Some parents are quite surprised at how quickly mild-mannered infants become willful and opinion-

# Focus On DIET AND HEALTH

## Overweight Children: A Growing Concern

The prevalence of children and adolescents in the United States who are overweight is on the rise. Health experts estimate that 15% of children (6 to 11 years of age) and 15% of adolescents (12 to 19 years of age) are at risk of being overweight (between the 85th and 95th percentile), and that another 15% are overweight (>95th percentile). Health conditions once only common in adults are now becoming more common in children and teens. For example, type 2 diabetes in our nation's youth is increasingly prevalent.[34] The American Diabetes Association reports that 8 to 45% of children with newly diagnosed diabetes have type 2 diabetes.[35] Being overweight as a child also increases the risk for cardiovascular disease later in life.

The health and social consequences experienced by overweight children and teens are of great concern to parents and health professionals. Furthermore, excess weight that persists throughout childhood and adolescence is likely to continue into adulthood. Many behaviors contribute to the increasing number of overweight children. However, the behaviors most strongly linked to weight gain in children are unhealthy eating patterns and physical inactivity. For example, watching television has become a national U.S. pastime and has replaced physical activity for many children. It is not surprising to find an association between obesity and the amount of time spent watching television.[36] Over one third of children in the United States watch three or more hours of television each day, and this does not include additional time playing computer and video games.[37]

Of additional concern is the fact that children often consume energy-dense foods while watching television. Many persuasive messages are directed at children encouraging them to consume snack foods. Approximately 90% of food commercials that air on Saturday morning children's television shows are for products with limited nutritional value.[38] Each year, the average American child is exposed to over 10,000 food advertisements on television alone. For these reasons, several health organizations such as the AAP encourage parents to limit the amount of time children spend watching television. It is recommended that children younger than 2 years of age watch no television at all, and that parents limit the number of hours older children view television to two hours per day.[39]

Restricting the number of hours that children watch television does not, by itself, prevent weight gain or facilitate weight loss. Children need to be physically active as well. It is important that parents and schools create an environment that encourages children to exercise. Unfortunately, some states have relaxed physical education requirements in public schools. Although the CDC recommends that children participate in 30 minutes of physical education class daily, only 6 to 8% of public schools meet this guideline. Based on a report released from the National Governors' Association, 25% of our nation's youth receive no physical education in school at all.[40]

In addition to exercise, providing healthy food choices at home and school is vitally important for children and adolescents. In the past, school cafeterias only served meals that met federal guidelines for nutritional standards. Today, many schools sell foods and beverages that are not part of the federal school meal program. Parents may also be surprised to discover the number of school cafeterias around the country operated by national fast-food chains. Money generated from selling "competitive foods" are often used to help support school-related activities and programs. A national survey revealed that over 70% of high schools have vending machines that sell snacks of poor nutritional quality such as soft drinks, candy, chips, and cookies.[41] In fact, very few vending machines are stocked with nutritional snacks such as milk or fruit. Some studies show that children consume, on average, 50 more cans of soda per year in schools where vending machines are available, compared to schools without vending machines.[42]

Preventing children from becoming overweight will take considerable effort at home and at school. Regardless of the many pressures and enticements, teaching children about the importance of physical activity, good nutrition, and providing nutritious foods are important steps in keeping them healthy. After all, there are as many healthy food choices available today as there are unhealthy ones.

ated toddlers. How parents respond to these feeding challenges can determine whether these behaviors persist or fade. Although forcing children to eat is never recommended, it is sometimes difficult for parents to remain calm when a child refuses to eat. Most experts agree that the child, not the parent, should be the one who determines if he or she eats, and how much.[43] Rather, it is the parent's job to make mealtime a pleasant experience and to provide nutritious, age-appropriate foods to choose from. Following some general guidelines for handling common childhood feeding problems can encourage healthy eating.

Some of these guidelines are listed below.

- *Avoid using food to control behavior.* Most experts agree that using food for rewards and/or punishments is not a good idea. Although this may modify behaviors in the short term, it will likely create food issues down the road. For example, giving a child a dessert to reward "good" behavior may establish a connection between sweet foods and approval. Rather, parents should teach children that food is pleasurable and nourishing and not something to turn to for approval, emotional comfort, or punishment.

- *Model good eating habits.* It is important for parents to serve as good role models. Studies show that if parents and/or siblings enjoy a particular food, the child will be more likely to enjoy it too.[44]

- *Use mealtime for family time.* Mealtime provides a time for the entire family to be together and share quality time. Studies show that families that eat meals together on a regular basis are more likely to have children who eat healthier diets than families that do not.[45]

- *Be patient.* Food preferences in children do not always appear rational to adults. That is, children often judge foods as acceptable or unacceptable based on attributes such as color, texture, and appearance rather than taste. Also, children likely experience strong flavors such as onions and certain spices more intensely than adults. Food likes and dislikes change over time, and with patience and encouragement, children are often willing to broaden their food preferences.

- *Encourage new foods.* Making new foods familiar to children is an important first step in food acceptance. This can sometimes be achieved by allowing children to help select and prepare new foods. Not only does this make a new food more familiar, it may also stimulate interest. It is important for parents to remember that accepting new foods takes time. Pressuring children to eat or to try new foods is not recommended.

- *Understand the importance of food rituals.* It is not uncommon for children to develop unusual food rituals. For example, some children become upset if certain foods are touching each other or if a sandwich is cut a certain way. Food rituals are a response to children's need for consistency and familiarity. A **food jag** is a type of food ritual in which the child wants to eat only one specific food everyday for an extended period of time. Parents often become concerned when this occurs. However, this common childhood behavior is usually temporary. Most experts agree that within reason, food rituals and food jags should be respected.[46]

- *Encourage nutritious snacking.* Children have small stomachs and therefore need to eat smaller portions than adults. They also need to eat more frequently. In fact, limiting food consumption to three meals a day is very difficult for children. This is why nutritious between-meal snacks are recommended. Although frequent snacking means that children are often not hungry at mealtime, this is not usually a problem as long as healthy, nutritious snacks are offered.

- *Encourage self-regulation.* It is important that children learn to self-regulate food intake based on internal cues of hunger and satiety. For this reason, serving sizes of food need to be age appropriate, allowing children to ask for more if desired. Although children often claim to be too "full" to eat certain foods, it is not recommended that they be forced to eat all the food on their plate. Rather than forcing children to eat, parents need to set limits. For example, if a child declines to eat his or her meal and then asks for dessert, it is reasonable to say no.

**CONNECTIONS** Recall that hunger is the physical desire for food, whereas satiety is the physical state when hunger has been satisfied (Chapter 9, page 322).

**food jag** In young children, a desire for a particular food.

## Recommended Energy Intakes for Children

During childhood, adequate energy is needed to both support total energy expenditure (TEE) and for the synthesis of new tissues. Thus, Estimated Energy Requirements (EERs) are equal to the sum of TEE plus the energy content of tissues associated with growth. Because the rate of growth slows during childhood, the amount of energy (kcal/day) needed to support growth also decreases. For example, the energy needed to support growth in infants (0 to 3 months of age) is approximately 175 kcal/day. By comparison, the energy needed to support growth in toddlers is only 20 kcal/day. Although energy needed for growth decreases with age, EERs increase. This is because some of the variables that affect TEE (such as age, height, weight, and physical activity level) increase. For example, the EER for a 1-month-old male infant weighing 9.7 pounds is 467 kcal/day. By 35 months of age (weight = 31.3 pounds), the EER increases to 1,184 kcal/day.

## Recommended Nutrient Intakes for Children

By 4 years of age, the AMDRs are close to those for adults. Although the amount of protein needed per kilogram of body weight decreases during childhood, recommended total protein intakes increase, overall. AMDRs and recommended protein intakes for toddlers and children (grams/day and grams/kilogram/day) are listed inside the front cover of this book. The RDA for protein for toddlers and children is 0.95 g of protein per kg of body weight per day. It is especially important to provide high-quality protein during this time, such as meat, yogurt, cheese, and eggs.

Although there are no RDAs for total fat for toddlers and young children, adequate fat intake (30 to 40% of energy for children 1 to 3 years of age and 25 to 35% for children 4 to 8 years of age) is important for growth and development. It is particularly important for children to consume adequate amounts of the essential fatty acids linoleic acid and linolenic acid. Carbohydrates are needed to provide the brain with glucose. After 1 year of age, the amount of glucose used by brain is about the same as adults. For this reason, carbohydrate recommendations for children and adults are the same. As for adults, dietary fiber is important for children's health. The AI for dietary fiber is 19 g/day for toddlers between 1 and 3 years of age. This increases to 25 g/day for young children between 4 and 8 years of age. By comparison, it is recommended that adults consume 25 to 38 g of dietary fiber/day.

### Micronutrients of Special Interest

Because height steadily increases throughout childhood, foods that provide nutrients related to bone health are particularly important during this time. For example, the AI for calcium increases from 500 mg/day for toddlers to 800 mg/day for young children. Milk and other dairy products are good sources of calcium. In fact, it would take 4 cups of broccoli to provide the same amount of calcium as 1 cup of milk. For this reason, the Dietary Guidelines for Americans recommends that children between the ages of 1 and 8 years consume 2 cups of low-fat milk or equivalent milk products each day. In general, 1 cup yogurt, 1½ ounces of natural cheese, or 2 ounces of processed cheese is equivalent to 1 cup of milk. Many children do not meet recommended intakes for calcium. Therefore calcium is a nutrient of concern in this population group.

CONNECTIONS ▶ Recall that an iron deficiency affects the ability to synthesize hemoglobin, resulting in anemia (Chapter 12, page 490).

Iron is another nutrient that is important for growing children. Toddlers and young children require 7 and 10 mg, respectively, of iron per day to meet their needs. Iron deficiency is one of the more common nutritional problems in childhood. Children who drink large amounts of milk and eat a limited variety of other foods are at greatest risk. A lack of iron-rich foods can lead to the development of iron deficiency anemia, causing children to become irritable and inattentive, and to have decreased appetites. To prevent iron deficiency, parents are encouraged to feed their children a variety of iron-rich foods such as meats, fish, poultry, eggs, legumes (peas and beans), enriched cereal products, whole-grain bread products, and iron-fortified foods.

Because children have unique eating patterns and nutritional needs, it is important to provide a variety of nutrient-dense foods. Table 14.7 surveys the amount of food from each food group needed to meet recommended nutrient and energy intakes based on the MyPyramid food guidance system. In addition to emphasizing a healthy diet, the Dietary Guidelines for Americans stresses the importance of regular physical activity to promote physical health and psychological well-being. Parents are encouraged to make sure that children engage in at least 60 minutes of physical activity daily or most days of the week.

## ESSENTIAL *Concepts*

Childhood is a stage of the life span characterized by growth and development. Growth is monitored using sex-specific growth charts. Children between the 85th and 95th percentile based on BMI-for-age are considered at risk for being overweight, and those at or above the 95th percentile are considered overweight. Children below the 5th percentile are considered underweight. Excess weight that persists throughout childhood is of concern and increases the risk for developing weight-related health problems. Developmental changes

## TABLE 14.7 Recommendations for Children Based on the MyPyramid Food Guidance System

| Food Group | Age: 3 years<br>Sex: Male<br>Physical Activity: 30–60 min./day<br>Estimated Energy Requirement:<br>1,400 kcal/day | Age: 3 years<br>Sex: Female<br>Physical Activity: 30–60 min./day<br>Estimated Energy Requirement:<br>1,200 kcal/day | Recommended Food Choices |
|---|---|---|---|
| Grains | 5-oz. equivalent | 4-oz. equivalent | Foods made from wheat, rice, oats, such as bread, pasta, oatmeal, breakfast cereals, and tortillas |
| Vegetables | 1½ cups | 2 cups | Includes all fresh, frozen, canned, and dried vegetables and vegetable juices |
| Fruits | 1½ cups | 1 cup | Includes all fresh, frozen, canned fruits, dried fruits, and fruit juices |
| Milk | 2 cups | 2 cups | Includes all fluid milk and foods made from milk such as yogurt and cheese |
| Meat and beans | 4-oz. equivalent | 3-oz. equivalent | Includes 1 oz. lean meat, poultry, or fish, 1 egg. Meat equivalents: 1 tbsp. peanut butter, ¼ cup cooked legumes, or ½ oz. of nuts or seeds |

SOURCE: United States Department of Agriculture. Available from: http://www.MyPyramid.gov.

can make feeding children challenging for parents. Parents should make mealtime pleasant and provide age-appropriate, healthy food choices. A diet that provides adequate nutrients and energy is important throughout childhood. Nutrients needed for bone health (such as calcium) and to support growth (such as iron) are particularly important. The Dietary Guidelines for Americans 2005 stress the importance of providing children with nutrient-dense foods and regular physical activity.

# Adolescence

Toward the end of childhood, hormonal changes begin to transform a child into an adolescent, marking the beginning of profound physical growth and psychological development. Hormones are responsible for changes in height, weight, skeletal mass, and body composition. In females, the onset of menstruation, known as **menarche,** also begins around this time. In fact, adolescence is unlike any other time in the human life cycle. As with any stage of growth, nutrition plays an important role, and this is certainly the case as the adolescent matures into a young adult. Unfortunately, this is also a time when unhealthy eating practices may develop. For example, weight dissatisfaction is common among teens and often leads to dieting and other destructive weight-loss behaviors.

## Growth and Development During Adolescence

Adolescence is the bridge between childhood and adulthood, and this transformation is signified by the onset of **puberty.** Defined as the maturation of the reproductive system and the capability of reproduction, puberty is brought on by changes in the concentration of certain hormones. During this time the adolescent begins to experience many physical, psychological, and social changes. Because these changes span a nine-year period, the DRIs for this stage of the life cycle are divided into two phases: 9 to 13 years and 14 to 18 years.

The onset of puberty varies, and adolescents of the same age can differ in terms of physical maturation. For example, some adolescents are "early" developers, whereas others are "late" developers. Consequently, nutritional needs of adolescents may depend more on the stage of physical maturation than chronological age. Also, females tend to experience puberty at an earlier age than do males. In general, puberty begins roughly around age 9 (years) in females and age 11 (years) in males.

Considerable growth takes place during adolescence. Before the onset of puberty, males and females have attained 84% of their adult height. During the adolescent growth spurt, females and males grow 6 and 8 inches in height, respectively. Bone mass also increases, with peak bone mass nearly attained by about 20 years of age. Because bone mass increases only slightly after adolescence, it is especially important that teens consume adequate intakes of nutrients for bone health, such as vitamin D, calcium, and phosphorous.

In addition to linear growth (height), changes in body weight and composition occur during adolescence. Overall, females and males gain an average of 35 and 45 pounds, respectively. However, changes in body composition differ quite a bit between females and males. For example, females experience a decrease in percent lean mass and a relative increase

**CONNECTIONS** Recall that peak bone mass is the maximum amount of minerals found in bones during the life span (Nutrition Matters, page 545).

**menarche** (men – ARCH – e) The first time a female menstruates.

**puberty** Maturation of the reproductive system.

in percent fat mass, whereas males experience an increase in percent lean mass and a relative decrease in percent fat mass. Although increased body fat in females is normal and healthy, it can contribute to weight dissatisfaction, which in turn can lead to unhealthy dieting and caloric restriction. Caloric restriction during adolescence can delay growth, development, and reproductive maturation. The topic of eating disorders and related health problems was addressed in the Nutrition Matters that followed Chapter 9.

The rapid changes in growth that take place during adolescence are accompanied by many developmental changes. For example, the desire for independence can often strain family relationships and lead to rebellious behaviors. Furthermore, there is a strong need to fit in and be accepted by peers, and healthy eating can become a low priority. In fact, peers are often more influential in determining food preferences than family members. Dieting, skipping meals, and increasing consumption of foods away from home are common occurrences among teenagers. However, with maturation can come the recognition that current health behaviors can have lasting impacts on long-term health.

## Nutritional Concerns and Recommendations During Adolescence

The rapid growth and development associated with adolescence increases the body's need for certain nutrients and energy. Table 14.8 provides an overview of the amount of food from each food group needed to meet recommended nutrient and energy intakes based on the MyPyramid food guidance system. An inadequate diet during this stage of life can compromise health and have lasting effects. For example, too little dietary calcium at a time when skeletal growth is increasing can compromise bone health later in life. For this reason, the AI for calcium during adolescence (9 to 18 years) is set at 1,300 mg/day. Getting enough calcium is best achieved by consuming dairy products such as milk, yogurt, and cheese.

Adolescents also require additional iron to support growth. And because of the iron loss associated with menstruation, the RDA for iron in females is higher than that of males (15 and 11 mg/day, respectively) during later adolescence. Reportedly, 10% of teenagers in the United States have impaired iron status or iron deficiency anemia.

Although growth is influenced by many factors, adequate intakes of protein and energy are particularly important. Younger adolescents require more protein per kilogram of body weight than do adults—0.95 g of protein per kilogram of body weight. For 14- to 18-year-olds, the RDA for protein requirements decreases to 0.85 g protein per kilogram of body weight per day. This is equivalent to about 46 and 52 g of protein per day for these age groups, respectively.

Estimated Energy Requirements (EERs) during adolescence take into account energy needed to maintain health, promote optimal growth, and support a desirable level of physical activity. Equations used to determine EERs in adolescents are the same as those for children except for the amount of energy needed for growth. For example, for boys and girls 9 through 18 years of age EER is calculated by summing the total energy expenditure plus 25 kcal/day for growth. Equations used to determine EERs during adolescence are shown in Appendix B.

**TABLE 14.8    Recommendations for Adolescents Based on MyPyramid Food Guidance System**

| Food Group | Age: 12 years<br>Sex: Female<br>Physical Activity:<br>30–60 min./day<br>Estimated Energy<br>Requirement:<br>2,000 kcal/day | Age: 12 years<br>Sex: Male<br>Physical Activity:<br>30–60 min./day<br>Estimated Energy<br>Requirement:<br>2,200 kcal/day | Age: 16 years<br>Sex: Female<br>Physical Activity:<br>30–60 min./day<br>Estimated Energy<br>Requirement:<br>2,000 kcal/day | Age: 16 years<br>Sex: Male<br>Physical Activity:<br>30–60 min./day<br>Estimated Energy<br>Requirement:<br>2,800 kcal/day |
|---|---|---|---|---|
| Grains | 6-oz. equivalent | 7-oz. equivalent | 6-oz. equivalent | 10-oz. equivalent |
| Vegetables | 2½ cups | 3 cups | 2½ cups | 3½ cups |
| Fruits | 2 cups | 2 cups | 2 cups | 2½ cups |
| Milk | 3 cups | 3 cups | 3 cups | 3 cups |
| Meat and beans | 5½-oz. equivalent | 6-oz. equivalent | 5½-oz. equivalent | 7-oz. equivalent |

SOURCE: United States Department of Agriculture. Available from: http://www.MyPyramid.gov.

# ESSENTIAL *Concepts*

Hormonal changes begin the transformation from childhood into adolescence, causing changes in height, weight, and body composition. Adolescence is also the beginning of reproductive maturity. The timing of these changes varies, and therefore adolescents of the same age can differ in terms of physical maturation and nutritional requirements. Linear growth is completed at the end of the adolescent growth spurt, although bone mass continues to increase into early adulthood. Thus it is important that adolescents have adequate intakes of nutrients related to bone health. Changes in body composition during adolescence are different for males and females. Overall, males experience an increase in percent lean mass relative to a decrease in percent body fat, whereas females decrease in percent lean mass relative to percent body fat. Although increased body fat in females is normal and healthy, it can contribute to weight dissatisfaction. The rapid growth and development associated with adolescence increases the body's need for certain nutrients and energy. Estimated Energy Requirements (EERs) take into account both energy expenditure and additional energy needed for growth.

# Adulthood and Age-Related Changes

Although physical maturity is achieved early in adulthood, many productive years still remain ahead. In the United States, life expectancy continues to increase, and people over 85 years old are becoming the fastest growing age group (Figure 14.10).[47] Life expectancy is different from **life span,** which is the maximum number of years an individual in a particular species has remained alive. In humans, both life expectancy and life span have been rising steadily. Undoubtedly there is an upper limit to the human life span, but as of now one of the oldest known people in the world lived 122 years. Although there is little doubt that genes play a major role in determining longevity, lifestyle is important as well.

The population trend dubbed the "graying of America" reflects a shift in the age structure of the United States. Never before have there been so many older people, and never before has life expectancy been so high. These changes

CONNECTIONS▶ Recall that life expectancy is the average number of years a person can expect to live; longevity is life expectancy at birth (Chapter 1, page 21).

**life span** Maximum number of years an individual in a particular species has remained alive.

## FIGURE 14.10 The Aging of the U.S. Population

The increased number of births between 1946 and 1964 resulted in a shift in the age structure of the U.S. population.

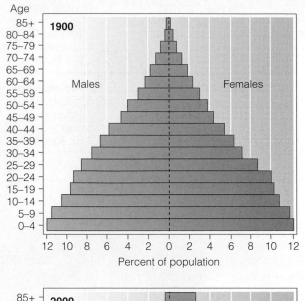

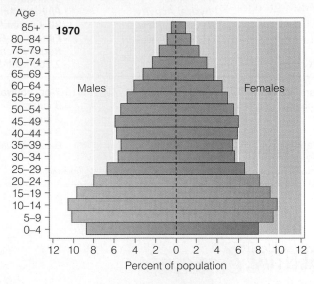

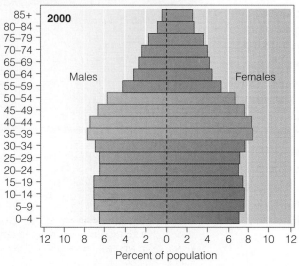

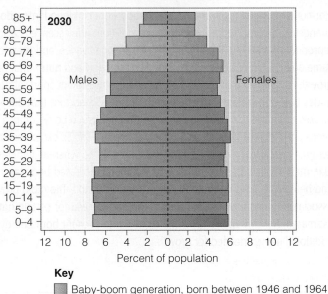

**Key**

Baby-boom generation, born between 1946 and 1964

SOURCE: Adapted from U.S. Census Bureau, Population Division.

© Pascal Parrot/Corbis Sygma

Jeanne Louise Calment (1875–1997), at the age of 122, was one of the oldest humans whose age was fully authenticated. She attributed her longevity to olive oil, port wine, and chocolate. Her genes probably contributed to her longevity; her father lived to the age of 94 years and her mother to 86.

are partly due to advances in medical technology and improved health care. The graying of America is also partly due to the increased number of births in the 1940s, 1950s, and 1960s. People born at this time are often labeled the "baby boom" generation. During the baby boom years many hospitals expanded their obstetric and gynecology units to accommodate the increase in births. Shortly thereafter, hundreds of schools were built throughout the United States to accommodate the growing number of school-age children. Today, there is an increased need for retirement communities and health care practitioners to care for the rising number of older "boomers."

## Nutrition During Adulthood

Adulthood is the period in the life span characterized by physical maturity. However, some males continue to grow in their early 20s, and bone mass increases slightly until age 30 for both males and females. In addition, basal metabolism gradually decreases with age during adulthood, reducing overall requirements for energy and nutrients. For these reasons, adulthood is divided into three groups—young adults (19 to 30 years), middle-aged adults (31 to 50 years), and older adults (>50 years). There is a wide spectrum of health and independence among older adults. Some people maintain active lifestyles, but others are frail and require a great deal of assistance. Functional status therefore is sometimes a better indicator of nutritional needs than chronological age. Because of this, the DRIs for older adults are subdivided further into two groups—51 to 70 years, and over 70 years.

## Theories of Aging

The branch of science and medicine dedicated to the social, behavioral, psychological, and health issues of aging is called **gerontology**. Although aging is inevitable, there is much a person can do to stay physically fit and healthy as long as possible. Target dietary behaviors identified in the national health objectives *Healthy People 2010* that promote healthy aging are summarized in Table 14.9. Gerontologists have many theories as to why we age. Some believe the body deteriorates because of daily wear and tear. Others believe that a combination of lifestyle, environmental, and genetic factors determines how quickly we age.

One of the greatest contributions in understanding biological aging came from experiments performed in the 1960s by the physiologist Leonard Hayflick. Using cell cultures, Hayflick demonstrated that the number of times cells can divide is limited.[48] Thus senescence, or the process of aging, may be programmed in cells, and factors that increase cell replication may speed up this process. For example, damage caused by unstable compounds called free radicals may increase cell replication and thus aging. In addition, free radicals extensively damage proteins, cell membranes, and DNA. Over time, damaged cells become unable to function fully, triggering a cascade of other age-related changes. Free radical damage has been implicated in several degenerative diseases such as cancer, atherosclerosis, and cataracts.[49] Free radicals are a by-product of metabolism, and lifestyle practices and environmental factors such as smoking, exposure to pollutants, and sun exposure increase their production as well.

The evidence that aging and many age-related chronic diseases may, in part, be caused by free radical damage raised hopes that antioxidant nutrients could slow the aging process by protecting the body from oxidative damage.

**CONNECTIONS** Recall that basal metabolism represents the energy expenditure associated with vital body functions such as respiration, beating of the heart, nervous function, and muscle tone (Chapter 9, page 324).

© C. Lyttle/zefa/Corbis

**Many older adults are physically fit and remain active.**

**CONNECTIONS** Recall that free radicals are compounds with unpaired electrons and can damage cell membranes, proteins, and DNA (Chapter 10, page 409).

---

**TABLE 14.9** *Healthy People 2010* **Nutrition Goals for Older People**

Increase the number of older adults who are at a healthy weight.
Decrease the number who are obese (BMI > 30 kg/m²).
Increase the proportion of older adults who
- Eat at least 2 servings of fruit/day.
- Eat at least 3 servings of vegetables/day, one third of which are dark green or deep yellow.
- Eat at least 6 servings of grain products/day, at least 3 of which are whole grains.
- Eat less than 10% of calories from saturated fats.
- Eat no more than 30% of total calories from fat.

SOURCE: *Healthy People 2010: National Health Promotion and Disease Prevention Objectives.* Washington, DC: U.S. Department of Health and Human Services; 2000.

**gerontology** (ger – on – TOL – o – gy) The study of aging.

**CONNECTIONS** Recall that antioxidant nutrients (such as vitamins C and E) prevent unstable free radicals from damaging cells by donating electrons to them (Chapter 10, page 409).

Although animal studies provide some supportive evidence, human data remain controversial.[50] Clinical trials are needed to clarify whether consumption of foods rich in antioxidants or antioxidant supplements can help slow the progression of age-related changes and help people live longer and healthier lives.

## Nutritional Issues of Adults

Although we cannot "turn back the hands of time," there is much adults can do to keep their bodies strong and healthy. Studies show that adults who adopt a healthy lifestyle that includes eating nutritious foods, maintaining a healthy body weight, regular physical activity, and not smoking are less likely to develop chronic diseases such as cardiovascular disease, type 2 diabetes, high blood pressure, and certain types of cancer. During this stage of life, recommended nutrient intakes are intended to reduce the risk of chronic disease while providing adequate amounts of essential nutrients. Table 14.10 provides an overview of the amount of food from each food group needed to meet recommended nutrient and energy intakes based on the MyPyramid food guidance system.

In general, older adults are considered an "at risk" population for developing many nutrition-related health problems. For example, issues related to food insecurity, social isolation, depression, illness, and the use of multiple medications can compromise nutritional status. The topics of hunger and insufficient food availability in the elderly are addressed in more detail in the Nutrition Matters following this chapter. Another reason why older adults are at increased risk for developing nutrition-related health problems is that physiological changes associated with aging can affect nutrient use. These changes are summarized in Table 14.11.

## Changes in Body Composition Affect Energy and Nutrient Requirements

As adults grow older, many experience a loss of lean mass and an increase in fat mass. Although genetics, physical activity, and nutritional status influence these shifts, in general, a 70-year-old male has approximately 22 more pounds of body fat and 24 pounds less muscle than a 20-year-old male.[51] A loss of lean mass, which in turn decreases energy requirements, explains why weight gain is often associated with aging. A key recommendation included in the Dietary

**TABLE 14.10 Recommendations for Older Adults Based on the MyPyramid Food Guidance System**

| Food Groups | Age: 60 years<br>Sex: Female<br>Physical Activity: 30 to 60 min./day<br>Estimated Energy Requirement: 1,800 kcal/day | Age: 60 years<br>Sex: Male<br>Physical Activity: 30 to 60 min./day<br>Estimated Energy Requirement: 2,400 kcal/day |
|---|---|---|
| Grains | 6 oz.-equivalent | 8 oz.-equivalent |
| Vegetables | 2½ cups | 3 cups |
| Fruits | 1½ cups | 2 cups |
| Milk | 3 cups | 3 cups |
| Meat and beans | 5 oz.-equivalent | 6.5 oz.-equivalent |

SOURCE: United States Department of Agriculture. Available from: http://www.MyPyramid.gov.

## TABLE 14.11  Physiological Changes of Aging

**Cardiovascular System**
- Elasticity in blood vessels is lost.
- Cardiac output decreases.
- Blood pressure increases.

**Endocrine System**
- Estrogen and testosterone levels fall.
- Secretion of growth hormone decreases.
- Glucose tolerance is reduced.
- Ability to convert provitamin D to active vitamin D diminishes.

**Gastrointestinal System**
- Secretion of saliva and mucus is reduced.
- Dentition quality declines and teeth are lost.
- Difficulty swallowing sets in.
- Secretion of gastric juice is reduced.
- Peristalsis decreases.
- Vitamin $B_{12}$ absorption is reduced.

**Musculoskeletal System**
- Bone mass is reduced.
- Lean mass is reduced.
- Fat mass increases.
- Resting metabolism decreases.
- Strength, flexibility, and agility are reduced.

**Nervous System**
- Appetite regulation alters.
- Thirst sensation is blunted.
- Ability to smell and taste decreases.
- Sleep patterns change.
- Visual acuity decreases.

**Renal System**
- Blood flow to kidneys diminishes.
- Kidney filtration rate decreases.
- Ability to clear blood of metabolic wastes decreases.

**Respiratory System**
- Respiratory rate decreases.

Guidelines for Americans is for adults to prevent gradual weight gain over time by decreasing energy intake and increasing physical activity.

In general, energy requirements decrease as a person gets older. Equations used to estimate energy requirements are the same for all adults. However, after the age of 30 total energy expenditure (TEE) decreases 7 kcal per year for women and 10 kcal per year for men. For example, all other factors being equal, a 70-year-old woman requires 280 fewer kcal/day than a 30-year-old woman. Likewise, a 70-year-old man requires 400 fewer kcal/day than a 30-year-old man. Of course, energy requirements depend on many factors such as physical activity, weight, and changes in the relative amounts of muscle and body fat.

In addition to weight gain, the loss of muscle mass can make older people less steady, increasing their risk for injury. According to the CDC, more than one third of adults 65 years of age and over experience fall-related injuries annually.[52] Not surprisingly, most head traumas and hip fractures in older adults are caused by falls. Exercise not only decreases fat mass and slows age-related bone loss, but it also helps strengthen muscles and improve coordination.

## Preventing Bone Disease

Age-related bone loss can lead to osteoporosis, which in turn can make bones fragile. Both men and women can develop osteoporosis, but it is by far more common in females. Although many factors influence bone density, it is especially important for older adults to have adequate intakes of nutrients that affect bone health. These include calcium, vitamin D, phosphorus, and magnesium. For adults over 50 years of age, the AI for calcium increases from 1,000 mg to 1,200 mg/day.

Considerable evidence also shows that older adults are at increased risk for vitamin D deficiency. This may in part be caused by limited exposure to sunlight. In addition, the ability to synthesize vitamin D from sunlight decreases as a person ages. For these reasons, the recommended intake of vitamin D increases after the age of 50 years from 5 to 10 μg a day. Recommended intakes of vitamin D are even higher (15 μg/day) for adults over 70 years of

**CONNECTIONS** Recall that osteoporosis is caused by a reduction in bone mass and can lead to fractures and loss of stature in older adults (Chapter 11, page 452; Chapter 13, page 520).

age. Although magnesium and phosphorus are also important for bone health, aging does not seem to affect dietary requirements for these nutrients.

## Hormonal Changes in Women Affect Nutritional Requirements

As women age, they experience a natural and progressive decline in their secretion of the hormone estrogen. This can cause irregular menstrual cycles and a variety of other physical changes. Unfortunately, the decline in estrogen during this **perimenopausal** stage of life increases the rate of bone loss. By the time a woman reaches **menopause,** around 50 to 60 years of age, her ovaries are producing very little estrogen and menstrual cycles completely stop. Declining levels of estrogen cause bone loss to accelerate, making some women's bones weak and fragile. Women who have dense bones before menopause have fewer problems.

Menopause is a normal part of aging and affects each woman differently. Some women experience profound emotional and physiological changes, whereas others experience no discomfort at all. In fact, some women go on hormone replacement therapy to get relief from menopausal symptoms. Although hormone replacement therapy also protects against bone loss, it is associated with increased risk for cancers of the breast and uterus.[53] It is important for women to discuss with their health care providers the benefits and risks associated with hormone replacement therapy.

As an alternative to hormone replacement therapy, some women find that consuming foods made from soybeans is helpful. This is because soybeans contain phytochemicals called **isoflavones** that are structurally similar to estrogen. However, isoflavones do not provide relief from menopausal symptoms in all women, and the reason for this is unclear. The consensus opinion of the North American Menopause Society is that some data support the effectiveness of isoflavones in treating menopausal symptoms, although many studies fail to show an effect.[54]

Because monthly blood loss associated with menstruation has stopped, menopause can improve iron status. In fact, the RDA for iron decreases for women during this life stage, from 18 to 8 mg/day—the same as recommended for adult men. Regardless, adequate iron intakes remain a concern, especially for older adults, who often limit their intake of meat, poultry, and fish.

## Changes in the Gastrointestinal Tract

Age-related changes in the gastrointestinal (GI) tract can also affect nutritional status in adults and especially in older adults. For example, aging muscles can become less responsive to neural signals, and this can slow the movement of food through the GI tract. Decreased GI motility is one reason why constipation is more common in older adults than in younger adults. When fecal material remains in the colon for prolonged periods, too much water resorption can occur, resulting in hard, compacted stool. Drinking adequate amounts of fluids and eating fiber-rich foods can improve GI function and help prevent this problem. Neuromuscular changes in the GI tract can also make swallowing more difficult. Difficulty in swallowing, called **dysphagia,** is a common cause of choking. This frightening experience can cause older adults to avoid eating certain foods. People who have difficulty swallowing should seek medical assistance. Preparing foods so that they are moist and have a soft texture is sometimes helpful.

CONNECTIONS Hormone replacement therapy is taken by women to restore estrogen levels that decline as a result of menopause or the surgical removal of the ovaries (Nutrition Matters, page 547).

**perimenopausal** Hormonal changes that precede menopause.

**menopause** The time in a woman's life when menstruation ceases, usually during the sixth decade of life.

**isoflavone** Compound that is structurally similar to the hormone estrogen; found primarily in soybeans.

**dysphagia** (dys – PHA – gia) Difficulty in swallowing.

With increased age there is also a decline in the number of cells that produce gastric secretions, such as hydrochloric acid and intrinsic factor. This condition, called **atrophic gastritis,** can decrease the bioavailability of nutrients such as calcium, iron, biotin, folate, vitamin $B_{12}$, and zinc. Without intrinsic factor vitamin $B_{12}$ cannot be absorbed, which can lead to pernicious anemia. Symptoms associated with pernicious anemia include dementia, memory loss, irritability, delusions, and personality changes, all of which can easily be overlooked in older adults. It is recommended that older adults take vitamin $B_{12}$ supplements or consume adequate amounts of foods fortified with vitamin $B_{12}$.

**CONNECTIONS** Recall that intrinsic factor is produced by the gastric cells, and is needed for the absorption of vitamin $B_{12}$; vitamin $B_{12}$ deficiency causes pernicious anemia (Chapter 10, page 404).

## Other Nutritional Issues in Older Adults

Other age-related physiological changes can contribute to nutritional deficiencies in older adults. For example, problems with oral health, missing teeth, or poorly fitting dentures can make food less enjoyable, often limiting the types of foods a person eats. Foods that require chewing such as meat, fruits, and vegetables can cause pain, embarrassment, and discomfort for older adults, and therefore tend to be avoided. It is important for older people who are experiencing problems with oral health to get proper dental care.

### Changes in Sensory Stimuli

Sensory changes in taste and smell can also affect food intake in older adults. With age, the ability to smell diminishes, making food tasteless and bland. Certain medications can also alter taste and diminish appetite. Older adults may find that adding spices to foods makes them more appealing, tasteful, and enjoyable. With advanced age, some people also experience a pattern of weight loss caused by a general decline in appetite or what is called **anorexia of aging.** Some health professionals believe this is caused by changes in hormones and neuropeptides that regulate food intake.[55] Anorexia of aging puts individuals at increased risk of protein-energy malnutrition.

### Inadequate Fluid Intake

With age, the sensation of thirst can also become blunted. Because of this, many older adults do not consume enough fluid and are at increased risk for dehydration. Dehydration can upset the balance of electrolytes in cells and tissues. A lack of fluid can also disrupt bowel function and can exacerbate constipation. In addition to inadequate fluid intakes, certain medications can also increase water loss from the body. Some older adults intentionally limit fluid intake because of embarrassment over loss of bladder control. Symptoms of dehydration, which are often overlooked in the elderly population, include headache, dizziness, fatigue, clumsiness, visual disturbances, and confusion. Adequate fluid consumption and early detection of dehydration in older adults is very important.

**CONNECTIONS** Recall that electrolytes dissociate into ions that carry a charge (Chapter 3, page 67).

### Drug–Nutrient Interactions

The elderly are at particularly high risk for adverse drug–nutrient interactions. This is, in part, because older adults often take multiple medications to treat a variety of chronic diseases. The most frequently prescribed medications for people over 50 years of age are for arthritis, hypertension, type 2 diabetes, cancer, and heart disease. The more medications a person takes, the greater the risk of experiencing a drug–nutrient interaction. For example, some drugs can cause loss of appetite, leading to inadequate food intake. Furthermore,

**atrophic gastritis** (gas – TRI – tis) Inflammation of the mucosal membrane that lines the stomach, reducing the number of cells that produce gastric secretions.

**anorexia of aging** Loss of appetite in the elderly that leads to weight loss and overall physiological decline.

some medications can alter taste, making food unpleasant to eat. Even nutrient absorption can be affected by some medications. It is important for elderly people to be aware of such problems and to receive instruction regarding the proper use of medications.

## Assessing Nutritional Risk in Older Adults

Clearly, many factors put older adults at increased nutritional risk. For this reason, several national health organizations jointly sponsored the Nutritional Screening Initiative (NSI), a collaborative effort to improve nutritional health in older adults. The NSI helped to identify risk factors closely associated with poor nutritional status in this group. These risk factors were compiled and used to develop a screening tool called the NSI DETERMINE checklist, shown in Figure 14.11. As you can see, each nutritional risk factor is represented in the DETERMINE acronym. Individuals with a nutritional score of 6 or more on the NSI DETERMINE checklist are considered to be at high nutritional risk.

**FIGURE 14.11  Nutritional Screening Initiative DETERMINE Checklist**

The Nutrition Screening Initiative was an effort by several organizations to develop screening tools to assess risk factors associated with poor nutritional health in older adults.

### DETERMINE YOUR NUTRITIONAL HEALTH

Circle the number in the "yes" column for those that apply to you or someone you know. Total your nutritional score.

| | YES |
|---|---|
| I have had an illness or condition that made me change the kind and/or amount of food I eat. | 2 |
| I eat fewer than 2 meals per day. | 3 |
| I eat few fruits or vegetables or milk products. | 2 |
| I have 3 or more drinks of beer, liquor or wine almost every day. | 2 |
| I have tooth or mouth problems that make it hard for me to eat. | 2 |
| I don't always have enough money to buy the food I need. | 4 |
| I eat alone most of the time. | 1 |
| I take 3 or more different prescribed or over-the-counter drugs a day. | 1 |
| Without wanting to, I have lost or gained 10 pounds in the last 6 months. | 2 |
| I am not always physically able to shop, cook and/or feed myself. | 2 |
| | TOTAL |

**Total Your Nutritional Score. If it's—**

| 0–2 | Good! Recheck your nutritional score in 6 months. |
|---|---|
| 3–5 | You are at moderate nutritional risk. See what can be done to improve your eating habits and lifestyle. |
| 6 or more | You are at high nutritional risk. Bring this Checklist the next time you see your doctor, dietitian or other qualified health or social service professional. |

### DETERMINE: Warning signs of poor nutritional health.

**Disease**
Any disease, illness or chronic condition which causes you to change the way you eat, or makes it hard for you to eat, puts your nutritional health at risk.

**Eating poorly**
Eating too little and eating too much both lead to poor health. Eating the same foods day after day or not eating fruit, vegetables, and milk products daily will also cause poor nutritional health.

**Tooth loss/mouth pain**
A healthy mouth, teeth and gums are needed to eat. Missing, loose or rotten teeth or dentures which don't fit well, or cause mouth sores, make it hard to eat.

**Economic hardship**
As many as 40% of older Americans have incomes of less than $6,000 per year. Having less—or choosing to spend less—than $25–30 per week for food makes it very hard to get the foods you need to stay healthy.

**Reducing social contact**
One-third of all older people live alone. Being with people daily has a positive effect on morale, well-being and eating.

**Multiple medicines**
Many older Americans must take medicines for health reasons. Almost half of older Americans take multiple medicines daily.

**Involuntary weight loss/gain**
Losing or gaining a lot of weight when you are not trying to do so is an important warning sign that must not be ignored. Being overweight or underweight also increases your chance of poor health.

**Needs assistance in self care**
Although most older people are able to eat, one of every five have trouble walking, shopping, buying food and cooking food, especially as they get older.

**Elder years above age 80**
Most older people lead full and productive lives. But as age increases, risk of frailty and health problems increase. Checking your nutritional health regularly makes good sense.

SOURCE: From the American Academy of Family Physicians, available at: http://www.aafp.org/x17367.xml.

Many services help improve overall health for older adults. For example, most communities have congregate meal programs where people can enjoy nutritious, low-cost meals in the company of others. Congregate meal programs are federally subsidized, and cost is based on a person's ability to pay. Another program that provides meals to senior citizens is the Meals on Wheels Association of America. This program, which relies heavily on volunteers, delivers low-cost meals to home-bound elderly adults and to others who are disabled. The mission of this program is to nourish and enrich the lives of people who are home bound and who need services that promote dignity and independent living. The federally funded Food Stamp Program also assists with food-related expenses of low-income adults who may be on fixed incomes.

A Meals on Wheels volunteer delivers meals to a home-bound elderly person.

## ESSENTIAL *Concepts*

The risk of developing diet-related chronic disease increases with age. Adults who remain physically active and maintain a nutritious diet tend to live longer and have fewer health problems. Physiological changes associated with aging can also affect nutritional status. As individuals grow older, they experience a relative loss of lean mass and increase in fat mass. For this reason, energy requirements decrease. The loss of muscle mass can make older people less steady and can increase the risk for injury. Exercise helps maintain a healthy weight, decreases fat mass, slows age-related bone loss, strengthens muscles, and improves coordination. Because age-related bone loss can make bones fragile, it is important for older adults to have adequate intakes of calcium, vitamin D, phosphorus, and magnesium. Many factors can contribute to inadequate food intake in older adults, including poor oral health, altered taste, and decreased ability to smell. Older adults may not be able to readily detect thirst and are at increased risk for dehydration. Age-related changes in the gastrointestinal tract can also affect nutritional status. A decrease in production of gastric secretions can impair absorption of iron, calcium, biotin, folate, vitamin $B_{12}$, and zinc. Of particular concern is the risk of pernicious anemia, which is especially a problem in this group. Services that provide food to older adults include congregate meal programs, meal delivery programs, and Food Stamps.

## Review Questions   Answers are found in Appendix H.

1. Which stage of the life span has the lowest rate of hyperplasia?
   a. Infancy
   b. Adolescence
   c. Pregnancy
   d. Senescence

2. Which of the following represents the correct sequence of prenatal development?
   a. embryo        fetus        zygote
   b. fetus         zygote       embryo
   c. zygote        blastocyst   fetus
   d. blastocyst    fetus        zygote

3. The first trimester for pregnancy corresponds to
   a. the embryonic period.
   b. the embryonic and fetal periods.
   c. the fetal period.
   d. the zygotic period.

4. Weight gain recommendations during pregnancy are based on
   a. prepregnancy BMI.
   b. the ratio of lean mass to fat mass.
   c. waist circumference.
   d. hip circumference.

5. Which of the following groups of women has the lowest recommended weight gain during pregnancy?
   a. Underweight women
   b. Overweight and obese women
   c. Women over 30 years of age
   d. Women between 25 and 30 years of age

6. Low _____ intake around the time of conception can increase the risk of having a child born with a neural tube defect.
   a. calcium
   b. linolenic acid

   **c.** iron

   **d.** folate

**7.** Milk production takes place in specialized structures called _____.

   **a.** alveoli

   **b.** mammary ducts

   **c.** the endometrium

   **d.** milk sinus

**8.** Colostrum is

   **a.** a brand of infant formula.

   **b.** the first milk released from the breast after a baby is born.

   **c.** mucus-like feces that collects in the large intestine in the fetus.

   **d.** a brand of nutrient supplements that are given to infants.

**9.** During which stage of life is menarche most likely to occur?

   **a.** Pregnancy

   **b.** Infancy

   **c.** Adolescence

   **d.** Adulthood

**10.** Calcium requirements during pregnancy are _____ a woman who is not pregnant (assume both women are 25 years of age).

   **a.** higher than

   **b.** lower than

   **c.** the same as that of

**11.** Which of the following changes in body composition is associated with aging in adults?

   **a.** Bone mass increases

   **b.** Lean mass increases

   **c.** Body fat increases

   **d.** All of the above

**12.** Describe growth patterns during infancy, childhood, and adolescence.

**13.** Describe the hormonal regulation of milk production.

**14.** List and describe three physiological changes associated with aging and how these changes can affect nutrient requirements.

**15.** Explain why older adults are considered "at risk" for developing impaired nutritional status.

## Media Links

A variety of study tools for this chapter are available at our website, www.thomsonedu.com/nutrition/mcguire.

Prepare for tests and deepen your understanding of chapter concepts with these online materials:

- Practice tests
- Flashcards
- Glossary
- Student lecture notebook
- Web links
- Animations
- Chapter summaries, learning objectives, and crossword puzzles

## Notes

**1.** Centers for Disease Control and Prevention. National Center for Health Statistics. Clinical Growth Charts. Available from: http://www.cdc.gov/nchs/about/major/nhanes/growthcharts/clinical_charts.htm.

**2.** American Academy of Pediatrics Policy Statement Committee on Fetus and Newborn. Pediatrics. 2004;114:1362–4.

**3.** National Center for Health Statistics. Data on Birth Defects. Available from: http://www.cdc.gov/nchs/data/factsheets/birthdefects.pdf.

**4.** Brent RL. Environmental causes of human congenital malformations: the pediatrician's role in dealing with these complex clinical problems caused by a multiplicity of environmental and genetic factors. Pediatrics. 2004;113:957–68.

**5.** Holmes LB. Teratogen-induced limb defects. American Journal of Medical Genetics. 2002;112:297–303.

**6.** Boyle RJ. Effects of certain prenatal drugs on the fetus and newborn. Pediatrics in Review. 2002;23:17–24.

**7.** Eustace LW, Kang DH, Coombs D. Fetal alcohol syndrome: a growing concern for health care professionals. Journal of Obstetrics, Gynecologic, and Neonatal Nursing. 2003;32:215–21.

**8.** Villar J, Merialdi M, Gulmezoglu AM, Abalos E, Carroli G, Kulier R, de Onis M. Characteristics of randomized controlled trials included in systematic reviews of nutritional interventions reporting maternal morbidity, mortality, preterm delivery, intrauterine growth restriction and small for gestational age and birth weight outcomes. Journal of Nutrition. 2003;133:1632S–9S.

**9.** Barker DJ. The foetal and infant origins of inequalities in health in Britain. Journal of Public Health Medicine. 1991;13:64–8.

10. Barker DJ. Fetal programming of coronary heart disease. Trends in Endocrinology and Metabolism. 2002;13:364–8.

11. Rasmussen KM. The "fetal origins" hypothesis: challenges and opportunities for maternal and child nutrition. Annual Review of Nutrition. 2000;21:73–95.

12. Institute of Medicine/National Academies of Science. Nutrition During Pregnancy. Washington, DC: National Academy Press; 1990.

13. U.S. Department of Health and Human Services and U.S. Environmental Protection Agency. What you need to know about mercury in fish and shellfish. March 2004. Available from: http://www.cfsan.fda.gov/~dms/admehg3.html.

14. Beard JL. Effectiveness and strategies of iron supplementation during pregnancy. American Journal of Clinical Nutrition. 2000;71:1288S–94S.

15. Feinleib M, Beresford SA, Bowman BA, Mills JL, Rader JI, Selhub J, Yetley EA. Folate fortification for the prevention of birth defects: case study. American Journal of Epidemiology. 2001;154:S60–9.

16. Misra DP, Astone N, Lynch CD. Maternal smoking and birth weight: interaction with parity and mother's own in utero exposure to smoking. Epidemiology. 2005;16:288–93.

17. Husgafvel-Pursiainen K. Genotoxicity of environmental tobacco smoke: a review. Mutation Research. 2004;567: 427–45.

18. American Academy of Pediatrics. Breastfeeding and the use of human milk. Pediatrics. 2005;115:496–501.

19. Agostoni C, Giovannini M. Cognitive and visual development: influence of differences in breast and formula fed infants. Nutrition and Health. 2001;15:183–8.

20. Dewey KG. Impact of breastfeeding on maternal nutritional status. Advances in Experimental Medical Biology. 2004;554:91–100. Winkvist A, Rasmussen KM. Impact of lactation on maternal body weight and body composition. Journal of Mammary Gland Biology and Neoplasia. 1999;4:309–18.

21. Taylor JS, Kacmar JE, Nothnagle M, Lawrence RA. A systematic review of the literature associating breastfeeding with type 2 diabetes and gestational diabetes. Journal of the American College of Nutrition. 2005;24:320–6.

22. Ryan AS, Wenjun Z, Acosta A. Breastfeeding continues to increase into the new millennium. Pediatrics. 2002;110:1103–9.

23. American Academy of Pediatrics. Breastfeeding and the use of human milk. Pediatrics. 2005;115:496–501.

24. Lind T, Hernell O, Lonnerdal B, Stenlund H, Domellof M, Persson LA. Dietary iron intake is positively associated with hemoglobin concentration during infancy but not during the second year of life. Journal of Nutrition. 2004;134:1064–70.

25. Gartner LM, Greer FR. Prevention of rickets and vitamin D deficiency: New guidelines for vitamin D intake. Pediatrics. 2003;111:908–10.

26. American Academy of Pediatrics. Breastfeeding and the use of human milk. Pediatrics. 2005;115:496–501.

27. Centers for Disease Control and Prevention. Recommendations to prevent and control iron deficiency in the United States. Morbidity and Mortality Weekly Reports. 1998;47(RR-3):1–36.

28. American Academy of Pediatrics Committee on Nutrition. The use and misuse of fruit juice in pediatrics. Pediatrics. 2001;107:1210–3.

29. American Academy of Pediatrics. Introduction of solid foods. Available from: http://www.aap.org/visit/nutrpolicies.htm.

30. Braganza SF. Food allergy. Pediatrics in Review. 2003;24:393–4.

31. American Academy of Pediatrics Committee on Nutrition. The use and misuse of fruit juice in pediatrics. Pediatrics. 2001;107:1210–3.

32. Tanzi MG, Gabay MP. Association between honey consumption and infant botulism. Pharmacotherapy. 2002;22: 1479–83.

33. Ogden CL, Flegal KM, Carroll MD, Johnson CL. Prevalence and trends in overweight among US children and adolescents. 1999–2000. JAMA (Journal of the American Medical Association). 2002;288:1728–32.

34. Botero D, Wolfsdorf JL. Diabetes mellitus in children and adolescents. Archives of Medical Research. 2005;36:281–90.

35. The American Diabetes Association. Diabetes statistics for youth. Available from: http://www.diabetes.org/diabetes-statistics/children.jsp.

36. Robinson TN. Television viewing and childhood obesity. Pediatric Clinics of North America. 2001;4:1017–25.

37. Crespo CJ, Smit E, Troiano RP, Bartlett SJ, Macera CA, Andersen RE. Television watching, energy intake, and obesity in US children: results from the third National Health and Nutrition Examination Survey, 1988–1994. Archives of Pediatric and Adolescent Medicine. 2001;155:360–5.

38. American Academy of Pediatrics Committee on Public Education. Children, adolescents, and television. Pediatrics. 2001;107:423–6.

39. American Academy of Pediatrics Committee on Public Education. Children, adolescents and television. Pediatrics. 2001;107:423–5.

40. Devlin LM. A call to action. North Carolina Medical Journal. 2002;63:302–3.

41. Institute of Medicine. Schools can play a role in preventing childhood obesity. Preventing childhood obesity: health in the balance. 2005. Available from: http://www.iom.edu.

42. American Academy of Pediatrics. Soft drinks in schools. Pediatrics. 2004;113:152–4.

43. Satter E. The feeding relationship: problems and interventions. Journal of Pediatrics. 1990;117:181–9.

44. Koivisto Hursti UK. Factors influencing children's food choice. Annals of Medicine. 1999;31:26–32.

45. Patrick H, Nicklas TA. A review of family and social determinants of children's eating patterns and diet quality. Journal of American College of Nutrition. 2005;2:83–92.

46. Fuller C, Keller L, Olson J, Plymale A. Helping preschoolers become healthy eaters. Journal of Pediatric Health Care. 2005;19:178–82.

47. Centers for Disease Control and Prevention. The state of aging and health in America. 2004. Available from: http://www.cdc.gov/aging/pdf/state_of_aging_and_health_in_america_2004.pdf.

48. Shay JW, Wright WE. Hayflick, his limit, and cellular ageing. National Review of Molecular Cellular Biology. 2000;1:72–6.

49. Willcox JK, Ash SL, Catignani GL. Antioxidants and prevention of chronic disease. Critical Reviews in Food Science and Nutrition. 2004;44:275–95.

50. Stanner SA, Hughes J, Kelly CN, Buttriss J. A review of the epidemiological evidence for the antioxidant hypothesis. Journal of Public Health Nutrition. 2004;7:407–22.

51. Chernoff R. Geriatric nutrition: The health professional's handbook, 3rd ed. Gaithersburg, MD: Aspen; 2006. Moretti C, Frajese GV, Guccione L, Wannenes F, De Martino MU, Fabbri A, Frajese G. Androgens and body composition in the aging male. Journal of Endocrinological Investigation. 2005;28:56–64.

52. Centers for Disease Control and Prevention. Reducing falls and resulting hip fractures among older women. Available from http://www.cdc.gov/ncipc/factsheets/falls.htm.

53. Beral V, Banks E, Reeves G. Evidence from randomised trials on the long-term effects of hormone replacement therapy. Lancet. 2002;360:942–4.

54. Consensus opinion of The North American Menopause Society. The role of isoflavones in menopausal health. Menopause. 2000;7:215–29.

55. Morley JE. Anorexia of aging: physiologic and pathologic. American Journal of Clinical Nutrition. 1997;66:760–73.

**Check out the following sources for additional information.**

56. Allen LH. Pregnancy and lactation. In: Present knowledge in nutrition, 8th ed. Bowman BA, Russell RM, editors. Washington, DC: ILSI Press; 2001.

57. Abrams SA. Nutritional rickets: an old disease returns. Nutrition Reviews. 2002;60:111–5.

58. Adair LS, Prentice AM. A critical evaluation of the fetal origins hypothesis and its implications for developing countries. Journal of Nutrition. 2004;134:191–3.

59. American Academy releases revised breastfeeding recommendations. Available from: http://www.aap.org/advocacy/releases/feb05breastfeeding.htm.

60. Barker DJP. Growth in utero and coronary heart disease. Nutrition Reviews. 1996;54:S1–S7.

61. Barker D. The midwife, the coincidence, and the hypothesis. British Medical Journal. 2003;327:1428–30.

62. Butte NF, Wong WW, Treuth MS, Ellis KJ, O'Brian Smith E. Energy requirements during pregnancy based on total energy expenditure and energy deposition. American Journal of Clinical Nutrition. 2004;79:1078–87.

63. Chen A, Rogan WJ. Breastfeeding and the risk of postneonatal death in the United States. Pediatrics. 2004;113:435–9.

64. Dewey KG. Nutrition, growth, and complementary feeding of the breastfed infant. Pediatric Clinics of North America. 2001;48:87–104.

65. Dwyer J. A vital sign: progress and prospects in nutrition screening of older Americans. Aging. 1993;5:13–21.

66. Obesity prevention in pediatric primary care. Four behaviors to target. (Editorial.) Archives of Pediatrics Adolescent Medicine. 2003;157:725–7.

67. Gregory K. Update on nutrition for preterm and full-term infants. Journal of Obstetric, Gynecologic, and Neonatal Nursing. 2005;34:98–108.

68. Hayflick L. Living forever and dying in the attempt. Experimental Gerontology. 2003;38:1231–41.

69. Hedley AA, Ogden CL, Johnson CL, Carroll MD, Curtin LR, Flegal KM. Overweight and obesity among US children, adolescents, and adults, 1999–2002. JAMA (Journal of the American Medical Association). 2004;291:2847–50.

70. Heird WC. Nutritional requirements during infancy. In: Present knowledge in nutrition, 8th ed. Bowman BA, Russell RM, editors. Washington, DC: ILSI Press; 2001.

71. Henderson A. Vitamin D and the breastfed infant. Journal of Obstetrics, Gynecologic, and Neonatal Nursing. 2005;34:367–72.

72. Hoelscher DM, Evans A, Parcel GS, Kelder SH. Designing effective nutrition interventions for adolescents. Journal of the American Dietetic Association. 2002;102:S52–63.

73. Jaakkola JJK, Gissler M. Maternal smoking in pregnancy, fetal development, and childhood asthma. American Journal of Public Health. 2004;94:136–41.

74. Kazal LA. Prevention of iron deficiency in infants and toddlers. American Family Physician. 2002;1:1217–24.

75. Kurzer MS. Phytoestrogen supplement use by women. Journal of Nutrition. 2003;133:1983S–6S.

76. Mascarenhas MR, Zemel BS, Tershakovvec AM, Stallings VA. Adolescence. In: Present knowledge in nutrition, 8th ed. Bowman BA, Russell RM, editors. Washington, DC: ILSI Press; 2001.

77. May PA, Gossage JP. Estimating the prevalence of fetal alcohol syndrome. A summary. Alcohol Research and Health. 2001;25:159–67.

78. McConahy KL, Picciano MF. How to grow a healthy toddler—12 to 24 months. Nutrition Today. 2003;38:156–63.

79. Motil KJ. Infant feeding: a critical look at infant formulas. Current Opinion in Pediatrics. 2000;12:469–76.

80. Murray-Kolb L, McConahy KL, Picciano MF, Smiciklas-Wright H, Birch LL, Mitchell DC. Nutritional guidance is needed during dietary transition in early childhood. Pediatrics. 2000;106:109–14.

81. Oddy WH. Breastfeeding protects against illness and infection in infants and children: a review of the evidence. Breastfeeding Reviews. 2001;9:11–8.

82. Ogden CL, Flegal KM, Carroll MD, Johnson CL. Prevalence and trends in overweight among US children and adolescents, 1999–2000. JAMA (Journal of the American Medical Association). 2002;288:1728–32.

83. Ortega RM. Dietary guidelines for pregnant women. Public Health Nutrition. 2001;4:1343–6.

84. Ozanne SE, Hales CN. Early programming of glucose-insulin metabolism. Trends in Endocrinology and Metabolism. 2002;13:368–73.

85. Raiten DJ, Picciano MF. Vitamin D and health in the 21st century: Bone and beyond. Executive summary. American Journal of Clinical Nutrition. 2004;80:1673S–7S.

86. Rajakumar K, Thomas SB. Reemerging nutritional rickets: A historical perspective. Archives of Pediatrics & Adolescent Medicine. 2005;159:335–41.

87. Tollefson L, Cordle F. Methyl mercury in fish: a review of residue levels, fish consumption and regulatory action in the United States. Environmental Health Perspectives 1986;68:203–8.

88. U.S. Department of Health and Human Services and U.S. Department of Agriculture. Dietary Guidelines for Americans 2005. Washington, DC: U.S. Government

Printing Office; 2005. Available from: http://www
.healthierus.gov/dietaryguidelines.

89. U.S. Department of Health and Human Services. Healthy
people 2010: understanding and improving health, 2nd ed.
Washington, DC: U.S. Government Printing Office; 2000.

90. Vellas BJ, Garry PJ. Aging. In: Present knowledge in nutrition, 8th ed. Bowman BA, Russell RM, editors. Washington,
DC: ILSI Press; 2001.

91. Wilson PR, Pugh LC. Promoting nutrition in breastfeeding
women. Journal of Obstetric, Gynecological, and Neonatal
Nursing. 2005;34:120–4.

# Food Insecurity, Hunger, and Malnutrition

*by Edward A. Frongillo and Claire M. Horan*

© Alexandra Winkler/Reuters/Corbis

© Tony Freeman/PhotoEdit

**Food insecurity results when households do not have sufficient food.**

country in the world—even wealthy countries such as the United States.

In this Nutrition Matters, you will learn about concerns related to global nutrition and the problems resulting from food insecurity, hunger, and malnutrition worldwide. You will also learn about domestic and international assistance programs designed to alleviate these devastating public health problems. A framework for understanding factors that contribute to food insecurity and hunger is also presented, as well as possible solutions.

## Overview of Food Insecurity

The term **food insecurity** refers to what households or individuals experience when food availability is limited or when access to food is uncertain. Currently there is enough food produced in the world to feed every person with at least 2,700 kcal per day.[2] In other words, food insecurity is not caused by insufficient food production. Rather, people become food insecure when they cannot grow enough of their own food or cannot purchase adequate amounts of food from other people. Other constraints such as limited physical or mental functioning can also contribute to food insecurity. In reality, the principal factors related to food insecurity in the world are poverty, war, and natural disaster.

*W*e have all experienced hunger—the physical sensation of not having enough to eat—at some time in our life. For most people hunger is easily remedied by eating. However, for some people this is not the case. When sufficient food is not available or accessible (a condition called food insecurity), hunger can result, leading to serious physical, social, and psychological consequences. The prevalence of food insecurity in the world is astonishing. The United Nations Food and Agriculture Organization (FAO) estimates that approximately 800 million people worldwide experience persistent hunger.[1]

Food insecurity and hunger are generally portrayed in the media by showing a very small, crying child with a belly swollen from starvation. Indeed, there are many starving children in the world today. Food insecurity and hunger are not limited to infants and children. Rather, these problems affect people of all ages and in every

**food insecurity** Having limited or uncertain access to adequate amounts of nutritious, safe foods.

Clearly, there are varying degrees of food insecurity, with chronic hunger representing one of the more severe consequences. As noted, the word *hunger* is often used to describe the physical discomfort experienced by individuals consuming insufficient food. However, on a global level the term *hunger* is more commonly used to describe a shortage of food. Because many people live with uncertainty as to whether they will have enough to eat, food insecurity and its resulting hunger are major social concerns in the world today.

## Effects of Food Insecurity

People living in food-insecure households respond to the threat of hunger in many different ways. Whereas some individuals take advantage of charitable organizations that assist people in need, others may resort to finding food by stealing, begging, or scavenging.[3] For many people, food insecurity can cause feelings of alienation, deprivation, and distress. In addition, it can adversely affect family dynamics and social interactions within the larger context of community. Of course, food insecurity that results in hunger can also lead to malnutrition and its associated health complications. Clearly, the short- and long-term consequences of food insecurity for each person in a household can be devastating.

## ESSENTIAL *Concepts*

Food insecurity occurs when individuals or households have limited or uncertain access to sufficient amounts of food. People in all stages of life and in many regions of the world experience food insecurity. In its more severe state, food insecurity can lead to hunger and malnutrition. Less severe food insecurity can adversely affect many other aspects of a person's life.

## Food Insecurity in the United States

Measuring food insecurity is a scientific challenge. Only when the extent and nature of the problem have been determined can effective strategies be developed and targeted to the appropriate populations. Just as there is no one cause of food insecurity, there is no one solution to the problem.

In wealthy countries, the prevalence of food insecurity is often difficult to assess because it is typically not associated with detectable signs of malnutrition. For this reason, clinical measurements of nutritional status such as anthropometry (for example, weight and height) are not always useful indicators of food insecurity. In the United States and other wealthy countries, there is a low occurrence of malnutrition in children because of lack of food. Most cases of malnutrition in this age group are not directly attributed to food insecurity but instead to poor parental care such as neglect or abuse.[4]

Because food insecurity is often not related to overt malnutrition in the United States, it is somewhat difficult to measure. Instead, health organizations collect and analyze data regarding food availability and access. The prevalence and incidence of food insecurity in U.S. households are estimated from this information.

## Extent of Food Insecurity in the United States

A survey developed by the U.S. Department of Agriculture (USDA) is used to assess the prevalence of food insecurity in the United States.[5] Household members are asked questions about behaviors related to food availability and access (Table 1). Depending on responses, a person is classified as being food secure, food insecure without hunger, food insecure with moderate hunger, or food insecure with severe hunger. Using this survey, researchers have estimated that 12 million people in the United States (11% of all households) cannot buy adequate amounts of food—at least some of the time. Of these, approximately 4 million households (3.5%) experience food insecurity with hunger. Alarmingly, the prevalence of food insecurity in the United States appears to be on the rise.

### TABLE 1    Challenges Faced by Food-Insecure Households

The Economic Research Service of the USDA surveys households to determine the presence and severity of food insecurity. The survey tracks the extent to which food-insecure households face the following challenges.

- Food supplies are exhausted before money is available to purchase more, or worry exists that food will run out before there is money to buy more.
- There is not enough money to purchase food for balanced meals.
- The size of meals is reduced, or meals are skipped, because of a lack of money to buy food.
- Adults and children may experience hunger, because there is not enough money to buy food.
- Adults and children may experience weight loss because there is not enough money to buy food.
- Adults and children may go without food for an entire day because there is not enough money to buy food.
- Children may be fed a limited variety of low-cost foods because there is not enough money to buy food.

SOURCE: U.S. Department of Agriculture. Available from: http://www.ers.usda.gov/publications/fanrr29/.

## Risk Factors for Food Insecurity and Hunger in the United States

Risk for experiencing food insecurity in the United States is associated with income, ethnicity, family structure, and location of the home.[6] Indeed, many of these risk factors are intertwined, making it difficult to determine how much each one alone contributes to food insecurity. Nonetheless, the link between income—specifically poverty—and food insecurity is indisputable, and is often the "common thread" among these factors. It is important to recognize that many people who live in poverty maintain steady employment. Statistics from the U.S. Department of Labor indicate that of the 31 million people who live below the poverty line, approximately 6 million are classified as the "working poor"—defined as having worked in the labor force for at least 27 weeks at the point of assessment with incomes below the official poverty level.[7] Many of the working poor have difficulties meeting monthly living expenses.

When money is limited, people are often forced to reduce food-related expenses in order to pay for such things as prescription medications, utilities, or health care.[8] It is not surprising to learn that one in three households living below the poverty line is food insecure. However, it is important to point out that not all households living below the poverty line are food insecure, and some households with incomes above the poverty line experience food insecurity. This is because an unexpected event such as a medical expense or a repair bill can cause some people, at least temporarily, to cross the line into poverty. Thus, many factors in addition to income can predispose a person or family to food insecurity.

In addition to income, food insecurity in the United States is more prevalent among certain ethnic groups. For example, black and Latino households are at higher risk for food insecurity than other ethnic groups. Households headed by a single woman are at even higher risk. In addition, people living in urban and rural areas are more likely to experience food insecurity than those living in suburban regions. However, it is important to recognize that all three of these factors (ethnicity, head of household, and location) are strongly associated with income, and thus poverty.[9]

## Consequences of Food Insecurity in the United States

Food insecurity in the United States does not typically lead to hunger or nutrient deficiencies. However, food insecurity still represents a major public health concern. There are many consequences of food insecurity, and these have been studied most extensively in women and children. These studies show that women often experience the negative consequences of food insecurity before their children do.[10] This is because mothers often protect their children from not having enough to eat by themselves consuming less food. For example, some studies have found that women in food-insecure households eat fewer fruits and vegetables themselves in order to better provide for their children.[11] Nonetheless, it is the children raised in food-insecure households that experience the most significant and lasting, long-term effects. For example, studies show that these children tend to have difficulties in school, lower scores on standardized tests, miss more days of school, exhibit more behavioral problems and depression, and have increased risk for suicide.[12]

It is also important to understand that the elderly are at increased risk for experiencing food insecurity.[13] However, they often experience it differently from both children and other adults. Elders with limited mobility and poor health, for example, may have food available to them, but have difficulty or anxiety associated with meal preparation. In addition, many older people have fixed incomes. This limited amount of money is often needed to pay for expenses associated with daily living. By spending less money on food, elders can lower their monthly expenses and have more money available for these important necessities. Not surprisingly, the latest USDA survey (2002) shows that the prevalence of food insecurity among the elderly is on the rise.[14]

## Providing Food-Based Assistance in the United States

In the United States, many programs and services are in place to alleviate food insecurity. Some of these are federally funded, whereas others are community efforts staffed by volunteers. A summary of these and other food assistance programs is provided in Table 2 and discussed briefly here.

One example of a food-based assistance program is the **Food Stamp Program**, which is often the first line of defense against hunger for low-income households in the United States. This program, administered by the USDA, serves over 20 million people per year. People who are eligible for the Food Stamp Program are given a monthly allotment of coupons or debit cards that are used to make food purchases. A household's monthly allotment depends on the number of people in the household and their income. On average, food stamp

**Food Stamp Program** A federally funded program that improves food availability to low-income households by providing coupons for the purchase of certain foods.

**TABLE 2    Examples of Food Assistance Programs in the United States**

| Program | Description and Website |
| --- | --- |
| Child and Adult Care Food Program (CACFP) | • Provides families with affordable, quality day care and nutrition for children and elderly adults.<br>• http://www.fns.usda.gov/cnd/CARE/ |
| Expanded Food and Nutrition Education Program (EFNEP) | • Assists low-income people to acquire knowledge, skills, attitudes, and behaviors necessary to maintain nutritionally balanced diets and to improve family health and nutritional well-being.<br>• http://www.csrees.usda.gov/nea/food/efnep/efnep.html |
| Food Stamp Program | • Enables low-income families to buy foods with coupons or debit cards at authorized grocery stores or farmers markets.<br>• http://www.fns.usda.gov/fsp/ |
| Head Start | • Provides nutrition and resources for children in low-income families. Head Start serves the nutritional and developmental needs of newborns and children through 5 years of age.<br>• http://www.acf.hhs.gov/programs/hsb/ |
| Meals on Wheels Association of America (MOWAA) | • Delivers meals to people who are elderly, homebound, disabled, frail, or at risk of malnutrition.<br>• http://www.mowaa.org/ |
| National School Lunch and School Breakfast Programs | • Provides low-income children with nutritious meals for free or at reduced cost.<br>• http://www.fns.usda.gov/cnd/Default.htm |
| Special Supplemental Nutrition Program for Women, Infants, and Children (WIC) | • Provides coupons to purchase nutritious foods, to low-income women, infants, and children who are at nutritional risk. WIC also provides nutrition education concerning breastfeeding, healthy food purchasing, and meal planning.<br>• http://www.fns.usda.gov/wic/ |

recipients receive about $80 worth of benefits per person per month. Food stamps can only be used to purchase food and cannot be used to buy tobacco, alcohol, paper products, or other nonfood items. Unfortunately, only half of the families that qualify for food stamps actually receive them.[15]

In addition to the Food Stamp Program, millions of pregnant (and lactating) women, infants, and children in the United States benefit from the **Special Supplemental Nutrition Program for Women, Infants,**

© Ken Hammond/USDA

WIC provides many important services to women, infants, and children.

**and Children,** known as WIC. This federally funded program, which is also administered through the USDA, assists families in making nutritious food purchases by providing coupons that are used to purchase a variety of WIC-approved foods in grocery stores. These foods are nutrient-dense, high-protein foods such items as peanut butter, milk, rice, beans, cereal, and canned tuna. Many farmers' markets also accept WIC coupons, allowing people to purchase a variety of locally grown fresh fruits and vegetables. In addition to assisting with food purchases, WIC provides health assistance and nutrition education to eligible pregnant women and mothers of young infants and children.

Other food-based assistance programs available in the Unites States are the **National School Lunch Program** and the **School Breakfast Program,** which provide nutritionally balanced meals to school-age children each school day, either free of charge or at a reduced cost. These programs, administered by the USDA,

**Special Supplemental Nutrition Program for Women, Infants, and Children (WIC)** A federally funded health and nutrition program that provides assistance to low income women, infants, and children.

**National School Lunch** and **School Breakfast Programs** Federally funded programs that provide free or subsidized nutritious meals to school-age children living in low-income households.

are available in all public schools, nonprofit private schools, residential child-care institutions, and after-school enrichment programs. The National School Lunch and School Breakfast Programs make it possible for many children living in food-insecure households to receive nutritious meals each school day. Because there is such a need for this service, many schools also provide breakfast, lunch, and snacks to children throughout summer vacation.

In addition to these and other government-funded programs, the private sector also provides services to make food more available to those in need. These non-profit organizations include food recovery programs, food banks and pantries, and food kitchens—many of which are staffed by members of the community. **Food recovery programs** work to collect and distribute food that otherwise would be wasted or discarded. The USDA estimates that more than 96 billion pounds of edible food are lost per year by food retailers and producers.[16] Food recovery programs such as that run by the America's Second Harvest organization work to minimize this loss by several methods, including the following.

- Field gleaning—gathering crops from farmers' fields that remain after harvest
- Perishable food rescue or salvage—collection of perishable produce from wholesale and retail sources such as grocery stores
- Food rescue—collection of prepared foods from restaurants, hotels, and caterers
- Nonperishable food collection—collection, from retail sources, of processed foods that have long shelf lives

Recovered foods are donated to food pantries, emergency kitchens, and homeless shelters, providing millions of people with food that would normally go to waste.

© Grant Halverson/AP Photos

Food recovery programs collect and distribute food that would otherwise go to waste. These volunteers help to bag more than 47,000 pounds of potatoes as part of the National Food Recovery Summit. The potatoes, which will provide more than 120,000 meals, are then delivered to food banks, churches, and the needy.

In addition to food recovery programs, community food banks, food pantries, and food kitchens also provide food to those in need. **Food banks** and **pantries** depend on community donations to stock their shelves with non-perishable items, which are then distributed to people. **Food kitchens** serve prepared meals to members of the community, but mostly to those who are homeless or living in shelters.

## ESSENTIAL *Concepts*

Food insecurity in the United States does not typically lead to hunger or nutrient deficiencies. Nonetheless, it is estimated that 11% of Americans experience food insecurity at some time during a year. Ethnicity, family structure, and location can all influence risk for becoming food insecure. Specifically, being black or Latino, being in a household headed by a single woman, and being of low income all increase risk. Children who experience food insecurity have more behavioral problems and do more poorly in school. The elderly are also at risk for food insecurity. There are many food assistance programs to help at-risk individuals obtain sufficient amounts of food. These include the Food Stamp Program, the Special Supplemental Nutrition Program for Women, Infants, and Children (WIC), the School Breakfast and School Lunch Programs, food recovery programs, and community food banks and kitchens.

## World Hunger and Malnutrition

Because poverty is more prevalent in underdeveloped countries, food insecurity tends to be widespread and severe in these regions of the world. Researchers estimate that about 18% of people in poor countries do not have enough to eat.[17] This amounts to nearly 800 million hungry people in the world, most of whom live in Asia. Although Asia has the largest *number* of hungry people, sub-Saharan Africa has the highest *percentage* of people who experience inadequate food intake—where one in every three people does not have enough food to eat. Causes of global food insecurity are complex. However, its consequences are often hunger, malnutrition, and other adverse societal outcomes.

**food recovery program** Organization that collects and distributes food that would otherwise be discarded.

**food bank** Agency that collects donated foods and distributes them to local food pantries, shelters, and soup kitchens.

**food pantry** Program that provides canned, boxed, and sometimes fresh foods directly to individuals in need.

**food kitchen** Program that prepares and serves meals to those in need.

## Causes of and Conditions Related to International Food Insecurity

Causes of food insecurity in poor countries are often different from those seen in the United States. Most experts agree that global food insecurity is not due to lack of available food on the international level. Rather, it is caused by diminished local food supply resulting from a variety of circumstances including famine, political instability, unavailability of land for growing crops, urbanization, population growth, and natural disasters. In addition, many poor countries also lack clean water—contributing further to poor health.

### Political Unrest

Availability of and access to food is often limited by civil strife, wars, and political unrest. Because of this turmoil, millions of people can be displaced from their homes, often relocating to crude facilities set up for refugees. In countries with large refugee populations, such as Sudan, Liberia, Rwanda, and Colombia, malnutrition is rampant. According to the United Nations, those living in refugee camps have the highest rates of disease and malnutrition of any group worldwide.[18] Because of political unrest, relief agencies are not always able to provide the aid that is needed.

### Urbanization and Use of Land

The use of land to produce food for export rather than to feed the region's people and support local economies can also greatly contribute to food shortages. Without land to grow crops, people cannot produce adequate food for themselves and their families. Because they hope to find new employment opportunities elsewhere, this situation often prompts relocation from rural to urban areas. This trend, called **urbanization,** is both a consequence and cause of food shortages in many parts of the world.

### Population Growth

Population growth in some of the poorest regions of the world has increased the challenge of providing adequate food and water to the world's inhabitants. Not surprisingly, countries with the fastest growing populations tend to be those already burdened with staggering rates of hunger and malnutrition.[19] Thus the ability to provide even the most basic of needs—food and shelter—may be compromised further. Because hunger and illness claim the lives of so many children, families living in poverty tend to have numerous children. This is because large families may be viewed as being essential for survival.

## Global Food Insecurity and Malnutrition

Although there are many consequences of food insecurity in poor countries, perhaps the most important is malnutrition. As you have learned, malnutrition is poor nutritional status resulting from inadequate or excessive intake of energy and/or nutrients. In cases of food shortages, malnutrition takes the form of undernutrition, which has both short-term and long-term negative effects on the health of individuals, families, and societies.

### Forms of Global Malnutrition

Malnutrition can take many forms. Whereas some people may consume enough energy but lack certain nutrients, others may lack both. The three most common micronutrient deficiencies in children around the world are iron, iodine, and vitamin A.[20] Approximately 3.5 billion infants and children worldwide suffer from iron and iodine deficiencies, both of which can impair growth and cognitive development. In addition, vitamin A deficiency affects 250 million preschool-age children in the world, causing blindness and many other consequences.[21] You can read more about these nutrients, and related deficiencies, in Chapters 10 through 13.

Protein-energy malnutrition (PEM) can also be a severe consequence of malnutrition. Inadequate intake of food providing protein and energy seriously impacts growth and development in infants and children. PEM makes infants and children less able to resist and fight off disease and infection. You can read more about the complications associated with PEM in Chapter 6.

### Consequences of Malnutrition for Individuals

Infants, children, and women are especially vulnerable to malnutrition. For example, malnutrition during pregnancy can deplete the mother's nutritional stores and increases the risk of having a low birth weight (LBW) baby. Poor maternal nutritional status can also increase risk of neonatal death. To make matters worse, nearly 60% of the deaths of infants and young children in the world are caused, in part, by malnutrition. This is because malnutrition strongly increases the likelihood of death when an infant or young child gets ill from diarrhea, malaria, respiratory infection, or other diseases common in poor countries.

Most infant deaths are not due to severe malnutrition but to mild and moderate forms. Nonetheless, the risk of death is greatly increased for severely malnourished children. For example, when a child is mildly underweight

**urbanization** A shift in a country's population from primarily rural to urban regions.

the risk of death increases 2.5 times. However, the risk of death increases to 8.4 times higher when the child is severely underweight. Because poor nutrition compromises the immune system, malnutrition can make the adverse effects of disease much stronger, leading prematurely to death.[22]

Malnutrition can also have serious consequences for a child's growth and development. As a consequence, 24% of children under 5 years of age—149 million children—are estimated to have **stunted growth.**[23] Children with stunted growth are proportionately small in terms of height and weight. Africa has the highest percentage (35%) of infants and young children with stunted growth, along with Asian, Latin American, and Caribbean countries. Even if the infant survives, stunted growth can compromise a person's well-being and ability to function later in life.

## Consequences of Food Insecurity for Societies

Aside from affecting an individual's health and well-being, malnutrition can harm whole societies. For example, extensive food insecurity and malnutrition can result in an entire nation of adults with reduced capacity for physical work and lower work productivity. These conse-

Children with stunted growth are proportionally small in height and weight.

© Priit J. Vesilind/National Geographic Society

quences can have profound long-term adverse effects on the economic growth and standard of living of a country. In this way, not only is poverty a cause of hunger, but hunger is also a cause of poverty. Therefore, addressing food insecurity and malnutrition is one important way to encourage economic progress in poor countries.

## Providing Food-Based Assistance to the Global Community

Unlike wealthy countries, developing countries often lack stable governments and have few programs in place to assist those in need. These countries typically depend on relief efforts provided by international organizations such as the World Health Organization (WHO), the United Nations Children's Fund (UNICEF), the Food and Agriculture Organization (FAO) of the United Nations, and the U.S. Peace Corps. These organizations try to assist countries in making lasting changes that will ultimately improve health and food security.

You may be wondering how one person can make a difference in the world, especially in terms of helping to alleviate world hunger. An example of how people become inspired to make a difference can be gleaned from John F. Kennedy, who, during a visit to the University of Michigan in 1961, challenged students to serve their country by working to improve the quality of life for others in developing countries. This inspirational speech was the beginning of a federally funded program called the **Peace Corps.** The mission of the Peace Corps is to promote world peace and friendship by the following means.

- Assisting countries in training men and women
- Bringing a better understanding of Americans to people in the countries served
- Bringing to Americans a better understanding of the people served

Today, over 182,000 people have served in the Peace Corps in over 138 different countries. These volunteers work to help others have a better life, by assisting farmers in growing crops, teaching mothers how to better care for their babies, and educating the entire community about health and disease prevention. Thus the Peace Corps offers the opportunity to make a difference in the lives of others, helping address the problems of food insecurity

**stunted growth** Delayed or slow growth resulting from chronic undernutrition.

**Peace Corps** A federally funded program whereby American volunteers live and work with people in underdeveloped countries.

and malnutrition in many parts of the world. Many other programs also offer opportunities for people to gain experience in and make contributions to other countries.

## ESSENTIAL *Concepts*

Food insecurity is more common and severe in nonindustrialized countries than more wealthy nations, often leading to widespread hunger and malnutrition. It is generally agreed that there is sufficient food in the world to provide nutrition to all its inhabitants. Worldwide food insecurity is instead usually due to a combination of poverty, inadequate food distribution, famine, political instability, land availability, urbanization, and natural disasters. Widespread consequences of food insecurity in poorer regions of the world include greater incidence of low birth weight, neonatal death, growth stunting, vitamin A deficiency, iron deficiency, and iodine deficiency. Decreased ability to engage in physical activity can also occur, resulting in a less efficient and productive work force. Although local governments often cannot provide assistance in the form of food and services, there are many international federally funded and private organizations whose goal is to alleviate hunger worldwide.

# Potential Solutions for Global Food Insecurity and Malnutrition

Experts generally agree that there is enough food in the world for everyone to have enough to eat.[24] So why is food insecurity and malnutrition so widespread—especially in the poorest countries? As you have learned, the causes of food insecurity and malnutrition vary by geographic zone, political unrest, national and local economic policy, and population growth. For this reason, it is important for nutritionists and policymakers to appreciate that the causes of food insecurity are varied and intertwined. Only by viewing the complexity of this issue can the relative importance of each contributing factor be addressed and effective solutions sought.

## Toward Alleviating Food Insecurity and Malnutrition

Malnutrition contributes to the suffering of millions of people in the world. In addition, it contributes to more than half of child deaths worldwide. Thus, expert committees agree that improving food availability and access must be a global priority. Such an effort has the potential to prevent approximately 6 million child deaths per year.[25] Although malnutrition is a direct consequence of insufficient dietary intake, its ultimate cause often has more to do with the economic and social circum-

stances of the poor. In addition, health conditions such as the high prevalence of HIV and other diseases can play an important role, as do larger societal problems such as violence and corruption. Thus, to remedy food insecurity and its related hunger, we must remedy these underlying contributing problems.

Many international organizations are committed to alleviating world hunger. One of these is the United Nations (UN), which is a multinational organization first established in 1945 to promote peace through international cooperation and collective security. Today, 191 countries are members of the UN, and combating international hunger is one of its current efforts. For example, the United Nations Children's Fund (UNICEF) has presented a conceptual framework that provides incentives to work toward the common goal of improving the quality of life (including relieving hunger) for people in the world's poorest countries. These incentives include providing financial reimbursement to families who make the effort to have their children attend school, or for starting their own small businesses. Financial incentives would also be given to nations for providing nutritional supplements for infants, children, and pregnant women, or promoting breastfeeding and family planning. Countries that implement these incentives will undoubtedly make significant progress toward improving the health of their people.

A plan called the Millennium Development Goal that has been put forth by the UN has pledged to halve the proportion of people who suffer from hunger in the world by the year 2015.[26] Endorsed by the majority of the countries in the world, this unprecedented effort addresses the needs of the world's poorest countries. To achieve this goal, efforts are currently underway to improve education, promote gender equality, reduce infant mortality, improve maternal health, combat HIV and other infectious diseases, promote sustainable agricultural practices, and develop global economic partnerships. Although this endeavor has been a major challenge requiring commitment and dedication from the worldwide community, its potential for easing the burden of food insecurity, hunger, and malnutrition in the world is substantial.

## ESSENTIAL *Concepts*

Solving the world's food insecurity and malnutrition problems continues to be a major challenge for both local and international communities. There is much interest in finding ways to combat world food insecurity. For example, the UN has proposed programs that provide incentives to improve food availability. These include giving families money if their children attend school, providing funds for families to start small businesses, promoting breastfeeding and family planning, and providing nutritional supplements for infants, children, and pregnant women. Another effort supported by the UN, called the Millennium Development Goal, has pledged to reduce substantially (by half) the proportion of people who suffer from hunger in the world by the year 2015. To achieve this goal, efforts are being made to improve education, promote gender equality, reduce infant mortality, improve maternal health, combat HIV and other infectious diseases, promote sustainable agricultural practices, and develop global partnerships in the world economy.

## Notes

1. Food and Agriculture Organization of the United Nations. The state of food insecurity in the world 2002. Available from: http://www.fao.org/documents/show_cdr.asp?url_file=/docrep/005/y7352e/y7352e00.htm.

2. Food and Agriculture Organization of the United Nations. The state of food insecurity in the world 2002. Available from: http://www.fao.org/documents/show_cdr.asp?url_file=/docrep/005/y7352e/y7352e00.htm.

3. Kempson KM, Palmer Keenan D, Sadani PS, Ridlen S, Scotto Rosato N. Food management practices used by people with limited resources to maintain food sufficiency as reported by nutrition educators. Journal of the American Dietetic Association. 2002;102:1795–79.

4. Kessler DB, Dawson P, editors. Failure to thrive and pediatric undernutrition. Baltimore: Paul H. Brooks; 1999.

5. United States Department of Agriculture. Household food security in the United States, 2002. Economic Research Service Food Assistance and Nutrition Research Report no. FANRR-35, October 2002. Available from: http://www.ers.usda.gov/publications/fanrr35/.

6. Pelletier DL, Olson CM, Frongillo EA. Food insecurity, hunger and undernutrition. In: Present knowledge in nutrition, 8th ed. Bowman B, Russell R, editors. Washington, DC: International Life Sciences Institute; 2001.

7. United States Department of Labor, Bureau of Labor Statistics. A profile of the working poor, 2000. Available from: http://www.bls.gov/cps/cpswp2000.htm.

8. Kim M, Ohls J, Cohen R. Hunger in America, 2001. National report prepared for America's Second Harvest. Princeton, NJ: Mathematica Policy Research Inc.; 2001. Available from: www.mathematica-mpr.com/pdfs/hunger2001.pdf.

9. Furness BW, Simon PA, Wold CM, Asarian-Anderson J. Prevalence and predictors of food insecurity among low-income households in Los Angeles County. Public Health Nutrition. 2004;7:791–4.

10. Olson CM. Nutrition and health outcomes associated with food insecurity and hunger. Journal of Nutrition. 1999;129:521S–4S.

11. Kendall A, Olson CM, Frongillo EA Jr. Relationship of hunger and food insecurity to food availability and consumption. Journal of the American Dietetic Association. 1996;96:1019–24.

12. Alaimo K, Olson CM, Frongillo EA. Family food insufficiency, but not low family income, is positively associated with dysthymia and suicide symptoms in adolescents. Journal of Nutrition. 2002;132:719–25.

13. Lee JS, Frongillo EA Jr. Nutritional and health consequences are associated with food insecurity among U.S. elderly persons. Journal of Nutrition. 2001;131:1503–9.

14. United States Department of Agriculture. Household food security in the United States, 2002. Economic Research Service Food Assistance and Nutrition Research Report no. FANRR-35, October 2002. Available from: http://www.ers.usda.gov/publications/fanrr35/.

15. Zedlewski SR. Leaving welfare often severs families' connections to the Food Stamp Program. Journal of the American Medical Women's Association. 2002;57:23–6.

16. U.S. Department of Agriculture. Food recovery and gleaning initiative: a citizen's guide to food recovery, 1996. Washington, DC: USDA; 1996.

17. Food and Agriculture Organization of the United Nations. The state of food insecurity in the world 2002. Available from: http://www.fao.org/documents/show_cdr.asp?url_file=/docrep/005/y7352e/y7352e00.htm.

18. United Nations Sub-Committee on Nutrition. Reports on the nutrition situation of refugees and displaced populations. 1998. Available from: http://www.unsystem.org/scn/Publications/RNIS/rnis25.pdf.

19. El-Ghannam AR. The global problems of child malnutrition and mortality in different world regions. Journal of Health and Social Policy. 2003;16:1–26.

20. Kapil U, Bhavna A. Adverse effects of poor micronutrient status during childhood and adolescence. Nutrition Reviews. 2002:S84–90.

21. Humphrey JH, West KP Jr, Sommer A. Vitamin A deficiency and attributable mortality among under-5-year-olds. Bulletin of the World Health Organization. 1992:225–32.

22. Caulfield LE, de Onis M, Blossner M, Black RE. Undernutrition as an underlying cause of child deaths associated with diarrhea, pneumonia, malaria, and measles. American Journal of Clinical Nutrition. 2004;80:193–8.

23. Milman A, Frongillo EA, de Onis M, Hwang JY. Differential improvement among countries in child stunting is associated with long-term development and specific interventions. Journal of Nutrition. 2005:135:1415–22.

24. Food and Agriculture Organization of the United Nations. The state of food insecurity in the world 2002. Available from: http://www.fao.org/documents/show_cdr.asp?url_file=/docrep/005/y7352e/y7352e00.htm.

25. Jones G, Steketee R, Bhutta Z, Morris S, the Bellagio Child Survival Study Group. How many child deaths can we prevent this year? Lancet. 2003:362:65–71.

26. UN Millennium Development Project. UN Millennium development goals. Millennium Development Project report 2005. Available from: http://www.un.org/millenniumgoals/.

# Appendixes

# Aids to Calculation

The study of nutrition often requires solving mathematical problems. The three most common calculations are conversions, percentages, and ratios.

## Conversions

It is often necessary to make conversions from one unit of measure to another. For example, pounds are converted to kilograms, inches are converted to centimeters, ounces are converted to grams, etc. Conversion factors convert 1 unit to another without changing the value of the measurement. For example, 2.2 lb. equals 1 kg. Thus the weight does not change, only the units used to express weight. Common conversion factors are listed next.

2.2 lb. = 1 kg
1 oz. = 28 g
1 inch = 2.54 cm

In addition, it is often necessary to determine how many calories (kcal) are in a given amount of food. This calculation is similar to a conversion.

1 g protein = 4 kcal protein
1 g fat = 9 kcal fat
1 g carbohydrate = 4 kcal carbohydrate

### Sample Conversions

***Example 1. Converting weight in lb. to weight in kg and weight in kg to weight in lb.***
To convert 150 lb. to kg, divide by 2.2, like this:

$$150 \text{ lb.} \div 2.2 = 68.2 \text{ kg}$$

To convert 68.2 kg to lb., multiply by 2.2, like this:

$$68.2 \text{ kg} \times 2.2 = 150 \text{ lb.}$$

***Example 2. Converting weight in oz. to weight in g and weight in g to weight in oz.***
To convert 4 oz. to g, multiply by 28, like this:

$$4 \text{ oz.} \times 28 = 112 \text{ g}$$

To convert 112 g to oz., divide 112 by 28, like this:

$$112 \text{ g} \div 28 = 4 \text{ oz.}$$

***Example 3. Converting height in inches to height in cm***
To convert 58 inches to centimeters, multiply 58 by 2.54, like this:

$$58 \text{ inches} \times 2.54 = 147.3 \text{ cm}$$

To convert 147.32 cm to inches, divide 147.32 cm by 2.54, like this:

$$147.3 \text{ cm} \div 2.54 = 58 \text{ inches}$$

***Example 4. Calculating kcal in food from weight in grams and weight in grams to kcal in food***
To calculate how many kcal are in 50 g of protein, multiply 50 by 4, like this:

$$50 \text{ g protein} \times 4 \text{ kcal/g} = 200 \text{ kcal}$$

To calculate how many g of protein it would take to get 200 kcal, divide 200 by 4, like this:

$$200 \text{ kcal protein} \div 4 \text{ kcal/g} = 50 \text{ g protein}$$

## Calculating Percentages

A percentage expresses the contribution of a part to a total. The total is always 100. To calculate a percentage, you must first determine "its part of the total," which is then expressed as a percentage by multiplying by 100. For example, what percentage of this meal's total kcal are from fat?

Total kcal in the meal = 400 kcal
Kcal from fat = 225 kcal

To solve this problem, divide the part (in this case, kcal from fat) by the total kcal (in this case, total kcal in the meal), then multiply by 100 to express the number as a percentage, like this:

$$(225 \text{ fat kcal} \div 400 \text{ total kcal}) \times 100 = 56\%$$

This meal also provides 87 kcal from carbohydrate and 87 kcal from protein. It may also be necessary to calculate percentages of total kcal from fat, carbohydrate, and protein. Calculate the percent of total kcal in this meal from fat, carbohydrate, and protein like this:

225 (fat) kcal ÷ 400 total kcal × 100 = 56%
87 (carbohydrate) kcal ÷ 400 total kcal × 100 = 22%
87 (protein) kcal ÷ 400 total kcal × 100 = 22%

Notice that when the percentages are added together (56% + 22% + 22%), they total 100%.

This type of calculation is very common. However, sometimes it is first necessary to determine how many kcal are in a specified weight of food. For example, nutrient content is often provided in g (50 g fat, 50 g carbohydrate, and 35 g protein). To calculate the percentages of total kcal from fat, carbohydrate, and protein, you must first calculate how many calories there are in these macronutrients.

This problem requires you to

- Calculate the kcal for each nutrient.
- Calculate total kcal in the food.
- Calculate percentages of total kcal from fat, carbohydrate, and protein.

*Step 1.* Calculate the kcal for each nutrient.

50 g fat (9 kcal/g) = 450 kcal from fat

50 g carbohydrate (4 kcal/g) = 200 kcal from carbohydrate

35 g protein (4 kcal/g) = 140 kcal from protein

*Step 2.* Calculate total kcal in the food.

450 kcal from fat + 200 kcal from carbohydrate + 140 kcal from protein = 790 kcal total

*Step 3.* Calculate percentages of total kcal from fat, carbohydrate, and protein.

450 kcal from fat ÷ 790 total kcal × 100 = 57% of kcal from fat

200 kcal from carbohydrate ÷ 790 total kcal × 100 = 25% of kcal from carbohydrate

140 kcal from protein ÷ 790 total kcal × 100 = 18% of kcal from protein

Notice that these percentages (57% + 25% + 18%) add up to 100% of total kcal.

## Ratios

A ratio reflects relative amounts of something by reducing one of the values to 1. Because the units being compared are always the same, a ratio is not expressed in terms of units. For example, suppose a diet provides 50 g of carbohydrate and 100 g of protein. The ratio is 1:2; that is, for every gram of carbohydrate there are 2 g of protein.

Here is another example. A diet provides 3,000 mg of sodium and 2,000 mg of potassium. The ratio here (calculated by dividing mg sodium by mg potassium) is 1.5:1. This means there are 1.5 mg of sodium in this diet for every 1 mg of potassium.

# Estimated Energy Requirement (EER) Calculations and Physical Activity (PA) Values

## Estimated Energy Requirement Calculations

| Age Group | Equations for Estimated Energy Requirement (EER; kcal/d)[a] |
|---|---|
| 0–3 months | $[89 \times \text{weight (kg)} - 100] + 175$ |
| 4–6 months | $[89 \times \text{weight (kg)} - 100] + 56$ |
| 7–12 months | $[89 \times \text{weight (kg)} - 100] + 22$ |
| 13–36 months | $[89 \times \text{weight (kg)} - 100] + 20$ |
| 3–8 years (male) | $88.5 - [61.9 \times \text{age (y)}] + PA \times [26.7 \times \text{weight (kg)} + 903 \times \text{height (m)}] + 20$ |
| 3–8 years (female) | $135.3 - [30.8 \times \text{age (y)}] + PA \times [10.0 \times \text{weight (kg)} + 934 \times \text{height (m)}] + 20$ |
| 9–18 years (male) | $88.5 - [61.9 \times \text{age (y)}] + PA \times [26.7 \times \text{weight (kg)} + 903 \times \text{height (m)}] + 25$ |
| 9–18 years (female) | $135.3 - [30.8 \times \text{age (y)}] + PA \times [10.0 \times \text{weight (kg)} + 934 \times \text{height (m)}] + 25$ |
| 19+ years (male) | $662 - [9.53 \times \text{age (y)}] + PA \times [15.91 \times \text{weight (kg)} + 539.6 \times \text{height (m)}]$ |
| 19+ years (female) | $354 - [6.91 \times \text{age (y)}] + PA \times [9.36 \times \text{weight (kg)} + 726 \times \text{height (m)}]$ |
| **Pregnancy** | |
| 14–18 years | |
|   1st trimester | Adolescent EER + 0 |
|   2nd trimester | Adolescent EER + 340 |
|   3rd trimester | Adolescent EER + 452 |
| 19–50 years | |
|   1st trimester | Adult EER + 0 |
|   2nd trimester | Adult EER + 340 |
|   3rd trimester | Adult EER + 452 |
| **Lactation** | |
| 4–18 years | |
|   1st six months | Adolescent EER + 330 |
|   2nd six months | Adolescent EER + 400 |
| 19–50 years | |
|   1st six months | Adult EER + 330 |
|   2nd six months | Adult EER + 400 |
| **Overweight or Obese[b]** | |
| 3–18 years (male) | $114 - [50.9 \times \text{age (y)}] + PA \times [19.5 \times \text{weight (kg)} + 1161.4 \times \text{height (m)}]$ |
| 3–18 years (female) | $389 - [41.2 \times \text{age (y)}] + PA \times [15.0 \times \text{weight (kg)} + 701.6 \times \text{height (m)}]$ |
| 19+ years (male) | $1086 - [10.1 \times \text{age (y)}] + PA \times [13.7 \times \text{weight (kg)} + 416 \times \text{height (m)}]$ |
| 19+ years (female) | $448 - [7.95 \times \text{age (y)}] + PA \times [11.4 \times \text{weight (kg)} + 619 \times \text{height (m)}]$ |

[a] "PA" stands for the "physical activity" value appropriate for the age and physiological state. These can be found in the next table.
[b] Body mass index (BMI) $\geq$ 25 kg/m²; values represent estimated total energy expenditure (TEE; kcal/d) for weight maintenance; weight loss can be achieved by a reduction in energy intake and/or an increase in energy expenditure.

# Physical Activity (PA) Values

| Age Group (sex) | Physical Activity Level [a] | Physical Activity (PA) Value |
|---|---|---|
| 3–8 years (male) | Sedentary | 1.00 |
| | Low active | 1.13 |
| | Active | 1.26 |
| | Very active | 1.42 |
| 3–8 years (female) | Sedentary | 1.00 |
| | Low active | 1.16 |
| | Active | 1.31 |
| | Very active | 1.56 |
| 3–18 years (overweight male)[b] | Sedentary | 1.00 |
| | Low active | 1.12 |
| | Active | 1.24 |
| | Very active | 1.45 |
| 3–18 years (overweight female)[b] | Sedentary | 1.00 |
| | Low active | 1.18 |
| | Active | 1.35 |
| | Very active | 1.60 |
| 9–18 years (male) | Sedentary | 1.00 |
| | Low active | 1.13 |
| | Active | 1.26 |
| | Very active | 1.42 |
| 9–18 years (female) | Sedentary | 1.00 |
| | Low active | 1.16 |
| | Active | 1.31 |
| | Very active | 1.56 |
| 19+ years (male) | Sedentary | 1.00 |
| | Low active | 1.11 |
| | Active | 1.25 |
| | Very active | 1.48 |
| 19+ years (female) | Sedentary | 1.00 |
| | Low active | 1.12 |
| | Active | 1.27 |
| | Very active | 1.45 |
| 19+ years (overweight/obese male)[b] | Sedentary | 1.00 |
| | Low active | 1.12 |
| | Active | 1.29 |
| | Very active | 1.59 |
| 19+ years (overweight/obese female)[b] | Sedentary | 1.00 |
| | Low active | 1.16 |
| | Active | 1.27 |
| | Very active | 1.44 |

[a] *Sedentary* activity level is characterized by no physical activity aside from that needed for independent living. *Low active* level is characterized by walking 1.5 to 3 miles/day at 2–4 mph (or equivalent) in addition to the light activity associated with typical day-to-day life. People who are *active* walk 3 to 10 miles/day at 2–4 mph (or equivalent) in addition to the light activity associated with typical day-to-day life. *Very active* individuals walk 10 or more miles/day at 2–4 mph (or equivalent) in addition to the light activity associated with typical day-to-day life.

[b] Body mass index (BMI) ≥ 25 kg/m².

SOURCE: Institute of Medicine. *Dietary Reference Intakes for Energy, Carbohydrate, Fiber, Fat, Fatty Acids, Cholesterol, Protein, and Amino Acids (Macronutrients).* Washington, DC: National Academies Press; 2005.

# Summary of the USDA Dietary Guidelines for Americans (2005)

| Guidelines | Key Recommendations | Special Recommendations |
| --- | --- | --- |
| **Adequate Nutrients Within Calorie Needs** | • Consume a variety of nutrient-dense foods and beverages within and among the basic food groups while choosing foods that limit the intake of saturated and *trans* fats, cholesterol, added sugars, salt, and alcohol.<br>• Meet recommended intakes within energy needs by adopting a balanced eating pattern, such as the USDA Food Guide or the DASH Eating Plan. | *People over age 50:* Consume vitamin $B_{12}$ in fortified foods or supplements.<br>*Women of childbearing age who may become pregnant:* Eat foods high in heme iron and/or consume iron-rich plant foods or iron-fortified foods with an enhancer of iron absorption, such as vitamin C–rich foods.<br>*Women of childbearing age who may become pregnant and those in the first trimester of pregnancy:* Consume adequate synthetic folic acid daily (from fortified foods or supplements) in addition to food forms of folate from a varied diet.<br>*Older adults, people with dark skin, and people exposed to insufficient ultraviolet band radiation (i.e., sunlight):* Consume extra vitamin D from vitamin D–fortified foods and/or supplements. |
| **Weight Management** | • To maintain body weight in a healthy range, balance calories from foods and beverages with calories expended.<br>• To prevent gradual weight gain over time, make small decreases in intake of food and beverage calories and increase physical activity. | *Those who need to lose weight:* Aim for a slow, steady weight loss by decreasing calorie intake while maintaining an adequate nutrient intake and increasing physical activity. For overweight children, reduce the rate of body weight gain while allowing growth and development. Those with chronic disease and/or on medication should consult a health care provider prior to starting a weight reduction program.<br>*Pregnant or breastfeeding women:* Ensure appropriate weight gain during pregnancy. Moderate weight reduction is safe during breastfeeding and does not compromise weight gain of the nursing infant. |
| **Physical Activity** | • Engage in regular physical activity and reduce sedentary activities to promote health, psychological well-being, and a healthy body weight.<br>   *Normal-weight adults:* Engage in at least 30 minutes (preferably 60 minutes) of moderate-intensity physical activity, above usual activity, at work or home on most days of the week.<br>   *Overweight adults:* Participate in at least 60 to 90 minutes of daily moderate-intensity physical activity while not exceeding caloric intake requirements.<br>• Achieve physical fitness by including cardiovascular conditioning, stretching exercises for flexibility, and resistance exercises or calisthenics for muscle strength and endurance. | *Children and adolescents:* Engage in at least 60 minutes of physical activity on most, preferably all, days of the week.<br>*Pregnant or breastfeeding women:* In the absence of medical or obstetric complications, pregnant women should incorporate at least 30 minutes of low-risk, moderate-intensity physical activity on most, if not all, days of the week. Be aware that neither acute nor regular exercise adversely affects a woman's ability to successfully breastfeed.<br>*Older adults:* Participate in regular physical activity to reduce functional declines associated with aging and to achieve other benefits of physical activity. |

| Guidelines | Key Recommendations | Special Recommendations |
|---|---|---|
| **Food Groups to Encourage** | • Consume a sufficient amount of fruits and vegetables while staying within energy needs. Two cups of fruit and 2½ cups of vegetables per day are recommended for a reference 2,000-kcal intake, with higher or lower amounts depending on the calorie level. <br> • Choose a variety of fruits and vegetables each day. In particular, select from all five vegetable subgroups (dark green, orange, legumes, starchy vegetables, and other vegetables) several times a week. <br> • Consume 3 or more ounce-equivalents of whole-grain products per day, with the rest of the recommended grains coming from enriched or whole-grain products. In general, at least half the grains should come from whole grains. <br> • Consume 3 cups per day of fat-free or low-fat milk or equivalent milk products. | *Children and adolescents:* Consume whole-grain products often; at least half the grains should be whole grains. Children 2 to 8 years should consume 2 cups per day of fat-free or low-fat milk or equivalent milk products. Children 9 years of age and older should consume 3 cups per day of fat-free or low-fat milk or equivalent milk products. |
| **Fats** | • Consume less than 10% of calories from saturated fatty acids and less than 300 mg/day of cholesterol, and keep *trans* fatty acid consumption as low as possible. <br> • Keep total fat intake between 20 to 35% of calories, with most fats coming from sources of poly- and monounsaturated fatty acids, such as fish, nuts, and vegetable oils. <br> • When selecting and preparing meat, poultry, dry beans, and milk or milk products, make choices that are lean, low fat, or fat free. <br> • Limit intake of fats and oils high in saturated and/or *trans* fatty acids, and choose products low in such fats and oils. | *Children and adolescents:* Keep total fat intake between 30 to 35% of calories for children 2 to 3 years of age and between 25 to 35% of calories for children and adolescents 4 to 18 years of age, with most fats coming from sources of polyunsaturated and monounsaturated fatty acids, such as fish, nuts, and vegetable oils. |
| **Carbohydrates** | • Choose fiber-rich fruits, vegetables, and whole grains often. <br> • Choose and prepare foods and beverages with little added sugars or caloric sweeteners, such as amounts suggested by the USDA Food Guide and the DASH Eating Plan. <br> • Reduce the incidence of dental caries by practicing good oral hygiene and consuming sugar- and starch-containing foods and beverages less frequently. | |
| **Sodium and Potassium** | • Consume less than 2,300 mg (approximately 1 teaspoon of salt) of sodium per day. <br> • Choose and prepare foods with little salt. At the same time, consume potassium-rich foods, such as fruits and vegetables. | *Individuals with hypertension, blacks, and middle-aged and older adults:* Consume no more than 1,500 mg of sodium per day, and meet the potassium recommendation (4,700 mg/day) with food. |
| **Alcoholic Beverages** | • Those who choose to drink alcoholic beverages should do so sensibly and in moderation—defined as the consumption of up to one drink per day for women and up to two drinks per day for men. <br> • Alcoholic beverages should not be consumed by some individuals, including those who cannot restrict their alcohol intake, women of childbearing age who may become pregnant, pregnant and lactating women, children and adolescents, individuals taking medications that can interact with alcohol, and those with specific medical conditions. <br> • Alcoholic beverages should be avoided by individuals engaging in activities that require attention, skill, or coordination. | |

*continued*

| Guidelines | Key Recommendations | Special Recommendations |
|---|---|---|
| **Food Safety** | • Clean hands, food contact surfaces, and fruits and vegetables. Meat and poultry should not be washed or rinsed.<br>• Separate raw, cooked, and ready-to-eat foods while shopping, preparing, or storing foods.<br>• Cook foods to a safe temperature to kill microorganisms.<br>• Chill (refrigerate) perishable food promptly and defrost foods properly.<br>• Avoid raw (unpasteurized) milk or any products made from unpasteurized milk, raw or partially cooked eggs or foods containing raw eggs, raw or undercooked meat and poultry, unpasteurized juices, and raw sprouts. | *Infants and young children, pregnant women, older adults, and those who are immuno-compromised:* Do not eat or drink raw (unpasteurized) milk or any products made from unpasteurized milk, raw or partially cooked eggs or foods containing raw eggs, raw or undercooked meat and poultry, raw or undercooked fish or shellfish, unpasteurized juices, and raw sprouts.<br><br>*Pregnant women, older adults, and those who are immunocompromised:* Only eat certain deli meats and frankfurters that have been reheated to steaming hot. |

SOURCE: U.S. Department of Health and Human Services; and U.S. Department of Agriculture. *Dietary Guidelines for Americans 2005.* Washington, DC: U.S. Government Printing Office; 2005. Available from: www.health.gov/dietaryguidelines.

# APPENDIX D

# List of FDA–Approved Qualified and Unqualified Health Claims

| Diet and Health Claims | Eligible Foods/Supplements | Model Claim Statements |
| --- | --- | --- |
| **Qualified Health Claims** | | |
| Selenium and Cancer | Dietary supplements containing selenium. | *Some scientific evidence suggests that consumption of selenium may reduce the risk of certain forms of cancer. However, FDA has determined that this evidence is limited and not conclusive.* |
| Antioxidant Vitamins and Cancer | Dietary supplements containing vitamin E and/or vitamin C. | *Some scientific evidence suggests that consumption of antioxidant vitamins may reduce the risk of certain forms of cancer. However, FDA has determined that this evidence is limited and not conclusive.* |
| Nuts and Heart Disease | Almonds, hazelnuts, peanuts, pecans, some pine nuts, pistachio nuts, and walnuts that are raw, blanched, roasted, salted, and/or lightly coated and/or flavored. <br><br> Nut-containing products other than whole or chopped nuts that contain at least 11 g of one or more of the nuts just listed. | *Scientific evidence suggests but does not prove that eating 1.5 oz. per day of most nuts as part of a diet low in saturated fat and cholesterol may reduce the risk of heart disease.* |
| Walnuts and Heart Disease | Whole or chopped walnuts. | *Supportive but not conclusive research shows that eating 1.5 oz. per day of walnuts, as part of a low-saturated-fat and low-cholesterol diet and not resulting in increased caloric intake, may reduce the risk of coronary heart disease.* |
| Omega-3 Fatty Acids and Coronary Heart Disease | Dietary supplements containing the omega-3 long-chain polyunsaturated fatty acids eicosapentaenoic acid (EPA) and/or docosahexaenoic acid (DHA). | *Consumption of omega-3 fatty acids may reduce the risk of coronary heart disease. FDA evaluated the data and determined that, although there is scientific evidence supporting the claim, the evidence is not conclusive.* |
| B Vitamins and Vascular Disease | Dietary supplements containing vitamin $B_6$, vitamin $B_{12}$, and/or folic acid. | *As part of a well-balanced diet that is low in saturated fat and cholesterol, folic acid, vitamin $B_6$, and vitamin $B_{12}$ may reduce the risk of vascular disease. FDA evaluated this claim and found that, the evidence in support of this claim is inconclusive.* |
| Monounsaturated Fatty Acids from Olive Oil, and Coronary Heart Disease | All products that are essentially pure olive oil or salad dressings that contain 6 g or more olive oil per serving that are low in cholesterol and saturated fat. | *Limited and not conclusive scientific evidence suggests that eating about 2 tablespoons (23 g) of olive oil daily may reduce the risk of coronary heart disease due to the monounsaturated fat in olive oil. To achieve this possible benefit, olive oil is to replace a similar amount of saturated fat and not increase the total number of calories you eat in a day.* |
| Phosphatidylserine, Cognitive Dysfunction, and Dementia | Dietary supplements containing soy-derived phosphatidylserine. | *Very limited and preliminary scientific research suggests that phosphatidylserine may reduce the risk of dementia in the elderly. FDA concludes that there is little scientific evidence supporting this claim.* |
| Folic Acid and Neural Tube Birth Defects (supplements) | Dietary supplements containing folic acid. | *Some studies suggest that 0.8 mg folic acid in a dietary supplement is more effective in reducing the risk of neural tube defects than a lower amount in foods. FDA does not endorse this claim. Public health authorities recommend that women of child-bearing age consume 0.4 mg folic acid daily from fortified foods or dietary supplements or both to reduce the risk of neural tube defects.* |

*continued*

| Diet and Health Claims | Eligible Foods/Supplements | Model Claim Statements |
|---|---|---|
| **Unqualified Health Claims** | | |
| Calcium and Osteoporosis | Foods and supplements high in bio-available calcium. Supplements must disintegrate and dissolve, and phosphorus content cannot exceed calcium content. | *Regular exercise and a healthy diet with enough calcium helps teens and young adult white and Asian women maintain good bone health and may reduce their high risk of osteoporosis later in life.* |
| Sodium and Hypertension | Foods and supplements low in sodium. | *Diets low in sodium may reduce the risk of high blood pressure, a disease associated with many factors.* |
| Dietary Fat and Cancer | Low-fat foods. | *Development of cancer depends on many factors. A diet low in total fat may reduce the risk of some cancers.* |
| Dietary Saturated Fat, Cholesterol, and Risk of Coronary Heart Disease | Foods that are low in saturated fat, low in cholesterol, and low in fat. | *Although many factors affect heart disease, diets low in saturated fat and cholesterol may reduce the risk of this disease.* |
| Fiber-Containing Grain Products, Fruits, and Vegetables, and Cancer | Grain products, fruits, or vegetables that are low in fat and good sources of dietary fiber (without fortification). | *Low-fat diets rich in fiber-containing grain products, fruits, and vegetables may reduce the risk of some types of cancer, a disease associated with many factors.* |
| Fruits, Vegetables, and Grain Products that Contain Fiber (particularly soluble fiber), and Risk of Coronary Heart Disease | A fruit, vegetable, or grain product that contains fiber; is low in saturated fat, low in cholesterol, low in fat, and contains at least 0.6 g of soluble fiber per serving (without fortification). | *Diets low in saturated fat and cholesterol and rich in fruits, vegetables, and grain products that contain some type of dietary fiber, particularly soluble fiber, may reduce the risk of heart disease, a disease associated with many factors.* |
| Fruits and Vegetables, and Cancer | A fruit or vegetable that is low in fat and is a good source (without fortification) of at least one of the following: vitamin A, vitamin C, or dietary fiber. | *Low-fat diets rich in fruits and vegetables (foods that are low in fat and may contain dietary fiber, vitamin A, or vitamin C) may reduce the risk of some types of cancer, a disease associated with many factors.* |
| Folate and Neural Tube Defects (foods) | Dietary supplements, or foods in that are naturally good sources of folate (i.e., only nonfortified food in conventional food form with at least 40 micrograms [μg] of folate per serving). | *Healthful diets with adequate folate may reduce a woman's risk of having a child with a brain or spinal cord defect.* |
| Dietary Sugar Alcohol and Dental Caries | Sugar-free foods containing a sugar alcohol such as xylitol or sorbitol. | *Frequent between-meal consumption of foods high in sugars and starches promotes tooth decay. The sugar alcohols in this food do not promote tooth decay.* |
| Soluble Fiber from Certain Foods, and Risk of Coronary Heart Disease | Foods low in saturated fat, cholesterol, and total fat that contain at least 0.75 g of soluble fiber from whole oats or barley or 1.7 g of psyllium husk soluble fiber per serving. | *Soluble fiber from foods, as part of a diet low in saturated fat and cholesterol, may reduce the risk of heart disease.* |
| Soy Protein and Risk of Coronary Heart Disease | Foods low in saturated fat, cholesterol, and total fat that contain at least 6.25 g of soy protein per serving. | *Diets low in saturated fat and cholesterol that include 25 g of soy protein a day may reduce the risk of heart disease. One serving of this food provides 30 g of soy protein.* |
| Plant Sterol/Stanol Esters and Risk of Coronary Heart Disease | Spreads, salad dressings, snack bars, and dietary supplements that are low in saturated fat and cholesterol and contain at least at least 1.7 g plant stanol esters per serving. | *Diets low in saturated fat and cholesterol that include two servings of foods that provide a daily total of at least 3.4 g of plant stanol esters in two meals may reduce the risk of heart disease. A serving of this food supplies 2.0 g of plant stanol esters.* |
| Whole-Grain Foods and Risk of Heart Disease and Certain Cancers | Low-fat, high-fiber foods that contain 51% or more whole-grain ingredients by weight per serving. | *Diets rich in whole-grain foods and other plant foods and low in total fat, saturated fat, and cholesterol may reduce the risk of heart disease and some cancers.* |
| Potassium and Risk of High Blood Pressure and Stroke | Good sources of potassium that are low in sodium, total fat, saturated fat and cholesterol. | *Diets containing foods that are a good source of potassium and that are low in sodium may reduce the risk of high blood pressure and stroke.* |

NOTES: Definitions of "good source of," "high in," and "low in" are defined by the U.S. Food and Drug Administration and are provided in Table 2.4, p. 53. "Qualified" health claims have less scientific backing than do "unqualified" health claims.

SOURCE: U.S. Food and Drug Administration (USFDA). A food labeling guide—Appendix B (June 1999). Available from: http://www.cfsan.fda.gov/~dms/flg-6b.html.USFDA. Summary of qualified health claims permitted (Sept. 2003). Available from: http://www.cfsan.fda.gov/~dms/qhc-sum.html.

# The Exchange System

The exchange system is a dietary plan created by the American Dietetic Association and the American Diabetes Association. This system provides people with a relatively easy approach to monitor portion size, caloric intake, and the amounts of calories from carbohydrates, proteins, and fats. This is done by selecting a certain number of "food exchanges" from the seven exchange food groups listed next.

## The Exchange System Food Groups

There are seven food groups in the exchange system, each providing a similar amount of calories, carbohydrates, protein, and fat. These groups are

- Starch (cereals, grains, pasta, breads, crackers, starch vegetables, and legumes)
- Fruit
- Milk (fat-free, reduced fat, and whole)
- Other carbohydrates (desserts and snacks with added sugars and fats)
- Nonstarchy vegetables
- Meat and meat substitutes (very lean, lean, medium-fat, and high-fat)
- Fats

There is an additional category referred to as free foods.

Each exchange food group lists a variety of foods that provide roughly the same amount of carbohydrate, fat, and protein. The following chart lists the amount of macronutrients and calories per food exchange from each exchange food group.

| Groups/Lists | Carbohydrate (grams) | Protein (grams) | Fat (grams) | Calories |
|---|---|---|---|---|
| **Carbohydrate Group** | | | | |
| Starch | 15 | 3 | 1 or less | 80 |
| Fruit | 15 | — | — | 60 |
| Milk | | | | |
|   Fat-free, low-fat | 12 | 8 | 0–3 | 90 |
|   Reduced-fat | 12 | 8 | 5 | 120 |
|   Whole | 12 | 8 | 8 | 150 |
| Other carbohydrates | 15 | Varies | Varies | Varies |
| Nonstarchy vegetables | 5 | 2 | — | 25 |
| **Meat and Meat Substitute Group** | | | | |
| Very lean | — | 7 | 0–1 | 35 |
| Lean | — | 7 | 3 | 55 |
| Medium-fat | — | 7 | 5 | 75 |
| High-fat | — | 7 | 8 | 100 |
| **Fat Group** | — | — | 5 | 45 |

SOURCE: © 1995 by the American Diabetes Association, Inc., and the American Dietetic Association.

## Meal Planning

It is important to first establish individual dietary requirements, including the optimal number of total calories and the correct proportion of carbohydrates, protein, and fats. The following chart lists various calorie levels and the allotted number of food exchanges per exchange group.

### Number of Exchanges per Day for Various Calorie Levels

| Calories | 1,200 kcal | 1,500 kcal | 1,800 kcal | 2,000 kcal | 2,200 kcal |
|---|---|---|---|---|---|
| **Exchange Groups** | | | | | |
| Starch/bread | 5 | 8 | 10 | 11 | 13 |
| Meat and meat substitutes | 4 | 5 | 7 | 8 | 8 |
| Nonstarchy vegetable | 2 | 3 | 3 | 4 | 4 |
| Fruit | 3 | 3 | 3 | 3 | 3 |
| Milk | 2 | 2 | 2 | 2 | 2 |
| Fat | 3 | 3 | 3 | 4 | 5 |

The exchange system is based on the three important principles of dietary planning—balance, moderation, and variety. However, it also recognizes the importance of individualized food menus. Therefore, the exchange system does not provide daily eating plans. Rather, a person using the exchange system creates his or her own daily menu by selecting foods from a list of foods provided with each exchange food group. The following diet plan is just one example of a daily menu using the exchange system for someone requiring 1,500 kcal/day.

| Exchange | Breakfast | Lunch | Snack | Dinner | Snack | Total |
|---|---|---|---|---|---|---|
| Starch | 2 | 2 | 1 | 2 | 1 | 8 |
| Lean meat | | 2 | | 3 | | 5 |
| Vegetable | | 1 | 1 | 1 | | 3 |
| Fruit | 1 | | 1 | | 1 | 3 |
| Milk | 1 | | | 1 | | 2 |
| Fat | 1 | | 1 | 1 | | 3 |

- *Breakfast:* 1 whole-wheat bagel, ½ cup orange juice, 1 cup reduced-fat milk, 1 teaspoon butter
- *Lunch:* 2 slices whole-wheat bread, 2 oz. canned tuna drained, 1 cup raw vegetables
- *Snacks:* 8 crackers, 1 cup raw vegetables, 1 apple, 1 tablespoon peanut butter, 1 orange
- *Dinner:* ⅔ cup rice, 3 oz. lean beef, 1 cup salad greens, 1 small whole-wheat dinner roll, 1 cup reduced-fat milk, 1 tablespoon salad dressing

# Exchange List: Starches

Each exchange under the starch list contains about 15 g of carbohydrates, 3 g of protein, and a trace of fat for a total of 80 calories. A half-cup of cooked cereal, grain, or pasta equals one exchange and 1 oz. of a bread product is one serving.

Choose starches with little fat and those that provide dietary fiber. This information is often provided on food labels. Starchy foods prepared with fat counts as one starch and one fat exchange.

| Food | Amount |
|---|---|
| **Bread** | |
| Bagel or English muffin | ½ (1 oz.) |
| Bread: whole-wheat, rye, white, pumpernickel | 1 slice |
| Bread (reduced-calorie) | 2 slices |
| Raisin bread (unfrosted) | 1 slice |
| Breadsticks (4 inches long by ½ inch across, crisp) | 4 |
| Dinner roll | 1 small (1 oz.) |
| Hamburger bun, hot dog bun | ½ (1 oz.) |
| Pita bread (6 inches across) | ½ |
| Tortilla (6 inches across) | 1 |
| **Cereal** | |
| Bran cereal | ½ cup |
| Hot cereal: oatmeal, Cream of Wheat | ½ cup |
| Grits | ½ cup |
| Grape-Nuts®, muesli, low-fat granola | ¼ cup |
| Other ready-to-eat cereals (unsweetened) | ¾ cup |
| Puffed cereal (unfrosted) | 1½ cups |
| Shredded Wheat® | 1 biscuit |
| Shredded Wheat® (spoon size) | ½ cup |
| Sugar frosted cereal | ½ cup |
| **Other Starches** | |
| Rice: white or brown (cooked) | ⅓ cup |
| Pasta: spaghetti, noodles, macaroni (cooked) | ⅓ cup |
| Wheat germ | 3 tbsp. |
| Barley, bulgur (cooked) | ½ cup |
| Couscous | ⅓ cup |
| **Peas, Beans, and Lentils** | |
| Baked beans | ⅓ cup |
| Dried beans, peas (cooked) | ½ cup |
| Lentils | ½ cup |
| Lima beans | ⅔ cup |
| **Starchy Vegetables** | |
| Corn | ½ cup |
| Corn on the cob (fresh or frozen) | 1 ear (5 oz.) |
| Mixed vegetables with corn, peas, or pasta | 1 cup |
| Parsnips | ½ cup |
| Peas (green) | ½ cup |
| Plantain | ½ cup |
| Potato (baked or boiled) | 1 small (3 oz.) |
| Potato (mashed) | ½ cup |
| Pumpkin | 1 cup |
| Winter squash: acorn, butternut, buttercup, Hubbard | 1 cup |
| Yam, sweet potato (fresh or without added sugar) | ½ cup |
| **Soups** | |
| Bean | ½ cup |
| Broth-based | 1 cup |
| Cream-based (low-fat or made with skim milk) | 1 cup |

| Food | Amount |
|---|---|
| **Crackers and Snacks** | |
| Animal crackers | 8 |
| Graham crackers (2½-inch squares) | 3 |
| Oyster crackers | 24 |
| Ry-Krisp® | 4 |
| Saltine crackers (2-inch squares) | 6 |
| Melba toast | 4 |
| Matzo | ¾ oz. |
| Popcorn (popped and no fat added or low-fat microwave) | 3 cups |
| Pretzel sticks | ¾ oz. |
| Rice cakes or popcorn cakes (4 inches across) | 2 |
| Rice cakes or popcorn minicakes | 5 |
| Snack chips: tortilla, potato (fat-free or baked) | 15 to 20 |

**Other Starches**

**The following foods are less nutritious. Use them occasionally as part of a planned meal or snack.**

*Foods higher in sugar but low in fat:*

| Food | Amount |
|---|---|
| Angel food cake | 1½-inch slice (1 oz.) |
| Frozen yogurt | ½ cup |
| Frozen yogurt (fat-free) | ⅓ cup |
| Gelatin (sugar-sweetened) | ½ cup |
| Gingersnaps | 3 |
| Ice cream (fat-free and no sugar added) | ½ cup |
| Pudding (sugar-free) | ½ cup |
| Pudding (sugar-sweetened) | ¼ cup |
| Sherbet, sorbet | ¼ cup |
| Vanilla wafers | 5 |

*Foods higher in fat but not in sugar:*

| Food | Amount |
|---|---|
| Biscuit (2½ inches across) | 1 small |
| Chow mein noodles | ½ cup |
| Cornbread (2-inch cube) | 1 piece (2 oz.) |
| Corn muffin (2 inches across) | 1 (2 oz.) |
| Croissant | 1 small |
| Croutons | ¾ cup |
| French fries | 16 to 25 (½ of a small order) |
| Muffin (cupcake size) | 1 small (1 oz.) |
| Pancake (4 inches across) | 1 |
| Quick bread: banana, pumpkin, zucchini | ⅜-inch slice (1 oz.) |
| Stuffing (bread) | ⅓ cup |
| Taco shells (6 inches across) | 2 |
| Waffle (4 inches across) | 1 |

*Foods higher in sugar and fat:*

| Food | Amount |
|---|---|
| Cake doughnut (plain) | 1 small |
| Cookies | 2 small (¾ oz. total) |
| Unfrosted cake | 2-inch square |

**NOTE:** The Exchange Lists are the basis of a meal planning system designed by a committee of the American Diabetes Association and the American Dietetic Association. While designed primarily for people with diabetes and others who must follow special diets, the Exchange Lists are based on principles of good nutrition that apply to everyone. Copyright © 2003 by the American Diabetes Association and the American Dietetic Association. Reprinted by permission.

## Exchange List: Nonstarchy Vegetables

Exchanges for nonstarchy vegetables are ½ cup cooked, 1 cup raw, and ½ cup juice. Each group contains 5 g of carbohydrates, 2 g of protein, and between 2 to 3 g of fiber. Fresh or frozen vegetables have less sodium than canned vegetables. Low-sodium canned vegetables are more desirable than high-sodium varieties. To reduce the amount of salt, drain the liquid from canned vegetables. Vegetables can be seasoned with spices, which adds flavor without added calories. Vegetables should be steamed or microwaved without added fat.

Alfalfa sprouts
Artichoke
Artichoke hearts
Asparagus
Bamboo shoots
Bean sprouts
Beans: green, Italian, yellow, or wax
Broccoli
Brussels sprouts
Cabbage
Carrots
Cauliflower

Celery
Chicory
Chinese cabbage
Cucumber
Eggplant
Greens: beet, collard, dandelion, kale, mustard, or turnip
Green onions or scallions
Jicama
Kohlrabi
Leeks
Lettuce: endive, escarole, leafy varieties, romaine, or iceberg

Mixed vegetables without corn, peas, or pasta
Mushrooms
Okra
Onions
Parsley
Snow peas or pea pods
Peppers (all varieties)
Radishes
Rhubarb, artificially sweetened
Rutabaga
Sauerkraut
Spinach

Summer squash
Swiss chard
Tomato, raw
Tomato, cherry
Tomato juice
Tomato paste
Tomato sauce
Turnips
Vegetable juice cocktail
Water chestnuts
Watercress

## Exchange List: Other Carbohydrates (sweets and desserts)

Being on a special diet does not mean totally eliminating sweets and desserts. Rather, dessert can be enjoyed as long as it is not overdone. Enjoy desserts on occasion and in moderation.

| Food | Amount | Exchanges (per Serving) |
|---|---|---|
| Angel food cake, without frosting | 1½-inch slice (1 oz.) | 1 carbohydrate |
| Brownie, without frosting | 2-inch square (1 oz.) | 1 carbohydrate and 1 fat |
| Cake, without frosting | 2-inch square (1 oz.) | 1 carbohydrate and 1 fat |
| Cake, with frosting | 2-inch square (2 oz.) | 2 carbohydrates and 1 fat |
| Cake doughnut, plain | 1 small | 1 carbohydrate |
| Chocolate mint patty, small | 1 piece (12 g) | 1 carbohydrate |
| Cookie or sandwich cookie with cream filling | 2 small (⅔ oz. total) | 1 carbohydrate and 1 fat |
| Frozen yogurt (fat-free) | ⅓ cup | 1 carbohydrate |
| Gelatin (sugar-sweetened) | ½ cup | 1 carbohydrate |
| Gingersnaps | 3 | 1 carbohydrate |
| Hershey's milk chocolate bar, plain | 1½ oz. | 2 carbohydrates and 2½ fats |
| Hershey's milk chocolate bar, with almonds | 1½ oz. | 1 carbohydrate, 2 fats and 1 meat |
| Hershey's chocolate kisses, plain | 6 pieces | 1 carbohydrate and 2 fats |
| Hershey's chocolate kisses, with almonds | 6 pieces | 1 carbohydrate and 2 fats |
| Ice cream (fat-free and no sugar added) | ½ cup | 1 carbohydrate |
| Jelly beans | 14 pieces (1 oz.) | 2 carbohydrates |
| Jelly beans (sugar-free) | 25 pieces (1 oz.) | 1 carbohydrate |
| Lindt truffles | 3 pieces | 1 carbohydrate and 3 fats |
| Low-carb chocolate-mint wafer bar | 1 oz. | ½ carbohydrate and 2 fats |
| M&Ms, plain | 1½ oz. | 2 carbohydrates and 2 fats |
| M&Ms, peanut | 1¾ oz. | 2 carbohydrates, 1½ fats, and 1 meat |
| Muffin (cupcake size) | 1 small (1 oz.) | 1 carbohydrate |
| Nestle's milk chocolate bar with crisped rice | 1½ oz. | 2 carbohydrates and 2 fats |
| Pancake (4 inches across) | 1 | 1 carbohydrate |
| Pudding (sugar-free) | ½ cup | 1 carbohydrate |
| Pudding (sugar-sweetened) | ¼ cup | 1 carbohydrate |
| Quick bread: banana, pumpkin, zucchini | ⅜ inch slice (1 oz.) | 1 carbohydrate |
| Reese's peanut butter cup (miniature) | 4 pieces | 1 carbohydrate and 1 fat |
| Sherbet, sorbet | ¼ cup | 1 carbohydrate |
| Snickers bar (fun size) | 1 bar (¾ oz.) | 1 carbohydrate and 1 fat |
| Strawberry twists | 2½ oz. | 3½ carbohydrates |
| Vanilla wafers | 5 | 1 carbohydrate |
| Waffle (4 inches across) | 1 | 1 carbohydrate |

## Exchange List: Milk and Milk Products

The milk and milk product exchange is categorized by fat content. A milk exchange is 1 cup or 8 oz. Skim and very low-fat milk is recommended, and whole milk and foods made with whole milk should be consumed sparingly. Fat-free or low-fat milk and milk products contain 12 g of carbohydrates, 8 g of protein, and 0 to 3 g of fat, and 90 kcal.

| Food | Amount |
| --- | --- |
| **Fat-free and Low-fat Milk Products** | |
| Milk, fat-free or low-fat (½% and 1%) | 1 cup (8 oz.) |
| Dry milk powder (fat-free) | ⅓ cup |
| Buttermilk (fat-free or low-fat) | 1 cup |
| Evaporated skim milk | ½ cup |
| Yogurt (plain, fat-free) | ⅔ cup (6 oz.) |
| Yogurt (fat-free, made with sugar substitute) | ⅔ cup (6 oz.) |
| Pudding (sugar-free, made with skim milk) | ½ cup |
| Hot chocolate mix (sugar-free, made with water) | 1 cup |

| Food | Amount |
| --- | --- |
| **Reduced-fat Milk Products** | |
| Milk, reduced-fat (2%) | 1 cup |
| Yogurt, plain, reduced-fat (2%) | ¾ cup |
| Soy milk (plain) | 1 cup |
| **Whole Milk Products** | |
| Milk (whole) | 1 cup |
| Evaporated whole milk | ½ cup |

## Exchange List: Fruits

Fruit exchanges contain about 15 g of carbohydrates for a total of 60 calories. Whole fruit is an important source of fiber. Select canned fruit and fruit juices that do not contain added sugar. Most importantly, avoid those packed in heavy syrup. Drain the juice from canned fruits prior to serving.

| Food | Amount |
| --- | --- |
| **Fresh and Dried Fruit** | |
| Apple | 1 small (2 inches across) |
| Apple, dried | 4 rings |
| Apricots | 4 medium |
| Apricots, dried | 8 halves |
| Banana | ½ (4 oz.) |
| Blackberries | ¾ cup |
| Blueberries | ¾ cup |
| Cantaloupe | ⅓ small (1 cup cubed) |
| Cherries | 12 large |
| Dates | 3 medium |
| Figs, dried | 1½ medium |
| Figs, fresh | 2 medium or 1½ large |
| Grapefruit | ½ large |
| Grapes | 17 small (3 oz.) |
| Guava | 1 medium |
| Honeydew melon | ⅛ medium (1 cup cubed) |
| Kiwi | 1 large |
| Kumquats | 5 medium |
| Mango | ½ small |
| Nectarine | 1 small |
| Orange | 1 small (2½ inches across, or 6½ oz.) |
| Papaya | ½ medium (1 cup) |
| Passion fruit | 3 medium |
| Peach | 1 medium |
| Pear | ½ large (4 oz.) |
| Persimmons | 2 medium |

| Food | Amount |
| --- | --- |
| **Fresh and Dried Fruit, *continued*** | |
| Pineapple, fresh | ¾ cup |
| Plums | 2 small (5 oz.) |
| Pomegranate | ½ medium |
| Prickly pear | 1 large |
| Prunes | 3 medium |
| Raisins | 2 tbsp. |
| Raspberries | 1 cup |
| Strawberries | 1¼ cup |
| Tangelo | 1 medium |
| Tangerines | 2 small (8 oz.) |
| Watermelon, cubed | 1¼ cup |
| **Canned or Frozen Fruit (unsweetened)** | |
| Applesauce, apricots, cherries, fruit cocktail, grapes, peaches, pears, pineapple, or plums | ½ cup |
| Grapefruit or mandarin oranges | ¾ cup |
| **Fruit Juice (unsweetened)** | |
| Apple cider, apple juice, apricot nectar, grapefruit juice, orange juice, peach nectar, pear nectar, pineapple juice, or tangerine juice | ½ cup |
| Cranberry juice cocktail, grape juice, prune juice, or fruit juice blends of 100% juice | ⅓ cup |
| Cranberry juice cocktail (reduced-calorie) | 1 cup |

## Exchange List: Fats

A fat exchange is based on a serving size that contains 5 grams of fat. There are three types of fats—saturated, polyunsaturated, and monounsaturated. Saturated should be consumed in moderation, substituting polyunsaturated or monounsaturated fats instead. Another type of fat, called *trans* fats, should also be consumed in minimal amounts.

| Food | Amount |
|---|---|
| **Monounsaturated Fats** | |
| Avocado | 2 tbsp. |
| Olives, black or ripe | 8 large |
| Olives, green | 10 large |
| Peanuts | 10 large |
| Peanut butter, smooth or crunchy | ½ tbsp. |
| Nuts: pecans, almonds, or cashews | 4 to 6 |
| Oil: canola, olive, peanut, or sesame | 1 tsp. |
| Sesame seeds | 1 tbsp. |
| Tahini or sesame paste | 2 tsp. |
| **Polyunsaturated Fats** | |
| Margarine | 1 tsp. |
| Margarine, reduced-fat or light | 1 tbsp. |
| Mayonnaise | 1 tsp. |
| Mayonnaise, reduced-fat | 1 tbsp. |
| Miracle Whip reduced-fat salad dressing | 1 tbsp. |
| Miracle Whip salad dressing | 2 tsp. |
| Nondairy cream substitute, liquid or powder | ¼ cup |
| Salad dressing, reduced-fat | 2 tbsp. |
| Salad dressing, regular | 1 tbsp. |
| Seeds: pumpkin, sunflower | 1 tbsp. |

| Food | Amount |
|---|---|
| **Polyunsaturated Fats, *continued*** | |
| Tartar sauce | 1 tbsp. |
| Tartar sauce, reduced-fat | 2 tbsp. |
| Walnuts | 4 halves |
| Bacon, crisp | 1 strip |
| Bacon fat | 1 tsp. |
| **Saturated Fats** | |
| Butter | 1 tsp. |
| Butter, reduced-fat | 1 tbsp. |
| Butter, whipped | 2 tsp. |
| Coconut, shredded | 2 tbsp. |
| Cream cheese | 1 tbsp. |
| Cram cheese, reduced-fat | 1½ tbsp. |
| Gravy | 2 tbsp. |
| Half-and-half (light cream) | 2 tbsp. |
| Heavy cream | 1 tbsp. |
| Salt pork | 1-inch cube |
| Shortening or lard | 1 tsp. |
| Sour cream | 2 tbsp. |
| Sour cream, reduced-fat | 3 tbsp. |

## Exchange List: Meat and Meat Substitutes

The exchange group for meat and meat substitutes, is categorized by fat content—lean meat, medium-fat meat, and high-fat meat. High-fat meat and meat substitute exchanges should be limited to three times per week. Fat should be removed before cooking. Legumes and other meat substitutes are also good sources of protein. Each exchange contains 7 g of protein.

| Food | Amount |
|---|---|
| **Lean Meats and Meat Substitutes** | |
| Poultry without skin (chicken, turkey, duck, goose, pheasant, Cornish hen) | 1 oz. |
| Wild game (venison, rabbit, elk, buffalo, ostrich) | 1 oz. |
| Dried bean, peas, lentils (cooked) | ½ cup |
| Fish (fresh or frozen) | 1 oz. |
| Herring | 1 oz. |
| Tuna, salmon, or mackerel (canned, drained) | 1 oz. |
| Sardines | 2 medium |
| Clams, crab, scallops, oysters, lobster, shrimp, imitation shellfish | 1 oz. |
| Beef, USDA select or choice, fat-trimmed (rib, chuck and rump roasts; ground round; round, sirloin, flank, T-bone, porterhouse steaks) | 1 oz. |
| Lamb (roast, chop, leg) | 1 oz. |
| Pork (tenderloin, center loin chop, ham) | 1 oz. |
| Veal (roast, lean chop) | 1 oz. |
| Cheese (less than 3 g of fat per ounce) | 1 oz. |
| Cottage cheese (fat-free, low-fat or regular) | ¼ cup |
| Parmesan cheese | 2 tbsp. |
| Egg substitute | ¼ cup |
| Egg whites | 2 |
| Hot dog, fat-free or low-fat (less than 3 g of fat per ounce) | 1 small |
| Luncheon meat, fat-free or low-fat (less than 3 g of fat per ounce) | 1 oz. |

| Food | Amount |
|---|---|
| **Medium-fat Meats and Meat Substitutes** | |
| Poultry with skin | 1 oz. |
| Fried fish | 1 oz. |
| Ground meat (beef, chicken, lamb, turkey) | 1 oz. |
| Beef (meatloaf, corned beef, short ribs, prime cuts trimmed of fat) | 1 oz. |
| Lamb (rib roast) | 1 oz. |
| Veal (cutlet) | 1 oz. |
| Sausage (less than 5 g of fat per ounce) | 1 oz. |
| Cheese (feta, mozzarella or others with less than 5 g of fat per ounce) | 1 oz. |
| Ricotta cheese | ¼ cup |
| Egg | 1 |
| Tempeh | ¼ cup |
| Tofu (soybean curd) | ½ cup (4 oz.) |
| **High-fat Meats and Meat Substitutes** | |
| Pork spareribs, ground pork | 1 oz. |
| Bacon | 3 slices |
| Sausage (Polish, bratwurst, kielbasa) | 1 oz. |
| Breakfast sausage | 1 patty or 2 links |
| Hot dog (turkey, chicken, beef, pork or combination) | 1 |
| Luncheon meats (bologna, salami) | 1 oz. |
| Organ meats (liver, heart) | 1 oz. |
| Cheese (American, cheddar, colby, Monterey jack, Swiss) | 1 oz. |
| Cheese spread | 2 tbsp. |
| Peanut butter | 1 tbsp. |

## Exchange List: Free Foods

Foods and drinks that contain less than 20 kcal or those with less than 5 g of carbohydrate per serving are considered free foods. Free foods without a specified serving size can be consumed in unlimited amounts. Those with specified serving sizes should be limited to three servings per day.

| Food | Amount |
| --- | --- |
| **Beverages** | |
| Water | — |
| Carbonated or flavored water (sugar-free) | — |
| Club soda | — |
| Coffee: regular or decaffeinated | — |
| Diet soft drinks (sugar-free) | — |
| Drink mixes, sugar-free | — |
| Mineral water | — |
| Tea | — |
| Tonic water (sugar-free) | — |
| **Seasonings** | |
| Butter flavoring (fat-free) | — |
| Garlic | — |
| Herbs | — |
| Pepper | — |
| Spices | — |
| Flavored extracts | — |
| Horseradish | — |
| Hot pepper sauce | — |
| Lemon juice | — |
| Lime juice | — |
| Nonstick pan spray | — |
| Pimento | — |
| Vinegar | — |
| Wine in cooking | — |
| Mustard | — |
| Worcestershire or soy sauce | — |
| **Miscellaneous** | |
| Bouillon or broth (fat-free) | — |
| Flavored gelatin (sugar-free) | — |
| Gum (sugar-free) | — |
| Sugar substitutes (aspartame, saccharin, or acesulfame-K) | — |
| Unflavored gelatin (plain) | — |

| Food | Amount |
| --- | --- |
| **Condiments** | |
| Barbecue sauce | 1 to 2 tbsp. |
| Cocktail sauce | 1 to 2 tbsp. |
| Dill pickles | 1½ large |
| Jam or jelly: low-sugar or light | 1 to 2 tbsp. |
| Ketchup | 1 to 2 tbsp. |
| Margarine, fat-free | 4 tbsp. |
| Mayonnaise, fat-free | 1 tbsp. |
| Miracle Whip salad dressing, fat-free | 1 tbsp. |
| Nondairy creamer | 2 tbsp. |
| Pancake syrup, sugar-free | 1 to 2 tbsp. |
| Pickle relish | 1 tbsp. |
| Salad dressing, fat-free | 1 tbsp. |
| Salsa | ¼ cup |
| Sour cream, fat-free | 1 tbsp. |
| Soy sauce, regular or light | 1 tbsp. |
| Sweet and sour sauce | 1 tbsp. |
| Sweet pickles, bread-and-butter pickles | 2 slices |
| Sweet pickles, gherkin | ¾ oz. |
| Teriyaki sauce | 1 tbsp. |
| **Miscellaneous** | |
| Cream cheese, fat-free | 1 tbsp. |
| Cocoa powder, unsweetened | 1 tbsp. |
| Cranberries, sweetened with sugar substitute | ½ cup |
| Hard candy, sugar-free | 2 to 3 pieces |
| Rhubarb, sweetened with sugar substitute | ½ cup |
| Whipped topping: low-fat or fat-free | 2 tbsp. |

# APPENDIX F

## Metropolitan Life Insurance Company Height and Weight Tables (1983)

| Height | Small Frame | Medium Frame | Large Frame |
|--------|-------------|--------------|-------------|
| | | Pounds | |
| **Men**[a] | | | |
| 5'2" | 128–134 | 131–141 | 138–150 |
| 5'3" | 130–136 | 133–143 | 140–153 |
| 5'4" | 132–138 | 135–145 | 142–156 |
| 5'5" | 134–140 | 137–148 | 144–160 |
| 5'6" | 136–142 | 139–151 | 146–164 |
| 5'7" | 138–145 | 142–154 | 149–168 |
| 5'8" | 140–148 | 145–157 | 152–172 |
| 5'9" | 142–151 | 148–160 | 155–176 |
| 5'10" | 144–154 | 151–163 | 158–180 |
| 5'11" | 146–157 | 154–166 | 161–184 |
| 6'0" | 149–160 | 157–170 | 164–188 |
| 6'1" | 152–164 | 160–174 | 168–192 |
| 6'2" | 155–168 | 164–178 | 172–197 |
| 6'3" | 158–172 | 167–182 | 176–202 |
| 6'4" | 162–176 | 171–187 | 181–207 |
| **Women**[b] | | | |
| 4'10" | 102–111 | 109–121 | 118–131 |
| 4'11" | 103–113 | 111–123 | 120–134 |
| 5'0" | 104–115 | 113–126 | 122–137 |
| 5'1" | 106–118 | 115–129 | 125–140 |
| 5'2" | 108–121 | 118–132 | 128–143 |
| 5'3" | 111–124 | 121–135 | 131–147 |
| 5'4" | 114–127 | 124–138 | 134–151 |
| 5'5" | 117–130 | 127–141 | 137–155 |
| 5'6" | 120–133 | 130–144 | 140–159 |
| 5'7" | 123–136 | 133–147 | 143–163 |
| 5'8" | 126–139 | 136–150 | 146–167 |
| 5'9" | 129–142 | 139–153 | 149–170 |
| 5'10" | 132–145 | 142–156 | 152–173 |
| 5'11" | 135–148 | 145–159 | 155–176 |
| 6'0" | 138–151 | 148–162 | 158–179 |

[a] Weights at ages 25 to 59 years based on lowest mortality. Weight in pounds according to frame (in indoor clothing weighing 5 lb., shoes with 1-inch heels).
[b] Weights at ages 25 to 59 years based on lowest mortality. Weight in pounds according to frame (in indoor clothing weighing 3 lb., shoes with 1-inch heels).

SOURCE: Courtesy of Metropolitan Life Insurance Company.

# Answers to Review Questions and Practice Calculations

## Chapter 1

### Review Questions

**1.** c **2.** d **3.** d **4.** a **5.** c **6.** d **7.** c
**8.** a **9.** a **10.** b

**11.** A nutrient is a substance found in foods can be used by the body to support health. Essential nutrients cannot be made in the body in sufficient amounts and are required in the diet. The nonessential nutrients can be made by the body in sufficient amounts. Thus we do not need to consume nonessential nutrients. Conditionally essential nutrients can be made by most people but by not others. People who cannot make enough of these compounds must get them from foods. **12.** A randomized, double-blind, placebo-controlled study is a type of intervention study in which participants are randomly assigned to receive a treatment (intervention) or placebo (control). In this way, each participant has an equal chance of being in each group. The term *double-blind* means that neither the investigator nor the participant knows to which group participants have been assigned. People in the placebo group are given a treatment that is similar to the treatment but does not contain or have the intervention of interest. This type of study is considered the "gold standard" in nutrition research, because it helps decrease the chance that researcher bias has occurred, helps account for placebo effect, and controls for confounding variables. It is also optimal if a researcher wishes to test for a cause-and-effect relationship. **13.** The *nutrition transition* refers to the shift from undernutrition to overnutrition (or unbalanced nutrition) that sometimes occurs when an agricultural society makes the transition to a more industrialized economy. Whereas nutrient deficiencies are common in less industrialized societies, chronic disease (such as heart attack and diabetes) is more common in industrialized societies.

### Practice Calculations

**1.** A serving of this food contains 200 kcal from carbohydrates, 100 kcal from protein, and 45 kcal from lipid. This translates to 58% of energy from carbohydrate, 29% from protein, and 13% from lipid. **2.** Answers vary depending on food choices. However, as a hypothetical example, consider a meal consisting of average serving sizes of, lima beans, grapes, a T-bone steak, and 2% milk. This meal would contribute 631 kcal, of which 152, 164, and 315 kcal would come from carbohydrates, proteins, and lipids, respectively. Thus 24, 26, and 50% of total calories would come from carbohydrates, proteins, and lipids, respectively.

## Chapter 2

### Review Questions

**1.** d **2.** d **3.** c **4.** b **5.** a **6.** b **7.** c
**8.** a **9.** b **10.** d

**11.** The four methods of assessing nutritional status are anthropometric measurements, biochemical measurements, clinical assessment, and dietary assessment. Anthropometric measurements are commonly used during infancy and childhood, whereas biochemical measurements are more frequently used during periods of suspected illness. Medical personnel use clinical assessment to ascertain signs and symptoms of malnutrition, and dietary assessment is frequently conducted by both clinicians and researchers. Although anthropometric measurements are often inexpensive to obtain, they cannot be used to determine specific nutrient deficiencies. Conversely, although biochemical and clinical assessment may be more costly, their results are more precise in that specific nutrient deficiencies can be diagnosed. Dietary assessment is useful in determining whether a diet has variety and balance, although it tends to be more time consuming than other methods. **12.** The four categories of Dietary Reference Intake (DRI) values are the Estimated Average Requirement (EAR), the Recommended Dietary Allowance (RDA), the Adequate Intake (AI) level, and the Tolerable Upper Intake Level (UL). Of these, the RDA and AI are meant to serve as dietary intake goals. Recommended Dietary Allowances (RDAs) were determined from information used to establish the EAR values. When EARs (and thus RDAs) could not be established, AIs were set. **13.** The Recommended Dietary Allowances (RDAs) and Adequate Intake (AI) levels can be used as dietary intake goals. Thus dietary planning should ensure that these amounts of nutrients are obtained from foods. The current Daily Values (DVs) are based on older nutrient intake guidelines but will likely be changed in the near future to reflect current RDA or AI values.

### Practice Calculations

**1.** Between 900 and 1,300 kcal should come from carbohydrates. One change that would decrease carbohydrate intake would be to vary the foods that are eaten at breakfast. Instead of having a bowl of cereal and a piece of toast each morning, alternate that with scrambled eggs or a yogurt. However, it would be unwise if this change decreased the amount of whole grains consumed all day. **2.** This person's EER is 3,025 kcal/day. **3.** She is consuming 333% of her RDA for vitamin E. However, because she is still consuming less than her UL for vitamin E, she need not be concerned with this.

## Chapter 3

### Review Questions

**1.** b **2.** a **3.** d **4.** b **5.** a **6.** b **7.** c
**8.** c **9.** b **10.** c

**11.** Oxidation is the loss of electrons, whereas reduction is the gain of electrons. When an atom is oxidized it loses an electron, which has a negative charge. Therefore the atom becomes more positive. **12.** Ionic bonds form when atoms with opposite charges are drawn together. That is, cations with positive charges are attracted to anions with negative charges. Covalent bonds form when atoms want to complete their outer valence shell. In this case, atoms share electrons. **13.** Electrons in nonpolar covalent bonds are shared equally between atoms. In a polar covalent bond, electrons are not shared equally. This makes one end of the molecule more negative and the other end of the molecule more positive. **14.** A solute dissolves in a solvent, making a solution. In this example, sodium chloride is the solute and water is the solvent. When sodium chloride dissolves in water, a solution of sodium chloride and water (salt water) forms. **15.** An acid releases hydrogen ions when dissolved in solution, whereas a base releases hydroxide ions. When an acid such as hydrochloric acid dissolves in water, hydrogen and chloride ions are released. The increased concentration of hydrogen ions causes the pH of the solution to decrease, becoming more acidic. When a base such as sodium hydroxide dissolves in water, sodium and hydroxide ions are released. The hydroxide ions combine with hydrogen ions to form water. This decreases the concentration of hydrogen ions in solution causing the pH to increase. **16.** During simple diffusion solutes cross cell membranes by moving from a higher concentration to a lower concentration, until equilibrium is reached. During facilitated diffusion, solutes require the assistance of a transport protein to cross cell membranes. Solutes move from a higher to a lower concentration until equilibrium is reached. Water crosses cell membranes by osmosis. The net movement of water is determined by the concentration of solutes. Water moves from a low solute concentration to a higher solute concentration, until the solute concentration is equal on both sides of the cell membrane. These three transport mechanisms are passive because energy is not required. Active transport mechanisms require energy to move solutes across cell membranes. These include carrier-mediated active transport and vesicular active transport. During carrier-mediated active transport, a solute moves from a region of lower to a region of higher concentration. This requires both energy and the assistance of a transport protein. There are two types of vesicular active transport— endocytosis and exocytosis. During endocytosis, substances outside the cell are enclosed in vesicles and released within the cell. During exocytosis, substances within cells are enclosed in vesicles and released outside the cell. **17.** A negative feedback system is a physiological response to a change in the internal environment that, when successful, is turned off when the initial stimulus is no longer present. The purpose of this system is to restore balance. Restored balance, or homeostasis, inhibits the response from continuing.

## Chapter 4
### Review Questions

**1.** c  **2.** a  **3.** b  **4.** b  **5.** c  **6.** c  **7.** c
**8.** a  **9.** d  **10.** c  **11.** d  **12.** d  **13.** a  **14.** b
**15.** d

**16.** The GI tract contains four tissue layers—the mucosa, submucosa, muscularis, and serosa. The mucosa is the innermost lining and consists of epithelial cells. These cells produce and release a variety of secretions needed for digestion. The submucosa is a layer of connective tissue that surrounds the mucosal layer. It contains blood vessels that provide nourishment to the mucosa and muscularis. It also contains lymphatic vessels and a network of nerves that control the release of secretions from the mucosa. The muscularis usually consists of two layers of smooth muscles. A network of nerves embedded between the layers of muscles control the contraction and relaxation of the muscularis. The function of the muscularis is to promote GI motility. The last layer is the serosa, which is also connective tissue, that encloses the GI tract. The serosa secretes a fluid that lubricates the digestive organs and anchors them within the abdominal cavity. **17.** Organs that produce and release secretions needed for digestion include the salivary glands, stomach, pancreas, gallbladder, and small intestine. The salivary glands release saliva, which contains water, salt, and digestive enzymes. Saliva moistens foods and facilitates chemical breakdown. The stomach releases gastric juice, which contains a mixture of water, hydrochloric acid, and enzymes. When food mixes with gastric juice, chyme is formed. The pancreas releases pancreatic juice, which contains water bicarbonate, and digestive enzymes. The bicarbonate neutralizes the acidic chyme as it passes from the stomach into the small intestine. The pancreatic enzymes are needed for the chemical breakdown of nutrients. The gallbladder releases bile (produced in the liver), which is an emulsifying agent. Bile enables fat to mix in the watery environment of the small intestine. Last, the small intestine releases a variety of enzymes, which are needed for the chemical breakdown of nutrients. **18.** Digestive events taking place in the mouth include the physical and chemical breakdown of food. Food mixes with saliva, which helps moisten the food. After swallowing, food moves into the stomach stimulating the release of the hormone gastrin. Gastrin stimulates the release of gastric juice and GI motility. Increased gastric motility causes food to mix with gastric juice. As it becomes more liquid, it is called chyme. The enzymes present in gastric juice chemically break down nutrients. Chyme passes into the small intestine. The small intestine releases enteric hormones—cholecystokinin (CCK), secretin, and gastric inhibitory protein. These hormones coordinate the release of secretions from the pancreas and gallbladder, the relaxation of sphincters, and GI motility. Enterocytes, which make up the brush border of the small intestine, produce enzymes for the final stage of digestion. When nutrients are completely broken down, they are absorbed into the enterocyte, the primary site of nutrient absorption. **19.** The lining of the small intestine has a large surface area, making it well suited for nutrient absorption. First, the mucosa is arranged in large, pleated folds that face toward the lumen.

These folds are covered with tiny projections called villi. Each villus is lined with absorptive epithelial cells called enterocytes. Each enterocyte is covered with even smaller projections called microvilli. The microvilli make up the absorptive surface, which is referred to as the brush border. These structures create an enormous surface area where nutrient absorption takes place. **20.** The enteric nervous system, the enteric endocrine system, and the central nervous system regulate GI motility (movement) and the release of GI secretions. The enteric nervous and endocrine systems are embedded in the various tissue layers of the GI tract. The enteric nervous system monitors and responds to changes related to digestive activities by communicating with muscles and glands. Neural connections maintain communication between the central nervous system and the GI tract. This is why emotional stimuli can affect GI function. The enteric endocrine system produces hormones that also regulate the process of digestion.

## Chapter 5
### Review Questions

**1.** c  **2.** c  **3.** d  **4.** d  **5.** d  **6.** b  **7.** c
**8.** b  **9.** c  **10.** d

**11.** Disaccharides consist of two monosaccharides bonded together. Sucrose is made up of glucose and fructose; lactose is made up of glucose and galactose; and maltose is made up of two glucose molecules. A glycosidic bond forms between monosaccharides by a condensation reaction. This occurs when a hydrogen atom from one monosaccharide interacts with a hydroxyl group from another monosaccharide. In the process, a water molecule is formed. **12.** The digestion of amylose and amylopectin involve many of the same enzymes. The exception is α-dextrinase, which is needed to hydrolyze α-1,6 glycosidic bonds present in amylopectin. The digestion of amylose and amylopectin begins in the mouth. The salivary glands release α-salivary amylase, which hydrolyzes α-1,4 glycosidic bonds present in both amylose and amylopectin. This results in a partial-breakdown product called dextrins. There is no starch digestion in the stomach. Dextrins then pass into the small intestine. The pancreas releases α-pancreatic amylase, which continues to hydrolyze α-1,4 glycosidic bonds. Dextrins from amylose are broken down to maltose, whereas dextrins from amylopectin result in maltose and limit-dextrins. Limit-dextrins contain α-1,6 glycosidic bonds that were located at branch points in the original amylopectin molecule. Maltose is hydrolyzed to glucose by the brush border enzyme maltase and limit-dextrins are hydrolyzed to glucose by α-dextrinase, also a brush border enzyme. **13.** The hormones insulin and glucagon play important roles in glucose homeostasis. Both of these hormones are released from the pancreas. When blood glucose increases, additional insulin is released into the blood. Some cells, called insulin-sensitive cells, require the presence of insulin to take up glucose. Insulin binds to receptors on the surface of insulin-sensitive cells. The binding of insulin to insulin receptors causes glucose transport molecules to relocate from the cell's cytoplasm to the surface of the cell membrane. Glucose transport proteins enable glucose to cross the cell membrane. As a result, blood glucose concentrations decrease. Insulin also facilitates glycogen synthesis. The hormone glucagon is released when blood glucose is low. Glucagon stimulates glycogenolysis in the liver. Glycogenolysis is the breakdown of glycogen and the release of glucose molecules. When liver glycogen is broken down, the glucose is released into the blood. As a result, blood glucose rises. **14.** The citric acid cycle both begins and ends with a molecule called oxaloacetate. Specifically, oxaloacetate combines with acetyl-CoA to produce citrate. Thus, adequate amounts of oxaloacetate are needed to keep the citric acid cycle fully functional. In addition, oxaloacetate is used to synthesize glucose by a process called gluconeogenesis. When glucose is limited, cells increase their use of fatty acids for energy. Fatty acid metabolism results in many molecules of acetyl-CoA, which normally enter the citric acid cycle by combining with oxaloacetate. However, when oxaloacetate is limited, molecules of acetyl-CoA are not able to readily enter the citric acid cycle. As a result, they enter a different metabolic pathway that forms ketones.

## Practice Calculations

**1.** 45% of 1800 kcal = 810 kcal
65% of 1800 kcal = 1170 kcal
810 kcal = 203 g of carbohydrate
1170 kcal = 293 g of carbohydrate
This person should consume 203–293 g carbohydrate per day

**2.** 25 g of added sugar = 100 kcal
10% of 1500 kcal = 150 kcal
Less than 10% of kcal are from added sugar, as recommended by the USDA Dietary Guidelines for Americans.

**3.** Recommendation 14 g dietary fiber per 1,000 kcal
For 2,500 kcal per day, a person should consume approximately 35 g of dietary fiber

# Chapter 6

## Review Questions

**1.** a **2.** b **3.** c **4.** d **5.** c **6.** b **7.** c
**8.** d **9.** c **10.** d

**11.** Sickle cell anemia is an inherited disease caused by a mutation in the DNA (gene) that codes for the protein hemoglobin. This alteration in the genetic code causes an incorrect amino acid to be inserted into the protein during translation, producing a hemoglobin molecule that cannot function properly. Signs and symptoms of sickle cell anemia include anemia, chest pain, swollen hands, and stunted growth. **12.** There are four levels of protein structure. These are called primary, secondary, tertiary, and quaternary structures. Primary structure refers to the number, types, and order of amino acids in the polypeptide chain. Secondary structure is how the protein's hydrogen bonding between amine and carboxylic acid groups causes folding—usually in the form of α-helices or β-folded sheets. Tertiary structure is caused by attractions between R-groups, and quaternary structure occurs when more than one polypeptide join together and/or a nonprotein subunit (a prosthetic group) are components of a single functional protein. **13.** Protein complementation is the process whereby incomplete proteins are consumed in combination so that the mix of foods provides sufficient amounts of all essential amino acids. Thus a "complete protein" is consumed. Protein complementation is especially important during periods of rapid growth and development (such as early childhood) when amino acid requirements are relatively high and overall food intake may be limited. **14.** Proteins serve many functions in the body. For example, they make up most of our muscle tissue and are therefore required for movement and basic physiological processes (such as cardiac function). The immune system also requires proteins for antibody production and for synthesizing and maintaining protective barriers such as the skin. Proteins also serve as important communicators in the body, making up many of the hormones that it produces. In addition, all enzymes are made from proteins. Thus all the metabolism required for basic body functions is protein dependent as well.

## Practice Calculations

**1.** Between 200 and 700 kcal should come from protein. Because most protein in the diet comes from meat and dairy products, one should substitute cereals, fruits, and vegetables for these foods when protein intake is higher than recommended. **2.** Answers can vary slightly depending on the food composition table used and the individual RDA for protein. If one uses the food composition table that accompanies this text, this meal provides 40 g of protein. For a person (for example, a 25-year-old female) who requires 46 g of protein daily, this would represent 87% of the RDA. Because animal products supply most of the protein in the diet, one might choose to consume milk instead of a soft drink to increase the protein content of this meal. **3.** The answer to this question is variable, depending on which MyPyramid is appropriate for you as well as your personal meat, bean, and dairy preferences. See your instructor for additional assistance with this problem.

# Chapter 7

## Review Questions

**1.** a **2.** b **3.** d **4.** c **5.** d **6.** a **7.** d
**8.** b **9.** d **10.** b

**11.** A phospholipid is composed of a glycerol molecule bonded to two fatty acids and a polar (phosphate-containing) head group. Of these components, the fatty acids are the most hydrophobic, and the polar head group is the most hydrophilic. Because it contains both polar (hydrophilic) and nonpolar (hydrophobic) regions, a phospholipid is said to be amphipathic. In general, phospholipids are important in the structure of cell membranes and are used to transport lipids in the body. **12.** Triglycerides, diglycerides, and monoglycerides are all types of lipids composed of one molecule of glycerol and variable numbers of fatty acids. Triglycerides contain three fatty acids, whereas diglycerides and monoglycerides contain two and one fatty acid(s), respectively. Of these, triglycerides are most abundant in the body. **13.** The Acceptable Macronutrient Distribution Ranges (AMDRs) suggest that we consume 20 to 35% of energy from lipids. The Dietary Guidelines for Americans recommend that we "choose fats wisely for good health." This means we should limit our intake of saturated fatty acids to less than 10% of calories and *trans* fatty acids to less than 1% of calories, while consuming greater amounts of polyunsaturated and monounsaturated fatty acids.

## Practice Calculations

**1.** A person requiring 2,000 kcal/day should consume between 100 and 200 kcal of linoleic acid daily. This translates to 11 to 22 g linoleic acid each day. Many foods contain linoleic acid including nuts, seeds, and a variety of vegetable oils. **2.** A person requiring 2,000 kcal/day should consume between 12 and 24 kcal of linolenic acid daily. This translates to 1.3 to 2.7 g linolenic acid each day. If linolenic acid intake is low, one might want to consume more foods containing canola oil or flaxseed oil. In addition, to assure intake of longer-chain ω-3 fatty acids, increased consumption of fatty fish and seafood may be beneficial. **3.** A person requiring 2,750 kcal/day should consume between 550 and 963 kcal of energy from lipids daily. This translates to 61 to 107 g lipid.

# Chapter 8

## Review Questions

**1.** d **2.** a **3.** c **4.** d **5.** c **6.** c **7.** a
**8.** c **9.** b **10.** c

**11.** The period shortly after a meal is referred to as the absorptive state. Absorbed nutrients enter the blood, stimulating the release of insulin. Insulin increases the activity of anabolic pathways involved in synthesizing proteins, fatty acids, triglycerides, and glycogen. During this time, glucose is the body's major source of energy (ATP). After 24 hours without food, blood glucose decreases and the body's glycogen stores are depleted. Glucagon is the dominant hormone during this time. Glucose comes from noncarbohydrate sources, mainly glucogenic amino acids, via gluconeogenesis. There is an increase in the mobilization of fatty acids for a source of energy. Ketogenesis also increases, and cells begin to use ketones as an energy source. **12.** Certain cells depend on glucose for a source of energy. When glucose is limited, cells synthesize glucose from noncarbohydrate sources, mainly glucogenic amino acids, via a metabolic process called gluconeogenesis. The rate of muscle breakdown is very rapid. Some cells decrease their glucose requirements by using an alternate energy source called ketones. In this way, ketones spare lean body mass. **13.** Carbohydrate catabolism begins with glycogenolysis, the breakdown of glycogen to glucose. Glucose enters the metabolic pathway called glycolysis. Glycolysis splits a glucose molecule, forming two molecules of pyruvate. If oxygen is readily available, each pyruvate combines with a molecule called coenzyme A, forming acetyl-CoA. If oxygen is not readily available, pyruvate is converted to lactate. Some ATP and NADH + H$^+$ are formed during glycolysis. Each molecule of acetyl-CoA enters the citric acid cycle by combining with oxaloacetate, forming citrate. Citrate is metabolized via the citric acid cycle, forming reduced coenzymes NADH + H$^+$ and FADH$_2$. ATP is also formed via GTP. The reduced coenzymes enter the electron transport chain and undergo oxidative phosphorylation. Each NADH + H$^+$ yields approximately 3 ATPs, whereas each FADH$_2$ yields approximately 2 ATPs. Protein catabolism begins with proteolysis, releasing amino acids. For amino acids to be used as an energy source, the amine group is removed by a two-step process—transamination and deamination. Transamination involves the transfer of the amine group from the amino acid to an α-ketoacid. Often this α-ketoacid is α-ketoglutarate, which forms the amino acid glutamate. The remaining carbon skeleton forms another α-ketoacid. Next, the newly formed glutamate is deaminated. This includes the removal of the amino group, which is used to synthesize ammonia. Ammonia is converted to urea, which is released into the blood. Urea is filtered out of the blood by the kidneys and excreted in the urine. The carbon skeleton remaining from the deaminated amino acid is used as an energy source. The structure of the α-ketoacid determines where it enters the citric acid cycle and how much ATP is produced. Lipid catabolism begins with lipolysis, the breakdown of triglycerides to glycerol and fatty acids. Glycerol can be used as a source of energy or used to synthesize glucose. Fatty acids are metabolized into numerous molecules of acetyl-CoA via a metabolic process called β-oxidation. For each acetyl-CoA formed, one NADH + H$^+$ and one FADH$_2$ is generated. Each molecule of acetyl-CoA enters the citric acid cycle by combining with oxaloacetate. On completion of this cycle, more NADH + H$^+$ and FADH$_2$ are generated. These reduced coenzymes enter the electron transport chain and undergo oxidative phosphorylation, producing ATP. **14.** Glycolysis is an anaerobic metabolic pathway that breaks down glucose to form two molecules of pyruvate. Small amounts of ATP are generated. The citric acid cycle is a metabolic pathway that begins when oxaloacetate combines with a molecule of acetyl-CoA to form citrate. This pathway generates NADH + H$^+$ and FADH$_2$. The metabolic pathway called β-oxidation metabolizes fatty acids to form numerous molecules of acetyl-CoA. For each molecule of acetyl-CoA formed, one molecule each of NADH + H$^+$ and FADH$_2$ are generated. **15.** Some ATP is generated by a process called substrate phosphorylation. This means that an inorganic phosphate is attached directly to a molecule of ADP to form ATP. Most ATP is generated via oxidative phosphorylation. This takes place in mitochondria and involves structures that make up the electron transport chain. Oxidative phosphorylation occurs when reduced coenzymes (NADH + H$^+$ and one FADH$_2$) become oxidized. The oxidation of the coenzymes provides the energy needed to phosphorylate ADP to generate ATP.

## Chapter 9
### Review Questions

**1.** a     **2.** a     **3.** a     **4.** b     **5.** a     **6.** a     **7.** b
**8.** c     **9.** b     **10.** d

**11.** Three methods used to assess body composition are hydrostatic weighing, DEXA, and skinfold thickness. Hydrostatic or underwater weighing requires a person to be submerged in water and compares a person's underwater weight to weight on land. DEXA uses x-ray beams to differentiate between fat mass and lean mass. Skinfold measures are made using calipers, an instrument that measures the thickness of fat folds. A mathematical formula is then used to estimate body fat. **12.** The set point theory suggests that body weight, in part, is regulated by internal mechanisms that help maintain a relatively stable body weight over an extended period of time. These internal mechanisms make adjustments in energy intake and energy expenditure to restore body weight to its desired weight or what is called the "set point." **13.** Adiposity is communicated to the brain in part by circulating concentrations of the hormones leptin and insulin. When adiposity increases, leptin and insulin concentrations increase, signaling neurons in the hypothalamus to release catabolic neuropeptides. Catabolic neuropeptides decrease energy intake and increase energy expenditure, resulting in negative energy balance and favoring weight loss. When adiposity decreases, leptin and insulin concentrations decrease, signaling neurons in the hypothalamus to release anabolic neuropeptides. These neuropeptides increase energy intake and decrease energy expenditure, resulting in positive energy balance and favoring weight gain. **14.** Advocates of low-carbohydrate weight-loss diets claim that the hormone insulin facilitates weight gain. The reduc-

tion of carbohydrate in the diet results in less insulin released from the pancreas. With lower concentrations of insulin in the blood, body fat is not as readily stored. **15.** Three signals originating from the GI tract that regulate hunger and satiety include gastric stretching, circulating levels of nutrients, and the hormone ghrelin. As the stomach fills with food, the stomach walls stretch, a process called gastric stretching. This signals the brain, which initiates the feeling of satiety. Increased blood concentrations of glucose, amino acids, and fatty acids trigger satiety, whereas low concentrations of glucose in the blood trigger hunger. The hormone ghrelin is produced by the stomach and released in response to hunger. Concentrations decrease after food intake. **16.** Total energy expenditure (TEE) refers to energy expenditure associated with basal metabolism, physical activity, and thermic effect of food. Basal metabolism, the largest component of TEE, is energy expenditure associated with vital body functions. Energy expenditure associated with physical activity accounts for 15 to 30% of TEE. The small component of TEE (10%) is the thermic effect of food. This refers to energy expenditure associated with processing food by the body.

### Practice Calculations

**1.** BMI = 25 kg/m$^2$; this person is at the higher end of the normal weight classification or at the lower end of the overweight classification. BMI does not take into account body composition. Therefore, physical activity must also be considered when making a final assessment. **2.** WHR = 1.05; in males, a WHR that exceeds 0.95 indicates android adiposity. Android adiposity is associated with increased risk for certain chronic diseases such as cardiovascular disease, hypertension, and type 2 diabetes.

## Chapter 10
### Review Questions

**1.** c     **2.** b     **3.** d     **4.** c     **5.** b     **6.** a     **7.** d
**8.** b     **9.** d     **10.** b

**11.** Thiamin is found in fish, legumes, and pork. Good sources of riboflavin include liver, meat, and dairy products. Niacin (or tryptophan, its precursor) is abundant in meat, tomatoes, and mushrooms. Good sources of pantothenic acid include mushrooms, organ meats, and sunflower seeds. Vitamin B$_6$ is plentiful in chickpeas, fish, and liver. Food sources of biotin include peanuts, tree nuts, and eggs. Folate is found in many fruits and vegetables such as spinach and oranges; it is also abundant in organ meats. Vitamin B$_{12}$ is found in shellfish, meat, and dairy products. Brightly colored fruits and vegetables such as citrus fruits, peppers, and papayas tend to be good sources of vitamin C, whereas choline is most abundant in eggs, liver, legumes, and pork. **12.** Vitamin B$_{12}$ deficiency results in the inability of the body to convert the inactive form of folate—5-methyl tetrahydrofolate (5-methyl THF)— to its active form of THF. Thus primary vitamin B$_{12}$ deficiency can cause secondary folate deficiency. Because the signs and symptoms of folate deficiency can lead a clinician to conclude that the person has primary folate deficiency, the true cause of malnutrition (vitamin B$_{12}$ deficiency) may be overlooked. Because of this, vitamin B$_{12}$ deficiency is said to be "masked" when it is accompanied by folate deficiency. **13.** Antioxidants donate electrons to other compounds, thus reducing them, and are important for a variety of reasons in the body. For example, antioxidants (such as vitamin C) protect the body from free radical damage. Free radicals are compounds that have unpaired electrons in their outer shells. These compounds seek to take electrons from other substances, thus oxidizing them. Vitamin C readily donates its own electrons to free radicals (stabilizing them), thus protecting DNA, proteins, and lipids from oxidative damage. Vitamin C also donates electrons to enzymes that need them in order to function. These enzymes tend to be metalloenzymes, containing iron or copper.

### Practice Calculations

**1.** Answers can vary slightly depending on food composition table used and individual RDA for folate. However, if one uses the food composition table that accompanies this text, this meal provides approximately 60 micrograms (μg) of folate. For a person (for example, a 25-year-old female) who requires 400 μg of folate daily, this would represent 15% of the RDA. To increase folate consumption, a variety of fruits and vegetables, legumes, and enriched cereal products should be added to the diet. For example, a salad could be ordered instead of the fries in this meal. **3.** The answer to this question is variable, depending on which MyPyramid is appropriate for you as well as your personal fruit and vegetable preferences. See your instructor for additional assistance with this problem.

## Chapter 11

### Review Questions

**1.** a  **2.** c  **3.** d  **4.** d  **5.** a  **6.** d  **7.** b
**8.** c  **9.** d  **10.** d

**11.** Preformed vitamin A is found in animal foods, being especially abundant in organ meats, fatty fish, and dairy products. Provitamin A carotenoids tend to be found in yellow, orange, and red fruits and vegetables such as cantaloupe and carrots. Egg yolks, foods containing dairy fat (such as butter), and fatty fish are good sources of vitamin D. Vitamin E is abundant in many vegetable oils, nuts, and seeds, and vitamin K is found in dark green vegetables (such as kale and spinach) as well as fish and legumes. All these vitamins can be destroyed by exposure to extreme temperature and oxygen, except for vitamin D which is relatively more stable. **12.** Several vitamins, including vitamins C and E, serve as antioxidants in the body. Vitamin E is especially important in protecting phospholipids from oxidative damage in cell membranes. Vitamin E donates its electrons to free radicals, thus stabilizing them and preventing them from taking electrons from the fatty acid components of the phospholipid bilayer. In addition, some carotenoids (such as β-carotene) are also potent antioxidants, protecting DNA, proteins, and fatty acids from oxidative damage. **13.** Vitamin D is called the "sunshine vitamin" because its synthesis in the body depends on exposure to sunlight. This involves the conversion of 7-dehydrocholesterol to previtamin $D_3$ (precalciferol) in the skin via a reaction that requires ultraviolet light found in sunlight—hence, the name "sunshine vitamin." Next, in the skin previtamin $D_3$ is converted to vitamin $D_3$ (cholecalciferol), which in turn is converted to 25-hydroxyvitamin D [25-(OH)$D_3$] in the liver and then [1,25-(OH$_2$)$D_3$ or calcitriol] in the kidneys.

### Practice Calculations

**1.** Answers can vary slightly depending on food composition table used and individual RDA for vitamin A. However, if one uses the food composition table that accompanies this text, this meal provides negligible amounts of vitamin A, representing 0% of any person's RDA. To increase vitamin A (or provitamin A carotenoid) consumption, milk or a milkshake could be substituted for the cola, and a mixed salad could be eaten as a side dish. **3.** The answer to this question is variable, depending on which MyPyramid is appropriate for you as well as your personal meat, bean, and dairy preferences. See your instructor for additional assistance with this problem.

## Chapter 12

### Review Questions

**1.** a  **2.** c  **3.** b  **4.** a  **5.** d  **6.** c  **7.** b
**8.** d  **9.** d  **10.** d

**11.** In some cases, the concentration of certain minerals in the soil and water can greatly affect the amount of these minerals in the plants and animals raised in a region. This is because plants and animals obtain their minerals from the soil and water available to them. **12.** Because the content and bioavailability of iron in animal products tends to be high, vegetarians are at greater risk of iron deficiency than omnivores. This is especially true for vegans who do not consume any animal products. Similarly, because bioavailability of zinc from animal sources is especially high, it is recommended that vegetarians (especially vegans) consume up to 50% more zinc than omnivores. To help maintain healthy iron status, vegetarians who do eat fish are advised to consume it along with iron-containing plant foods; iron supplements may be necessary.

In addition, zinc-containing foods such as legumes and fortified cereal products should be emphasized. **13.** Both cretinism and goiter are caused by iodine deficiency. However, cretinism occurs in infants of severely iodine-deficient mothers, whereas goiter occurs later in life. Although this is not well understood, it is likely that babies with cretinism are not born with goiters because goiters have had insufficient time to form. It is also possible that the immature thyroid gland is unable to respond to iodine deficiency by increasing its growth and thus developing a goiter. There may also be genetic factors predisposing some fetuses to developing cretinism without goiters *in utero.*

### Practice Calculations

**1.** Answers can vary slightly depending on food composition table used and individual RDA for iron. However, if one uses the food composition table that accompanies this text, this meal provides approximately 4.9 milligrams (mg) of iron. For a person (for example, a 25-year-old male) who requires 8 mg of iron daily, this would represent approximately 61% of the RDA. To increase iron consumption, a cheeseburger could be substituted for the hamburger, or milk could be consumed instead of the cola. **3.** The answer to this question is variable, depending on which MyPyramid is appropriate for you as well as your personal meat, beans, and dairy preferences. See your instructor for additional assistance with this problem.

## Chapter 13

### Review Questions

**1.** b  **2.** d  **3.** d  **4.** b  **5.** b  **6.** d  **7.** d
**8.** a  **9.** c  **10.** c

**11.** Reducing salt intake can be accomplished in a variety of ways, including choosing foods labeled as being "low-sodium" or "sodium-free," preparing foods without additional salt, adding less salt at the table, limiting the amount of sodium-containing condiments used, and replacing high-salt snacks (such as chips) with fresh fruits and vegetables. **12.** Osmosis is the phenomenon by which water moves from a region of low solute concentration to a region of high solute concentration. Using the principle of osmosis, the body can actively transport sodium and chloride ions across cell membranes in such a way that water passively follows. For example, for water to move from an intracellular space to an extracellular compartment, chloride can be actively pumped from the cell's cytosol out to the surrounding interstitial space. In this way, fluids can become part of a cell's secretions. **13.** Phosphorus is needed for energy metabolism because it is a critical component of adenosine triphosphate (ATP), and it is required for fatty acid transport via lipoproteins. Phosphorus deficiency can cause fatigue and weakness for many reasons. For example, without sufficient phosphorus to generate ATP, energy-requiring reactions (such as those needed for muscle function) cannot occur. In addition, because fatty acids provide an important source of energy to cells, phosphorus deficiency can decrease the availability of these energy-yielding nutrients to the body by inhibiting lipoprotein synthesis.

### Practice Calculations

**1.** Answers can vary slightly depending on food composition table used and individual AI for sodium. However, if one uses the food composition table that accompanies this text, this meal provides approximately 1,006 mg of sodium. For a person (for example, a 25-year-old male) who requires 1,500 mg (1.5 g) of sodium daily, this would represent approximately 67% of the AI and 44% of the maximum sodium intake recommended in the Dietary Guidelines for Americans (2,300 mg/day). As most of the sodium in this meal was from the fries, sodium intake could be decreased by substituting a baked potato, salad, or fruit salad for this component. **2.** The answer to this question is variable, depending on which MyPyramid is appropriate for you as well as your personal fruit and vegetable preferences. See your instructor for additional assistance with this problem.

# Chapter 14

## Review Questions

**1.** d      **2.** b      **3.** a      **4.** d      **5.** b      **6.** d      **7.** a
**8.** b      **9.** c      **10.** c      **11.** c

**12.** Infancy has one of the highest rates of growth and development during the life cycle. During the first year of life, weight triples and length doubles. The rate of growth decreases during childhood. During adolescence, hormonal changes cause the reproductive organs to mature. The onset of puberty varies, but females tend to experience puberty at an earlier age than males. During the adolescent growth spurt, female height increases by approximately 6 inches, whereas males gain an additional 8 inches. In addition to increased height, weight increases and body composition changes. **13.** Infant suckling causes neural signals to stimulate the hypothalamus. In turn, the hypothalamus signals the pituitary gland to release the hormones prolactin and oxytocin. Prolactin stimulates the secretory cells in the mammary gland to synthesize milk, whereas oxytocin causes small muscles, called myoepithelial cells, to contract. This forces milk out of the alveoli and into the mammary ducts. The release of milk is called milk let-down. **14.** Three physical changes associated with aging include changes in body composition (loss of lean mass and increased fat mass), a decline in the production of gastric secretions, and age-related bone loss. These changes can affect nutrient requirements. For example, the decline in lean mass can reduce caloric requirements. Decreased ability to produce gastric secretions can interfere with digestion and nutrient absorption. Requirements for calcium increase because age-related bone loss can lead to osteoporosis. **15.** There are many reasons why elderly people are at increased risk for nutrient deficiencies. Many elderly people are on fixed incomes, limiting money available for food. Some elderly people become isolated and depressed, which can lead to inadequate food intake. Also, loss of teeth can limit the variety and types of food consumed. This too can lead to nutritional problems.

# Glossary

**1,25-dihydroxyvitamin D ([1,25-(OH)₂ D₃], calcitriol)** The active form of vitamin D in the body produced in the kidneys from 25-(OH) D₃.

**24-hour recall** A retrospective dietary assessment method that analyzes each food and drink consumed over the previous 24 hours.

**25-hydroxyvitamin D [25-(OH) D₃]** An inactive form of vitamin D that is made from cholecalciferol in the liver.

**5-methyltetrahydrofolate (5-methyl THF)** An inactive form of folate.

**7-dehydrocholesterol** A metabolite of cholesterol that is converted to cholecalciferol (vitamin D₃) in the skin.

## A

**α-dextrinase** Intestinal enzyme that hydrolyzes α-1,6 glycosidic bonds.

**α-helix** A common configuration that makes up many proteins' secondary structures.

**α-ketoacid** The structure remaining after the amino group has been removed from an amino acid.

**α-tocopherol** The most active form of vitamin E.

**"ABCDs" of nutritional status assessment** Four components of assessing nutritional status: anthropometric measurements, biochemical measurements, clinical assessment, and dietary assessment.

**absorption** The passage of nutrients through the lining of the GI tract into the blood or lymphatic circulation.

**absorptive state** (postprandial period) The first four hours after a meal.

**Acceptable Macronutrient Distribution Ranges (AMDRs)** Recommendations concerning the distribution or percentages of energy from each of the macronutrient classes.

**acetaldehyde dehydrogenase (ALDH)** An enzyme that converts acetaldehyde to acetic acid.

**acid** A substance that releases hydrogen ions (H⁺) when dissolved.

**acidic** Having a pH less than 7.

**acidosis** A condition resulting from the accumulation of acids in body fluids.

**acrodermatitis enteropathica** A genetic abnormality resulting in decreased absorption of dietary zinc.

**acrylamide** A compound that is formed in starchy foods (such as potatoes) when heated to high temperatures.

**active site** An area on an enzyme that binds substrates in a chemical reaction.

**acute starvation** Early stages of starvation defined as the first five days of fasting or minimal food intake, beginning 24 hours after the last meal.

**adaptation response** Physiological changes that result from exercise that help improve physical fitness.

**adaptive thermogenesis** Energy expended to adapt to changes in the environment or to physiological conditions.

**adenosine triphosphate (ATP)** A chemical used by the body when it needs to perform work.

**Adequate Intake (AI) level** Nutrient intake of healthy populations that appears to support adequate nutritional status set as goals of dietary assessment and planning when RDAs cannot be established.

**adipocyte** A specialized cell that makes up adipose tissue.

**aflatoxin** A toxic compound produced by certain molds that grow on peanuts, some grains, and soybeans.

**Al-Anon** An organization dedicated to helping people cope with alcoholic family members and friends.

**Alateen** An organization dedicated to helping children cope with an alcoholic parent.

**albumin** A protein important in regulating fluid balance between intravascular and interstitial spaces.

**alcohol** An organic compound containing one or more hydroxyl (–OH) groups attached to carbon atoms.

**alcohol dehydrogenase (ADH)** An enzyme found mostly in the liver that metabolizes ethanol to acetaldehyde.

**alcohol dehydrogenase pathway** The primary metabolic pathway that chemically breaks down alcohol in the liver.

**alcoholic cardiomyopathy** Condition that results when the heart muscle weakens in reponse to heavy alcohol consumption.

**alcoholic hepatitis** Inflammation of the liver caused by chronic alcohol abuse.

**Alcoholics Anonymous (AA)** An organization dedicated to helping people achieve and maintain sobriety.

**aldosterone** A hormone produced by the adrenal glands in response to low blood sodium concentration (hyponatremia) and angiotensin II.

**alkalosis** A condition resulting from excess base in body fluids.

**alpha (α) end** The end of a fatty acid with the carboxylic acid (–COOH) group.

**alveolus** (plural, *alveoli*) A cluster of milk-producing cells that make up the mammary glands.

**amino acid** Nutrient composed of a central carbon bonded to an amino group, carboxylic acid group, and a side-chain group (R-group).

**amino group** (–NH₂) The nitrogen-containing component of an amino acid.

**amphibolic** pathway Metabolic pathway that generates intermediate products that can be used for both catabolism and anabolism.

**amphipathic** Having both nonpolar (noncharged) and polar (charged) portions.

**amylopectin** A type of starch consisting of a highly branched arrangement of glucose molecules.

**amylose** A type of starch consisting of a linear chain of glucose molecules.

**anabolic neuropeptide** A protein released by nerve cells that stimulates hunger and/or decreases energy expenditure.

**anabolic pathway** A series of metabolic reactions that require energy to make complex molecules from simpler ones.

**anal sphincters** Internal and external sphincters that regulate the passage of feces through the anal canal.

**android adiposity** Body fat mostly in the intra-abdominal cavity; also referred to as visceral fat.

**aneurysm** The outward bulging of a blood vessel.

**angina pectoris** Pain in the region of the heart, caused by a portion of the heart muscle receiving inadequate amounts of blood.

**angiotensinogen** An inactive protein made by the liver that is converted by renin into angiotensin I.

**angiotensin I** The precursor of angio-tensin II.

**angiotensin II** A protein derived from angiotensin I in the lungs; stimulates aldosterone release.

**animal study** The use of experimental animal subjects such as mice, rats, or primates.

**anion** An ion that has a net negative charge.

**anorexia of aging** Loss of appetite in the elderly that leads to weight loss and overall physiological decline.

**anorexia nervosa (AN)** An eating disorder characterized by an irrational fear of gaining weight or becoming obese.

**anorexia nervosa, binge-eating/purging type** An eating disorder characterized by food restriction as well as bingeing and/or purging.

**anorexia nervosa, restricting type** An eating disorder characterized by food restriction.

**anthropometric measurements** Measurements or estimates of physical aspects of the body such as height, weight, circumferences, and body composition.

**antibody** Protein produced by the immune system; responds to the presence of foreign proteins in the body; helps fight infection.

**antidiuretic hormone (ADH)** (also called vasopressin) A hormone produced in the pituitary gland and released during periods of low blood volume; stimulates the kidneys to decrease urine production, thus conserving water.

**antioxidant** A compound that readily gives up electrons (and hydrogen ions) to other substances.

**aorta** The main artery that initially carries blood from the heart to all areas of the body except the lungs.

**apoptosis** The normal process by which a cell leaves the cell cycle and die; programmed cell death.

**appendix** A small, finger-like appendage attached to the cecum.

**appetite** A psychological desire for food.

**appropriate for gestational age (AGA)** An infant that weighs between the 10th and the 90th percentile for weight for gestational age.

**arachidonic acid** A long-chain, polyunsaturated ω-6 fatty acid produced from linoleic acid.

**arcuate nucleus** A collection of neurons in the hypothalamus that release anabolic and catabolic neuropeptides, which influence food intake and/or energy expenditure.

**ariboflavinosis** A disease caused by riboflavin deficiency.

**arterioles** Small blood vessels that branch off from arteries.

**artery** A blood vessel that carries blood away from the heart.

**ascites** Edema that occurs in the abdominal cavity.

**atherosclerosis** The hardening and narrowing of blood vessels caused by buildup of fatty deposits and inflammation in the vessel walls.

**atom** The smallest portion that an element can be divided into and still retain its properties.

**ATP creatine phosphate (ATP-CP) pathway** An anaerobic metabolic pathway that uses ADP and creatine phosphate to generate ATP.

**ATP synthase** A mitochondrial enzyme that adds a phosphate to ADP to form ATP.

**atrophic gastritis** Inflammation of the mucosal membrane that lines the stomach, reducing the number of cells that produce gastric secretions.

**autoantibodies** Antibodies produced by the immune system; attack the body's own cells.

**autoimmune disease** An immune response that results in the destruction of normal body cells.

**autoimmune disorder** Condition that occurs when the immune system produces antibodies that attack and destroy tissues in the body.

**avidin** A protein present in egg whites that binds biotin, making it unavailable for absorption.

# B

**β-carotene** A provitamin A carotenoid.

**B-complex vitamins** A term used to describe all the B vitamins.

**β-folded sheet** A common configuration that makes up many proteins' secondary structures.

**β-oxidation** The series of chemical reactions that breaks down fatty acids to molecules of acetyl-CoA.

**balance** (or **proportionality**) A concept that emphasizes eating appropriate relative amounts of foods from each food group.

**bariatrics** The branch of medicine concerned with the treatment of obesity.

**basal energy expenditure (BEE)** Energy expended for basal metabolism over a 24-hour period.

**basal metabolic rate (BMR)** Energy expended for basal metabolism per hour (expressed as kcal/hour).

**basal metabolism** Energy expended to sustain metabolic activities related to basic vital body functions such as respiration, muscle tone, and nerve function.

**base** A substance that releases hydroxide ions (OH⁻) when dissolved in water.

**basic** or **alkaline** Having a pH greater than 7.

**basolateral membrane** The cell membrane that faces away from the lumen of the GI tract.

**benign tumor** A growth that has minimal physiological consequences and does not metastasize.

**beriberi** A disease that results from thiamin deficiency.

**bile** A fluid, made by the liver and stored and released from the gallbladder, that contains bile salts, cholesterol, water, and bile pigments.

**bile acids** Amphipathic substances made from cholesterol in the liver; a component of bile important for lipid digestion and absorption.

**bile salt–dependent cholesteryl ester hydrolase** An enzyme produced in the pancreas that cleaves fatty acids from cholesteryl esters.

**bingeing** Uncontrolled consumption of large quantities of food in a relatively short period of time.

**binge drinking** Consumption of five or more drinks in males and four or more drinks in females with the intent to become intoxicated.

**binge eating disorder (BED)** Recurring episodes of consuming large amounts of food within a short period of time not followed by purging.

**bioavailability** The extent to which nutrients are absorbed into the blood or lymphatic system.

**biochemical measurement** Laboratory analysis of biological samples, such as blood and urine, used in nutritional assessment.

**bioelectrical impedance** A method used to assess body composition based on measuring the body's electrical conductivity.

**biological marker (biomarker)** A measurement in a biological sample such as blood or urine that reflects a nutrient's function.

**Bioterrorism Act** Federal legislation aimed to ensure the continued safety of the U.S. food supply from intentional harm by terrorists.

**biotin** (vitamin B₇) A water-soluble vitamin involved in energy metabolism and regulation of gene expression.

**bisphosphonates** A class of drugs sometimes taken to help reduce bone loss.

**Bitot's spots** A sign of vitamin A deficiency characterized by white spots on the eye; caused by buildup of dead cells and secretions.

**blastocyst** Early stage of embryonic development.

**blood alcohol concentration (BAC)** A unit of measurement that describes the level of alcohol in the blood.

**blood clot** (also called thrombosis) A small, insoluble particle made of clotted blood and clotting factors.

**body composition** The distribution of fat, lean mass (muscle), and minerals in the body.

**body mass index (BMI)** An indicator of body fatness calculated as weight (kg) divided by height squared (m²).

**bolus** A soft, rounded mass of chewed food.

**bomb calorimeter** A device used to measure the amount of energy in a food.

**bone marrow** The soft, spongy, inner part of bone that makes red blood cells, white blood cells, and platelets.

**bone mass** The total amount of bone mineral in the body.

**bone mineral density** The amount of bone mineral per unit area.

**bone remodeling** (bone turnover) The process by which older and damaged bone is removed and replaced by new bone.

**bovine somatotropin (bST; bovine growth hormone)** A protein hormone produced by cattle and used in the dairy industry to enhance milk production.

**bovine spongiform encephalopathy (BSE; mad cow disease)** A fatal disease in cattle caused by ingesting prions.

**brush border** The absorptive surface of the small intestine made up of thousands of microvilli that line enterocytes.

**buffer** A substance that releases or binds hydrogen ions in order to resist changes in pH.

**bulimia nervosa (BN)** An eating disorder characterized by repeated cycles of bingeing and purging.

# C

**cachexia** A condition in which a person loses lean body mass (muscle tissue).

**calbindin** A transport protein made in enterocytes that assists in calcium absorption; synthesis is stimulated by vitamin D.

**calcitonin** A hormone produced in the thyroid gland in response to high blood calcium levels.

**calcium (Ca)** A major mineral found in the skeleton and needed for blood clotting, muscle and nerve function, and energy metabolism.

**calorie** A unit of measure used to express the amount of energy in a food.

**cancer** A condition characterized by unregulated cell division.

**capillaries** Blood vessels with thin walls, which allow for the exchange of materials between blood and tissues.

**carbohydrate** Organic compound made up of varying numbers of monosaccharides.

**carboxylation reaction** A metabolic reaction in which a bicarbonate subunit (HCO₃) is added to a molecule.

**carcinogen** Compound or condition that causes cancer.

**cardiac arrhythmia** Irregular heartbeat caused by high intakes of alcohol.

**cardiovascular disease** A disease of the heart or vascular system.

**carnitine** A molecule found in muscle and liver cells that transports fatty acids across the mitochondrial membrane.

**carotenoids** Brightly colored compounds found in some foods; structures similar to that of vitamin A.

**carrier-mediated active transport** An energy-requiring mechanism whereby a substance moves from a region of lower concentration to a region of higher concentration.

**cartilage** The soft, nonmineralized precursor of bone.

**catabolic neuropeptide** A protein released by nerve cells that inhibits hunger and/or stimulates energy expenditure.

**catabolic pathway** A series of metabolic reactions that break down complex molecules into simpler ones, often releasing energy in the process.

**cataract** A cloudy growth that develops on the lens of the eye, causing impaired vision.

**cation** An ion that has a net positive charge.

**cause-and-effect relationship** (also called causal relationship) When an alteration in one variable causes a change in another variable.

**cecum** The pouch that marks the first section of the large intestine.

**celiac disease** An autoimmune response, to the protein gluten, that damages the absorptive surface of the small intestine; also called gluten-sensitive enteropathy.

**cell culture system** Specific type of cells that can be grown in the laboratory and used for research purposes.

**cell cycle** The process by which cells grow, mature, replicate their DNA, and divide.

**cell differentiation** The process in which an immature cell becomes a specific type of mature cell.

**cell signaling** The first step in protein synthesis in which the cell receives a signal to produce a protein.

**cell turnover** The cycle of cell formation and cell breakdown.

**central nervous system** The part of the nervous system made up of the brain and spinal cord.

**cephalic phase** The response of the central nervous system to sensory stimuli, such as smell, sight, and taste that occurs before food enters the GI tract, characterized by increased GI motility and release of GI secretions.

**cerebral beriberi** (Wernicke-Korsakoff syndrome) A form of thiamin deficiency characterized by poor muscle control and paralysis of the eye muscles.

**ceruloplasmin** The protein that transports copper in the blood.

**chain length** The number of carbons in a fatty acid's backbone.

**cheilosis** Sores occurring on the outsides and corners of the lips.

**chelator** A substance that binds compounds in the gastrointestinal tract, making them unavailable for absorption.

**chemical bonds** Electric forces that hold atoms together in a molecule.

**chemoreceptor** A sensory receptor that responds to a chemical stimulus.

**chemotherapy** The use of drugs to stop the growth of cancer.

**chief cells** Exocrine cells in the gastric mucosa that produce the protein-digesting enzyme pepsin.

**Chinese restaurant syndrome** Severe headaches and nausea reportedly caused by consuming large amounts of monosodium glutamate, in people who are sensitive to it.

**chloride (Cl)** A major mineral important for regulating fluid balance, protein digestion in the stomach (via HCl), and carbon dioxide removal by the lungs.

**cholecalciferol** (vitamin $D_3$) The form of vitamin D in animal foods and made by the human body.

**cholecystokinin (CCK)** A hormone, produced by the duodenum, that stimulates the release of enzymes from the pancreas and stimulates the gallbladder to contract and release bile.

**cholesterol** A sterol found in animal foods and made in the body; required for bile acid and steroid hormone synthesis.

**cholesterol ratio** The mathematical ratio of total blood cholesterol to high-density lipoprotein (HDL) cholesterol.

**cholesteryl ester** A sterol ester made of a cholesterol molecule bonded to a fatty acid via an ester linkage.

**choline** A water-soluble compound used by the body to synthesize acetylcholine (a neurotransmitter) and a variety of phospholipids needed for cell membrane structure; considered a conditionally essential nutrient.

**chromium (Cr)** An essential trace mineral needed for proper insulin function.

**chromium picolinate** A form of chromium taken as an ergogenic aid by some athletes.

**chromosome** A strand of DNA in a cell's nucleus.

**chronic disease** A noninfectious disease that develops slowly and persists over time.

**chylomicron** A lipoprotein made in the enterocyte that transports large lipids away from the small intestine in the lymph.

**chylomicron remnant** The lipoprotein particle that remains after a chylomicron has lost most of its fatty acids.

**chyme** The thick fluid resulting from the mixing of food with gastric secretions in the stomach.

**cirrhosis** The formation of scar tissue in the liver caused by chronic alcohol abuse.

**cis double bond** A carbon–carbon double bond in which the hydrogen atoms are arranged on the same side of the double bond.

**citrate** The first intermediate product in the citric acid cycle formed when acetyl-CoA joins with oxaloacetate.

**citric acid cycle** An amphibolic metabolic pathway that oxidizes acetyl-CoA to yield carbon dioxide, NADH + H⁺, FADH₂, and ATP via substrate phosphorylation.

**coagulation** The process by which blood clots are formed.

**coenzyme** Organic molecule, often derived from vitamins, needed for enzymes to function.

**cofactor** A nonprotein component of an enzyme, often a mineral, needed for its activity.

**collagen** The main protein found in connective tissue, including skin, bones, teeth, cartilage, and tendons.

**colon** The portion of the large intestine that carries material from the cecum to the rectum.

**colostrum** The first secretion from the breasts after birth that provides nourishment and immunological protection to newborns.

**common bile duct** The duct that transports secretions from the liver, pancreas, and gallbladder into the duodenum.

**complete protein source** A food that contains all the essential amino acids in relative amounts needed by the body.

**complex carbohydrates** Category of carbohydrate that includes oligosaccharides and polysaccharides.

**compound** A molecule made up of two or more different types of atoms.

**computerized nutrient database** Software that provides information concerning the nutrient and energy contents of many foods.

**condensation** A chemical reaction that results in the formation of water.

**conditionally essential nutrient** Normally non-essential nutrient that, under certain circumstances, becomes essential.

**cones and rods** Cells in the retina that are needed for vision.

**confounding variable** A factor, other than the one of interest, that might influence the outcome of an experiment.

**connective tissue** Tissue that supports, connects, and anchors body structures.

**control group** A group of people, animals, or cells in an intervention study that does not receive the experimental treatment.

**copper (Cu)** An essential trace mineral that acts as a cofactor for nine enzymes involved in redox reactions.

**Cori cycle** The metabolic pathway that regenerates glucose by circulating lactate from muscle to the liver, where it undergoes gluconeogenesis.

**cornea** The outermost layer of tissue covering the front of the eye.

**correlation** (also called association) When a change in one variable is related to a change in another variable.

**cortical bone** (compact bone) The dense, hard layer of bone found directly beneath the periosteum.

**cortisol** Hormone secreted by the adrenal glands in response to stress that helps increase blood glucose availability via gluconeogenesis and glycogenolysis.

**coupled reactions** Chemical reactions that take place simultaneously often involving the oxidation of one molecule and the reduction of another.

**covalent bond** A chemical bond created by the sharing of one or more electrons.

**C-reactive protein (CRP)** A protein produced in the liver and smooth muscles that participates in the immune system's response to a noxious stimulus, injury, or infection; elevated circulating concentrations are related to increased risk for cardiovascular disease.

**creatine phosphate (CP)** A high-energy compound consisting of creatine and phosphate used to generate ATP.

**creatine phosphokinase** An enzyme that splits creatine phosphate to generate ATP.

**cretinism** A form of IDD that affects babies born to iodine-deficient mothers.

**Creutzfeldt-Jakob disease** A fatal disease in humans caused by a genetic mutation or surgical contamination with prions.

**critical period** Period in development when cells and tissue rapidly grow and differentiate to form body structures.

**Crohn's disease** A chronic inflammatory condition that usually affects the ileum and/or first portion of the large intestine.

**cross-contamination** The transfer of microorganisms from one food to another or from one surface or utensil to another.

**cross-tolerance** Tolerance to one drug that causes tolerance to other similar drugs.

**cupric ion (Cu²⁺)** The more oxidized form of copper.

**cuprous ion (Cu⁺)** The more reduced form of copper.

**cystic fibrosis** A genetic (inherited) disease in which a defective chloride transporter results in the inability of the body to transport chloride out of cells.

**cyst** A stage of the life cycle of some parasites.

**cytochrome** A heme protein complex that is part of the electron transport chain in mitochondria.

**cytochromes** Iron-containing protein complexes that combine electrons, hydrogen ions, and oxygen to form water.

**cytochrome c oxidase** A copper-containing enzyme needed in the electron transport chain.

**cytochrome P450** An iron-containing enzyme that helps stabilize free radicals.

**cytoplasm** (also called **cytosol**) The gel-like matrix inside cells but outside of cell organelles.

## D

**Daily Value (DV)** Recommended intake of a nutrient based on either a 2,000- or 2,500-kcal diet.

**danger zone** The temperature range between 40 and 140°F in which pathogenic organisms grow most readily.

**DASH (Dietary Approaches to Stop Hypertension) diet** A dietary pattern emphasizing fruits, vegetables, and low-fat dairy products designed to lower blood pressure.

**db gene** The gene that codes for the leptin receptor.

**db/db mouse** Obese mouse with a mutation in the gene that codes for the leptin receptor.

**deamination** The removal of an amino group from an amino acid.

**defecation** The expulsion of feces from the body through the rectum and anal canal.

**dehydration** A condition in which the body has an insufficient amount of water.

**dehydrogenase** A type of enzyme that catalyzes the removal and transfer of electrons and hydrogen atoms between organic compounds.

**denaturation** The alteration of a protein's three-dimensional structure by heat, acid, enzymes, or agitation.

**dental fluorosis** Discoloration and pitting of teeth caused by excessive fluoride intake.

**development** Change in the complexity and/or attainment of a function.

**dextrin** A partial breakdown product formed during starch digestion consisting of varying numbers of glucose units.

**diabetes mellitus** Medical condition characterized by a lack of insulin or impaired insulin utilization that results in elevated blood glucose levels.

**diet record** A prospective dietary assessment method that requires the individual to write down detailed information about foods and drinks consumed over a specified period of time.

**dietary assessment** The evaluation of a person's dietary intake.

**dietary fiber** Polysaccharide found in plants that is not digested or absorbed in the human small intestine.

**dietary folate equivalent (DFE)** A unit of measure used to describe the amount of bioavailable folate in a food or supplement.

**Dietary Guidelines for Americans** Dietary recommendations, developed by the USDA and DHHS, that give specific nutritional guidance to individuals as well as advice about physical activity, alcohol intake. and food safety.

**Dietary Reference Intakes (DRIs)** A set of four types of nutrient intake reference standards used to assess and plan dietary intake; these include the Estimated Average Requirements (EARs), Recommended Dietary Allowances (RDAs), Adequate Intake levels (AIs), and the Tolerable Upper Intake Levels (ULs).

**dietary supplements** Products intended to supplement the diet that contain vitamins, minerals, amino acids, herbs or other plant-derived substances, and/or a multitude of other food-derived compounds.

**dietitian** A nutritionist who works as a clinician, assisting people in making healthy dietary choices.

**digestion** The physical and chemical breakdown of food by the digestive system into a form that allows nutrients to be absorbed.

**digestive enzymes** Biological catalysts that facilitate chemical reactions that break chemical bonds by the addition of water (hydrolysis), resulting in the breakdown of large molecules into smaller components.

**diglyceride** (also called **diacylglycerol**) A lipid made of glycerol bonded to two fatty acids.

**direct calorimetry** A measurement of energy expenditure obtained by assessing heat loss.

**disaccharidase** Intestinal enzyme that hydrolyzes glycosidic bonds in disaccharides.

**disaccharide** Carbohydrate consisting of two monosaccharides bonded together.

**discretionary calorie allowance** The amount of calories that can still be eaten, without promoting weight gain, after all the essential nutrients have been consumed.

**disengaged families** Family interaction characterized by a lack of cohesiveness and little parental involvement.

**disinhibition** A loss of inhibition.

**disordered eating** Eating pattern that includes irregular eating, consistent undereating, and/or consistent overeating.

**distillation** A process used to make a concentrated alcohol beverage by condensing and collecting alcohol vapors.

**diuretic** A substance or drug that causes water loss from the body.

**diverticular disease, or diverticulosis** Condition in the large intestine; characterized by the presence of pouches that form along the intestinal wall.

**diverticulitis** Inflammation of diverticula (pouches) in the lining of the large intestine.

**docosahexaenoic acid (DHA)** A long-chain, polyunsaturated ω-3 fatty acid produced from linolenic acid.

**double-blind study** A human experiment in which neither the participants nor the scientists know to which group the participants have been assigned.

**doubly labeled water** Water that contains stable isotopes of hydrogen and oxygen atoms.

**down-regulation** The "turning off" of protein synthesis.

**dry beriberi** A form of thiamin deficiency characterized by muscle loss and leg cramps.

**Dual-Energy X-Ray Absorptiometry (DEXA)** A method used to assess body composition by passing x-ray beams at different energy levels through the body.

**dual-energy x-ray absorptiometry (DEXA or DXA)** A method of measuring body composition.

**duodenum** The first segment of the small intestine.

**dysphagia** Difficulty in swallowing.

## E

**eating disorder** Extreme disturbance in eating behaviors that can result in serious medical conditions, psychological consequences, and dangerous weight loss.

**eating disorders not otherwise specified (EDNOS)** A category of eating disorders that includes some, but not all, of the diagnostic criteria for anorexia nervosa and/or bulimia nervosa.

**edema** The buildup of fluid in the interstitial spaces.

**eicosanoids** Biologically active compounds synthesized from arachidonic acid and EPA.

**eicosapentaenoic acid (EPA)** A long-chain, polyunsaturated ω-3 fatty acid produced from linolenic acid.

**electrolyte** A chemical compound that separates into ions when in solution.

**electron** A subatomic particle that orbits around the nucleus of an atom that carries a negative charge.

**electron sharing** The sharing of 1 or more valence electrons between atoms.

**electron transfer** The transfer of 1 or more electrons between atoms.

**electron transport chain** A series of chemical reactions that transfer electrons and hydrogen ions from $NADH + H^+$ and $FADH_2$ along protein complexes in the inner mitochondrial membrane, ultimately producing ATP.

**element** A substance made up of only one type of atom.

**elimination** The process whereby solid waste is formed and expelled from the body.

**embryonic period** The stage of development that extends through the eighth week of pregnancy.

**emulsification** The process by which large lipid globules are broken down and stabilized into smaller lipid droplets.

**end product** The final product in a metabolic pathway.

**endocytosis** A form of vesicular active transport whereby the cell membrane surrounds substances and releases them to the cytoplasm.

**endoplasmic reticulum** A cell organelle involved in the synthesis and transport of materials within and from cells.

**endurance training** Physical training that improves pulmonary and cardiovascular function.

**energy** The capacity to do work.

**energy balance** A state in which energy intake equals energy expenditure.

**energy metabolism** Chemical reactions that enable cells to store and use energy from nutrients.

**energy-yielding nutrient** A nutrient that the body can use to produce ATP.

**enmeshment** Family interaction whereby family members are overly involved with one another and have little autonomy.

**enrichment** The fortification of a select group of foods (rice, flour, bread or rolls, farina, pasta, corn meal, and corn grits) with FDA-specified levels of thiamin, niacin, riboflavin, folate, and iron.

**enteric endocrine system** Hormones secreted by the mucosal lining of the GI tract that regulate GI motility and secretion.

**enteric (intestinal) toxin** A toxic agent produced by an organism after it enters the gastrointestinal tract.

**enteric nervous system** Neurons located within the submucosa and muscularis layers of the digestive tract.

**enterocytes** Epithelial cells that make up the surface of each villus.

**environmental factor** An element or variable in our surroundings that we may or may not have control over (such as pollution and temperature).

**enzyme-substrate complex** A substrate attached to an enzyme's active site.

**epidemiologic study** A study in which data are collected from a group of individuals who are not asked to change their behaviors in any way.

**epiglottis** A cartilage flap that covers the trachea while swallowing.

**epinephrine** Hormone released from the adrenal glands in response to stress; helps increase blood glucose levels by promoting glycogenolysis.

**epithelial tissue** Tissue that forms a protective layer on bodily surfaces and lines internal organs, ducts, and cavities.

**equilibrium** The state when substances separated by a membrane are in equal concentration.

**ergocalciferol** (vitamin $D_2$) The form of vitamin D found in plant foods and vitamin-D fortified foods.

**esophagus** The passageway that begins at the pharynx and ends at the stomach.

**essential nutrient** A substance that must be obtained from the diet, because the body needs it and cannot make it in required amounts.

**ester bond** The type of chemical bond that holds fatty acids onto triglycerides, phospholipids, and cholesteryl esters.

**Estimated Average Requirement (EAR)** The amount of a nutrient that meets the physiological requirements of half (50%) the healthy population of similar individuals.

**Estimated Energy Requirement (EER)** Average energy intake required to maintain energy balance in healthy individuals based on sex, age, physical activity level, weight, and height.

**ethanol** An alcohol produced by the chemical breakdown of sugar by yeast.

**exocytosis** A form of vesicular active transport whereby cell products are enclosed in a vesicle and the contents of the vesicle are released to the outside of the cell.

**extracellular** Situated outside of a cell or cells.

## F

**facilitated diffusion** A passive transport mechanism whereby a substance crosses a cell membrane with the assistance of a transport protein.

**fasting hypoglycemia** Low blood glucose that occurs when the pancreas releases excess insulin regardless of food intake.

**fat** A lipid that is solid at room temperature.

**fatty acid** A lipid consisting of a chain of carbons with a methyl ($-CH_3$) group on one end and a carboxylic acid group ($-COOH$) on the other.

**fatty liver** A condition caused by excess alcohol consumption characterized by the accumulation of triglycerides in the liver.

**female athlete triad** A combination of interrelated conditions: disordered eating, amenorrhea, and osteopenia.

**fermentable fiber** Fiber that can be fermented by bacteria in the large intestine.

**fermentation** Metabolism, by bacteria, that occurs under relatively anaerobic conditions; the process whereby yeast chemically breaks down sugar to produce ethanol and carbon dioxide.

**ferric iron** ($Fe^{3+}$) The oxidized form of iron.

**ferritin** A protein important for iron absorption and storage in the body.

**ferrous iron** ($Fe^{2+}$) The reduced form of iron.

**fetal origins hypothesis** A theory suggesting that conditions during gestational development can alter risk for chronic diseases later in life.

**fetal period** Period of gestation after week 8 through birth.

**fibrinogen** A water-soluble protein that is converted to the water-insoluble protein fibrin.

**fibrin** An insoluble protein that forms blood clots.

**FightBAC!®** A public education program developed to reduce foodborne bacterial illness.

**filtration** The process of selective removal of metabolic waste products from the blood.

**flavin adenine dinucleotide (FAD)** The oxidized form of the coenzyme that is able to accept 2 electrons and 2 hydrogen ions, forming $FADH_2$; a coenzyme form of riboflavin.

**flavin mononucleotide (FMN)** A coenzyme form of riboflavin.

**fluoride** ($F^-$) A nonessential trace mineral that strengthens bones and teeth.

**folate** (also called folacin) A water-soluble vitamin involved in single-carbon transfer reactions; needed for amino acid metabolism and DNA synthesis.

**folic acid** The form of folate commonly used in vitamin supplements and food fortification.

**food allergy** A condition in which the body's immune system reacts against a ptotein in food.

**food bank** Agency that collects donated foods and distributes them to local food pantries, shelters, and soup kitchens.

**food biosecurity** Measures aimed at preventing the food supply from falling victim to planned contamination.

**food composition table** Tabulated information concerning the nutrient and energy contents of foods.

**food craving** A strong psychological desire for a particular food or foods.

**food frequency questionnaire** A retrospective dietary assessment method that assesses food selection patterns over an extended period of time.

**food insecurity** Having limited or uncertain access to adequate amounts of nutritious, safe foods.

**food jag** In young children, a desire for a particular food.

**food kitchen** Program that prepares and serves meals to those in need.

**food pantry** Program that provides canned, boxed, and sometimes fresh foods directly to individuals in need.

**food preoccupied** Spending an inordinate amount of time thinking about food.

**food recovery program** Organization that collects and distributes food that would otherwise be discarded.

**Food Stamp Program** A federally funded program that improves food availability to low-income households by providing coupons for the purchase of certain foods.

**foodborne illness** A disease caused by ingesting food.

**fortified food** A food to which nutrients have been added.

**free radical** A reactive molecule with 1 or more unpaired electrons; free radicals are destructive to cell membranes, DNA, and proteins.

**fructose** A 6-carbon monosaccharide found in fruits and vegetables; also called levulose.

**full-term infants** Babies born with gestational ages between 37 and 42 weeks.

**functional food** A food that contains an essential nutrient, phytochemical, or zoonutrient and that is thought to benefit human health.

## G

**galactose** A 6-carbon monosaccharide found mainly bonded with glucose to form the milk sugar lactose.

**gastric banding** A band around the upper portion of the stomach, dividing it into a small upper pouch and a larger lower pouch.

**gastric bypass** A surgical procedure that reduces the size of the stomach and bypasses a segment of the small intestine so that fewer calories are absorbed.

**gastric emptying** The process by which food leaves the stomach and enters the small intestine.

**gastric inhibitory protein (GIP)** A hormone produced by endocrine cells lining the small intestine that controls GI motility; also called gastric inhibitory peptide.

**gastric juice** Digestive secretions produced by exocrine cells that make up gastric pits.

**gastric lipase** Enzyme produced in the stomach that hydrolyzes ester bonds between fatty acids and glycerol molecules.

**gastric phase** The phase of digestion stimulated by the arrival of food into the stomach characterized by increased GI motility and release of GI secretions.

**gastric pits** Invaginations of the mucosal lining of the stomach that contain specialized endocrine and exocrine cells.

**gastrin** A hormone secreted by endocrine cells that stimulates the production and release of gastric juice.

**gastroesophageal reflux disease (GERD)** A condition caused by the weakening of the gastroesophageal sphincter, which enables gastric juices to reflux into the esophagus, causing irritation to the mucosal lining.

**gastroesophageal sphincter (GES)** A circular muscle that regulates the flow of food between the esophagus and the stomach; also called lower esophageal sphincter or cardiac sphincter.

**gastrointestinal (GI) tract** A tubular passage that runs from the mouth to the anus that includes several organs that participate in the process of digestion; also called the digestive tract.

**genetic factor** An inherited element or variable in our lives that cannot be altered.

**genetic makeup (genome)** The particular DNA contained in a person's cells.

**genetically modified organism (GMO)** An organism (plant or animal) made by genetic engineering.

**gene** A portion of a chromosome that codes for the primary structure of a polypeptide.

**gene therapy** The use of altered genes to enhance health.

**gerontology** The study of aging.

**gestational age** Length of pregnancy, determined by counting the number of weeks between the first day of a woman's last normal menstrual period and birth.

**gestational diabetes** Type of diabetes; characterized by insulin resistance that develops in pregnant women.

**GI motility** Mixing and propulsive movements of the gastrointestinal tract caused by contraction and relaxation of the muscularis.

**GI secretions, digestive juices** Substances released by organs that make up the digestive system that facilitate the process of digestion; also called digestive juices.

**glossitis** Inflamed tongue.

**glucagon** Hormone secreted by the pancreatic α-cells in response to decreased blood glucose.

**glucogenic amino acid** An amino acid that can be converted to glucose via gluconeogenesis.

**gluconeogenesis** Synthesis of glucose from non-carbohydrate sources.

**glucose** A 6-carbon monosaccharide produced in plants by photosynthesis.

**glucose transporter** Intracellular transport proteins that assist in the transport of glucose molecules across cell membranes.

**glutathione peroxidases** A group of selenoprotein enzymes that have redox functions in the body.

**gluten** A protein found in cereal grains such as wheat, rye, barley, and possibly oats.

**glycemic index (GI)** Measure based on the extent to which a food containing 50 g of carbohydrates increases blood glucose concentrations.

**glycemic load (GL)** Measure that assesses the effect of a food on blood glucose that takes into account both the glycemic index of a food as well as its carbohydrate content.

**glycemic response** Effect of a food on the extent and/or duration of the rise in blood glucose levels.

**glycogen** Polysaccharide consisting of a highly branched arrangement of glucose molecules found primarily in liver and skeletal muscle.

**glycogenesis** Formation of glycogen.

**glycogenolysis** The breakdown of liver and muscle glycogen into glucose.

**glycolysis** The metabolic pathway that splits glucose into two 3-carbon molecules called pyruvate.

**glycosidic bond** Chemical bond formed by condensation of two monosaccharides.

**goiter** A form of IDD that affects children and adults; characterized by an enlarged thyroid gland.

**goitrogens** Compounds found in some vegetables that decrease iodine utilization by the thyroid gland.

**Golgi apparatus** An organelle that packages macromolecules and other substances in vesicles.

**gout** A condition caused by the accumulation of uric acid in the joints.

**growth** An increase in size and/or number of cells.

**guanosine triphosphate (GTP)** A high-energy compound similar to ATP.

**gynoid adiposity** Subcutaneous body fat in the lower part of the body, typically the thighs and hips.

## H

**haustral contractions** Slow muscular movements that move the colonic contents back and forth and that help compact the feces.

**Hawthorne effect** Phenomenon in which study results are influenced by alteration of something that is not related to the actual study intervention.

**Hazard Analysis Critical Control Points (HACCP) system** A food safety protocol used to decrease contamination of foods during processing.

*Healthy People 2010* A publication that sets long-term national goals for improving overall health of Americans.

**health claims** FDA-approved statements relating a food or food component to health benefits.

**heart attack** (also called myocardial infarction) An often life-threatening condition in which blood flow to some or all of the heart muscle is completely blocked.

**heart disease** (also called coronary heart disease) A condition that occurs when the heart muscle does not receive enough blood.

**heat exhaustion** Rise in body temperature that occurs when the body has difficulty dissipating heat.

**heat stroke** Rise in body temperature due to excess loss of body fluids associated with sweating.

**heme iron** Iron that is a component of a heme group; heme iron includes hemoglobin in blood, myoglobin in muscles, and cytochromes in mitochondria.

**hemodilution** A decrease in the number of red blood cells per volume of plasma caused by plasma volume expansion.

**hemoglobin** A complex protein composed of four iron-containing heme groups and four protein subunits needed for oxygen and carbon dioxide transport in the body.

**hemolytic anemia** Decreased ability of the blood to carry oxygen and carbon dioxide due to rupturing of red blood cells.

**hemosiderin** An iron-storing protein.

**hepatic portal circulation** A circulatory route that delivers nutrient-rich blood from the small intestine to the liver.

**hepatic portal vein** A blood vessel that circulates blood to the liver from the GI tract.

**hereditary hemochromatosis** A genetic abnormality resulting in the increased absorption of iron in the intestine.

**heterocyclic amines** Cancer-causing compounds that can be formed when meat is cooked at high temperatures.

**high-density lipoprotein (HDL)** A lipoprotein made by the liver that circulates in the blood to collect excess cholesterol from cells.

**high-quality protein source** A complete protein source with adequate amino-acid bioavailability.

**high osmotic pressure** When a solution has a large amount of solutes dissolved in it.

**homeostasis** A state of balance or equilibrium.

**homocysteine** A compound that is converted to methionine in a folate and vitamin $B_{12}$-requiring, coupled reaction.

**hormone-sensitive lipase** An enzyme that catalyzes the hydrolysis of ester bonds that attach fatty acids to the glycerol molecule; mobilizes fatty acids from adipose tissue.

**hormone replacement therapy (HRT)** Medication containing estrogen and progesterone, sometimes taken by women after having their ovaries removed or after menopause.

**hormone** A chemical substance produced by an endocrine gland that is released into the blood and stimulates a response elsewhere in the body.

**hunger** The physiological drive to consume food.

**hydrogen bond** A force of attraction between hydrogen atoms and atoms such as oxygen or nitrogen.

**hydrolysis** A chemical reaction whereby compounds react with water.

**hydrophilic** A substance that dissolves or mixes with water.

**hydrophobic** A substance that does not dissolve or mix with water.

**hydrostatic weighing** (underwater weighing) Method for estimating body composition that compares weight on land to weight underwater.

**hydroxyapatite** $[Ca_{10}(PO_4)_6(OH)_2]$ The mineral matrix of bones and teeth.

**hypercarotenemia** A condition in which carotenoids accumulate in the skin, causing it to become yellow-orange.

**hypercholesterolemia** Elevated levels of cholesterol in the blood.

**hyperglycemia** Abnormally high level of glucose in the blood.

**hyperinsulinemia** Condition characterized by high blood insulin.

**hyperkeratosis** A symptom of vitamin A deficiency in which immature skin cells overproduce the protein keratin, causing rough and scaly skin.

**hyperlipidemia** Elevated levels of lipids in the blood.

**hypernatremia** High blood sodium concentration.

**hyperplastic growth** Growth associated with an increase in cell number.

**hypertriglyceridemia** Elevated levels of triglycerides in the blood.

**hypertrophic growth** Growth associated with an increase in cell size.

**hypervitaminosis A** A condition in which elevated circulating vitamin A levels result in signs and symptoms of toxicity.

**hypoglycemia** Abnormally low level of glucose in the blood.

**hypokalemia** Low blood potassium concentration.

**hyponatremia** Low blood sodium concentration.

**hypothalamus** An area of the brain that controls many involuntary functions by the release of hormones.

**hypothesis** A prediction about the relationship between variables.

# I

**ileocecal sphincter** The sphincter that separates the ileum from the cecum and regulates the flow of material between the small and large intestines.

**ileum** The last segment of the small intestine that comes after the jejunum.

**impaired glucose regulation** Condition characterized by elevated levels of glucose in the blood.

**inborn error of metabolism** A disease that is caused by the absence of an enzyme needed for metabolism.

**incidence** The number of people who are newly diagnosed with a condition in a given period of time.

**incomplete protein source** A food that lacks or contains very low amounts of one or more essential amino acids.

**indirect calorimetry** A measurement of energy expenditure obtained by assessing oxygen consumption and carbon dioxide production.

**infantile beriberi** A form of thiamin deficiency that occurs in infants breastfed by thiamin-deficient mothers.

**infant mortality rate** The number of infant deaths ($<$ 1 year of age) per 1,000 live births in a given year.

**infectious agent of foodborne illness** A pathogen or compound in food that causes illness and can be passed or transmitted from one infected animal or person to another.

**infectious disease** A contagious illness caused by a pathogen such as a bacteria, virus, or parasite.

**inflammation** A response to cellular injury that is characterized by capillary dilation, white blood cell infiltration, release of immune factors, redness, heat, and pain.

**inflammatory bowel diseases (IBDs)** Chronic conditions such as ulcerative colitis and Crohn's disease that cause inflammation of the lower GI tract.

**initiation** The first stage of cancer, in which a normal gene is transformed into a cancer-forming gene (oncogene).

**inorganic compound** A substance that does not contain carbon–carbon bonds or carbon–hydrogen bonds.

**insoluble fiber** Dietary fiber that is incapable of being dissolved in water.

**insulin** Hormone secreted by the pancreatic β-cells in response to increased blood glucose.

**insulin receptor** Protein found on the surface of certain cell membranes that bind insulin.

**insulin resistance** Condition characterized by the inability of insulin receptors to respond to the hormone insulin.

**interaction** When the relationship between two factors is influenced or modified by another factor.

**intermediate-density lipoprotein (IDL)** A lipoprotein that results from the loss of fatty acids from a VLDL; IDLs are ultimately converted to LDLs.

**intermediate product** A product formed before a metabolic pathway reaches completion, often serving as a substrate in the next chemical reaction.

**intermembrane space** The space between the inner and outer mitochondrial membranes.

**interstitial fluid** Fluid that surrounds cells.

**interval training** Athletic training that alternates between fast bursts of intensive exercise and slower, less demanding activity.

**intervention study** An experiment in which something is altered or changed to determine its effect on something else.

**intestinal phase** The phase of digestion when chyme enters the small intestine, characterized by both a decrease in gastric motility and secretion of gastric juice.

**intra-abdominal (visceral) fat** Body fat associated with internal organs within the abdominal cavity.

**intracellular** Situated within a cell.

**intrauterine growth retardation (IUGR)** Slow or delayed growth in utero.

**intrinsic factor** A protein produced by exocrine cells in the gastric pits of the stomach, needed for vitamin $B_{12}$ absorption.

**in vitro** Involving the use of cells or environments that are not part of a living organism.

**in vivo** Involving the study of natural phenomena in a living organism.

**iodide (I⁻)** The most abundant form of iodine in the body.

**iodine (I)** An essential trace mineral that is a component of the thyroid hormones.

**iodine deficiency disorders (IDDs)** A broad spectrum of conditions caused by inadequate iodine.

**ion** An atom that has acquired an electrical charge by gaining or losing one or more electrons.

**ionic bond** The force of attraction between ions with opposite charges.

**ionic compound** A substance, such as table salt, composed of ions.

**ionization** The dissociation of a compound into ions.

**iron (Fe)** A trace mineral needed for oxygen and carbon dioxide transport, energy metabolism, removal of free radicals, and synthesis of DNA.

**irradiation** A food preservation process that applies radiant energy to foods, to kill bacteria.

**irritable bowel syndrome (IBS)** A condition that typically affects the lower GI tract, causing abdominal pain, muscle spasms, diarrhea, and/or constipation.

**isoflavone** Compound that is structurally similar to the hormone estrogen; found primarily in soybeans.

# J

**jejunum** The midsection of the small intestine, located between the duodenum and the ileum.

# K

**Keshan disease** A disease resulting from selenium deficiency.

**ketoacidosis** Severe metabolic condition resulting from the accumulation of ketones; a rise in ketone levels in the blood, causing the pH of the blood to decrease.

**ketogenesis** Metabolic pathway that leads to the production of ketones.

**ketogenic amino acids** Amino acids that can be used to make ketones.

**ketogenic diets** Diets that stimulate ketone production.

**ketone** Organic compound used as an energy source during starvation, fasting, low-carbohydrate diets, or uncontrolled diabetes.

**ketosis** Condition resulting from overproduction of ketone bodies; an accumulation of ketones in body tissues and fluids.

**kidneys** The organs that filter metabolic waste products from the blood and play a role in maintaining blood volume.

**kilocalorie (kcal, or Calorie)** 1,000 calories.

**kwashiorkor** A form of PEM often characterized by edema in the extremities (hands, feet).

**kyphosis (dowager's hump)** A curvature of the upper spine, caused by osteoporosis.

# L

**labile amino acid pool** In the body, amino acids that are immediately available to cells for protein synthesis and other purposes.

**lactase** Intestinal enzyme that hydrolyzes lactose into glucose and galactose.

**lactation** The production and release of milk.

**lacteal** A lymphatic vessel found in an intestinal villus.

**lacto-ovo-vegetarian** A type of vegetarian who consumes dairy products and eggs in an otherwise plant-based diet.

**lactogenesis** The onset of milk production.

**lactose** Disaccharide consisting of glucose and galactose; produced by mammary glands.

**lactose intolerance** Inability to digest the milk sugar lactose, caused by a lack of the enzyme lactase.

**lactovegetarian** A type of vegetarian who consumes dairy products (but not eggs) in an otherwise plant-based diet.

**large for gestational age (LGA)** An infant that weighs more than the 90th percentile for weight for gestational age.

**LDL receptor** Membrane-bound protein that binds LDLs, causing them to be taken up and dismantled.

**leptin** A hormone produced mainly by adipose tissue that helps regulate body weight.

**let-down** The movement of milk through the mammary ducts toward the nipple.

**life expectancy** A statistical prediction of the average number of years of life remaining to a person at a specific age.

**life span** Maximum number of years an individual in a particular species has remained alive.

**lifestyle factor** Behavioral component of our lives over which we may or may not have control (such as diet and tobacco use).

**limiting amino acid** An essential amino acid in the lowest concentration in an incomplete protein source.

**limit dextrin** A partial breakdown product formed during amylopectin digestion that contain three to four glucose molecules and an α-1,6 glycosidic bond.

**lingual lipase** An enzyme produced in the salivary glands that hydrolyzes ester bonds between fatty acids and glycerol molecules.

**linoleic acid** An essential ω-6 fatty acid.

**linolenic acid** An essential ω-3 fatty acid.

**lipid** Organic substance that is relatively insoluble in water and soluble in organic solvents.

**lipogenesis** The metabolic processes that result in fatty acid and, ultimately, triglyceride synthesis.

**lipolysis** The breakdown of triglycerides into fatty acids and glycerol.

**lipoprotein** A spherical particle made of varying amounts of triglycerides, cholesterol, cholesteryl esters, phospholipids, and proteins.

**lipoprotein lipase** An enzyme that hydrolyzes the ester bond between a fatty acid and glycerol in a triglyceride molecule.

**longevity** Life expectancy at birth.

**low birth weight (LBW)** An infant that weighs less than 2,500 g at birth.

**low osmotic pressure** When a solution has a small amount of solutes dissolved in it.

**low-density lipoprotein (LDL)** A lipoprotein that delivers cholesterol to cells.

**low-quality protein source** A food that is either an incomplete protein source or one that has low amino acid bioavailability.

**lumen** The cavity inside a tubular structure in the body.

**lymph** A fluid, found in lymphatic vessels, that is derived from tissue fluids.

**lymphatic duct** An enclosed canal that circulates lymph.

**lymphatic system** A circulatory system made up of vessels and lymph that flows from organs and tissues, drains excess fluid from spaces that surround cells, picks up dietary fats from the digestive tract, and plays a role in immune function.

**lymphocyte** A type of cell in the immune system that produces antibodies that attack foreign cells.

**lysophospholipid** A lipid composed of a glycerol bonded to a polar head group and a fatty acid; final product of phospholipid digestion.

**lysosome** An organelle that contains enzymes that degrade molecules.

## M

**macrocytic anemia** A condition in which red blood cells are large, caused by inability of the cell to mature and divide appropriately; can be due to folate deficiency and/or vitamin $B_{12}$ deficiency.

**macromolecules** Large molecules made by cells such as proteins, that are made up of smaller subunits.

**macronutrients** The class of nutrients that we need to consume in relatively large quantities.

**macrophage** A white blood cell that is part of the body's immune defense.

**macula** A portion of the retina important for sight.

**macular degeneration** A chronic disease that results from deterioration of the retina.

**magnesium (Mg)** A major mineral needed for stabilizing enzymes and ATP and as a cofactor for many enzymes.

**malignant tumor** A cancerous growth.

**malnutrition** Poor nutritional status caused by either undernutrition or overnutrition.

**maltase** Enzyme that hydrolyzes maltose into two glucose molecules.

**maltose** Disaccharide consisting of two glucose molecules bonded together; formed during the chemical breakdown of starch.

**mammary duct** Structure that transports milk from the secretory cells toward the nipple.

**manganese (Mn)** An essential trace mineral that is a cofactor for enzymes needed for bone

formation and the metabolism of carbohydrates, proteins, and cholesterol.

**marasmus** A form of PEM characterized by extreme wasting of muscle and adipose tissue.

**marine toxin** A poison produced by ocean algae.

**mastication** Chewing and grinding of food by the teeth to prepare for swallowing.

**matter** Material that has mass and occupies space.

**maximal oxygen consumption ($VO_2$ max)** Maximum volume of oxygen that can be delivered per unit of time.

**meat factor** An unidentified compound, found in meat, that increases the absorption of non-heme iron.

**mechanoreceptor** A sensory receptor that responds to pressure, stretching, or mechanical stimulus.

**meconium** Dark green, feces-like matter that collects in the intestine of an unborn baby.

**medical history** Questions asked to assess overall health.

**Mediterranean diet** A dietary pattern originating from the region surrounding the Mediterranean Sea that includes relatively high intakes of fruits and vegetables, nuts, seeds, fish, and certain oils (such as olive oil) and low intake of meat.

**menadione** A form of vitamin K produced commercially.

**menaquinone** A form of vitamin K produced by bacteria.

**menarche** The first time a female menstruates.

**menopause** The time in a woman's life when menstruation ceases, usually during the sixth decade of life.

**messenger ribonucleic acid (mRNA)** A form of RNA involved in gene transcription.

**metabolic pathway** A series of interrelated enzyme-catalyzed chemical reactions that take place in cells.

**metabolic syndrome** Condition characterized by an abnormal metabolic profile, abdominal body fat, and insulin resistance.

**metabolism** Chemical reactions that take place in the body.

**metalloenzyme** An enzyme that contains a mineral cofactor.

**metallothionine** A protein in the enterocyte that regulates zinc absorption and elimination.

**metastasis** The spreading of cancer cells to other parts of the body via the blood or lymph.

**micelle** A water-soluble, spherical structure formed in the small intestine via emulsification.

**microcytic hypochromic anemia** A condition in which red blood cells are small and light in color due to inadequate hemoglobin synthesis; can be due to vitamin $B_6$ deficiency.

**microflora, microbiota** Bacteria that reside in the large intestine.

**micronutrients** The class of nutrients that we need to consume in relatively small quantities.

**microsomal ethanol-oxidizing system (MEOS)** A pathway used to metabolize alcohol when it is present in high amounts.

**microsomes** Cell organelles associated with endoplasmic reticula

**microvilli** Hairlike projections on the surface of enterocytes.

**mineral** Inorganic substance, other than water, that is required by the body in small amounts.

**miscarriage** Death of a fetus during the first 20 weeks of pregnancy.

**mitochondrial matrix** The inner compartment of the mitochondrion.

**mitochondria** Cellular organelles involved in generating energy (ATP).

**moderation** A concept that emphasizes not consuming too much of a particular type of food.

**molecular formula** Indicates the number and types of atoms in a molecule.

**molecule** A substance held together by chemical bonds.

**molybdenum (Mo)** An essential trace mineral that is a cofactor for enzymes needed for amino acid and purine metabolism.

**monoglyceride (also called monoacylglycerol)** A lipid made of a glycerol bonded to a single fatty acid.

**monomer** Small molecules that join together to form a polymer.

**monosaccharide** Carbohydrate consisting of a single sugar.

**monounsaturated fatty acid (MUFA)** A fatty acid that contains one carbon–carbon double bond in its backbone.

**morbidity rate** The number of illnesses in a given period of time.

**mortality rate** The number of deaths in a given period of time.

**mucosa** The lining of the gastrointestinal tract that is made up of epithelial cells; also called mucosal lining.

**mucus** A substance that coats and protects mucous membranes.

**muscle dysmorphia** Pathological preoccupation with increasing muscularity.

**muscle tissue** Tissue that specializes in movement.

**muscularis** The layer of tissue in the gastrointestinal tract that consists of at least two layers of smooth muscle.

**mutation** The alteration of a gene.

**myoepithelial cells** Muscle cells that surround alveoli and that contract, forcing milk into the mammary ducts.

**myoglobin** A heme protein found in muscle.

**MyPyramid** A graphic and accompanying interactive website developed by the USDA to illustrate the recommendations put forth in the Dietary Guidelines for Americans.

## N

**National Health and Nutrition Examination Survey (NHANES)** A federally funded epidemiologic study begun in the 1970s to assess trends in diet and health in the U.S. population.

**National School Lunch** and **School Breakfast Programs** Federally funded programs that provide free or subsidized nutritious meals to school-age children living in low-income households.

**near-infrared interactance** A method used to assess body composition, based on the scattering and absorbance of light in the near-infrared spectrum.

**negative energy balance** A state in which energy intake is less than energy expenditure.

**negative feedback system** Physiological response that works to restore a system to normal.

**negative nitrogen balance** The condition in which protein (nitrogen) intake is less than protein (nitrogen) loss by the body.

**nephrons** Tubules in the kidneys that filter waste materials from the blood that are later excreted in the urine.

**neural tissue** Tissue that specializes in communication via neurons.

**neural tube defect** A malformation in which the neural tissue does not form properly during fetal development.

**neuropeptide** A hormone-like protein released by nerve cells.

**neutron** A subatomic particle in the nucleus of an atom with no electrical charge.

**niacin** (vitamin B₃) An essential water-soluble vitamin involved in energy metabolism, electron transport chain, synthesis of fatty acids and proteins, metabolism of vitamin C and folate, glucose homeostasis, and cholesterol metabolism.

**niacin equivalent (NE)** A unit of measure that describes the niacin content and/or tryptophan in food.

**nicotinamide adenine dinucleotide (NAD⁺)** The oxidized form of the coenzyme that is able to accept 2 electrons and 2 hydrogen ions, forming NADH + H⁺.

**nicotinamide adenine dinucleotide phosphate (NHDP⁺)** The oxidized form of the coenzyme that is able to accept 2 electrons and 2 hydrogen ions, forming NADPH + H⁺.

**night blindness** A condition characterized by impaired ability to see in the dark.

**nitrites** Nitrogen-containing compounds that are often added to processed meats to enhance color and flavor.

**nitrogen balance** The condition in which protein (nitrogen) intake equals protein (nitrogen) loss by the body.

**nitrosamines** Nitrogen-containing chemical carcinogens that are produced from nitrites.

**nonessential nutrient** A substance found in food and used by the body to promote health but not required to be consumed in the diet.

**nonexercise activity thermogenesis (NEAT)** Energy expended for movement such as fidgeting and maintaining posture.

**nonheme iron** Iron that is not attached to a heme group.

**noninfectious agent of foodborne illness** An inert (nonliving) substance, in food, that causes illness.

**nonpolar** Molecule that are held together by covalent bonds that share electrons equally.

**nonprovitamin A carotenoid** A carotenoid that cannot be converted to vitamin A.

**nucleus** A membrane-enclosed organelle that contains the genetic material DNA.

**nutrient content claims** FDA-regulated phrases and words that can be included on a food's packaging to describe its nutrient content.

**nutrient density** The relative ratio of nutrients in a food in comparison to total calories.

**nutrient requirement** The lowest chronic intake level of a nutrient that supports a defined level of nutritional status in a particular individual.

**nutrient** A substance in foods used by the body for energy, maintenance of body structures, or regulation of chemical processes.

**nutrigenomics** The science of how genetics and nutrition together influence health.

**nutrition** The science of how living organisms obtain and use food to support processes required for life.

**nutritional adequacy** The situation in which a person consumes the required amount of a nutrient to meet physiological needs.

**nutritional sciences** A broad spectrum of academic and social disciplines related to nutrition.

**nutritional status** The health of a person as it relates to how well his or her diet meets that person's individual nutrient requirements.

**nutritional toxicity** Overconsumption of a nutrient resulting in dangerous (toxic) effects.

**Nutrition Facts panel** A required component of food packaging that contains information about the nutrient content of the food.

**nutrition transition** The shift from undernutrition to overnutrition or unbalanced nutrition that often occurs simultaneously with the industrialization of a society.

## O

**ob gene** The gene that codes for the protein leptin.

**ob/ob mouse** Obese mouse with a mutation in the gene that produces the hormone leptin.

**obesity** Excess body fat.

**octet rule** The "desire" of an atom to have 8 electrons in its outer valence shell.

**oil** A lipid that is liquid at room temperature.

**oligosaccharide** Carbohydrate made of relatively few (3 to 10) monosaccharides.

**omega (ω) end** The end of a fatty acid with the methyl (−CH₃) group.

**omega-6 (ω-6) fatty acid** A fatty acid in which the first double bond is located between the sixth and seventh carbons from the methyl or omega (ω) end.

**omega-3 (ω-3) fatty acid** A fatty acid in which the first double bond is located between the third and fourth carbons from the methyl or omega (ω) end.

**oncogene** An abnormal gene that transforms normal cells to cancer cells.

**opsin** The protein component of rhodopsin.

**organ** A group of tissues that combine to carry out coordinated functions.

**organelle** Within a cell, a structure that has a particular function.

**organic compound** A substance that contains carbon–carbon bonds or carbon–hydrogen bonds.

**organic foods** Plant and animal foods that have been grown, harvested, and processed without conventional pesticides, fertilizers, growth promoters, bioengineering, or ionizing radiation.

**organ system** Organs that work collectively to carry out related functions.

**osmosis** A passive transport mechanism whereby water moves from a region of lower solute concentration to that of a higher solute concentration.

**ossification** The process by which minerals are added to cartilage, ultimately resulting in bone formation.

**osteoblast** A bone cell that promotes bone formation.

**osteoclast** A bone cell that promotes bone breakdown.

**osteomalacia** Softening of the bones in adults that can be due to vitamin D deficiency.

**osteopenia** A disease characterized by moderate bone demineralization.

**osteoporosis** A serious bone disease resulting in weak, porous bones.

**overnutrition** Overconsumption of one or more nutrients and/or energy.

**overweight** Excess weight for a given height.

**oxaloacetate** The final product of the citric acid cycle, which becomes the substrate for the first reaction in this pathway.

**oxidation** The loss of one or more electrons.

**oxidative phosphorylation** The chemical reactions that link the oxidation of NADH + H⁺ and FADH₂ to the phosphorylation of ADP to form ATP and water.

**oxytocin** A hormone produced in the pituitary gland that stimulates the movement of milk into the mammary ducts.

## P

**pancreatic α-amylase** Enzyme released from the pancreas that digests starch by hydrolyzing α-1,4 glycosidic bonds.

**pancreatic juice** Pancreatic secretions that contain bicarbonate and enzymes needed for digestion.

**pancreatic lipase** An enzyme produced in the pancreas that hydrolyzes ester bonds between fatty acids and glycerol molecules.

**pancreatitis** Inflammation of the pancreas.

**pantothenic acid** (vitamin B₅) A water-soluble vitamin involved in energy metabolism, hemoglobin synthesis, and phospholipid synthesis.

**parasite** An organism that, during part of its life cycle, must live within or on another organism without benefiting its host.

**parathyroid hormone (PTH)** A hormone produced in the parathyroid glands that stimulates the conversion of 25-(OH) D₃ to 1,25-(OH)₂ D₃ in the kidneys.

**parietal cells** Exocrine cells within the gastric mucosa that secrete hydrochloric acid and intrinsic factor.

**partial hydrogenation** A process by which some carbon–carbon double bonds found in PUFAs are converted to carbon–carbon single bonds, resulting in the production of *trans* fatty acids.

**passive transport** A non–energy-requiring mechanism whereby a substance moves from a region of higher concentration to that of a lower concentration.

**pasteurization** A food preservation process that subjects foods to heat to kill bacteria, yeasts, and molds.

**pathogenic** Something that causes disease.

**Peace Corps** A federally funded program whereby American volunteers live and work with people in underdeveloped countries.

**peak bone mass** The greatest amount of bone mineral that a person has during his or her life.

**peer-reviewed journal** A publication that requires a group of scientists to read and approve a study before it is published.

**pellagra** A disease caused by niacin and/or tryptophan deficiency.

**pepsin** An enzyme needed for protein digestion.

**pepsinogen** The inactive form of pepsin.

**peptic ulcer** The presence of irritation and/or erosion of the mucosal lining in the stomach, duodenum, or esophagus.

**peptide bond** A chemical bond that joins amino acids.

**Percent Daily Value (% DV)** The percentage of the recommended intake (DV) of a nutrient provided by a single serving of a food.

**periconceptional period** Prenatal stage of development beginning with conception through the formation of the blastocyst.

**perimenopausal** Hormonal changes that precede menopause.

**periosteum** The outer covering of bone, consisting of blood vessels, nerves, and connective tissue.

**peristalsis** Waves of muscular contractions that move materials in the GI tract in a forward direction.

**pernicious anemia** A condition caused by vitamin $B_{12}$ deficiency due to lack of intrinsic factor.

**peroxisome** An organelle that contains enzymes that break down amino acids, fatty acids, and toxic substances in cells.

**pH scale** A scale, ranging from 0 to 14, that signifies the acidity or alkalinity of a solution.

**phenylketonuria (PKU)** An inborn error of metabolism in which phenylalanine cannot be converted to tyrosine.

**phosphatidylcholine** (also called lecithin) A phospholipid that contains choline as its polar head group; commonly added to foods as an emulsifying agent.

**phospholipase A2** An enzyme produced in the pancreas that hydrolyzes fatty acids from phospholipids.

**phospholipid** A lipid composed of a glycerol bonded to two fatty acids and a polar head group.

**phosphorus (P)** A major mineral needed for cell membranes, bone and tooth structure, DNA, RNA, ATP, lipid transport, and a variety of reactions in the body that require phosphorylation.

**photosynthesis** Process whereby plants trap energy from the sun to produce glucose from carbon dioxide and water.

**phylloquinone** A form of vitamin K found in foods.

**phytates** Phosphorus-containing compounds often found in the outer coating of a kernel of grain, vegetables, and legumes.

**phytochemical** A substance found in plants and thought to benefit human health (above and beyond the provision of essential nutrients and energy).

**phytoestrogen** An estrogen-like phytochemical.

**phytosterol** Sterol made by plants.

**pica** An abnormal eating behavior that involves consuming nonfood substances such as dirt or clay.

**placebo** A "fake" treatment given to the control group that cannot be distinguished from the actual treatment.

**placebo effect** The phenomenon in which there is an apparent effect of the treatment because the individual expects or believes that it will work.

**placenta** An organ made up of fetal and maternal tissue that supplies nutrients and oxygen to the fetus.

**plaque** A complex of cholesterol, fatty acids, cells, cellular debris, and calcium that can form inside blood vessels and within vessel walls.

**plasma** The fluid portion of the blood.

**polar head group** A phosphate-containing charged chemical structure that is a component of a phospholipid.

**polar molecule** Molecules that have both partial positively and negatively charged portions.

**polycyclic aromatic hydrocarbons** Cancer-causing compounds that can be formed when meat is grilled.

**polymer** Large molecules made up of repeating subunits.

**polyphenols** Organic compounds found in some foods.

**polysaccharide** Complex carbohydrate made of many monosaccharides.

**polyunsaturated fatty acid (PUFA)** A fatty acid that contains more than one carbon–carbon double bond in its backbone.

**positive energy balance** A state in which energy intake is greater than energy expenditure.

**positive nitrogen balance** The condition in which protein (nitrogen) intake is greater than protein (nitrogen) loss by the body.

**postabsorptive state** The period of time (4 to 24 hours after a meal) when no dietary nutrients are being absorbed.

**postpartum amenorrhea** The period after giving birth and before the return of menses.

**potassium (K)** A major mineral important in fluid balance, muscle and nerve function, and energy metabolism.

**prebiotic foods** Indigestible foods that stimulate the growth of bacteria that naturally reside in the large intestine.

**preformed toxin** Poisonous substance produced by microbes while they are in a food (prior to ingestion).

**preterm (premature) infant** A baby born at 37 weeks or less gestation.

**prevalence** The total number of people who have a condition in a given period of time.

**previtamin $D_3$ (precalciferol)** An intermediate product made during the conversion of 7-dehydrocholesterol to cholecalciferol (vitamin $D_3$) in the skin.

**primary (information) source** The publication in which a scientific finding was first published.

**primary malnutrition** Poor nutritional status caused strictly by inadequate diet.

**primary structure** The sequence of amino acids that make up a single polypeptide chain.

**prion** A misshapen protein that causes other proteins to also become distorted, damaging nervous tissue.

**probiotic foods** Foods and dietary supplements that contain live bacteria.

**product** A molecule produced in a chemical reaction.

**proenzyme** An inactive precursor of an enzyme.

**progression** The third stage of cancer in which tumor cells rapidly divide and invade surrounding tissues.

**prolactin** A hormone produced in the pituitary gland that stimulates the production of milk in alveoli.

**prolonged starvation** Food deprivation lasting longer than one week.

**promotion** The second stage of cancer in which the initiated cell begins to replicate itself, forming a tumor.

**proof** A measure of the alcohol content of distilled liquor.

**prospective dietary assessment** Type of dietary assessment that assesses present food and beverage intake.

**prosthetic group** A nonprotein component of a protein that is part of the quaternary structure.

**protease** An enzyme that cleaves peptide bonds.

**protein** Nitrogen-containing macronutrient made from amino acids.

**protein complementation** Combining incomplete protein sources to provide all essential amino acids in relatively adequate amounts.

**protein turnover** The balance between protein degradation and protein synthesis in the body.

**protein-energy malnutrition (PEM)** Protein deficiency accompanied by inadequate intake of energy and often of other essential nutrients as well.

**proteolysis** The breakdown of proteins into peptides or amino acids.

**prothrombin** A clotting factor (protein) that is converted to the enzyme thrombin.

**proton** A subatomic particle in the nucleus of an atom that carries a positive charge.

**protozoa** Very small (single-cell) organisms that are sometimes parasites.

**provitamin A carotenoid** A carotenoid that can be converted to vitamin A.

**puberty** Maturation of the reproductive system.

**PubMed** A computerized database that allows access to approximately 11 million biomedical journals.

**pulmonary arteries** Blood vessels that transport oxygen-poor blood from the right side of the heart to the lungs.

**pulmonary circulation** The division of the cardiovascular system that circulates deoxygenated blood from the heart to the lungs, and oxygenated blood from the lungs back to the heart.

**pulmonary veins** Blood vessels that transport oxygen-rich blood from the lungs to the heart.

**purging** Self-induced vomiting and/or misuse of laxatives, diuretics, and/or enemas.

**purine** A compound that makes up DNA and RNA.

**pyloric sphincter** A circular muscle that regulates the flow of food between the stomach and the duodenum.

**pyridoxal phosphate (PLP)** The coenzyme form of vitamin $B_6$.

**pyrimidine** A compound that makes up DNA and RNA.

**pyruvate** An intermediate product formed during metabolism of carbohydrates and some amino acids.

# Q

**qualified health claims** Statements concerning less well established health benefits that have been ascribed to a particular food or food component.

**quaternary structure** The combining of polypeptide chains with other polypeptide chains and/or prosthetic groups.

# R

**randomized, double-blind, placebo-controlled study** The type of experiment that is considered to be the ideal research design for testing a cause-and-effect hypothesis.

**random assignment** When study participants have equal chance of being assigned to each experimental group.

**rate** A measure of something within a specific period of time.

**reactive hypoglycemia** Low blood glucose that occurs after eating carbohydrate-rich foods, caused by the release of too much insulin.

**Recommended Dietary Allowance (RDA)** The average chronic intake level of a nutrient thought to meet the nutrient requirements of nearly all (97%) healthy people in a particular physiological state and age.

**rectum** The lower portion of the large intestine between the sigmoid colon and the anal canal.

**reduction** The gain of one or more electrons.

**reduction-oxidation (redox) reactions** Chemical reactions that take place simultaneously whereby one molecule gives up 1 or more electrons (is oxidized) while the other molecule receives 1 or more electrons (is reduced).

**reference standard** A value that represents what would be expected in a healthy population of similar age and sex.

**renin** An enzyme produced in the kidneys in response to low blood pressure; converts angiotensinogen to angiotensin I.

**researcher bias** When the researcher influences the results of a study.

**resorption** The cycle of reuptake and return of previously removed materials to the blood.

**resting energy expenditure (REE)** Energy expended for resting metabolism over a 24-hour period.

**resting heart rate** Number of heartbeats per minute measured at rest.

**resting metabolic rate (RMR)** A measure of energy expenditure assessed under less stringent conditions than is BMR.

**restrained eaters** People who experience cycles of fasting followed by bingeing.

**retina** The inner lining of the back of the eye.

**retinoid (preformed vitamin A)** A term used to describe all forms of vitamin A.

**retinol activity equivalent (RAE)** A unit of measure used to describe the combined amount of preformed vitamin A and provitamin A carotenoids in foods.

**retinol-binding protein** and **transthyretin** Proteins that carry retinol in the blood.

**retrospective dietary assessment** Type of dietary assessment that assesses previously consumed foods and beverages.

**reverse cholesterol transport** Process by which HDLs remove cholesterol from nonhepatic (nonliver) tissue for transport back to the liver.

**R-group** The portion of an amino acid's structure that distinguishes it from other amino acids.

**rhodopsin** A compound, in the retina, that consists of the protein opsin and the vitamin *cis*-retinal; needed for night vision.

**riboflavin (vitamin B$_2$)** An essential water-soluble vitamin involved in energy metabolism, the synthesis of a variety of vitamins, nerve function, and protection of lipids.

**ribosome** A cellular organelle involved in protein synthesis; a particle associated with the endoplasmic reticulum in the cytoplasm, involved in gene translation.

**rickets** A symptom of vitamin D deficiency in young children characterized by deformed bones, especially in the legs.

**risk factor** A lifestyle, environmental, or genetic factor related to a person's chances of developing a disease.

**rough endoplasmic reticulum (RER)** An organelle studded with ribosomes that synthesizes proteins.

**R protein** A protein produced in the stomach that binds to vitamin B$_{12}$.

**rugae** Folds that line the stomach wall.

# S

**saliva** A secretion released into the mouth by the salivary glands; moistens food and starts the process of digestion.

**salivary α-amylase** Enzyme released from the salivary glands, that digests starch by hydrolyzing α-1,4 glycosidic bonds.

**satiety** The state in which hunger is satisfied and a person feels he or she has had enough to eat.

**saturated fatty acid (SFA)** A fatty acid that contains only carbon–carbon single bonds in its backbone.

**scientific method** Steps used by scientists to explain observations.

**scurvy** A condition caused by vitamin C deficiency; symptoms include bleeding gums, bruising, poor wound healing, and skin irritations.

**secondary diabetes** Diabetes that results from other diseases, medical conditions, or medication.

**secondary malnutrition** Poor nutritional status caused by factors such as illness.

**secondary structure** Folding of a protein because of hydrogen bonds that form between elements of the amino acid backbone (not R-groups).

**secretin** A hormone, secreted by the duodenum, that stimulates the release of sodium bicarbonate and enzymes from the pancreas.

**segmentation** A muscular movement in the gastrointestinal tract that moves the contents back and forth within a small region.

**selective estrogen receptor modulator (SERM)** A drug that does not contain the hormone estrogen but causes estrogen-like effects in the body.

**selenium (Se)** An essential trace mineral that is important for redox reactions, thyroid function, and activation of vitamin C.

**selenomethionine** The amino acid methionine that has been altered to contain selenium instead of sulfur.

**selenoprotein** A protein that contains selenomethionine instead of sulfur-containing methionine.

**semipermeable membrane** A barrier that allows passage of some, but not all, molecules across it.

**senescence** The process of aging, during which function diminishes.

**serosa** Connective tissue that encloses the gastrointestinal tract.

**serotype** A specific strain of a larger class of organism.

**set point theory** A theory that suggests changes in energy intake and energy expenditure help maintain a stable body weight.

**shellfish poisoning** A group of foodborne illnesses caused by consuming shellfish that contain marine toxins.

**sickle cell anemia** A disease in which an alteration in the amino acid sequence of hemoglobin causes red blood cells to become misshapen and decreases the ability of the blood to carry oxygen and carbon dioxide.

**signs** Physical indicators of disease that can be seen by others, such as pale skin and skin rashes.

**simple carbohydrate,** or **simple sugar** Category of carbohydrates consisting of mono- or disaccharides.

**simple diffusion** A passive transport mechanism whereby a substance crosses a cell membrane without the assistance of a transport protein.

**simple relationship** A relationship between two factors that is not influenced or modified by another factor.

**single-blind study** A human experiment in which the participants do not know to which group they have been assigned.

**skeletal fluorosis** Weakening of the bones caused by excessive fluoride intake.

**skinfold caliper** An instrument used to measure the thickness of subcutaneous fat.

**skinfold thickness** An anthropometric measurement of body fatness.

**small for gestational age (SGA)** An infant that is below the 10th percentile for weight for gestational age.

**smooth endoplasmic reticulum (SER)** An organelle that is involved in lipid synthesis.

**sodium (Na)** A major mineral important for regulating fluid balance, nerve function, and muscle contraction.

**soluble fiber** Dietary fiber that is capable of being dissolved in water.

**solute** A substance that dissolves in a solvent.

**solution** A mixture of two or more substances that are uniformly dispersed.

**solvent** The component of a solution in which a solute dissolves.

**Special Supplemental Nutrition Program for Women, Infants, and Children (WIC)** A federally funded health and nutrition program that provides assistance to low income women, infants, and children.

**specific heat** The energy required to raise the temperature of a substance.

**sphincter** A muscular band that narrows an opening between organs in the GI tract.

**sphincter of Oddi** The sphincter that regulates the passage of secretions from the common bile duct into the duodenum.

**spina bifida** A form of neural tube defect in which the spine does not properly form.

**sports anemia** A physiological response to training caused by a disproportionate increase in plasma volume relative to the number of red blood cells.

**stable isotope** A form of an element that contains additional neutrons.

**steroid hormone** A hormone made from cholesterol.

**sterol** A type of lipid with a distinctive multiring structure; a common example is cholesterol.

**sterol ester** A chemical compound consisting of a sterol molecule bonded to a fatty acid via an ester linkage.

**stomatitis** Swollen mouth.

**strength training** Physical training that increases muscle growth and strength.

**stroke** A condition that occurs when a portion of the brain does not receive enough blood.

**stroke volume** The amount of blood pumped out of the heart with each contraction.

**stunted growth** Delayed or slow growth resulting from chronic undernutrition.

**subcutaneous adipose tissue** Adipose tissue found directly under the skin.

**submucosa** A layer of tissue that lies between the mucosa and muscularis tissue layers.

**substrate (reactant)** A molecule that enters a chemical reaction.

**substrate phosphorylation** The transfer of an inorganic phosphate group to ADP to form ATP.

**sucrase** Intestinal enzyme that hydrolyzes sucrose into glucose and fructose.

**sucrose** Disaccharide consisting of glucose and fructose; found primarily in fruits and vegetables.

**sulfite** A naturally occurring compound, in some foods, that can be used as a food additive to prevent discoloration and bacterial growth.

**superoxide dismutase** A copper-containing enzyme that reduces the superoxide free radical to form hydrogen peroxide.

**symptoms** Manifestations of disease that cannot be seen by others, such as stomach pain or loss of appetite.

**systemic circulation** The division of the cardiovascular system that begins and ends at the heart and delivers blood to all the organs except the lungs.

## T

**Tangier disease** A genetic disorder resulting in the production of faulty HDL particles that cannot take up cholesterol from cells.

**teratogen** Environmental agent that can alter normal cell growth and development, causing a birth defect.

**tertiary structure** Folding of a polypeptide chain because of interactions among the R-groups of the amino acids.

**tetany** A condition in which muscles tighten and are unable to relax.

**tetrahydrofolate (THF)** The active form of folate.

**thermic effect of food (TEF)** Energy expended for the digestion, absorption, and metabolism of nutrients.

**the female athlete triad** A combination of interrelated conditions: an eating disorder, amenorrhea, and osteopenia.

**thiamin (vitamin B$_1$)** An essential water-soluble vitamin involved in energy metabolism; synthesis of DNA, RNA, and NADPH+H$^+$; and nerve function.

**thiamin pyrophosphate (TPP)** The coenzyme form of thiamin that has two phosphate groups.

**thiamin triphosphate (TTP)** A form of thiamin with three phosphate groups.

**thoracic duct** The major duct of the lymphatic system; releases lymph into the blood at the subclavian vein.

**thrifty gene hypothesis** A theory that suggests people who are most likely to survive starvation are those most likely to gain weight when food is abundant.

**thrombin** The enzyme that catalyzes the conversion of fibrinogen to fibrin.

**thyroid-stimulating hormone (TSH)** A hormone produced in the pituitary gland that stimulates uptake of iodine by the thyroid gland.

**thyroxine (T$_4$)** The less active form of thyroid hormone; contains 4 atoms of iodine.

**tissue** An aggregation of specialized cell types that are similar in form and function.

**Tolerable Upper Intake Level (UL)** The highest level of chronic intake of a nutrient thought to be not detrimental to health.

**tolerance** Response to high and repeated drug exposure that results in a reduced effect.

**total energy expenditure (TEE)** Total energy expended or used by the body.

**trabecular bone** The inner, less dense layer of bone that contains the bone marrow.

**trace mineral** An essential mineral that is required in amounts less than 100 mg daily.

**transamination** The transfer of an amino group from one amino acid to another organic compound to form a different amino acid.

**transcobalamin** The protein that transports vitamin B$_{12}$ in the blood.

**transcription** The process by which mRNA is made using DNA as a template.

**transferrin** A protein important for iron transport in the blood.

**transferrin receptor** Protein found on cell membranes that binds transferrin, allowing the cell to take up iron.

**transfer ribonucleic acid (tRNA)** A form of RNA in the cytoplasm involved in gene translation.

**transient ischemic attack (TIA)** A "ministroke" that is caused by a temporary decrease in blood flow to the brain.

**translation** The process by which amino acids are linked together via peptide bonds on ribosomes, using mRNA and tRNA.

***trans* double bond** A carbon–carbon double bond in which the hydrogen atoms are arranged on opposite sides of the double bond.

***trans* fatty acid** A fatty acid containing at least one *trans* double bond.

**triglyceride (also called triacylglycerol)** A lipid composed of a glycerol bonded to three fatty acids.

**triiodothyronine (T$_3$)** The more active form of thyroid hormone; contains 3 atoms of iodine.

**trypsin, chymotrypsin, elastase,** and **carboxypeptidase** Active enzymes involved in protein digestion in the small intestine.

**trypsinogen, chymotrypsinogen, proelastase,** and **procarboxypeptidase** Inactive proenzymes produced in the pancreas and released into the small intestine in response to CCK.

**tumor** A growth sometimes caused by cancer.

**Type 1 osteoporosis** The form of osteoporosis that occurs in women, caused by hormone-related bone loss.

**Type 2 osteoporosis** The form of osteoporosis that occurs in men and women caused by age-related and lifestyle factors.

## U

**ulcerative colitis** A type of inflammatory bowel disease (IBD) that causes chronic inflammation of the colon.

**umami** A taste, in addition to the four basic taste components, that imparts a savory or meat-like taste.

**undernutrition (or nutritional deficiency)** Inadequate intake of one or more nutrients and/or energy.

**unqualified health claims** Health claims that can be included on a food's packaging to describe a specific, scientifically supported health benefit.

**unsaturated fatty acid** A fatty acid that contains at least one carbon–carbon double bond in its backbone.

**up-regulation** The "turning on" of protein synthesis.

**urbanization** A shift in a country's population from primarily rural to urban regions.

**urea** A relatively nontoxic, nitrogen-containing compound that is produced from ammonia after deamination.

**ureters** Ducts that carry urine from the kidneys to the bladder.

**urethra** The duct that carries urine from the bladder to the outside of the body.

**urinary bladder** The organ that collects urine.

**USDA Food Guides** Dietary recommendations developed by the USDA based on categorizing foods into "food groups."

## V

**valence shell** The outermost orbital of an atom.

**variant Creutzfeldt-Jakob disease** A form of Creutzfeldt-Jakob disease that may be caused by consuming BSE-contaminated foods.

**variety** A concept that emphasizes the importance of eating different varieties of a food type.

**vasodilation** Relaxation of the smooth muscles inside a blood vessel; increases the diameter of the vessel.

**vegan** A type of vegetarian who consumes no animal products.

**vegetarian** A person who does not consume any or selected foods and beverages made from animal products.

**vein** A blood vessel that carries blood toward the heart.

**ventilation rate** Rate of breathing.

**venules** Small blood vessels that branch off from veins.

**very low density lipoprotein (VLDL)** A lipoprotein made by the liver that contains a large amount of triglyceride; its major function is to deliver fatty acids to cells.

**vesicular active transport** An energy-requiring mechanism whereby large molecules move into or out of cells by an enclosed vesicle.

**villi (plural of *villus*)** Small, finger-like projections that cover the inner surface of the small intestine.

**visceral adipose tissue** Adipose tissue surrounding the vital organs.

**vitamin A deficiency disorder (VADD)** A multifaceted disease resulting from vitamin A deficiency.

**vitamin B$_6$** A water-soluble vitamin involved in the metabolism of proteins and amino acids; the synthesis of neurotransmitters and hemoglobin; glycogenolysis; and regulation of steroid hormone function.

**vitamin B$_{12}$ (cobalamin)** A water-soluble vitamin involved in energy metabolism and the conversion of homocysteine to methionine.

**vitamin C (ascorbic acid)** A water-soluble vitamin that has antioxidant functions in the body.

**vitamin K deficiency bleeding (VKDB)** A disease that occurs in newborn infants; characterized by uncontrollable internal bleeding (hemorrhage) from inadequate vitamin K.

**W**

**waist-to-hip ratio (WHR)** The ratio of waist circumference to hip circumference used to determine body fat distribution pattern.

**wet beriberi** A form of thiamin deficiency characterized by severe edema.

**X**

**xerophthalmia** A condition characterized by serious damage to the cornea due to vitamin A deficiency that can lead to blindness.

**Z**

**zinc (Zn)** An essential trace mineral involved in gene expression, immune function, and cell growth.

**zinc finger** Zinc containing three-dimensional structure of some proteins that allows them to regulate gene expression.

**zoonutrient** A substance found in animal foods and thought to benefit human health (above and beyond the provision of essential nutrients and energy).

**zygote** An ovum that has been fertilized by a sperm.

# Credits

**Chapter 1**

1 © Matthew Farruggio
2 © Matthew Farruggio
5 Food Collection/Index Stock Imagery
6 © Polara Studios
7 (top) © Nathan Benn/Corbis; (center) © Jerry Alexander/The Image Bank/Getty Images; (bottom) © Polara Studios
13 (top left) © Digital Vision/Getty Images; (center left) © John Giustina Photography/Iconica/Getty Images; (bottom left) © GK Hart/Vikki Hart/Photodisc/Getty Images; (right) © Maximilian Stock LTD/Phototake
23 © Dylan Ellis/Digital Vision/Getty Images

**Chapter 2**

28 © Matthew Farruggio
29 © Matthew Farruggio
43 (all) USDA
45 © Myrna Engler
47 USDA
48 (both) USDA
52 © Myrna Engler

**Chapter 3**

60 © Matthew Farruggio
61 © Astrid & Hanns-Frieder Michler/Photo Researchers, Inc.

**Chapter 4**

88 © Matthew Farruggio
89 © Matthew Farruggio
99 Dr. Fred Hossler/Visuals Unlimited
103 Gary Gaugler/Visuals Unlimited

**Chapter 5**

124 © Matthew Farruggio
125 © Matthew Farruggio
132 © Andrew Syred/Photo Researchers, Inc.
133 © Biophoto Associates/Photo Researchers, Inc.
140 Scott Bauer/USDA
151 © Doug Mazell/Index Stock Imagery
160 © Yoav Levy/Phototake
161 (all) Courtesy of the Clendening History of Medicine Library, University of Kansas Medical Center

162 © BSIP/Phototake
169 © Kent Foster/Visuals Unlimited

**Chapter 6**

172 © Matthew Farruggio
173 © Matthew Farruggio
176 © Ulrich Kerth/StockFood/Getty Images
177 © Polara Studios
181 © Dr. Stanley Flegler/Visuals Unlimited
191 © Dr. Marazzi/Photo Researchers, Inc.
195 © Kevin Dodge/Corbis
196 © Chabruken/The Image Bank/Getty Images
201 © The Wellcome Photo Library
207 © Dynamic Graphics Group/Creatas/Alamy
209 © Robert Harding Picture Library Ltd./Alamy
210 © Dr. Dennis Kunkel/Visuals Unlimited
211 © Medioimages/Getty Images

**Chapter 7**

220 © Matthew Farruggio
221 © Matthew Farruggio
229 © Michael Malovlich/Masterfile
237 © Jose Luis Pelaez/Corbis
240 © Royalty-free/Corbis
252 © Rita Maas/FoodPix/Jupiter Images
259 © Charles Thatcher/Stone/Getty Images

**Chapter 8**

268 © Matthew Farruggio
269 © Matthew Farruggio
299 Courtesy of Richard Weindruch, University of Wisconsin
300 © AP/Wide World Photos
304 © David Madison/Allsport Concepts/Getty Images
306 (left) © Photodisc/Getty Images; (right) © Duomo/Corbis
307 Courtesy of Kathy Beerman
311 © Mark Thiessen/National Geographic Society, Medical Imaging by Ken Eward/Biografx
316 © David Young-Wolff/PhotoEdit

**Chapter 9**

318 © Matthew Farruggio
319 USDA
320 Courtesy John H. Campbell, UCLA
327 Courtesy of Korr Medical Technologies
333 (bottom left) Courtesy of Bodystat Ltd.; (bottom center) © Wellcome Photo Library; (bottom right) Courtesy of Futrex, Inc.; (top left) © David Young-Wolff/PhotoEdit; (top center) Courtesy of Life Measurement, Inc.; (top right) © Wellcome Photo Library

334 (both) Courtesy of Kevin Davy/Ph.D., Virginia Polytechnic and State University
337 (top) © Felicia Martinez/PhotoEdit; (bottom) © Michael Newman/PhotoEdit
340 Courtesy Gokhan Hotamisligil, M.D., Ph.D., Harvard School of Public Health
350 Scott Bauer/ARS/USDA
351 © Dennis MacDonald/PhotoEdit
361 © Photodisc/Getty Images
363 © David Young Wolff/PhotoEdit
367 © Wellcome Photo Library
369 © Robert Brenner/PhotoEdit
370 (left) © John Lamb/The Image Bank/Getty Images; (right) © Kathy Willens/AP Photos
372 © Robin Sachs/PhotoEdit

**Chapter 10**

376 © Matthew Farruggio
377 © Matthew Farruggio
385 © Riccardo Marcialis/StockFood Munich/Stockfood America
388 © Rick Mariani/Stockfood America
390 (all) © Wellcome Photo Library
391 © Buntrock/StockFood Munich/Stockfood America
392 (left) © Stockfood America; (center) © Stockfood America; (right) © Rancho Gordo
393 © Wellcome Photo Library
394 © Smith-StockFood Munich/Stockfood America
396 © Amos Schliack/Stockfood
397 (left) © Carolina Biological Supply/Phototake, Inc.; (center) © Dr. Gladden Willis/Visuals Unlimited; (right) © American Society of Hematology. All Rights Reserved.
400 © Polara Studios
404 © Eising Food Photography/Stockfood America
407 © Polara Studios
410 © Dr. Marazzi/Science Photo Library/Photo Researchers, Inc.
419 (left) © Photodisc/Getty Images; (right) © A. Inden/zefa/Corbis
429 © Arthur Glauberman/Photo Researchers, Inc.

**Chapter 11**

434 © Matthew Farruggio
435 © Matthew Farruggio
438 © MaXx Images/Stockfood America
439 Courtesy Surfasonline.com
444 (both) © Wellcome Photo Library
445 (left) © Frans Lemmens/zefa/Corbis; (right) Courtesy Syngenta
446 © Wellcome Photo Library
452 © Wellcome Photo Library

454 © Pete A. Eising/StockFood Munich/ Stockfood America

457 © Eising Food Photography/Stockfood America

459 © Dr. David M. Phillips/Visuals Unlimited

467 © Photodisc/Getty Images

473 © Jose Luis Pelaez, Inc./Corbis

476 Peggy Greb/ARS/USDA

477 © Tim Hill/Stockfood America

## Chapter 12

480 © Matthew Farruggio

481 © Matthew Farruggio

485 © Cazals/StockFood Munich/Stockfood America

486 © Eising Food Photography/Stockfood America

491 (both) © Martin Chillmaid/Photo Researchers, Inc.

496 © Food Collection/Stockfood America

497 © Aurora Photos/StockFood Munich/ Stockfood America

498 © Wellcome Photo Library

499 © Phototake, Inc.

501 © Rubberball Productions/Getty Images

502 © Photodisc/Getty Images

505 © A. Hbrkova/StockFood Munich/ Stockfood America

506 © Wellcome Photo Library

508 © Wellcome Photo Library

## Chapter 13

514 © Matthew Farruggio

515 © Matthew Farruggio

518 © Stockbyte Platinum/Getty Images

527 © Maximilian Stock LTD/Stockfood

530 © Schnare & Stief/Stockfood America

538 Peggy Greb/ARS/USDA

544 © Yoav Levy/Phototake, Inc.

548 (left) © Topham/The Image Works; (right) © Martin Rotker/Phototake, Inc.

549 Courtesy Neurodiagnostics, Inc.

## Chapter 14

554 © Matthew Farruggio

555 © Matthew Farruggio

558 (all) © Lennart Nilsson/Albert Bonniers Förlag AB, from "A Child Is Born," Dell Publishing Company

561 © Jonathan Nourok/PhotoEdit

562 © Leonard McCombe/Time Life Pictures/ Getty Images

564 (both) Courtesy MRC Epidemiology Resource Centre, Southampton General Hospital

571 © David Young-Wolff/PhotoEdit

578 © Michael Newman/PhotoEdit

580 © Myrleen Ferguson Cate/PhotoEdit

590 © Pascal Parrot/Corbis Sygma

591 © C. Lyttle/zefa/Corbis

597 © Karen Preuss/The Image Works

602 (left) © Alexandra Winkler/Reuters/Corbis; (right) © Tony Freeman/PhotoEdit

605 Ken Hammond/USDA

606 © Grant Halverson/AP Photos

608 © Priit J. Vesilind/National Geographic Society

611 (both) © Photodisc/Getty Images

# Index

Boldface page references indicate definitions of terms.

## A

"ABCDs" of nutritional status assessment, **31**–36
Abdominal obesity, 167
Absorption, **91**, 108–109. *See also* Bioavailability; Digestion; *specific nutrients*
  of alcohol, 420–421
  of carbohydrates, 140–143
  of fat-soluble vitamins, 250, 436
  of fluids and electrolytes in large intestine, 112
  of lipids, 244–245
  of major minerals, 516
  of protein, 188
  in small intestine, 104, 105, 108, 109
Absorptive state, **297**–298
Acceptable Macronutrient Distribution Ranges (AMDRs), 37, **41**–42, 351–352
  for carbohydrates, 41, 152
  for children, 585
  for essential fatty acids, 229
  for lipids, 252
  for protein, 197, 203, 567–568
Accessory organs, 90–91, 104
Accutane™ (isotreinoin), 562
Acesulfame K, 155
Acetaldehyde, 424, 425
Acetaldehyde dehydrogenase (ALDH), **423**, 424
Acetic acid (vinegar), 73
Acetylcholine, synthesis of, 412
Acetyl-CoA synthase, 423
Acetyl coenzyme A (acetyl-CoA)
  from beta-oxidation, 294
  citric acid cycle and, 284
  conversion of pyruvate to, 282, 283, 386, 425, 426
  formation, 423–424
  ketones from, 294
  lipogenesis and, 292
  synthesis, 289, 290
Achiote paste, 439
Acidic solution, **72**, 73
Acidosis, **74**, 75
Acids, **73**–74
Acrodermatitis enteropathica, **506**
Acrylamide, **212**
Actin, 189
Active transport, 80, 81
  absorption by, 108
  of monosaccharides, 140, 141
  for movement of electrolytes, 533, 534
Acute starvation, 297, **298**–299

Adaptation response, **310**
Adaptive thermogenesis, **324**
Adenine, 276, 277
Adenosine, 276, 277
Adenosine diphosphate (ADP), 277, 306, 489–490
Adenosine triphosphate. *See* ATP (adenosine triphosphate)
Adequate Intake (AI) level, 36, **39**
  for biotin, 399
  for chloride, 529
  for chromium, 502
  dietary assessment and, 39, 40
  for dietary fiber, 151
  for essential fatty acids, 229
  for fluoride, 508
  for manganese, 503
  for pantothenic acid, 395
  for potassium, 531
  for sodium, 526, 529
  for vitamin D, 453
  for vitamin K, 460
ADH (alcohol dehydrogenase), **423**, 424
  ADH pathway, **423**–424
ADH (antidiuretic hormone), **537**
  blood volume/pressure regulated by, 537–538
Adipocytes, **233**, 320, 321
Adipose tissue, 320, 321
  amino acids stored as fat in, 193
  basal metabolic rate of, 326
  calories in one pound of, 320
  energy balance signals from, 342
  energy stored as, 233–234
  as hormone-producing endocrine "gland," 165–166, 320
  leptin produced by, 347
  methods to estimate percent of, 331–333
  set point theory and, **339**
  subcutaneous, **233**, 332–333, 334
  visceral, **233**
Adiposity. *See also* Fat, body
  android, **334**
  BMI and, 329
  gynoid, **334**
  hormones signaling, 347
Adolescence, 587–589
  eating disorders in, 368–369
  growth and development during, 587–588
  nutritional concerns and recommendations during, 588–589
  overweight and obesity in, 336, 338, 583
  physical inactivity in, 338
  teen pregnancy, 565
Adoption studies of body weight, 339
ADP (adenosine diphosphate), 277, 306, 489–490

Adrenal glands, blood sodium regulation and, 527, 528
Adulthood, 589–597. *See also* Older adults
  nutritional issues in, 592–597
  nutrition during, 590–591
  protein deficiency in, 201
  subdivisions of, 591
Adult-onset diabetes mellitus. *See* Diabetes, type 2
Advertising, impact of, 338, 583
Aerobic metabolism/metabolic pathways, 278, 282–285, 305–306, 308–309
Aflatoxin, 208, **209**
African Americans. *See also* Race and ethnicity
  hypertension among, 529
  obesity rates among, 338
  salt sensitivity among, 529
AGA (appropriate for gestational age), **562**, 563
Age. *See also* Life cycle nutrition
  age structure of U.S. population, 589–590
  basal metabolic rate and, 325, 326
  bone mineral density and, **546**
  calcium bioavailability and, 517–518
  Estimated Energy Requirements (EERs) by, A-4
  gestational, 559, 560, 561–563
  osteoporosis and, 546
Aging
  anorexia of, **595**
  hormonal changes in women and, 594
  moderate caloric restriction and, 298, 299
  physiological changes of, 593, 594–595
  theories of, 591–592
  trend, 22
  of U.S. population, 590
AI. *See* Adequate Intake (AI) level
Air, preventing vitamin loss from, 384
Alanine aminotransferase (ALT), 274
Al-Anon®, **431**
Alateen®, **431**
Albumin, 189, **190**, 191
  fatty acid-albumin complexes, 245
Alcohol, 419, **420**–433
  absorption, 420–421
  binge drinking, **430**–431, 432
  blood alcohol concentration, **421**, 422, 423
  blood lipid levels and, 264
  blood pressure and, 262
  bone health and, 550
  cancer risk and, 477
  cardiovascular disease and, 262, 429–431

# Anatomy of MyPyramid

**One size doesn't fit all**

USDA's new MyPyramid symbolizes a personalized approach to healthy eating and physical activity. The symbol has been designed to be simple. It has been developed to remind consumers to make healthy food choices and to be active every day. The different parts of the symbol are described below.

**Activity**

Activity is represented by the steps and the person climbing them, as a reminder of the importance of daily physical activity.

**Moderation**

Moderation is represented by the narrowing of each food group from bottom to top. The wider base stands for foods with little or no solid fats or added sugars. These should be selected more often. The narrower top area stands for foods containing more added sugars and solid fats. The more active you are, the more of these foods can fit into your diet.

**Personalization**

Personalization is shown by the person on the steps, the slogan, and the URL. Find the kinds and amounts of food to eat each day at MyPyramid.gov.

**Proportionality**

Proportionality is shown by the different widths of the food group bands. The widths suggest how much food a person should choose from each group. The widths are just a general guide, not exact proportions. Check the Web site for how much is right for you.

**Variety**

Variety is symbolized by the 6 color bands representing the 5 food groups of the Pyramid and oils. This illustrates that foods from all groups are needed each day for good health.

**Gradual Improvement**

Gradual improvement is encouraged by the slogan. It suggests that individuals can benefit from taking small steps to improve their diet and lifestyle each day.

**MyPyramid.gov**
**STEPS TO A HEALTHIER YOU**

USDA U.S. Department of Agriculture Center for Nutrition Policy and Promotion April 2005 CNPP-16

USDA is an equal opportunity provider and employer

GRAINS | VEGETABLES | FRUITS | OILS | MILK | MEAT & BEANS

---

## Summary of the USDA MyPyramid Food Guidance System

### Key Recommendations[a]

#### Nutrient-Dense Foods
- Maximize the amount of nutrient-dense foods that you consume by choosing and preparing low-fat meats and dairy products, limiting intake of foods with added sugars, and drinking alcohol only in moderation (if at all).

#### Discretionary Calories
- To maintain a healthy weight, do not exceed your discretionary calorie allowance.
- Increase your discretionary calorie allowance by increasing physical activity.

#### Physical Activity
- Make physical activity a regular part of the day.
- At a minimum, do moderate-intensity activity for 30 minutes most days, or preferably every day.

#### Grains
- Consume 3–10 oz. per day (or equivalent).
- Whole-grain and folate-fortified products are recommended.

#### Fruits and Vegetables
- Consume 1–2½ cups of fruit per day.
- Consume 1–4 cups of vegetables per day.
- Choose a variety of fruits and vegetables each day.

#### Dairy Products
- Consume 2–3 cups per day.
- Low-fat or fat-free dairy foods are recommended.

#### Meats, Poultry, Eggs, and Fish
- Consume 2–7 oz. per day (or equivalent)
- Choose fish more often and avoid foods high in saturated fats, *trans* fats, and cholesterol.

#### Oils
- Consume 3–11 tsp. per day.
- Amount recommended depends greatly on caloric needs.

---

[a]Amounts recommended by MyPyramid are dependent on age, sex, and physical activity level. Personalized recommendations can be generated by going to the MyPyramid website (http://www.mypyramid.gov).

SOURCE: Adapted from U.S. Department of Health and Human Services and U.S. Department of Agriculture, *Dietary Guidelines for Americans 2005.* Government Printing Office, Washington, D.C., 2005 (http://www.healthierus.gov/dietaryguidelines) and U.S. Department of Agriculture MyPyramid website (http://www.mypyramid.gov).

## Summary of the 2005 USDA Dietary Guidelines

### Adequate Nutrients Within Calorie Needs
- Consume a variety of nutrient-dense foods and beverages while choosing foods that limit the intake of saturated and *trans* fats, cholesterol, added sugars, salt, and alcohol.

### Weight Management
- Balance calories from foods and beverages with calories expended.

### Physical Activity
- Engage in regular physical activity and reduce sedentary activities to promote health, psychological well-being, and a healthy body weight.

### Foods to Encourage
- *Fruits and Vegetables:* Choose a variety of fruits and vegetables each day.
- *Whole Grains:* Consume 3 or more ounce-equivalents of whole-grain products per day.
- *Low-Fat Dairy Products:* Consume 3 cups per day of fat-free or low-fat milk or equivalent milk products.

### Fats
- Keep total fat intake between 20 to 35% of calories, with most fats coming from sources of polyunsaturated and monounsaturated fatty acids.
- Limit intake of fats and oils high in saturated and/or *trans* fatty acids.

### Carbohydrates
- Choose fiber-rich fruits, vegetables, and whole grains often.
- Choose and prepare foods and beverages with little added sugars or caloric sweeteners.

### Sodium and Potassium
- Choose and prepare foods with little salt. At the same time, consume potassium-rich foods, such as fruits and vegetables.

### Alcoholic Beverages
- Those who choose to drink alcoholic beverages should do so sensibly and in moderation.

### Food Safety
- Clean hands, food contact surfaces, and fruits and vegetables.
- Separate raw, cooked, and ready-to-eat foods.
- Cook foods to a safe temperature.
- Chill perishable food promptly and defrost foods properly.
- Avoid raw milk or any products made from unpasteurized milk, raw eggs, unpasteurized juice, or raw sprouts.

SOURCE: Adapted from U.S. Department of Health and Human Services and U.S. Department of Agriculture, *Dietary Guidelines for Americans 2005*. Government Printing Office, Washington, D.C., 2005 (http://www.healthierus.gov/dietaryguidelines) and U.S. Department of Agriculture MyPyramid website (http://www.mypyramid.gov).

## Commonly Used Weights, Measures, and Metric Conversion Factors

### Length

1 meter (m) = 39 in, 3.28 ft, or 100 cm.
1 centimeter (cm) = 0.39 in, 0.032 ft, or 0.01 m.
1 inch (in) = 2.54 cm, 0.083 ft, or 0.025 m.
1 foot (ft) = 30 cm, 0.30 m, or 12 in.

### Temperature

| | Celsius$^a$ | | Fahrenheit | |
|---|---|---|---|---|
| Boiling point | 100°C | | 212°F | Boiling point |
| Body temperature | 37°C | | 98.6°F | Body temperature |
| Melting point | 0°C | | 32°F | Melting point |

- To find degrees Fahrenheit (°F) when you know degrees Celsius (°C), multiply by 1.8 and then add 32.
- To find degrees Celsius (°C) when you know degrees Fahrenheit (°F), subtract 32 and then multiply by 0.56.

### Volume

1 liter (L) = 1000 mL, 0.26 gal, 1.06 qt, 2.11 pt, or 34 oz.
1 milliliter (mL) = 1/1000 L or 0.03 fluid oz.
1 gallon (gal) = 128 oz., 16 c, 3.78 L, 4 qt, or 8 pt.
1 quart (qt) = 32 oz., 4 c, 0.95 L, or 2 pt.
1 pint (pt) = 16 oz., 2 c, 0.47 L, or 0.5 qt.
1 cup (c) = 8 oz., 16 tbsp., 237 mL, or 0.24 L.
1 ounce (oz.) = 30 mL, 2 tbsp., or 6 tsp.
1 tablespoon (tbsp.) = 3 tsp., 15 mL, or 0.5 oz.
1 teaspoon (tsp.) = 5 mL or 0.17 oz.

### Weight

1 kilogram (kg) = 1000 g, 2.2 lb., or 35 oz.
1 gram (g) = 1/1000 kg, 1000 mg, or 0.035 oz.
1 milligram (mg) = 1/1000 g or 1000 µg.
1 microgram (µg) = 1/1000 mg.
1 pound (lb.) = 16 oz., 454 g, or 0.45 kg.
1 ounce (oz.) = 28 g or 0.062 lb.

### Energy

1 kilojoule (kJ) = 0.24 kcal, 240 calories, or 0.24 Calories.
1 kcalorie (kcal) = 4.18 kJ, 1000 calories, or 1 Calorie.

$^a$Also known as *centigrade*.